Nonlinear Systems and Their Remarkable Mathematical Structures

Nonlinear Systems and Their Remarkable Mathematical Structures

Volume 3, Contributions from China

Norbert Euler
Centro Internacional de Ciencias A.C., Cuernavaca

Da-jun Zhang
Shanghai University, Shanghai

CRC Press
Taylor & Francis Group
Boca Raton London New York

CRC Press is an imprint of the
Taylor & Francis Group, an **informa** business

First edition published 2022
by CRC Press
6000 Broken Sound Parkway NW, Suite 300, Boca Raton, FL 33487-2742

and by CRC Press
4 Park Square, Milton Park, Abingdon, Oxon OX14 4RN

CRC Press is an imprint of Taylor & Francis Group, an Informa business

Library of Congress Cataloging-in-Publication Data

Names: Euler, Norbert, editor.
Title: Nonlinear systems and their remarkable mathematical structures / edited by Norbert Euler.
Description: Boca Raton : CRC Press, [2019]- | Includes bibliographical references and index.
Identifiers: LCCN 2018038379 (print) | LCCN 2019724820 (ebook) | ISBN 9780429893810 epub | ISBN 9780429470462 ebook | ISBN 9781138601000 hardback : alk. paper : vol. 1 | ISBN 9781138601000 (hardback : alk. paper : vol. 1) | ISBN 9780429470462 (ebook)
Subjects: LCSH: Nonlinear theories. | Differential equations, Nonlinear. | Nonlinear systems.
Classification: LCC QA427 (ebook) | LCC QA427 .N672 2019 (print) | DDC 515/.355--dc23
LC record available at https://lccn.loc.gov/2018038379

ISBN 13: 978-0-367-54108-8 (hbk)
ISBN 13: 978-0-367-54112-5 (pbk)
ISBN 13: 978-1-00-308767-0 (ebk)

Typeset in CMR10 font
by KnowledgeWorks Global Ltd.

Contents

Preface

We are delighted to present Volume 3 in this series that aims to provide a comprehensive collection of works that report the most recent results in the mathematical description of nonlinear phenomena for dynamical processes in nature. An important difference between Volume 3 and the previous two volumes is that, in the current volume, we focus on contributions by Chinese mathematicians working in China. For that reason, Volume 3 consists exclusively of contributions by Chinese researchers, whereby some works are in collaboration with colleagues from outside of China. The book will give the reader a good overview of the current research activities in the subject of nonlinear systems in China. We should however emphasize that, due to the obvious restrictions of physical space in the form of pages and the fact that some professors have unfortunately been too busy to contribute to this book when the call was made, the book does not contain contributions from all the active groups of researchers currently working in China in the subject of nonlinear mathematical system.

The book consists of 18 chapters and invited contributions, presented in two parts, namely, **Part A**: *Integrability and Symmetry*, and **Part B**: *Algebraic, Analytic and Geometric Methods*. We now give a short description of each chapter.

Part A consists of nine chapters, numbered **A1** to **A9**. In **A1** the author, *J P Cheng*, gives a detailed review on the BKP hierarchy constructed by using the neutral free Fermions, and further discusses the constrained BKP hierarchy and modified BKP hierarchy. In **A2** the authors, *W Fu and F W Nijhoff*, give an elementary introduction to the direct linearisation approach of integrable systems. In **A3** the authors, *J Hietarinta and D J Zhang*, present a comprehensive review of the discrete Boussinesq equations based on their three-component forms on an elementary quadrilateral. In **A4** the authors, *J Kang, X C Liu, P J Olver and C Z Qu*, provide a survey on Liouville correspondences between integrable hierarchies and also provide prototypical examples. In **A5** the authors, *Q P Liu and L Xue*, give a brief review concerning Darboux transformations for supersymmetric integrable systems and also provide a list of supersymmetric integrable systems. In **A6** the author, *S Y Lou*, first reviews four types of methods to find nonlocal symmetries of integrable systems and then discusses applications of nonlocal symmetries in generating new integrable equations and finding solutions. In **A7** the authors, *H J Zhou and Y Chen*, consider an extended nonlinear Schrödinger equation with third-order term and fourth-order term and present high-order soliton solutions in the framework of the Riemann-Hilbert problem. In **A8** the author, *Z X Zhou*, first reviews the general construction of Darboux transformations for systems without symmetries and then discusses the Darboux transformation for systems with unitary symmetry and more complicated symmetries. In **A9** the author, *D Zuo*, reviews recent developments about Frobenius manifold structures on the orbit spaces of reflection groups and their extensions and proposes a conjecture for general cases.

Part B consists of nine chapters, numbered **B1** to **B9**. In **B1** the author, *X K Chang*, briefly reviews the developments pertaining to the relation between Toda lattices and peakon solutions with emphasis on their isospectral structures in terms of orthogonal functions. In **B2** the authors, *M M Chen, X G Geng, K D Wang and B Xue*, develop spectral analysis of 3×3 matrix spectral problems and obtain long-time asymptotics of the initial value problem for the generalized coupled derivative nonlinear Schrödinger equation in the situation of solitonless by using the nonlinear steepest descent method. In **B3** the authors, *X B Hu and G F Yu*, by using recursion operators of integrable nonlinear evolution equations, construct so-called unified bilinear forms for several well known or less known integrable hierarchies of equations. In **B4** the authors *L Ling and L-C Zhao* discuss rogue wave patterns and modulational instability in nonlinear Schrödinger hierarchy and present some novel results for breather dynamics and high-order rogue wave decomposition. In **B5** the authors, *W Liu, X G Geng and B Xue*, derive algebro-geometric solutions to the modified Blaszak-Marciniak lattice hierarchy related to a 3×3 matrix spectral problem. In **B6** the authors, *Y L Yang and E G Fan*, consider the Cauchy problem for the modified Schrödinger equation and obtain the long time asymptotics of the equation using nonlinear steepest descent method and the $\bar{\partial}$-analysis. In **B7** the authors, *R X Yao, W Wang and Y Li*, obtain two new hierarchies of multiple solitons of (2+1)-dimensional Sawada-Kotera type equation and also investigate soliton molecules of the equation. In **B8** the author, *C Zhang*, gives a review of the notions of integrable boundary for integrable PDEs, and provides an effective means, called dressing the boundary, to construct soliton solutions of the associated models on the half-line. In **B9** the authors, *H Q Zhao and Z N Zhu*, review the connection between integrability of integrable spatial discrete hierarchy and integrable nonlinear PDE hierarchy.

Norbert Euler (Bad Ems, 30 January 2021)

Da-jun Zhang (Shanghai, 30 January 2021)

The Authors

1. X K Chang, *Academy of Mathematics and Systems Science, Chinese Academy of Sciences; University of Chinese Academy of Sciences, P.R. China* [**B1**]
2. M M Chen, *Zhengzhou University, P.R. China* [**B2**]
3. Y Chen, *East China Normal University, P.R. China* [**A7**]
4. J P Cheng, *China University of Mining and Technology, P.R. China* [**A1**]
5. E G Fan, *Fudan University, P.R. China* [**B6**]
6. W Fu, *East China Normal University, P.R. China* [**A2**]
7. X G Geng, *Zhengzhou University, P.R. China* [**B2**], [**B5**]
8. J Hietarinta, *University of Turku, Finland* [**A3**]
9. X B Hu, *Academy of Mathematics and Systems Science, Chinese Academy of Sciences, P.R. China* [**B3**]
10. J Kang, *Northwest University, P.R. China* [**A4**]
11. Y Li, *Shaanxi Normal University, P.R. China* [**B7**]
12. L Ling, *South China University of Technology, P.R. China* [**B4**]
13. Q P Liu, *China University of Mining and Technology (Beijing), P.R. China* [**A5**]
14. W Liu, *Zhengzhou University; Shijiazhuang Tiedao University, P.R. China* [**B5**]
15. X C Liu, *Xi'an Jiaotong University, P.R. China* [**A4**]
16. S Y Lou, *Ningbo University, P.R. China* [**A6**]
17. F W Nijhoff, *University of Leeds, United Kingdom* [**A2**]
18. P J Olver, *University of Minnesota, USA* [**A4**]
19. C Z Qu, *Ningbo University, P.R. China* [**A4**]
20. K D Wang, *Zhengzhou University, P.R. China* [**B2**]
21. W Wang, *Shaanxi Normal University; Xi'an University of Finance and Economics, P.R. China* [**B7**]
22. B Xue, *Zhengzhou University, P.R. China* [**B2**], [**B5**]

23. L Xue, *Ningbo University, P.R. China* [**A5**]

24. Y L Yang, *Fudan University, P.R. China* [**B6**]

25. R X Yao, *Shaanxi Normal University, P.R. China* [**B7**]

26. G F Yu, *Shanghai Jiao Tong University, P.R. China* [**B3**]

27. C Zhang, *Shanghai University, P.R. China* [**B8**]

28. D J Zhang, *Shanghai University, P.R. China* [**A3**]

29. H Q Zhao, *Shanghai University of International Business and Economics, P.R. China* [**B9**]

30. L-C Zhao, *Northwest University, Xi'an; Shaanxi Key Laboratory for Theoretical Physics Frontiers, P.R. China* [**B4**]

31. H J Zhou, *East China Normal University, P.R. China* [**A7**]

32. Z X Zhou, *Fudan University, P.R.China* [**A8**]

33. Z N Zhu, *Shanghai Jiao Tong University, P.R. China* [**B9**]

34. D Zuo, *University of Science and Technology of China, P.R. China* [**A9**]

A1. BKP hierarchy and modified BKP hierarchy

Jipeng Cheng

School of Mathematics, China University of Mining and Technology, Xuzhou, Jiangsu 221116, P.R. China

Abstract

The highest weight representation of the infinite Lie algebras is used to construct various soliton equations by the famous Kyoto school. In this chapter, we firstly review the construction of the BKP hierarchy corresponding to $\mathfrak{o}(\infty)$ by using the neutral free Fermions. As an important reduction of the BKP hierarchy, the constrained BKP hierarchy is further investigated in the aspect of the bilinear equations. It is shown that the bilinear equations in the form of tau functions can fully determine the constrained BKP hierarchy. Then the bilinear equations of the constrained BKP hierarchy are given in the Fermionic picture, and further interpretations of these bilinear equations are obtained. Particularly, there is the structure of the modified BKP hierarchy in the constrained BKP hierarchy, which is presented in the forms of the bilinear equations. Last, we investigate the dressing structure and the Lax structure of the modified BKP hierarchy. Note that the soliton equations obtained by the highest weight representation are usually given in the forms of the bilinear equations; therefore, the work here may be one try in further study of the integrable properties for these soliton equations.

1 Introduction

In the famous work of the Kyoto school, Sato *et al*, connected the soliton equations with the infinite dimensional Lie algebra and its highest weight representation [6, 11, 12, 19]. The corresponding group orbit of the highest weight vector, called the tau function, forms an infinite Grassmannian, and the defining equations, i.e., the Plücker relations, are just the soliton equations [6, 11, 19]. The most basic example is the KP hierarchy, whose corresponding Lie algebra is $gl(\infty)$ [6, 11, 12, 19]. Their work has gained great success in the study of the soliton theory. One can refer to [6, 11, 12, 13, 19] and the references in them for more details. But the soliton equations obtained in this way are usually presented in the form of bilinear equations, and for some of them, the corresponding Lax equations are unknown, for example, the modified BKP hierarchy [11, 28].

The BKP hierarchy is an important integrable system, corresponding to the Lie algebra $\mathfrak{o}(\infty)$ [6, 11, 13, 14, 28, 29]. The BKP hierarchy shares many important properties with the KP hierarchy, such as owning one tau function. The neutral free Fermions can be used to construct the tau function of the BKP hierarchy [6, 29], which can be viewed as the $O(\infty)$ orbit of the vacuum vector in the Fermionic picture. By Boson-Fermion correspondence, the tau function can be written into the usual form (see Section 2 for more details). Recently, there are many important results in the BKP hierarchy; for example, the additional symmetries and the quantum torus symmetries [16, 17, 25, 26]; various symmetry function solutions, including Schur Q functions [9, 22, 24] and polynomial tau functions [14]; relations with Bures ensemble and Pfaffian point process [28, 10]; the Grassmannian that corresponds to the total descendant potential [4]; the applications of Pfaffians [1, 3]; and the projective Hurwitz numbers [20].

The constrained BKP hierarchy [2, 5, 18, 23] is an important reduction of the BKP hierarchy, which is defined by imposing special conditions on the Lax operator (see (3.1)). In [23], the bilinear equations of the constrained BKP hierarchy are investigated in the forms of wave functions [2, 18] and tau functions. They have proved the equivalence of the bilinear equations in the form of the wave functions with the constrained BKP hierarchy, but they did not show that the bilinear equations in the forms of the tau functions can also lead to the constrained BKP hierarchy. Here in this chapter, we fill this gap, so that the bilinear equations in the forms of the tau functions can completely determine the constrained BKP hierarchy. Based upon this fact, we give a Fermionic description of the constrained BKP hierarchy, which may be helpful for understanding the essential properties of the constrained BKP hierarchy. Some further illustrations of the bilinear equations in the Fermionic picture are also given.

In the bilinear equations of the constrained BKP hierarchy, the structure of the modified BKP hierarchy [11, 28] is found, which is another important research object in this chapter. In the KP case, there is the $(l-l')$-th modified KP hierarchy [11], but there is only the 1st modified BKP hierarchy [11, 28], since it is usually very special for the structure of the Fermionic Fock space generated by neutral free Fermions [11, 28] (see Section 2 for more details). Here, we will concentrate on the construction of the dressing structures and the Lax structures. Also, the connection with the BKP hierarchy is considered. Based upon this, the corresponding modified BKP equation is given. These results are just one try in better understanding the Lax structures of the soliton equations derived by the methods in [6, 11, 12, 13, 19] (which are usually presented in the form of the bilinear equations).

This chapter is organized as follows. In Section 2, we give a very detailed review on the BKP hierarchy constructed by using the neutral free Fermions. Then we give a further discussion of the bilinear equations of the constrained BKP hierarchy, and also the corresponding Fermionic descriptions are given in Section 3. Last, in Section 4, we construct the dressing structures and the Lax structures of the modified BKP hierarchy and also investigate the connections with the BKP hierarchy.

2 BKP hierarchy and neutral Fermions

In this section, we will give a detailed review of the BKP hierarchy by using the neutral free Fermions. One can refer to [6, 11, 13, 28, 29] for more details. Firstly, the $O(\infty)$ orbit of the vacuum can be seen as the tau function of the BKP hierarchy, which satisfies the bilinear equations in the Fermionic form. Then by using the

Boson-Fermion correspondence, the corresponding bilinear equations are written into the Bosonic form. Based upon the bilinear equation, the wave function, dressing operator and Lax operator are introduced. At last, the dressing equations and the Lax equations are obtained.

2.1 BKP hierarchy in the Fermionic picture

Firstly, consider a vector space $V = \oplus_{j\in\mathbb{Z}}\mathbb{C}\phi_j$ with a symmetric bilinear forms,

$$(\phi_i, \phi_j) = (-1)^i\delta_{i+j,0}. \tag{2.1}$$

Then the Clifford algebra $\mathcal{A}$ can be defined by the quotient of the tensor algebra $T(V) = \mathbb{C} \oplus \bigoplus_{j\geq 1} V^{\otimes j}$ by its ideal I generated by

$$[\phi_i, \phi_j]_+ \triangleq \phi_i\phi_j + \phi_j\phi_i = (\phi_i, \phi_j)1, \tag{2.2}$$

that is $\mathcal{A} = T(V)/I$. Here ϕ_i is called the neutral Fermion.

The Fock space $\mathcal{F}$ for neutral Fermions and its dual space are defined by

$$\mathcal{F} = \mathcal{A}/\mathcal{A}V^-, \quad \mathcal{F}^* = \mathcal{A}/\mathcal{A}V^+, \tag{2.3}$$

where $V^\pm = \bigoplus_{\pm j>0}\mathbb{C}\phi_j \oplus \mathbb{C}(\phi_0 - \frac{1}{\sqrt{2}})$. If we denote $|0\rangle$ and $\langle 0|$ as the residue class of 1 in $\mathcal{F}$ and $\mathcal{F}^*$, respectively, then one can find that

$$\begin{aligned} &\phi_{-j}|0\rangle = 0, \quad \langle 0|\phi_j = 0, \quad j > 0, \\ &\phi_0|0\rangle = \frac{1}{\sqrt{2}}|0\rangle, \quad \langle 0|\phi_0 = \langle 0|\frac{1}{\sqrt{2}}. \end{aligned} \tag{2.4}$$

Therefore,

$$\begin{aligned} \mathcal{F} =&\text{span}\{\phi_{n_1}\cdots\phi_{n_k}|0\rangle\}, \quad \mathcal{F}^* = \text{span}\{\langle 0|\phi_{-n_k}\cdots\phi_{-n_1}\}, \\ &\text{with} \quad n_1 > n_2\cdots > n_k > 0. \end{aligned} \tag{2.5}$$

By using (2.2) and (2.4), one can define the nondegenerate bilinear pairing

$$\begin{aligned} &\mathcal{F}^* \times \mathcal{F} \longrightarrow \mathbb{C} \\ &(\langle 0|a, b|0\rangle) \longrightarrow \langle 0|ab|0\rangle = \langle ab\rangle, \end{aligned}$$

with $\langle 0|0\rangle = 1$. Particularly,

$$\langle 0|\phi_i\phi_j|0\rangle = \begin{cases} (-1)^i\delta_{i+j,0}, & j > 0; \\ \frac{1}{2}\delta_{i,0}, & j = 0; \\ 0, & j < 0. \end{cases} \tag{2.6}$$

For a general case, one can compute $\langle 0|v_1v_2\cdots v_r|0\rangle$ by the so-called Wick theorem, with $v_1, v_2, \cdots, v_r \in V' = \oplus_{j\neq 0}\mathbb{C}\phi_j$, that is

$$\langle v_1v_2\cdots v_r\rangle = \begin{cases} 0, & \text{if } r \text{ is odd;} \\ \sum_\eta \text{sign}(\eta)\langle v_{\eta(1)}v_{\eta(2)}\rangle\cdots\langle v_{\eta(r-1)}v_{\eta(r)}\rangle, & \text{if } r \text{ is even.} \end{cases} \tag{2.7}$$

Here η is the permutation such that $\eta(1) < \eta(2), \cdots, \eta(r-1) < \eta(r)$ and $\eta(1) < \eta(3) < \cdots < \eta(r-1)$. If we introduce $\phi(z) = \sum_{i\in\mathbb{Z}} \phi_i z^i$, then by Wick's Theorem

$$\langle \phi(z_1)\phi(z_2)\cdots\phi(z_m)\rangle = \frac{1}{2^{\frac{m}{2}}} \prod_{1\le i<j\le m} \frac{z_i - z_j}{z_i + z_j}. \tag{2.8}$$

Remark 1. Here the vacuum states $|0\rangle$ and $\langle 0|$ are given by Kac and van de Leur in [13], which are different from the ones in [6, 11]. If we denote the corresponding vacuum states in [6, 11] by $|0\rangle'$ and $\langle 0|'$, then they can be related by

$$|0\rangle = \frac{1}{\sqrt{2}}|0\rangle' + \phi_0|0\rangle', \quad \langle 0| = \frac{1}{\sqrt{2}}\langle 0|' + \langle 0|'\phi_0. \tag{2.9}$$

Particularly,

$$\langle 0|v_1v_2\cdots v_{2r}|0\rangle = \langle 0|'v_1v_2\cdots v_{2r}|0\rangle', \tag{2.10}$$

where $v_i \in V$. What's more, the corresponding Fock spaces are different for [13] and [6, 11]. The vacuum states $|0\rangle'$ and $\langle 0|'$ are more familiar in physics, since the expectation values of an odd number of Ferminons are zero, while $\langle 0|\phi_0|0\rangle = \frac{1}{\sqrt{2}}$. But $|0\rangle$ and $\langle 0|$ will be more convenient when discussing the BKP hierarchy, see [27]'s arXiv version for more details.

Denote $E_{i,j} = (\delta_{\mu,i}\delta_{\nu,j})_{\mu,\nu\in\mathbb{Z}}$, then one can define

$$o(\infty) = \Big\{ \sum_{i,j\in\mathbb{Z}} a_{i,j}E_{i,j} | a_{i,j} = (-1)^{i+j+1}a_{-j,-i}, \quad a_{i,j} = 0 \text{ for } |i-j| \gg 0 \Big\}.$$

Obviously, any $A \in o(\infty)$ can be written into

$$A = \sum_{i,j\in\mathbb{Z}} a_{i,j}Z_{i,j}, \quad Z_{i,j} = (-1)^j E_{i,-j} - (-1)^i E_{j,-i}. \tag{2.11}$$

Note that

$$\begin{aligned}[Z_{i,j}, Z_{i',j'}] =& (-1)^j\delta_{i'+j,0}Z_{i,j'} - (-1)^j\delta_{j+j',0}Z_{i,i'} \\ &+ (-1)^i\delta_{i+i',0}Z_{j',j} - (-1)^i\delta_{i+j',0}Z_{i',j},\end{aligned} \tag{2.12}$$

and by using the formula

$$[AB, C] = A[B,C]_+ - [A,C]_+B = A[B,C] + [A,C]B, \tag{2.13}$$

one can obtain

$$\begin{aligned}&[\phi_i\phi_j, \phi_{i'}\phi_{j'}] = \phi_i[\phi_j, \phi_{i'}\phi_{j'}] + [\phi_i, \phi_{i'}\phi_{j'}]\phi_j \\ =&\phi_i[\phi_j,\phi_{i'}]_+\phi_{j'} - \phi_i\phi_{i'}[\phi_j,\phi_{j'}]_+ + [\phi_i,\phi_{i'}]_+\phi_{j'}\phi_j - \phi_{i'}[\phi_i,\phi_{j'}]_+\phi_j \\ =&(-1)^j\delta_{i'+j,0}\phi_i\phi_{j'} - (-1)^j\delta_{j+j',0}\phi_i\phi_{i'} \\ &+ (-1)^i\delta_{i+i',0}\phi_{j'}\phi_j - (-1)^i\delta_{i+j',0}\phi_{i'}\phi_j.\end{aligned} \tag{2.14}$$

By comparing (2.12) with (2.14), one can find that $\phi_i\phi_j$ shares the same commutation relations with the matrices Z_{ij}. So we can identify $\phi_i\phi_j$ with Z_{ij}. Next, we will consider a larger algebra. For any $A = \sum_{i,j\in\mathbb{Z}} a_{ij}Z_{ij} \in o(\infty)$, let us consider

$$X_A = \sum_{i,j\in\mathbb{Z}} a_{i,j} : \phi_i\phi_j :, \tag{2.15}$$

where $:\phi_i\phi_j := \phi_i\phi_j - \langle\phi_i\phi_j\rangle$. According to the relations below,

$$\begin{aligned}
&[:\phi_i\phi_j :, :\phi_{i'}\phi_{j'}:]\\
=&(-1)^j\delta_{i'+j,0} : \phi_i\phi_{j'} : -(-1)^j\delta_{j+j',0} : \phi_i\phi_{i'} :\\
&+ (-1)^i\delta_{i+i',0} : \phi_{j'}\phi_j : -(-1)^i\delta_{i+j',0} : \phi_{i'}\phi_j :\\
&+ (-1)^{i+j}(\delta_{i+j',0}\delta_{i'+j,0} - \delta_{i+i',0}\delta_{j+j',0})(Y_B(-i) - Y_B(j)),
\end{aligned} \tag{2.16}$$

where $Y_B(n) = \begin{cases} 1, & n > 0; \\ \frac{1}{2}, & n = 0; \\ 0, & n < 0. \end{cases}$, the central extension of $o(\infty)$ can be defined as,

$$\mathfrak{o}(\infty) = \{X_A | A \in o(\infty)\} \oplus \mathbb{C}1. \tag{2.17}$$

Then the corresponding group of $\mathfrak{o}(\infty)$ is given by

$$O(\infty) = \{g | g = e^{X_1} \cdots e^{X_l}, X_i \in \mathfrak{o}(\infty)\}. \tag{2.18}$$

Lemma 1. *Assume $g \in O(\infty)$, then*

$$g\phi_n g^{-1} = \sum_m a_{m,n}\phi_m, \tag{2.19}$$

where a_{mn} satisfies

$$\sum_l (-1)^l a_{m,-l}a_{n,l} = (-1)^n\delta_{m+n,0}. \tag{2.20}$$

Proof. Firstly for any $B = (b_{i,j})_{i,j\in\mathbb{Z}}$,

$$[X_B, \phi_n] = \sum_m (-1)^n(b_{m,-n} - b_{-n,m})\phi_m. \tag{2.21}$$

Assume $g = e^{X_B}$, then by $e^A Be^{-A} = e^{adA}(B)$,

$$g\phi_n g^{-1} = \sum_m a_{m,n}\phi_m, \tag{2.22}$$

where $a = e^{\hat{b}}$ and $\hat{b}_{m,n} = (-1)^n(b_{m,-n} - b_{-n,m})$. Particularly, $\hat{b} \in o(\infty)$, which is equivalent to

$$J^{-1}\hat{b}^T J = -\hat{b}, \tag{2.23}$$

where $J=((-1)^i\delta_{i+j,0})_{i,j\in\mathbb{Z}}$ and $\hat{b}^T$ means the transpose of $\hat{b}$. Thus

$$J^{-1}a^TJ=\sum_{l=0}^{\infty}\frac{1}{l!}J^{-1}(\hat{b}^T)^lJ=\sum_{l=0}^{\infty}\frac{1}{l!}(-\hat{b})^l=e^{-\hat{b}}=a^{-1}, \tag{2.24}$$

which is just (2.20). ■

Consider the following operator on $\mathcal{F}\otimes\mathcal{F}$:

$$S=\sum_{j\in\mathbb{Z}}(-1)^j\phi_j\otimes\phi_{-j}, \tag{2.25}$$

Then we have the lemma below

Lemma 2. *If $g\in O(\infty)$, then*

$$[S,g\otimes g]=0. \tag{2.26}$$

Proof. By Lemma 1, there exists $(a_{ij})_{i,j\in\mathbb{Z}}$ satisfying (2.20) such that

$$\begin{aligned}(g\otimes g)\cdot S=&\sum_{j\in\mathbb{Z}}(-1)^jg\phi_j\otimes g\phi_{-j}\\ =&\sum_{jlm}(-1)^ja_{l,j}a_{m,-j}\phi_lg\otimes\phi_mg=\sum_{lm}(-1)^m\delta_{m+l,0}\phi_lg\otimes\phi_mg\\ =&\sum_{m}(-1)^m\phi_{-m}g\otimes\phi_mg=S\cdot(g\otimes g)\end{aligned}$$

■

Proposition 1. *[13, 6, 11, 29] If $|u\rangle\in\mathcal{F}$ and $|u\rangle\neq0$, then $|u\rangle\in O(\infty)|0\rangle$ if and only if*

$$S(|u\rangle\otimes|u\rangle)=\frac{1}{2}|u\rangle\otimes|u\rangle. \tag{2.27}$$

Proof. Firstly, it is obvious that (2.27) is correct for $|u\rangle=|0\rangle$ by (2.4). Then if $|u\rangle\in O(\infty)|0\rangle$, then there exists $g\in O(\infty)$ such that $|u\rangle=g|0\rangle$, then (2.27) can be proved by Lemma 2. As for the proof of the converse, the readers can refer to [13] for more details. ■

Equation (2.27) is called the bilinear equation of the Fermionic BKP hierarchy.

2.2 Boson-Fermion correspondence

Next we will rewrite (2.27) into the Bosonic forms. For this, let us introduce

$$\sum_{n\in\mathbb{Z}}H_nz^{-n-1}=\frac{1}{2z}:\phi(z)\phi(-z):. \tag{2.28}$$

By comparing the coefficients of the z^{-n-1},

$$H_n = \frac{1}{2}\sum_{i\in\mathbb{Z}}(-1)^i : \phi_{-i-n}\phi_i : . \tag{2.29}$$

It should be noted that $H_{2n} = 0$, since when $n \neq 0$,

$$\begin{aligned} H_{2n} =& \frac{1}{2}\sum_{i\in\mathbb{Z}}(-1)^i\phi_{-i-2n}\phi_i = -\frac{1}{2}\sum_{i\in\mathbb{Z}}(-1)^i\phi_i\phi_{-i-2n} \\ =& -\frac{1}{2}\sum_{j\in\mathbb{Z}}(-1)^{j+2n}\phi_{-j-2n}\phi_j = -H_{2n}, \end{aligned} \tag{2.30}$$

and

$$H_0 = -\frac{1}{2}\sum_{i>0}(-1)^i\phi_i\phi_{-i} + \frac{1}{2}\sum_{i<0}(-1)^i\phi_{-i}\phi_i = 0, \tag{2.31}$$

where we have set $j = -i - 2n$. Therefore, here we only discuss H_n with n odd. By (2.4), one can obtain

$$H_n|0\rangle = 0, \quad \langle 0|H_{-n} = 0, \quad n > 0. \tag{2.32}$$

Further, we can find that $\{H_{2n+1}\}_{n\in\mathbb{Z}}$ form a Heisenberg algebra. Actually, by using (2.16), one can find that for n and m odd (assume $n > 0$ firstly and $n < 0$ is similar),

$$\begin{aligned} &[H_n, H_m] = \frac{1}{4}\sum_{ij}(-1)^{i+j}[: \phi_{-i-n}\phi_i :, : \phi_{-j-m}\phi_j :] \\ =& -\frac{1}{2}\sum_{j}(-1)^j(: \phi_{-j-m}\phi_{j-n} : + : \phi_{-j-m-n}\phi_j :) \\ &+ \frac{1}{2}\delta_{m+n,0}\sum_{i}(Y_B(i+n) - Y_B(i)) \\ =& \frac{1}{2}\delta_{m+n,0}\cdot\left(Y_B(0) - Y_B(-n) + Y_B(n) - Y_B(0) + \sum_{i=-n+1}^{-1} 1\right) = \frac{1}{2}n\delta_{m+n,0}, \end{aligned} \tag{2.33}$$

Here we have changed j into $n + j$ in the term $: \phi_{-j-m}\phi_{j-n} :$. The reason we can do this is that $: \phi_{-j-m}\phi_{j-n} : a|0\rangle \neq 0$ holds only for finite j with $a \in \mathcal{A}$.

Further if for $t = (t_1 = x, t_3, t_5, \cdots)$ denote

$$H(t) = \sum_{n=0}^{\infty} t_{2n+1}H_{2n+1}, \tag{2.34}$$

then one can find

$$e^{H(t)}\phi(z)e^{-H(t)} = e^{\xi(t,z)}\phi(z), \tag{2.35}$$

where $\xi(t,z)=\sum_{n=0}^{\infty}t_{2n+1}z^{2n+1}$. In fact, according to

$$\begin{aligned}[H(t),\phi(z)]=&\frac{1}{2}\sum_{n\geq 0}\sum_{lm}t_{2n+1}z^m(-1)^l[:\phi_{-l-2n-1}\phi_l:,\phi_m]\\=&\frac{1}{2}\sum_{n\geq 0}\sum_{lm}t_{2n+1}z^m(-1)^l\Big(\phi_{-l-2n-1}[\phi_l,\phi_m]_+-[\phi_{-l-2n-1},\phi_m]_+\phi_l\Big)\\=&\sum_{n\geq 0}\sum_{m}t_{2n+1}z^m\phi_{m-2n-1}=\xi(t,z)\phi(z).\end{aligned}\tag{2.36}$$

Then (2.35) can be proved by $e^ABe^{-A}=e^{adA}(B)$. After the preparation above, now we give an isomorphism between $\mathcal{F}$ and $\mathcal{B}=\mathbb{C}[t_1,t_3,\cdots]$ in the theorem below.

Proposition 2. *The correspondence*

$$\sigma:\ \mathcal{F}\to\mathcal{B},\quad |u\rangle\mapsto\sigma(|u\rangle)=\langle 0|e^{H(t)}|u\rangle,\tag{2.37}$$

is an isomorphism. Further,

$$\sigma H_n\sigma^{-1}=\frac{\partial}{\partial t_n},\quad \sigma H_{-n}\sigma^{-1}=\frac{n}{2}t_n,\tag{2.38}$$

where $n>0$ and n is odd.

Proof. Firstly, because $[H_n,H_m]=0$ for $n,m>0$,

$$\frac{\partial}{\partial t_n}\sigma(|u\rangle)=\frac{\partial}{\partial t_n}\langle 0|e^{H(t)}|u\rangle=\langle 0|e^{H(t)}H_n|u\rangle=\sigma(H_n|u\rangle),$$

which is just the first relation of (2.38). As for the second relation of (2.38),

$$\sigma(H_{-n}|u\rangle)=\langle 0|e^{H(t)}H_{-n}|u\rangle=\langle 0|(H_{-n}+\frac{n}{2}t_n)e^{H(t)}|u\rangle=\frac{n}{2}t_n\sigma(|u\rangle),$$

where we have used (2.32) and $e^{H(t)}H_{-n}e^{-H(t)}=e^{adH(t)}(H_{-n})=H_{-n}+\frac{n}{2}t_n$.

Next, we try to prove σ is bijective. Note that $\sigma(|0\rangle)=1$. Thus the successive applications of H_{-n} $(n>0)$ and the corresponding images can give rise to any monomial in $\mathcal{B}$. So σ is obviously surjective. In order to show σ is injective, we introduce the degree of $\mathcal{F}$ in the way below,

$$\deg|0\rangle=0,\quad \deg\phi_j=j.$$

Then there is a direct sum decomposition of $\mathcal{F}$, that is,

$$\mathcal{F}=\bigoplus_{l\in\mathbb{Z}_{\geq 0}}\mathcal{F}_l,\quad \mathcal{F}_l=\{f\in\mathcal{F}|\deg f=l\}.\tag{2.39}$$

Note that $\mathcal{F}_l=span\{\phi_{n_1}\cdots\phi_{n_k}|0\rangle|n_1>\cdots>n_k>0,\quad n_1+\cdots n_k=l\}$. So the dimension of $\mathcal{F}_l$ is the number of the partition of l into distinct positive integers. If we define the formal character of $\mathcal{F}$ as

$$\dim_q\mathcal{F}=\sum_{l\in\mathbb{Z}_{\geq 0}}\dim\mathcal{F}_l\cdot q^l.\tag{2.40}$$

Since ϕ_j contributes q^j to the character of $\mathcal{F}$,

$$\dim_q \mathcal{F} = \prod_{j=1}^{\infty}(1+q^j) = \prod_{j=1}^{\infty}\frac{1-q^{2j}}{1-q^j}$$
$$= \frac{\prod_{j=1}^{\infty}(1-q^{2j})}{\prod_{j=1}^{\infty}(1-q^{2j})\prod_{j=0}^{\infty}(1-q^{2j+1})} = \prod_{j=0}^{\infty}(1-q^{2j+1})^{-1}. \quad (2.41)$$

Similarly, if we define $\deg t_j = j$, then

$$\mathcal{B} = \bigoplus_{l\in\mathbb{Z}} \mathcal{B}_l, \quad \mathcal{B}_l = \{u \in \mathcal{B} | \deg u = l\}. \quad (2.42)$$

The corresponding formal character of $\mathcal{B}$ is given by

$$\dim_q \mathcal{B} = \sum_{l\in\mathbb{Z}_{\geq 0}} \dim \mathcal{B}_l \cdot q^l = \prod_{j=0}^{\infty}(1-q^{2j+1})^{-1}, \quad (2.43)$$

where we have used $\dim \mathcal{B}_l$ is the number of the partitions of l into odd positive integers.

From (2.41) and (2.43), one can know $\dim \mathcal{F}_l = \dim \mathcal{B}_l$. So σ is injective by further considering the fact that σ is surjective. ■

Then the Boson-Fermion correspondence is given by the proposition below.

Proposition 3. *[6, 11, 13, 29]*

$$\phi(z) = \frac{1}{\sqrt{2}}\exp\left(\sum_{n>0,odd}\frac{2H_{-n}}{n}z^n\right)\exp\left(-\sum_{n>0,odd}\frac{2H_n}{n}z^{-n}\right). \quad (2.44)$$

To prove this proposition, we need the lemma below.

Lemma 3. *The following relation holds.*

$$\sqrt{2}\langle 0|\phi(z) = \langle 0|e^{-H(2\varepsilon(z^{-1}))}, \quad (2.45)$$

where $\varepsilon(z) = (z, \frac{z^3}{3}, \frac{z^5}{5}, \cdots)$.

Proof. In fact, one only needs to show

$$\sqrt{2}\langle 0|\phi(z)\phi(z_1)\cdots\phi(z_{m-1})|0\rangle = \langle 0|e^{-H(2\varepsilon(z^{-1}))}\phi(z_1)\cdots\phi(z_{m-1})|0\rangle. \quad (2.46)$$

According (2.8),

$$\text{LHS of (2.46)} = \frac{1}{2^{\frac{m-1}{2}}}\prod_{i=1}^{m-1}\frac{z-z_i}{z+z_i}\prod_{1\leq j<l\leq m-1}\frac{z_j-z_l}{z_j+z_l}.$$

While

$$\begin{aligned}\text{RHS of (2.46)} =& \prod_{i=1}^{m-1} e^{-\xi(2\varepsilon(z^{-1}),z_i)}\langle 0|\phi(z_1)\cdots\phi(z_{m-1})e^{-H(2\varepsilon(z^{-1}))}|0\rangle \\ =& \prod_{i=1}^{m-1} e^{-\xi(2\varepsilon(z^{-1}),z_i)}\langle 0|\phi(z_1)\cdots\phi(z_{m-1})|0\rangle \\ =&\frac{1}{2^{\frac{m-1}{2}}}\prod_{i=1}^{m-1}\frac{z-z_i}{z+z_i}\prod_{1\le j<l\le m-1}\frac{z_j-z_l}{z_j+z_l},\end{aligned}$$

which coincides with LHS of (2.46). Here we have used $e^{-\xi(2\varepsilon(z^{-1}),z')} = \frac{z-z'}{z+z'}$. ■

Now we give the proof of Proposition 3.

Proof. For any $|u\rangle \in \mathcal{F}$, by using (2.35), Proposition 2 and Lemma 3

$$\begin{aligned}&\sigma(\phi(z)|u\rangle) = \langle 0|e^{H(t)}\phi(z)|u\rangle \\ =&e^{\xi(t,z)}\langle 0|\phi(z)e^{H(t)}|u\rangle = \frac{1}{\sqrt{2}}e^{\xi(t,z)}e^{-2\xi(\tilde{\partial},z^{-1})}\langle 0|e^{H(t)}|u\rangle \\ =&\frac{1}{\sqrt{2}}\sigma\exp\left(\sum_{n>0,odd}\frac{2H_{-n}}{n}z^n\right)\exp\left(-\sum_{n>0,odd}\frac{2H_n}{n}z^{-n}\right)\sigma^{-1}\cdot\sigma(|u\rangle),\end{aligned}$$

where $\tilde{\partial} = (\frac{\partial}{\partial t_1}, \frac{1}{3}\frac{\partial}{\partial t_3}, \cdots)$. ■

By now, we have established the Boson-Fermion correspondence. Particularly,

$$\sigma\phi(z)\sigma^{-1} = \frac{1}{\sqrt{2}}e^{\xi(t,z)}e^{-2\xi(\tilde{\partial},z^{-1})}. \tag{2.47}$$

If apply $\langle 0|e^{H(t)} \otimes \langle 0|e^{H(t')}$ to the bilinear equations (2.27) and denote $\tau(t) = \sigma(|u\rangle) = \langle 0|e^{H(t)}|u\rangle$ with $|u\rangle = g|0\rangle$ and $g \in O(\infty)$, we will obtain by (2.35) and Lemma 3,

$$\begin{aligned}\frac{1}{2}\tau(t)\tau(t') =&\sum_j(-1)^j\langle 0|e^{H(t)}\phi_j g|0\rangle\langle 0|e^{H(t')}\phi_{-j}g|0\rangle \\ =&\text{Res}_z\frac{1}{z}\langle 0|e^{H(t)}\phi(z)g|0\rangle\langle 0|e^{H(t')}\phi(-z)g|0\rangle \\ =&\text{Res}_z\frac{1}{2z}e^{\xi(t-t',z)}\tau(t-2\varepsilon(z^{-1}))\tau(t'+2\varepsilon(z^{-1})).\end{aligned}$$

Here $\text{Res}_z\sum_i a_i z^i = a_{-1}$. Thus one has the proposition below.

Proposition 4. *If denote* $\tau(t) = \langle 0|e^{H(t)}g|0\rangle$ *with* $g \in O(\infty)$, *then*

$$\text{Res}_z\frac{1}{z}e^{\xi(t-t',z)}\tau(t-2\varepsilon(z^{-1}))\tau(t'+2\varepsilon(z^{-1})) = \tau(t)\tau(t'), \tag{2.48}$$

which is the bilinear equation of the BKP hierarchy in the Bosonic form.

2.3 Lax structure of BKP hierarchy

Next we will discuss the wave functions, dressing operators and the Lax equations of the BKP hierarchy. In this subsection, all n are odd.

For this, denote

$$w(t,z) = \frac{\tau(t - 2\varepsilon(z^{-1}))}{\tau(t)} e^{\xi(t,z)}, \tag{2.49}$$

then (2.48) can be converted into

$$\mathrm{Res}_z \frac{1}{z} w(t,z) w(t',z) = 1, \tag{2.50}$$

where $w(t,z)$ is called the wave function of the BKP hierarchy. Further, if we introduce a pseudo-differential operator

$$W = 1 + w_1\partial + w_2\partial^{-2} + w_3\partial^{-3} + \cdots, \quad \partial = \partial_x \tag{2.51}$$

such that $w(t,z) = W(e^{\xi(t,z)})$. W is called the dressing operator. Before further discussion, the lemma below is needed.

Lemma 4. *[8] If we let $A(x,\partial_x) = \sum_i a_i(x)\partial_x^i$ and $B(x',\partial_{x'}) = \sum_j b_j(x')\partial_{x'}^j$ be two pseudo-differential operators, then*

$$\mathrm{Res}_z A(x,\partial_x)(e^{xz}) \cdot B(x',\partial_{x'})(e^{-x'z}) = A(x,\partial_x)B^*(x,\partial_x)\partial_x(\Delta^0), \tag{2.52}$$

where $B^*(x,\partial_x) = \sum_j (-\partial_x)^j b_j(x)$, $\Delta^0 = (x-x')^0$ and

$$\partial_x^{-a}(\Delta^0) = \begin{cases} 0, & a < 0, \\ \frac{(x-x')^a}{a!}, & a \geq 0. \end{cases} \tag{2.53}$$

Denote $\hat{t} = (t_3, t_5, \cdots)$ and let $\hat{t}' = \hat{t}$, then by Lemma 4, (2.50) will become into

$$\begin{aligned} 1 =& \mathrm{Res}_z W(x,\hat{t},\partial_x)\partial_x^{-1}(e^{xz}) \cdot W(x',\hat{t},\partial_{x'})(e^{-x'z}) \\ =& W(t,\partial)\partial^{-1}W(t,\partial)^*\partial((x-x')^0), \end{aligned} \tag{2.54}$$

which tells us $(W\partial^{-1}W^*\partial)_{\leq 0} = 1$. Here $(\sum_i a_i\partial^i)_{\leq 0} = \sum_{i\leq 0} a_i\partial^i$. On the other hand, from the forms of W (see (2.51)), we know that $W\partial^{-1}W^*\partial = 1 + \mathcal{O}(\partial^{-1})$. Therefore,

$$W\partial^{-1}W^* = \partial^{-1}, \tag{2.55}$$

which is the BKP constraint on the dressing operator W.

If we further apply ∂_{t_n} (n is odd) on the (2.50) and let $\hat{t}' = \hat{t}$, then

$$\begin{aligned} 0 =& \mathrm{Res}_z \left(W(x,\hat{t},\partial_x)_{t_n} + W(x,\hat{t},\partial_x)\partial_x^n\right)\partial_x^{-1}(e^{xz}) \cdot W(x',\hat{t},\partial_{x'})(e^{-x'z}) \\ =& \left(W(t,\partial)_{t_n} + W(t,\partial)\partial^n\right)\partial^{-1}W(t,\partial)^*\partial((x-x')^0). \end{aligned}$$

Thus by (2.55) and (2.51)

$$0 = \left(\left(W_{t_n} + W\partial^n\right)\partial^{-1}W^*\partial\right)_{\leq 0} = W_{t_n}W^{-1} + \left(W\partial^nW^{-1}\right)_{\leq 0}.$$

Therefore, one can obtain the dressing equation below,

$$W_{t_n} = -\left(W\partial^nW^{-1}\right)_{\leq 0}W = -\left(W\partial^nW^{-1}\right)_{<0}W, \quad n = 1, 3, 5, \cdots, \tag{2.56}$$

where we have used the lemma below.

Lemma 5. *For odd number $n > 0$,*

$$(W\partial^nW^{-1})_{[0]} = 0, \tag{2.57}$$

where $(\sum_i a_i\partial^i)_{[0]} = a_0$.

Proof. From (2.55),

$$(W\partial^nW^{-1})^* = -(W^*)^{-1}\partial^nW^* = -\partial W\partial^nW^{-1}\partial^{-1}. \tag{2.58}$$

If assume $W\partial^nW^{-1} = \sum_{i\leq n} a_i\partial^i$, one can rewrite the above relation into

$$\sum_{i\leq n} a_i\partial^{i-1} = -\partial^{-1}\sum_{i\leq n}(-\partial)^i a_i. \tag{2.59}$$

By comparing the coefficients of ∂^{-1} of both sides in the above relation,

$$a_0 = -a_0 \Rightarrow a_0 = 0. \tag{2.60}$$

■

Remark 2. It should be noted that the evolution equation (2.56) of the dressing operator is consistent with (2.55). That is to say

$$0 = W_{t_n}\partial^{-1}W^* + W\partial^{-1}W^*_{t_n} = W_{t_n}W^{-1}\partial^{-1} + \left(W_{t_n}W^{-1}\partial^{-1}\right)^* \tag{2.61}$$

should still hold after substitution of (2.55), which is equivalent to

$$(W\partial^nW^{-1})^*_{\geq 0} = -\partial(W\partial^nW^{-1})_{\geq 0}\partial^{-1}. \tag{2.62}$$

From (2.55), one can obtain

$$(W\partial^nW^{-1})^* = -\partial(W\partial^nW^{-1})\partial^{-1}$$

whose positive order part with respect to ∂ is just (2.62) by Lemma 5.

Now we can introduce the Lax operator

$$L = W\partial W^{-1} = \partial + u_1\partial^{-1} + u_2\partial^{-2} + \cdots, \tag{2.63}$$

satisfying the following BKP constraint

$$L^* = -\partial L \partial^{-1}, \tag{2.64}$$

which can be obtained by (2.55). From this constraint, one can express $u_{2n} = f(u_1, u_3, \cdots, u_{2n-1})$ with f is the differential polynomial of u_i. For example,

$$u_2 = -u_{1x}, \quad u_4 = u_{1xxx} - 2u_{3x}, \cdots. \tag{2.65}$$

The Lax equation can be derived from the evolution equation (2.56) of the dressing operator W, which is

$$\begin{aligned} L_{t_n} =& W_{t_n} \partial W^{-1} - W \partial W^{-1} W_{t_n} W^{-1} \\ =& \left[W_{t_n} W^{-1}, W \partial W^{-1}\right] = -\left[\left(W \partial^n W^{-1}\right)_{<0}, W \partial W^{-1}\right] \\ =& -\left[(L^n)_{<0}, L\right] = \left[(L^n)_{\geq 0}, L\right], \end{aligned}$$

that is,

$$L_{t_n} = \left[(L^n)_{\geq 0}, L\right], \tag{2.66}$$

which is also consistent with the BKP constraint (2.64)(see Remark 2 for reference). By comparing the coefficients of ∂^{-i} $(i \geq 1)$, one can obtain many different differential equations. For example,

$$\begin{aligned} u_{1t_3} =& -2u_{1xxx} + 3u_{3x} + 6u_1 u_{1x}, \\ u_{3t_3} =& 3u_{1xxxxx} - 5u_{3xxx} + 3u_1 u_{1xxx} + 3u_1 u_{3x} \\ & + 3u_{5x} + 9u_3 u_{1x} + 9u_{1x} u_{1xx}, \\ u_{1t_5} =& 6u_{1xxxxx} - 10u_{3xxx} + 5u_{5x} - 10u_1 u_{1xxx} + 20u_1 u_{3x} \\ & + 20u_3 u_{1x} + 30u_1^2 u_{1x}. \end{aligned}$$

If eliminate u_3 and u_5, and set $u_1 = u$, $y = t_3$, $t = \frac{5}{9} t_5$, then

$$\begin{aligned} u_t =& \int u_{yy} dx + 3u_x \int u_y dx - \frac{1}{5} u_{xxxxx} - 3uu_{xxx} \\ & - 3u_x u_{xx} - 9u^2 u_x + u_{xxy} + 3uu_y, \end{aligned} \tag{2.67}$$

which is the BKP equation, and also called the $2+1$-dimensional Sawada-Kotera equation [15].

3 Constrained BKP hierarchy

In this section, we will further discuss the bilinear equations of the constrained BKP hierarchy and give the Fermionic description of the constrained BKP hierarchy.

3.1 Background on constrained BKP hierarchy

One important reduction of the BKP hierarchy is symmetry constraint [2, 5, 18, 23], defined by the following constraint on the Lax operator

$$L^k = (L^k)_{\geq 0} + q_1\partial^{-1}q_{2x} - q_2\partial^{-1}q_{1x}. \tag{3.1}$$

Here k is a fixed odd positive integer and $q_i(i = 1, 2)$ are two independent functions. In order to be consistent with the Lax equation (2.66), that is

$$((L^k)_{t_n})_{<0} = [(L^n)_{\geq 0}, L^k]_{<0},$$

q_1 and q_2 should satisfy

$$q_{i,t_n} = (L^n)_{\geq 0}(q_i), \tag{3.2}$$

since if L has the form (3.1),

$$((L^k)_{t_n})_{<0} = q_{1t_n}\cdot\partial^{-1}q_{2x} + q_1\partial^{-1}\cdot(q_{2t_n})_x - q_{2t_n}\cdot\partial^{-1}q_{1x} - q_2\partial^{-1}\cdot(q_{1t_n})_x$$

and

$$\begin{aligned}
[(L^n)_{\geq 0}, L^k]_{<0} =& [(L^n)_{\geq 0}, q_1\partial^{-1}q_{2x} - q_2\partial^{-1}q_{1x}] \\
=& (L^n)_{\geq 0}(q_1)\cdot\partial^{-1}q_{2x} + q_1\partial^{-1}\cdot((L^n)_{\geq 0}(q_2))_x \\
&- (L^n)_{\geq 0}(q_2)\cdot\partial^{-1}q_{1x} - q_2\partial^{-1}\cdot((L^n)_{\geq 0}(q_1))_x,
\end{aligned}$$

where we have used (2.62) and the lemma below.

Lemma 6. *[21] Given any differential operator A and any function $f(x)$,*

$$(Af\partial^{-1})_{<0} = A(f)\cdot\partial^{-1}, \quad (\partial^{-1}fA)_{<0} = \partial^{-1}\cdot A^*(f). \tag{3.3}$$

If one function q satisfies (3.2), then q is called the eigenfunction. The system of (2.66), (3.1) and (3.2) is called the constrained BKP hierarchy. Here we give some examples of the constrained BKP hierarchy.

$$\begin{aligned}
q_{1t_3} =& 3q_1q_{2x}q_{1x} - 3q_2q_{1x}^2 + q_{1xxx}, \\
q_{2t_3} =& 3q_1q_{2x}^2 - 3q_2q_{2x}q_{1x} + q_{2xxx}.
\end{aligned}$$

and

$$\begin{aligned}
q_{1t_5} =& 10q_1^2q_{2x}^2q_{1x} - 20q_1q_{2x}q_2q_{1x}^2 + 10q_2^2q_{1x}^3 + 5q_{2xx}q_{1xx}q_1 - 5q_{1xx}^2q_2 \\
&+ 5q_{1xx}q_1q_{2x} - 10q_{1xxx}q_2q_{1x} + 5q_1q_{1x}q_{2xxx} + q_{1xxxxx}, \\
q_{2t_5} =& 10q_1^2q_{2x}^3 - 20q_1q_{2x}^2q_2q_{1x} + 10q_{2x}q_2^2q_{1x}^2 + 5q_{2xx}^2q_1 - 5q_{2xx}q_{1xx}q_2 \\
&- 5q_{1xxx}q_{2x}q_2 + 10q_1q_{2x}q_{2xxx} - 5q_2q_{1x}q_{2xxx} + q_{2xxxxx}.
\end{aligned}$$

The bilinear equations of the constrained BKP hierarchy are given in the proposition below.

Proposition 5. *[2, 18, 23] The constrained BKP hierarchy, i.e., the system of (2.66), (3.1) and (3.2), is equivalent to the following bilinear equations:*

$$\mathrm{Res}_z z^{k-1} w(t,z) w(t',-z) = q_1(t) q_2(t') - q_2(t) q_1(t') \tag{3.4}$$

$$\mathrm{Res}_z z^{-1} w(t,z) \Omega(q_i(t'), w(t',-z)_{x'}) = q_i(t), \quad i = 1, 2, \tag{3.5}$$

where $\Omega(q_i(t), w(t,-z)_x) = \int q_i(t) w(t,-z)_x dx$.

Further, if we denote $\rho_i(t) = q_i(t)\tau(t)$, one can obtain the proposition [23] below.

Proposition 6. *[23]* τ, ρ_1 *and* ρ_2 *satisfy the following bilinear equations:*

$$\mathrm{Res}_z z^{k-1} \tau(t - 2\varepsilon(z^{-1})) \tau(t' + 2\varepsilon(z^{-1})) e^{\xi(t-t',z)} = \rho_1(t)\rho_2(t') - \rho_2(t)\rho_1(t'), \tag{3.6}$$

$$\mathrm{Res}_z z^{-1} \tau(t - 2\varepsilon(z^{-1})) \rho_i(t' + 2\varepsilon(z^{-1})) e^{\xi(t-t',z)} = 2\rho_i(t)\tau(t') - \rho_i(t')\tau(t), \quad i = 1, 2. \tag{3.7}$$

In [23], the authors have obtained that for the constrained BKP hierarchy (2.66), (3.1) and (3.2), the bilinear equations (3.6) and (3.7) of τ, ρ_1 and ρ_2 are always correct. However, if (3.6) and (3.7) are known, could we obain the constrained BKP hierarchy (2.66), (3.1) and (3.2)? So next we will investigate whether (3.6) and (3.7) are equivalent to the constrained BKP hierarchy.

3.2 Further discussions on bilinear equations of constrained BKP hierarchy

Lemma 7. *Given two functions* $\rho(t)$ *and* $\tau(t)$*, if they satisfy the following bilinear equation*

$$\mathrm{Res}_z z^{-1} \tau(t - 2\varepsilon(z^{-1})) \rho(t' + 2\varepsilon(z^{-1})) e^{\xi(t-t',z)} = 2\rho(t)\tau(t') - \rho(t')\tau(t), \tag{3.8}$$

then

$$\left(\frac{\rho(t \pm 2\varepsilon(\lambda^{-1}))}{\tau(t)} e^{\mp\xi(t,\lambda)}\right)_x = \frac{\rho(t)^2}{\tau(t)^2} \left(\frac{\tau(t \pm 2\varepsilon(\lambda^{-1}))}{\rho(t)} e^{\mp\xi(t,\lambda)}\right)_x. \tag{3.9}$$

Further

$$\left(\frac{\rho(t)}{\tau(t)}\right)_x \frac{\tau(t - 2\varepsilon(\lambda^{-1}))}{\tau(t)} e^{\xi(t,\lambda)} = \left(\frac{\rho(t)\tau(t - 2\varepsilon(\lambda^{-1})) - \rho(t - 2\varepsilon(\lambda^{-1}))\tau(t)}{2\tau(t)^2} e^{\xi(t,\lambda)}\right)_x, \tag{3.10}$$

$$\frac{\rho(t)}{\tau(t)} \left(\frac{\tau(t + 2\varepsilon(\lambda^{-1}))}{\tau(t)} e^{-\xi(t,\lambda)}\right)_x = \left(\frac{\rho(t + 2\varepsilon(\lambda^{-1}))\tau(t) + \rho(t)\tau(t + 2\varepsilon(\lambda^{-1}))}{2\tau(t)^2} e^{-\xi(t,\lambda)}\right)_x. \tag{3.11}$$

Proof. Firstly, apply ∂_x on (3.8) and set $t-t'=2\varepsilon(\lambda^{-1})$,

$$\begin{aligned}&\mathrm{Res}_z z^{-1}\left(\tau(t-2\varepsilon(z^{-1}))_x+z\tau(t-2\varepsilon(z^{-1}))z\right)\\&\quad\times\rho(t-2\varepsilon(\lambda^{-1})+2\varepsilon(z^{-1}))\frac{1+z/\lambda}{1-z/\lambda}\\&=2\rho(t)_x\tau(t-2\varepsilon(\lambda^{-1}))-\rho(t-2\varepsilon(\lambda^{-1}))\tau(t)_x.\end{aligned}\tag{3.12}$$

Note that the residue should be computed in the circle satisfying $|z|<|\lambda|$ with clockwise. So if we denote LHS of (3.12) as $\mathrm{Res}_z f(z)$, then

$$\mathrm{Res}_z f(z)=-\mathrm{Res}_{z=\lambda}f(z)-\mathrm{Res}_{z=\infty}f(z).$$

To compute $\mathrm{Res}_{z=\infty}f(z)$, one needs

$$\begin{aligned}-\frac{1}{\mu^2}f(\mu^{-1})=&\frac{1}{\mu}\Big(\tau(t-2\varepsilon(\mu))_x+\frac{1}{\mu}\tau(t-2\varepsilon(\mu))\Big)\\&\times\rho(t-2\varepsilon(\lambda^{-1})+2\varepsilon(\mu))\frac{1+\lambda\mu}{1-\lambda\mu}\triangleq\frac{1}{\mu}g_1(\mu)+\frac{1}{\mu^2}g_2(\mu)\end{aligned}$$

Therefore

$$\begin{aligned}&\mathrm{Res}_{z=\infty}f(z)=-\mathrm{Res}_{\mu=0}\frac{1}{\mu^2}f(\mu^{-1})\\&=\mathrm{Res}_{\mu=0}\left(\frac{1}{\mu}g_1(\mu)+\frac{1}{\mu^2}g_2(\mu)\right)=g_1(0)+\lim_{\mu\to0}\frac{dg_2(\mu)}{d\mu}\\&=-\tau(t)_x\rho(t-2\varepsilon(\lambda^{-1}))+2\tau(t)\rho(t-2\varepsilon(\lambda^{-1}))_x+2\lambda\tau(t)\rho(t-2\varepsilon(\lambda^{-1}))\end{aligned}$$

and

$$\mathrm{Res}_{z=\lambda}f(z)=-2\left(\tau(t-2\varepsilon(\lambda^{-1}))_x+\lambda\tau(t-2\varepsilon(\lambda^{-1}))\right)\rho(t)$$

By summarizing the results above, one can obtain

$$\begin{aligned}0=&\rho(t)_x\tau(t-2\varepsilon(\lambda^{-1}))-\tau(t-2\varepsilon(\lambda^{-1}))_x\rho(t)\\&+\rho(t-2\varepsilon(\lambda^{-1}))_x\tau(t)-\rho(t-2\varepsilon(\lambda^{-1}))\tau(t)_x\\&+\lambda\Big(\rho(t-2\varepsilon(\lambda^{-1}))\tau(t)-\rho(t)\tau(t-2\varepsilon(\lambda^{-1}))\Big),\end{aligned}$$

which leads to (3.9). As for (3.10), they are just the equivalent forms of (3.9) by direct computations. (3.11) can be obtained by (3.10). ■

Proposition 7. *Under the same conditions of Lemma 7, $\rho(t)$ and $\tau(t)$ are the tau functions of the BKP hierarchy, that is*

$$\begin{aligned}&\mathrm{Res}_z z^{-1}\rho(t-2\varepsilon(z^{-1}))\rho(t'+2\varepsilon(z^{-1}))e^{\xi(t-t',z)}=\rho(t)\rho(t'),\\&\mathrm{Res}_z z^{-1}\tau(t-2\varepsilon(z^{-1}))\tau(t'+2\varepsilon(z^{-1}))e^{\xi(t-t',z)}=\tau(t)\tau(t').\end{aligned}$$

Proof. Firstly, divide (3.8) by $\rho(t)\rho(t')$ and apply ∂_x on both sides,

$$\mathrm{Res}_z z^{-1}\left(\frac{\tau(t-2\varepsilon(z^{-1}))}{\rho(t)}e^{\xi(t,z)}\right)_x \frac{\rho(t'+2\varepsilon(z^{-1}))}{\rho(t')}e^{-\xi(t',z)} = -\left(\frac{\tau(t)}{\rho(t)}\right)_x.$$

By using (3.9),

$$\mathrm{Res}_z z^{-1}\left(\frac{\rho(t-2\varepsilon(z^{-1}))}{\tau(t)}e^{\xi(t,z)}\right)_x \frac{\rho(t'+2\varepsilon(z^{-1}))}{\rho(t')}e^{-\xi(t',z)} = \left(\frac{\rho(t)}{\tau(t)}\right)_x. \quad (3.13)$$

If we denote $q(t) = \frac{\rho(t)}{\tau(t)}$, $W = \frac{\rho(t-2\varepsilon(\partial^{-1}))}{\rho(t)}$ and let $\hat{t} = \hat{t}'$, one can further obtain by Lemma 4,

$$\begin{aligned} q(t)_x =& \mathrm{Res}_z\left(\partial_x q(t)W(t,\partial_x)\partial_x^{-1}\right)(e^{xz}) \cdot \left(W(t',\partial_{x'})\right)(e^{-x'z}) \\ =& \Big(\partial_x q(t)W(t,\partial_x)\partial_x^{-1}W(t,\partial_x)^*\partial_x\Big)\Big((x-x')^0\Big). \end{aligned}$$

Therefore by using $W = 1 + \mathcal{O}(\partial^{-1})$,

$$(\partial qW\partial^{-1}W^*\partial)_{\leq 0} = q_x \Rightarrow \partial qW\partial^{-1}W^*\partial = q_x + q\partial = \partial q,$$

which is just the BKP constraint on the dressing operator, i.e., $W\partial^{-1}W^* = \partial^{-1}$. Next apply ∂_{t_n} (n is odd) on the (3.13), let $\hat{t}' = \hat{t}$ and use Lemma 4

$$q_{xt_n} = \Big((\partial qW)_{t_n} + \partial qW\partial^n\Big)\Big(\partial^{-1}W^*\partial\Big)\Big((x-x')^0\Big).$$

Therefore, by the BKP constraint (2.55),

$$\begin{aligned} q_{xt_n} =& (\partial qW)_{t_n}W^{-1} - \Big((\partial qW)_{t_n}W^{-1}\Big)_{>0} + \Big(\partial qW\partial^n W^{-1}\Big)_{\leq 0} \\ =& (\partial qW)_{t_n}W^{-1} - q_{t_n}\partial + \Big(\partial qW\partial^n W^{-1}\Big)_{\leq 0}. \end{aligned}$$

Further by Lemma 5,

$$\begin{aligned} \partial qW_{t_n}W^{-1} =& -\Big(\partial qW\partial^n W^{-1}\Big)_{\leq 0} \\ =& -\Big(\partial q\Big(W\partial^n W^{-1}\Big)_{<0}\Big)_{\leq 0} = -\partial q\Big(W\partial^n W^{-1}\Big)_{<0}. \end{aligned}$$

Therefore

$$W_{t_n} = -\Big(W\partial^n W^{-1}\Big)_{<0}W, \quad (3.14)$$

which is the evolution equation of the dressing operator. Thus we have obtained the complete data of the BKP hierarchy (2.55) and (2.56).

But in order to prove $\rho(t)$ is the tau function of the BKP hierarchy, one needs to further discuss. For this, define the wave function $w(t,z) = W(e^{\xi(t,z)}) = \frac{\rho(t-2\varepsilon(z^{-1}))}{\rho(t)}e^{\xi(t,z)}$, then

$$w(t,\lambda)_{t_n} = (W\partial^n W^{-1})_{\geq 0}(w(t,\lambda)) = (W\partial^n W^{-1})_{\geq 1}(w(t,\lambda)). \quad (3.15)$$

Therefore, $\partial_t^\alpha w(x',\hat{t},-\lambda)$ can be written in the following way

$$\partial_t^\alpha w(x',\hat{t},-\lambda)=P_\alpha(x',\hat{t})\big(w(x',\hat{t},-\lambda)\big), \tag{3.16}$$

where $\partial_t^\alpha=\prod_{n=1}^{\infty}\partial_{t_{2n+1}}^{\alpha_{2n+1}}$, $\alpha=(\alpha_1,\alpha_3,\cdots)$ and $P_\alpha(x',\hat{t})=\sum_{i\geq 1}a_{\alpha,i}(x',\hat{t})\partial_{x'}^i$ is a differential operator. Particularly, $P_0=1$. Next by considering the Taylor expansion of w with respect to $\hat{t}'$ at $\hat{t}$

$$w(x',\hat{t}',-\lambda)=\sum_{\alpha=(\alpha_3,\alpha_5,\cdots)\geq 0}\frac{(\hat{t}'-\hat{t})^\alpha}{\alpha!}\partial_t^\alpha w(x',\hat{t},-\lambda), \tag{3.17}$$

with

$$(\hat{t}'-\hat{t})^\alpha=\prod_{n=1}^{\infty}(t'_{2n+1}-t_{2n+1})^{\alpha_{2n+1}},\quad \alpha!=\prod_{n=1}^{\infty}\alpha_{2n+1}!,$$
$$\alpha\geq 0\quad\Leftrightarrow\quad \alpha_{2n+1}\geq 0,\ n=1,2,\cdots.$$

Now by Lemma 4 and (2.55),

$$\begin{aligned}
&\mathrm{Res}_\lambda\frac{1}{\lambda}w(t,\lambda)w(t',-\lambda)\\
=&\mathrm{Res}_\lambda\frac{1}{\lambda}\sum_{\alpha\geq 0}\frac{(\hat{t}'-\hat{t})^\alpha}{\alpha!}w(t,\lambda)P_\alpha(x',\hat{t})(w(x',\hat{t},-\lambda))\\
=&\mathrm{Res}_\lambda\sum_{\alpha\geq 0}\frac{(\hat{t}'-\hat{t})^\alpha}{\alpha!}W(x,\hat{t},\partial_x)\partial_x^{-1}(e^{\lambda x})\cdot P_\alpha(x',\hat{t})W(x',\hat{t},\partial_x')(e^{-\lambda x'})\\
=&\sum_{\alpha\geq 0}\frac{(\hat{t}'-\hat{t})^\alpha}{\alpha!}W\partial^{-1}W^*P_\alpha(t)^*\partial\Big((x-x')^0\Big)\\
=&\sum_{\alpha\geq 0}\frac{(\hat{t}'-\hat{t})^\alpha}{\alpha!}\partial^{-1}P_\alpha(t)^*\partial\Big((x-x')^0\Big)=1,
\end{aligned}$$

where one should note that $\partial^{-1}P_\alpha(t)^*$ is a differential operator, except $\alpha=0$.

At last, by introducing one form

$$\omega=\sum_{n\geq 1}dt_{2n-1}\cdot\mathrm{Res}_z z^{2n-1}\Big(-\sum_{m=1}z^{-2m}\partial_{t_{2m-1}}+\frac{1}{2}\partial_z\Big)\log\hat{w}(t,z), \tag{3.18}$$

where $\hat{w}(t,z)=w(t,z)e^{-\xi(t,z)}$. It can be proved that ω is closed (see [6, 7] for more details). Thus there exists a function $f(t)$, called the tau function of the BKP hierarchy, such that

$$d\log f(t)=\omega,\quad w(t,z)=\frac{f(t-2\varepsilon(z^{-1}))}{f(t)}e^{\xi(t,z)}. \tag{3.19}$$

Recall that $w(t,z)=\frac{\rho(t-2\varepsilon(z^{-1}))}{\rho(t)}e^{\xi(t,z)}$, thus $\rho(t)=cf(t)$ with c constant. So $\rho(t)$ is a tau function of the BKP hierarchy.

By (3.8) and $\mathrm{Res}_z h(-z) = -\mathrm{Res}_z h(z)$, one can find

$$\mathrm{Res}_z z^{-1}\rho(t-2\varepsilon(z^{-1}))\tau(t'+2\varepsilon(z^{-1}))e^{\xi(t-t',z)} = 2\tau(t)\rho(t') - \tau(t')\rho(t). \quad (3.20)$$

So by similar methods, one can also show $\tau(t)$ is still a tau function of the BKP hierarchy. ■

By Lemma 7 and Proposition 7, one can show the inverse of Proposition 6 is also correct, that is the theorem below.

Theorem 1. *If we assume that $\tau(t)$, $\rho_1(t)$ and $\rho_2(t)$ satisfy (3.6)(3.7), and denote $\psi(t,\lambda) = \frac{\tau(t-2\varepsilon(\lambda^{-1}))}{\tau(t)}e^{-\xi(t,\lambda)}$, $q_i(t) = \frac{\rho_i(t)}{\tau(t)}$ with $i = 1,2$, then (3.4) and (3.5) still hold, which implies the constrained BKP hierarchy (2.66), (3.2) and (3.1).*

Proof. (3.4) is obvious by (3.6). From (3.11), one can know that

$$\begin{aligned} S(q_i(t), w(t,-z)_x) =& \frac{\rho_i(t+2\varepsilon(\lambda^{-1}))\tau(t) + \rho_i(t)\tau(t+2\varepsilon(\lambda^{-1}))}{2\tau(t)^2}e^{-\xi(t,\lambda)} \\ =& \frac{\rho_i(t+2\varepsilon(\lambda^{-1}))}{2\tau(t)}e^{-\xi(t,\lambda)} + \frac{1}{2}q_i(t)w(t,-z) \end{aligned}$$

Then divide (3.7) by $\tau(t)\tau(t')$,

$$\mathrm{Res}_z z^{-1}w(t,z)\Big(2\Omega(q_i(t'), w(t',-z)_{x'}) - q_i(t')w(t',-z)\Big) = 2q_i(t) - q_i(t').$$

Further according to Proposition 7, one can obtain (3.5). ■

3.3 Constrained BKP hierarchy in the Fermionic picture

From the discussion above, (3.6) and (3.7) can fully determine the constrained BKP hierarchy. Here in this section, we will give a Fermionic description of the constrained BKP hierarchy based upon (3.6) and (3.7) in the theorem below.

Theorem 2. *For the constrained BKP hierarchy (3.6) and (3.7), if denote $|u\rangle = \sigma^{-1}(\tau(t))$ and $|f_i\rangle = \sigma^{-1}(\rho_i)$ with $i = 1,2$, then*

$$\sum_{j\in\mathbb{Z}}(-1)^j\phi_j|u\rangle\otimes\phi_{-j}|u\rangle = \frac{1}{2}|u\rangle\otimes|u\rangle, \quad (3.21)$$

$$\sum_{j\in\mathbb{Z}}(-1)^j\phi_j|u\rangle\otimes\phi_{-j}|f_i\rangle = |f_i\rangle\otimes|u\rangle - \frac{1}{2}|u\rangle\otimes|f_i\rangle, \quad (3.22)$$

$$\sum_{j\in\mathbb{Z}}(-1)^{j+k}\phi_j|u\rangle\otimes\phi_{-k-j}|u\rangle = \frac{1}{2}|f_1\rangle\otimes|f_2\rangle - \frac{1}{2}|f_2\rangle\otimes|f_1\rangle. \quad (3.23)$$

Proof. Firstly, by Proposition 2 and Lemma 3

$$\begin{aligned} &\tau(t\pm 2\varepsilon(z^{-1}))e^{\mp\xi(t,z)} = \sqrt{2}\sigma(\phi(\mp z)|u\rangle), \\ &\rho_i(t'\pm 2\varepsilon(z^{-1}))e^{\mp\xi(t,z)} = \sqrt{2}\sigma(\phi(\mp z)|f_i\rangle). \end{aligned}$$

Then (3.6) and (3.7) can be rewritten into (3.23) and (3.22), respectively. (3.21) can be obtained from the fact τ is a tau function of BKP hierarchy, according to Proposition 7. ■

From Proposition 1, we can know that (3.21) is equivalent to $|u\rangle \in O(\infty)|0\rangle$. The interpretations of (3.22) and (3.23) are given in the theorem below.

Theorem 3. *Let $|u\rangle \in O(\infty)|0\rangle$, then*

$$(3.22) \Leftrightarrow (1) \text{ there exists } \alpha_i \in V \text{ such that } |f_i\rangle = \alpha_i|u\rangle.$$

$$\Leftrightarrow (2) \sum_{j\in\mathbb{Z}}(-1)^j\phi_j|f_i\rangle \otimes \phi_{-j}|u\rangle = |u\rangle \otimes |f_i\rangle - \frac{1}{2}|f_i\rangle \otimes |u\rangle.$$

Proof. (3.22)⇒(1). Firstly assume $|u\rangle = |0\rangle$. If (3.22) holds, then

$$\sum_{j\geq 0}(-1)^j\phi_j|0\rangle \otimes \phi_{-j}|f_i\rangle = |f_i\rangle \otimes |0\rangle - \frac{1}{2}|0\rangle \otimes |f_i\rangle. \tag{3.24}$$

Since all $\phi_j|0\rangle$ $(j \geq 0)$ are independent, thus there exists $\alpha_i = \sum_{j\geq 0} c_j\phi_j \in V$ such that $|f_i\rangle = \alpha_i|0\rangle$.

When $|u\rangle = g|0\rangle$, by considering that $S = \sum_{>0}(-1)^j\phi_j \otimes \phi_{-j}$ can commute with $g \otimes g$ (see Lemma 2), one can obtain

$$\sum_{j\geq 0}(-1)^j\phi_j|0\rangle \otimes \phi_{-j}g^{-1}|f_i\rangle = g^{-1}|f_i\rangle \otimes |0\rangle - \frac{1}{2}|0\rangle \otimes g^{-1}|f_i\rangle. \tag{3.25}$$

Therefore, there exists $\beta_i \in V$ such that $g^{-1}|f_i\rangle = \beta_i|0\rangle$. Further by Lemma 1,

$$|f_i\rangle = g\beta_i|0\rangle = \alpha_i g|0\rangle,$$

for some $\alpha_i \in V$.

(1)⇒(3.22). If $|f_i\rangle = \alpha_i|u\rangle$ with $\alpha_i = \sum_l a_l\phi_l$, then

$$\sum_j(-1)^j\phi_j|u\rangle \otimes \phi_{-j}|f_i\rangle = \sum_j(-1)^j\phi_j|u\rangle \otimes \Big((\alpha_i, \phi_{-j}) - \alpha_i\phi_{-j}\Big)|u\rangle$$

$$= \alpha_i|u\rangle \otimes |u\rangle - \Big(1 \otimes \alpha_i\Big) \cdot \sum_j(-1)^j\phi_j|u\rangle \otimes \phi_{-j}|u\rangle$$

$$= \alpha_i|u\rangle \otimes |u\rangle - \frac{1}{2}|u\rangle \otimes \alpha_i|u\rangle = |f_i\rangle \otimes |u\rangle - \frac{1}{2}|u\rangle \otimes |f_i\rangle$$

(2)⇒(1) is the same as (3.22)⇒(1), while (1)⇒(2) is similar to (1)⇒(3.22). ■

Corollary 1. For two functions $\tau(t)$ and $\rho(t)$,

$$\mathrm{Res}_z z^{-1}\tau(t - 2\varepsilon(z^{-1}))\rho(t' + 2\varepsilon(z^{-1}))e^{\xi(t-t',z)} = 2\rho(t)\tau(t') - \rho(t')\tau(t),$$

is equivalent to

$$\mathrm{Res}_z z^{-1}\rho(t - 2\varepsilon(z^{-1}))\tau(t' + 2\varepsilon(z^{-1}))e^{\xi(t-t',z)} = 2\tau(t)\rho(t') - \tau(t')\rho(t).$$

Remark 3. This corollary can also be easily proved by the fact $\mathrm{Res}_z f(z) = -\mathrm{Res}_z f(-z)$.

Theorem 4. *Given $g \in O(\infty)$ satisfying (2.19) and two independent vectors $\alpha_i \in V$ with $i = 1, 2$, assume $|u\rangle = g|0\rangle$ and $f_i = \alpha_i|u\rangle$, then*

$$(3.23) \Leftrightarrow \text{ there exist constants } c_{1l} \text{ and } c_{2l} \text{ such that } \tilde{\phi}_{-l} - (c_{1l}\alpha_1 + c_{2l}\alpha_2) \in Ann(|u\rangle), \quad l \geq 0,$$

where $\tilde{\phi}_{-l} = \sum_j a_{k+j,-l}\phi_j$ with the matrix (a_{ij}) given in (2.19) and $Ann(|u\rangle) = \{\alpha|\alpha \in V, \ \alpha|u\rangle = 0\}$.

Proof. Firstly, if (3.23) holds, then by (2.19) and $a^{-1} = J^{-1}a^T J$ (i.e., $(a^{-1})_{i,j} = (-1)^{i+j}a_{-j,-i}$),

$$\begin{aligned}\sum_{j\in\mathbb{Z}}(-1)^{j+k}\phi_j|u\rangle \otimes \phi_{-k-j}|u\rangle &= \sum_{j,l\in\mathbb{Z}}(-1)^{j+k}(a^{-1})_{l,-k-j}\phi_j g|0\rangle \otimes g\phi_l|0\rangle \\ &= \sum_{j,l\in\mathbb{Z}}(-1)^l a_{k+j,-l}\phi_j g|0\rangle \otimes g\phi_l|0\rangle \\ &= \sum_{l\geq 0}(-1)^l\tilde{\phi}_{-l}g|0\rangle \otimes g\phi_l|0\rangle \\ &= \frac{1}{2}|f_1\rangle \otimes |f_2\rangle - \frac{1}{2}|f_2\rangle \otimes |f_1\rangle.\end{aligned}$$

By considering the vectors $g\phi_l|0\rangle$ with $l \geq 0$ are linearly independent and $f_i = \alpha_i|u\rangle$, we can conclude that there exist constants c_{1l} and c_{2l} such that

$$\tilde{\phi}_{-l}g|0\rangle = c_{1l}|f_1\rangle + c_{2l}|f_2\rangle \Rightarrow \tilde{\phi}_{-l} - (c_{1l}\alpha_1 + c_{2l}\alpha_2) \in Ann(|u\rangle), \quad l \geq 0.$$

Conversely, assume $\tilde{\phi}_{-l} - (c_{1l}\alpha_1 + c_{2l}\alpha_2) \in Ann(|u\rangle)$ for $l \geq 0$. Here c_{1l} and c_{2l} will be determined later. Then by (2.19)

$$\begin{aligned}&\sum_{j\in\mathbb{Z}}(-1)^{j+k}\phi_j|u\rangle \otimes \phi_{-k-j}|u\rangle = \sum_{l\geq 0}(-1)^l\tilde{\phi}_{-l}g|0\rangle \otimes g\phi_l|0\rangle \\ &= |f_1\rangle \otimes \sum_{l\geq 0}\sum_{m\in\mathbb{Z}}(-1)^l c_{1l}a_{m,l}\phi_m g|0\rangle + |f_2\rangle \otimes \sum_{l\geq 0}\sum_{m\in\mathbb{Z}}(-1)^l c_{2l}a_{m,l}\phi_m g|0\rangle.\end{aligned}$$

If we assume $\alpha_1 = \sum_n A_n\phi_n$ and $\alpha_2 = \sum_n B_n\phi_n$, then by comparing $\frac{1}{2}|f_1\rangle \otimes |f_2\rangle - \frac{1}{2}|f_2\rangle \otimes |f_1\rangle$, one can obtain

$$\sum_{l\geq 0}(-1)^l c_{1l}a_{m,l} = \frac{1}{2}B_m, \quad \sum_{l\geq 0}(-1)^l c_{2l}a_{m,l} = -\frac{1}{2}A_m,$$

which implies to

$$c_{1l} = \frac{(-1)^l}{2}\sum_{m\in\mathbb{Z}}(a^{-1})_{l,m}B_m, \quad c_{2l} = -\frac{(-1)^l}{2}\sum_{m\in\mathbb{Z}}(a^{-1})_{l,m}A_m.$$

■

4 Modified BKP hierarchy

In Section 3, we have discussed the bilinear equations of the constrained BKP hierarchy. In fact, among the corresponding bilinear equations (3.6) and (3.7), we would like to point out that (3.7) is the bilinear equation of the modified BKP hierarchy [11, 13, 28]. Next in this section, we will investigate the dressing operators, the dressing equations and the Lax equations of the modified BKP hierarchy, and relations between the modified BKP hierarchy and the BKP hierarchy. In this section, all n are odd.

4.1 Dressing structure and Lax equation

For this, our starting point is

$$\mathrm{Res}_z z^{-1}\tau(t-2\varepsilon(z^{-1}))\rho(t'+2\varepsilon(z^{-1}))e^{\xi(t-t',z)} = 2\rho(t)\tau(t') - \rho(t')\tau(t). \quad (4.1)$$

Proposition 8. *(4.1) is equivalent to*

$$\mathrm{Res}_z z^{-1}\frac{\tau(t'-2\varepsilon(z^{-1}))}{\rho(t')}e^{\xi(t',z)}\int\frac{\rho(t)}{\tau(t)}\Big(\frac{\tau(t+2\varepsilon(z^{-1}))}{\tau(t)}e^{-\xi(t,z)}\Big)_x dx = 1. \quad (4.2)$$

Proof. Firstly, if $\tau(t)$ and $\rho(t)$ satisfy (4.1), by using (3.11) and Proposition 7, one can obtain

$$\begin{aligned}\text{LHS of (4.2)} =& \frac{1}{2\rho(t')\tau(t)^2}\mathrm{Res}_z z^{-1}e^{\xi(t'-t,z)}\tau(t'-2\varepsilon(z^{-1}))\\ &\times\Big(\rho(t+2\varepsilon(z^{-1}))\tau(t)+\rho(t)\tau(t+2\varepsilon(z^{-1}))\Big)\\ =&\frac{\Big(2\rho(t')\tau(t)-\tau(t')\rho(t)\Big)\tau(t)+\rho(t)\tau(t)\tau(t')}{2\rho(t')\tau(t)^2}=1.\end{aligned}$$

Conversely if (4.2) is correct, then we try to prove (4.1). For this, we firstly show $\tau(t)$ is the tau function of the BKP hierarchy. In fact, if we differentiate both sides of (4.2) with respect to x, one can obtain

$$\mathrm{Res}_z z^{-1}\frac{\tau(t'-2\varepsilon(z^{-1}))}{\tau(t')}e^{\xi(t',z)}\Big(\frac{\tau(t+2\varepsilon(z^{-1}))}{\tau(t)}e^{-\xi(t,z)}\Big)_x = 0. \quad (4.3)$$

If we denote $W = \frac{\tau(t-2\varepsilon(\partial^{-1}))}{\tau(t)}$, then by the similar methods in Proposition 7, one can obtain

$$(W\partial^{-1}W^*\partial^2)_{\leq 0} = 0, \quad (4.4)$$

$$\Big((W_{t_n}+W\partial^n)\partial^{-1}W^*\partial^2\Big)_{\leq 0} = 0, \quad n \text{ is odd}. \quad (4.5)$$

(4.4) means $W\partial^{-1}W^* = \partial^{-1}$, which is just the BKP constraint. Inserting the relation $W\partial^{-1}W^* = \partial^{-1}$ into (4.5), one can obtain

$$W_{t_n} = -(W\partial^n W^{-1})_{\leq -1}W,$$

which completely determines the data of the BKP hierarchy, combining with the relation $W\partial^{-1}W^* = \partial^{-1}$. Further by similar methods in Proposition 7, one can at last prove that $\tau(t)$ is the tau function of the BKP hierarchy.

After the preparation above, now let us try to prove (4.1). Denote

$$\psi(t,z) = \hat{\psi}(t,z)e^{\xi(t,z)} = \frac{\tau(t-2\varepsilon(z^{-1}))}{\rho(t)}e^{\xi(t,z)} = \Big(\frac{\tau(t)}{\rho(t)} + \mathcal{O}(z^{-1})\Big)e^{\xi(t,z)}, \tag{4.6}$$

$$\begin{aligned}\psi^*(t,z) &= \hat{\psi}^*(t,z)e^{\xi(t,z)} = z^{-1}\int \frac{\rho(t)}{\tau(t)}\Big(\frac{\tau(t+2\varepsilon(z^{-1}))}{\tau(t)}e^{-\xi(t,z)}\Big)_x dx \\ &= \Big(\frac{\rho(t)}{\tau(t)} + \mathcal{O}(z^{-1})\Big)z^{-1}e^{-\xi(t,z)}.\end{aligned} \tag{4.7}$$

Then (4.2) can be rewritten into

$$\mathrm{Res}_z\psi(t',z)\psi^*(t,z) = 1. \tag{4.8}$$

Further, if set $t' = t + 2\varepsilon(\lambda^{-1}))$,

$$\mathrm{Res}_z\hat{\psi}(t+2\varepsilon(\lambda^{-1}),z)\hat{\psi}^*(t,z)\frac{1+z/\lambda}{1-z/\lambda} = 1. \tag{4.9}$$

By the similar methods in Lemma 7,

$$\hat{\psi}^*(t,z) = \frac{\tau(t+2\varepsilon(z^{-1}))\rho(t)}{2z\tau(t)^2} + \frac{\rho(t+2\varepsilon(z^{-1}))}{2z\tau(t)}. \tag{4.10}$$

One can finally obtain (4.1), after inserting this relation (4.10) into (4.8) and using the fact that $\tau(t)$ is the tau function of the BKP hierarchy. ■

Because of the equivalence of (4.1) and (4.2), now we can discuss the dressing structure of the modified BKP hierarchy from (4.2). Firstly, denote

$$q(t) = \frac{\rho(t)}{\tau(t)}, \quad Z = \frac{\tau(t-2\varepsilon(\partial^{-1}))}{\rho(t)} = q(t)^{-1} + \mathcal{O}(\partial^{-1}), \tag{4.11}$$

then

$$\psi(t,z) = Z(e^{-\xi(t,z)}), \quad \psi^*(t,z) = -\Big(\partial^{-1}q\partial qZ\partial^{-1}\Big)(e^{-\xi(t,z)}). \tag{4.12}$$

So by similar way in Proposition 7,

$$(Z\partial^{-1}Z^*q\partial q)_{\leq 0} = 1, \quad ((Z_{t_n} + Z\partial^n)\partial^{-1}Z^*q\partial q)_{\leq 0} = 0, \quad n \text{ is odd}. \tag{4.13}$$

Thus one can obtain the theorem below.

Theorem 5. *For $Z = z_0 + z_1\partial^{-1} + z_2\partial^{-2} + \cdots$ defined in (4.11),*

$$Z\partial^{-1}Z^* = z_0\partial^{-1}z_0, \quad Z_{t_n} = -(Z\partial^nZ^{-1})_{\leq 0}Z, \quad n \text{ is odd}. \tag{4.14}$$

Here Z is called the dressing operator of the modified BKP hierarchy, and (4.14) is the corresponding constraint and the evolution equation. Note that the constraint $Z\partial^{-1}Z^* = z_0\partial^{-1}z_0$ can be shown in the components, that is

$$\sum_{j=0}^{l}(-1)^j z_j z_{l-j} + \sum_{i=1}^{l-1}\sum_{j=0}^{l-i}(-1)^{i+j}\binom{-l+i-1}{i} z_j z_{l-i-j}^{(i)} = 0, \quad l \geq 1. \tag{4.15}$$

Therefore,

$$\begin{aligned}
z_2 =& \frac{1}{2z_0}(z_1^2 - 2z_0 z_{1x} + 2z_1 z_{0x}),\\
z_4 =& -\frac{1}{8z_0^3}\Big(z_1^4 + 12z_1^3 z_{0x} - 12z_1^2 z_0 z_{1x} - 16z_1^2 z_0 z_{0xx} + 44z_1^2 z_{0x}^2\\
&- 8z_1 z_3 z_0^2 + 16z_1 z_{1xx} z_0^2 + 8z_1 z_{0xxx} z_0^2 - 56z_1 z_0 z_{1x} z_{0x} - 48z_1 z_0 z_{0x} z_{0xx}\\
&+ 48z_1 z_{0x}^3 + 16z_0^3 z_{3x} - 8z_0^3 z_{1xxx} + 12z_0^2 z_{1x}^2 + 24z_0^2 z_{1x} z_{0xx} - 16z_3 z_{0x} z_0^2\\
&+ 24z_0^2 z_{0x} z_{1xx} - 48z_0 z_{1x} z_{0x}^2\Big).
\end{aligned}$$

Further if define the Lax operator

$$\mathcal{L} = Z\partial Z^{-1} = \partial + v_0 + v_1\partial^{-1} + v_2\partial^{-2} + \cdots, \tag{4.16}$$

with $v_0 = -(\log z_0)_x$, then the corresponding constraints on the Lax operator and the Lax equation are given as follows,

$$\mathcal{L}^* = -e^{\int v_0 dx}\partial e^{\int v_0 dx}\cdot\mathcal{L}\cdot e^{-\int v_0 dx}\partial^{-1}e^{-\int v_0 dx}, \quad \mathcal{L}_{t_n} = [(\mathcal{L}^n)_{\geq 1}, \mathcal{L}]. \tag{4.17}$$

The constraints on the Lax operator $\mathcal{L}$ in the component form are given by

$$\begin{aligned}
(1-(-1)^l)v_{1+l} =& z_0^2\sum_{i=0}^{l}(-1)^{i+1}\left(\binom{-i}{l+1-i}(v_i z_0^{-2})^{(l-i+1)}\right.\\
&\left.+\binom{-i}{l-i}(v_i z_0^{-1}(z_0^{-1})_x)^{(l-i)}\right) - z_0(v_l z_0^{-1})_x,
\end{aligned} \tag{4.18}$$

that is,

$$\begin{aligned}
v_2 =& -v_0 v_1 - v_{1x},\\
v_4 =& 2v_0^3 v_1 + 4v_{0x}v_0 v_1 + 5v_0^2 v_{1x} + 3v_{0x}v_{1x}\\
&- 3v_0 v_3 + 4v_0 v_{1xx} + v_{0xx}v_1 - 2v_{3x} + v_{1xxx},
\end{aligned}$$

The corresponding Lax equations can lead to

$$v_{0t_3} = 3v_{0x}v_0^2 + 3v_0 v_{1x} + 3v_0 v_{0xx} + 3v_{0x}v_1 + 3v_{0x}^2 + v_{0xxx}, \tag{4.19}$$

$$\begin{aligned}
v_{1t_3} =& -3v_{1x}v_0^2 - 6v_{0x}v_1 v_0 - 9v_{1x}v_{0x} + 6v_1 v_{1x} - 6v_0 v_{1xx}\\
&- 3v_1 v_{0xx} + 3v_{3x} - 2v_{1xxx},
\end{aligned} \tag{4.20}$$

$$\begin{aligned} v_{0t_5} =& 5v_{0x}v_0^4 + 30v_{0x}^2v_0^2 + 10v_{0xx}v_0^3 - 5v_{1xx}v_0^2 + 15v_{0x}^3 \\ &+ 10v_1v_{0x}^2 + 50v_{0xx}v_{0x}v_0 + 10v_1^2v_{0x} + 20v_{1x}v_1v_0 \\ &+ 10v_{0xx}v_1v_0 + 10v_0^2v_{0xxx} + 5v_{1xx}v_{0x} + 15v_{0xxx}v_{0x} \\ &+ 5v_3v_{0x} + 10v_{1x}v_{0xx} + 10v_{0xx}^2 + 5v_0v_{0xxxx} + 5v_0v_{3x} \\ &+ 5v_1v_{0xxx} + v_{0xxxxx}. \end{aligned} \tag{4.21}$$

One can eliminate v_1 and v_3, and obtain the differential equations of v_0.

$$\begin{aligned} v_{0t} =& v_{0y}v_{0xx}v_0^{-1} - 2v_{0x}v_{0xx}^2v_0^{-2} + 2v_{0x}^3v_{0xx}v_0^{-3} - 2v_{0x}^2v_{0xxx}v_0^{-2} \\ &+ v_{0xxxx}v_{0x}v_0^{-1} + v_{0xxx}v_{0xx}v_0^{-1} - 2v_0^2v_{0y} - v_{0x}^3 - v_{0xxy} \\ &- 3v_{0x}v_{0y} - 3v_0v_{0xy} + v_{0xxx}v_0^2 - v_{0x}v_0^4 - \frac{1}{5}v_{0xxxxx} \\ &+ \Big(3v_{0x}v_{0xx}v_0^{-2} - v_{0y}v_0^{-1} - v_{0x}v_0^{-2}\int v_{0y}dx + 2v_0v_{0x} \\ &- 2v_{0x}^3v_0^{-3} - v_{0xxx}v_0^{-1}\Big)\int v_{0y}dx + \Big(3v_{0x} + 2v_0^{-1}\int v_{0y}dx \\ &- 2v_{0xx}v_0^{-1} + 2v_{0x}^2v_0^{-2} + v_0^2 + 1\Big)\int v_{0yy}dx + \Big(3v_0 \\ &- v_{0x}v_0^{-1}\Big)\int v_{0yyy}dx + 2\int v_{0yyyy}dx \\ &+ v_{0x}\Big(-\int v_{0y}v_0^{-2}(\int v_{0y}dx)dx + \int v_0^{-1}(\int v_{0yy}dx)dx \\ &- 2\int v_0v_{0y}dx + \int v_{0xx}v_{0y}v_0^{-2} - \int v_{0xxy}v_0^{-1}dx\Big). \end{aligned} \tag{4.22}$$

Here we have set $y = t_3$ and $t = \frac{5}{9}t_5$. This equation is too cumbersome. In next subsection, a short differential equation of $\exp(\int v_0)$ will be given.

4.2 Relations with BKP hierarchy

In this subsection, we will investigate the relations between the modified KP hierarchy and the BKP hierarchy.

Proposition 9. *If q is an eigenfunction of the BKP hierarchy and L is the corresponding Lax operator, then*

$$\mathcal{L} = q^{-1}Lq \tag{4.23}$$

is the Lax operator of the modified BKP hierarchy.

Proof. By using the formula in [21],

$$(q^{-1}Aq)_{\geq 1} = q^{-1}\cdot A_{\geq 0}\cdot q - q^{-1}\cdot A_{\geq 0}(q), \tag{4.24}$$

one can easily find

$$\mathcal{L}_{t_n} = [q^{-1}(L^n)_{\geq 0}q + q^{-1}q_{t_n}, \mathcal{L}] = [(\mathcal{L}^n)_{\geq 1}, \mathcal{L}]. \tag{4.25}$$

Further, it is obvious that $\mathcal{L}^* = -q\partial q\mathcal{L}q^{-1}\partial^{-1}q^{-1}$. ■

Proposition 10. *Assume that $\mathcal{L}$ is the Lax operator of the modified BKP hierarchy and z_0 is the ∂^0-terms in the dressing operator Z, then*

$$L = z_0^{-1}\mathcal{L}z_0 \tag{4.26}$$

is the Lax operator of the BKP hierarchy and $q \triangleq z_0^{-1}$ is the eigenfunction of the BKP hierarchy with respect to L.

Proof. We firstly show $q_{t_n} = (L^n)_{\geq 0}(q)$. In fact, this can be obtained by comparing the ∂^0-terms in the dressing equation

$$Z_{t_n} = -(Z\partial^n Z^{-1})_{\leq 0}Z = -(z_0 L^n z_0^{-1})_{\leq 0}Z,$$

that is

$$z_{0t_n} = -(z_0 L^n z_0^{-1})_{[0]}z_0 = -(z_0(L^n)_{\geq 0}z_0^{-1})_{[0]}z_0 = -z_0^2 \cdot (L^n)_{\geq 0}(z_0^{-1}). \tag{4.27}$$

Then the evolution of L is given by

$$\begin{aligned} L_{t_n} =& [q_{t_n}q^{-1} + q(\mathcal{L}^n)_{\geq 1}q^{-1}, L] \\ =& \Big[q_{t_n}q^{-1} + q(q^{-1}\cdot(L^n)_{\geq 0}\cdot q - q^{-1}(L^n)(q))q^{-1}, L\Big] = [(L^n)_{\geq 0}, L]. \end{aligned}$$

The BKP constraint can be easily obtained from $\mathcal{L} = q\partial q\mathcal{L}q^{-1}\partial^{-1}q^{-1}$. ∎

By considering the relations between the modified BKP hierarchy and the BKP hierarchy, one can express the modified BKP equation in the way below. Note that given the Lax operator $\mathcal{L} = \partial + v_0 + v_1\partial^{-1} + \cdots$ of the modified BKP hierarchy, one can find $v_0 = q^{-1}q_x$, where q is the eigenfunction of the BKP hierarchy with the Lax operator $L = q\mathcal{L}q^{-1} = \partial + u_1\partial^{-1} + \cdots$. Therefore, the evolution equation of v_0 can be expressed by q. By $q_{t_n} = (L^n)_{\geq 0}(q)$ and $L_{t_n} = [(L^n)_{\geq 0}, L]$,

$$\begin{aligned} &q_{t_3} = q_{xxx} + 3u_1 q_x, \quad q_{t_5} = q_{xxxxx} + 5u_1 q_{xxx} + (5u_3 + 10u_1^2)q_x, \\ &u_{1t_3} = -2u_{1xxx} + 3u_{3x} + 6u_1 u_{1x}. \end{aligned}$$

Then by eliminating u_1, u_3 and setting $y = t_3$, $t = \frac{5}{9}t_5$, one can obtain

$$\begin{aligned} q_t =& -\frac{1}{5}q_{xxxxx} + 2q_{xxy} + \Big(4q_{xx}q_{xxxx} - 4q_{xx}q_{xy} - q_{xxx}q_y + q_y^2\Big)q_x^{-1} \\ &+ \Big(-4q_{xx}^2 q_{xxx} + 4q_{xx}^2 q_y\Big)q_x^{-2} + q_x\Big(\int (q_{yy} - q_{xxxy})q_x^{-1}dx \\ &+ \int (q_{xy}q_{xxx} - q_{xy}q_y)q_x^{-2}dx\Big), \end{aligned} \tag{4.28}$$

which is equivalent to (4.22).

Acknowledgements

Many thanks to Professor Jingsong He (Shenzhen University, China) for the long-term encouragement and support.

References

[1] Chang X K, Hu X B, Li S H and Zhao J X, An application of Pfaffians to multipeakons of the Novikov equation and the finite Toda lattice of BKP type, *Adv. Math.* **338**, 1077–1118, 2018.

[2] Cheng J P, He J S and Hu S, The "ghost" symmetry of the BKP hierarchy, *J. Math. Phys.* **51**, 053514, 2010.

[3] Cheng J P and He J S, The applications of the gauge transformation for the BKP hierarchy, *J. Math. Anal. Appl.* **410**, 989–1001, 2014.

[4] Cheng J P and Milanov T, The 2-component BKP Grassmannian and simple singularities of type D, *Int. Math. Res. Not.*, rnz325, 2019.

[5] Cheng Y, Constraints of the Kadomtsev-Petviashvili hierarchy, *J. Math. Phys.* **33**, 3774–3782, 1992.

[6] Date E, Kashiwara M, Jimbo M and Miwa T, Transformation groups for soliton equations, in *Proceedings of RIMS symposium on non-linear integrable systems-classical theory and quantum theory (Kyoto, Janpan, 1981)*, World Scientific, Singapore, 39–119, 1983.

[7] Dickey L A, *Soliton equations and Hamiltonian systems*, World Scientific, Singapore, 2003.

[8] Geng L M, Chen H Z, Li N and Cheng J P, Bilinear identities and squared eigenfunctions symmetries of the BC_r-KP hierarchy, *J. Nonlin. Math. Phys.* **26**, 404–419, 2019.

[9] Harnad J and Orlov A Yu, Fermionic approach to bilinear expansions of Schur functions in Schur Q-functions, arXiv: 2008.13734.

[10] Hu X B, Li S H, The partition function of the Bures ensemble as the τ-function of BKP and DKP hierarchies: continuous and discrete, *J. Phys. A* **50**, 285201, 2017.

[11] Jimbo M and Miwa T, Solitons and infinite dimensional Lie algebras, *Publ. RIMS, Kyoto Univ.* **19**, 943–1001, 1983.

[12] Kac V G, Raina A K and Rozhkovskaya N, *Bombay lectures on highest weight representations of infinite dimensional Lie algebras (second edition)*, World Scientific, Singapore, 2013.

[13] Kac V G and van de Leur J W, The geometry of spinors and the multicomponent BKP and DKP hierarchies, in *The bispectral problem (Montreal, 1997)*, *CRM Proc. Lect. Notes* **14**, *AMS, Providence, RI*, 159–202, 1998.

[14] Kac V G and van de Leur J W, Polynomial tau-functions of BKP and DKP hierarchies, *J. Math. Phys.* **60**, 071702, 2019.

[15] Konopelchenko B G and Dubrovsky V G, Some new integrable nonlinear evolution equations in $2+1$ dimensions, *Phys. Lett. A* **102**, 15–17, 1984.

[16] Li C Z, Dispersionless and multicomponent BKP hierarchies with quantum torus symmetries, *J. Geom. Phys.* **119**, 103–111, 2017.

[17] Li C Z and He J S, Quantum torus symmetry of the KP, KdV and BKP hierarchies, *Lett. Math. Phys.* **104**, 1407–1423, 2014.

[18] Loris I and Willox R, Symmetry reductions of the BKP hierarchy, *J. Math. Phys.* **40**, 1420–1431, 1999.

[19] Miwa T, Jimbo M and Date E, *Solitons: differential equations, symmetryies and infinite dimensional algebras*, Cambridge Univesity Press, Cambridge, 2000.

[20] Natanzon S M, Orlov A Yu, BKP and projective Hurwitz numbers, *Lett. Math. Phys.* **107**, 1065–1109, 2017.

[21] Oevel W, Darboux theorems and Wronskian formulas for integrable system I: constrained KP flows, *Physica A* **195**, 533–576, 1993.

[22] Rozhkovskaya N, Multiparameter Schur Q-functions are solutions of the BKP hierarchy, *SIGMA* **15**, 065, 2019.

[23] Shen H F and Tu M H, On the constrained B-type Kadomtsev-Petviashvili hierarchy: Hirota bilinear equations and Virasoro symmetry, *J. Math. Phys.* *52*, 032704, 2011.

[24] Shigyo Y, On the expansion coefficients of tau-function of the BKP hierarchy, *J. Phys. A* **49**, 295201, 2016.

[25] Tian K L, He J S, Cheng J P and Cheng Y, Additional symmetries of constrained CKP and BKP hierarchies, *Sci. China Math.* **54**, 257–268, 2011.

[26] Tian K L, He J S and Foerster A, Negative generators of the Virasoro constraints for BKP hierarchy, *Rom. Rep. Phys.* **72**, 101, 2020.

[27] van de Leur J W and Orlov A Yu, Random turn walk on a half line with creation of particles at the origin, *Phys. Lett. A* **373**, 2675–2681, 2009 (arXiv: 0801.0066).

[28] Wang Z L and Li S H, BKP hierarchy and Pfaffian point process, *Nucl. Phys. B* **939**, 447–464, 2019.

[29] You Y, Polynomial solutions of the BKP hierarchy and projective representations of symmetric groups, in *Infinite-dimensional Lie algebras and groups (Luminy-Marseille, 1988)*, *Adv. Ser. Math. Phys.* **7**, 449–464, 1989.

$A2.$ Elementary introduction to direct linearisation of integrable systems

Wei Fu [a] *and Frank W. Nijhoff* [b]

[a] *School of Mathematical Sciences and Shanghai Key Laboratory of Pure Mathematics and Mathematical Practice, East China Normal University, 500 Dongchuan Road, Shanghai 200241, P.R. China*

[b] *School of Mathematics, University of Leeds, Leeds LS2 9JT, United Kingdom*

Abstract

We give an elementary introduction to the direct linearisation approach of integrable systems, which provides a unified framework to understand nonlinear discrete and continuous equations and their underlying integrability structures. As the most fundamental integrable models, the discrete and continuous Kadomtsev–Petviashvili hierarchies are taken as examples to illustrate the direct linearisation method.

1 Introduction

The notion of integrability came from classical mechanics, and it means that a finite-dimensional Hamiltonian equation possesses a sufficient number of commuting first integrals, resulting in the fact that the system is exactly solvable. In the modern theory of integrable systems, such a notion was generalised to the infinite-dimensional case, in order to deal with the solvability problem for nonlinear partial differential equations (PDEs). The integrability of a nonlinear PDE is normally understood from the following two aspects: One is that a PDE has an infinite number of first integrals (which are in involution). This is a very natural generalisation of the integrability of finite-dimensional models, and it is equivalent to discussing the existence of infinitely many commuting symmetries for a given nonlinear PDE, or alternatively, searching for a recursion operator which generates the whole hierarchy for the nonlinear PDE, by starting from a seed symmetry [16]. The other one is the existence of explicit solutions allowing an infinite number of degrees of freedom for a nonlinear PDE, motivated by the pioneer work of Gardner, Greene, Kruskal and Miura [12], who proposed the inverse scattering transform when trying to solve the initial value problem with zero boundary condition for the well-known Korteweg–de Vries (KdV) equation, and constructed a particular class of solutions containing an arbitrary number of free parameters (i.e., solutions depending on discrete spectral variables), which is now called 'soliton'. In addition to the inverse scattering method, many mathematical techniques were established to solve nonlinear PDEs, including Hirota's bilinear method [14], Darboux transformation [17], and especially, the algebro-geometric approach for solving nonlinear integrable equations with periodic boundary conditions, giving rise to the so-called finite gap solutions relying on continuous spectral variables, see [24].

Although the above two properties are widely accepted by the integrable community, and are considered as the criteria to judge whether a nonlinear differential equation is integrable, there is still no rigorous definition of integrability (i.e., no precise axiom) at the current stage. A milestone to answer this question is attributed to Sato and his Kyoto school. In [25], Sato explained that the existing integrability properties such as exact solutions, Lax pairs, bilinear forms, etc. are actually interconnected through an object entitled the tau function, by studying the remarkable Kadomtsev–Petviashvili (KP) hierarchy whose solution describes an orbit on an infinite-dimensional Grassmannian. Motivated by Sato's observation on the KP hierarchy, the Kyoto school (see e.g., [15] and references therein) further established the relationship between nonlinear PDEs and infinite-dimensional Lie algebras (cf. Drinfel'd and Sokolov [3]), and proposed a unified framework for soliton hierarchies. The result of the Kyoto school in a sense proves that different criteria to test the integrability of nonlinear equations are actually equivalent.

Parallel to the theory of continuous integrable systems (i.e., integrable PDEs), the study of discrete integrable systems (namely integrable difference equations) also has arisen people's attention in the past decades. One effective way of dealing with integrable discretisation of a nonlinear PDE is to consider its Bäcklund transformation, in which case the Bäcklund transformation itself and the associated superposition formula (see e.g., [26]) are thought of as the resulting semi-discrete and fully discrete equations, respectively, and the Bäcklund parameters play the role of the corresponding lattice parameters [13]. Discrete equations arising from such a procedure not only take elegant forms (that often appear in discrete geometry) in their own right, but also possess much richer structure compared with their continuous counterparts, since a single discrete equation encodes the whole hierarchy of the corresponding continuous equation. The integrability of these equations is realised through a concept called multi-dimensional consistency (MDC), see e.g., [1, 23], namely a discrete equation is consistently embedded into a higher-dimensional lattice space, within which it is compatible with its analogues with regard to different lattice directions and parameters. In other words, the MDC is the discrete version of the existence of infinitely many commuting symmetries. More importantly, the MDC property also guarantees that a nonlinear difference equation is integrable, as it potentially produces key integrability characteristics such as Lax pairs and exact solutions, etc.

In contrast to continuous integrable systems, many classical methods are not applicable to the discrete integrable systems, mainly due to the nonlocality of discrete shift operation (to be more precise, the Leibnitz rule is broken). For this reason, generalisations of existing techniques were developed to study integrable difference equations particularly, including the discrete Sato theory [2] and the direct linearisation (DL) theory [19–21], as the two systematical frameworks for integrable discrete models. The former generalises the bilinear identities in the continuous Sato theory to the discrete situation, and successfully establishes an effective method to construct discrete bilinear equations and discrete tau functions; while the idea of the latter methodology is to investigate nonlinear discrete systems based on linear integral equations, from the aspect of the inverse scattering transform.

In this chapter, we aim to give an elementary introduction to the DL theory of both discrete and continuous integrable systems. The DL approach was proposed by Fokas and Ablowitz [4, 5], as a generalisation of the Riemann–Hilbert method, to solve initial value problems of integrable PDEs, and it was then developed by Nijhoff, Quispel, Capel and their collaborators to study the underlying integrability structures of both discrete and continuous equations, including examples such as discrete and continuous KdV, Boussinesq (BSQ) and Kadomtsev–Petviashvili (KP) systems, see e.g., [19–21]. Recently, the authors further established the links between the DL and the root systems of Lie algebras, and included more examples of discrete and continuous integrable systems within the DL picture, see [6, 8–11]. As an elementary introduction towards graduate students, we only take the discrete and continuous KP hierarchies as examples to explain how the DL approach works in the theory of integrable systems, since they are the most fundamental models among the theory of integrable systems.

The organisation is as follows. In Section 2, we introduce the notions of infinite matrices and vectors, as well as their operations. Based on the infinite matrices and vectors, the general theory of the DL approach is given in Section 3. Section 4 is concerned with the direct linearising framework of the discrete and continuous KP hierarchies.

2 Infinite matrices and vectors

2.1 Infinite matrices

In the finite-dimensional case, there are a finite number of indices labelling entries in a matrix/vector. In order to define an infinite matrix, we need a "centred" matrix/vector together with the so-called index-raising operators to label all the entries. This involves the use of the objects (generators) $\mathbf{\Lambda}$, ${}^t\mathbf{\Lambda}$ and $\boldsymbol{O}$ in an associative algebra (with unit) $\mathcal{A}$ over a field $\mathcal{F}$, obeying the relation

$$\boldsymbol{O}\,{}^t\mathbf{\Lambda}^{-j}\mathbf{\Lambda}^{-i}\boldsymbol{O} = \delta_{j,i}\boldsymbol{O}, \quad i, j \in \mathbb{Z}, \tag{2.1}$$

where the powers ${}^t\mathbf{\Lambda}^i$ and $\mathbf{\Lambda}^j$ are the ith and jth compositions of ${}^t\mathbf{\Lambda}$ and $\mathbf{\Lambda}$ (and their inverse), respectively. In general, $\mathbf{\Lambda}$, ${}^t\mathbf{\Lambda}$ and $\boldsymbol{O}$ do not commute, and we require that $\mathbf{\Lambda}$ and ${}^t\mathbf{\Lambda}$ act as each other's transpose. From (2.1) it is also easy to see that $\boldsymbol{O}$ is a projector satisfying $\boldsymbol{O}^2 = \boldsymbol{O}$ by setting $i = j = 0$.

Following the idea in the case of finite-dimensional matrices, we define

$$\boldsymbol{E}^{(i,j)} = \mathbf{\Lambda}^{-i}\boldsymbol{O}\,{}^t\mathbf{\Lambda}^{-j}, \quad \forall i, j \in \mathbb{Z}, \tag{2.2}$$

which implies $\boldsymbol{O} = \boldsymbol{E}^{(0,0)}$.

A direct calculation shows that

$$\begin{aligned} \boldsymbol{E}^{(i_1,j_1)}\boldsymbol{E}^{(i_2,j_2)} &= \mathbf{\Lambda}^{-i_1}\boldsymbol{O}\,{}^t\mathbf{\Lambda}^{-j_1}\mathbf{\Lambda}^{-i_2}\boldsymbol{O}\,{}^t\mathbf{\Lambda}^{-j_2} \\ &= \delta_{j_1,i_2}\mathbf{\Lambda}^{-i_1}\boldsymbol{O}\,{}^t\mathbf{\Lambda}^{-j_2} = \delta_{j_1,i_2}\boldsymbol{E}^{(i_1,j_2)}, \end{aligned} \tag{2.3}$$

where we have used the rule (2.1) for the second equality. The above relation will later govern the multiplication of two infinite matrices.

An, in general, infinite matrix $\boldsymbol{U}$ is defined as

$$\boldsymbol{U} = \sum_{i,j\in\mathbb{Z}} U_{i,j}\boldsymbol{E}^{(i,j)}, \tag{2.4}$$

where the coefficients $U_{i,j}$ are the (i,j)-entries in the infinite matrix, taking values in the field $\mathcal{F}$. From the definition, we can observe that members in $\{\boldsymbol{E}^{(i,j)}|i,j\in\mathbb{Z}\}$ form a basis of the linear space of infinite matrices. The transpose operation on $\boldsymbol{U}$ is defined by

$${}^{t}\boldsymbol{U} = \sum_{i,j\in\mathbb{Z}} U_{j,i}\boldsymbol{E}^{(i,j)}.$$

In particular, we can easily prove that ${}^{t}\boldsymbol{O} = \boldsymbol{O}$ and ${}^{t}\boldsymbol{E}^{(i,j)} = \boldsymbol{E}^{(j,i)}$, following the definition (note that $\boldsymbol{O}$ and $\boldsymbol{E}^{(i,j)}$ have their respective (i',j')-entries $\delta_{i',0}\delta_{j',0}$ and $\delta_{i',i}\delta_{j',j}$).

Suppose that $\boldsymbol{U}$ and $\boldsymbol{V}$ are two infinite matrices having entries $U_{i,j}$ and $V_{i,j}$, and $\mathfrak{p}$ is an element taken from the field $\mathcal{F}$. We can show that operations of these infinite matrices such as addition, multiplication of two infinite matrices as well as scalar multiplication of an infinite matrix by $\mathfrak{p}\in\mathcal{F}$ obey the following rules:

$$\boldsymbol{U}+\boldsymbol{V} = \sum_{i,j\in\mathbb{Z}} (U_{i,j}+V_{i,j})\boldsymbol{E}^{(i,j)},$$

$$\mathfrak{p}\boldsymbol{U} = \sum_{i,j\in\mathbb{Z}} (\mathfrak{p}\,U_{i,j})\boldsymbol{E}^{(i,j)},$$

$$\boldsymbol{U}\cdot\boldsymbol{V} = \sum_{i,j\in\mathbb{Z}} \left(\sum_{i'\in\mathbb{Z}} U_{i,i'}V_{i',j}\right)\boldsymbol{E}^{(i,j)}.$$

The first two equations are proven directly as they follow from the definition (2.4). The proof of the third one requires (2.3). In fact,

$$\begin{aligned}
\boldsymbol{U}\boldsymbol{V} &= \sum_{i_1,j_1\in\mathbb{Z}}\sum_{i_2,j_2\in\mathbb{Z}} U_{i_1,j_1}V_{i_2,j_2}\boldsymbol{E}^{(i_1,j_1)}\boldsymbol{E}^{(i_2,j_2)}\\
&= \sum_{i_1,j_1\in\mathbb{Z}}\sum_{i_2,j_2\in\mathbb{Z}} U_{i_1,j_1}V_{i_2,j_2}\delta_{j_1,i_2}\boldsymbol{E}^{(i_1,j_2)}\\
&= \sum_{i_1,j_2\in\mathbb{Z}}\left(\sum_{i_2\in\mathbb{Z}} U_{i_1,i_2}V_{i_2,j_2}\right)\boldsymbol{E}^{(i_1,j_2)}.
\end{aligned}$$

We point out that the definition (2.4) also covers finite-dimensional matrices by restricting the number of non-zero coefficients to a finite number, i.e., $U_{i,j} = 0$ for $i \neq 1,2,\cdots,M$ or $j\neq 1,2,\cdots,N$ for some given positive integers M and N, resulting in an $M\times N$ finite-dimensional matrix given by $\boldsymbol{U} = \sum_{i=1,j=1}^{M,N} U_{i,j}\boldsymbol{E}_{M\times N}^{(i,j)}$.

An alternative way to understand the objects $\boldsymbol{O}$, $\boldsymbol{\Lambda}$ and ${}^t\boldsymbol{\Lambda}$ is considering them as infinite matrices of size $\infty \times \infty$ having their respective (i,j)-entry

$$(\boldsymbol{O})_{i,j} = \delta_{i,0}\delta_{0,j}, \quad (\boldsymbol{\Lambda})_{i,j} = \delta_{i+1,j}, \quad \text{and} \quad ({}^t\boldsymbol{\Lambda})_{i,j} = \delta_{i,j+1}.$$

Such a realisation is actually compatible with the statement above, see Subsection 2.3. The infinite matrices $\boldsymbol{\Lambda}$ and ${}^t\boldsymbol{\Lambda}$ play the role of the index-raising operators; to be more precise, $\boldsymbol{\Lambda}$ (resp. ${}^t\boldsymbol{\Lambda}$) raises the row (resp. column) index of an infinite matrix by left (resp. right) multiplication.

2.2 Infinite-dimensional vectors

Following the same idea, we also introduce infinite-dimensional vectors as follows. Suppose that $\boldsymbol{o}$ and ${}^t\boldsymbol{o}$ are two objects obeying the relations

$${}^t\boldsymbol{o}{}^t\boldsymbol{\Lambda}^{-j}\boldsymbol{\Lambda}^{-i}\boldsymbol{o} = \delta_{j,i} \quad \text{and} \quad \boldsymbol{o}{}^t\boldsymbol{o} = \boldsymbol{O}, \tag{2.5}$$

where $i,j \in \mathbb{Z}$ and $\boldsymbol{o}$, ${}^t\boldsymbol{o}$, $\boldsymbol{\Lambda}$ and ${}^t\boldsymbol{\Lambda}$ do not commute with each other in general.

The objects $\boldsymbol{o}$ and ${}^t\boldsymbol{o}$, together with the index-raising operators $\boldsymbol{\Lambda}$ and ${}^t\boldsymbol{\Lambda}$, are the ingredients to construct infinite-dimensional vectors. We define

$$\boldsymbol{e}^{(i)} = \boldsymbol{\Lambda}^{-i}\boldsymbol{o}, \quad {}^t\boldsymbol{e}^{(i)} = {}^t\boldsymbol{o}{}^t\boldsymbol{\Lambda}^{-i}, \quad \forall i \in \mathbb{Z},$$

which play the roles of bases of an arbitrary column vector and an arbitrary row vector. Meanwhile, with the help of (2.5), we can show that

$$\boldsymbol{e}^{(i)}{}^t\boldsymbol{e}^{(j)} = \boldsymbol{\Lambda}^{-i}\boldsymbol{o}{}^t\boldsymbol{o}{}^t\boldsymbol{\Lambda}^{-j} = \boldsymbol{\Lambda}^{-i}\boldsymbol{O}{}^t\boldsymbol{\Lambda}^{-j} = \boldsymbol{E}^{(i,j)} \tag{2.6a}$$

as well as

$${}^t\boldsymbol{e}^{(j)}\boldsymbol{e}^{(i)} = {}^t\boldsymbol{o}{}^t\boldsymbol{\Lambda}^{-j}\boldsymbol{\Lambda}^{-i}\boldsymbol{o} = \delta_{j,i}. \tag{2.6b}$$

An infinite-dimensional column vector $\boldsymbol{u}$ and its transpose ${}^t\boldsymbol{u}$ (i.e., an infinite-dimensional row vector) are defined as

$$\boldsymbol{u} = \sum_{i\in\mathbb{Z}} u_i \boldsymbol{e}^{(i)} \quad \text{and} \quad {}^t\boldsymbol{u} = \sum_{i\in\mathbb{Z}} u_i {}^t\boldsymbol{e}^{(i)}, \tag{2.7}$$

respectively, where u_i (as the corresponding components) are elements in the same field $\mathcal{F}$. Elements in the sets $\{\boldsymbol{e}^{(i)}|i \in \mathbb{Z}\}$ and $\{{}^t\boldsymbol{e}^{(i)}|i \in \mathbb{Z}\}$ form the respective bases for an arbitrary infinite-dimensional vector and its transpose, respectively.

For arbitrary infinite-dimensional vectors $\boldsymbol{u} = \sum_{i\in\mathbb{Z}} u_i \boldsymbol{e}^{(i)}$ and $\boldsymbol{v} = \sum_{j\in\mathbb{Z}} v_j \boldsymbol{e}^{(j)}$ and their transpose, where u_i, v_j are elements from the field $\mathcal{F}$, and for arbitrary $\mathfrak{p}$ also from $\mathcal{F}$, we have the basic operations as follows:

$$\boldsymbol{u} + \boldsymbol{v} = \sum_{i\in\mathbb{Z}} (u_i + v_i)\boldsymbol{e}^{(i)}, \quad {}^t\boldsymbol{u} + {}^t\boldsymbol{v} = \sum_{i\in\mathbb{Z}} (u_i + v_i){}^t\boldsymbol{e}^{(i)},$$

$$\mathfrak{p}\,\boldsymbol{u} = \sum_{i\in\mathbb{Z}} (\mathfrak{p}\,u_i)\boldsymbol{e}^{(i)}, \quad \mathfrak{p}\,{}^t\boldsymbol{u} = \sum_{i\in\mathbb{Z}} (\mathfrak{p}\,u_i){}^t\boldsymbol{e}^{(i)},$$

$$\boldsymbol{u}{}^t\boldsymbol{v} = \sum_{i,j\in\mathbb{Z}} u_i v_j \boldsymbol{E}^{(i,j)}, \quad {}^t\boldsymbol{v}\boldsymbol{u} = \sum_{i\in\mathbb{Z}} u_i v_i.$$

The derivations of the first four relations are trivial, as they can be observed from the definition of infinite-dimensional vectors immediately; while the last two equations follow from (2.6) and we give their derivations below:

$$\boldsymbol{u}^t\boldsymbol{v} = \sum_{i\in\mathbb{Z}}\sum_{j\in\mathbb{Z}} u_i v_j \boldsymbol{e}^{(i)}\,{}^t\boldsymbol{e}^{(j)} = \sum_{i\in\mathbb{Z}}\sum_{j\in\mathbb{Z}} u_i v_j \boldsymbol{E}^{(i,j)},$$

$${}^t\boldsymbol{v}\boldsymbol{u} = \sum_{i\in\mathbb{Z}}\sum_{j\in\mathbb{Z}} u_i v_j\,{}^t\boldsymbol{e}^{(j)}\boldsymbol{e}^{(i)} = \sum_{i\in\mathbb{Z}}\sum_{j\in\mathbb{Z}} u_i v_j \delta_{i,j} = \sum_{i\in\mathbb{Z}} u_i v_i.$$

In other words, $\boldsymbol{u}^t\boldsymbol{v}$ is an infinite matrix and ${}^t\boldsymbol{v}\boldsymbol{u}$ is a scalar quantity, which obey the same rules (except that the summation is over $\mathbb{Z}$) as those for finite-dimensional vectors.

The case of N-component (column and row) vectors is obtained by the restriction $u_i = 0$ for $i \neq 1, 2, \cdots, N$, leading to $\boldsymbol{u} = \sum_{i=1}^N u_i \boldsymbol{e}_N^{(i)}$ and ${}^t\boldsymbol{u} = \sum_{i=1}^N u_i\,{}^t\boldsymbol{e}_N^{(i)}$, namely a finite-dimensional vector can be considered as a degeneration of an infinite-dimensional vector.

Making use of (2.6), we can also show that $\boldsymbol{E}^{(i,j)}$, $\boldsymbol{e}^{(i)}$ and ${}^t\boldsymbol{e}^{(i)}$ satisfy the following relations:

$$\boldsymbol{E}^{(i,j)}\boldsymbol{e}^{(i')} = \delta_{i',j}\boldsymbol{e}^{(i)}, \quad {}^t\boldsymbol{e}^{(j')}\boldsymbol{E}^{(i,j)} = \delta_{i,j'}\,{}^t\boldsymbol{e}^{(j)}. \tag{2.8}$$

The relation (2.8) allows the multiplication of an infinite matrix and an infinite-dimensional vector. For instance, for arbitrary $\boldsymbol{U}$ taking the form of (2.4) and $\boldsymbol{u}$ and ${}^t\boldsymbol{u}$ given by (2.7), we have

$$\boldsymbol{U}\boldsymbol{u} = \sum_{i,j\in\mathbb{Z}}\sum_{i'\in\mathbb{Z}} U_{i,j} u_{i'} \boldsymbol{E}^{(i,j)}\boldsymbol{e}^{(i')} = \sum_{i\in\mathbb{Z}}\left(\sum_{j\in\mathbb{Z}} U_{i,j}u_j\right)\boldsymbol{e}^{(i)},$$

$${}^t\boldsymbol{u}\boldsymbol{U} = \sum_{i'\in\mathbb{Z}}\sum_{i,j\in\mathbb{Z}} u_{i'} U_{i,j}\,{}^t\boldsymbol{e}^{(i')}\boldsymbol{E}^{(i,j)} = \sum_{j\in\mathbb{Z}}\left(\sum_{i\in\mathbb{Z}} u_i U_{i,j}\right){}^t\boldsymbol{e}^{(j)},$$

namely $\boldsymbol{U}\boldsymbol{u}$ is a column vector and ${}^t\boldsymbol{u}\boldsymbol{U}$ is a row vector.

2.3 Visualisations

The above subsection gives the definitions of infinite matrices and vectors in an algebraic way. In fact, these abstract notations have their respective visualisation. In this subsection, we give visualisations of the projector $\boldsymbol{O}$ (an "centred" matrix), the index-raising operators $\boldsymbol{\Lambda}$ and ${}^t\boldsymbol{\Lambda}$, as well as the "centred" vectors $\boldsymbol{o}$ and ${}^t\boldsymbol{o}$.

The projector $\boldsymbol{O}$ is a particular infinite matrix when we take $U_{i,j} = \delta_{i,0}\delta_{j,0}$ in (2.4), namely $\boldsymbol{O}$ is a "centred" matrix that has only nonzero value in the (0, 0)-entry. Similarly, the unit infinite matrix[1] $\boldsymbol{I}$ is a particular case of (2.4) when $U_{i,j} = \delta_{i,j}$.

[1]Sometimes we also use 1 (instead of $\boldsymbol{I}$) to denote a unit infinite matrix.

Both $\boldsymbol{O}$ and $\boldsymbol{I}$ have their respective visualisation as follows:

$$\boldsymbol{O} = \begin{pmatrix} \ddots & & & \\ & 0 & & \\ & & \boxed{1} & \\ & & & 0 \\ & & & & \ddots \end{pmatrix}, \quad \boldsymbol{I} = \begin{pmatrix} \ddots & & & \\ & 1 & & \\ & & \boxed{1} & \\ & & & 1 \\ & & & & \ddots \end{pmatrix},$$

where the boxes denote the "centres" (i.e., the $(0,0)$-entries) of the two matrices.

The index-raising operators $\boldsymbol{\Lambda}$ and ${}^t\boldsymbol{\Lambda}$ themselves are not infinite matrices; however, we can visualise them by considering their operations on the unit infinite matrix $\boldsymbol{I}$, namely $\boldsymbol{\Lambda}\cdot\boldsymbol{I}$ and $\boldsymbol{I}\cdot{}^t\boldsymbol{\Lambda}$. Observing that the index-raising operators acting on $\boldsymbol{E}^{(i,j)}$ gives rise to relations as follows:

$$\boldsymbol{\Lambda}\boldsymbol{E}^{(i,j)} = \boldsymbol{\Lambda}^{-i+1}\boldsymbol{O}{}^t\boldsymbol{\Lambda}^{-j} = \boldsymbol{E}^{(i-1,j)},$$
$$\boldsymbol{E}^{(i,j)}{}^t\boldsymbol{\Lambda} = \boldsymbol{\Lambda}^{-i}\boldsymbol{O}{}^t\boldsymbol{\Lambda}^{-j+1} = \boldsymbol{E}^{(i,j-1)},$$

according to the definition of $\boldsymbol{E}^{(i,j)}$, we consider $\boldsymbol{\Lambda}\boldsymbol{I}$ and $\boldsymbol{I}{}^t\boldsymbol{\Lambda}$, respectively, and obtain the following expressions:

$$\boldsymbol{\Lambda}\cdot\boldsymbol{I} = \sum_{i,j\in\mathbb{Z}} \delta_{i,j}\boldsymbol{E}^{(i-1,j)} = \sum_{i,j\in\mathbb{Z}} \delta_{i+1,j}\boldsymbol{E}^{(i,j)},$$
$$\boldsymbol{I}\cdot{}^t\boldsymbol{\Lambda} = \sum_{i,j\in\mathbb{Z}} \delta_{i,j}\boldsymbol{E}^{(i,j-1)} = \sum_{i,j\in\mathbb{Z}} \delta_{i,j+1}\boldsymbol{E}^{(i,j)}.$$

This implies that $\boldsymbol{\Lambda}\boldsymbol{I}$ and $\boldsymbol{I}{}^t\boldsymbol{\Lambda}$ have their respective (i,j)-entries $\delta_{i+1,j}$ and $\delta_{i,j+1}$. Therefore, the visualizations of them are given by

$$\boldsymbol{\Lambda}\cdot\boldsymbol{I} = \begin{pmatrix} \ddots & \ddots & & & \\ & 0 & 1 & & \\ & & \boxed{0} & 1 & \\ & & & 0 & \ddots \\ & & & & \ddots \end{pmatrix} \quad \text{and} \quad \boldsymbol{I}\cdot{}^t\boldsymbol{\Lambda} = \begin{pmatrix} \ddots & & & & \\ \ddots & 0 & & & \\ & 1 & \boxed{0} & & \\ & & 1 & 0 & \\ & & & \ddots & \ddots \end{pmatrix},$$

respectively.

Similarly, $\boldsymbol{o}$ and ${}^t\boldsymbol{o}$ are degenerations of $\boldsymbol{u}$ and ${}^t\boldsymbol{u}$ defined in (2.7), respectively, when $u_i = \delta_{i,0}$. Thus, we can conclude that $\boldsymbol{o}$ and ${}^t\boldsymbol{o}$ are the two "centred" vectors, and their visualisations are

$$\boldsymbol{o} = \left(\cdots,0,\boxed{1},0,\cdots\right)^{\mathrm{T}}, \quad {}^t\boldsymbol{o} = \left(\cdots,0,\boxed{1},0,\cdots\right),$$

respectively.

The matrix $\boldsymbol{E}^{(i,j)}$ and the vectors $\boldsymbol{e}^{(i)}$ and ${}^t\boldsymbol{e}^{(i)}$ can also be visualised in the same way. By observing their corresponding coefficients, we can conclude that $\boldsymbol{E}^{(i,j)}$ is an infinite matrix having the (i,j)-entry 1 and other entries zero; $\boldsymbol{e}^{(i)}$ is an infinite-dimensional column vector having the ith-component 1 and other components zero, and ${}^t\boldsymbol{e}^{(i)}$ is its transpose (therefore a row vector).

2.4 Some useful operations

In this subsection, we explain some operations on infinite matrices and vectors, and give some concrete examples which would be used in the DL framework.

For an infinite matrix $\boldsymbol{U}$, its (i,j)-entry $(\boldsymbol{U})_{i,j}$ is defined as the coefficient of $\boldsymbol{E}^{(i,j)}$ in the expansion (2.4). Similarly, the ith-components of $\boldsymbol{u}$ and ${}^t\boldsymbol{u}$ are defined as the coefficients of $\boldsymbol{e}^{(i)}$ and ${}^t\boldsymbol{e}^{(i)}$ according to (2.7). We also point out that the action $(\,\cdot\,)_{i,j}$ can all be expressed by $(\,\cdot\,)_{0,0}$ through the relation

$$(\boldsymbol{U})_{i,j} = U_{i,j} = (\boldsymbol{\Lambda}^i \boldsymbol{U} {}^t\boldsymbol{\Lambda}^j)_{0,0}.$$

Similarly, for an infinite-dimensional vector $\boldsymbol{u}$ and its transpose ${}^t\boldsymbol{u}$ as defined in (2.7), we have

$$(\boldsymbol{\Lambda}^i \boldsymbol{u})_0 = u_i, \quad ({}^t\boldsymbol{u}\,{}^t\boldsymbol{\Lambda}^j)_0 = u_j.$$

Therefore, we only need the action $(\,\cdot\,)_{0,0}$ for convenience in the future. Sometimes, there also exist operations involving the index-raising operators $\boldsymbol{\Lambda}$ and ${}^t\boldsymbol{\Lambda}$ in their rational form acting on $\boldsymbol{U}$, or $\boldsymbol{u}$ (as well as ${}^t\boldsymbol{u}$). For instance[2],

$$\left(\frac{1}{1-\boldsymbol{\Lambda}}\boldsymbol{U}\right)_{0,0} = \left(\sum_{i=0}^{\infty}\boldsymbol{\Lambda}^i\boldsymbol{U}\right)_{0,0} = \sum_{i=0}^{\infty}(\boldsymbol{\Lambda}^i\boldsymbol{U})_{0,0} = \sum_{i=0}^{\infty}U_{i,0}.$$

Next, we give some examples of the action $(\,\cdot\,)$ on multiplications of infinite matrices and vectors, especially for the case when the projector $\boldsymbol{O}$ is involved. Suppose that $\boldsymbol{U} = \sum_{i,j\in\mathbb{Z}} U_{i,j}\boldsymbol{E}^{(i,j)}$ and $\boldsymbol{V} = \sum_{i,j\in\mathbb{Z}} V_{i,j}\boldsymbol{E}^{(i,j)}$. We have

$$(\boldsymbol{\Lambda}^{i_1}\boldsymbol{U}{}^t\boldsymbol{\Lambda}^{j_1}\boldsymbol{O}\boldsymbol{\Lambda}^{i_2}\boldsymbol{V}{}^t\boldsymbol{\Lambda}^{j_2})_{0,0} = U_{i_1,j_1}V_{i_2,j_2}.$$

This relation can be proven directly. Notice $\boldsymbol{O} = \boldsymbol{O}^2$ (a particular case of (2.1)). We can prove that

$$\begin{aligned}
&\boldsymbol{\Lambda}^{i_1}\boldsymbol{E}^{(i,j)}{}^t\boldsymbol{\Lambda}^{j_1}\boldsymbol{O}\boldsymbol{\Lambda}^{i_2}\boldsymbol{E}^{(i',j')}{}^t\boldsymbol{\Lambda}^{j_2} \\
&= \boldsymbol{\Lambda}^{i_1-i}\boldsymbol{O}{}^t\boldsymbol{\Lambda}^{j_1-j}\boldsymbol{O}\boldsymbol{O}\boldsymbol{\Lambda}^{i_2-i'}\boldsymbol{O}{}^t\boldsymbol{\Lambda}^{j_2-j'} \\
&= \delta_{j_1-j,0}\delta_{0,i_2-i'}\boldsymbol{\Lambda}^{i_1-i}\boldsymbol{O}{}^t\boldsymbol{\Lambda}^{j_2-j'} = \delta_{j_1-j,0}\delta_{0,i_2-i'}\boldsymbol{E}^{(i-i_1,j'-j_2)},
\end{aligned}$$

where the relation (2.1) is used. Then, it follows from the above relation that

$$\begin{aligned}
&\boldsymbol{\Lambda}^{i_1}\boldsymbol{U}{}^t\boldsymbol{\Lambda}^{j_1}\boldsymbol{O}\boldsymbol{\Lambda}^{i_2}\boldsymbol{V}{}^t\boldsymbol{\Lambda}^{j_2} \\
&= \sum_{i,j,i',j'\in\mathbb{Z}} U_{i,j}V_{i',j'}\boldsymbol{\Lambda}^{i_1}\boldsymbol{E}^{(i,j)}{}^t\boldsymbol{\Lambda}^{j_1}\boldsymbol{O}\boldsymbol{\Lambda}^{i_2}\boldsymbol{E}^{(i',j')}{}^t\boldsymbol{\Lambda}^{j_2} \\
&= \sum_{i,j,i',j'\in\mathbb{Z}} U_{i,j}V_{i',j'}\delta_{j_1-j,0}\delta_{0,i_2-i'}\boldsymbol{E}^{(i-i_1,j'-j_2)} \\
&= \sum_{i,j'\in\mathbb{Z}} U_{i,j_1}V_{i_2,j'}\boldsymbol{E}^{(i-i_1,j'-j_2)} = \sum_{i,j'\in\mathbb{Z}} U_{i+i_1,j_1}V_{i_2,j'+j_2}\boldsymbol{E}^{(i,j')},
\end{aligned}$$

[2]Here 1 denotes the identity.

which implies that its $(0,0)$-entry is $U_{i_1,j_1}V_{i_2,j_2}$. From this example, we observe that the projector $\boldsymbol{O}$ separates the multiplication of infinite matrices, namely the right-hand side of the above equation can be expressed as $(\boldsymbol{\Lambda}^{i_1}\boldsymbol{U}{}^t\boldsymbol{\Lambda}^{j_1})_{0,0}(\boldsymbol{\Lambda}^{i_2}\boldsymbol{U}{}^t\boldsymbol{\Lambda}^{j_2})_{0,0}$.

We also have a similar relation for the multiplication of an infinite matrix $\boldsymbol{U}$ and an infinite-dimensional column vector $\boldsymbol{u}$ when $\boldsymbol{O}$ is involved. For example, by considering $\boldsymbol{U}=\sum_{i,j\in\mathbb{Z}}U_{i,j}\boldsymbol{E}^{(i,j)}$ and $\boldsymbol{u}=\sum_i u_i\boldsymbol{e}^{(i)}$, the following relation holds:

$$(\boldsymbol{\Lambda}^{i_1}\boldsymbol{U}{}^t\boldsymbol{\Lambda}^{j_1}\boldsymbol{O}\boldsymbol{\Lambda}^{i_2}\boldsymbol{u})_0=U_{i_1,j_1}u_{i_2},$$

which still tells us that the projector $\boldsymbol{O}$ separates the multiplication, namely the right-hand side can be written as $(\boldsymbol{\Lambda}^{i_1}\boldsymbol{U}\boldsymbol{\Lambda}^{j_1})_{0,0}(\boldsymbol{\Lambda}^{i_2}\boldsymbol{u})_0$. The proof of the relation is quite similar to the previous proof. The key point is that one needs to use $\boldsymbol{O}=\boldsymbol{O}^2=\boldsymbol{O}\boldsymbol{o}{}^t\boldsymbol{o}$ as well as the rules given in (2.4) and (2.5).

Furthermore, we introduce the following two important infinite-dimensional vectors which will be quite important in the DL framework as follows:

$$\begin{aligned}\boldsymbol{c}_k&=\sum_{i\in\mathbb{Z}}k^i\,\boldsymbol{e}^{(i)}=(\cdots,k^{-1},\boxed{1},k,\cdots)^{\mathrm{T}},\\ {}^t\boldsymbol{c}_{k'}&=\sum_{i\in\mathbb{Z}}k'^i\,{}^t\boldsymbol{e}^{(i)}=(\cdots,k'^{-1},\boxed{1},k',\cdots).\end{aligned}$$

It is easy to verify that the vectors $\boldsymbol{c}_k$ and ${}^t\boldsymbol{c}_{k'}$ obviously satisfy

$$\boldsymbol{\Lambda}\boldsymbol{c}_k=k\boldsymbol{c}_k,\quad {}^t\boldsymbol{c}_{k'}{}^t\boldsymbol{\Lambda}=k'{}^t\boldsymbol{c}_{k'}\quad\text{and}\quad(\boldsymbol{c}_k)_0=1,\quad({}^t\boldsymbol{c}_{k'})_0=1,\tag{2.9}$$

namely $\boldsymbol{c}_k$ and ${}^t\boldsymbol{c}_{k'}$ are eigenvectors of the index-raising operators $\boldsymbol{\Lambda}$ and ${}^t\boldsymbol{\Lambda}$, respectively, with their corresponding eigenvalues k and k'. For instance, the infinite-dimensional vectors $\boldsymbol{c}_k$ and ${}^t\boldsymbol{c}_{k'}$ obey the following relations:

$$\boldsymbol{\Lambda}^i\boldsymbol{c}_k{}^t\boldsymbol{c}_{k'}{}^t\boldsymbol{\Lambda}^j=k^ik'^j\boldsymbol{c}_k{}^t\boldsymbol{c}_{k'},\quad {}^t\boldsymbol{c}_{k'}{}^t\boldsymbol{\Lambda}^j\boldsymbol{O}\boldsymbol{\Lambda}^i\boldsymbol{c}_k=k^ik'^j.$$

The first relation follows from (2.9) directly. For the second one, by direct calculation we have

$$\begin{aligned}{}^t\boldsymbol{c}_{k'}{}^t\boldsymbol{\Lambda}^j\boldsymbol{O}\boldsymbol{\Lambda}^i\boldsymbol{c}_k&=\sum_{i',j'\in\mathbb{Z}}k^{i'}k'^{j'}\,{}^t\boldsymbol{e}^{(j')}{}^t\boldsymbol{\Lambda}^j\boldsymbol{O}\boldsymbol{\Lambda}^i\boldsymbol{e}^{(i')}\\ &=\sum_{i',j'\in\mathbb{Z}}k^{i'}k'^{j'}\,{}^t\boldsymbol{o}{}^t\boldsymbol{\Lambda}^{j-j'}\boldsymbol{o}{}^t\boldsymbol{o}\boldsymbol{\Lambda}^{i-i'}\boldsymbol{o}\\ &=\sum_{i',j'\in\mathbb{Z}}k^{i'}k'^{j'}\delta_{j-j',0}\delta_{0,i-i'}=k^ik'^j,\end{aligned}$$

where we use $\boldsymbol{O}=\boldsymbol{o}{}^t\boldsymbol{o}$ the second equality and (2.5) for the third equality. Based on the second relation, there is one more example which is needed in the later sections. For the infinite matrix $\boldsymbol{\Omega}=-\sum_{i=0}^{\infty}(-{}^t\boldsymbol{\Lambda})^{-i-1}\boldsymbol{O}\boldsymbol{\Lambda}^i$, we have

$${}^t\boldsymbol{c}_{k'}\boldsymbol{\Omega}\boldsymbol{c}_k=\frac{1}{k+k'}\doteq\Omega_{k,k'}.$$

Thus, $\boldsymbol{\Omega}$ is an infinite matrix representation of $\Omega_{k,k'}$. This is because some direct calculation shows that

$$ {}^t\boldsymbol{c}_{k'}\boldsymbol{\Omega}\boldsymbol{c}_k = -\sum_{i=0}^{\infty}(-k')^{-i-1}k^i = \frac{1}{k'}\sum_{i=0}^{\infty}\left(-\frac{k}{k'}\right)^i = \frac{1}{k'}\frac{1}{1+\frac{k}{k'}} = \frac{1}{k+k'}. $$

In other words, the infinite matrix $\boldsymbol{\Omega}$ obeys the relation

$$ \boldsymbol{\Omega}\boldsymbol{\Lambda} + {}^t\boldsymbol{\Lambda}\boldsymbol{\Omega} = \boldsymbol{O}. $$

2.5 Trace and determinant

In the approach we also need the notion of trace and determinant of an infinite matrix, but only in the case when they involve the projector $\boldsymbol{O}$.

The trace of an infinite matrix $\boldsymbol{U}$ taking the form of (2.4) is defined as

$$ \operatorname{tr}\boldsymbol{U} = \sum_{i\in\mathbb{Z}} U_{i,i}. \tag{2.10} $$

Unlike the finite case, this definition may raise an issue about divergence due to the infinite summation. However, in our approach we only consider the case when $\boldsymbol{O}$ is involved, which avoids the divergence issue. The following identity for the trace holds:

$$ \operatorname{tr}(\boldsymbol{O}\boldsymbol{U}) = \operatorname{tr}(\boldsymbol{U}\boldsymbol{O}) = U_{0,0} = (\boldsymbol{U})_{0,0}, \tag{2.11} $$

which allows us to transfer the trace of an infinite matrix to a scalar quantity. We only prove $\operatorname{tr}(\boldsymbol{O}\boldsymbol{U}) = U_{0,0}$, and the proof of the other one is similar. A particular case (when $\boldsymbol{E}^{(i_1,j_1)} = \boldsymbol{E}^{(0,0)} = \boldsymbol{O}$) of (2.3) implies that $\boldsymbol{O}\boldsymbol{E}^{(i,j)} = \delta_{0,i}\boldsymbol{E}^{(0,j)}$. Thus, we have

$$ \boldsymbol{O}\boldsymbol{U} = \sum_{i,j\in\mathbb{Z}} U_{i,j}\boldsymbol{O}\boldsymbol{E}^{(i,j)} = \sum_{i,j\in\mathbb{Z}} U_{i,j}\delta_{0,i}\boldsymbol{E}^{(0,j)} = \sum_{j\in\mathbb{Z}} U_{0,j}\boldsymbol{E}^{(0,j)}, $$

and consequently the (i,j)-entry of $\boldsymbol{U}$ is expressed by $\delta_{0,i}U_{0,j}$. According to the definition (2.10), we have $\operatorname{tr}(\boldsymbol{O}\boldsymbol{U}) = \sum_{i\in\mathbb{Z}}\delta_{0,i}U_{0,i} = U_{0,0}$. The trace also has the properties

$$ \operatorname{tr}(\boldsymbol{U}+\boldsymbol{V}) = \operatorname{tr}\boldsymbol{U} + \operatorname{tr}\boldsymbol{V} \quad \text{and} \quad \operatorname{tr}(\boldsymbol{U}\boldsymbol{V}) = \operatorname{tr}(\boldsymbol{V}\boldsymbol{U}), $$

where $\boldsymbol{U}$ and $\boldsymbol{V}$ are two arbitrary infinite matrices involving $\boldsymbol{O}$.

These properties are exactly the same as the ones in the finite-dimensional matrix theory. The proof is similar, namely it is proven by using the definition. We give a concrete example as follows: Suppose that $\boldsymbol{U}$ is an arbitrary infinite matrix, $\boldsymbol{\Lambda}$ and ${}^t\boldsymbol{\Lambda}$ are the index-raising operators, and $\boldsymbol{O}$ is the projector. We have the following identity:

$$ \operatorname{tr}((\boldsymbol{O}\boldsymbol{\Lambda} - {}^t\boldsymbol{\Lambda}\boldsymbol{O})\boldsymbol{U}) = (\boldsymbol{\Lambda}\boldsymbol{U} - \boldsymbol{U}{}^t\boldsymbol{\Lambda})_{0,0} = U_{1,0} - U_{0,1}. $$

This is a consequence of the above two propositions. By using the linearity of the trace, we can prove

$$\mathrm{tr}((\boldsymbol{O}\boldsymbol{\Lambda} - {}^t\boldsymbol{\Lambda}\boldsymbol{O})\boldsymbol{U}) = \mathrm{tr}(\boldsymbol{O}\boldsymbol{\Lambda}\boldsymbol{U}) - \mathrm{tr}({}^t\boldsymbol{\Lambda}\boldsymbol{O}\boldsymbol{U}).$$

Next, using the cyclic permutation, the right-hand side can be written as

$$\mathrm{tr}(\boldsymbol{O}(\boldsymbol{\Lambda}\boldsymbol{U})) - \mathrm{tr}(\boldsymbol{O}(\boldsymbol{U}{}^t\boldsymbol{\Lambda})).$$

The identity is proven in virtue of (2.11).

The determinant of a finite-dimensional non-zero square matrix $\boldsymbol{A}$ obeys the following important identity:

$$\ln(\det \boldsymbol{A}) = \mathrm{tr}(\ln \boldsymbol{A}),$$

which provides us with a way to generalise the notion to the infinite-dimensional case.

The determinant of an infinite matrix $\boldsymbol{U}$ is defined as

$$\det \boldsymbol{U} = \exp\left(\mathrm{tr}(\ln \boldsymbol{U})\right).$$

The right-hand side should be understand as a series expansion. In our framework, we only deal with a particular class of determinants taking the form of $\det(1+*)$, where $\boldsymbol{O}$ is involved in $*$. Under such an assumption, a determinant can be evaluated in terms of a scalar quantity; thus, we avoid the issue of divergence in the expansion.

In applications, we need the well-known Weinstein–Aronszajn formula to evaluate the determinant of an infinite matrix. As examples, we list the rank 1 and rank 2 cases. The determinant of infinite matrices obeys the following identities:

$$\det(1+\boldsymbol{U}\boldsymbol{O}\boldsymbol{V}) = 1 + (\boldsymbol{V}\boldsymbol{U})_{0,0},$$
$$\det(1+\boldsymbol{U}(\boldsymbol{O}\boldsymbol{\Lambda} - {}^t\boldsymbol{\Lambda}\boldsymbol{O})\boldsymbol{V}) = \det\begin{pmatrix} 1+(\boldsymbol{\Lambda}\boldsymbol{V}\boldsymbol{U})_{0,0} & -(\boldsymbol{\Lambda}\boldsymbol{V}{}^t\boldsymbol{\Lambda})_{0,0} \\ (\boldsymbol{V}\boldsymbol{U})_{0,0} & 1-(\boldsymbol{V}\boldsymbol{U}{}^t\boldsymbol{\Lambda})_{0,0} \end{pmatrix}.$$

The left-hand sides of the above equalities should be understood as

$$\det(1+(\boldsymbol{U}\boldsymbol{o})({}^t\boldsymbol{o}\boldsymbol{V})) \quad \text{and} \quad \det\left(1+(\boldsymbol{U}\boldsymbol{o}, -\boldsymbol{U}{}^t\boldsymbol{\Lambda}\boldsymbol{o})\begin{pmatrix} {}^t\boldsymbol{o}\boldsymbol{\Lambda}\boldsymbol{V} \\ {}^t\boldsymbol{o}\boldsymbol{V} \end{pmatrix}\right),$$

respectively, where $\boldsymbol{U}\boldsymbol{o}$ and $\boldsymbol{U}{}^t\boldsymbol{\Lambda}\boldsymbol{o}$ are two infinite-dimensional column vectors, and ${}^t\boldsymbol{o}\boldsymbol{V}$ and ${}^t\boldsymbol{o}\boldsymbol{\Lambda}\boldsymbol{V}$ are two infinite-dimensional row vectors. Therefore, the above equalities are nothing but the infinite matrix analogues of the rank 1 and rank 2 cases of the identity

$$\det(1+\boldsymbol{A}\boldsymbol{B}) = \det(1+\boldsymbol{B}\boldsymbol{A}),$$

where $\boldsymbol{A}$ and $\boldsymbol{B}$ are finite-dimensional matrices of sizes $M\times N$ and $N\times M$, respectively.

3 General theory of the direct linearisation

3.1 Notations

Before we discuss the DL framework, we introduce some notations in terms of continuous and discrete dynamics. We consider two infinite sets of variables

$$\boldsymbol{x} = (x_1, x_2, \cdots) \quad \text{and} \quad \boldsymbol{n} = (n_1, n_2, \cdots),$$

in which x_j are the continuous flow variables, and n_j are the discrete flow variables associated with their corresponding lattice parameters p_j. We also note that the indices j for $j \in \mathbb{Z}$ of x and n denote the positive flows (when $j > 0$) and negative flows (when $j < 0$), respectively. Consider a smooth function $f = f_n \doteq f(\boldsymbol{x}; \boldsymbol{n})$, where n is an extra discrete variable associated with its corresponding lattice parameter zero.

We introduce the notation ∂_j denoting the partial derivative with respect to the continuous variable x_j, namely

$$\partial_j f(\boldsymbol{x}, \boldsymbol{n}) \doteq \frac{\partial}{\partial x_j} f(\boldsymbol{x}; \boldsymbol{n}).$$

Based on the partial derivative, we further give the definition of Hirota's bilinear derivative with the notation D_j as follows:

$$D_j f \cdot g \doteq \left(\frac{\partial}{\partial x_j} - \frac{\partial}{\partial x'_j} \right) f(\cdots, x_j, \cdots) g(\cdots, x'_j, \cdots) \bigg|_{x'_j = x_j},$$

for smooth functions f and g.

We also introduce the discrete forward and backward shift operators T_j and T_j^{-1} in terms of n_j defined by

$$\begin{aligned} \mathrm{T}_j f(\cdots, n_j, \cdots) &= f(\cdots, n_j + 1, \cdots), \\ \mathrm{T}_j^{-1} f(\cdots, n_j, \cdots) &= f(\cdots, n_j - 1, \cdots). \end{aligned}$$

3.2 Linear integral equation

The idea of the direct linearising method is to solve a nonlinear integrable equation by considering its associated linear integral equation, namely a solution to the linear integral equation can help to express a solution to the nonlinear equation by nonlinearisation. Reversely, considering the structure of a linear integral equation also provides a way to construct its related nonlinear equation. For the most general structure of the DL framework, we consider a linear integral equation taking the form of

$$\boldsymbol{u}_k + \iint_D \mathrm{d}\zeta(l, l') \rho_k \Omega_{k,l'} \sigma_{l'} \boldsymbol{u}_l = \rho_k \boldsymbol{c}_k, \tag{3.1}$$

where the wave function $\boldsymbol{u}_k$ is an infinite column vector having its ith-component $u_k^{(i)}$ as a smooth function of the continuous independent variables $\boldsymbol{x}$, the discrete independent variables $\boldsymbol{n}$ (and also their associated lattice parameters p_j), the extra discrete variable n, as well as the spectral parameter k; the Cauchy kernel $\Omega_{k,l'}$ is an algebraic expression of the spectral parameters k and l', independent of the independent variables $\boldsymbol{x}$, $\boldsymbol{n}$, n and lattice parameters $\{p_j|j \in \mathbb{Z}\}$; the plane wave functions ρ_k and $\sigma_{l'}$ are expressions of the discrete and continuous independent variables (i.e., flow variables) $\boldsymbol{x}$, $\boldsymbol{n}$, n as well as the lattice parameters $\{p_j|j \in \mathbb{Z}\}$, also depending on the spectral parameters k and l', respectively; the infinite column vector $\boldsymbol{c}_k$ takes the form of $\boldsymbol{c}_k = (\cdots, k^{-1}, 1, k, \cdots)^{\mathrm{T}}$; the measure $\mathrm{d}\zeta(l,l')$ depending on the spectral variables l and l', and the integration domain D can be determined later for particular classes of solutions.

Although the wave function $\boldsymbol{u}_k$ in the linear integral equation (3.1) is a multi-component (i.e., infinite-component) vector, it is still suitable to think of it as a scalar equation since each component $u_k^{(i)}$ solves an independent linear integral equation. The reason why we introduce an infinite number of components is because this brings a possibility to study all the gauge equivalent integrable equations in the same class simultaneously.

The existence of two spectral variables l and l' in the measure $\mathrm{d}\zeta(l,l')$ on the domain D is crucial as they govern the general solution for 3D integrable hierarchies. Such a structure is also a reflection of an underlying nonlocal Riemann–Hilbert (or $\bar{\partial}$-) problem. In fact, the core step in a nonlocal Riemann–Hilbert problem is solving a linear integral equation, which is a special reduction of (3.1). When the measure collapses, namely l and l' become algebraically dependent, the double integral turns out to be a single integral, which results in 2D integrable equations. This is how a dimensional reduction of a 3D equation is performed.

The plane wave factors ρ_k and $\sigma_{l'}$ describe the dispersion of a nonlinear integrable equation, namely its linear structure. The Cauchy kernel, instead, reflects the nonlinear structure of the nonlinear equation; for instance, it produces the phase factor in a two-soliton solution, which is a characteristic helping to understand the nonlinear interaction between two solitons.

3.3 Infinite matrix representation

We have explained the linear integral equation in the DL. The next step is to give the infinite matrix representation of the integral equation and the related quantities. Before we deal with this, we need to introduce some essential infinite matrices, which reflect the structure of the linear integral equation.

We introduce an infinite matrix $\boldsymbol{\Omega}$ defined by the following relation:

$$\Omega_{k,k'} = {}^{t}\boldsymbol{c}_{k'}\boldsymbol{\Omega}\,\boldsymbol{c}_k. \tag{3.2}$$

This relation implies that $\boldsymbol{\Omega}$ is the infinite matrix representation of the Cauchy kernel $\Omega_{k,k'}$. We also define an infinite matrix $\boldsymbol{C}$ as follows:

$$\boldsymbol{C} = \iint_D \mathrm{d}\zeta(k,k')\rho_k \boldsymbol{c}_k {}^{t}\boldsymbol{c}_{k'}\sigma_{k'}. \tag{3.3}$$

The core part of this infinite matrix is $\rho_k\sigma_{k'}$, namely the effective dispersion in the linear integral equation. Thus, $\boldsymbol{C}$ is the infinite matrix representation of the plane wave factor of the linear integral equation. Moreover, we define an infinite matrix $\boldsymbol{U}$ as follows:

$$\boldsymbol{U} = \iint_D \mathrm{d}\zeta(k,k')\boldsymbol{u}_k{}^t\boldsymbol{c}_{k'}\sigma_{k'}. \tag{3.4}$$

The infinite matrix $\boldsymbol{U}$ is a non-linearisation of the wave function – the wave function $\boldsymbol{u}_k$ solves the linear integral equation and we complement it with the plane wave factor $\sigma_{k'}$.

From the above definitions, we can conclude that the main idea to construct the infinite matrix representations of the certain quantities in the linear integral equation is to get the infinite-dimensional vectors $\boldsymbol{c}_k$ and ${}^t\boldsymbol{c}_{k'}$ involved. Actually, such a treatment brings us all the powers of the spectral variables k and k' in the structure, which will later help us with construct closed-form nonlinear equations.

We can now express the linear integral equation by the above infinite matrices. The linear integral (3.1) has the infinite matrix representation

$$\boldsymbol{u}_k = (1 - \boldsymbol{U\Omega})\rho_k\boldsymbol{c}_k. \tag{3.5}$$

According to the definition of $\boldsymbol{\Omega}$, we can replace the Cauchy kernel and rewrite the linear integral equation (3.1) as

$$\begin{aligned}\rho_k\boldsymbol{c}_k &= \boldsymbol{u}_k + \iint_D \mathrm{d}\zeta(l,l')\rho_k\sigma_{l'}\boldsymbol{u}_l{}^t\boldsymbol{c}_{l'}\boldsymbol{\Omega}\boldsymbol{c}_k \\ &= \boldsymbol{u}_k + \left(\iint_D \mathrm{d}\zeta(l,l')\boldsymbol{u}_l{}^t\boldsymbol{c}_{l'}\sigma_{l'}\right)\boldsymbol{\Omega}\rho_k\boldsymbol{c}_k,\end{aligned}$$

which implies that $\boldsymbol{u}_k = (1 - \boldsymbol{U\Omega})\rho_k\boldsymbol{c}_k$. As a consequence of (3.5), the infinite matrix $\boldsymbol{U}$ satisfies the following relation:

$$\boldsymbol{U} = (1 - \boldsymbol{U\Omega})\boldsymbol{C}, \quad \text{or alternatively} \quad \boldsymbol{U} = \boldsymbol{C}(1 + \boldsymbol{\Omega C})^{-1}. \tag{3.6}$$

This follows from (3.5) directly. One can act $\iint_D \mathrm{d}\zeta(k,k')^t\boldsymbol{c}_{k'}\sigma_{k'}$ on (3.5), which gives rise to the above relation.

In addition, we also need the tau function in the framework. The tau function in the DL framework is defined as

$$\tau = \det(1 + \boldsymbol{\Omega C}). \tag{3.7}$$

The tau function contains the main structure of the linear integral equation, namely, the Cauchy kernel (in $\boldsymbol{\Omega}$), the plane wave factors and the measure (in $\boldsymbol{C}$). Thus, it can be thought of as a fundamental quantity in a sense. Alternatively, one can also define the function by $\tau = \det(1 + \boldsymbol{C\Omega})$. This is because

$$\det(1 + \boldsymbol{C\Omega}) = \exp\{\mathrm{tr}[\ln(1 + \boldsymbol{C\Omega})]\} = \exp\{\mathrm{tr}[\ln(1 + \boldsymbol{\Omega C})]\} = \det(1 + \boldsymbol{\Omega C}),$$

which is exactly the same as the definition of the tau function (3.7).

4 Discrete and continuous KP hierarchies

4.1 Infinite matrix representation of KP

As we have seen in the above section, the key ingredients in the DL framework are the Cauchy kernel, the plane wave factors, as well as the integration measure. For the discrete and continuous KP hierarchies, we select the kernel

$$\Omega_{k,k'} = \frac{1}{k+k'}, \quad \text{with the measure} \quad \mathrm{d}\zeta(k,k') \quad \text{being arbitrary,} \tag{4.1}$$

and the plane wave factors as follows:

$$\rho_k = \prod_{j=1}^{\infty}(p_j+k)^{n_j}\left(\sum_{j=1}^{\infty}k^j x_j\right), \tag{4.2a}$$

$$\sigma_{k'} = \prod_{j=1}^{\infty}(p_j-k')^{-n_j}\left(-\sum_{j=1}^{\infty}(-k')^j x_j\right). \tag{4.2b}$$

Once these objects are given, the linear integral equation is fixed, and therefore, the DL structure is fully determined.

Following the definition of $\boldsymbol{C}$ in Section 3, one derives the dynamical evolutions of the infinite matrix $\boldsymbol{C}$

$$(\mathrm{T}_j\boldsymbol{C})(p_j-{}^t\boldsymbol{\Lambda}) = (p_j+\boldsymbol{\Lambda})\boldsymbol{C} \quad \text{and} \quad \partial_j\boldsymbol{C} = \boldsymbol{\Lambda}^j\boldsymbol{C} - \boldsymbol{C}(-{}^t\boldsymbol{\Lambda})^j, \tag{4.3}$$

for $j=1,2,\cdots$. There relations are natural consequences if one considers the dynamical evolutions of the effective plane wave factor $\rho_k\sigma_{k'}$, with the help of Equation (2.9). Notice that the kernel given in (4.1), we can deduce that its corresponding infinite matrix $\boldsymbol{\Omega}$ satisfies $\boldsymbol{\Omega}\boldsymbol{\Lambda} + {}^t\boldsymbol{\Lambda}\boldsymbol{\Omega} = \boldsymbol{O}$ according the example in Section 3, and subsequently one can further derive

$$\boldsymbol{\Omega}(p_j+\boldsymbol{\Lambda}) - (p_j-{}^t\boldsymbol{\Lambda})\boldsymbol{\Omega} = \boldsymbol{O} \quad \text{as well as} \quad \boldsymbol{\Omega}\boldsymbol{\Lambda}^j - (-{}^t\boldsymbol{\Lambda})^j\boldsymbol{\Omega} = \boldsymbol{O}_j, \tag{4.4}$$

where

$$\boldsymbol{O}_j = \sum_{i=0}^{j-1}(-{}^t\boldsymbol{\Lambda})^{j-1-i}\boldsymbol{O}\,\boldsymbol{\Lambda}^i,$$

for $j=1,2,\cdots$.

Now it is possible to construct the infinite matrix representation of the discrete and continuous KP hierarchies. The key object in the formulism is the matrix $\boldsymbol{U}$ introduced in (3.4). If we shift (3.6) with respect to n_j, or differentiate this equation with respect to x_j, the following equations arise:

$$(\mathrm{T}_j\boldsymbol{U})(p_j-{}^t\boldsymbol{\Lambda}) = (p_j+\boldsymbol{\Lambda})\boldsymbol{U} - (\mathrm{T}_j\boldsymbol{U})\boldsymbol{O}\boldsymbol{U}, \tag{4.5a}$$

$$\partial_j\boldsymbol{U} = \boldsymbol{\Lambda}^j\boldsymbol{U} - \boldsymbol{U}(-{}^t\boldsymbol{\Lambda})^j - \boldsymbol{U}\boldsymbol{O}_j\boldsymbol{U}. \tag{4.5b}$$

Below we provide the derivations of both equations. Shifting Equation (3.6) with respect to n_j and multiplying it by $p_j - {}^t\boldsymbol{\Lambda}$, we obtain

$$\begin{aligned}(\mathrm{T}_j\boldsymbol{U})(p_j - {}^t\boldsymbol{\Lambda}) &= [1 - (\mathrm{T}_j\boldsymbol{U})\boldsymbol{\Omega}](\mathrm{T}_j\boldsymbol{C})(p_j - {}^t\boldsymbol{\Lambda}) \\ &= [1 - (\mathrm{T}_j\boldsymbol{U})\boldsymbol{\Omega}](p_j + \boldsymbol{\Lambda})\boldsymbol{C},\end{aligned}$$

in which the first equation in (4.3) is used. Now we make use of the first equation in (4.4), the above equation is further reformulated as

$$(\mathrm{T}_j\boldsymbol{U})(p_j - {}^t\boldsymbol{\Lambda}) = (p_j + \boldsymbol{\Lambda})\boldsymbol{C} - (\mathrm{T}_j\boldsymbol{U})[\boldsymbol{O} + (p_j - {}^t\boldsymbol{\Lambda})\boldsymbol{\Omega}]\boldsymbol{C},$$

or equivalently,

$$(\mathrm{T}_j\boldsymbol{U})(p_j - {}^t\boldsymbol{\Lambda})(1 + \boldsymbol{\Omega}\boldsymbol{C}) = (p_j + \boldsymbol{\Lambda})\boldsymbol{C} - (\mathrm{T}_j\boldsymbol{U})\boldsymbol{O}\boldsymbol{C},$$

which is nothing but (4.5a) if one multiplies it by $(1 + \boldsymbol{\Omega}\boldsymbol{C})^{-1}$. As for (4.5b), direct calculation shows that

$$\partial_j\boldsymbol{U} = (1 - \boldsymbol{U}\boldsymbol{\Omega})(\partial_j\boldsymbol{C}) - (\partial_j\boldsymbol{U})\boldsymbol{\Omega}\boldsymbol{C},$$

and thus, we have

$$\begin{aligned}(\partial_j\boldsymbol{U})(1 + \boldsymbol{\Omega}\boldsymbol{C}) &= (1 - \boldsymbol{U}\boldsymbol{\Omega})[\boldsymbol{\Lambda}^j\boldsymbol{C} - \boldsymbol{C}(-{}^t\boldsymbol{\Lambda})^j] \\ &= \boldsymbol{\Lambda}^j\boldsymbol{C} - \boldsymbol{U}\boldsymbol{\Omega}\boldsymbol{\Lambda}^j\boldsymbol{C} - \boldsymbol{U}(-{}^t\boldsymbol{\Lambda})^j \\ &= \boldsymbol{\Lambda}^j\boldsymbol{C} - \boldsymbol{U}(-{}^t\boldsymbol{\Lambda})^j - \boldsymbol{U}[\boldsymbol{O}_j + (-{}^t\boldsymbol{\Lambda})^j\boldsymbol{\Omega}]\boldsymbol{C} \\ &= \boldsymbol{\Lambda}^j\boldsymbol{C} - \boldsymbol{U}(-{}^t\boldsymbol{\Lambda})^j(1 + \boldsymbol{\Omega}\boldsymbol{C}) - \boldsymbol{U}\boldsymbol{O}_j\boldsymbol{C},\end{aligned}$$

where we have made use of the second relation in (4.4). This is nothing but the dynamical equation of $\boldsymbol{U}$ with respect to x_j, if we multiply $(1 + \boldsymbol{\Omega}\boldsymbol{C})^{-1}$ from the right.

Equation (4.5) is the infinite matrix representation of the discrete and continuous KP hierarchies in the DL framework, in which the index j is chosen from the set of positive integers, and this guarantees that KP has an infinite number of degrees of freedom, in both discrete and continuous spaces.

4.2 Closed-form nonlinear equations

Equations in (4.5) provides a formal representation of the discrete and continuous KP equations. Since it is wildly known that the KP hierarchy is composed of infinitely many $(2+1)$-dimensional equations, in this subsection, we construct $(2+1)$-dimensional closed-form discrete and continuous equations, based on a single variable $u = U_{0,0}$, from Equation (4.5).

We first construct the fully discrete KP hierarchy, which was derived from (4.5a). Taking the $(0,0)$-entry of (4.5a), the following equations arise:

$$p_j\mathrm{T}_ju - \mathrm{T}_jU_{0,1} = p_ju + U_{1,0} - (\mathrm{T}_ju)u. \tag{4.6}$$

for $j = 1, 2, \cdots$. Now if we arbitrarily select three distinct directions, for example p_h, p_i and p_j, we have

$$
\begin{aligned}
&p_h \mathrm{T}_h u - \mathrm{T}_h U_{0,1} = p_h u + U_{1,0} - (\mathrm{T}_h u)u, \\
&p_i \mathrm{T}_i u - \mathrm{T}_i U_{0,1} = p_i u + U_{1,0} - (\mathrm{T}_i u)u,
\end{aligned}
$$

in addition to (4.6). From these three equations, it is possible to eliminate $U_{1,0}$ and obtain the following equations:

$$
\begin{aligned}
&p_h \mathrm{T}_h u - p_i \mathrm{T}_i u + \mathrm{T}_i U_{0,1} - \mathrm{T}_h U_{0,1} = (p_h - p_i)u + u(\mathrm{T}_i u - \mathrm{T}_h u), \\
&p_i \mathrm{T}_i u - p_j \mathrm{T}_j u + \mathrm{T}_j U_{0,1} - \mathrm{T}_i U_{0,1} = (p_i - p_j)u + u(\mathrm{T}_j u - \mathrm{T}_i u), \\
&p_j \mathrm{T}_j u - p_h \mathrm{T}_h u + \mathrm{T}_h U_{0,1} - \mathrm{T}_j U_{0,1} = (p_j - p_h)u + u(\mathrm{T}_h u - \mathrm{T}_j u).
\end{aligned}
$$

Now if we perform T_j, T_h and T_i on the above three equations, respectively, and eliminate $U_{0,1}$, the following closed-form equation is derived:

$$
\begin{aligned}
&(p_h - \mathrm{T}_h u)(p_i - p_j + \mathrm{T}_h \mathrm{T}_j u - \mathrm{T}_h \mathrm{T}_i u) \\
&\quad + (p_i - \mathrm{T}_i u)(p_j - p_h + \mathrm{T}_i \mathrm{T}_h u - \mathrm{T}_i \mathrm{T}_j u) \\
&\qquad + (p_j - \mathrm{T}_j u)(p_h - p_i + \mathrm{T}_j \mathrm{T}_i u - \mathrm{T}_j \mathrm{T}_h u) = 0,
\end{aligned}
\tag{4.7}
$$

which is the fully discrete KP equation. Notice that p_h, p_i and p_j are arbitrary, all the equations in the discrete KP hierarchy take the same form with regard to different lattice directions and parameters. This is the MDC property of the discrete KP equation.

Next, we consider the semi-discrete KP hierarchy (the hierarchy with one continuous variables and two discrete variables). In this case, we need to combine simultaneously (4.5a) and (4.5b), and make use of the following relations:

$$
\begin{aligned}
&(\mathrm{T}_h \boldsymbol{U})(p_h - {}^t\boldsymbol{\Lambda}) = (p_h + \boldsymbol{\Lambda})\boldsymbol{U} - (\mathrm{T}_h \boldsymbol{U})\boldsymbol{O}\boldsymbol{U}, \\
&(\mathrm{T}_i \boldsymbol{U})(p_i - {}^t\boldsymbol{\Lambda}) = (p_i + \boldsymbol{\Lambda})\boldsymbol{U} - (\mathrm{T}_i \boldsymbol{U})\boldsymbol{O}\boldsymbol{U}, \\
&\partial_j \boldsymbol{U} = \boldsymbol{\Lambda}^j \boldsymbol{U} - \boldsymbol{U}(-{}^t\boldsymbol{\Lambda})^j - \boldsymbol{U}\boldsymbol{O}_j \boldsymbol{U}.
\end{aligned}
$$

By eliminating $\boldsymbol{\Lambda}\boldsymbol{U}$ and $\boldsymbol{U}{}^t\boldsymbol{\Lambda}$ from the first two equations, respectively, the following two equations are derived:

$$
\begin{aligned}
(\mathrm{T}_i - \mathrm{T}_h)\boldsymbol{U}{}^t\boldsymbol{\Lambda} &= (p_h - p_i)\boldsymbol{U} - (p_h \mathrm{T}_h - p_i \mathrm{T}_i)\boldsymbol{U} - (\mathrm{T}_h \boldsymbol{U} - \mathrm{T}_i \boldsymbol{U})\boldsymbol{O}\boldsymbol{U}, \\
(\mathrm{T}_i - \mathrm{T}_h)\boldsymbol{\Lambda}\boldsymbol{U} &= (p_h - p_i)(\mathrm{T}_h \mathrm{T}_i \boldsymbol{U}) \\
&\quad - (p_h \mathrm{T}_i - p_i \mathrm{T}_h)\boldsymbol{U} - (\mathrm{T}_h \mathrm{T}_i \boldsymbol{U})\boldsymbol{O}(\mathrm{T}_h - \mathrm{T}_i)\boldsymbol{U}.
\end{aligned}
$$

With the help of these two equations, one is able to eliminate $\boldsymbol{\Lambda}$ and ${}^t\boldsymbol{\Lambda}$ in the continuous equations, and consequently obtains the semi-discrete KP hierarchy by taking the $(0,0)$-entry. For example, the $j = 1$ case gives rise to the closed-form equation

$$
\partial_1(\mathrm{T}_i u - \mathrm{T}_h u) = (p_h - p_i + \mathrm{T}_i u - \mathrm{T}_h u)(u - \mathrm{T}_h u - \mathrm{T}_i u + \mathrm{T}_h \mathrm{T}_i u), \tag{4.8}
$$

which is the semi-discrete KP equation.

Furthermore, by combining the dynamical relations

$$\begin{aligned}
&(\mathrm{T}_h\boldsymbol{U})(p_h - {}^t\boldsymbol{\Lambda}) = (p_h + \boldsymbol{\Lambda})\boldsymbol{U} - (\mathrm{T}_h\boldsymbol{U})\boldsymbol{O}\boldsymbol{U},\\
&\partial_1\boldsymbol{U} = \boldsymbol{\Lambda}\boldsymbol{U} + \boldsymbol{U}{}^t\boldsymbol{\Lambda} - \boldsymbol{U}\boldsymbol{O}\boldsymbol{U},\\
&\partial_j\boldsymbol{U} = \boldsymbol{\Lambda}^j\boldsymbol{U} - \boldsymbol{U}(-{}^t\boldsymbol{\Lambda})^j - \boldsymbol{U}\boldsymbol{O}_j\boldsymbol{U},
\end{aligned}$$

we are able to construct the semi-continuous KP hierarchy (with two continuous and one discrete independent variables), namely a hierarchy having n_h, x_1 and x_j, in which x_j is the flow variable. From the first two relations, we have

$$\begin{aligned}
&(\mathrm{T}_h - 1)\boldsymbol{\Lambda}\boldsymbol{U} = \partial_1\mathrm{T}_h\boldsymbol{U} - p_h(\mathrm{T}_h\boldsymbol{U} - \boldsymbol{U}) + (\mathrm{T}_h\boldsymbol{U})\boldsymbol{O}(\mathrm{T}_h\boldsymbol{U} - \boldsymbol{U}),\\
&(\mathrm{T}_h - 1)\boldsymbol{U}{}^t\boldsymbol{\Lambda} = p_h(\mathrm{T}_h\boldsymbol{U} - \boldsymbol{U}) - \partial_1\boldsymbol{U} + (\mathrm{T}_h\boldsymbol{U} - \boldsymbol{U})\boldsymbol{O}\boldsymbol{U},
\end{aligned}$$

which can help us with reducing the powers of $\boldsymbol{\Lambda}$ and ${}^t\boldsymbol{\Lambda}$ in the relation for $\partial_j\boldsymbol{U}$. Then the $(0,0)$-entry gives us all the equations in the semi-continuous KP hierarchy, in which the first nontrivial equation is the x_2-flow

$$\partial_2(\mathrm{T}_h u - u) = \partial_1^2(\mathrm{T}_h u + u) + \partial_1(\mathrm{T}_h u - u)^2 - 2p_h\partial_1(\mathrm{T}_h u - u), \tag{4.9}$$

namely the semi-continuous KP equation, which in the literature is often referred to as the differential-difference KP equation.

Finally, we present the continuous KP hierarchy, in which case the continuous variables x_1 and x_2 are fixed as base variables, and x_j is the flow variable. For this reason, we select the following equations from (4.5b):

$$\begin{aligned}
&\partial_1\boldsymbol{U} = \boldsymbol{\Lambda}\boldsymbol{U} + \boldsymbol{U}{}^t\boldsymbol{\Lambda} - \boldsymbol{U}\boldsymbol{O}\boldsymbol{U},\\
&\partial_2\boldsymbol{U} = \boldsymbol{\Lambda}^2\boldsymbol{U} - \boldsymbol{U}{}^t\boldsymbol{\Lambda}^2 - \boldsymbol{U}(\boldsymbol{O}\boldsymbol{\Lambda} - {}^t\boldsymbol{\Lambda}\boldsymbol{O})\boldsymbol{U},\\
&\partial_j\boldsymbol{U} = \boldsymbol{\Lambda}^j\boldsymbol{U} - \boldsymbol{U}(-{}^t\boldsymbol{\Lambda})^j - \boldsymbol{U}\boldsymbol{O}_j\boldsymbol{U}.
\end{aligned}$$

Similarly, the first two equations provide us with the following:

$$\begin{aligned}
&\boldsymbol{\Lambda}\boldsymbol{U} = \frac{1}{2}\partial_1^{-1}[\partial_2\boldsymbol{U} + \partial_1^2\boldsymbol{U} + 2\boldsymbol{U}\boldsymbol{O}(\partial_1\boldsymbol{U})],\\
&\boldsymbol{U}(-{}^t\boldsymbol{\Lambda}) = \frac{1}{2}\partial_1^{-1}[\partial_2\boldsymbol{U} - \partial_1^2\boldsymbol{U} - 2(\partial_1\boldsymbol{U})\boldsymbol{O}\boldsymbol{U}],
\end{aligned}$$

from which one is able to reduce the order of $\boldsymbol{\Lambda}$ and ${}^t\boldsymbol{\Lambda}$ in the third equation and obtain an equation depending on only $\boldsymbol{U}$ and $\boldsymbol{O}$, namely the jth flow in the continuous KP hierarchy. The KP equation arises as the x_3-flow in this hierarchy, taking the form of

$$\partial_3 u = \frac{1}{4}\partial_1^3 u + \frac{3}{2}(\partial_1 u)^2 + \frac{3}{4}\partial_1^{-1}\partial_2^2 u, \tag{4.10}$$

as the equation from the $(0,0)$-entry.

Equations (4.7), (4.8), (4.9) and (4.10) are the fully discrete, semi-discrete, semi-continuous and fully continuous KP equations. There are two different viewpoints to understand these equations as follows: One can start from the discrete equation (4.7), by taking the continuum limit, one can reach to (4.8), (4.9) and (4.10) step by step. Alternatively, we can think of the continuous equation (4.10) as our starting point, and then verify that (4.9) is its Bäcklund transformation associated with the Bäcklund parameter p_h, which further implies that (4.8) is the nonlinear superposition formula, and (4.7) is the closed-form algebraic relations for six solutions of the continuous KP equation.

4.3 Associated linear problems

We consider the associated linear problems for the discrete and continuous KP hierarchies in this subsection, which is done through the infinite matrix representation of the linear integral equation, namely Equation (3.5). By following the similar derivation of (4.5), we can derive from (3.5) that the infinite-dimensional vector $\boldsymbol{u}_k$ satisfies the following dynamical relations:

$$\mathrm{T}_j\boldsymbol{u}_k = (p_j+\boldsymbol{\Lambda})\boldsymbol{u}_k - (\mathrm{T}_j\boldsymbol{U})\boldsymbol{O}\,\boldsymbol{u}_k, \tag{4.11a}$$

$$\partial_j\boldsymbol{u}_k = \boldsymbol{\Lambda}^j\boldsymbol{u}_k - \boldsymbol{U}\,\boldsymbol{O}_j\boldsymbol{u}_k, \tag{4.11b}$$

for $j=1,2,\cdots$. Below we give the derivation of these equations. Forward-shifting Equation (3.5) with respect to n_j by one unit gives us

$$\mathrm{T}_j\boldsymbol{u}_k = [1-(\mathrm{T}_j\boldsymbol{U})\boldsymbol{\Omega}](\mathrm{T}_j\rho_k)\boldsymbol{c}_k = [1-(\mathrm{T}_j\boldsymbol{U})\boldsymbol{\Omega}](p_j+\boldsymbol{\Lambda})\rho_k\boldsymbol{c}_k.$$

In virtue of (4.4), we further obtain

$$\begin{aligned}\mathrm{T}_j\boldsymbol{u}_k &= (p_j+\boldsymbol{\Lambda})\rho_k\boldsymbol{c}_k - (\mathrm{T}_j\boldsymbol{U})\boldsymbol{\Omega}(p_j+\boldsymbol{\Lambda})\rho_k\boldsymbol{c}_k \\ &= (p_j+\boldsymbol{\Lambda})\rho_k\boldsymbol{c}_k - (\mathrm{T}_j\boldsymbol{U})\left[\boldsymbol{O}+(p_j-{}^t\boldsymbol{\Lambda})\boldsymbol{\Omega}\right]\rho_k\boldsymbol{c}_k.\end{aligned}$$

With the help of (4.5a), this can be written as

$$\mathrm{T}_j\boldsymbol{u}_k = (p_j+\boldsymbol{\Lambda})\rho_k\boldsymbol{c}_k - (\mathrm{T}_j\boldsymbol{U})\boldsymbol{O}\rho_k\boldsymbol{c}_k - [(p_j+\boldsymbol{\Lambda})\boldsymbol{U}-(\mathrm{T}_j\boldsymbol{U})\boldsymbol{O}\boldsymbol{U}]\,\boldsymbol{\Omega}\rho_k\boldsymbol{c}_k,$$

which is nothing but (4.11a). Next, by taking the derivative of $\boldsymbol{u}_k$ in (3.5) with respect to x_j, we have

$$\begin{aligned}\partial_j\boldsymbol{u}_k &= (1-\boldsymbol{U}\boldsymbol{\Omega})(\partial_j\rho_k)\boldsymbol{c}_k - (\partial_j\boldsymbol{U})\boldsymbol{\Omega}\rho_k\boldsymbol{c}_k \\ &= (1-\boldsymbol{U}\boldsymbol{\Omega})\boldsymbol{\Lambda}^j\rho_k\boldsymbol{c}_k - [\boldsymbol{\Lambda}^j\boldsymbol{U}-\boldsymbol{U}(-{}^t\boldsymbol{\Lambda})^j-\boldsymbol{U}\boldsymbol{O}_j\boldsymbol{U}]\boldsymbol{\Omega}\rho_k\boldsymbol{c}_k \\ &= \boldsymbol{\Lambda}^j(1-\boldsymbol{U}\boldsymbol{\Omega})\rho_k\boldsymbol{c}_k - \boldsymbol{U}[\boldsymbol{\Omega}\boldsymbol{\Lambda}^j-(-{}^t\boldsymbol{\Lambda})^j\boldsymbol{\Omega}]\rho_k\boldsymbol{c}_k + \boldsymbol{U}\boldsymbol{O}_j\boldsymbol{U}\boldsymbol{\Omega}\rho_k\boldsymbol{c}_k,\end{aligned}$$

which results in (4.11b) if one makes use of the second equation in (4.4).

Equations listed in (4.11) form the infinite vector representation of the linear problems for the discrete and continuous KP hierarchies. In order to write down the linear problems for the discrete and continuous KP equations, we introduce the wave function $\phi \doteq (\boldsymbol{u}_k)_0$. We first consider the fully discrete linear equations. By selecting p_h and p_i for (4.11a), respectively, it is possible to eliminate $\boldsymbol{\Lambda u}_k$, and as a consequence we obtain

$$(\mathrm{T}_h - \mathrm{T}_i)\boldsymbol{u}_k = [p_h - p_i + (\mathrm{T}_i\boldsymbol{UO} - (\mathrm{T}_h\boldsymbol{U})\boldsymbol{O}]\boldsymbol{u}_k,$$

whose 0th-component gives rise to

$$(\mathrm{T}_i - \mathrm{T}_j)\phi = (p_i - p_j + \mathrm{T}_j u - \mathrm{T}_i u)\phi. \tag{4.12}$$

Secondly, Equation (4.11a) can be written as

$$\boldsymbol{\Lambda u}_k = [\mathrm{T}_h - p_h + (\mathrm{T}_h\boldsymbol{U})\boldsymbol{O}]\boldsymbol{u}_k,$$

which is the key formula to reduce the order of $\boldsymbol{\Lambda}$ in (4.11b), and therefore, we are able to write down a closed-form equation for $\boldsymbol{u}_k$ with T_h and ∂_j operations, whose simplest case provides us with

$$(\mathrm{T}_h - \partial_1)\boldsymbol{u}_k = [p_h + \boldsymbol{UO} - (\mathrm{T}_h\boldsymbol{U})\boldsymbol{O}]\boldsymbol{u}_k.$$

Taking the 0th-component of this equation, we derive

$$(\mathrm{T}_h - \partial_1)\phi = [p_h + u - (\mathrm{T}_h u)]\phi. \tag{4.13}$$

Likewise, if we concentrate on only the continuous equations from (4.11b), the powers of $\boldsymbol{\Lambda}$ can be reduced through

$$\boldsymbol{\Lambda u}_k = \partial_1\boldsymbol{u}_k + \boldsymbol{UOu}_k,$$

namely the first flow in the sequence, and thus $\partial_j\boldsymbol{u}_k$ is expressed by only $\boldsymbol{u}_k$ and its higher-order derivatives with respective to x_1. And then the action of taking $(0,0)$-entry helps us to write down the continuous linear problems, including

$$\partial_2\phi = [\partial_1^2 + 2(\partial_1 u)]\phi, \tag{4.14}$$

$$\partial_3\phi = \left[\partial_1^3 + 3(\partial_1 u)\partial_1 + \frac{3}{2}(\partial_1^2 u + \partial_2 u)\right]\phi \tag{4.15}$$

as the first two equations. Equations (4.12), (4.13) and (4.14) together with their their analogues with regard to different directions form the linear problems for the whole discrete and continuous KP hierarchies.

4.4 Bilinearisation

Following the definition of the tau function (3.7), we can also study the bilinear formalism in the DL framework. This is done by computing dynamical evolutions of the tau function. Applying the forward shift operation on τ yields

$$\begin{aligned}\mathrm{T}_j\tau &= \det[1 + \boldsymbol{\Omega}(\mathrm{T}_j\boldsymbol{C})] = \det\left[1 + \boldsymbol{\Omega}(p_j + \boldsymbol{\Lambda})\boldsymbol{C}(p_j - {}^t\boldsymbol{\Lambda})^{-1}\right] \\ &= \det[1 + \boldsymbol{\Omega C} + (p_j - {}^t\boldsymbol{\Lambda})^{-1}\boldsymbol{OC}] = \tau\left[1 + \left(\boldsymbol{U}\frac{1}{p_j - {}^t\boldsymbol{\Lambda}}\right)_{0,0}\right],\end{aligned}$$

where in the last step the property of the determinant given in Section 2 is used, and thus the following dynamical relation is obtained:

$$\frac{\mathrm{T}_j\tau}{\tau}=V_{-p_j},\quad \text{and similarly,}\quad \frac{\mathrm{T}_j^{-1}\tau}{\tau}=W_{p_j}, \tag{4.16}$$

where $V_\cdot$ and $W_\cdot$ are defined as

$$V_a\doteq 1-\left(\boldsymbol{U}\frac{1}{a+{}^t\boldsymbol{\Lambda}}\right)_{0,0}\quad\text{and}\quad W_a\doteq 1-\left(\frac{1}{a-\boldsymbol{\Lambda}}\boldsymbol{U}\right)_{0,0},$$

respectively. Now we consider $[(4.5\mathrm{a})(p_j-\boldsymbol{\Lambda})^{-1}]_{0,0}$ and its p_i counterpart. By eliminating the redundant variables and reserving τ and u, we have

$$(p_i-p_j)\frac{\tau(\mathrm{T}_i\mathrm{T}_j\tau)}{(\mathrm{T}_i\tau)(\mathrm{T}_j\tau)}=p_i-p_j+\mathrm{T}_ju-\mathrm{T}_iu. \tag{4.17a}$$

Next we consider the evolutions of the tau function with respect to x_j. Direct computation shows that

$$\begin{aligned}\partial_j\ln\tau&=\partial_j\ln[\det(1+\boldsymbol{\Omega}\boldsymbol{C})]=\partial_j\operatorname{tr}[\ln(1+\boldsymbol{\Omega}\boldsymbol{C})]\\&=\operatorname{tr}[(1+\boldsymbol{\Omega}\boldsymbol{C})^{-1}\boldsymbol{\Omega}(\boldsymbol{\Lambda}^j\boldsymbol{C}-\boldsymbol{C}(-{}^t\boldsymbol{\Lambda})^j)]=\operatorname{tr}(\boldsymbol{O}_j\boldsymbol{U}),\end{aligned}$$

for $j=1,2,\cdots$, in virtue of the second equation in (4.4). In particular, the formula for the x_1-flow is nothing but the important bilinear transformation for the KP equation, which is in the form of

$$\partial_1\ln\tau=u. \tag{4.17b}$$

Equations (4.17a) and (4.17b) establish the connection between u and τ, for both discrete and continuous cases. Thanks to them, the bilinear discrete and continuous KP hierarchies are easily constructed within the DL scheme, in which the simplest equations are as follows:

$$\begin{aligned}&(p_i-p_j)(\mathrm{T}_h\tau)(\mathrm{T}_i\mathrm{T}_j\tau)\\&\qquad+(p_j-p_h)(\mathrm{T}_i\tau)(\mathrm{T}_j\mathrm{T}_h\tau)+(p_h-p_i)(\mathrm{T}_j\tau)(\mathrm{T}_h\mathrm{T}_i\tau)=0, &&(4.18\mathrm{a})\\&D_1(\mathrm{T}_i\tau)\cdot(\mathrm{T}_h\tau)=(p_h-p_i)[\tau(\mathrm{T}_h\mathrm{T}_i\tau)-(\mathrm{T}_h\tau)(\mathrm{T}_i\tau)], &&(4.18\mathrm{b})\\&(D_1^2-D_2+2p_hD_1)\tau\cdot(\mathrm{T}_h\tau)=0, &&(4.18\mathrm{c})\\&(D_1^4-4D_1D_3+3D_2^2)\tau\cdot\tau=0, &&(4.18\mathrm{d})\end{aligned}$$

namely the bilinear forms of Equations (4.7), (4.8), (4.9) and (4.10).

5 Concluding remarks

In this chapter, we present an introduction to the DL approach to integrable systems. By introducing the notions of infinite matrix and infinite-dimensional vector, the general theory of the DL is established on the infinite matrix representation of a linear integral equation, which provides a very effective path to investigate integrable equations within a single framework. The discrete and continuous KP hierarchies are taken as examples to illustrate the DL scheme, from which their underlying integrability structures are explained, from different perspectives such as exact solutions, Lax pairs and tau functions. Since this is an elementary introduction, there are a certain number of details in the DL which are not covered in this paper. We list these in the following paragraphs for those who would like to see the whole picture of the DL framework.

When we construct closed-form equations, we only consider the $(0,0)$-entry of the infinite matrix. In fact, other entries also form closed-form equations in the DL scheme, among which the two typical entries are the ones corresponding to the modified equations and Schwarzian equations. These equations are not independent in their own rights, and they are deep down related to each other through differential/difference transforms. For example, we refer the reader to [8, 10] for the other possible nonlinear forms of the discrete and continuous KP equations.

The linear integral equation also has its adjoint equation, with the same kernel and plane wave factors. The nonlinearisation of the adjoint wave function will result in the same potential, namely the infinite matrix $\boldsymbol{U}$. In this sense, the obtained nonlinear equations are still the same; this is the reason why we omit this part here in this chapter. However, the adjoint linear integral equation itself will lead to the so-called adjoint linear problems. This is rather important for higher-dimensional equations such as the KP hierarchy, as the linear problems and their adjoints govern the two spectral parameters k and k' separately, forming the binary Darboux transformations. While on the lower-dimensional level, the adjoint linear problems are not necessary as they degenerate and become the same as the standard linear problems.

In our scheme, the discrete, semi-discrete and continuous equations arise from the same infinite matrix $\boldsymbol{U}$ once the kernel and the plane wave factors are fixed. This implies that all these equations are compatible with each other; in other words, they act as symmetries for each other, i.e., the MDC property. In addition to symmetries, the DL framework also allows us to construct master symmetries of discrete equations, which is realised by considering the lattice parameters p_j as the independent variables, see [22] and [11] for the discrete KdV case and the discrete KP case, respectively.

The measure in the linear integral equation determines the nonlinear structure of a certain class of integrable equations. In the class of KP, there is no restriction on the measure, which means the KP hierarchy is the most general model within the DL framework. By performing reductions on the measure, namely imposing constraints on the spectral variables k and k', we restrict the solution space of the KP hierarchy, and subsequently, integrable sub-equations arise as reduced equations of KP. Examples include other KP-type equations, 2D integrable equations, as well as Painlevé equations, etc., see e.g., [8, 9, 22, 29] and references therein.

The infinite matrix $\boldsymbol{U}$ defined as (3.4) provides the most general solution (which we refer to as the direct linearising solution) to the resulting nonlinear equations, in the sense that it relies on the continuous spectral variables, from the viewpoint of the DL. By taking measures and contours of particular forms, the linear integral equation degenerates, and as a consequence, special classes of solutions such as soliton solution and multipole solutions can be constructed, see e.g., [8,28]. Furthermore, the DL approach has a generalisation called direct linearising transform, from which one can even find more general solutions to discrete integrable systems, such as exact solutions under periodic boundary conditions, see [18,27].

Acknowledgements

This project was supported by the National Natural Science Foundation of China (grant no. 11901198) and Shanghai Pujiang Program (grant no. 19PJ1403200). WF was also partially sponsored by the Science and Technology Commission of Shanghai Municipality (grant no. 18dz2271000) as well as the Fundamental Research Funds for the Central Universities.

References

[1] Bobenko A I and Suri Yu B, Integrable systems on quad-graphs, *Int. Math. Res. Not.* **2002** 573–611, 2002.

[2] Date E, Jimbo M and Miwa T, Method for generating discrete soliton equations I–V, *J. Phys. Soc. Jpn.* **51** 5116–4224, 4125–4131, 1982, **52** 388–393, 761–765, 766–771, 1983.

[3] Drinfel'd V G and Sokolov V V, Lie algebras and equations of Korteweg–de Vries type, *J. Sov. Math.* **30** 1975–2036, 1985.

[4] Fokas A S and Ablowitz M J, Linearization of the Korteweg–de Vries and Painlevé II equations, *Phys. Rev. Lett.* **47** 1096–1110, 1981.

[5] Fokas A S and Ablowitz M J, On the inverse scattering and direct linearizing transforms for the Kadomtsev–Petviashvili equation, *Phys. Lett. A* **94** 67–70, 1983.

[6] Fu W, Direct linearisation of the discrete-time two-dimensional Toda lattices, *J. Phys. A: Math. Theor.* **51** 334001, 2018.

[7] Fu W, Direct linearization approach to discrete integrable systems associated with $\mathbb{Z}_{\mathcal{N}}$ graded Lax pairs, *Proc. R. Soc. A* **476** 20200036, 2020.

[8] Fu W and Nijhoff F W, Direct linearizing transform for three-dimensional discrete integrable systems: the lattice AKP, BKP and CKP equations, *Proc. R. Soc. A* **473** 20160905, 2017.

[9] Fu W and Nijhoff F W, On reductions of the discrete Kadomtsev–Petviashvili-type equations, *J. Phys. A: Math. Theor.* **50** 505203, 2017.

[10] Fu W and Nijhoff F W, Linear integral equations, infinite matrices, and soliton hierarchies, *J. Math. Phys.* **59** 071101, 2018.

[11] Fu W and Nijhoff F W, On nonautonomous differential-difference AKP, BKP and CKP equations, *Proc. R. Soc. A* **477** 20200717, 2021.

[12] Gardner C S, Greene J M, Kruskal M D and Miura R M, Method for solving the Korteweg–de Vries equation, *Phys. Rev. Lett.* **19** 1095–1097, 1967.

[13] Hietarinta J, Joshi N and Nijhoff F W, *Discrete systems and integrability*, Cambridge University Press, Cambridge, 2016.

[14] Hirota R, *The direct method in soliton theory*, Cambridge University Press, Cambridge, 2004.

[15] Jimbo M and T. Miwa, Solitons and infinite dimensional Lie algebras, *Publ. RIMS* **19** 943–1001, 1983.

[16] Magri F, A simple model of the integrable Hamiltonian equation, *J. Math. Phys.* **19** 1156–1162, 1978.

[17] Matveev V B and Salle M A, *Darboux transformations and solitons*, Springer, Berlin, 1991.

[18] Nijhoff F W and Atkinson J, Elliptic N-soliton solutions of ABS lattice equations, *Int. Math. Res. Not.* **2010** 3837–3895, 2010.

[19] Nijhoff F W, Capel H W, Wiersma G L and Quispel G R W, Bäcklund transformations and three-dimensional lattice equations, *Phys. Lett. A* **105** 267–272, 1984.

[20] Nijhoff F W, Papageorgiou V G, Capel H W and Quispel G R W, The lattice Gel'fand–Dikii hierarchy, *Inverse Probl.* **8** 597–621, 1992.

[21] Nijhoff F W, Quispel G R W and Capel H W, Direct linearisation of difference-difference equations, *Phys. Lett. A* **97** 125–128, 1983.

[22] Nijhoff F W, Ramani A, Grammaticos B, and Ohta Y, On Discrete Painlevé equations associated with the lattice KdV systems and the Painlevé VI equation, *Stud. Appl. Math.* **106** 261–314, 2001.

[23] Nijhoff F W and Walker A J, The discrete and continuous Painlevé VI hierarchy and the Garnier systems, *Glasgow Math. J.* **43A** 109–123, 2001.

[24] Novikov S P, Manakov S V, Pitaevskii L P and Zakharov V E, *Theory of solitons: the inverse scattering method*, Springer, Berlin, 1984.

[25] Sato M, Soliton equations as dynamical systems on an infinite dimensional Grassmann manifolds, *RIMS Kôkyûroku* **439** 30–46, 1981.

[26] Wahlquist H D and Estabrook F B, Bäcklund transformation for solutions of the Korteweg–de Vries equation, *Phys. Rev. Lett.* **31** 1386–1390, 1973.

[27] Yoo-kong S and Nijhoff F W, Elliptic (N, N')-soliton solutions of the lattice Kadomtsev–Petviashvili equation, *J. Math. Phys.* **54** 043511, 2013.

[28] Zhang D J and Zhao S L, Solutions to ABS lattice equations via generalized Cauchy matrix approach, *Stud. Appl. Math.* **131** 72–103, 2013.

[29] Zhang D J, Zhao S L and Nijhoff F W, Direct linearization of extended lattice BSQ systems, *Stud. Appl. Math.* **129** 220–248, 2012.

A3. Discrete Boussinesq-type equations

Jarmo Hietarinta [a] *and Da-jun Zhang* [b]

[a] *Department of Physics and Astronomy, University of Turku, FIN-20014 Turku, Finland*
E-mail: jarmo.hietarinta@utu.fi

[b] *Department of Mathematics, Shanghai University, Shanghai 200444, P.R. China*
E-mail: djzhang@staff.shu.edu.cn

Abstract

We present a comprehensive review of the discrete Boussinesq equations based on their three-component forms on an elementary quadrilateral. These equations were originally found by Nijhoff et al. using the direct linearization method and later generalized by Hietarinta using a search method based on multidimensional consistency. We derive from these three-component equations their two- and one-component variants. From the one-component form we derive two different semi-continuous limits as well as their fully continuous limits, which turn out to be PDE's for the regular, modified and Schwarzian Boussinesq equations. Several kinds of Lax pairs are also provided. Finally we give their Hirota bilinear forms and multi-soliton solutions in terms of Casoratians.

1 Introduction

Among the $1+1$ dimensional soliton equations there are evolution equations, such as the Korteweg–de Vries (KdV) equation, in which time derivatives appear in first order, but there are also important equations with higher order time derivatives, such as the Boussinesq (BSQ) equation. An essential difference between these equations is in the initial data required: For KdV it would be enough to give, e.g., $u(x,t=0)$, while for the second order BSQ equation we would need $u(x,t=0)$ and $\partial_t u(x,t=0)$, or something similar.

The difference between the first and second order time evolution is reflected also in the integrable discretizations of these equations. For first order equations a well-defined evolution is obtained from a staircase-like initial data together with an equation defined on the elementary square of the lattice. For higher order time evolutions, one would then need either initial data on a number of parallel staircases with an equation on a larger stencil or alternatively, multi-component initial data with a larger number of equations on the small stencil.

The recent rapid advances in the study of integrable partial difference equations (PΔE) are to a large extent due to the efficient use of the particular integrability property of *multi-dimensional consistency* (MDC), which is related to the existence of hierarchies in the continuous case [16]. In its simplest form it involves dimensions 2 and 3 and is called Consistency-Around-a-Cube (CAC). The MDC property was discussed already in [2,25,30] but in full force it was applied in [1] (with some further technical assumptions), and this provided a classification of first order equations defined on an elementary lattice square of the Cartesian 2D lattice, the so-called Adler-Bobenko-Suris (ABS) list. The requirement of MDC can also be applied on multi-component equations on the elementary plaquette. A partial classification of three-component equations was done in [14] on the basis of CAC and most of the results turned out to be discrete versions BSQ equations (DBSQ).

Multicomponent equations were also studied from the perspective of direct linearization approach (DLA) and several equations were found [11, 27, 28, 33, 37, 41]. In addition to the CAC and DLA approaches, DBSQ equations have been derived also by applying a three-reduction on the three-term Hirota-Miwa equation [6, 22], or on the four-term Miwa equation [20]. Still further results have been obtained using the Cauchy matrix approach [7] or graded Lax pairs [9].

In this chapter we will discuss in detail the multi-component DBSQ equations. In Section 2 we compare the various three-component forms that have appeared in the literature and their connections by gauge transformations or by Möbius transformations. Their symmetries are also briefly discussed. In fact, all the DBSQ-type equations found in [14] can be viewed as extensions of some known lattice equations found in the 1990's. In Section 3 we discuss how the dynamics of the first order three-component equations can be represented by two- or one-component forms on a larger stencil. We also discuss the continuum limits of the one-component forms in Section 4 and show that the limits really are BSQ-like equations.

In Section 5 we present the Lax pairs of the various discrete forms. As for the solutions of DBSQ-type equations, besides the results from direct linearisation and Cauchy matrix approach (see [7, 27, 36, 37, 41]), equation B2 has been bilinearized in [18] and solutions were given in terms of Casoratians; in Section 6 we investigate bilinear forms for A2 and C3 equations.

2 DBSQ-type equations

2.1 Basic concepts and definitions

The discrete equations that we discuss here are all defined on the Cartesian $\mathbb{Z} \times \mathbb{Z}$ lattice. Most of the time the equations are defined on a single quadrilateral but larger stencils are sometimes needed. The independent variables live on the vertices of the lattice and are therefore labeled by the vertex coordinates, see Figure 1.

Sometimes the equations have a simpler look if we replace the subscript with a tilde or a hat, or use some other simplified notation, for example

$$\mathbf{u} = \mathbf{u}_{n,m} = \mathbf{u}_{0,0},\ \widetilde{\mathbf{u}} = \mathbf{u}_{n+1,m} = \mathbf{u}_{1,0},\ \widehat{\mathbf{u}} = \mathbf{u}_{n,m+1} = \mathbf{u}_{0,1},\ \widehat{\widetilde{\mathbf{u}}} = \mathbf{u}_{n+1,m+1} = \mathbf{u}_{1,1}. \tag{2.1}$$

The equation(s) on the quadrilateral are given by $\mathbf{Q}(\mathbf{u}, \widetilde{\mathbf{u}}, \widehat{\mathbf{u}}, \widehat{\widetilde{\mathbf{u}}}) = 0$, where $\mathbf{Q}$ are affine multi-linear polynomials, and one may then ask whether the system of equations is integrable according to some definition. We will use the MDC criterion which means that the equation defined on the 2D-lattice can be extended consistently into higher dimensions.

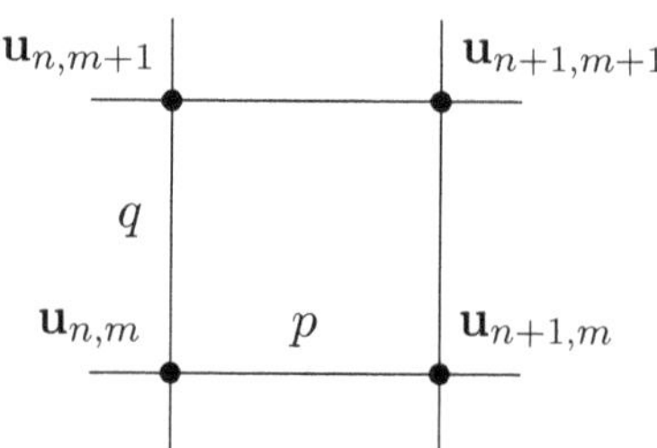

Figure 1. Elementary quadrilateral of the lattice, with possibly multi-component corner variables $\mathbf{u}$. The parameters p, q characterize the distance between lattice points.

As an example consider the simple case of the 1-component lattice potential KdV equation (H1 in the ABS list):

$$Q(u,\widetilde{u},\widehat{u},\widehat{\widetilde{u}};p,q) := (u-\widehat{\widetilde{u}})(\widehat{u}-\widetilde{u}) - p^2 + q^2 = 0.$$

Since u is defined on the 2D-lattice, it naturally depends only on the coordinates n, m, but in order to extend the equations into a 3D setting, which is the simplest requirement of MDC, we introduce a third variable k and denote $u_{n,m,k+1} = \overline{u}$; the associated lattice parameter is r. For each of the six sides of the cube we have an equation:

$$\text{bottom:}\quad Q(u,\widetilde{u},\widehat{u},\widehat{\widetilde{u}};p,q) = 0, \qquad \text{top:}\quad Q(\overline{u},\overline{\widetilde{u}},\overline{\widehat{u}},\overline{\widehat{\widetilde{u}}};p,q) = 0, \tag{2.2a}$$

$$\text{back:}\quad Q(u,\widehat{u},\overline{u},\overline{\widehat{u}};\alpha,\beta) = 0, \qquad \text{front:}\quad Q(\widetilde{u},\widehat{\widetilde{u}},\overline{\widetilde{u}},\overline{\widehat{\widetilde{u}}};\alpha,\beta) = 0, \tag{2.2b}$$

$$\text{left:}\quad Q(u,\overline{u},\widetilde{u},\overline{\widetilde{u}};\gamma,\delta) = 0, \qquad \text{right:}\quad Q(\widehat{u},\overline{\widehat{u}},\widehat{\widetilde{u}},\overline{\widehat{\widetilde{u}}};\gamma,\delta) = 0. \tag{2.2c}$$

(Note that we used arbitrary parameters in this example and hope to determine them by consistency.) In order to study CAC we choose $u, \widetilde{u}, \widehat{u}, \overline{u}$ as initial values and then from the equations on the LHS we can compute $\widehat{\widetilde{u}}$, $\overline{\widehat{u}}$, $\overline{\widetilde{u}}$. When these are used on the RHS we have three ways to compute $\overline{\widehat{\widetilde{u}}}$, but they must all yield the same value. In this case we find the condition

$$(\alpha^2-\beta^2)+(\gamma^2-\delta^2)+(p^2-q^2) = 0.$$

Since back and front equations do not depend on p we find $(\alpha^2-\beta^2) = (q^2-r^2)$ and $(\gamma^2-\delta^2) = r^2-p^2$ for some r and then all RHS equations yield the very symmetric form

$$\overline{\widehat{\widetilde{u}}} = \frac{\widehat{u}\widetilde{u}(q^2-p^2)+\widetilde{u}\overline{u}(p^2-r^2)+\overline{u}\widehat{u}(r^2-q^2)}{\overline{u}(p^2-q^2)+\widehat{u}(r^2-p^2)+\widetilde{u}(q^2-r^2)}.$$

Note that the RHS of this expression does not depend on u, this is called the tetrahedron property.

The application of MDC on one-component equations resulted in the ABS-list [1]. For multi-component equations there are no equally comprehensive classifications. It should also be noted that passing the CAC test is necessary but not sufficient [15] for integrability.

2.2 Hietarinta's list of equations

In [14] Hietarinta made a partial classification of Boussinesq (BSQ) type lattice equations using CAC. Since BSQ equations are of second order in time, their discrete analogues are either multi-component on a quadrilateral, or defined on a larger stencil. In [14] three-component approach was used and some equations were defined on the links and one equation on the full quadrilateral. Using CAC, the following three-component DBSQ-type equations were found:

B2:

$$\widetilde{y} = x\widetilde{x} - z, \tag{2.3a}$$

$$\widehat{y} = x\widehat{x} - z, \tag{2.3b}$$

$$y = x\widehat{\widetilde{x}} - \widehat{\widetilde{z}} + b_0(\widehat{\widetilde{x}} - x) + b_1 + \frac{P - Q}{\widetilde{x} - \widehat{x}}, \tag{2.3c}$$

A2:

$$\widetilde{y} = z\widetilde{x} - x, \tag{2.4a}$$

$$\widehat{y} = z\widehat{x} - x, \tag{2.4b}$$

$$y = x\widehat{\widetilde{z}} - b_0 x + \frac{P\widetilde{x} - Q\widehat{x}}{\widehat{z} - \widetilde{z}}, \tag{2.4c}$$

C3:

$$\widetilde{y}\, z = \widetilde{x} - x, \tag{2.5a}$$

$$\widehat{y}\, z = \widehat{x} - x, \tag{2.5b}$$

$$\widehat{\widetilde{z}}\, y = b_0\, x + b_1 + z\frac{P\,\widetilde{y}\,\widehat{z} - Q\,\widehat{y}\,\widetilde{z}}{\widetilde{z} - \widehat{z}}, \tag{2.5c}$$

and

C4:

$$\widetilde{y}\, z = \widetilde{x} - x, \tag{2.6a}$$

$$\widehat{y}\, z = \widehat{x} - x, \tag{2.6b}$$

$$\widehat{\widetilde{z}}\, y = x\,\widehat{\widetilde{x}} + b_2 + z\,\frac{P\,\widetilde{y}\,\widehat{z} - Q\,\widehat{y}\,\widetilde{z}}{\widetilde{z} - \widehat{z}}. \tag{2.6c}$$

Here the parameters P and Q are related to lattice spacing parameters p, q in n and m-directions, respectively.

The convention for naming the variables was designed for MDC and for analyzing the evolution (Section 2.5): The quasilinear equations are defined on the edges of the quadrilateral; "a" equations depend always on x, $\widetilde{x}$, z, and $\widetilde{y}$, and the "b" equations on x, $\widehat{x}$, z, and $\widehat{y}$, i.e., the dependence is always the same, only the algebra is different. Also, we list the equations in the order B2, A2, C3,4 because, as we will see later, they correspond to regular, modified and Schwarzian BSQ equations, respectively.

The coupling constants b_i are arbitrary and generalize some previous results. Note, however, that b_1 in B2 can be removed with the transformation

$$(x, y, z) \mapsto (x, y - \frac{b_1}{3}(n + m - 1), z + \frac{b_1}{3}(n + m)), \tag{2.7}$$

and b_0 in A2 can be removed using

$$(x, y, z) \mapsto (x, y + \frac{b_0 x}{3}(n+m), z + \frac{b_0}{3}(n+m+1)). \tag{2.8}$$

In the following we do not keep these removable parameters.

In addition some two-component forms were found in [14]:

C2-1:

$$\widehat{\widetilde{x}} = \frac{\widehat{x}\widetilde{z} - \widetilde{x}\widehat{z}}{\widetilde{z} - \widehat{z}}, \tag{2.9a}$$

$$\widehat{\widetilde{z}} = -b_0 z\,\widehat{\widetilde{x}} + z\,\frac{P\,\widehat{z} - Q\,\widetilde{z}}{\widetilde{z} - \widehat{z}}, \tag{2.9b}$$

and

C2-2:

$$\widehat{\widetilde{x}} = \frac{\widehat{x}\widetilde{z} - \widetilde{x}\widehat{z}}{\widetilde{z} - \widehat{z}}, \tag{2.10a}$$

$$x\widehat{\widetilde{z}} = -b_0 z + z\,\frac{P\,\widetilde{x}\,\widehat{z} - Q\,\widehat{x}\,\widetilde{z}}{\widetilde{z} - \widehat{z}}. \tag{2.10b}$$

These are actually discrete versions of KdV so we will not discuss them further here.

2.3 Relations between the C-equations

Let us first note that lattice equations are classified only up to local rational-linear (i.e., Möbius) transformations, and that equations related by them are considered same. However, for some purposes a particular form may be better in practice.

Note that for the C-equations one can derive relation (2.9a) by eliminating y. After the transformations discussed below it is often useful to use this relation when comparing results.

First note that if $b_0 \neq 0$, then by the transformation

$$x \to x - \frac{b_1}{b_0}, \tag{2.11}$$

one can remove from the C3 equation (2.5c) the parameter b_1 and then we can consider the following form:

$$\widetilde{y}\,z = \widetilde{x} - x, \tag{2.12a}$$

$$\widehat{y}\,z = \widehat{x} - x, \tag{2.12b}$$

$$\widehat{\widetilde{z}}\,y = b_0\,x + z\frac{P\,\widetilde{y}\,\widehat{z} - Q\,\widehat{y}\,\widetilde{z}}{\widetilde{z} - \widehat{z}}, \tag{2.12c}$$

which we call $C3_{b_0}$.

Since the transformation (2.11) fails when $b_0 = 0$, the following equation

$$\widetilde{y}\,z = \widetilde{x} - x, \tag{2.13a}$$

$$\widehat{y}\,z = \widehat{x} - x, \tag{2.13b}$$

$$\widehat{\widetilde{z}}\,y = b_1 + z\frac{P\,\widetilde{y}\,\widehat{z} - Q\,\widehat{y}\,\widetilde{z}}{\widetilde{z} - \widehat{z}} \tag{2.13c}$$

is not a trivial subcase of C3 equation (2.5). Since (2.13c) does not contain x, we get a two-component form after eliminating x from (2.13a, 2.13b) and their shifts, this results in

$$\widehat{\widetilde{y}} = -z\frac{\widetilde{y} - \widehat{y}}{\widetilde{z} - \widehat{z}}. \tag{2.13d}$$

Thus (2.13c,2.13d) is a two-component form, let us call it $C3_{b_1}$.

We will next show that C4 can be obtained from C3 by a Möbius transformation. As the first step we note that from equations (2.5a,2.5b) and their shifts one can derive

$$x = \widehat{\widetilde{x}} + z\,\frac{\widetilde{y}\widehat{z} - \widehat{y}\widetilde{z}}{\widetilde{z} - \widehat{z}}. \tag{2.14}$$

Using it to replace $\frac{1}{2}b_0 x$ in (2.12c) we get the following alternative form for $C3_{b_0}$:

$$\widetilde{y}\,z = \widetilde{x} - x, \tag{2.15a}$$

$$\widehat{y}\,z = \widehat{x} - x, \tag{2.15b}$$

$$\widehat{\widetilde{z}}\,y = z\,\frac{(P - c_2)\,\widetilde{y}\,\widehat{z} - (Q - c_2)\,\widehat{y}\,\widetilde{z}}{\widetilde{z} - \widehat{z}} + c_2(x + \widehat{\widetilde{x}}), \quad (c_2 = \frac{b_0}{2}). \tag{2.15c}$$

Now, inserting the (mixed) Möbius transformation [41]

$$x = \frac{x_1 - c_2}{2c_2(x_1 + c_2)}, \quad y = \frac{y_1}{x_1 + c_2}, \quad z = \frac{z_1}{x_1 + c_2}, \tag{2.16}$$

into (2.15) we get

$$\widetilde{y}_1\,z_1 = \widetilde{x}_1 - x_1, \tag{2.17a}$$

$$\widehat{y}_1\,z_1 = \widehat{x}_1 - x_1, \tag{2.17b}$$

$$\widehat{\widetilde{z}}_1\,y_1 = z_1\,\frac{(P - c_2)\,\widetilde{y}_1\,\widehat{z}_1 - (Q - c_2)\,\widehat{y}_1\,\widetilde{z}_1}{\widetilde{z}_1 - \widehat{z}_1} + x_1\,\widehat{\widetilde{x}}_1 - c_2^2, \tag{2.17c}$$

which is $C4_{b_2}$ equation (2.6), after redefining

$$P \to P - c_2, \quad Q \to Q - c_2, \quad b_2 = -c_2^2. \tag{2.18}$$

The above transformation fails if $b_0 = 0$ in C3, i.e., if $b_2 = 0$ in C4, but that special case can be obtained from $C3_{b_1=1}$ by the following transformation:

$$x = -1/x_1, \quad y = y_1/x_1, \quad z = z_1/x_1. \tag{2.19}$$

But since $C3_{b_1=1}$ depends only on 2 variables, the transformation (2.19) in fact eliminates the x variable.

In summary, among the C-equations we only need to consider the three-component equation $C3_{b_0}$ (2.12) for $b_0 \neq 0$, and the two-component equations $C3_{b_1}$ (2.13c,2.13d) (for arbitrary b_1).

2.4 Symmetries

2.4.1 $n \leftrightarrow m$ reflection symmetry

As can be easily seen, all the equations are invariant under the $n \leftrightarrow m$ reflection, i.e.,$\,\tilde{}\leftrightarrow\hat{}$, accompanied by $P \leftrightarrow Q$ parameter change.

2.4.2 Reversal symmetry

By reversal symmetry we mean symmetry under changing all tildes to undertildes and hats to underhats. More precisely, the indices change sign, and then the generic point is renamed:

$$x_{n+\nu,m+\mu} \mapsto x_{-n-\nu,-m-\mu} = x_{n'-\nu,m'-\mu},$$

after which we can drop the primes. This reversal is then with respect to the lattice point (n, m). In the notation where only shifts relative to (n, m) are indicated (such as $x_{0,1}$) we have $x_{\nu,\mu} \mapsto x_{-\nu,-\mu}$, after which we usually shift the whole equation.

B2: If we apply this reversal to B2 equation (2.3a) we have

$$\widetilde{y} = x\widetilde{x} - z \quad \xmapsto{\text{reversal}} \quad \utilde{y} = x\utilde{x} - z \quad \xmapsto{\text{shift}} \quad y = \widetilde{x}x - \widetilde{z}.$$

Thus we have reversal symmetry of (2.3a) (and (2.3b)) if we add the exchange $y \leftrightarrow z$. As for (2.3c), it turns out that we should also take $b_0 \leftrightarrow -b_0$ and $(P, Q) \leftrightarrow (-P, -Q)$ which is natural for a reversal of direction. In summary, the B2 equations are reversal invariant if accompanied with

$$(x, y, z, P, Q, b_0) \mapsto (x, z, y, -P, -Q, -b_0). \tag{2.20}$$

C: Similarly for the C3 equations we have reversal symmetry if we include variable changes

$$(x, y, z) \mapsto (-x, z, y). \tag{2.21}$$

For (2.5a, 2.5b) this is manifest, and also for most terms in (2.5c) but some terms need more computations. For example we have

$$z\frac{P\widetilde{y}\widehat{z} - Q\widehat{y}\widetilde{z}}{\widetilde{z} - \widehat{z}} \mapsto z\frac{P\utilde{y}\underset{\wedge}{z} - Q\underset{\wedge}{y}\utilde{z}}{\utilde{z} - \underset{\wedge}{z}} \mapsto \widehat{\widetilde{z}}\frac{P\widehat{y}\widetilde{z} - Q\widetilde{y}\widehat{z}}{\widehat{z} - \widetilde{z}} \mapsto \widehat{\widetilde{y}}\frac{P\widehat{z}\widetilde{y} - Q\widetilde{z}\widehat{y}}{\widehat{y} - \widetilde{y}},$$

and to finish the computation we should still show that $z/(\widetilde{z} - \widehat{z}) = \widehat{\widetilde{y}}/(\widehat{y} - \widetilde{y})$ but this follows by taking suitable linear combination of (2.5a), (2.5b) and their shifts.

The only other special term in C3 is $b_0\, x$ in (2.5c) which in this process changes to $-b_0\widehat{\widetilde{x}}$. However, by taking again a suitable combination of the (2.5a, 2.5b) and their shifts we can derive

$$\widehat{\widetilde{x}} = x - z\frac{\widetilde{y}\widehat{z} - \widehat{z}\widetilde{y}}{\widetilde{z} - \widehat{z}},$$

which combines with the P, Q term. Thus if $b_0 \neq 0$, we have reversal symmetry if we do the further parameter replacements

$$(P, Q, b_0, b_1) \mapsto (P + b_0, Q + b_0, -b_0). \tag{2.22}$$

The two-component equation $\mathrm{C3}_{b_1}$ of (2.13c,2.13d) is also reversal symmetric.

A2: The case of A2 is a bit more complicated: From (2.4a) we have

$$\widetilde{y} = z\widetilde{x} - x \quad \mapsto \quad \underset{\sim}{y} = z\underset{\sim}{x} - x \quad \mapsto \quad y = \widetilde{z}x - \widetilde{x}$$

but the usual map (2.21) does not take this to the original form. Dividing the last equation by $x\widetilde{x}$ yields $y/(x\widetilde{x}) = \widetilde{z}/\widetilde{x} - 1/x$. Now it can be seen that we get the original form if instead of (2.21) we have

$$(y, z, x) \mapsto (-z/x, -y/x, 1/x). \tag{2.23}$$

For the P, Q term we only need to show that $\widehat{\widetilde{x}}x/(\widehat{y}\widetilde{x} - \widetilde{y}\widehat{x}) = 1/(\widehat{z} - \widetilde{z})$, which follows from (2.4a,2.4b) and their shifts. The form (2.23) suggests that the transformation can be simplified if we use another variable $w := y/x$. Then we get the alternate form

A2-alt:

$$z - \widetilde{w} = \frac{x}{\widetilde{x}}, \tag{2.24a}$$

$$z - \widehat{w} = \frac{x}{\widehat{x}}, \tag{2.24b}$$

$$(\widetilde{z} - \widehat{z})(\widehat{\widetilde{z}} - w) = P\,\frac{\widetilde{x}}{x} - Q\,\frac{\widehat{x}}{x}. \tag{2.24c}$$

These are reversal invariant with

$$(w, z, x) \mapsto (-z, -w, 1/x). \tag{2.25}$$

This is easy to show if one uses the formula

$$\frac{\widehat{\widetilde{x}}\,x}{\widetilde{x}\,\widehat{x}} = \frac{\widetilde{w} - \widehat{w}}{\widetilde{z} - \widehat{z}}$$

which follows from (2.24a,2.24b).

2.5 Initial values and evolution

We will consider evolution starting from initial values given on a staircase, on which the inside corner points are given by $n + m = 0$ and the outside corner points by $n + m = 1$, see Figure 2 (a). Another possible initial staircase is given by $n - m = 0$, $n - m = -1$, see Figure 2 (b). We have three sets of variables and several equations, so we must study carefully which kind of initial values are necessary and make sense.

Let us first consider a staircase in the NW-SE direction, Figure 2 (a). We assume that x is given on all points of the staircase and to get started, also $z_{0,0}$. Then it

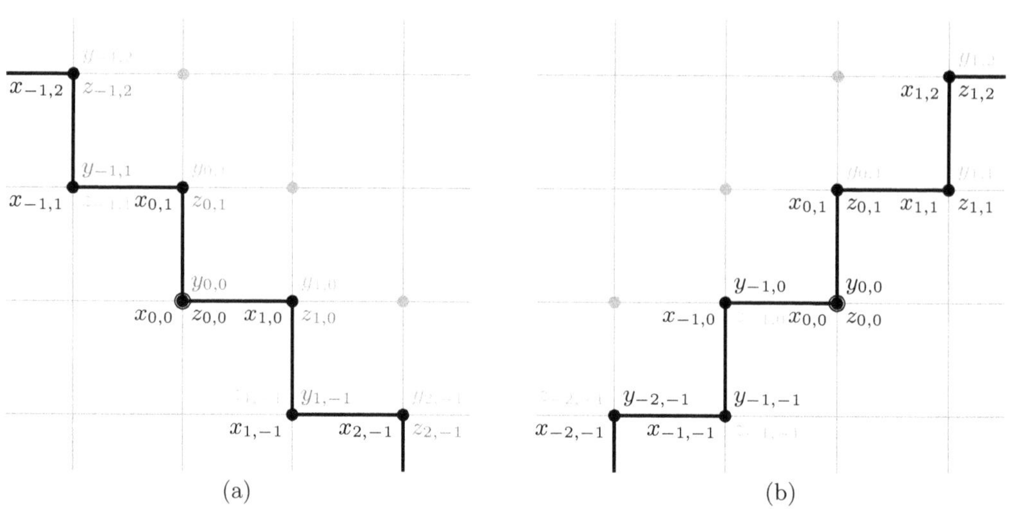

Figure 2. The initial values given on the staircase. The black variables must be given in order to compute the red variables on the staircase using the quasilinear equations. In order to compute the values at the red dots (one step in the evolution) it is also necessary to give the blue variables in case (a).

is possible to compute, step by step, the red y and z values using the quasilinear equations, i.e., we can compute $y_{n,m}$ for $n+m=1$ and $z_{n,m}$ for $n+m=0$, $n \neq 0$. If the staircase is in the NE-SW direction, as in Figure 2 (b), it is necessary to give $z_{n,m}$ for $n-m=0,-1$, $n \geq 0$ and $y_{n,m}$ for $n-m=0,-1$, $n \leq 0$.

After the above we still have three equations left for each square, by which we should be able to compute values at points where $n+m=2$ (red dots in Figure 2). But before that can be done for the staircase (a) we need more initial values: In order to use the remaining quasilinear equations we also need $z_{n,m}$, $n+m=1$, and for the fully nonlinear equation, $y_{n,m}$, $n+m=0$. These additional necessary initial values are given in blue in Figure 2. For staircase (b) all necessary initial values were needed already for filling in the staircase.

Since the systems are reversal symmetric, the same initial values work for evolution in the opposite direction.

2.6 Connection with the direct-linearization results

2.6.1 The direct-linearization approach

Many of the discrete Boussinesq equations discussed here were derived earlier (see [24,27,36,37]) using the direct linearization scheme (DLA) of Capel, Nijhoff, Quispel et al. The results of Hietarinta [14] provided generalizations to the early results, but they were subsequently also derived from the DLA point of view in [41] by generalizing the dispersion relation.

The DLA was first proposed by Fokas and Ablowitz [8], and soon was developed to the study of discrete integrable systems [27, 28, 33]. In this approach, an infinite matrix is introduced via a linear integral equation with certain plane wave factors and discrete equations arise as closed forms of the shift relations of the elements of the matrix.

For the DBSQ equations, one first introduces an integral equation for infinite order column vector $\boldsymbol{u}(k)$:

$$\boldsymbol{u}(k)+\rho(k)\sum_{j=1}^{2}\int_{\Gamma_j} d\mu_j(l)\,\boldsymbol{u}(l)\sigma(-\omega_j(l))=\rho(k)\boldsymbol{c}_k^T,$$

where Γ_j and $d\mu_j(l)$ are contours and measures that need to be suitably chosen, $\boldsymbol{c}_k$ is an infinite order constant column vector $(\cdots,k^{-2},k^{-1},1,k,k^2,\cdots)^T$, plane wave factors are

$$\rho(k)=(p+k)^n(q+k)^m\rho^{(0)}(k),\quad \sigma(k')=(p-k')^{-n}(q-k')^{-m}\sigma^{(0)}(k'),$$

$\rho^{(0)},\sigma^{(0)}$ are constants, $\omega_j(k)$ are defined through

$$p^3+\alpha_2p^2-(k^3+\alpha_2k^2)=(p-k)(p-\omega_1(k))(p-\omega_2(k)).$$

Then, introduce an $\infty\times\infty$ matrix $\boldsymbol{U}$ by

$$\boldsymbol{U}=\sum_{j=1}^{2}\int_{\Gamma_j} d\mu_j(l)\,\sigma(-\omega_j(l))\boldsymbol{u}(l)\,\boldsymbol{c}_{-\omega_j(l)}^T.$$

After that one can define scalar functions

$$S_{(a,b)}^{(i,j)}=\boldsymbol{e}^T(a+\Lambda)^{-1}\boldsymbol{U}(b-\Lambda)^{-1}\boldsymbol{e},\quad i,j\in\mathbb{Z},\quad a,b\in\mathbb{C},$$

where $\boldsymbol{e}=(\cdots,0,0,1,0,0,\cdots)^T$ in which only the center element is 1, and $\Lambda=(\lambda_{i,j})_{\infty\times\infty}$ in which $\lambda_{i,j}=\delta_{i,j-1}$. More explicitly we have

$$\begin{aligned}
&u^{(i,j)}=(-1)^jS_{(0,0)}^{(i,j)},\quad s_{a,b}=S_{(a,b)}^{(-1,-1)},\quad v_a=1-S_{(a,0)}^{(-1,0)},\quad w_b=1+S_{(0,b)}^{(0,-1)},\\
&s_a=a+S_{(a,0)}^{(-1,1)},\quad t_b=-b+S_{(0,b)}^{(1,-1)},\quad r_a=a^2-S_{(a,0)}^{(-1,2)},\quad z_b=b^2+S_{(0,b)}^{(2,-1)}.
\end{aligned}$$

DBSQ equations arise from closed forms of the shift relations of the above elements (cf. [41]).

Due to the different origin, the equations from DLA appear in a different gauge and in this section we elaborate the connections.

2.6.2 B2

Let us focus on B2 equation (2.3), in which we may assume that $b_1=0$, as mentioned before. If we use on (2.3) the transformation [41]

$$x=u^{(0,0)}-x_0, \tag{2.26a}$$

$$y=u^{(0,1)}-x_0u^{(0,0)}+y_0, \tag{2.26b}$$

$$z=u^{(1,0)}-x_0u^{(0,0)}+z_0, \tag{2.26c}$$

where

$$x_0 = np + mq + c_1, \tag{2.27a}$$
$$y_0 = \frac{1}{2}(np + mq + c_1)^2 - \frac{1}{2}(np^2 + mq^2 + c_2) - c_3, \tag{2.27b}$$
$$z_0 = \frac{1}{2}(np + mq + c_1)^2 + \frac{1}{2}(np^2 + mq^2 + c_2) + c_3 \tag{2.27c}$$

and $c_j (j = 1, 2, 3)$ are constants, we obtain

$$p\widetilde{u}^{(0,0)} - \widetilde{u}^{(0,1)} = pu^{(0,0)} + u^{(1,0)} - \widetilde{u}^{(0,0)}u^{(0,0)}, \tag{2.28a}$$
$$q\widehat{u}^{(0,0)} - \widehat{u}^{(0,1)} = qu^{(0,0)} + u^{(1,0)} - \widehat{u}^{(0,0)}u^{(0,0)}, \tag{2.28b}$$
$$-\frac{P-Q}{p-q+\widehat{u}^{(0,0)}-\widetilde{u}^{(0,0)}} = \frac{G_3(-p,-q)}{q-p} + (p+q+b_0)(u^{(0,0)} - \widehat{\widetilde{u}}^{(0,0)}) - u^{(0,0)}\widehat{\widetilde{u}}^{(0,0)} + \widehat{\widetilde{u}}^{(1,0)} + u^{(0,1)}, \tag{2.28c}$$

where

$$G_3(a, b) = g_3(a) - g_3(b), \tag{2.29a}$$
$$g_3(a) = a^3 + \alpha_2 a^2, \tag{2.29b}$$

and $b_0 = -\alpha_2$. This agrees with the DLA result (Eqs.(32a,33) in Ref. [41]) provided that we parametrise P, Q in terms of p, q as follows:

$$P = g_3(-p), \quad Q = g_3(-q), \tag{2.30}$$

with $b_0 = -\alpha_2$. Note that one can always replace $-p$ and $-q$ by a and b, respectively, and then get the parametrisation used in [19] for getting soliton solutions.

The above parametrisation of P, Q provides also a connection to the lattice potential KdV equation [26,28]: Using (2.30, 2.29b) and then taking the singular limit $b_0 \to \infty$ we get from (2.3c)

$$(x - \widehat{\widetilde{x}})(\widetilde{x} - \widehat{x}) = -p^2 + q^2. \tag{2.31}$$

This one-component equation appears as H1 equation in the ABS list [1].

Here is a second alternative form of (2.3) in the special case $b_0 = 0$:

$$\omega(\widehat{u}^{(0,1)} - \widetilde{u}^{(0,1)}) = p\widetilde{u}^{(0,0)} - q\widehat{u}^{(0,0)} - u^{(0,0)}(p - q + \widehat{u}^{(0,0)} - \widetilde{u}^{(0,0)}), \ (\omega = e^{\frac{2\pi i}{3}}), \tag{2.32a}$$
$$\widehat{u}^{(1,0)} - \widetilde{u}^{(1,0)} = q\widetilde{u}^{(0,0)} - p\widehat{u}^{(0,0)} + \widehat{\widetilde{u}}^{(0,0)}(p - q + \widehat{u}^{(0,0)} - \widetilde{u}^{(0,0)}), \tag{2.32b}$$
$$\widehat{\widetilde{u}}^{(1,0)} - \omega u^{(0,1)} = pq - (p + q + u^{(0,0)})(p + q - \widehat{\widetilde{u}}^{(0,0)}) + \frac{p^3 - q^3}{p - q + \widehat{u}^{(0,0)} - \widetilde{u}^{(0,0)}}. \tag{2.32c}$$

This is Eq. (5.3.12) in Ref. [37] as well as Eqs.(2.15a-c) with $N = 3$ in Ref. [27]. In the above system, (2.32a) and (2.32b) can be obtained through (2.28a) − (2.28b) and $\widehat{(2.28a)} - \widetilde{(2.28b)}$, and (2.32c) is (2.28c) with $b_0 = 0$, in addition, $u^{(0,1)} \to -\omega u^{(0,1)}$.

Note that the alternative forms (2.28) with (2.29) and (2.32) have the "background solution" $u^{(k,l)} = 0$, corresponding to (2.27) for (2.3). Thus these alternative forms will be useful once we start to construct soliton solutions in Section 6 and continuum limits in Section 4.

2.6.3 A2

For the A2 equation (2.4) (without the removable parameter b_0 (cf. (2.8))) several alternative forms have been presented in the literature.

The form

$$\widetilde{s}_a = (p + u^{(0,0)})\widetilde{v}_a - (p-a)v_a, \tag{2.33a}$$
$$\widehat{s}_a = (q + u^{(0,0)})\widehat{v}_a - (q-a)v_a, \tag{2.33b}$$
$$(p + q - \widehat{\widetilde{u}}^{(0,0)} + \frac{s_a}{v_a} - \alpha_2)(p - q + \widehat{u}^{(0,0)} - \widetilde{u}^{(0,0)}) = p_a\frac{\widetilde{v}_a}{v_a} - q_a\frac{\widehat{v}_a}{v_a}, \tag{2.33c}$$

was derived from direct linearisation approach (see Eq. (30) in Ref. [41]), here p_a and q_a are defined as

$$p_a = \frac{-G_3(-p,-a)}{p-a}, \quad q_a = \frac{-G_3(-q,-a)}{p-a}. \tag{2.34}$$

The transformation between Eq.(2.4) and Eq.(2.33) is given by

$$v_a = \frac{x}{x_a}, \quad u^{(0,0)} = z - z_0, \quad s_a = \frac{1}{x_a}(y - v_a y_a), \tag{2.35}$$

where

$$x_a = (p-a)^{-n}(q-a)^{-m}c_1, \tag{2.36a}$$
$$z_0 = (c_3 - p)n + (c_3 - q)m + c_2, \tag{2.36b}$$
$$y_a = x_a(z_0 - c_3), \tag{2.36c}$$

and where c_1, c_2 are constants, $c_3 = \alpha_2/3$ (if $b_0 = 0$) and

$$P = -G_3(-p,-a), \quad Q = -G_3(-q,-a). \tag{2.37}$$

Note that the above P, Q can be equivalently reparametrised as

$$P = G_3(p,a), \ Q = G_3(q,a) \tag{2.38}$$

if we take $c_3 = -\alpha_2/3$ in (2.36). Either convention can be adopted.

Another system related to A2 was given in Eqs.(4.22a, 4.21b) of Ref. [24], i.e.,

$$\widetilde{s}\widetilde{v} = (p+u)\widetilde{v} - pv, \tag{2.39a}$$
$$\widehat{s}\widehat{v} = (p+u)\widehat{v} - pv, \tag{2.39b}$$
$$(p + q + s - \widehat{\widetilde{u}})(p - q + \widehat{u} - \widetilde{u}) = \frac{p^2\widetilde{v} - q^2\widehat{v}}{v}, \tag{2.39c}$$

which can be derived from (2.33) by taking

$$s_a = sv, \ u_0 = u, \ v_a = v, \ a = 0, \ \alpha_2 = 0. \tag{2.40}$$

Then there is the system

$$p - q + \widehat{u} - \widetilde{u} = \frac{(p-a)\widehat{v} - (q-a)\widetilde{v}}{\widehat{\widetilde{v}}}, \tag{2.41a}$$

$$p - q + \widehat{s} - \widetilde{s} = (p-a)\frac{v}{\widetilde{v}} - (q-a)\frac{v}{\widehat{v}}, \tag{2.41b}$$

$$(p + q + s - \widehat{\widetilde{u}})(p - q + \widehat{u} - \widetilde{u}) = \frac{p_a\widetilde{v} - q_a\widehat{v}}{v}, \tag{2.41c}$$

which is given by Eqs.(A.4a,b,c) in [27], and named as the "Toda-MBSQ equation". In fact, (2.41a) and (2.41b) can be derived from (2.33a) and (2.33b), by using

$$s_a = sv, \ v_a = v, \ u^{(0,0)} = u, \ \alpha_2 = 0, \tag{2.42}$$

and then eliminating s and u, respectively.

Finally, eliminating s_a from (2.33a) and (2.33b) yields

$$p - q + \widehat{u}_0 - \widetilde{u}_0 = \frac{(p-a)\widehat{v}_a - (q-a)\widetilde{v}_a}{\widehat{\widetilde{v}}_a}, \tag{2.43}$$

which, together with (2.33a) and (2.33c) with $\alpha_2 = 0$, gives the system {(5.3.7a), (5.3.14), (5.3.15)} in Ref. [37].

2.6.4 C3

The C3 equation is related to the following equation derived by DLA [41]

$$(p-a)\, S_{a,b} - (p-b)\, \widetilde{S}_{a,b} = \widetilde{v}_a\, w_b, \tag{2.44a}$$

$$(q-a)\, S_{a,b} - (q-b)\, \widehat{S}_{a,b} = \widehat{v}_a\, w_b, \tag{2.44b}$$

$$v_a\, \widehat{\widetilde{w}}_b = w_b\, \frac{\frac{p_a}{p-b}\, \widehat{w}_b\, \widetilde{v}_a - \frac{q_a}{q-b}\, \widetilde{w}_b\, \widehat{v}_a}{(p-b)\, \widetilde{w}_b - (q-b)\, \widehat{w}_b} + \frac{G_3(-a,-b)}{(p-b)\,(q-b)}\, S_{a,b}, \tag{2.44c}$$

$S_{a,b} = s_{a,b} - 1/(a-b)$ and the connection between the two equations is [41]

$$S_{a,b} = \left(\frac{p-a}{p-b}\right)^n \left(\frac{q-a}{q-b}\right)^m x, \tag{2.45a}$$

$$v_a = -(p-a)^n\, (q-a)^m y, \tag{2.45b}$$

$$w_b = (p-b)^{-n}\, (q-b)^{-m} z, \tag{2.45c}$$

where

$$P = -G_3(-p,-a), \quad Q = -G_3(-q,-b), \quad b_0 = G_3(-a,-b), \tag{2.46}$$

and $G_3(a,b)$, p_a and q_a are defined as before.

Note that since α_2 is arbitrary, we can replace α_2 by $-\alpha_2$ and thus in C3 (2.12) P, Q and b_0 can be reparametrised as

$$P = G_3(p,a), \quad Q = G_3(q,b), \quad b_0 = -G_3(a,b), \tag{2.47}$$

where $G_3(a,b)$ is defined as (2.29).

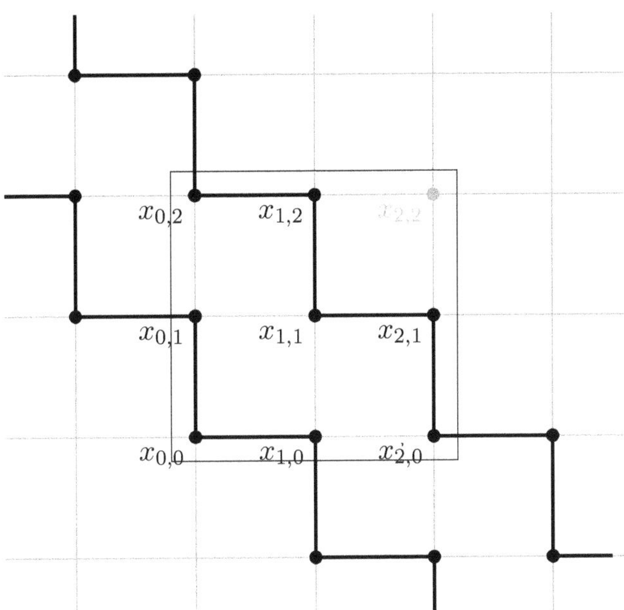

Figure 3. For one-component second order time evolution, initial values can be given on two consecutive staircases. The evolution equation will then be given on the indicated 9-point stencil.

3 Two- and one-component forms

So far we have discussed the BSQ equations in their three-component forms, e.g., in terms of x, y, z, or $u^{(0,0)}, u^{(1,0)}, u^{(0,1)}$, etc. As explained in Section 2.5, Figure 2, for the discrete BSQ equations it is then necessary to give the initial values for two components on a staircase-like configuration before the next step in the evolution can be computed. An alternative formulation of second order time evolution is to use only one component and give initial values on two consecutive staircases as in Figure 3. In that case the equation usually involves the points within a 3×3 stencil. (Note that in the two-component case, it is not necessary that the equations are defined on a square, only that the next step in the evolution can be calculated once values on the staircase are given.)

In this section we will derive two- and one-component forms of these equations. The process of variable elimination can also be interpreted as a Bäcklund transformation (BT). Assuming we have the following situation:

$$G[x] = 0 \quad \xleftarrow{\text{eliminate } y} \quad \begin{cases} A[x,y] = 0 \\ B[x,y] = 0 \end{cases} \quad \xrightarrow{\text{eliminate } x} \quad H[y] = 0$$

Then we say that the pair $\{A[x,y] = 0, B[x,y] = 0\}$ provides a BT between $G[x] = 0$ and $H[y] = 0$. Another way of looking at the above is to consider the pair $\{A[x,y] = 0, B[x,y] = 0\}$ as two equations for one variable x, which can be solved, provided that the other variable y satisfies some "integrability condition". In the context of PDE's this is a familiar situation. For example, from the pair $\partial_x \psi = A(x, y)$, $\partial_y \psi = B(x, y)$ we can solve ψ only if $\partial_x B = \partial_y A$. In more general cases this problem of "formal integrability" or "involutivity" of a set of PDE's can become quite complicated,[1] and the problem has been analysed at length in the mathematics literature (the reader is referred to a recent monograph [35]).

Now we have a partial *difference* version of the same thing: we can integrate one of the variables, say x if the other variable, say y, satisfies some PΔE (condition for integrability), which we should find. To do that, we may need several shifted consequences of the original equations, which provide not only new equations but also new variables. The hope is that we eventually have a sufficient number of equations to solve for all shifts of x and still have one more equation that gives a condition on the other variable.[2]

Since we will need several shifts in both directions we will use the notation where we give as subscripts just the shifts with respect to the basic position (n, m), for example $\widetilde{y} = y_{1,0}$.

3.1 Generalities about the elimination process

Before studying the specific equations we can make some general observations. Our starting point is the set of three equations for three variables, A2, B2, or C3. Since we always have an equation with un-shifted z it is easy to solve for that variable and use it in the remaining ones. Similarly for y, we just need to apply shifts, although due to reversal symmetry we do not have to consider this case. Thus we can easily construct a two-component pair of equations in y and x or in z and x.

The situation with x is different, because the first two equations contain $x_{0,0}$, $x_{1,0}$, $x_{0,1}$. We can use these to eliminate all shifted x and the resulting equation would then be a polynomial in $x_{0,0}$. In the optimal case the $x_{0,0}$-dependent part would factor out, but this does not always happen. Furthermore, it turns out that sometimes it is beneficial to absorb some x-dependence into y by writing the equation in terms of $w := y/x$.

As discussed in Section 2.4.2, all of our equations are reversal symmetric, which in particular exchanges z and y. We will therefore only need to construct two-component forms in terms of (y, z) and (y, x) or (z, x).

[1]The idea is to compute differential consequences of the initial equations and try to find an "involutive completion" after which the new differential consequences are just prolongations and produce no genuine new equations. One of the difficult problems is to decide when one can stop this process (this is apparently based on Spencer cohomology).

[2]After this, it remains to prove that all the remaining equations are satisfied due to this one condition (although this is never done in practice).

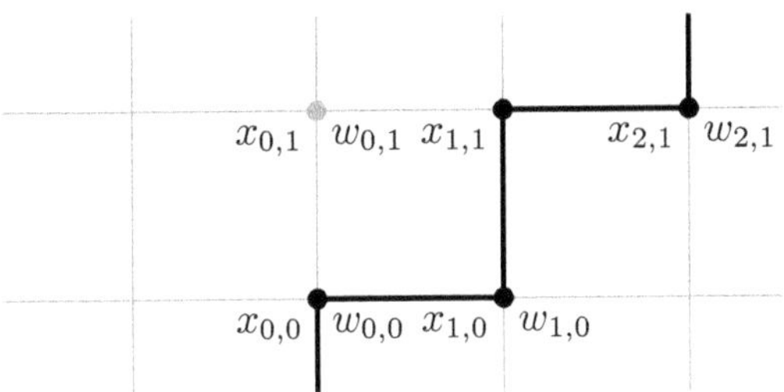

Figure 4. Equations (3.2a) and (3.2c) allow evolution in the NW direction, even if initial values outside the square are needed.

3.2 A2

We take the A2 equation in the alternate form (2.24) with $w := y/x$:

$$z_{0,0} - w_{1,0} = \frac{x_{0,0}}{x_{1,0}}, \tag{3.1a}$$

$$z_{0,0} - w_{0,1} = \frac{x_{0,0}}{x_{0,1}}, \tag{3.1b}$$

$$(z_{1,0} - z_{0,1})(z_{1,1} - w_{0,0}) = P\,\frac{x_{1,0}}{x_{0,0}} - Q\,\frac{x_{0,1}}{x_{0,0}}, \tag{3.1c}$$

because it is reversal symmetric, and because in this form x appears homogeneously and is therefore easier to eliminate.

A2 in terms of x and $w := y/x$: Solving for $z_{0,0}$ from (3.1a) we get from (3.1b) and (3.1c)

$$w_{1,0} - w_{0,1} = \frac{x_{0,0}}{x_{0,1}} - \frac{x_{0,0}}{x_{1,0}}, \tag{3.2a}$$

$$w_{2,1} - w_{0,0} = \frac{x_{1,1}}{x_{0,0}}\,\frac{P\,x_{1,0} - Q\,x_{0,1}}{x_{1,0} - x_{0,1}} - \frac{x_{1,1}}{x_{2,1}}. \tag{3.2b}$$

This is suitable for next eliminating w, but if we want to eliminate x, then it is best to first eliminate $x_{0,0}$ from (3.2b) using (3.2a) which gives the alternative form

$$\left(w_{2,1} - w_{0,0} + \frac{x_{1,1}}{x_{2,1}}\right)(w_{1,0} - w_{0,1}) = P\,\frac{x_{1,1}}{x_{0,1}} - Q\,\frac{x_{1,1}}{x_{1,0}}. \tag{3.2c}$$

The corresponding equation for z, x can be obtained by reversal symmetry, together with (2.25). Note that Equations (3.2b) and (3.2c) are not defined on the basic square but contain an extra point at $(2, 1)$. Nevertheless, they allow evolution from a staircase initial conditions, as shown in Figure 4.

A2 in terms of z and $w := y/x$: In order to eliminate x-dependence from (3.1a) and (3.1b) we need to take shifts and ratios, leading to [9, 31]

$$\frac{w_{1,0}-z_{0,0}}{w_{0,1}-z_{0,0}}=\frac{w_{1,1}-z_{0,1}}{w_{1,1}-z_{1,0}}, \tag{3.3a}$$

$$(w_{0,0}-z_{1,1})(z_{1,0}-z_{0,1})=\frac{P}{w_{1,0}-z_{0,0}}-\frac{Q}{w_{0,1}-z_{0,0}}. \tag{3.3b}$$

The second equation has the alternative form obtained by eliminating the $z_{0,0}$-dependency

$$(w_{0,0}-z_{1,1})(w_{1,0}-w_{0,1})=\frac{P}{w_{1,1}-z_{0,1}}-\frac{Q}{w_{1,1}-z_{1,0}}. \tag{3.3c}$$

Equations (3.3b) and (3.3c) are connected by

$$(P,Q,n,m,w,z)\to(-P,-Q,-n,-m,z,w).$$

The equations (3.3) are defined on the basic quadrilateral and are 3-dimensionally consistent; the triply shifted quantities are given in the Appendix.

A2 in terms of x only from (3.2a) **and** (3.2b)**:** From equations (3.2a),(3.2b) it is easy to derive an equation for x alone. These equations are of the following type

$$w_{1,0}-w_{0,1}=A_{0,0},\quad w_{2,1}-w_{0,0}=B_{0,0}, \tag{3.4a}$$

and by taking suitable shifts, one can eliminate w and derive the "integrability condition"

$$B_{1,0}-B_{0,1}=A_{2,1}-A_{0,0}. \tag{3.4b}$$

When this is calculated, we get an equation on a 3×3 stencil (cf. Figure 3) (see Eq. (A.5) in [27], Eq. (5.2) in [23], Eq. (4.9) in [24] and Eq. (5.7.6) in [37]).

$$\left(\frac{P\,x_{1,1}-Q\,x_{0,2}}{x_{0,2}-x_{1,1}}\right)\frac{x_{1,2}}{x_{0,1}}-\left(\frac{P\,x_{2,0}-Q\,x_{1,1}}{x_{1,1}-x_{2,0}}\right)\frac{x_{2,1}}{x_{1,0}}=\frac{x_{0,0}}{x_{1,0}}-\frac{x_{0,0}}{x_{0,1}}-\frac{x_{1,2}}{x_{2,2}}+\frac{x_{2,1}}{x_{2,2}}. \tag{3.5}$$

This equation is reversal symmetric with (2.25) and only changes sign.

A2 in terms of w only from (3.3a) **and** (3.3c)**:** In order to derive other one-component equations, one needs a slightly more complicated sequence of elimination steps. In order to eliminate z from (3.3a) and (3.3c), we observe that the $z_{n,m}$ that appear in these equations are located on the lattice as given in Figure 5.1, where (a) corresponds to (3.3c) and (b) to (3.3a).

The elimination process then goes on as follows: Assume that the z-values at points 1 and 2 of Figure 5.2 are arbitrary (say, $z_{0,0}$ and $z_{1,0}$), then using (b) we can

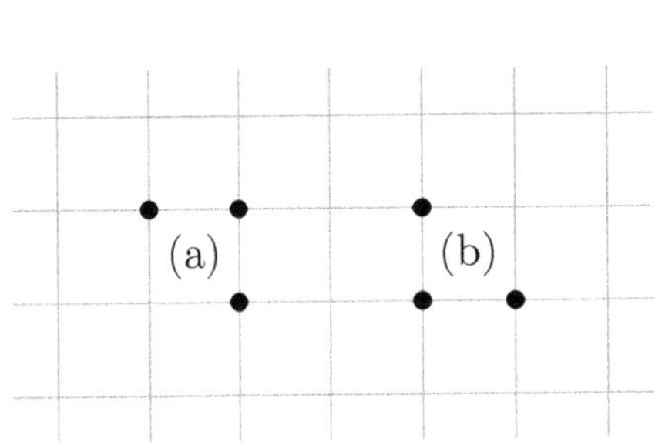

Figure 5.1

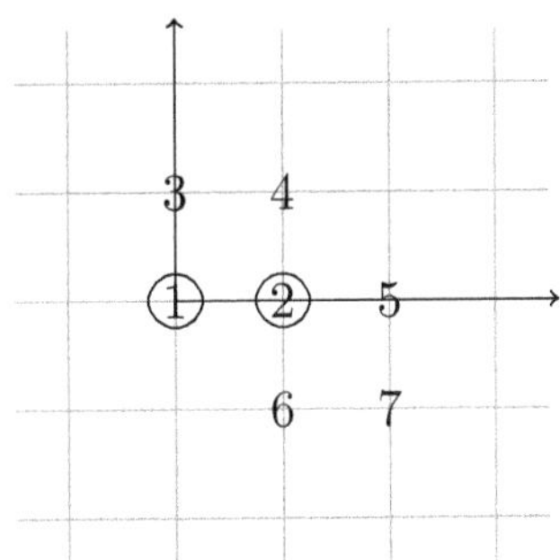

Figure 5.2

Figure 5. If equations (a) and (b) have variables located in the lattice as in Figure 5.1 then in the elimination process one needs to consider 7 points as given in Figure 5.2.

compute the value at point 3; we denote this process by $(1;2) \xrightarrow{(b)} 3$. There are two routes to compute the z-value at point 5:

$$
\begin{aligned}
&(1;2) \xrightarrow{(b)} 3, \quad (2;3) \xrightarrow{(a)} 4, \quad (4;2) \xrightarrow{(b)} 5,\\
&(1;2) \xrightarrow{(a)} 6, \quad (2;6) \xrightarrow{(b)} 7, \quad (2;7) \xrightarrow{(a)} 5.
\end{aligned}
$$

The resulting two values for z at 5 must be the same, for the arbitrary values of z at points 1 and 2. Equating the two computed values yields a rational expression in the arbitrary initial values $z_{0,0}$, $z_{1,0}$. In this case the numerator does not contain the initial values and gives the equation. The necessary polynomial algebra is straightforward but tedious and is best done using a computer algebra system (such as REDUCE [13] or Mathematica). The result is (see Eq. (1.3) in [27], Eq. (5.3) in [23], Eq. (4.18) in [24] and Eq. (5.7.3) in [37].)

$$
\frac{P-Q}{w_{2,0}-w_{1,1}} - \frac{P-Q}{w_{1,1}-w_{0,2}} - (w_{2,2}-w_{0,1})(w_{2,1}-w_{1,2}) - (w_{0,0}-w_{2,1})(w_{1,0}-w_{0,1}) = 0. \tag{3.6}
$$

The equation for z alone can be obtained by reversal symmetry and its form is identical to (3.6) except for sign changes (cf.(2.25)).

A2 in terms of w only from (3.2a) **and** (3.2c)**:** We could also eliminate x from (3.2) in order to obtain an equation in w. In the present A2 case, it is not necessary because we did already derive the w equation using another sequence of eliminations. However, in some later cases we need this different kind of elimination process, so we will do it here as an exercise with guaranteed success.

First note that the x variables appear in lattice positions as illustrated in Figure 6, with Figure 6.1(a) corresponding to (3.2a) and Figure 6.1(b) to (3.2c). The

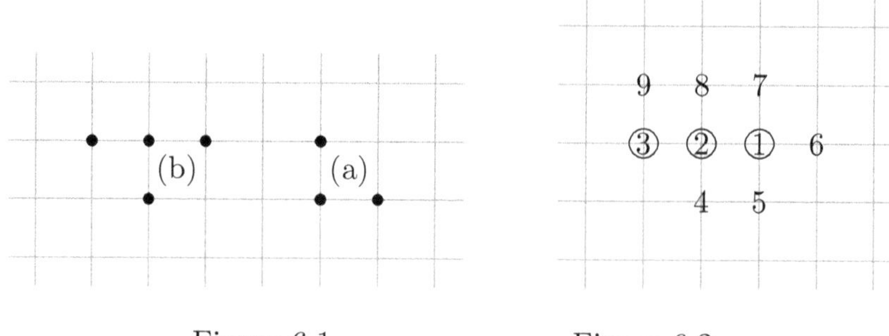

Figure 6.1 Figure 6.2

Figure 6. If equations (a) and (b) have variables located in the lattice as in Figure 6.1 then in the elimination process one needs to consider 9 points as given in Figure 6.2.

elimination process is now as follows: We assume that x values at circled points 1,2,3 are given and then compute the other x values as follows

$$(1;2;3) \xrightarrow{(b)} 4, \quad (2;4) \xrightarrow{(a)} 5, \quad (1;2;5) \xrightarrow{(b)} 6, \quad (6;1) \xrightarrow{(a)} 7,$$
$$(1,2) \xrightarrow{(a)} 8, \quad (2;3) \xrightarrow{(a)} 9, \quad (2;8;9) \xrightarrow{(b)} 7.$$

The two values of x at point 7 must be the same, which gives us an equation which should hold for arbitrary values of x at 1,2,3. When the computations are done, the result is (3.6), as expected.

3.3 B2

The B2 equation is given by

$$y_{1,0} + z_{0,0} = x_{0,0}x_{1,0}\ , \tag{3.7a}$$
$$y_{0,1} + z_{0,0} = x_{0,0}x_{0,1}\ , \tag{3.7b}$$
$$y_{0,0} + z_{1,1} = x_{0,0}x_{1,1} + b_0(x_{1,1} - x_{0,0}) + \frac{P-Q}{x_{1,0} - x_{0,1}}. \tag{3.7c}$$

Note that Equations (3.7a,3.7b) are similar to (3.1a,3.1b), and therefore the derivation of two-component forms is similar.

B2 in terms of y and x: Solving $z_{0,0}$ from (3.7a) and using in the other equations yields

$$y_{1,0} - y_{0,1} = x_{0,0}(x_{1,0} - x_{0,1}), \tag{3.8a}$$
$$y_{0,0} - y_{2,1} = (x_{0,0} - x_{2,1})x_{1,1} + b_0(x_{1,1} - x_{0,0}) + \frac{P-Q}{x_{1,0} - x_{0,1}}. \tag{3.8b}$$

Another two-component form is obtained if we replace y by $w := y + z$ and after that eliminate z. The result is

$$\begin{aligned} w_{0,1} - w_{1,0} + (x_{0,0} + x_{1,1})(x_{0,1} - x_{1,0}) &= 0, \qquad (3.9a) \\ \frac{P - Q - x_{1,0}\, w_{1,0} + x_{0,1}\, w_{0,1}}{x_{1,0} - x_{0,1}} - w_{0,0} - w_{1,1} + b_0(x_{1,1} - x_{0,0}) & \\ +(x_{0,0} + x_{1,1})(x_{1,0} + x_{0,1}) + x_{0,0}\, x_{1,1} &= 0. \qquad (3.9b) \end{aligned}$$

A notable difference with (3.8) is that this is defined on an elementary quadrilateral; it has the CAC property. This form was presented in [34], but its relation to B2 was left open.

B2 in terms of x only from (3.8a) **and** (3.8b)**:** The structure of equations (3.8a) and (3.8b) is as in (3.4a) so that we get the one-component equation using (3.4b):

$$\begin{aligned} (P - Q)\left(\frac{1}{x_{2,0} - x_{1,1}} - \frac{1}{x_{1,1} - x_{0,2}}\right) + b_0(x_{0,1} - x_{1,0} + x_{2,1} - x_{1,2}) \\ -(x_{2,2} - x_{0,1})(x_{2,1} - x_{1,2}) - (x_{0,0} - x_{2,1})(x_{1,0} - x_{0,1}) = 0. \end{aligned} \qquad (3.10)$$

This equation is a generalization (b_0-terms) of (3.6) derived for A2. Attempts to eliminate x seem to lead into very complicated equations.

3.4 C3

As noted in Section 2.3 we only need to consider C3 equation

$$\begin{aligned} y_{1,0}\, z_{0,0} &= x_{1,0} - x_{0,0}, && (3.11a) \\ y_{0,1}\, z_{0,0} &= x_{0,1} - x_{0,0}, && (3.11b) \\ z_{1,1}\, y_{0,0} &= b_0\, x_{0,0} + b_1 + z_{0,0} \frac{P\, y_{1,0}\, z_{0,1} - Q\, y_{0,1}\, z_{1,0}}{z_{1,0} - z_{0,1}}, && (3.11c) \end{aligned}$$

for the three-component case $b_0 \neq 0, b_1 = 0$ and for the two-component case $b_0 = 0$, b_1 arbitrary. We can treat them together for some computations.

C3 in terms of x, y: After solving for z from (3.11a) and using it in (3.11b) and (3.11c), we get the two-component form

$$y_{1,0}(x_{0,1} - x_{0,0}) = y_{0,1}(x_{1,0} - x_{0,0}), \qquad (3.12a)$$

$$\begin{aligned} y_{0,0}(x_{2,1} - x_{1,1}) =& y_{2,1}(b_0\, x_{0,0} + b_1) \\ &+ y_{2,1} \frac{(x_{1,1} - x_{0,1})(x_{0,0} - x_{1,0})\, P - (x_{1,1} - x_{1,0})(x_{0,0} - x_{0,1})\, Q}{x_{1,0} - x_{0,1}}. \end{aligned} \qquad (3.12b)$$

This form is well suited for eliminating y next, because it is linear in y, but if we would like to eliminate x the following alternative would be useful:

$$(x_{2,1}-x_{1,1})\frac{y_{0,0}}{y_{2,1}}=(b_0\,x_{0,0}+b_1)-\frac{(x_{1,1}-x_{0,1})\,y_{1,0}\,P-(x_{1,1}-x_{1,0})\,y_{0,1}\,Q}{y_{1,0}-y_{0,1}},\tag{3.12c}$$

because it is linear in x. These equations are defined on a configuration given in Figure 4 ($w\leftrightarrow y$). The equations for x,z are the same as above, and can be obtained by reversal, cf.(2.21,2.22).

C3 in terms of x only: Equations (3.12a, 3.12b) have the form

$$\frac{y_{1,0}}{y_{0,1}}=A_{0,0},\quad \frac{y_{0,0}}{y_{2,1}}=B_{0,0}.$$

The dependence is similar to (3.4), except that instead of an additive case we now have a multiplicative case. The integrability condition is

$$\frac{A_{2,1}}{A_{0,0}}=\frac{B_{0,1}}{B_{1,0}},$$

which yields

$$\frac{(x_{2,2}-x_{1,2})(x_{0,2}-x_{1,1})(x_{0,1}-x_{0,0})}{(x_{2,2}-x_{2,1})(x_{1,1}-x_{2,0})(x_{1,0}-x_{0,0})}=\frac{(x_{1,1}-x_{0,2})(b_0\,x_{0,1}+b_1)+(x_{1,2}-x_{0,2})(x_{0,1}-x_{1,1})P-(x_{1,2}-x_{1,1})(x_{0,1}-x_{0,2})Q}{(x_{2,0}-x_{1,1})(b_0\,x_{1,0}+b_1)+(x_{2,1}-x_{1,1})(x_{1,0}-x_{2,0})P-(x_{2,1}-x_{2,0})(x_{1,0}-x_{1,1})Q}.\tag{3.13}$$

Note that this is invariant under $n\leftrightarrow m$, $P\leftrightarrow Q$, and reversal symmetric with (2.22).

C3 in terms of y only: The dependence on x in (3.12a) and (3.12c) is linear and of the form given in Figure 6, and using that method we can eliminate x. The resulting equation in terms of y is the same as (3.5) in terms of x. Note in particular that dependence on the parameters b_0,b_1 drops out from the y-equation, while it remains in the x-equation.

C3 in terms of y and z: The method of eliminating x depends sensitively on the additional x-dependent term. We need to consider separately the two cases.

- b_1 arbitrary, $b_0=0$. In this case Equation (3.11c) does not depend on x at all, while form (3.11a) and (3.11b) x can be easily eliminated, leaving

$$y_{1,1}(z_{1,0}-z_{0,1})+z_{0,0}(y_{1,0}-y_{0,1})=0,\tag{3.14a}$$

$$y_{0,0}z_{1,1}=b_1+z_{0,0}\frac{Py_{1,0}z_{0,1}-Qy_{0,1}z_{1,0}}{z_{1,0}-z_{0,1}}.\tag{3.14b}$$

Note that by the gauge transformation

$$z = w\,p^n\,q^m, \quad y = -v\,p^{-n}\,q^{-m}, \tag{3.15}$$

with $P = p^3$, $Q = q^3$ and $b_1 = b_1' pq$, we will get from (3.14)

$$\frac{p\,\widetilde{w} - q\,\widehat{w}}{w} = \frac{p\,\widehat{v} - q\,\widetilde{v}}{\widehat{\widetilde{v}}} = \frac{p\,\widetilde{v}\,\widehat{w} - q\,\widehat{v}\,\widetilde{w}}{b_1' + v\,\widehat{\widetilde{w}}}. \tag{3.16}$$

When $b_1' = 0$ we get

$$\frac{p\,\widetilde{w} - q\,\widehat{w}}{w} = \frac{p\,\widehat{v} - q\,\widetilde{v}}{\widehat{\widetilde{v}}} = \frac{p\,\widetilde{v}\,\widehat{w} - q\,\widehat{v}\,\widetilde{w}}{v\,\widehat{\widetilde{w}}}, \tag{3.17}$$

which was already presented in [24].

- $b_1 = 0$, $b_0 \neq 0$. In this case it is best to absorb some x into y by defining $w := y/x$. Then (3.11a) and (3.11b) yield

$$w_{1,1}z_{0,0}(w_{1,0}z_{1,0} - w_{0,1}z_{0,1}) - w_{1,1}(z_{1,0} - z_{0,1}) - z_{0,0}(w_{1,0} - w_{0,1}) = 0, \tag{3.18a}$$

while the other equation becomes

$$b_0 + \frac{w_{1,1}}{w_{1,0} - w_{0,1}}\left(\frac{Pw_{1,0}z_{0,1}}{w_{1,1}z_{0,1} - 1} - \frac{Qw_{0,1}z_{1,0}}{w_{1,1}z_{1,0} - 1}\right) - w_{0,0}z_{1,1} = 0. \tag{3.18b}$$

Note that in both cases the equation pair is still quadrilateral.

In order to derive one-component forms, one can use the method of Figure 5, after some modifications in the equations.

C3 in terms of z or y from (3.14): Eliminating variable y leads to (3.5) but in terms of z and n, m reversed.

In order to eliminate z one should first eliminate $z_{0,0}$ from (3.14b) using (3.14a), after which the method of Figure 5 works and yields (3.5) in terms of y.

C3 in terms of w or z from (3.18): After eliminating z one gets (3.5), but now in terms of w and with $(p, q) \mapsto (p - b_0, p - b_0)$.

In order to eliminate w one should first eliminate $w_{1,1}$ from (3.14b) using (3.14a), and then one gets (3.5) in terms of z but n, m reversed.

Summary of one-component forms:

- A2 has two one-component forms: (3.6) in $w = y/x$ and in z, and (3.5) in x. They correspond to regular and modified BSQ, respectively.
- C3 has two one-component forms: (3.13) in x and (3.5) in y or $w = y/x$ or z, corresponding to Schwarzian and modified BSQ, respectively.

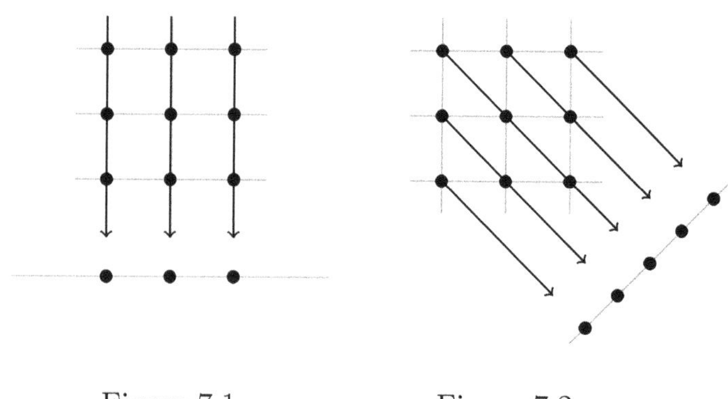

Figure 7.1 Figure 7.2

Figure 7. Two ways to project the 3×3 stencil to a line: Figure 7.1 gives the straight limit, Figure 7.2 the skew limit. A 90° rotation is also possible.

Thus we can say that the three-component form of A2 contains both regular and modified BSQ and the three-component form of C3 contains both Schwarzian and modified BSQ. B2 on the other hand contains only regular BSQ, but in a generalized form in comparison to A2. It is an interesting open question whether there is a three-component version which contains all three different BSQ equations in full generality.

4 Continuum limits

We will now consider (semi-)continuous limits of the derived one-component equations: (3.5) for A2, (3.10) for B2, and (3.13) for $C3_{b_0}$. These are defined on a 3×3 stencil, see Figure 7.

The technical aspects of taking semi-continuous and fully continuous limits involve several choices, including the gauge (or background solution) and the way the lattice parameters behave under these limits. For example, we know that p, q are related to lattice spacing, but should we let, for example, $p \to 0$ or $p \to \infty$? Before going into specific equations we are going to discuss some aspects that apply to all cases.

4.1 Common features

4.1.1 Approaching continuum

The semi-continuous limit means that in some direction the lattice points approach each other to form a continuum. There are two simple ways to squeeze the 3×3 stencil onto a line, the straight and the skewed way, see Figure 7.1 and 7.2, respectively. (Squeezing in other directions is also possible, but the result would depend on still more points and probably would not be useful for applications.)

Before taking the limit, we must anchor the discrete and continuous variables. First of all we shift the equations so that the center of the 3×3 stencil is at $(0, 0)$.

For the *straight limit* in the m direction, we can take $m = 0$ to correspond to $\xi = 0$ and then generically the discrete variable m corresponds to continuous variable ξ and we can write

$$u_{n,\mu} = u_n(\xi + \epsilon\mu), \tag{4.1}$$

where ϵ measures the distance between two lattice points in the m direction. In practice $\mu \in \{-1, 0, 1\}$. For the *skew limit* we must take instead

$$x_{\nu,\mu} = u_{\nu+\mu}(\tau + \tfrac{1}{2}(\nu - \mu)\epsilon), \tag{4.2}$$

where $\nu, \mu \in \{-1, 0, 1\}$.

Now that the dependence on ϵ is given we can expand in ϵ,

$$u_n(\xi + \epsilon s) = u_n(\xi) + \epsilon s\, u_n'(\xi) + \frac{1}{2}\epsilon^2 s^2\, u_n''(\xi) + \cdots. \tag{4.3}$$

4.1.2 Parameter relations

The above discussion was about limits in general, but in concrete cases we must first determine the connections between the various parameters in the equation and the lattice spacing. One method to get information about this is to linearize the discrete equation and study its discrete plane wave solutions or plane wave factors (PWF) (cf. [26, 33]) Once the PWF is known one can figure out what should be done in order to get, as a limit, the continuous PWF. This is described in Chapter 5 of [16]. In general terms the discrete-continuous relation is an application of

$$\lim_{n\to\infty}\left(1 + \frac{\alpha}{n}\right)^n = e^\alpha. \tag{4.4}$$

Before one can linearize a nonlinear equation, it is necessary to choose a background solution (or gauge) around which to expand. For Equation (3.5) we must choose a multiplicative gauge

$$x_{n,m} = (p-a)^{-n+n_0}(q-b)^{-m+m_0}c_1(1 + \epsilon v_{n,m} + \cdots), \tag{4.5}$$

while for (3.10) we must choose additive gauge, such as

$$x_{n,m} = \sigma(n - n_0)(p - a) + \sigma(m - m_0)(q - b) + \epsilon u_{n,m} + \cdots, \tag{4.6}$$

but for (3.13) no gauge is necessary.

We have used here (n_0, m_0) as origin but in the expansion these should drop out.

For each of the considered DBSQ equations one finds for $u_{n,m}$ or $v_{n,m}$ the PWF solution:

$$\text{PWF:} \qquad \left(\frac{p - \omega(k)}{p - k}\right)^n \left(\frac{q - \omega(k)}{q - k}\right)^m, \tag{4.7}$$

where $\omega(k) \neq k$ is one of the roots of $g_3(\omega) - g_3(k) = 0$, where g_3 is a cubic polynomial, such as (2.29b).

Straight limit: In order to use (4.4) for a straight limit in the m-direction, we write

$$\left(\frac{q-\omega(k)}{q-k}\right)^m = \left(1+\frac{k-\omega(k)}{q-k}\right)^m$$

and therefore q must approach infinity as m, i.e., $m/q=\xi$. Then

$$\left(\frac{q-\omega(k)}{q-k}\right)^m = \left(1+\frac{\xi(k-\omega(k))}{m-\xi k}\right)^m \longrightarrow \exp[\xi(k-\omega(k))], \quad \text{as } m\to\infty.$$

Sometimes we may need higher order corrections to the above as follows:

$$\left(\frac{q-\omega(k)}{q-k}\right)^m = \exp\left[\xi(k-\omega(k))+\tfrac{1}{2}(k^2-\omega(k)^2)\xi/q+O(1/q^2)\right].$$

These results suggest that in a straight limit in m direction we should take

$$m=\xi\, q, \quad q\to\infty. \tag{4.8}$$

Skew limit: In the skew limit, as in Figure 7.2, we first need to make a 45° degree rotation, and after defining $N=n+m$, $m'=\frac{1}{2}(n-m)$, i.e., $n=\frac{1}{2}N+m'$, $m=\frac{1}{2}N-m'$, we get[3]

$$\begin{aligned}
&\left(\frac{p-\omega(k)}{p-k}\right)^n\left(\frac{q-\omega(k)}{q-k}\right)^m\\
&=\left(\frac{p-\omega(k)}{p-k}\cdot\frac{q-\omega(k)}{q-k}\right)^{N/2}\left(\frac{p-k}{p-\omega(k)}\cdot\frac{q-\omega(k)}{q-k}\right)^{m'}\\
&=\left(\frac{p-\omega(k)}{p-k}\cdot\frac{q-\omega(k)}{q-k}\right)^{N/2}\left(1+\frac{(k-\omega(k))(p-q)}{(p-\omega(k))(q-k)}\right)^{m'}.
\end{aligned}$$

This suggests that $p-q$ should approach 0, thus we take

$$q=p-\delta, \quad \tau=\delta m', \quad \delta\to 0 \tag{4.9}$$

and then the expansion yields

$$\left(1+\frac{(k-\omega(k))\delta}{(p-\omega(k))(p-k-\delta)}\right)^{\tau/\delta} = \exp\left[\frac{(k-\omega(k))\tau}{(p-\omega(k))(p-k)}+O(\delta^2)\right].$$

Thus the main results we have so far obtained from PWF analysis are the limit behaviors (4.8) and (4.9) for straight and skew limits, respectively. Expanding in higher orders will be useful in deriving double limits.

Finally we must relate the lattice parameters p,q used in the limits to the parameters P,Q appearing in the equation, we use

$$P=g_3(p) \text{ or } P=g_3(p)-g_3(a), \quad \text{where } g_3(x):=x^3+\alpha_2x^2+\alpha_1x+\alpha_0, \tag{4.10}$$

[3]Note that one can also take $N=n+m$, $m'=m$, cf. [16].

and similarly for Q. (This g_3 is a generalized version of (2.29b) with additional α_j.) The cubic term is a signature of Boussinesq equations. It should also be noted that as far as the CAC test is concerned, the parameters P, Q can be arbitrary. However, if we want the discrete equation to have a continuum limit to a (semi-)continuous Boussinesq equation, then the cubic form is necessary; furthermore P, Q may depend on some other parameters of the equation.

4.2 B2

As the first case we take equation (3.10), but centered at the origin

$$(P-Q)\left(\frac{1}{x_{1,-1}-x_{0,0}}-\frac{1}{x_{0,0}-x_{-1,1}}\right)+b_0(x_{-1,0}-x_{0,-1}+x_{1,0}-x_{0,1})$$
$$-(x_{1,1}-x_{-1,0})(x_{1,0}-x_{0,1})-(x_{-1,-1}-x_{1,0})(x_{0,-1}-x_{-1,0})=0.$$

4.2.1 Straight limit

We use the linear background $x=pn+qm+\epsilon\, u$ and take $\epsilon=1/q\to 0$ with initially arbitrary P, Q as in (4.10). Then using (4.1) and expanding we find from order 0 that $\alpha_3=1$, and at order 1 that $\alpha_2=-b_0$.[4] Finally (if $\alpha_1=0$, in agreement with (2.29b)) at order 2 we get the semi-discrete three-point equation

$$\begin{aligned}&u_1''+u_0''+u_{-1}''-3u_1'(p-u_0+u_1)+3u_{-1}'(p-u_{-1}+u_0)\\&+(p-u_0+u_1)^3-(p-u_{-1}+u_0)^3\\&-b_0[u_{-1}'-u_1'+(2p-u_{-1}+u_1)(u_{-1}-2u_0+u_1)]=0.\end{aligned}\tag{4.11}$$

where the primes refer to derivatives with respect to ξ, cf. (4.1). Note that when $b_0=0$, the above equation is Eq. (5.9.13) in [37] (with $u\to -u$). The b_0 term is actually a discrete derivative of the semi-discrete lpKdV equation

$$u_0'+u_1'+(2p-u_0+u_1)(u_0-u_1)=0.$$

4.2.2 Skew limit

As discussed before, we should make a 45° rotation and then take the continuous limit in the m' direction. We use the same linear background as before, and the form of P, Q found for the straight limit, but instead of (4.8) and (4.1) we use (4.9) and (4.2).

Then at order ϵ^3 we get the following five-point equation

$$\begin{aligned}&(3p^2-2pb_0)\,\ddot{u}_0+(\dot{u}_0+1)^2[(u_2-u_{-1}-b_0+3p)(\dot{u}_1+1)\\&\qquad+(u_{-2}-u_1+b_0-3p)(\dot{u}_{-1}+1)]=0,\end{aligned}\tag{4.12}$$

where the dot refers to derivatives with respect to τ in (4.2). For $b_0=0$ this agrees with (5.9.4) in [37], up to $p\to -p$.

[4]This means in particular that although P, Q and b_0 were independent in the CAC analysis, they must be related in the indicated way before we have a reasonable continuum limit.

4.2.3 Double limit

Now we take Equation (4.11) and do the straight continuum limit on the remaining discrete variable using $p = 1/\delta$. We set $u_\nu(\xi) = v(\tau + \nu\delta, \xi)$ and apply Taylor expansion as in (4.3) and the leading term in the expansion would then be $(\partial_\xi - \partial_\tau)v(\tau, \xi) = 0$, which means that the naive limit does not work. In order to get a better understanding of the situation, we return to the PWF (4.7). With $q \to \infty$ limit already taken we expand in $1/p$:

$$\left(\frac{p-\omega(k)}{p-k}\right)^n e^{(k-\omega(k))\xi} = \exp\Big[(k-\omega(k))(\xi+\tau) + \tfrac{1}{2}(k^2-\omega^2(k))\tfrac{\tau}{p} + O(\tfrac{1}{p^2})\Big], \; \tau = \frac{n}{p}. \tag{4.13}$$

This suggests that we should introduce new variables

$$x = \xi + \tau, \quad t = \tau/p, \tag{4.14}$$

which means

$$\partial_\xi = \partial_x, \quad \partial_\tau = \partial_x + \tfrac{1}{p}\partial_t, \tag{4.15}$$

and apply these in the expansion, before actually taking the limit $p \to \infty$. In other words, the variables ξ and τ are infinitesimally close and therefore the leading term combines with some higher order terms. With this modification the leading term becomes

$$v_{xxxx} - 12v_{xx}v_x - 8b_0 v_{xt} + 12v_{tt} = 0.$$

The v_{xt} term can be converted into v_{xx} by the translation $\partial_t \mapsto \partial_t + \frac{1}{3}b_0\partial_x$ (which could have been be added into (4.15)) which yields

$$v_{xxxx} - 12v_{xx}v_x - \tfrac{4}{3}b_0^2 v_{xx} + 12v_{tt} = 0. \tag{4.16}$$

This is the standard Boussinesq equation with a mass term.

We can also take the double continuum limit by starting from (4.12), where the continuous variable is τ and $p = 1/\delta$. Thus we expand $u_\nu(\tau) = v(\xi + \nu\delta, \tau)$, but a change of the continuous variables is also necessary. If we define x, t by

$$\partial_\tau = [1 + \delta\tfrac{1}{3}b_0]\delta^2\partial_x + \delta^3\partial_t, \quad \partial_\xi = \partial_x,$$

we get (4.16).

Finally, we can take a double limit directly from the fully discrete form (3.10). For that purpose we take the linear gauge as before, and the limit by $p = p_0/\epsilon$, $q = q_0/\epsilon$, $\epsilon \to 0$ with

$$u_{n,m} = -v(x + nA + mB, y + (n+m)\epsilon^2/(p_0 q_0)), \tag{4.17}$$

where

$$A := -\epsilon/p_0 - (\epsilon^2 b_0)/(3p_0^2), \quad B := -\epsilon/q_0 - (\epsilon^2 b_0)/(3q_0^2). \tag{4.18}$$

After expanding in ϵ, the leading term is again (4.16).

4.3 A2

Now we consider equation (3.5), centered at the origin:

$$\left(\frac{P\,x_{0,0}-Q\,x_{-1,1}}{x_{-1,1}-x_{0,0}}\right)\frac{x_{0,1}}{x_{-1,0}}-\left(\frac{P\,x_{1,-1}-Q\,x_{0,0}}{x_{0,0}-x_{1-1}}\right)\frac{x_{1,0}}{x_{0,-1}}=\frac{x_{-1,-1}}{x_{0,-1}}-\frac{x_{-1,-1}}{x_{-1,0}}-\frac{x_{0,1}}{x_{1,1}}+\frac{x_{1,0}}{x_{1,1}}.$$

In this case we use the gauge transformation (4.5) and the cubic parametrisation (4.10).

4.3.1 Straight limit

As in Section 4.2.1 we use (4.8), (4.10) and (4.1) but now with a multiplicative gauge (4.5) and take $\alpha_3=\sigma=1$, $b=a$ after which the leading term in the expansion will yield a second order three-point equation:

$$\begin{aligned}&\frac{v_1''}{v_1}+\frac{v_0''}{v_0}+\frac{v_{-1}''}{v_{-1}}-2\left(\frac{v_1'}{v_1}\right)^2-\left(\frac{v_0'}{v_0}\right)^2-\frac{v_1'v_0'}{v_1v_0}+\frac{v_{-1}'v_0'}{v_{-1}v_0}\\&\quad+3(p-a)\left(\frac{v_{-1}'}{v_0}-\frac{v_1'v_0}{v_1^2}\right)+(\alpha_2+3a)\left(\frac{v_{-1}'}{v_{-1}}-\frac{v_1'}{v_1}\right)-\left(\frac{v_1}{v_0}-\frac{v_0}{v_{-1}}\right)\frac{P(p)}{a-p}\\&\quad-\left(\frac{v_{-1}}{v_0}-\frac{v_0}{v_1}\right)(\alpha_2+3a)(a-p)+\left(\frac{v_{-1}^2}{v_0^2}-\frac{v_0^2}{v_1^2}\right)(a-p)^2=0.\end{aligned}\tag{4.19}$$

This can also be written as

$$\begin{aligned}&[\ln(v_{-1}v_0v_1)]''+(\ln v_{-1}-\ln v_1)'[\alpha_2+3a+(\ln(v_{-1}v_0v_1))']\\&\quad+(p-a)\left[\frac{v_{-1}}{v_0}(\alpha_2+3a+3(\ln v_{-1})')-\frac{v_0}{v_1}(\alpha_2+3a+3(\ln v_1)')\right]\\&\quad+\frac{P(p)}{p-a}\left(\frac{v_1}{v_0}-\frac{v_0}{v_{-1}}\right)+(p-a)^2\left(\frac{v_{-1}^2}{v_0^2}-\frac{v_0^2}{v_1^2}\right)=0.\end{aligned}\tag{4.20}$$

If $\alpha_2=a=0$ this is the same as (5.9.14) in [37].

4.3.2 Skew limit

For the skew limit, we also use multiplicative gauge (4.5) and expansion (4.2) with (4.9) and cubic parametrisation. The semi-continuous limit is then

$$\partial_\tau\left(\frac{v_1}{v_{-1}}\frac{\Pi_1\,\dot v_0+\Pi_2\,v_0}{(p-a)\dot v_0-v_0}\right)-\frac{v_{-2}}{v_{-1}}\left((p-a)\frac{\dot v_{-1}}{v_{-1}}-1\right)+\frac{v_1}{v_2}\left((p-a)\frac{\dot v_1}{v_1}-1\right)=0,\tag{4.21}$$

where

$$\begin{aligned}\Pi_1&=(g_3(p)-g_3(a))/(a-p),\\\Pi_2&=-(a+2p+\alpha_2).\end{aligned}$$

4.3.3 Double limit

Again we can derive the fully continuous limit from either the straight semi-continuous Equation (4.19) or from the skew semi-continuous limit (4.21) or taking a double limit directly from (3.5). For A2 there is the special feature that in order to reach the modified BSQ equation we must also change the dependent variables by $v = e^V$.

Starting with the straight semi-continuous limit (4.19) we use again $p = 1/\delta$ and $u_\nu = v(y + \nu\delta, x)$, $v = e^V$ and furthermore we use the change of variables[5]

$$\partial_y = (1 - \tfrac{1}{3}\alpha_2\delta)\partial_x + \tfrac{1}{2}\delta\partial_t,$$

and then we obtain

$$V_{xxxx} - 6V_{xx}V_x^2 + 6V_{xx}V_t - 4(\alpha_2 + 3a)V_{xx}V_x + 4(\alpha_1 - \tfrac{1}{3}\alpha_2^2)V_{xx} + 3V_{tt} = 0. \quad (4.22)$$

This is the modified Boussinesq equation (cf. [21], Equation (3.5)).

From the skew limit (4.21) we use $p = 1/\delta$, $\alpha_3 = 1$ and the variable change

$$\partial_y = \delta^2(-1 + \tfrac{1}{3}\alpha_2\delta)\partial_x - \delta^3\partial_t,$$

and get the same result (4.22).

Finally the double limit directly from (3.5) is taken with (4.17) except now

$$A := \epsilon/p_0 - (\epsilon/p_0)^2\tfrac{1}{3}\alpha_2, \quad B := \epsilon/q_0 - (\epsilon/q_0)^2\tfrac{1}{3}\alpha_2. \quad (4.23)$$

The result is again the same as in (4.22).

4.4 C3

We consider next the continuum limits of (3.13), centered at $(0,0)$:

$$\frac{(x_{1,1} - x_{0,1})(x_{-1,1} - x_{0,0})(x_{-1,0} - x_{-1,-1})}{(x_{1,1} - x_{1,0})(x_{0,0} - x_{1,-1})(x_{0,-1} - x_{-1,-1})} =$$
$$\frac{(x_{0,0} - x_{-1,1})(b_0\, x_{-1,0} + b_1) + (x_{0,1} - x_{-1,1})(x_{-1,0} - x_{0,0})P - (x_{0,1} - x_{0,0})(x_{-1,0} - x_{-1,1})Q}{(x_{1,-1} - x_{0,0})(b_0\, x_{0,-1} + b_1) + (x_{1,0} - x_{0,0})(x_{0,-1} - x_{1,-1})P - (x_{1,0} - x_{1,-1})(x_{0,-1} - x_{0,0})Q}.$$

4.4.1 Straight limit

We use the usual straight limit approach (4.1), (4.8), and (4.10) but without any gauge transformation. The leading term in the expansion is then

$$\partial_x \log\left[P(p)(u_1 - u_{-1}) - b_1 - b_0 u_{-1} + \frac{u_1' u_0' u_{-1}'}{(u_1 - u_0)(u_0 - u_{-1})}\right]$$
$$= \frac{u_1'}{u_1 - u_0} - \frac{u_{-1}'}{u_0 - u_{-1}}. \quad (4.24)$$

A special case of this appears in [37] (5.9.15).

[5]This transformation is needed in order to eliminate the cross-term V_{xt} in favor of the mass term V_{xx}.

4.4.2 Skew limit

For the skew limit we use (4.2), (4.9) and (4.10), without gauge transform, and the leading term in the expansion yields

$$\partial_\tau \log\Big[P(p)(u_1-u_{-1})-b_1-b_0u_{-1} \\ +\frac{(3p^2+2\alpha_2 p+\alpha_1)(u_1-u_0)(u_0-u_{-1})}{\dot u_0}\Big] = \frac{\dot u_{-1}}{u_{-1}-u_{-2}}-\frac{\dot u_1}{u_2-u_1}. \tag{4.25}$$

This has some similarity to the straight limit. When $P(p)=p^3$, $b_1=b_0=0$, one will get (5.9.9) in [37] (up to signs).

4.4.3 Double limit

Starting from (4.24) and using $p=1/\delta$ and $u_\nu=v(y+\nu\delta,x)$ with the additional variable change defined by

$$\partial_y=(1-\tfrac{1}{3}\alpha_2\delta)\partial_x+\tfrac{1}{2}\delta\partial_t,$$

we obtain

$$3\partial_t\left(\frac{v_t}{v_x}\right)+\partial_x\left(\frac{v_{xxx}}{v_x}+\frac{3}{2}\frac{v_t^2-v_{xx}^2}{v_x^2}-4\frac{b_1+b_0v}{v_x}\right)=0. \tag{4.26}$$

The double limit obtained by starting from the skew semi-continuous limit (4.25), together with variable change

$$\partial_y=(-1+\tfrac{1}{3}\alpha_2\delta)\delta^2\partial_x+\tfrac{1}{2}\delta^3\partial_t,$$

again yields (4.26).

The direct double limit from (3.13) using (4.17) with (4.23), but with the second variable in (4.17) given by $y-(n+m)\epsilon^2/(2p_0q_0)$, leads again to (4.26), except for $b_1\mapsto -b_1$.

If $b_0=b_1=0$ in (4.26) it is the Schwarzian BSQ (see, e.g., [38] Eq. (4.9)), however the additional b_0,b_1 terms break the Möbius invariance.

5 Lax pairs

In general Lax pairs can be generated from CAC: One takes the side equations and interprets the bar-shifted variables as linear Lax variables. However, the equations defined on an edge only are not convenient, thus one usually takes some linear combinations of them. The construction of Lax matrices using CAC can be automatized to some extent, see [3–5].

One important requirement for the Lax matrices is that they should contain a spectral parameter, in the following it will be R.

5.1 B2

The B2 equations were given (2.3). On the left side of the cube we then have

$$\widetilde{y} = x\widetilde{x} - z, \quad \overline{y} = x\overline{x} - z, \tag{5.1}$$
$$\widetilde{\overline{y}} = \overline{x}\widetilde{\overline{x}} - \overline{z}, \quad \widetilde{\overline{y}} = \widetilde{x}\widetilde{\overline{x}} - \widetilde{z}, \tag{5.2}$$

and if from the second set we eliminate $\widetilde{\overline{y}}$ or $\widetilde{\overline{x}}$ we get

$$\widetilde{\overline{x}} = \frac{\widetilde{z} - \overline{z}}{\widetilde{x} - \overline{x}}, \qquad \widetilde{\overline{y}} = \frac{\widetilde{z}\,\overline{x} - \overline{z}\,\widetilde{x}}{\widetilde{x} - \overline{x}}. \tag{5.3}$$

Now we introduce [41]

$$\overline{x} = \frac{\phi_1}{\phi_0}, \quad \overline{z} = \frac{\phi_2}{\phi_0}, \quad \overline{y} = \frac{\phi_3}{\phi_0}, \tag{5.4}$$

and then the equations in (5.3) can be written as

$$\widetilde{\phi}_0 \;=\; \gamma(\widetilde{x}\phi_0 - \phi_1), \tag{5.5a}$$
$$\widetilde{\phi}_1 \;=\; \gamma(\widetilde{z}\phi_0 - \phi_2), \tag{5.5b}$$
$$\widetilde{\phi}_3 \;=\; \gamma(\widetilde{z}\phi_1 - \widetilde{x}\phi_2), \tag{5.5c}$$

where γ is the separation factor. For the $\widetilde{\overline{z}}$ equation we take tilde-bar version of (2.3c) (with also $Q \to R$) and eliminate from it $\widetilde{\overline{x}}$ using (5.3), this results with

$$\widetilde{\phi}_2 \;=\; \gamma(A_p\,\phi_0 + (b_0\,x - b_1 + y)\phi_1 - (b_0 + x)\phi_2), \tag{5.5d}$$
$$\text{where } A_p \;:=\; (b_1 - b_0 x - y)\widetilde{x} + (b_0 + x)\widetilde{z} + P - R. \tag{5.5e}$$

The above can also be done for the hat-bar version. We then get the matrix equations

$$\widetilde{\phi} = \boldsymbol{L}\phi\,, \quad \widehat{\phi} = \boldsymbol{M}\phi\,, \tag{5.6}$$

in which [41]

$$\phi = \begin{pmatrix} \phi_0 \\ \phi_1 \\ \phi_2 \\ \phi_3 \end{pmatrix}, \quad \boldsymbol{L}^{B2}_{4\times 4} = \gamma \begin{pmatrix} \widetilde{x} & -1 & 0 & 0 \\ \widetilde{z} & 0 & -1 & 0 \\ A_p & y + b_0 x - b_1 & -b_0 - x & 0 \\ 0 & \widetilde{z} & -\widetilde{x} & 0 \end{pmatrix}. \tag{5.7}$$

The matrix $\boldsymbol{M}$ is the hat-q version of $\boldsymbol{L}$. Now the last column of $\boldsymbol{L}, \boldsymbol{M}$ is a null column and therefore we can reduce the system to a 3×3 matrix problem with

$$\phi = \begin{pmatrix} \phi_0 \\ \phi_1 \\ \phi_2 \end{pmatrix}, \quad \boldsymbol{L}^{B2}_{3\times 3} = \gamma \begin{pmatrix} \widetilde{x} & -1 & 0 \\ \widetilde{z} & 0 & -1 \\ A_p & y + b_0 x - b_1 & -b_0 - x \end{pmatrix}. \tag{5.8}$$

The compatibility condition following from (5.6) is

$$\widehat{\boldsymbol{L}}\boldsymbol{M} = \widetilde{\boldsymbol{M}}\boldsymbol{L}. \tag{5.9}$$

For this equation it would be best to choose the separation factor γ so that det $\boldsymbol{L}$ =const., because then part of the matrix equation is immediately satisfied by taking determinants of both sides. In this case we can take $\gamma = 1$ because det $\boldsymbol{L} = P - R$. With this choice (5.9) yields three equations:

$$\widehat{\widetilde{x}} = \frac{\widetilde{z} - \widehat{z}}{\widetilde{x} - \widehat{x}}, \quad x = \frac{\widetilde{y} - \widehat{y}}{\widetilde{x} - \widehat{x}}, \tag{5.10a}$$

$$\widehat{\widetilde{z}} = b_0(\widehat{\widetilde{x}} - x) + x\widehat{\widetilde{x}} + b_1 - y + \frac{P - Q}{\widetilde{x} - \widehat{x}}. \tag{5.10b}$$

This is not completely equivalent to (2.3) because we do not have un-shifted z, nor $\widetilde{z}$ and $\widehat{z}$ separately.

In the language of DLA version of B2 (2.28) (with $b_1 = 0$), the 3×3 Lax matrices are as follows,

$$\boldsymbol{L}^{B2D} = \begin{pmatrix} p - \widetilde{u}^{(0,0)} & 1 & 0 \\ -\widetilde{u}^{(1,0)} & p - b_0 & 1 \\ * & -u^{(0,1)} - 2b_0u^{(0,0)} - 2b_0^2 & p + 2b_0 + u^{(0,0)} \end{pmatrix}, \tag{5.11}$$

in which $* = R - P - (p - \widetilde{u}^{(0,0)})[(p + b_0)(p + u^{(0,0)}) + u^{(0,1)}] - (p + u^{(0,0)} + 2b_0)\widetilde{u}^{(1,0)}$, and where $\boldsymbol{M}^{B2D}$ is obtained from (5.11) by replacing p by q and $\widetilde{\ }$ by $\widehat{\ }$, furthermore $\det[\boldsymbol{L}^{B2D}] = R - P$. The compatibility leads to the equations

$$\widehat{u}^{(1,0)} - \widetilde{u}^{(1,0)} = (p - q + \widehat{u}^{(0,0)} - \widetilde{u}^{(0,0)})\widehat{\widetilde{u}}^{(0,0)} - p\widehat{u}^{(0,0)} + q\widetilde{u}^{(0,0)} \tag{5.12a}$$

and

$$\widehat{u}^{(0,1)} - \widetilde{u}^{(0,1)} = (p - q + \widehat{u}^{(0,0)} - \widetilde{u}^{(0,0)})u^{(0,0)} - p\widetilde{u}^{(0,0)} + q\widehat{u}^{(0,0)}, \tag{5.12b}$$

together with an equation that is equivalent to equation (2.28c), if we use $\widehat{\widetilde{u}}^{(0,0)}$ as solved from the first equation and $u^{(0,0)}$ solved from the second.

The B2 Lax pair was first given in [27] with $\alpha_1 = \alpha_2 = b_0 = b_1 = 0$ and in [41] with full parametrisation.

5.2 A2

From CAC: The A2 equation was given in (2.4), where we can take $b_0 = 0$. Using (2.4a) and the bar-version of (2.4b) and their shifts, we can easily derive

$$\overline{\widetilde{x}} = \frac{\widetilde{x} - \overline{x}}{\widetilde{z} - \overline{z}}, \qquad \overline{\widetilde{y}} = \frac{\widetilde{x}\,\overline{z} - \overline{x}\,\widetilde{z}}{\widetilde{z} - \overline{z}}. \tag{5.13}$$

To these and to the tilde-bar version of (2.4c) we use (5.4) and obtain the Lax matrix [41]

$$\boldsymbol{L} = \gamma(x, \widetilde{x}) \begin{pmatrix} \widetilde{z} & 0 & -1 & 0 \\ \widetilde{x} & -1 & 0 & 0 \\ \frac{y\widetilde{z}}{x} - \frac{P\widetilde{x}}{x} & \frac{R}{x} & -\frac{y}{x} & 0 \\ 0 & -\widetilde{z} & \widetilde{x} & 0 \end{pmatrix}. \tag{5.14}$$

Again the last column vanishes and we use instead an invertible 3×3 matrix [41]

$$\boldsymbol{L}^{A2} = \gamma(x, \widetilde{x}) \begin{pmatrix} \widetilde{z} & 0 & -1 \\ \widetilde{x} & -1 & 0 \\ \frac{y\widetilde{z}}{x} + \frac{P\widetilde{x}}{x} & -\frac{R}{x} & -\frac{y}{x} \end{pmatrix}, \tag{5.15}$$

and the matrix $\boldsymbol{M}^{A2}$ is the hat-q version of $\boldsymbol{L}^{A2}$. If we again normalize the Lax matrix by the condition det$[\boldsymbol{L}]$ =const, we should take $\gamma(a, b) = (a/b)^{1/3}$. Then the compatibility condition (5.9) yields

$$\widehat{\widetilde{x}} = \frac{\widetilde{x} - \widehat{x}}{\widetilde{z} - \widehat{z}}, \qquad x = \frac{\widehat{x}\widetilde{y} - \widetilde{x}\widehat{y}}{\widetilde{x} - \widehat{x}}, \tag{5.16}$$

together with eq. (2.4c).

From DLA: In the DLA the A2 equation is given in (2.33) and the Lax matrix given there is [41]

$$\boldsymbol{L}^{A2} = \begin{pmatrix} p - a & \widetilde{v}_a & 0 \\ 0 & p - \widetilde{u}^{(0,0)} & 1 \\ \frac{G_3(R, -a)}{v_a} & * & p - \alpha_2 + \frac{s_a}{v_a} \end{pmatrix}, \tag{5.17}$$

in which $* = (p - \widetilde{u}^{(0,0)})(p - \alpha_2 + s_a/v_a) - p_a\widetilde{v}_a/v_a$, R stands for the spectral parameter, and $\boldsymbol{M}^{A2}$ is obtained from (5.17) by replacing p by q and $\widetilde{}$ by $\widehat{}$. (Recall that v_a is related to x, $u^{(0,0)}$ to z and s_a to y.) The Lax compatibility condition leads to the equations (2.33c) and

$$p - q + \widehat{u}^{(0,0)} - \widetilde{u}^{(0,0)} = (p - a)\frac{\widehat{v}_a}{\widehat{\widetilde{v}}_a} - (q - a)\frac{\widetilde{v}_a}{\widehat{\widetilde{v}}_a}, \tag{5.18a}$$

$$p - q + \frac{\widehat{s}_a}{\widehat{v}_a} - \frac{\widetilde{s}_a}{\widetilde{v}_a} = (p - a)\frac{v_a}{\widetilde{v}_a} - (q - a)\frac{v_a}{\widehat{v}_a}. \tag{5.18b}$$

From the two-component form: The A2 equation has also a two-component form,

$$\frac{z - \widetilde{w}}{z - \widehat{w}} = \frac{\widehat{z} - \widehat{\widetilde{w}}}{\widetilde{z} - \widehat{\widetilde{w}}}, \tag{5.19a}$$

$$(\widetilde{z} - \widehat{z})(\widehat{\widetilde{z}} - w) = \frac{P}{z - \widetilde{w}} - \frac{Q}{z - \widehat{w}}. \tag{5.19b}$$

This pair is still defined on the elementary square and is 3-dimensionally consistent, but it is not possible to construct a Lax pair using the sides of the consistency cube, because it leads to an expression that is quadratic in the auxiliary field ϕ. However, a 2×2 Lax matrix was given in [31] (see Eq. (2.4.5)) (although this does not contain spectral parameter):

$$\boldsymbol{L}^{A2}_{2\times 2} = \begin{pmatrix} \widetilde{z} & -1 \\ w\widetilde{z} + \frac{P}{z - \widetilde{w}} & -w \end{pmatrix}, \tag{5.20a}$$

$$\boldsymbol{M}^{A2}_{2\times 2} = \begin{pmatrix} \widehat{z} & -1 \\ w\widehat{z} + \frac{Q}{z - \widehat{w}} & -w \end{pmatrix}, \tag{5.20b}$$

which directly yields (5.19) from the compatibility $\widehat{\boldsymbol{L}}\boldsymbol{M} = \widetilde{\boldsymbol{M}}\boldsymbol{L}$.

Still another Lax matrix generating (5.19) is given in [9] (see page 16)

$$\boldsymbol{L}^{FX} = \begin{pmatrix} \widetilde{z} - w & 0 & -1 \\ -1 & \widetilde{w} - z & 0 \\ \frac{P}{\widetilde{w}-z} - (\widetilde{w} - w)(\widetilde{z} - w) & R & \widetilde{w} - w \end{pmatrix}, \tag{5.21}$$

with corresponding $\boldsymbol{M}^{FX}$. As mentioned above, this Lax pair cannot arise from CAC analysis of the type that worked for the other equations. However, since this Lax matrix resembles some of the other CAC-generated Lax matrices, we could try to reverse-engineer and see where it could come from. If we write out the 3×3 version of $\widetilde{\phi} = \boldsymbol{L}^{FX}\phi$ and then divide the second and third equation by the first we get

$$\frac{\widetilde{\phi}_1}{\widetilde{\phi}_0} = \frac{1 + (\widetilde{z} - w)\frac{\phi_1}{\phi_0}}{w - \widetilde{z} + \frac{\phi_2}{\phi_0}}, \tag{5.22a}$$

$$\frac{\widetilde{\phi}_2}{\widetilde{\phi}_0} = \frac{R\frac{\phi_1}{\phi_0} + (\widetilde{w} - w)\frac{\phi_1}{\phi_0} + \frac{P}{\widetilde{w}-z} - (\widetilde{w} - w)(\widetilde{z} - w)}{w - \widetilde{z} + \frac{\phi_2}{\phi_0}}. \tag{5.22b}$$

Previously we associated bar-quantities to ϕ ratios as in (5.4) but it does not work here. We must instead take

$$\frac{\phi_1}{\phi_0} = \frac{1}{\overline{w} - z}, \quad \frac{\phi_2}{\phi_0} = \overline{z} - w, \tag{5.23}$$

and this choice yields (5.19) from (5.22).

5.3 C3

The C3 equation (2.5) is

$$x - \widetilde{x} = \widetilde{y}z\ , \quad x - \widehat{x} = \widehat{y}z\ , \tag{5.24a}$$

$$\widehat{\widetilde{z}}\, y = z\frac{P\widetilde{z}\widetilde{y} - Q\widehat{z}\widehat{y}}{\widetilde{z} - \widehat{z}} + b_1 + b_0\, x\ . \tag{5.24b}$$

Several other equations can be derived from (5.24a), for example

$$\widehat{\widetilde{x}} = \frac{\widehat{x}\widetilde{z} - \widetilde{x}\widehat{z}}{\widetilde{z} - \widehat{z}}, \quad x = \frac{\widehat{x}\widetilde{y} - \widetilde{x}\widehat{y}}{\widetilde{y} - \widehat{y}}, \quad z = -\frac{\widetilde{x} - \widehat{x}}{\widetilde{y} - \widehat{y}}, \quad \widehat{\widetilde{y}} = -z\frac{\widetilde{y} - \widehat{y}}{\widetilde{z} - \widehat{z}}. \tag{5.25}$$

If we consider the tilde-bar versions of these and take equation for $\overline{\widetilde{x}}$ from (5.25) and equation (5.24b) in which $\widetilde{y}$ and $\overline{y}$ have been eliminated using (5.24a), then we can use the CAC method and construct the 3×3 Lax matrix

$$\boldsymbol{L}^{C3}_{3\times 3} = \frac{1}{z}\begin{pmatrix} \widetilde{z} & 0 & -1 \\ 0 & \widetilde{z} & -\widetilde{x} \\ \frac{\widetilde{z}}{y}(b_1 + (b_0 - R)\,x) & \frac{\widetilde{z}}{y}R & \frac{1}{y}(P(\widetilde{x} - x) - b_1 - b_0\,x) \end{pmatrix}. \tag{5.26}$$

The compatibility conditions arising from this Lax matrix and its hat-Q companion yield the equations for $\widehat{\widetilde{x}}$ and x in (5.25). However, for $\widehat{\widetilde{z}}$ it produces an equation that agrees with (5.24b) only after we also use the equation for z in (5.25), which does not follow from Lax compatibility.

There is also the 4×4 Lax matrix that can be obtained from CAC by adding the $\widehat{\widetilde{y}}$ equation from (5.25):

$$\boldsymbol{L}^{C3}_{4\times 4} = \frac{1}{z}\begin{pmatrix} \widetilde{z} & 0 & 0 & -1 \\ 0 & \widetilde{z} & 0 & -\widetilde{x} \\ -z\widetilde{y} & 0 & z & 0 \\ \frac{\widetilde{z}}{y}(b_1+b_0\,x) & 0 & -\frac{z\widetilde{z}}{y}R & \frac{1}{y}(Pz\widetilde{y}-b_1-b_0\,x) \end{pmatrix}, \tag{5.27}$$

and the matrix $\boldsymbol{M}^{C3}_{4\times 4}$ is the hat-Q version. In this case the compatibility conditions yield equation (5.24b) and equations for $\widehat{\widetilde{x}}$, x, $\widehat{\widetilde{y}}$ in (5.25).

If $b_0 = 0$ we also have the two-component version (2.13c,2.13d) containing z and y only. Using the CAC method we can construct the 3×3 Lax matrix

$$\boldsymbol{L}^{C3b_1}_{3\times 3} = \frac{1}{z}\begin{pmatrix} \widetilde{z} & 0 & -1 \\ -z\widetilde{y} & z & 0 \\ b_1\frac{\widetilde{z}}{y} & -R\frac{z\widetilde{z}}{y} & \frac{1}{y}(Pz\widetilde{y}-b_1) \end{pmatrix} \tag{5.28}$$

and with corresponding $\boldsymbol{M}^{C3b_1}_{3\times 3}$ the compatibility condition exactly returns (2.13c, 2.13d).

6 Bilinear structures of DBSQ-type equations

6.1 Preliminary

6.1.1 Discrete Hirota's bilinear form and Casoratians

Suppose $f_j(n,m)$ and $g_j(n,m)$ are functions defined on $\mathbb{Z}\times\mathbb{Z}$. Then a one-component discrete Hirota bilinear equation has the following form [17, 20]

$$\sum_j c_j\, f_j(n+\nu_j^+, m+\mu_j^+)\, g_j(n+\nu_j^-, m+\mu_j^-) = 0,$$

where it is essential that the index sums $\mu_j^+ + \mu_j^- = \mu_s$, $\nu_j^+ + \nu_j^- = \nu_s$ do not depend on j.

The solutions to a discrete Hirota bilinear equation will be given by a **Casoratian**, which is a determinant of a matrix composed of different shifts of a vector. For example, given functions $\psi_i(n,m,l)$ we define the column vector

$$\psi(n,m,l) = \big(\psi_1(n,m,l), \psi_2(n,m,l), \cdots, \psi_N(n,m,l)\big)^T, \tag{6.1}$$

and then the N-th order Casoratian reads

$$\big|\psi(n,m,l_1), \psi(n,m,l_2), \cdots, \psi(n,m,l_N)\big|. \tag{6.2}$$

For such a determinant we use a shorthand notation $|l_1, l_2, \cdots, l_N|$. Furthermore, if the Casoratian contains consecutive columns we use condensed notation such as (cf. [10]),

$$|0, 1, \cdots, N-1| = |\widehat{N-1}|, \quad |0, 1, \cdots, N-2, N| = |\widehat{N-2}, N|.$$

For the DBSQ-type equations discussed in this chapter the solutions can be expressed through a Casoratian $f = |\widehat{N-1}|$ which is composed by the *entry function* Ψ:

$$\begin{aligned}\psi_j(l, n, m, \alpha, \beta) \quad := \quad & \sum_{s=0}^{2} (-\omega_s(k_j))^l (p - \omega_s(k_j))^n (q - \omega_s(k_j))^m \\ & \times (a - \omega_s(k_j))^\alpha (b - \omega_s(k_j))^\beta \, \varrho_{j,s}^{(0)},\end{aligned} \tag{6.3}$$

which contains 5 independent variables n, m, l, α and β. In the following we do not mention those variables that are obvious from context. Here $\omega_s(k_j),\ s = 1, 2$ are roots of

$$g_3(\omega(k_j)) = g_3(k_j), \tag{6.4}$$

where g_3 defined in (2.29b) and $\omega_0(k_j) \equiv k_j$.

For shifts in the index variables (l, n, m, α, β) we often use shorthand notation: in addition to tilde and hat for shifts in the n- and m-direction (as in (2.1)), we introduce bar, circle and dot for the shifts in the l-, α- and β-directions, respectively, i.e.,

$$\overline{f}(n, m, l, \alpha, \beta) = f(n, m, l+1, \alpha, \beta), \tag{6.5a}$$

$$\mathring{f}(n, m, l, \alpha, \beta) = f(n, m, l, \alpha+1, \beta), \tag{6.5b}$$

$$\dot{f}(n, m, l, \alpha, \beta) = f(n, m, l, \alpha, \beta+1). \tag{6.5c}$$

When the symbol is below the variable it means backward shift: e.g., $\underset{\sim}{f}(n, m, l, \alpha, \beta) = f(n-1, m, l, \alpha, \beta)$.

A more general form than (6.3) is

$$\psi_j(l) = \sum_{s=0}^{2} (\delta - \omega_s(k_j))^l (p - \omega_s(k_j))^n (q - \omega_s(k_j))^m (a - \omega_s(k_j))^\alpha (b - \omega_s(k_j))^\beta \, \varrho_{j,s}^{(0)}, \tag{6.6}$$

which is referred to as the δ-extension of (6.3). Thus (6.6) can be considered as a function containing symmetrically five dimensions, corresponding to the direction coordinates (n, m, l, α, β) and their lattice spacing parameters (p, q, a, b, δ). Here for $\psi_j(l)$ in (6.6) we already omitted n, m, α, β for convenience, we will often do this if it does not cause any confusion.

Finally we note that a Casoratian can sometimes be written as a Wronskian. Suppose we define

$$\phi_j(l) = \sum_{s=0}^{2} e^{(\delta - \omega_s(k_j))l} (p - \omega_s(k_j))^n (q - \omega_s(k_j))^m (a - \omega_s(k_j))^\alpha (b - \omega_s(k_j))^\beta \, \varrho_{j,s}^{(0)}, \tag{6.7}$$

and the corresponding vector as in (6.1). Then the following Casoratian and Wronskian are equal to each other,

$$|\psi(l_1), \psi(l_2), \cdots, \psi(l_N)|_{C[\psi]} = |\partial_l^{l_1}\phi(l), \partial_l^{l_2}\phi(l), \cdots, \partial_l^{l_N}\phi(l)|_{W[\phi]}.$$

6.1.2 Laplace expansion

The proof that a Casoratian solves a bilinear equation is usually given by reducing the problem to a three-term Laplace expansion of a zero determinant. We will first give a generic result.

Lemma 1. *Suppose that* $\mathbf{P}$ *is a* $N \times (N-1)$ *matrix, and* $\mathbf{Q}$ *its* $N \times (N-k+1)$ *sub-matrix obtained by removing arbitrary* $(k-2)$ *columns from* $\mathbf{P}$ *where* $k \geq 3$*. Let* $\mathbf{a}_i (i = 1, 2, \cdots, k)$ *be some* N*-th order column vectors. Then we have*

$$\sum_{i=1}^{k} (-1)^{i-1} |\mathbf{P}, \mathbf{a}_i|\, |\mathbf{Q}, \mathbf{a}_1, \cdots, \mathbf{a}_{i-1}, \mathbf{a}_{i+1}, \cdots, \mathbf{a}_k| = 0. \tag{6.8}$$

This is a special case of Plüker relations.

In fact, let $\mathbf{B}$ be the $N \times (k-2)$ matrix consisting of those $(k-2)$ column vectors that are removed from $\mathbf{P}$ so that $|\mathbf{QB}| = |\mathbf{P}|$. Then it is easy to see that the following $2N \times 2N$ determinant vanishes:

$$\begin{vmatrix} \mathbf{Q} & \mathbf{0} & \mathbf{B} & \mathbf{a}_1 & \cdots & \mathbf{a}_k \\ \mathbf{0} & \mathbf{Q} & \mathbf{0} & \mathbf{a}_1 & \cdots & \mathbf{a}_k \end{vmatrix} = 0. \tag{6.9}$$

The LHS of (6.8) is actually the Laplace expansion of the LHS of (6.9).

Equation (6.8) is quite useful in proving solutions. When $k = 3$, (6.8) yields

$$|\mathbf{P}, \mathbf{a}_1|\, |\mathbf{Q}, \mathbf{a}_2, \mathbf{a}_3| - |\mathbf{P}, \mathbf{a}_2|\, |\mathbf{Q}, \mathbf{a}_1, \mathbf{a}_3| + |\mathbf{P}, \mathbf{a}_3|\, |\mathbf{Q}, \mathbf{a}_1, \mathbf{a}_2| = 0. \tag{6.10}$$

If we denote $\mathbf{P} = (\mathbf{Q}, \mathbf{a}_0)$, it reads

$$|\mathbf{Q}, \mathbf{a}_0, \mathbf{a}_1|\, |\mathbf{Q}, \mathbf{a}_2, \mathbf{a}_3| - |\mathbf{Q}, \mathbf{a}_0, \mathbf{a}_2|\, |\mathbf{Q}, \mathbf{a}_1, \mathbf{a}_3| + |\mathbf{Q}, \mathbf{a}_0, \mathbf{a}_3|\, |\mathbf{Q}, \mathbf{a}_1, \mathbf{a}_2| = 0, \tag{6.11}$$

which can be viewed as an expression of Plüker relation, and were used to verify Wronskian (or Casoratian) solutions to bilinear equations [10] and also to prove many determinantal identities (see [12]). When $k = 4$, (6.8) yields

$$\begin{aligned} &|\mathbf{P}, \mathbf{a}_1|\, |\mathbf{Q}, \mathbf{a}_2, \mathbf{a}_3, \mathbf{a}_4| - |\mathbf{P}, \mathbf{a}_2|\, |\mathbf{Q}, \mathbf{a}_1, \mathbf{a}_3, \mathbf{a}_4| + |\mathbf{P}, \mathbf{a}_3|\, |\mathbf{Q}, \mathbf{a}_1, \mathbf{a}_2, \mathbf{a}_4| \\ &-|\mathbf{P}, \mathbf{a}_4|\, |\mathbf{Q}, \mathbf{a}_1, \mathbf{a}_2, \mathbf{a}_3| = 0, \end{aligned} \tag{6.12}$$

which is also useful in solution verification (see [43]). In this chapter, besides (6.10) and (6.12), we need also $k = 5$ case, which is

$$\begin{aligned} &|\mathbf{P}, \mathbf{a}_1|\, |\mathbf{Q}, \mathbf{a}_2, \mathbf{a}_3, \mathbf{a}_4, \mathbf{a}_5| - |\mathbf{P}, \mathbf{a}_2|\, |\mathbf{Q}, \mathbf{a}_1, \mathbf{a}_3, \mathbf{a}_4, \mathbf{a}_5| + |\mathbf{P}, \mathbf{a}_3|\, |\mathbf{Q}, \mathbf{a}_1, \mathbf{a}_2, \mathbf{a}_4, \mathbf{a}_5| \\ &-|\mathbf{P}, \mathbf{a}_4|\, |\mathbf{Q}, \mathbf{a}_1, \mathbf{a}_2, \mathbf{a}_3, \mathbf{a}_5| + |\mathbf{P}, \mathbf{a}_5|\, |\mathbf{Q}, \mathbf{a}_1, \mathbf{a}_2, \mathbf{a}_3, \mathbf{a}_4| = 0. \end{aligned} \tag{6.13}$$

Here is another determinantal property which is often used in Wronskian/Casoratian verification.

Lemma 2. *[10]*

$$\sum_{j=1}^{N} |\mathbf{a}_1, \cdots, \mathbf{a}_{j-1}, \mathbf{b}\mathbf{a}_j, \mathbf{a}_{j+1}, \cdots, \mathbf{a}_N| = \left(\sum_{j=1}^{N} b_j\right) |\mathbf{a}_1, \cdots, \mathbf{a}_N|, \tag{6.14}$$

where $\mathbf{a}_j = (a_{1j}, \cdots, a_{Nj})^T$ *and* $\mathbf{b} = (b_1, \cdots, b_N)^T$ *are* N*-th order column vectors and* $\mathbf{b}\mathbf{a}_j$ *stands for* $(b_1 a_{1j}, \cdots, b_N a_{Nj})^T$.

A generalized version of this lemma can be found in [39, 42]. For further details see [40].

6.2 B2

6.2.1 B2

Equation B2 ((2.3) with $b_1 = 0$) is given by

$$B_1 := \quad \widetilde{y} - x\widetilde{x} + z = 0, \tag{6.15a}$$

$$B_2 := \quad \widehat{y} - x\widehat{x} + z = 0, \tag{6.15b}$$

$$B_3 := \quad y - b_0(\widehat{\widetilde{x}} - x) - x\widehat{\widetilde{x}} + \widehat{\widetilde{z}} - \frac{P - Q}{\widetilde{x} - \widehat{x}} = 0, \tag{6.15c}$$

where

$$P = p^3 - b_0 p^2 + R, \quad Q = q^3 - b_0 q^2 + R. \tag{6.15d}$$

It has background solution [18]

$$x_0 = pn + qm + c_1, \tag{6.16a}$$

$$z_0 = \tfrac{1}{2}x_0^2 + \tfrac{1}{2}(p^2 n + q^2 m + c_2) + c_3, \tag{6.16b}$$

$$y_0 = \tfrac{1}{2}x_0^2 - \tfrac{1}{2}(p^2 n + q^2 m + c_2) - c_3, \tag{6.16c}$$

where c_1, c_2, c_3 are arbitrary constants.

By the dependent variable transformation [18]

$$x = x_0 - \frac{g}{f}, \quad z = z_0 - x_0\frac{g}{f} + \frac{h}{f}, \quad y = y_0 - x_0\frac{g}{f} + \frac{s}{f}, \tag{6.17}$$

we can bilinearize the B2 lattice consisting of (6.15) and their shifts as follows

$$B_1 = \frac{\mathcal{B}_1}{f\widetilde{f}}, \; B_2 = \frac{\mathcal{B}_2}{f\widehat{f}}, \; B_3 = \frac{\mathcal{B}_3\mathcal{B}_4 + (p-q)f\widehat{\widetilde{f}}\mathcal{B}_4 + [p^2 + pq + q^2 - b_0(p+q)]\widetilde{f}\widehat{f}\mathcal{B}_3}{(\widetilde{x} - \widehat{x})f\widetilde{f}\widehat{f}\widehat{\widetilde{f}}}, \tag{6.18}$$

where the bilinear equations are

$$\mathcal{B}_1 := \widetilde{f}(h+pg) - \widetilde{g}(g+pf) + f\widetilde{s} = 0, \tag{6.19a}$$

$$\mathcal{B}_2 := \widehat{f}(h+qg) - \widehat{g}(g+qf) + f\widehat{s} = 0, \tag{6.19b}$$

$$\mathcal{B}_3 := \widetilde{f}\widehat{g} - \widehat{f}\widetilde{g} + (p-q)(\widetilde{f}\widehat{f} - f\widehat{\widetilde{f}}) = 0, \tag{6.19c}$$

$$\mathcal{B}_4 := [p^2+pq+q^2-b_0(p+q)](f\widehat{\widetilde{f}}-\widetilde{f}\widehat{f})+(p+q-b_0)(\widehat{\widetilde{f}}g-f\widehat{\widetilde{g}})+\widehat{\widetilde{f}}s+f\widehat{\widetilde{h}}-g\widehat{\widetilde{g}}=0. \tag{6.19d}$$

We have also used the parametrisation (6.15d).

The set of bilinear equations (6.19) admits N-soliton solutions in the following Casoratian form,

$$f = |\widehat{N-1}|,\ g = |\widehat{N-2}, N|,\ h = |\widehat{N-2}, N+1|,\ s = |\widehat{N-3}, N-1, N|, \tag{6.20}$$

composed of $\psi = (\psi_1, \psi_2, \cdots, \psi_N)^T$ with (6.3). A proof for these Casoratian solutions can be found in [19], where the meaning of the parameter b_0 is also discussed.

An alternate bilinearization was given in [22], where they first transformed the 9-point equation (3.10) without b_0 into a pair of equations living on a 2×4 stencil (Eqs. (53, 54)), which were then bilinearized using only f and g (see Eqs. (56, 57, 65, 66)).

6.2.2 B2-δ

In the above derivation we used Casoratians with entries (6.3). However, there is a natural generalization of the entry function with a new parameter δ as given in (6.6). We will now derive the corresponding generalized equations.

As the first step we compute the difference of f, g, h, s (as defined in (6.20)) for $\delta = 0$ and for $\delta \neq 0$. Using binomial expansion on $(\delta - \omega_s(k_j))^l$ we see that only the right-most columns contribute and find [18] (here $f \equiv f(0)$, $f' \equiv f(\delta)$)

$$f' = f, \tag{6.21a}$$

$$g' = g + N\delta\, f, \tag{6.21b}$$

$$h' = h + (N+1)\delta\, g + \tfrac{1}{2}N(N+1)\,\delta^2 f, \tag{6.21c}$$

$$s' = s + (N-1)\delta\, g + \tfrac{1}{2}N(N-1)\,\delta^2 f, \tag{6.21d}$$

where N is the size of the determinant. Now inverting this for f, g, h, s and using them on (6.19) it follows that the primed quantities solve the δ-modified bilinear equations

$$\mathcal{B}_1^\delta := \widetilde{f'}\,[h' + (p-\delta)g'\,] - \widetilde{g'}\,[g' + (p-\delta)f'\,] + f'\,\widetilde{s'} = 0, \tag{6.22a}$$

$$\mathcal{B}_2^\delta := \widehat{f'}\,[h' + (q-\delta)g'\,] - \widehat{g'}\,[g' + (q-\delta)f'\,] + f'\,\widehat{s'} = 0, \tag{6.22b}$$

$$\mathcal{B}_3^\delta := \widetilde{f'}\,\widehat{g'} - \widehat{f'}\,\widetilde{g'} + (p-q)\left(\widetilde{f'}\,\widehat{f'} - f'\,\widehat{\widetilde{f'}}\right) = 0, \tag{6.22c}$$

$$\begin{aligned}\mathcal{B}_4^\delta &:= (p^2+pq+q^2-b_0(p+q))\left(f'\,\widehat{\widetilde{f'}} - \widetilde{f'}\,\widehat{f'}\right)\\ &\quad - (p+q-b_0+\delta)\left(f'\,\widehat{\widetilde{g'}} - \widehat{\widetilde{f'}}g'\right) + \widehat{\widetilde{f'}}\,s' + f'\,\widehat{\widetilde{h'}} - g'\,\widehat{\widetilde{g'}} = 0.\end{aligned} \tag{6.22d}$$

Furthermore, from (6.17) we get

$$x' = x_0' - \frac{g'}{f'} = (x_0' - N\delta) - \frac{g}{f} = x, \tag{6.23a}$$

provided that $x_0' = x_0 + N\delta$, which can be accommodated by a change in the constant: $c_1' = c_1 + N\delta$. Similarly

$$\begin{aligned} z' &= z_0' - x_0'\frac{g'}{f'} + \frac{h'}{f'}\\ &= z_0' - (x_0+N\delta)\frac{g}{f} - x_0'N\delta + \frac{h}{f} + (N+1)\delta\frac{g}{f} + \tfrac{1}{2}N(N+1)\,\delta^2\\ &= z - \delta(x - x_0), \end{aligned} \tag{6.23b}$$

$$y' = y + \delta(x - x_0), \tag{6.23c}$$

provided that the constants in z_0, y_0 are adjusted so that $\frac{1}{2}(c_2'-c_2)+c_3'-c_3+\frac{1}{2}\delta^2 N = 0$. Using these we get the δ-modified nonlinear equations:

$$B_1^\delta := \widetilde{y'} - x'\widetilde{x'} + z' - \delta\left(\widetilde{x'} - x' - p\right) = 0, \tag{6.24a}$$

$$B_2^\delta := \widehat{y'} - x'\widehat{x'} + z' - \delta\left(\widehat{x'} - x' - q\right) = 0, \tag{6.24b}$$

$$B_3^\delta := y' - b_0\left(\widehat{\widetilde{x'}} - x'\right) - x'\widehat{\widetilde{x'}} + \widehat{\widetilde{z'}} + \frac{p^3-q^3-b_0(p^2-q^2)}{\widehat{x'} - \widetilde{x'}} + \delta\left(\widehat{\widetilde{x'}} - x' - p - q\right) = 0. \tag{6.24c}$$

Equations (6.22) and (6.24) were given for the $b_0 = 0$ case in [18] and for generic b_0 in [32].

6.3 A2

6.3.1 A2

The A2 equation (after removing parameter b_0) is given by

$$A_1 := \quad \widetilde{x}z - \widetilde{y} - x = 0, \tag{6.25a}$$

$$A_2 := \quad \widehat{x}z - \widehat{y} - x = 0, \tag{6.25b}$$

$$A_3 := \quad x\widehat{\widetilde{z}} - y - \frac{P\widetilde{x} - Q\widehat{x}}{\widetilde{z} - \widehat{z}} = 0, \tag{6.25c}$$

where P, Q will be parametrised by (2.37). It has background or seed solution (2.36)

$$x_a = (p-a)^{-n}(q-a)^{-m}c_1, \tag{6.26a}$$
$$z_0 = (c_3 - p)n + (c_3 - q)m + c_2, \tag{6.26b}$$
$$y_a = x_a(z_0 + a - c_3), \tag{6.26c}$$

where $c_3 = \alpha_2/3$, and c_1, c_2, α_i are arbitrary constants.

By the dependent variable transformation

$$x = x_a\frac{\underset{\circ}{f}}{f}, \quad z = z_0 + \frac{g}{f}, \quad y = y_a\frac{\underset{\circ}{f}}{f} + x_a\frac{\underset{\circ}{g}}{f}, \tag{6.27}$$

A2 is bilinearized into

$$\mathcal{A}_1 := \widetilde{\underset{\circ}{f}}(g + pf) - (p-a)\underset{\circ}{f}\widetilde{f} - f(a\widetilde{\underset{\circ}{f}} + \widetilde{\underset{\circ}{g}}) = 0, \tag{6.28a}$$
$$\mathcal{A}_2 := \widehat{\underset{\circ}{f}}(g + qf) - (q-a)\underset{\circ}{f}\widehat{f} - f(a\widehat{\underset{\circ}{f}} + \widehat{\underset{\circ}{g}}) = 0, \tag{6.28b}$$
$$\mathcal{A}_3 := \widetilde{f}\widehat{g} - \widehat{f}\widetilde{g} + (p-q)(\widetilde{f}\widehat{f} - f\widehat{\widetilde{f}}) = 0, \tag{6.28c}$$
$$\mathcal{A}_4 := (p-q)[((p+q+a-\alpha_2)\underset{\circ}{f} + \underset{\circ}{g})\widehat{\widetilde{f}} - \widehat{\widetilde{g}}\underset{\circ}{f}] - p_a\widetilde{\widetilde{f}}\underset{\circ}{f} + q_a\widehat{\widehat{f}}\underset{\circ}{f} = 0, \tag{6.28d}$$

where p_a, q_a are defined as (2.34) and the circle shift was defined in (6.5). The connection to (6.25) is by

$$A_1 = \frac{\widetilde{x}_a\mathcal{A}_1}{f\widetilde{f}}, \; A_2 = \frac{\widehat{x}_a\mathcal{A}_2}{f\widehat{f}}, \; A_3 = x_a\frac{\mathcal{A}_3\mathcal{A}_4 + (p-q)f\widehat{\widetilde{f}}\mathcal{A}_4 + (p_a\widetilde{\widetilde{f}}\underset{\circ}{f} - q_a\widehat{\widehat{f}}\underset{\circ}{f})\mathcal{A}_3}{(p-q)f\widetilde{f}\widehat{f}\widehat{\widetilde{f}}(\widehat{z} - \widetilde{z})}. \tag{6.29}$$

Multisoliton solutions to (6.28) are given through

$$f = |\widehat{N-1}|, \; g = |\widehat{N-2}, N|, \tag{6.30}$$

composed of ψ given in (6.6) with $\delta = 0$.

6.3.2 A2-δ

The δ deformation discussed in Section 6.2.2 can also be applied on A2. In this case the definitions of f, g are as in (6.20), and therefore the δ-dependence is as in (6.21), i.e., $f' = f$, $g' = g + N\delta\, f$. From this and (6.27) it follows

$$x = x', \quad y = y', \quad z = z' - \delta, \tag{6.31}$$

provided that we choose $c_2' - c_2 + N\delta = 0$. Then we find A2-δ as

$$A_1^\delta := \widetilde{x}'z' - \widetilde{y}' - x' - \delta\widetilde{x}' = 0, \tag{6.32a}$$
$$A_2^\delta := \widehat{x}'z' - \widehat{y}' - x' - \delta\widehat{x}' = 0, \tag{6.32b}$$
$$A_3^\delta := x'\widehat{\widetilde{z}}' - y' - \delta x' + \frac{P\widetilde{x}' - Q\widehat{x}'}{\widetilde{z}' - \widehat{z}'} = 0. \tag{6.32c}$$

On the other hand, it can be easily seen that equations (6.28) are invariant under $g \mapsto g + cf$.

6.4 $\mathrm{C3}_{b_0}$

The equation $\mathrm{C3}_{b_0}$ is

$$C_1 := \widetilde{x} - \widetilde{y}\,z - x = 0, \tag{6.33a}$$
$$C_2 := \widehat{x} - \widehat{y}\,z - x = 0, \tag{6.33b}$$
$$C_3 := \widehat{\widetilde{z}}\,y - d_2\,x - z\frac{P\,\widetilde{y}\,\widehat{z} - Q\,\widehat{y}\,\widetilde{z}}{\widetilde{z} - \widehat{z}} = 0. \tag{6.33c}$$

It has 0SS

$$x_{a,b} = \frac{1}{b-a}\left(\frac{p-b}{p-a}\right)^n\left(\frac{p-b}{q-a}\right)^m, \tag{6.34a}$$
$$z_b = (p-b)^n(q-b)^m, \tag{6.34b}$$
$$y_a = -(p-a)^{-n}(q-a)^{-m}, \tag{6.34c}$$

where we have used parametrisation (2.46). By the transformation (c.f., (6.5))

$$x = x_{a,b}\frac{\dot{f}}{\underset{\circ}{f}}, \quad z = z_b\frac{\dot{f}}{f}, \quad y = y_a\frac{f}{\underset{\circ}{f}}, \tag{6.35}$$

from (6.33) we have its bilinear form

$$\mathcal{C}_1 := (p-b)\underset{\circ}{\dot{\widetilde{f}}}f - (p-a)\underset{\circ}{\dot{f}}\widetilde{f} + (b-a)\dot{f}\underset{\circ}{\widetilde{f}} = 0, \tag{6.36a}$$
$$\mathcal{C}_2 := (q-b)\underset{\circ}{\dot{\widehat{f}}}f - (q-a)\underset{\circ}{\dot{f}}\widehat{f} + (b-a)\dot{f}\underset{\circ}{\widehat{f}} = 0, \tag{6.36b}$$
$$\mathcal{C}_3 := (p-b)\widehat{f}\dot{\widetilde{f}} - (q-b)\widetilde{f}\dot{\widehat{f}} + (p-q)\widehat{\widetilde{f}}\dot{f} = 0, \tag{6.36c}$$
$$\mathcal{C}_4 := (p-q)\dot{\widehat{\widetilde{f}}}\underset{\circ}{f} + \frac{(p-q)a_b}{(p-b)(q-b)}\widehat{\widetilde{f}}\underset{\circ}{\dot{f}} - \frac{p_a}{p-b}\dot{\widetilde{f}}\underset{\circ}{\widetilde{f}} + \frac{q_a}{q-b}\dot{\widetilde{f}}\underset{\circ}{\widehat{f}} = 0, \tag{6.36d}$$

where p_a, q_a are defined as (2.34) and here we also use $a_b = b_a = p_a|_{p=b}$, and the connection with (6.33) is

$$C_1 = \frac{-x_{a,b}\mathcal{C}_1}{(p-a)f\widetilde{f}}, \quad C_2 = \frac{-x_{a,b}\mathcal{C}_2}{(q-a)f\widehat{f}},$$
$$C_3 = \frac{(p-b)(q-b)(b-a)x_{a,b}z_b}{(p-q)f\widetilde{f}\widehat{f}\widehat{\widetilde{f}}(\widetilde{z}-\widehat{z})}\left[\mathcal{C}_3\mathcal{C}_4 + \left(\frac{p_a}{p-b}\dot{\widetilde{f}}\underset{\circ}{\widetilde{f}} - \frac{q_a}{q-b}\dot{\widetilde{f}}\underset{\circ}{\widehat{f}}\right)\mathcal{C}_3 - (p-q)\widehat{\widetilde{f}}\dot{f}\mathcal{C}_4\right]. \tag{6.37}$$

Casoratian solution of (6.36) is given by $f = |\widehat{N-1}|$ with ψ composed by (6.6). Note that C3-δ is the same as (6.33).

The bilinearization of $\mathrm{C3}_{b_1}$ is still open.

Equation name	A2	B2	$C3_{b_0}$	$C3_{b_1}$
3 component versions	$\boxed{(2.4)}$, $\boxed{(2.24)}$	$\boxed{(2.3)}$	$\boxed{(2.12)}$	$\leftarrow$
2 component versions	(3.2), $\boxed{(3.3)}$	(3.8) $\boxed{(3.9)}$	(3.12), $\boxed{(3.18)}$	$\boxed{(3.14)}$
1 component versions	(3.5), (3.6)	(3.10)	(3.13), (3.5)	$\leftarrow$
Straight limit	(4.19)	(4.11)	(4.24)	$\leftarrow$
Skew limit	(4.21)	(4.12)	(4.25)	$\leftarrow$
Double limit	(4.22)	(4.16)	(4.26)	$\leftarrow$
Lax Pairs	(5.15)+others	(5.8), (5.11)	(5.26)	(5.28)
Bilinear form	(6.28)	(6.19)	(6.36)	?

Table 1. A summary of the relevant equations. Those defining equations that live on the elementary quadrilateral are boxed. The "$\leftarrow$" in the last column means the result is included in the $C3_{b_0}$ case.

7 Conclusions

In this review we have discussed the fully discrete versions Boussinesq equations given as three-component equations on the basic quadrilateral. Their derivation using CAC and DLA was compared. From the three-component equation we derived two- and one-component versions on a larger stencil. Then we derived two semi-continuous limits and the fully continuous limits for the one-component versions. We discussed also several versions of Lax pairs. Finally we gave their Hirota bilinear forms, which are important for constructing solutions. The basic results are summarized in the adjoining Table 1 which points to the relevant equations.

We also note that recently an elliptic scheme of DLA has been developed, in which the spacing parameters P and Q can be parametrised by Weierstrass elliptic functions [29].

In addition to the question of bilinearizing equation $C3_{b_1}$, there are some interesting open questions: For example, the classification of higher order CAC lattice equations (cf. the ABS list containing 9 equations), higher genus solutions of the DBSQ-type equations, reductions of the hierarchy of lattice KP equations, etc.

Acknowledgments

We would like to thank J. Schiff for bringing Equation (3.9) to our attention. This work was supported by the NSF of China (grant numbers 11875040 and 11631007). All computations were done with REDUCE [13].

Appendix A Triply shifted variables from CAC

Since the equations are integrable, their triply shifted forms are the same independent of the sides used in the computation, and are therefore tilde-hat-bar symmetric. They can be computed if the equations are defined on a quadrilateral. This holds for all three-component forms and for some two-component forms, but the one-component forms are all defined on a bigger stencil.

A.1 Three-component forms

Three-component forms all live in the elementary quadrilateral and were in fact derived using CAC. The triply shifted quantities are as follows [14]:

For A2

$$\overline{\widehat{\widetilde{x}}} = x\,\frac{\widetilde{x}(\widehat{z}-\overline{z})+\widehat{x}(\overline{z}-\widetilde{z})+\overline{x}(\widetilde{z}-\widehat{z})}{p\widetilde{x}(\widehat{z}-\overline{z})+q\widehat{x}(\overline{z}-\widetilde{z})+r\overline{x}(\widetilde{z}-\widehat{z})}, \tag{A.1a}$$

$$\overline{\widehat{\widetilde{y}}} = \frac{y}{x}\overline{\widehat{\widetilde{x}}}+\frac{p\widetilde{x}(\widehat{x}-\overline{x})+q\widehat{x}(\overline{x}-\widetilde{x})+r\overline{x}(\widetilde{x}-\widehat{x})}{p\widetilde{x}(\widehat{z}-\overline{z})+q\widehat{x}(\overline{z}-\widetilde{z})+r\overline{x}(\widetilde{z}-\widehat{z})}, \tag{A.1b}$$

$$\overline{\widehat{\widetilde{z}}} = z-x\,\frac{p(\widehat{z}-\overline{z})+q(\overline{z}-\widetilde{z})+r(\widetilde{z}-\widehat{z})}{p\widetilde{x}(\widehat{z}-\overline{z})+q\widehat{x}(\overline{z}-\widetilde{z})+r\overline{x}(\widetilde{z}-\widehat{z})}. \tag{A.1c}$$

For B2

$$\overline{\widehat{\widetilde{x}}} = b_0+x+\frac{(q-r)\widetilde{x}+(r-p)\widehat{x}+(p-q)\overline{x}}{\widetilde{x}(\widehat{z}-\overline{z})+\widehat{x}(\overline{z}-\widetilde{z})+\overline{x}(\widetilde{z}-\widehat{z})}, \tag{A.2a}$$

$$\overline{\widehat{\widetilde{y}}} = b_0x+y+\frac{(q-r)\widetilde{z}+(r-p)\widehat{z}+(p-q)\overline{z}}{\widetilde{x}(\widehat{z}-\overline{z})+\widehat{x}(\overline{z}-\widetilde{z})+\overline{x}(\widetilde{z}-\widehat{z})}, \tag{A.2b}$$

$$\overline{\widehat{\widetilde{z}}} = z+b_0\overline{\widehat{\widetilde{x}}}-\frac{(q-r)\widehat{x}\overline{x}+(r-p)\overline{x}\widetilde{x}+(p-q)\widetilde{x}\widehat{x}}{\widetilde{x}(\widehat{z}-\overline{z})+\widehat{x}(\overline{z}-\widetilde{z})+\overline{x}(\widetilde{z}-\widehat{z})}. \tag{A.2c}$$

For C3

$$\overline{\widehat{\widetilde{x}}} = \frac{b_0x+b_1\overline{\widehat{\widetilde{y}}}}{y}+x+z\,\frac{\widetilde{z}\widehat{y}\overline{y}(q-r)+\widehat{z}\overline{y}\widetilde{y}(r-p)+\overline{z}\widetilde{y}\widehat{y}(p-q)}{\widetilde{z}(q\widehat{y}-r\overline{y})+\widehat{z}(r\overline{y}-p\widetilde{y})+\overline{z}(p\widetilde{y}-q\widehat{y})}, \tag{A.3a}$$

$$\overline{\widehat{\widetilde{y}}} = y\,\frac{\widetilde{z}(\widehat{y}-\overline{y})+\widehat{z}(\overline{y}-\widetilde{y})+\overline{z}(\widetilde{y}-\overline{y})}{\widetilde{z}(q\widehat{y}-r\overline{y})+\widehat{z}(r\overline{y}-p\widetilde{y})+\overline{z}(p\widetilde{y}-q\widehat{y})}, \tag{A.3b}$$

$$\begin{aligned}\overline{\widehat{\widetilde{z}}} = {} & b_0z+(b_1+b_0x)\frac{(q-r)\widetilde{z}+(r-p)\widehat{z}+(p-q)\overline{z}}{\widetilde{z}(q\widehat{y}-r\overline{y})+\widehat{z}(r\overline{y}-p\widetilde{y})+\overline{z}(p\widetilde{y}-q\widehat{y})} \\ & +zy\frac{qr\widetilde{z}(\widehat{y}-\overline{y})+rp\widehat{z}(\overline{y}-\widetilde{y})+pq\overline{z}(\widetilde{y}-\overline{y})}{\widetilde{z}(q\widehat{y}-r\overline{y})+\widehat{z}(r\overline{y}-p\widetilde{y})+\overline{z}(p\widetilde{y}-q\widehat{y})}.\end{aligned} \tag{A.3c}$$

A.2 Two-component forms

When the two-component form is defined on a quadrilateral, we can compute the triply shifted quantities. This is true for (3.3),(3.14),(3.18).

For A2 (3.3) where $w := y/x$

$$\widehat{\widetilde{\overline{w}}} = w + \frac{P(\overline{w}-\widehat{w}) + Q(\widetilde{w}-\overline{w}) + R(\widehat{w}-\widetilde{w})}{\overline{z}(\widetilde{w}-z)(\widehat{w}-\overline{w}) + \widehat{z}(\widehat{w}-z)(\overline{w}-\widetilde{w}) + \overline{z}(\overline{w}-z)(\widetilde{w}-\widehat{w})}, \tag{A.4a}$$

$$\widehat{\widetilde{\overline{z}}} = \tag{A.4b}$$
$$\frac{P\widetilde{w}(\widehat{w}-z)(\overline{w}-z)(\widehat{z}-\overline{z}) + Q\widehat{w}(\overline{w}-z)(\widetilde{w}-z)(\overline{z}-\widetilde{z}) + R\overline{w}(\widetilde{w}-z)(\widehat{w}-z)(\widetilde{z}-\widehat{z})}{P(\widehat{w}-z)(\overline{w}-z)(\widehat{z}-\overline{z}) + Q(\overline{w}-z)(\widetilde{w}-z)(\overline{z}-\widetilde{z}) + R(\widetilde{w}-z)(\widehat{w}-z)(\widetilde{z}-\widehat{z})}.$$

For B2 (3.9) where $w := y + z$

$$\widehat{\widetilde{\overline{x}}} = b_0 + \frac{P(\overline{x}-\widehat{x}) + Q(\widetilde{x}-\overline{x}) + R(\widehat{x}-\widetilde{x})}{\widetilde{w}(\overline{x}-\widehat{x}) + \widehat{w}(\widetilde{x}-\overline{x}) + \overline{w}(\widehat{x}-\widetilde{x})}, \tag{A.5a}$$

$$\widehat{\widetilde{\overline{w}}} = \widehat{\widetilde{\overline{x}}}(b_0 - x) + w + x^2 +$$
$$\frac{P(\overline{w}-\widehat{w}+(\overline{x}-\widehat{x})\widetilde{x}) + Q(\widetilde{w}-\overline{w}+(\widetilde{x}-\overline{x})\widehat{x}) + R(\widehat{w}-\widetilde{w}+(\widehat{x}-\widetilde{x})\overline{x})}{\widetilde{w}(\overline{x}-\widehat{x}) + \widehat{w}(\widetilde{x}-\overline{x}) + \overline{w}(\widehat{x}-\widetilde{x})}. \tag{A.5b}$$

For $C3_{b_1}$ (3.14):

$$\widehat{\widetilde{\overline{y}}} = -\,y\frac{\widetilde{y}(\widehat{z}-\overline{z}) + \widehat{y}(\overline{z}-\widetilde{z}) + \overline{y}(\widetilde{z}-\widehat{z})}{P\widetilde{y}(\widehat{z}-\overline{z}) + Q\widehat{y}(\overline{z}-\widetilde{z}) + R\overline{y}(\widetilde{z}-\widehat{z})}, \tag{A.6a}$$

$$\widehat{\widetilde{\overline{z}}} = -\,z\frac{PQ\overline{z}(\widetilde{y}-\widehat{y}) + RP\widehat{z}(\overline{y}-\widetilde{y}) + QR\widetilde{z}(\widehat{y}-\overline{y})}{P\widetilde{y}(\widehat{z}-\overline{z}) + Q\widehat{y}(\overline{z}-\widetilde{z}) + R\overline{y}(\widetilde{z}-\widehat{z})}$$
$$+\,b_1\frac{P(\widehat{z}-\overline{z}) + Q(\overline{z}-\widetilde{z}) + R(\widetilde{z}-\widehat{z})}{P\widetilde{y}(\widehat{z}-\overline{z}) + Q\widehat{y}(\overline{z}-\widetilde{z}) + R\overline{y}(\widetilde{z}-\widehat{z})}. \tag{A.6b}$$

For $C3_{b_0}$ (3.18):

$$\widehat{\widetilde{\overline{w}}} = -w\frac{(\widetilde{w}\widehat{w}z+\overline{w})(\widetilde{z}-\widehat{z}) + (\widehat{w}\overline{w}z+\widetilde{w})(\widehat{z}-\overline{z}) + (\overline{w}\widetilde{w}z+\widehat{w})(\overline{z}-\widetilde{z})}{\mathcal{D}_w}, \tag{A.7a}$$

$$\mathcal{D}_w = \widetilde{w}(P+b_0)(z(\widehat{w}\widehat{z}-\overline{w}\overline{z}) - \widehat{z}+\overline{z}) + \widehat{w}(Q+b_0)(z(\overline{w}\overline{z}-\widetilde{w}\widetilde{z}) - \overline{z}+\widetilde{z})$$
$$+\,\overline{w}(R+b_0)(z(\widetilde{w}\widetilde{z}-\widehat{w}\widehat{z}) - \widetilde{z}+\widehat{z}), \tag{A.7b}$$

$$\widehat{\widetilde{\overline{z}}} = z\frac{\mathcal{N}_z}{\mathcal{D}_z}, \tag{A.7c}$$

$$\mathcal{N}_z = PQ(\widetilde{w}-\widehat{w})(\overline{w}z-1)\overline{z} + QR(\widehat{w}-\overline{w})(\widetilde{w}z-1)\widetilde{z} + RP(\overline{w}-\widetilde{w})(\widehat{w}z-1)\widehat{z}$$
$$+\,b_0[P(\widehat{w}z-1)(\overline{w}z-1)(\widehat{z}-\overline{z}) + Q(\overline{w}z-1)(\widetilde{w}z-1)(\overline{z}-\widetilde{z})$$
$$+\,R(\widetilde{w}z-1)(\widehat{w}z-1)(\widetilde{z}-\widehat{z})], \tag{A.7d}$$

$$\mathcal{D}_z = P(\widehat{w}z-1)(\overline{w}z-1)(\widehat{z}-\overline{z})\widetilde{w} + Q(\overline{w}z-1)(\widetilde{w}z-1)(\overline{z}-\widetilde{z})\widehat{w}$$
$$+\,R(\widetilde{w}z-1)(\widehat{w}z-1)(\widetilde{z}-\widehat{z})\overline{w}. \tag{A.7e}$$

Note that none of these results have the tetrahedron property.

References

[1] Adler V E, Bobenko A I and Suris Yu B, Classification of integrable equations on quad-graphs. The consistency approach, *Commun. Math. Phys.* **233**, 513-543, 2003.

[2] Bobenko A I and Suris Yu B, Integrable systems on quad-graphs, *Int. Math. Res. Not.* **11**, 573-611, 2002.

[3] Bridgman T J, LaxPairPartialDifferenceEquations.m: A Mathematica package for the symbolic computation of Lax pairs of nonlinear partial difference equations defined on quadrilaterals. `http://inside.mines.edu/~whereman/software/LaxPairPartialDifferenceEquations/V2`, 2017.

[4] Bridgman T, Hereman W, Quispel G R W and van der Kamp P H, Symbolic computation of Lax pairs of partial difference equations using consistency around the cube, *Found. Comput. Math.* **13**, 517-544, 2013.

[5] Bridgman T and Hereman W, Lax pairs for edge-constrained Boussinesq systems of partial difference equations, in *Nonlinear Systems and Their Remarkable Mathematical Structures,* Vol.2, Eds. N. Euler, M.C. Nucci, Taylor & Francis, Boca Raton, 59-88, 2020.

[6] Date E, Jimbo M and Miwa T, Method for generating discrete soliton equations. III *J. Phys. Soc. Jpn.* **52** 388-393, 1983.

[7] Feng W, Zhao S L and Zhang D J, Exact solutions to lattice Boussinesq-type equations, *J. Nonlin. Math. Phys.* **19**, No.1250032 (15pp), 2012.

[8] Fokas A S and Ablowitz M J, Linearization of the Korteweg-de Vries and Painlevé II equations. *Phys. Rev. Lett.* **47**, 1096-1110, 1981.

[9] Fordy A and Xenitidis P, $\mathbb{Z}_N$ graded discrete Lax pairs and integrable difference equations, *J. Phys. A: Math. Theor.* **50**, No.165205 (30pp), 2017.

[10] Freeman N C and Nimmo J J C, Soliton solutions of the KdV and KP equations: the Wronskian technique, *Phys. Lett. A* **95**, 1-3, 1983.

[11] Fu W and Nijhoff F W, Direct linearizing transform for three-dimensional discrete integrable systems: the lattice AKP, BKP and CKP equations, *Proc. R. Soc. A* **473**, No.20160915 (22pp), 2017.

[12] Gragg W B, The Padé table and its relation to certain algorithms of numerical analysis, *SIAM Review* **14**, 1-62, 1972.

[13] Hearn A C and Schöpf R, REDUCE User Manual `https://reduce-algebra.sourceforge.io/`, 2019.

[14] Hietarinta J, Boussinesq-like multi-component lattice equations and multi-dimensional consistency, *J. Phys. A: Math. Theor.* **44**, No.165204 (22pp), 2011.

[15] Hietarinta J, Search for CAC-integrable homogeneous quadratic triplets of quad equations and their classification by BT and Lax, *J. Nonlin. Math. Phys.* **26**, 358-389, 2019.

[16] Hietarinta J, Joshi N and Nijhoff F W, *Discrete Systems and Integrability*, Camb. Univ. Press, Cambridge, 2016.

[17] Hietarinta J and Zhang D J, Soliton solutions for ABS lattice equations: II: Casoratians and bilinearization, *J. Phys. A: Math. Theor.* **42**, No.404006 (30pp), 2009.

[18] Hietarinta J and Zhang D J, Multisoliton solutions to the lattice Boussinesq equation, *J. Math. Phys.* **51**, No.033505 (12pp), 2010.

[19] Hietarinta J and Zhang D J, Soliton taxonomy for a modification of the lattice Boussinesq equation, *SIGMA* **7**, No.061 (14pp), 2011.

[20] Hietarinta J and Zhang D J, Hirota's method and the search for integrable partial difference equations. 1. Equations on a 3×3 stencil, *J. Difference Equa. Appl.* **19**, 1292-1316, 2013.

[21] Hirota R, and Satsuma J, Nonlinear evolution equations generated from the Bäcklund transformation for the Boussinesq equation, *Prog. Theore. Phys.* **57**, 797-807, 1977.

[22] Maruno K, and Kajiwara K, The discrete potential Boussinesq equation and its multisoliton solutions, *Applicable Analysis* **89**, 593-609, 2010.

[23] Nijhoff F W, On some "Schwarzian Equations" and their discrete analogues, in: Eds. A.S. Fokas and I.M. Gel'fand, *Algebraic Aspects of Integrable Systems: In memory of Irene Dorfman*, Birkhäuser Boston, 237-260, 1997.

[24] Nijhoff F W, Discrete Painlevé equations and symmetry reduction on the lattice, in: Eds. A.I. Bobenko and R. Seiler, *Discrete Integrable Geometry and Physics*, Oxford Univ. Press, 209-234, 1999.

[25] Nijhoff F W, Lax pair for the Adler (lattice Krichever-Novikov) system, *Phys. Lett. A* **297**, 49-58, 2002.

[26] Nijhoff F W and Capel H W, The discrete Korteweg-de Vries equation, *Acta Appl. Math.* **39**, 133-158, 1995.

[27] Nijhoff F W, Papageorgiou V G, Capel H W and Quispel G R W, The lattice Gel'fand-Dikii hierarchy, *Inverse Problems* **8**, 597-621, 1992.

[28] Nijhoff F W, Quispel G R W and Capel H W, Direct linearization of nonlinear difference-difference equations, *Phys. Lett. A* **97**, 125-128, 1983.

[29] Nijhoff F W, Sun Y Y and Zhang D J, Elliptic solutions of Boussinesq type lattice equations and the elliptic Nth root of unity, arXiv: 1909.02948.

[30] Nijhoff F W and Walker A J, The discrete and continuous Painlevé VI hierarchy and the Garnier systems, *Glasgow Math. J.* **43A**, 109-123, 2001.

[31] Nong L J, *Solutions to Discrete Boussinesq-type Systems*, PhD Thesis, Shanghai Univ., 2014.

[32] Nong L J, Zhang D J, Shi Y and Zhang W Y, Parameter extension and the quasi-rational solution of a lattice Boussinesq equation, *Chin. Phys. Lett.* **30**, No.040201 (4pp), 2013.

[33] Quispel G R W, Nijhoff F W, Capel H W and van der Linden J, Linear integral equations and nonlinear difference-difference equations, *Physica* **125A**, 344-380, 1984.

[34] Rasin A G and Schiff J, Bäcklund transformations for the Boussinesq equation and merging solitons, *J. Phys. A: Math. Theor.* **50**, No.325202 (21pp), 2017.

[35] Seiler W M, *Involution*, Springer-Verlag Berlin Heidelberg, 2010.

[36] Tongas A S and Nijhoff F W, The Boussinesq integrable system. Compatible lattice and continuum structures, *Glasgow Math. J.* **47A**, 205-219, 2005.

[37] Walker A J, *Similarity Reductions and Integrable Lattice Equations*, PhD Thesis, Univ. Leeds, 2001.

[38] Weiss J, The Painlevé property for partial differential equations. II: Bäcklund transformation, Lax pairs, and the Schwarzian derivative, *J. Math. Phys.* **24**, 1405-1413, 1983.

[39] Zhang D J, Notes on solutions in Wronskian form to soliton equations: KdV-type, arXiv:0603008v3 [nlin.SI], preprint, 2006.

[40] Zhang D J, Wronskian solutions of integrable systems, in *Nonlinear Systems and Their Remarkable Mathematical Structures,* Vol.2, Eds. N. Euler, M.C. Nucci, Taylor & Francis, Boca Raton, 415-444, 2020.

[41] Zhang D J, Zhao S L and Nijhoff F W, Direct linearization of an extended lattice BSQ system, *Stud. Appl. Math.* **129**, 220-248, 2012.

[42] Zhang D J, Zhao S L, Sun Y Y and Zhou J, Solutions to the modified Korteweg-de Vries equation, *Rev. Math. Phys.* **26**, No.14300064 (42pp), 2014.

[43] Zhou J, Zhang D J and Zhao S L, Breathers and limit solutions of the nonlinear lumped self-dual network equation, *Phys. Lett. A* **373**, 3248-3258, 2009.

$A4$. Liouville correspondences for integrable hierarchies

Jing Kang[a], *Xiaochuan Liu*[b], *Peter J. Olver*[c] *and Changzheng Qu*[d]

[a] *Center for Nonlinear Studies and School of Mathematics, Northwest University, Xi'an 710069, P.R. China*
jingkang@nwu.edu.cn

[b] *School of Mathematics and Statistics, Xi'an Jiaotong University, Xi'an 710049, P.R. China*
liuxiaochuan@mail.xjtu.edu.cn

[c] *School of Mathematics, University of Minnesota, Minneapolis, MN 55455, USA*
olver@umn.edu

[d] *School of Mathematics and Statistics, Ningbo University, Ningbo 315211, P.R. China*
quchangzheng@nbu.edu.cn

Abstract

This chapter contains a survey on Liouville correspondences between integrable hierarchies. A Liouville transformation between the corresponding isospectral problems induces a Liouville correspondence between their flows and Hamiltonian functionals. As prototypical examples, we construct Liouville correspondences for the modified Camassa-Holm, the Novikov, the Degasperis-Procesi, the two-component Camassa-Holm and the two-component Novikov (Geng-Xue) hierarchies. In addition, a new Liouville correspondence for a certain dual Schrödinger integrable hierarchy is presented.

1 Introduction

Due to the pioneering works [1] and [43], it is known that the spatial isospectral problem in the Lax-pair formulation of an integrable system plays a fundamental role when constructing the soliton solutions using the inverse scattering transform, as well as analyzing the long-time behavior of solutions based on the Riemann-Hilbert approach. The transition from one isospectral problem to another via a change of variables can usually be identified as a form of *Liouville transformation*; see [61] and [64] for this terminology. It is then expected that such a correspondence based on a Liouville transformation, which we call a *Liouville correspondence* for brevity, can be used to establish an inherent correspondence between associated integrability properties including symmetries, conserved quantities, soliton solutions, Hamiltonian structures, etc. Indeed, this is a basic idea for investigating the integrability of a new system by establishing its relation to a known integrable system through some kind of transformation, which, besides Liouville transformations, can include Bäcklund transformations, Miura transformations, gauge transformations, Darboux transformations, hodograph transformations, etc. [12, 15, 41, 59, 62, 66, 70, 74]. Applying an appropriate transformation enables one to adapt known solutions and integrable structures in order to derive explicit solutions and investigate the integrability properties for the transformed system.

In recent years, a great deal of attention has been devoted to integrable systems of the Camassa-Holm type, following the discovery of their novel properties, including the structure of nonlinear dispersion, which (as a rule) supports non-smooth soliton solutions, such as peakons, cuspons, compactons, etc. [10, 52], and the ability of such systems to model wave-breaking phenomena. Previous investigations demonstrate that many integrable hierarchies of Camassa-Holm type admit a Liouville correspondence with certain classical integrable hierarchies. In the following, recent advances in the study of Liouville correspondences for the integrable hierarchies of Camassa-Holm type and non-Camassa-Holm type are summarized.

Among integrable systems of Camassa-Holm type, the best-studied example is the Camassa-Holm (CH) equation

$$n_t + 2v_x n + v n_x = 0, \qquad n = v - v_{xx}, \tag{1.1}$$

that has a quadratic nonlinearity [8, 9, 17, 27, 35], while the modified Camassa-Holm (mCH) equation [67]

$$m_t + \left(\left(u^2 - u_x^2\right) m\right)_x = 0, \qquad m = u - u_{xx}, \tag{1.2}$$

is a prototypical integrable model with cubic nonlinearity, which presents several novel properties, as described, for instance, in [7, 30, 53, 54, 58, 69]. The CH equation (1.1) appears as an integrable generalization of the Korteweg–de Vries (KdV) equation possessing infinitely many symmetries [27], and is shown to correspond to the first negative flow of the KdV integrable hierarchy using the Liouville transformation [26]; see also [5, 6]. A novel link between the mCH equation (1.2) and the modified KdV (mKdV) equation was found and used to obtain the multisoliton solutions of (1.2) from the known multisoliton solutions of the mKdV equation in [58]. In addition, all the equations in the CH, mCH, KdV, and mKdV hierarchies have a classical bi-Hamiltonian form (see [67] for example). The interesting feature here is that the two Hamiltonians for the CH and mCH integrable hierarchies can be constructed from those of the KdV and mKdV hierarchies, respectively, using the approach of *tri-Hamiltonian duality* [24, 26, 67]. This approach is based on the observation that, by applying an appropriate scaling argument, many standard integrable equations that possess a bi-Hamiltonian structure in fact admit a compatible triple of Hamiltonian operators. Different combinations of the members of the compatible Hamiltonian triple can generate different types of integrable bi-Hamiltonian systems, which admit a *dual* relation. In [72], another kind of duality by exploiting the zero curvature formulations of the CH and mCH hierarchies and the KdV and mKdV hierarchies was also discussed.

As a consequence of these connections between the CH equation (1.1) and the KdV equation, and between the mCH equation (1.2) and the mKdV equation, it is anticipated that the KdV and mKdV hierarchies should be related to their respective dual counterparts, the CH and mCH hierarchies, in a certain manner. A relationship between the KdV and CH hierarchies provided by the approach of loop groups is explored in [73]. Note that the Lax-pair formulations of equations in the CH, mCH, KdV, and mKdV hierarchies are all based on a second order isospectral problem. Using the Liouville transformation between the spatial isospectral problems of the CH hierarchy and the KdV hierarchy, the Liouville correspondence between these two integrable hierarchies is established in [44] and [60], and in addition gives rise to a correspondence between the Hamiltonian functionals of the two hierarchies. The analysis for the KdV-CH setting depends strongly on the subtle relation between one of the original KdV Hamiltonian operators and one of the dual CH Hamiltonian operators. Furthermore, the relation between the smooth traveling-wave solutions of the CH equation and the KdV equation under the Liouville transformation was investigated in [45]. In [36], we established the Liouville correspondence between the integrable mCH hierarchy induced by (1.2) and the mKdV hierarchy, including the explicit relations between their equations and Hamiltonian functionals. In contrast to the CH-KdV setting, the analysis in [36] was based on the interrelation between the respective recursion operators and the conservative structure of all the equations in the mCH hierarchy. It is worth pointing out that the tri-Hamiltonian duality relationships helps establish corresponding Liouville correspondences in both the CH-KdV and mCH-mKdV cases as shown in [44] and [36]. As a by-product, we constructed in [36] a novel transformation mapping the mCH equation (1.2) to the CH equation (1.1) in terms of the respective Liouville correspondences between the CH-KdV hierarchies and between the mCH-mKdV hierarchies.

In [67], the nonlinear Schrödinger equation

$$u_t = \mathrm{i}\left(u_{xx} + u|u|^2\right),$$

was investigated using the method of tri-Hamiltonian duality, and the dual integrable version

$$u_t + \mathrm{i}\, u_{xt} = |u|^2\,(\mathrm{i}\, u - u_x), \tag{1.3}$$

was derived (see also [21]).The dual Schrödinger equation (1.3) has attracted much interest in recent years for its soliton solutions, well-posedness, physical relevance, etc. [23, 46, 47, 48, 75]. In this article, we present a new result of Liouville correspondence for the integrable hierarchy induced by equation (1.3). We prove that this dual integrable hierarchy can be related to the mKdV integrable hierarchy by a different Liouville transformation compared to the transformation used to establish the Liouville correspondence between the mCH hierarchy and the mKdV hierarchy.

The tri-Hamiltonian duality theory is a fruitful approach for deriving new dual integrable hierarchies from known integrable soliton hierarchies, provided that the bi-Hamiltonian formulation of the latter hierarchy admits a compatible triple of Hamiltonian operators through appropriate rescalings. For integrable soliton hierarchies with generalized bi-Hamiltonian structures, i.e., compatible pairs of Dirac structures [20], the corresponding Hamiltonian operators usually do not support a decomposition as linear combination of different parts using the scaling argument, and so constructing the associated dual integrable hierarchy in this case is unclear. On the other hand, other interesting CH-type integrable hierarchies without tri-Hamiltonian duality structure can be found using some particular classification procedure. Two representative examples are the Degasperis-Procesi (DP) and Novikov integrable hierarchies.

The DP integrable equation with quadratic nonlinearity

$$n_t = 3v_x n + v n_x, \qquad n = v - v_{xx}, \tag{1.4}$$

was derived in [19] as a result of the asymptotic integrability method for classifying (a class of) third-order nonlinear dispersive evolution equations. In such a classification framework, the CH equation and the KdV equation are two only other integrable candidates. The DP equation (1.4) is integrable with a Lax pair involving a 3×3 isospectral problem as well as a bi-Hamiltonian structure [18]. In [18, 31], it was pointed out that using the Liouville transformation, equation (1.4) is related to the first negative flow of the Kaup-Kupershmidt (KK) hierarchy, which is initiated from the classical KK integrable equation [40, 42]

$$P_\tau + P_{yyyyy} + 10PP_{yyy} + 25P_yP_{yy} + 80P^2P_y = 0. \tag{1.5}$$

The Novikov integrable equation with cubic nonlinearity

$$m_t = 3uu_x m + u^2 m_x, \qquad m = u - u_{xx}, \tag{1.6}$$

was discovered as a consequence of the symmetry classification of nonlocal partial differential equations involving both cubic and quadratic nonlinearities in [63]. The Lax pair formulation with 3×3 isospectral problem and bi-Hamiltonian structure were established in [32]. It was shown in [32] that the Novikov equation (1.6) is related by the Liouville transformation to the first negative flow of the Sawada-Kotera (SK) hierarchy, which is initiated from the classical SK integrable equation [11, 71]

$$Q_\tau + Q_{yyyyy} + 5QQ_{yyy} + 5Q_yQ_{yy} + 5Q^2Q_y = 0. \tag{1.7}$$

Although the DP and Novikov hierarchies are bi-Hamiltonian, the corresponding Hamiltonian operators are not amenable to the tri-Hamiltonian duality construction, in particular one that is related to the Hamiltonian operators of the KK and SK equations. In addition, the KK equation (1.5) and the SK equation (1.7) both have generalized bi-Hamiltonian formulations with the corresponding hierarchy generated by the respective recursion operators, which are obtained by composing symplectic and implectic operators that fail to satisfy the conditions of non-degeneracy or invertibility [28]. Therefore, studying a Liouville correspondence for the DP or Novikov integrable hierarchies requires a more delicate analysis. In [37], using the Liouville transformations and conservative structures in both the DP and Novikov settings, we found the corresponding operator identities relating the recursion operator of the DP/Novikov hierarchy and the adjoint operator of the recursion operator of the KK/SK hierarchy, and then were able to establish the Liouville correspondences between the DP and KK integrable hierarchies, as well as the Novikov and SK integrable hierarchies. In particular, a nontrivial operator factorization for the recursion operator of the SK equation discovered in [14] plays a key role in constructing the Liouville correspondence for the Novikov hierarchy. Note that the SK equation (1.5) and the KK equation (1.7) are related to the so-called Fordy-Gibbons-Jimbo-Miwa equation via certain Miura transformations [25]. Exploiting such a relation, we also obtained in [37] a nontrivial link between the Novikov equation (1.6) and the DP equation (1.4).

One can also investigate Liouville correspondences for multi-component integrable hierarchies. The integrable two-component CH hierarchy and the integrable two-component Novikov (Geng-Xue) hierarchy are two typical examples. The well-studied two-component CH (2CH) system

$$m_t + 2u_x m + u m_x + \rho\rho_x = 0, \qquad \rho_t + (\rho u)_x = 0, \qquad m = u - u_{xx}, \tag{1.8}$$

arises as the dual version for the integrable two-component Ito system

$$u_t = u_{xxx} + 3uu_x + vv_x, \qquad v_t = (uv)_x,$$

introduced in [33] using the method of tri-Hamiltonian duality [67]. Such a duality structure ensures that the system (1.8) is integrable, with bi-Hamiltonian formulation and compatible Hamiltonians, which thereby recursively generate the entire 2CH integrable hierarchy, with (1.8) forming the second flow in the positive direction.

In [16], the 2CH system (1.8) was derived as a model describing shallow water wave propagation. In [13], the Lax pair of system (1.8) was converted into a Lax pair of the integrable system

$$P_\tau = \rho_y, \quad Q_\tau = \frac{1}{2}\rho\, P_y + \rho_y P, \quad \rho_{yyy} + 2\rho_y Q + 2(\rho\, Q)_y = 0, \tag{1.9}$$

by a Liouville-type transformation introduced in [2, 3, 4]. In [13], this system was found to be the first negative flow of the Ablowitz-Kaup-Newell-Segur hierarchy [1] in terms of the spectral structure of its Lax-pair formulation. Although the 2CH and Ito integrable systems are related by tri-Hamiltonian duality, in contrast to the CH-KdV and mCH-mKdV cases, the Liouville correspondence between these two hierarchies is unexpected because the transformation between the corresponding isospectral problems is not obvious.

It is anticipated that one can establish a Liouville correspondence between the 2CH hierarchy and a second integrable hierarchy involving the integrable system (1.9) as one of flows in the negative direction. Nevertheless, the integrable structures including the recursion operator and Hamiltonians for such a hierarchy are not clear. The required integrability information was also not presented in [13]. In [38], we elucidated this entire integrable hierarchy, which we call the hierarchy associated with system (1.9) or the *associated 2CH (a2CH) hierarchy* for brevity. We show that it has a bi-Hamiltonian structure and establish a Liouville correspondence between the 2CH and a2CH hierarchies. As in the scalar case, verifying the Liouville correspondence relies on analyzing the underlying operators, which have a matrix form in the multi-component case, and a more careful calculation of the nonlinear interplay among the various components is hence required. Furthermore, we find in [38] that the second positive flow of the a2CH integrable hierarchy is closely related to shallow water models studied in [34] and [39].

The Novikov equation (1.6) has the following two-component integrable generalization

$$m_t+3vu_xm+uvm_x=0,\ n_t+3uv_xn+uvn_x=0,\ m=u-u_{xx},\ n=v-v_{xx},\quad (1.10)$$

which was introduced by Geng and Xue [29], and is referred to as the GX system; see [55] and references therein. As a prototypical multi-component integrable system with cubic nonlinearity, the GX system (1.10) supports special multi-peakon dynamics and has recently attracted much attention [49, 50, 51, 55, 56]. In [49], it was shown that there exists a certain Liouville transformation converting the Lax pair of GX system (1.10) into the Lax pair of the integrable system

$$\begin{aligned} Q_\tau &= \frac{3}{2}(q_y+p_y)-(q-p)P, \qquad & p_{yy}+2p_yP+pP_y+pP^2-pQ+1=0,\\ P_\tau &= \frac{3}{2}(q-p), & q_{yy}-2q_yP-qP_y+qP^2-qQ+1=0, \end{aligned} \quad (1.11)$$

where $q=v\,m^{2/3}n^{-1/3}$ and $p=u\,m^{-1/3}n^{2/3}$. The system (1.11) is bi-Hamiltonian, whose bi-Hamiltonian structure is derived in [49]. In [38], the entire *associated GX (aGX)* integrable hierarchy is investigated, in which (1.11) is the first negative flow. The Liouville correspondence between the integrable GX hierarchy generated by (1.10) and the aGX hierarchy is also established.

We conclude this section by outlining the rest of the survey. Section 2 is devoted to the Liouville correspondence for the mCH integrable hierarchy. The Liouville correspondences between the Novikov and SK hierarchies, and between the DP and KK hierarchies are discussed in Section 3. In Section 4, multi-component cases are discussed. Finally, we present the new result for the Liouville correspondence of the dual Schrödinger hierarchy generated by (1.3).

2 Liouville correspondences for the CH and mCH integrable hierarchies

In this section, we present our procedure to establish the Liouville correspondence between the mCH hierarchy and the (defocusing) mKdV hierarchy in details. A novel transformation mapping the mCH equation (1.2) to the CH equation (1.1) in terms of the respective Liouville correspondences between the CH-KdV hierarchies and between the mCH-mKdV hierarchies found in [36] is also addressed.

2.1 Liouville correspondence for mCH integrable hierarchy

Let us begin by presenting the basic CH-KdV case. For the CH equation (1.1), the bi-Hamiltonian structure takes the following form

$$n_t=\mathcal{J}\,\frac{\delta E_2}{\delta n}=\mathcal{L}\,\frac{\delta E_1}{\delta n},$$

where $E_1 = E_1(n)$, $E_2 = E_2(n)$ are Hamiltonian functionals, and the compatible Hamiltonian operators are given by

$$\mathcal{J} = -\left(\partial_x - \partial_x^3\right), \qquad \mathcal{L} = -\left(\partial_x n + n\,\partial_x\right).$$

These are related to the Hamiltonian pair

$$\overline{\mathcal{D}} = \partial_y, \qquad \overline{\mathcal{L}} = \tfrac{1}{4}\partial_y^3 - \tfrac{1}{2}\left(P\partial_y + \partial_y P\right) \tag{2.1}$$

of the KdV equation

$$P_\tau + P_{yyy} - 6\,PP_y = 0. \tag{2.2}$$

It was proved in [44] and [60] that the corresponding Liouville transformation relating their isospectral problems transforms between the CH and KdV hierarchies. The following identities

$$\mathcal{L}^{-1} = -\frac{1}{2\sqrt{n}}\,\overline{\mathcal{D}}^{-1}\frac{1}{n}, \qquad \overline{\mathcal{L}} = \frac{1}{4n}\,\mathcal{J}\frac{1}{\sqrt{n}}$$

relating the Hamiltonian operators under the Liouville transformation play an important role in the analysis used in [44] and [60].

The mCH equation (1.2) can be written in bi-Hamiltonian form [67]

$$m_t = \mathcal{J}\,\frac{\delta \mathcal{H}_2}{\delta m} = \mathcal{K}\,\frac{\delta \mathcal{H}_1}{\delta m}, \qquad m = u - u_{xx},$$

where

$$\mathcal{J} = -\left(\partial_x - \partial_x^3\right), \qquad \mathcal{K} = -\partial_x\, m\, \partial_x^{-1}\, m\, \partial_x \tag{2.3}$$

are compatible Hamiltonian operators, while the corresponding Hamiltonian functionals are given by

$$\mathcal{H}_1(m) = \int m\,u\,\mathrm{d}x, \qquad \mathcal{H}_2(m) = \frac{1}{4}\int \left(u^4 + 2u^2u_x^2 - \frac{1}{3}u_x^4\right)\mathrm{d}x.$$

In general, for an integrable bi-Hamiltonian equation with two compatible Hamiltonian operators $\mathcal{K}$ and $\mathcal{J}$, Magri's theorem [57, 66] establishes the formal existence of an infinite hierarchy

$$m_t = K_n(m) = \mathcal{J}\,\frac{\delta \mathcal{H}_n}{\delta m} = \mathcal{K}\,\frac{\delta \mathcal{H}_{n-1}}{\delta m}, \qquad n = 1, 2, \ldots, \tag{2.4}$$

of higher-order commuting bi-Hamiltonian systems, based on the higher-order Hamiltonian functionals $\mathcal{H}_n = \mathcal{H}_n(m)$, $n = 0, 1, 2, \ldots$, common to all members of the hierarchy. The members in the hierarchy (2.4) are obtained by applying successively the recursion operator $\mathcal{R} = \mathcal{K}\,\mathcal{J}^{-1}$ to a seed symmetry [65], which in the mCH setting takes the following form:

$$m_t = K_1(m) = -2m_x, \qquad \text{with} \qquad \mathcal{H}_0(m) = \int m\,\mathrm{d}x.$$

The positive flows in the mCH hierarchy (2.4) are hence

$$m_t + \left(\mathcal{K}\mathcal{J}^{-1}\right)^n (2\,m_x) = 0, \qquad n = 0, 1, \ldots. \tag{2.5}$$

Clearly, the mCH equation (1.2) appears in this hierarchy as

$$m_t = K_2(m) = -\left((u^2 - u_x^2)\,m\right)_x = \mathcal{R}K_1(m).$$

Similarly, one obtains an infinite number of higher-order commutative bi-Hamiltonian systems in the negative direction:

$$m_t = K_{-n}(m) = \mathcal{J}\,\frac{\delta\mathcal{H}_{-n}}{\delta m} = \mathcal{K}\,\frac{\delta\mathcal{H}_{-(n+1)}}{\delta m}, \qquad n = 1, 2, \ldots,$$

starting from the Casimir functional $\mathcal{H}_C(m) = \int 1/m\,dx$ of $\mathcal{K}$. Then, the first equation $m_t = K_{-1}(m)$ in the negative direction of the mCH hierarchy is

$$m_t = \mathcal{J}\,\frac{\delta\mathcal{H}_{-1}}{\delta m} = \mathcal{J}\,\frac{\delta\mathcal{H}_C}{\delta m} = \left(\frac{1}{m^2}\right)_x - \left(\frac{1}{m^2}\right)_{xxx}, \tag{2.6}$$

which is called the Casimir equation in [67]. It was noted in [67] that equation (2.6) has the form of a Lagrange transformation, modulo an appropriate complex transformation, of the mKdV equation. Successively applying $\mathcal{J}\mathcal{K}^{-1}$ produces the hierarchy of negative flows, in which the n-th member is

$$m_t = K_{-n}(m) = -\left(\mathcal{J}\mathcal{K}^{-1}\right)^{n-1}\mathcal{J}\frac{1}{m^2}, \quad n = 1, 2, \ldots. \tag{2.7}$$

As the original soliton equation in the duality relationship with the mCH equation (1.2), the (defocusing) mKdV equation

$$Q_\tau + Q_{yyy} - 6\,Q^2 Q_y = 0 \tag{2.8}$$

also admits a hierarchy consisting of an infinite number of integrable equations in both the positive and negative directions. Each member in the positive direction takes the form

$$Q_\tau = \overline{K}_n(Q) = \overline{\mathcal{J}}\,\frac{\delta\overline{\mathcal{H}}_n}{\delta Q} = \overline{\mathcal{K}}\,\frac{\delta\overline{\mathcal{H}}_{n-1}}{\delta Q}, \qquad n = 1, 2, \ldots, \tag{2.9}$$

where

$$\overline{\mathcal{K}} = -\frac{1}{4}\partial_y^3 + \partial_y Q \partial_y^{-1} Q \partial_y, \qquad \overline{\mathcal{J}} = -\partial_y, \tag{2.10}$$

are the compatible Hamiltonian operators, and $\overline{\mathcal{H}}_n = \overline{\mathcal{H}}_n(Q)$, $n = 0, 1, 2, \ldots$, are the corresponding Hamiltonian functionals. Using the recursion operator $\overline{\mathcal{R}} = \overline{\mathcal{K}}\,\overline{\mathcal{J}}^{-1}$, the positive flows in (2.9) are

$$Q_\tau + \left(\overline{\mathcal{K}}\,\overline{\mathcal{J}}^{-1}\right)^n (4\,Q_y) = 0, \qquad n = 0, 1, \ldots. \tag{2.11}$$

The negative flow in the following form

$$\overline{\mathcal{R}}^n Q_\tau = 0, \qquad n = 1, 2, \ldots,$$

can be rewritten as

$$\partial_y \left(\tfrac{1}{4}\partial_y - Q\partial_y^{-1}Q\right) \left(\overline{\mathcal{K}}\,\overline{\mathcal{J}}^{-1}\right)^{n-1} Q_\tau = 0,$$

due to the forms of the Hamiltonian operators (2.10), and thus, for each $n \geq 1$,

$$\left(\tfrac{1}{4}\partial_y - Q\partial_y^{-1}Q\right) \left(\overline{\mathcal{K}}\,\overline{\mathcal{J}}^{-1}\right)^{n-1} Q_\tau = \overline{C}_{-n}, \tag{2.12}$$

with $\overline{C}_{-n}$ being the corresponding constant of integration.

The zero curvature formulation [68, 72] for the mCH equation (1.2) is

$$\mathbf{\Psi}_x = \begin{pmatrix} -\frac{1}{2} & \frac{1}{2}\lambda m \\ -\frac{1}{2}\lambda m & \frac{1}{2} \end{pmatrix} \mathbf{\Psi}, \qquad \mathbf{\Psi} = \begin{pmatrix} \psi_1 \\ \psi_2 \end{pmatrix} \tag{2.13}$$

and

$$\mathbf{\Psi}_t = \begin{pmatrix} \lambda^{-2} + \frac{1}{2}(u^2 - u_x^2) & -\lambda^{-1}(u - u_x) - \frac{1}{2}\lambda m(u^2 - u_x^2) \\ \lambda^{-1}(u + u_x) + \frac{1}{2}\lambda m(u^2 - u_x^2) & -\lambda^{-2} - \frac{1}{2}(u^2 - u_x^2) \end{pmatrix} \mathbf{\Psi}.$$

On the other hand, the zero curvature formulation for (2.8) comes from the compatibility condition between

$$\mathbf{\Phi}_y = \begin{pmatrix} -\mathrm{i}\,\mu & \mathrm{i}\,Q \\ -\mathrm{i}\,Q & \mathrm{i}\,\mu \end{pmatrix} \mathbf{\Phi}, \qquad \mathbf{\Phi} = \begin{pmatrix} \phi_1 \\ \phi_2 \end{pmatrix} \tag{2.14}$$

and

$$\mathbf{\Phi}_\tau = \begin{pmatrix} -4\,\mathrm{i}\,\mu^3 - 2\,\mathrm{i}\,\mu Q^2 & 4\mu^2\,\mathrm{i}\,Q + 2\,\mathrm{i}\,Q^3 - 2\mu Q_y - iQ_{yy} \\ -4\mu^2\,\mathrm{i}\,Q - 2\,\mathrm{i}\,Q^3 - 2\mu Q_y + iQ_{yy} & 4\,\mathrm{i}\,\mu^3 + 2\,\mathrm{i}\,\mu Q^2 \end{pmatrix} \mathbf{\Phi}.$$

One can verify that the following Liouville transformation

$$\mathbf{\Phi} = \begin{pmatrix} -1 & \mathrm{i} \\ -\mathrm{i} & 1 \end{pmatrix} \mathbf{\Psi}, \qquad y = \int^x m(\xi)\,\mathrm{d}\xi,$$

will convert the isospectral problem (2.13) into the isospectral problem (2.14), with

$$Q = \frac{1}{2m}, \qquad \lambda = -2\mu.$$

We hence introduce the following transformation:

$$y = \int^x m(t, \xi)\,\mathrm{d}\xi, \qquad \tau = t, \qquad Q(\tau, y) = \frac{1}{2\,m(t, x)}, \tag{2.15}$$

and investigate how it affects the underlying correspondence between the flows of the mCH and (defocusing) mKdV hierarchies.

Hereafter, for a non-negative integer n, we denote the n-th equation in the positive and negative directions of the mCH hierarchy by $(\mathrm{mCH})_n$ and $(\mathrm{mCH})_{-n}$, respectively, while the n-th positive and negative flows in the (defocusing) mKdV hierarchy are denoted by $(\mathrm{mKdV})_n$ and $(\mathrm{mKdV})_{-n}$, respectively. Applying the Lemmas 3.2 and 3.3 in [36], we are able to establish a Liouville correspondence between the (defocusing) mKdV and mCH hierarchies.

Theorem 1. *Under the transformation* (2.15), *for each* $n \in \mathbb{Z}$, *the* $(\mathrm{mCH})_{n+1}$ *equation is related to the* $(\mathrm{mKdV})_{-n}$ *equation. More precisely, for each integer* $n \geq 0$,

(i). *m solves equation* (2.5) *if and only if Q satisfies $Q_\tau = 0$ for $n = 0$ or* (2.12) *for $n \geq 1$, with $\overline{C}_{-n} = 1/(-4)^n$;*

(ii). *For $n \geq 1$, the function m is a solution of the following rescaled version of* (2.7)*:*

$$m_t = K_{-n}(m) = \frac{(-1)^{n+1}}{2^{2n-1}} \left(\mathcal{J}\mathcal{K}^{-1}\right)^{n-1} \mathcal{J}\frac{1}{m^2}, \quad n = 1, 2, \ldots,$$

if and only if Q satisfies equation (2.11)*. In addition, for $n = 0$, the corresponding equation $m_t = 0$ is equivalent to $Q_\tau + 4Q_y = 0$.*

Magri's bi-Hamiltonian scheme enables one to recursively construct an infinite hierarchy of Hamiltonian functionals of the mCH equation (1.2). The Hamiltonian functionals $\mathcal{H}_n = \mathcal{H}_n(m)$ satisfy the recursive formula

$$\mathcal{J}\frac{\delta \mathcal{H}_n}{\delta m} = \mathcal{K}\frac{\delta \mathcal{H}_{n-1}}{\delta m}, \qquad n \in \mathbb{Z},$$

where $\mathcal{K}$ and $\mathcal{J}$ are the two compatible Hamiltonian operators (2.3) admitted by the mCH equation. On the other hand, the recursive formula

$$\overline{\mathcal{J}}\frac{\delta \overline{\mathcal{H}}_n}{\delta Q} = \overline{\mathcal{K}}\frac{\delta \overline{\mathcal{H}}_{n-1}}{\delta Q}, \qquad n \in \mathbb{Z}$$

can be used to obtain Hamiltonian functionals of the (defocusing) mKdV equation (2.8). Applying the Lemmas 4.1 and 4.2 in [36], we proved the following relation between the sequences of the Hamiltonian functionals admitted by the mCH and the (defocusing) mKdV equations.

Theorem 2. *For any non-zero integer n, each Hamiltonian functionals $\overline{\mathcal{H}}_n(Q)$ of the (defocusing) mKdV equation yields the Hamiltonian functionals $\mathcal{H}_{-n}(m)$ of the mCH equation, under the Liouville transformation* (2.15)*, according to the following identity*

$$\mathcal{H}_{-n}(m) = (-1)^n 2^{2n-1}\, \overline{\mathcal{H}}_n(Q), \qquad 0 \neq n \in \mathbb{Z}.$$

2.2 Relationship between the mCH and CH equations

It is wellknown that the KdV equation and the mKdV equation are linked by the Miura transformation. This leads to a question whether there exists a transformation relating their respective dual counterparts, in other words, a transformation between the CH equation (1.1) and the mCH equation (1.2). From the viewpoint of tri-Hamiltonian duality, the CH equation (1.1) is regarded as the dual integrable counterpart of the KdV equation (2.2). The KdV equation (2.2) is related to the (defocusing) mKdV equation (2.8) via the Miura transformation

$$\mathcal{B}(P,Q) \equiv P - Q^2 + Q_y = 0. \tag{2.16}$$

Furthermore, Fokas and Fuchssteiner [22] proved that all the positive members of the KdV hierarchy admit the same Miura transformation. In addition, we have the following result.

Proposition 1. *Assume that Q satisfies the equation*

$$\left(\overline{\mathcal{K}}\,\overline{\mathcal{J}}^{-1}\right) Q_\tau = 0, \tag{2.17}$$

where $\overline{\mathcal{K}}$ and $\overline{\mathcal{J}}$ are given in (2.10). *Then $P = Q^2 - Q_y$ satisfies*

$$\left(\overline{\mathcal{L}}\,\overline{\mathcal{D}}^{-1}\right) P_\tau = 0, \tag{2.18}$$

where $\overline{\mathcal{L}}$ and $\overline{\mathcal{D}}$ are defined by (2.1).

Using Proposition 1, we are able to construct a transformation from the mCH equation (1.2) to the CH equation (1.1). First, it was shown in [26, 44, 60] that the following Liouville transformation

$$P(\tau,y) = \frac{1}{n(t,x)}\Big(\frac{1}{4} - \frac{(n(t,x)^{-1/4})_{xx}}{n(t,x)^{-1/4}}\Big), \; y = \int^x \sqrt{n}\mathrm{d}\xi, \; n = v - v_{xx}, \; \tau = t, \tag{2.19}$$

relating the respective isospectral problems for the CH hierarchy and the KdV hierarchy, gives rise to the one-to-one correspondence between the CH equation (1.1) and the first negative flow (2.18). On the other hand, from Theorem 1, $m(t,x)$ satisfies the mCH equation (1.2) if and only if

$$Q(\tau,y) = \frac{1}{2m(t,x)}, \qquad y = \int^x m(t,\xi)\,\mathrm{d}\xi, \qquad \tau = t, \tag{2.20}$$

is the solution of equation (2.17). We deduce that the composite transformation including (2.16), (2.19), and (2.20) defines a map from the mCH equation (1.2) to the CH equation (1.1), albeit not one-to-one.

Proposition 2. *Assume $m(t,x)$ is the solution of the mCH equation* (1.2). *Then, $n(t,x)$ satisfies the CH equation* (1.1), *where $n(t,x)$ is determined by the relation* (2.19) *with $P(\tau,y) = Q^2(\tau,y) - Q_y(\tau,y)$ and $Q(\tau,y)$ defined by* (2.20).

3 Liouville correspondences for the Novikov and DP integrable hierarchies

In this section, we first survey the main results on Liouville correspondences for the Novikov and Degasperis-Procesi (DP) hierarchies, and then show the implicit relationship which associates the Novikov and DP equations.

3.1 Liouville correspondences for the Novikov and DP hierarchies

The Novikov equation (1.6) can be expressed in bi-Hamiltonian form [32]

$$m_t = K_1(m) = \mathcal{J}\,\frac{\delta \mathcal{H}_1}{\delta m} = \mathcal{K}\,\frac{\delta \mathcal{H}_0}{\delta m}, \qquad m = u - u_{xx},$$

where

$$\mathcal{K} = \frac{1}{2} m^{\frac{1}{3}}\,\partial_x\, m^{\frac{2}{3}}\,(4\partial_x - \partial_x^3)^{-1}\, m^{\frac{2}{3}}\,\partial_x\, m^{\frac{1}{3}}, \quad \mathcal{J} = (1-\partial_x^2)\, m^{-1}\,\partial_x\, m^{-1}\,(1-\partial_x^2)$$

are the compatible Hamiltonian operators. The corresponding Hamiltonian functionals are given by

$$\mathcal{H}_0(m) = 9\int (u^2 + u_x^2)\; \mathrm{d}x, \quad \mathcal{H}_1(m) = \frac{1}{6}\int um\partial_x^{-1} m(1-\partial_x^2)^{-1}(u^2 m_x + 3uu_x m)\, \mathrm{d}x.$$

As for the Novikov integrable hierarchy

$$m_t = K_n(m) = \mathcal{J}\,\frac{\delta \mathcal{H}_n}{\delta m} = \mathcal{K}\,\frac{\delta \mathcal{H}_{n-1}}{\delta m}, \qquad n \in \mathbb{Z}, \tag{3.1}$$

the Novikov equation (1.6) serves as the first member in the positive direction of (3.1). While, in the opposite direction, note that

$$K_0(m) = \mathcal{J}\,\frac{\delta \mathcal{H}_0}{\delta m} = 0.$$

The first negative flow is the Casimir equation

$$m_t = K_{-1}(m) = 3\,\mathcal{J}\, m^{-\frac{1}{3}}.$$

In addition, the Lax pair for the Novikov equation (1.6) consists of [32]

$$\mathbf{\Psi}_x = \begin{pmatrix} 0 & \lambda m & 1 \\ 0 & 0 & \lambda m \\ 1 & 0 & 0 \end{pmatrix} \mathbf{\Psi}, \qquad \mathbf{\Psi} = \begin{pmatrix} \psi_1 \\ \psi_2 \\ \psi_3 \end{pmatrix}, \tag{3.2}$$

and

$$\mathbf{\Psi}_t = \begin{pmatrix} \frac{1}{3}\lambda^{-2} - uu_x & \lambda^{-1}u_x - \lambda u^2 m & u_x^2 \\ \lambda^{-1}u & -\frac{2}{3}\lambda^{-2} & -\lambda^{-1}u_x - \lambda u^2 m \\ -u^2 & \lambda^{-1}u & \frac{1}{3}\lambda^{-2} + uu_x \end{pmatrix} \mathbf{\Psi}.$$

It was proved in [32] that by the Liouville transformation

$$y = \int^x m^{\frac{2}{3}}(t,\xi)\,\mathrm{d}\xi, \quad \tau = t, \quad Q = \frac{4}{9}\,m^{-\frac{10}{3}}m_x^2 - \frac{1}{3}\,m^{-\frac{7}{3}}m_{xx} - m^{-\frac{4}{3}}, \tag{3.3}$$

the isospectral problem (3.2) is transformed into

$$\Phi_{yyy} + Q\,\Phi_y = \mu\,\Phi, \tag{3.4}$$

with $\Phi = \psi_2$ and $\mu = \lambda^2$. The linear system (3.4) is a third-order spectral problem for the SK equation (1.7), which together with the corresponding time evolution of $\Phi = \psi_2$ yields

$$Q_\tau = W_y, \qquad W_{yy} + Q\,W = T, \qquad T_y = 0. \tag{3.5}$$

As noted in [32], the system (3.5) is the first negative flow of the SK hierarchy. Based on the Liouville transformation between the isospectral problems of Novikov equation and the first negative flow of the SK hierarchy, we are inspired to establish the Liouville correspondence for their entire hierarchies.

As far as the SK equation (1.7) is concerned, it exhibits a generalized bi-Hamiltonian structure, whose corresponding integrable hierarchy is generated by a recursion operator $\overline{\mathcal{R}} = \overline{\mathcal{K}}\,\overline{\mathcal{J}}$, with an implectic (Hamiltonian) operator

$$\overline{\mathcal{K}} = -\left(\partial_y^3 + 2\,Q\,\partial_y + 2\partial_y\,Q\right)$$

and a symplectic operator

$$\overline{\mathcal{J}} = 2\partial_y^3 + 2\partial_y^2\,Q\,\partial_y^{-1} + 2\partial_y^{-1}\,Q\,\partial_y^2 + Q^2\,\partial_y^{-1} + \partial_y^{-1}\,Q^2.$$

More precisely, the SK equation (1.7) can be written as

$$Q_\tau = \overline{K}_1(Q) = \overline{\mathcal{K}}\,\frac{\delta \overline{\mathcal{H}}_0}{\delta Q},$$

where the Hamiltonian functional is

$$\overline{\mathcal{H}}_0(Q) = \frac{1}{6}\int \left(Q^3 - 3Q_y^2\right)\,\mathrm{d}y.$$

In conclusion, the positive flows of the SK hierarchy are given by

$$Q_\tau = \overline{K}_n(Q) = \left(\overline{\mathcal{K}}\,\overline{\mathcal{J}}\right)^{n-1}\overline{K}_1, \qquad n = 1, 2, \ldots.$$

On the other hand, in the negative direction, in view of the fact that the trivial function $f \equiv 0$ satisfies the equation

$$\overline{\mathcal{J}}\cdot f = \frac{\delta \overline{\mathcal{H}}_0}{\delta Q},$$

as proposed in [37], the n-th negative flow has the form

$$\overline{\mathcal{R}}^n Q_\tau = 0, \qquad n = 1, 2, \ldots.$$

The following theorem is taken from [37] to illustrate the underlying one-to-one correspondence between the flows in the Novikov and SK hierarchies. In this section, for a positive integer n, the n-th equation in the positive and negative directions of the Novikov hierarchy are denoted by $(\mathrm{Nov})_n$ and $(\mathrm{Nov})_{-n}$, respectively, while the n-th positive and negative flows of the SK hierarchy are denoted by $(\mathrm{SK})_n$ and $(\mathrm{SK})_{-n}$, respectively. Based on Lemmas 2.3 and 2.4 in [37], we are able to establish the following result.

Theorem 3. *Under the Liouville transformation* (3.3), *for each positive integer* $n \in \mathbb{Z}^+$, *the n-th positive flow* $(\mathrm{Nov})_n$ *and negative flow* $(\mathrm{Nov})_{-n}$ *of the Novikov hierarchy are mapped into the n-th negative flow* $(\mathrm{SK})_{-n}$ *and positive flow* $(\mathrm{SK})_n$*, respectively.*

In addition, in the Novikov-SK setting, in order to establish the explicit relationship between the flows in the positive Novikov hierarchy and the flows in the negative SK hierarchy, the following factorization of the recursion operator $\overline{\mathcal{R}} = \overline{\mathcal{K}}\,\overline{\mathcal{J}}$ of the SK equation is necessary to identify the equations transformed from the positive flows in the Novikov hierarchy as the corresponding negative flows in the SK hierarchy exactly. The factorization is based on the following operator identity [14]:

$$\overline{\mathcal{R}} = -2\left(\partial_y^4 + 5Q\partial_y^2 + 4Q_y\partial_y + Q_{yy} + 4Q^2 + 2Q_y\partial_y^{-1}Q\right)\left(\partial_y^2 + Q + Q_y\partial_y^{-1}\right).$$

Based on the Liouville correspondence, as shown in [37], there also exists a one-to-one correspondence between the sequences of the Hamiltonian functionals $\{\mathcal{H}_n(m)\}$ of the Novikov equation and $\{\overline{\mathcal{H}}_n(Q)\}$ of the SK equation. In particular, with their Hamiltonian pairs $\mathcal{K}$, $\mathcal{J}$ and $\overline{\mathcal{K}}$, $\overline{\mathcal{J}}$ in hand, the corresponding Hamiltonian functionals $\{\mathcal{H}_n(m)\}$ and $\{\overline{\mathcal{H}}_n(Q)\}$ are determined by the following two recursive formulae:

$$\mathcal{J}\,\frac{\delta\mathcal{H}_n}{\delta m} = \mathcal{K}\,\frac{\delta\mathcal{H}_{n-1}}{\delta m}, \qquad \overline{\mathcal{J}}\,\overline{\mathcal{K}}\,\frac{\delta\overline{\mathcal{H}}_{n-1}}{\delta Q} = \frac{\delta\overline{\mathcal{H}}_n}{\delta Q}, \qquad n \in \mathbb{Z},$$

respectively. Indeed, in [37], the relationship between the two hierarchies and the effect of the Liouville transformation on the variational derivatives were investigated. Applying Lemmas 2.5 and 2.6 in [37], we can prove the following theorem.

Theorem 4. *Under the Liouville transformation* (3.3), *for each $n \in \mathbb{Z}$, the Hamiltonian functionals $\overline{\mathcal{H}}_n(Q)$ of the SK equation are related to the Hamiltonian functionals $\mathcal{H}_{-n}(m)$ of the Novikov equation, according to the following identity*

$$\mathcal{H}_n(m) = 18\,\overline{\mathcal{H}}_{-(n+2)}(Q), \qquad n \in \mathbb{Z}.$$

In analogy with the Liouville correspondence between the Novikov and SK hierarchies, there exists a similar correspondence between the DP and KK hierarchies, as well as their respective hierarchies of the Hamiltonian functionals. The main results are presented in the following two theorems. We refer the interested reader to [37] for further details.

Theorem 5. *Under the Liouville transformation*

$$y = \int^{x} n^{\frac{1}{3}}(t,\xi)\,\mathrm{d}\xi, \quad \tau = t, \quad P = \frac{1}{4}\left(\frac{7}{9}n^{-\frac{8}{3}}n_x^2 - \frac{2}{3}n^{-\frac{5}{3}}n_{xx} - n^{-\frac{2}{3}}\right), \tag{3.6}$$

for each positive integer $l \in \mathbb{Z}^+$, the l-th positive flow $(\mathrm{DP})_l$ *and negative flow* $(\mathrm{DP})_{-l}$ *of the DP hierarchy are mapped into the l-th negative flow* $(\mathrm{KK})_{-l}$ *and positive flow* $(\mathrm{KK})_l$ *of the KK hierarchy, respectively.*

Theorem 6. *Under the Liouville transformations* (3.6)*, for each $l \in \mathbb{Z}$, the Hamiltonian functional $\overline{\mathcal{E}}_l(P)$ of the KK equation* (1.5) *is related to that $\mathcal{E}_l(n)$ of the DP equation* (1.4)*, according to the following identity*

$$\mathcal{E}_l(n) = 36\,\overline{\mathcal{E}}_{-(l+2)}(P), \qquad l \in \mathbb{Z}.$$

3.2 Relationship between the Novikov and DP equations

As proposed in [37], a further significant application of the Liouville correspondence between the Novikov-SK and DP-KK hierarchies is to establish the relationship between the Novikov and DP equations, which is motivated by the following issues. Firstly, it has been shown in [25] that under the Miura transformations

$$Q - V_y + V^2 = 0, \qquad P + V_y + \frac{1}{2}V^2 = 0, \tag{3.7}$$

the SK equation (1.7) and the KK equation (1.5) are respectively transformed into the *Fordy-Gibbons-Jimbo-Miwa equation*

$$V_\tau + V_{yyyyy} - 5\left(V_yV_{yyy} + V_{yy}^2 + V_y^3 + 4VV_yV_{yy} + V^2V_{yyy} - V^4V_y\right) = 0. \tag{3.8}$$

In [37], this relationship was generalized to first negative flows of the SK and KK hierarchies. Indeed, in view of (3.7), the recursion operator $\overline{\mathcal{R}}$ of the SK equation and the recursion operator $\mathcal{R}^*$ of equation (3.8) satisfy

$$\mathcal{R}^* = T_1\overline{\mathcal{R}}T_1^{-1}, \qquad \text{with} \qquad T_1 = (2V - \partial_y)^{-1}.$$

Similarly, the recursion operator $\widehat{\mathcal{R}}$ of the KK equation is linked with the recursion operator $\mathcal{R}^*$ according to the identity

$$\mathcal{R}^* = T_2\widehat{\mathcal{R}}T_2^{-1}, \qquad \text{with} \qquad T_2 = (V + \partial_y)^{-1}.$$

Based on this, one has the following result.

Lemma 1. *Assume that V satisfies the equation*

$$\mathcal{R}^*\,V_\tau = 0. \tag{3.9}$$

Then $Q = V_y - V^2$ and $P = -V_y - V^2/2$ satisfy the first negative flow of the SK hierarchy $\overline{\mathcal{R}}\,Q_\tau = 0$ and the first negative flow of the KK hierarchy $\widehat{\mathcal{R}}\,P_\tau = 0$, respectively.

Finally, using Lemma 1, combined with the Liouville correspondences between the Novikov-SK and DP-KK hierarchies, we establish the relationship between the Novikov equation and the DP equation, which is summarized in the following theorem.

Theorem 7. *Both the Novikov equation* (1.6) *and the DP equation* (1.4) *are linked with equation* (3.9) *in the following sense. If* $V(\tau, y)$ *is a solution of equation* (3.9), *then the function* $m(t,x)$ *determined implicitly by the relation*

$$V_y - V^2 = -m^{-1}\,(1-\partial_x^2)\,m^{-\frac{1}{3}}, \qquad y = \int^x m^{\frac{2}{3}}(t,\,\xi)\,\mathrm{d}\xi, \qquad \tau = t,$$

satisfies the Novikov equation (1.6), *while the function* $n(t,x)$ *determined by*

$$V_y + \frac{1}{2}V^2 = \frac{1}{4}n^{-\frac{1}{2}}\,(1-4\partial_x^2)\,n^{-\frac{1}{6}}, \qquad y = \int^x n^{\frac{1}{3}}(t,\,\xi)\,\mathrm{d}\xi, \qquad \tau = t,$$

satisfies the DP equation (1.4).

4 Liouville correspondences for multi-component integrable hierarchies

In this section, we shall survey the main results concerning Liouville correspondences for the 2CH and GX hierarchies.

4.1 Liouville correspondence for 2CH hierarchy

The 2CH system (1.8) is a bi-Hamiltonian integrable system [67]

$$\begin{pmatrix} m \\ \rho \end{pmatrix}_t = \mathcal{J}\,\delta\mathcal{H}_2(m,\,\rho) = \mathcal{K}\,\delta\mathcal{H}_1(m,\,\rho), \quad \delta\mathcal{H}_n(m,\rho) = \left(\frac{\delta\mathcal{H}_n}{\delta m},\,\frac{\delta\mathcal{H}_n}{\delta\rho}\right)^T, \; n = 1,\,2,$$

with compatible Hamiltonian operators

$$\mathcal{K} = \begin{pmatrix} m\partial_x + \partial_x m & \rho\partial_x \\ \partial_x\rho & 0 \end{pmatrix}, \qquad \mathcal{J} = \begin{pmatrix} \partial_x - \partial_x^3 & 0 \\ 0 & \partial_x \end{pmatrix}, \tag{4.1}$$

and the associated Hamiltonian functionals

$$\mathcal{H}_1(m,\,\rho) = -\frac{1}{2}\int (u^2+u_x^2+\rho^2)\,\mathrm{d}x, \qquad \mathcal{H}_2(m,\,\rho) = -\frac{1}{2}\int u(u^2+u_x^2+\rho^2)\,\mathrm{d}x.$$

The Hamiltonian pair (4.1) induces the hierarchy

$$\begin{pmatrix} m \\ \rho \end{pmatrix}_t = \mathbf{K}_n = \mathcal{J}\delta\mathcal{H}_n = \mathcal{K}\delta\mathcal{H}_{n-1}, \quad \delta\mathcal{H}_n = \left(\frac{\delta\mathcal{H}_n}{\delta m},\,\frac{\delta\mathcal{H}_n}{\delta\rho}\right)^T, \quad n\in\mathbb{Z}, \tag{4.2}$$

of commutative bi-Hamiltonian systems, based on the corresponding Hamiltonian functionals $\mathcal{H}_n = \mathcal{H}_n(m,\rho)$. The positive flows of (4.2) begin with the seed system

$$\begin{pmatrix} m \\ \rho \end{pmatrix}_t = \mathbf{K}_1 = -\begin{pmatrix} m \\ \rho \end{pmatrix}_x,$$

and the 2CH system (1.8) is the second member. On the other hand, the negative flows start from the *Casimir system*

$$\begin{pmatrix} m \\ \rho \end{pmatrix}_t = \mathbf{K}_{-1} = \mathcal{J}\delta\mathcal{H}_C,$$

with the associated Casimir functional $\mathcal{H}_C(m,\rho) = \int m/\rho \, dx$ for the Hamiltonian operator $\mathcal{K}$. The first negative flow for the hierarchy (4.2) has the explicit form

$$m_t = (\partial_x - \partial_x^3)\left(\frac{1}{\rho}\right), \qquad \rho_t = -\left(\frac{m}{\rho^2}\right)_x, \qquad m = u - u_{xx},$$

which, together with the inverse recursion operator $\mathcal{R}^{-1} = \mathcal{J}\,\mathcal{K}^{-1}$ produces the members in the negative direction of (4.2), namely

$$\begin{pmatrix} m \\ \rho \end{pmatrix}_t = \mathbf{K}_{-n} = (\mathcal{J}\,\mathcal{K}^{-1})^{n-1}\,\mathcal{J}\begin{pmatrix} \rho^{-1} \\ -m\rho^{-2} \end{pmatrix}, \qquad n = 1, 2, \ldots.$$

In [38], we established the Liouville correspondence between the 2CH hierarchy and another integrable hierarchy, called the associated 2CH (a2CH) hierarchy. The transformation relating these hierarchies is motivated by the Liouville transformation [13]

$$\begin{gathered} \tau = t, \qquad y = \int^x \rho(t,\xi)\, d\xi, \qquad P(\tau,y) = -m(t,x)\,\rho(t,x)^{-2}, \\ Q(\tau,y) = -\frac{1}{4}\rho(t,x)^{-2} + \frac{3}{4}\rho(t,x)^{-4}\rho_x^2(t,x) - \frac{1}{2}\rho(t,x)^{-3}\rho_{xx}(t,x), \end{gathered} \tag{4.3}$$

which converts the isospectral problem

$$\Psi_{xx} + \left(-\frac{1}{4} - \lambda m + \lambda^2\rho^2\right)\Psi = 0, \quad \Psi_t = \left(\frac{1}{2\lambda} - u\right)\Psi_x + \frac{u_x}{2}\Psi$$

of the 2CH system into

$$\Phi_{yy} + (Q + \lambda P + \lambda^2)\Phi = 0, \qquad \Phi_\tau - \frac{1}{2\lambda}\rho\,\Phi_y + \frac{1}{4\lambda}\rho_y\Phi = 0, \tag{4.4}$$

with $\Phi = \sqrt{\rho}\,\Psi$.

We clarified in [38] some integrability properties of the a2CH hierarchy. This hierarchy is generated by the recursion operator

$$\overline{\mathcal{R}} = \frac{1}{2}\begin{pmatrix} 0 & \partial_y^2 + 4Q + 2Q_y\partial_y^{-1} \\ -4 & 4P + 2P_y\partial_y^{-1} \end{pmatrix},$$

and the corresponding positive flows and the negative flows are given by

$$\begin{pmatrix} Q \\ P \end{pmatrix}_\tau = \overline{\mathbf{K}}_n = \overline{\mathcal{R}}^{n-1}\overline{\mathbf{K}}_1, \qquad \overline{\mathcal{R}}^n \begin{pmatrix} Q \\ P \end{pmatrix}_\tau = \overline{\mathbf{K}}_0, \qquad n = 1, 2, \ldots,$$

respectively, where $\overline{\mathbf{K}}_1 = (-Q_y, -P_y)^T$ is the usual seed symmetry and the trivial symmetry $\overline{\mathbf{K}}_0 = (0, 0)^T$ is determined by $\overline{\mathcal{R}}\,\overline{\mathbf{K}}_0 = \overline{\mathbf{K}}_1$. Furthermore, the a2CH hierarchy can be written in bi-Hamiltonian form

$$\begin{pmatrix} Q \\ P \end{pmatrix}_\tau = \overline{\mathbf{K}}_n = \overline{\mathcal{K}}\delta\overline{\mathcal{H}}_{n-1} = \overline{\mathcal{J}}\delta\overline{\mathcal{H}}_n, \quad \delta\overline{\mathcal{H}}_n = \left(\frac{\delta\overline{\mathcal{H}}_n}{\delta Q}, \frac{\delta\overline{\mathcal{H}}_n}{\delta P}\right)^T, \quad n \in \mathbb{Z}, \tag{4.5}$$

using the compatible Hamiltonian operators

$$\overline{\mathcal{K}} = \frac{1}{4}\begin{pmatrix} \mathcal{L}\,\partial_y^{-1}\mathcal{L} & 2\mathcal{L}\,\partial_y^{-1}(P\partial_y + \partial_y P) \\ 2(P\partial_y + \partial_y P)\partial_y^{-1}\mathcal{L} & 4(P\partial_y + \partial_y P)\partial_y^{-1}(P\partial_y + \partial_y P) + 2\mathcal{L} \end{pmatrix},$$

$$\overline{\mathcal{J}} = \frac{1}{2}\begin{pmatrix} 0 & \mathcal{L} \\ \mathcal{L} & 2(P\partial_y + \partial_y P) \end{pmatrix}, \qquad \mathcal{L} = \partial_y^3 + 2Q\partial_y + 2\partial_y Q.$$

In particular, as noted in [38], the second positive flow of the 2CH hierarchy takes the explicit form

$$Q_\tau = -\frac{1}{2}P_{yyy} - 2QP_y - Q_yP, \qquad P_\tau = 2Q_y - 3PP_y, \tag{4.6}$$

which can be written as the bi-Hamiltonian form (4.5) with

$$\overline{\mathcal{H}}_1 = -\int P\,\mathrm{d}y, \qquad \overline{\mathcal{H}}_2 = -\int \left(\frac{1}{2}P^2 + 2Q\right)\mathrm{d}y.$$

Moreover, the system (4.6) can be obtained from the y-component of the Lax-pair formulation (4.4) together with

$$\Phi_\tau + (2\lambda + P)\Phi_y - \frac{1}{2}P_y\Phi = 0.$$

Based on this, system (4.6) is shown to be equivalent to the Kaup-Boussinesq system [39].

The scheme of the Liouville correspondence between the 2CH and a2CH hierarchies as proposed in [38] is as follows, where, for a positive integer n, the $(\mathrm{2CH})_n$, $(\mathrm{2CH})_{-n}$ and $(\mathrm{a2CH})_n$, $(\mathrm{a2CH})_{-n}$ denote the n-th positive and negative flows of 2CH hierarchy and the a2CH hierarchy, respectively.

Theorem 8. *Under the Liouville transformation* (4.3), *for each integer n, the* $(\mathrm{2CH})_{n+1}$ *equation is mapped into the* $(\mathrm{a2CH})_{-n}$ *equation.*

Furthermore, as a consequence of the Liouville transformation, we are led to the one-to-one correspondence between the Hamiltonian functionals in the 2CH and a2CH hierarchies.

Theorem 9. *Under the Liouville transformation* (4.3), *for each nonzero integer n, the Hamiltonian functionals $\mathcal{H}_n(m,\rho)$ of the 2CH hierarchy are related to the Hamiltonian functionals $\overline{\mathcal{H}}_n(Q,P)$ of the a2CH hierarchy, according to*

$$\mathcal{H}_n(m,\rho) = \overline{\mathcal{H}}_{-n}(Q,P), \qquad 0 \neq n \in \mathbb{Z}.$$

4.2 Liouville correspondence for the Geng-Xue hierarchy

The GX system (1.10) can be written in a bi-Hamiltonian form [50]

$$\begin{pmatrix} m \\ n \end{pmatrix}_t = \mathcal{K}\delta\mathcal{H}_1(m,\,n) = \mathcal{J}\delta\mathcal{H}_2(m,\,n),$$

where the compatible Hamiltonian operators are

$$\mathcal{K} = \frac{3}{2}\begin{pmatrix} 3m^{\frac{1}{3}}\partial_x m^{\frac{2}{3}}\Omega^{-1}m^{\frac{2}{3}}\partial_x m^{\frac{1}{3}} + m\partial_x^{-1}m & 3m^{\frac{1}{3}}\partial_x m^{\frac{2}{3}}\Omega^{-1}n^{\frac{2}{3}}\partial_x n^{\frac{1}{3}} - m\partial_x^{-1}n \\ 3n^{\frac{1}{3}}\partial_x n^{\frac{2}{3}}\Omega^{-1}m^{\frac{2}{3}}\partial_x m^{\frac{1}{3}} - n\partial_x^{-1}m & 3n^{\frac{1}{3}}\partial_x n^{\frac{2}{3}}\Omega^{-1}n^{\frac{2}{3}}\partial_x n^{\frac{1}{3}} + 3n\partial_x^{-1}n \end{pmatrix},$$

$$\mathcal{J} = \begin{pmatrix} 0 & \partial_x^2 - 1 \\ 1 - \partial_x^2 & 0 \end{pmatrix}, \qquad \Omega = \partial_x^3 - 4\,\partial_x,$$

while

$$\mathcal{H}_1(m,\,n) = \int un\,\mathrm{d}x, \qquad \mathcal{H}_2(m,\,n) = \int (u_x v - u v_x)un\,\mathrm{d}x$$

are the initial Hamiltonian functionals. The GX integrable hierarchy can be obtained by applying the resulting hereditary recursion operator $\mathcal{R} = \mathcal{K}\,\mathcal{J}^{-1}$ to the particular seed system

$$\begin{pmatrix} m \\ n \end{pmatrix}_t = \mathbf{G}_1(m,n) = \begin{pmatrix} -m \\ n \end{pmatrix}.$$

Hence, the l-th member in the positive direction of the GX hierarchy takes the form

$$\begin{pmatrix} m \\ n \end{pmatrix}_t = \mathbf{G}_l(m,n) = \mathcal{R}^{l-1}\,\mathbf{G}_1(m,n), \quad l = 1,2,\ldots, \tag{4.7}$$

and the GX system (1.10) is exactly the second positive flow. While, the l-th negative flow of the GX hierarchy is

$$\begin{pmatrix} m \\ n \end{pmatrix}_t = \mathbf{G}_{-l}(m,n) = (\mathcal{J}\,\mathcal{K}^{-1})^{l-1}\,\mathcal{J}\delta\mathcal{H}_C, \qquad l = 1,\,2,\ldots, \tag{4.8}$$

where

$$\mathcal{H}_C(m,\,n) = 3\int \Delta^{\frac{1}{3}}\,\mathrm{d}x, \quad \text{with} \quad \delta\mathcal{H}_C = \left(m^{-\frac{2}{3}}n^{\frac{1}{3}},\, m^{\frac{1}{3}}n^{-\frac{2}{3}}\right)^T,$$

is the Casimir functional for the Hamiltonian operator $\mathcal{K}$, and we set $\Delta = mn$ throughout this subsection.

The Lax-pair formulation for the GX system (1.10) takes the form [29]

$$\begin{aligned} \mathbf{\Psi}_x &= \begin{pmatrix} 0 & \lambda m & 1 \\ 0 & 0 & \lambda n \\ 1 & 0 & 0 \end{pmatrix}\mathbf{\Psi}, \qquad \mathbf{\Psi} = \begin{pmatrix} \psi_1 \\ \psi_2 \\ \psi_3 \end{pmatrix}, \\ \mathbf{\Psi}_t &= \begin{pmatrix} -u_x v & \lambda^{-1}u_x - \lambda uvm & u_x v_x \\ \lambda^{-1}v & -\lambda^{-2} + u_x v - u v_x & -\lambda uvn - \lambda^{-1}v_x \\ -uv & \lambda^{-1}u & uv_x \end{pmatrix}\mathbf{\Psi}. \end{aligned} \tag{4.9}$$

As is shown in [49], by the Liouville transformation

$$y = \int^x \Delta^{\frac{1}{3}} \mathrm{d}\xi, \ \tau = t, \ Q = \frac{1}{\Delta^{\frac{2}{3}}} + \frac{1}{6}\frac{\Delta_{xx}}{\Delta^{\frac{5}{3}}} - \frac{7}{36}\frac{\Delta_x^2}{\Delta^{\frac{8}{3}}}, \ P = \frac{1}{2}\frac{n^{\frac{2}{3}}}{m^{\frac{4}{3}}}\left(\frac{m}{n}\right)_x, \tag{4.10}$$

the isospectral problem (4.9) is converted into

$$\mathbf{\Phi}_y = \begin{pmatrix} 0 & \lambda & Q \\ 0 & P & \lambda \\ 1 & 0 & 0 \end{pmatrix} \mathbf{\Phi}, \qquad \mathbf{\Phi} = \begin{pmatrix} \phi_1 \\ \phi_2 \\ \phi_3 \end{pmatrix},$$

$$\mathbf{\Phi}_\tau = \frac{1}{2}\begin{pmatrix} A & 2\lambda^{-1}(p_y + pP) & p+q \\ 2\lambda^{-1}q & A - 2\lambda^{-2} & 2\lambda^{-1}(Pq - q_y) \\ 0 & 2\lambda^{-1}p & A \end{pmatrix} \mathbf{\Phi},$$

where

$$A = q_y p - q p_y - 2pqP, \qquad q = vm^{\frac{2}{3}}n^{-\frac{1}{3}}, \qquad p = um^{-\frac{1}{3}}n^{\frac{2}{3}}.$$

The compatibility condition $\mathbf{\Phi}_{y\tau} = \mathbf{\Phi}_{\tau y}$ gives rise to the following integrable system

$$\begin{aligned} Q_\tau &= \frac{3}{2}(q_y + p_y) - (q-p)P, \qquad & p_{yy} + 2p_yP + pP_y + pP^2 - pQ + 1 = 0, \\ P_\tau &= \frac{3}{2}(q-p), & q_{yy} - 2q_yP - qP_y + qP^2 - qQ + 1 = 0, \end{aligned}$$

which can be viewed as a negative flow of an integrable hierarchy, namely the associated Geng-Xue (aGX) integrable hierarchy. In addition, the Hamiltonian pair admitted by the aGX hierarchy are

$$\overline{\mathcal{K}} = \mathbf{\Gamma}\begin{pmatrix} 0 & \Theta \\ -\Theta^* & 0 \end{pmatrix}\mathbf{\Gamma}^* \quad \text{and} \quad \overline{\mathcal{J}} = \frac{1}{2}\begin{pmatrix} \mathcal{E} & 0 \\ 0 & -3\partial_y \end{pmatrix}, \tag{4.11}$$

where the matrix operator $\mathbf{\Gamma}$, and operators Θ, $\mathcal{E}$ are defined by

$$\mathbf{\Gamma} = -\frac{1}{6}\begin{pmatrix} \mathcal{E}\partial_y^{-1} & \mathcal{E}\partial_y^{-1} \\ (3\partial_y^2 - 2\partial_y P)\partial_y^{-1} & -(3\partial_y^2 + 2\partial_y P)\partial_y^{-1} \end{pmatrix}, \quad \begin{aligned} \Theta &= \partial_y^2 + P\partial_y + \partial_y P + P^2 - Q, \\ \mathcal{E} &= \partial_y^3 - 2Q\partial_y - 2\partial_y Q. \end{aligned}$$

Consequently, the l-th positive flow and negative flow of the aGX integrable hierarchy have the form

$$\begin{pmatrix} Q \\ P \end{pmatrix}_\tau = \overline{\mathbf{G}}_l = -\overline{\mathcal{R}}^{l-1}\begin{pmatrix} Q \\ P \end{pmatrix}_y, \qquad \overline{\mathcal{R}}^l\begin{pmatrix} Q \\ P \end{pmatrix}_\tau = \overline{\mathbf{G}}_0 = \begin{pmatrix} 0 \\ 0 \end{pmatrix}, \qquad l = 1, 2, \ldots, \tag{4.12}$$

respectively.

The main result on the Liouville correspondence between the GX and aGX integrable hierarchies given in [38] is described in the following theorem. Adopting a similar notation as above, the l-th positive and negative flows of the GX and aGX hierarchies are denoted by $(\mathrm{GX})_l$ and $(\mathrm{GX})_{-l}$, and by $(\mathrm{aGX})_l$ and $(\mathrm{aGX})_{-l}$, respectively.

Theorem 10. *Under the Liouville transformation* (4.10), *for each integer* $l \geq 1$,

(i). *If* $\big(m(t,x),\, n(t,x)\big)$ *is a solution of the* $(\mathrm{GX})_l$ *system* (4.7), *then the corresponding* $\big(Q(\tau,y),\, P(\tau,y)\big)$ *satisfies the* $(\mathrm{aGX})_{-l}$ *system* (4.12);

(ii). *If* $\big(m(t,x),\, n(t,x)\big)$ *is a solution of the* $(\mathrm{GX})_{-l}$ *system* (4.8), *then the corresponding* $\big(Q(\tau,y),\, P(\tau,y)\big)$ *satisfies the* $(\mathrm{aGX})_{l+1}$ *system* (4.12).

Considering the correspondence between the two hierarchies of Hamiltonian functionals admitted by these two integrable hierarchies, we have the following theorem.

Theorem 11. *For any nonzero integer* l, *each Hamiltonian functionals* $\mathcal{H}_l(m,n)$ *of the GX hierarchy relates the Hamiltonian functionals* $\overline{\mathcal{H}}_l(Q,P)$, *under the Liouville transformation* (4.10), *according to the following identity*

$$\mathcal{H}_l(m,n) = 6(-1)^{l+1}\overline{\mathcal{H}}_{-(l+1)}(Q,P), \qquad 0 \neq l \in \mathbb{Z}.$$

Remark 1. Notably, as claimed in [38] that the recursion operator $\overline{\mathcal{R}}$ for the aGX hierarchy satisfies the following composition identity

$$\overline{\mathcal{R}} = \mathcal{U}\,\mathcal{V}, \tag{4.13}$$

where $\mathcal{U}$ and $\mathcal{V}$ are the matrix operators defined by

$$\mathcal{U} = \begin{pmatrix} \mathcal{E} & \mathcal{E} \\ \mathcal{F} + 3\,\partial_y^2 & \mathcal{F} - 3\partial_y^2 \end{pmatrix}, \qquad \mathcal{F} = -(2P\partial_y + 2P_y),$$
$$\mathcal{V} = \frac{1}{54}\begin{pmatrix} 3\,\partial_y^{-1}\,\Theta\,\partial_y^{-1} & \partial_y^{-1}\,\Theta\,\partial_y^{-1}\,(2P - 3\,\partial_y) \\ -3\,\partial_y^{-1}\,\Theta^*\,\partial_y^{-1} & -\partial_y^{-1}\,\Theta^*\,\partial_y^{-1}\,(2P + 3\,\partial_y) \end{pmatrix}.$$

Formula (4.13) can be viewed as a new operator factorization for $\overline{\mathcal{R}}$, which is different with the decomposition of $\overline{\mathcal{R}} = \overline{\mathcal{K}}\,\overline{\mathcal{J}}^{-1}$ using the Hamiltonian pair given in (4.11). It is worth mentioning that such a novel factorization is a key issue in the proof of Theorem 10, especially in identifying the systems transformed from the negative (positive) flows of the GX hierarchy to be the corresponding positive (negative) flows of the aGX hierarchy.

5 Liouville correspondences for the dual Schrödinger and (defocusing) mKdV hierarchies

The nonlinear Schrödinger (NLS) equation

$$\mathrm{i}\,u_t + u_{xx} + \sigma u\,|u|^2 = 0, \qquad \sigma = \pm 1 \tag{5.1}$$

is a reduction of a bi-Hamiltonian system

$$\begin{pmatrix} u \\ v \end{pmatrix}_t = \mathcal{L}\mathcal{D}^{-1}\begin{pmatrix} u \\ v \end{pmatrix}_x, \tag{5.2}$$

where

$$\mathcal{L} = \begin{pmatrix} \partial_x + u\partial_x^{-1}v & -u\partial_x^{-1}u \\ -v\partial_x^{-1}v & \partial_x + v\partial_x^{-1}u \end{pmatrix}, \qquad \mathcal{D} = \begin{pmatrix} -\mathrm{i} & 0 \\ 0 & \mathrm{i} \end{pmatrix}$$

are compatible Hamiltonian operators. System (5.2) reduces to the NLS equation (5.1) when $v = \sigma\overline{u}$. In spirit of the general approach of tri-Hamiltonian duality, we introduce the following Hamiltonian pair

$$\mathcal{K} = \begin{pmatrix} m\partial_x^{-1} m & -m\partial_x^{-1} n \\ -n\partial_x^{-1} m & n\partial_x^{-1} n \end{pmatrix}, \qquad \mathcal{J} = \begin{pmatrix} 0 & \mathrm{i} - \partial_x \\ -(\mathrm{i} + \partial_x) & 0 \end{pmatrix}, \tag{5.3}$$

and define $m = u + \mathrm{i}\, u_x$, $n = v - \mathrm{i}\, v_x$, which leads to the integrable system

$$m_t = \mathrm{i}\, muv, \qquad n_t = -\,\mathrm{i}\, nuv. \tag{5.4}$$

If we let $v = \overline{u}$ and then $n = \overline{m}$, system (5.4) reduces to

$$m_t = u_t + \mathrm{i}\, u_{xt} = |u|^2\, (\mathrm{i}\, u - u_x), \tag{5.5}$$

which is exactly the dual version (1.3) derived in [67]. Hence, we call system (5.4) the dual Schrödinger equation and the integrable hierarchy generated by (5.4) the dual Schrödinger hierarchy.

In general, the bi-Hamiltonian integrable hierarchy initiated with the dual NLS (dNLS) equation takes the form

$$\begin{pmatrix} m \\ n \end{pmatrix}_t = \mathbf{F}_l = \mathcal{K}\, \delta\mathcal{H}_{l-1}(m,\, n) = \mathcal{J}\, \delta\mathcal{H}_l(m,\, n), \quad n = \overline{m}, \quad l = 1, 2, \ldots, \tag{5.6}$$

with $\delta\mathcal{H}_l(m,\, n) = \big(\delta\mathcal{H}_l/\delta m,\ \delta\mathcal{H}_l/\delta n\big)^T$, which is governed by the usual recursion procedure using the resulting hereditary recursion operator $\mathcal{R} = \mathcal{K}\,\mathcal{J}^{-1}$. More precisely, the dNLS equation serves as the second member corresponding to $l = 2$ in (5.6), where the Hamiltonian functionals are

$$\mathcal{H}_1 = \mathrm{i} \int m\, \overline{u}_x \, \mathrm{d}x, \qquad \mathcal{H}_2 = \frac{1}{2} \int m\, u |u|^2 \, \mathrm{d}x,$$

and the seed Hamiltonian system is

$$\begin{pmatrix} m \\ n \end{pmatrix}_t = \mathbf{F}_1 = \mathcal{J}\, \delta\mathcal{H}_1(m,\, n) = \begin{pmatrix} m \\ n \end{pmatrix}_x.$$

To obtain the negative hierarchy, note that (5.4) admits a conserved functional

$$\mathcal{H}_{-1} = \int (m\, n)^{\frac{1}{2}}\, \mathrm{d}x,$$

which satisfies

$$\mathbf{F}_0 = \mathcal{K}\, \delta\mathcal{H}_{-1}(m,\, n) = \big(cm,\ -cn\big)^T,$$

where c is the integration constant. This means that one can take

$$\begin{pmatrix} m \\ n \end{pmatrix}_t = \mathbf{F}_0 = \mathcal{J}\, \delta\mathcal{H}_0(m,\, n), \qquad \mathcal{H}_0 = -c\,\mathrm{i} \int (uv - \mathrm{i}\, u_x v)\, \mathrm{d}x,$$

as the initial equation. Then the l-th negative flow of the dNLS hierarchy is given by

$$\mathcal{R}^l \begin{pmatrix} m \\ n \end{pmatrix}_t = \mathbf{F}_0, \qquad l = 1, 2, \dots. \tag{5.7}$$

We now focus our attention on the isospectral problem associated with the dNLS equation in [47]:

$$\begin{aligned} \mathbf{\Psi}_x &= \begin{pmatrix} \frac{\mathrm{i}}{2} & \lambda m \\ \lambda n & -\frac{\mathrm{i}}{2} \end{pmatrix} \mathbf{\Psi}, \qquad \mathbf{\Psi} = \begin{pmatrix} \psi_1 \\ \psi_2 \end{pmatrix}, \\ \mathbf{\Psi}_t &= \begin{pmatrix} \frac{\mathrm{i}}{2}(uv + \frac{1}{2\lambda^2}) & \frac{u}{2\lambda} \\ \frac{v}{2\lambda} & -\frac{\mathrm{i}}{2}(uv + \frac{1}{2\lambda^2}) \end{pmatrix} \mathbf{\Psi}. \end{aligned} \tag{5.8}$$

One can verify that the following Liouville transformation

$$\mathbf{\Phi} = \begin{pmatrix} (n/m)^{\frac{1}{4}} & (n/m)^{-\frac{1}{4}} \\ (n/m)^{\frac{1}{4}} & -(n/m)^{-\frac{1}{4}} \end{pmatrix} \mathbf{\Psi}, \qquad y = \int^x (mn)^{\frac{1}{2}}(\xi)\, \mathrm{d}\xi$$

will convert the isospectral problem (5.8) into the isospectral problem

$$\mathbf{\Phi}_y = \begin{pmatrix} -\mathrm{i}\,\mu & Q \\ Q & \mathrm{i}\,\mu \end{pmatrix} \mathbf{\Phi}, \qquad \mathbf{\Phi} = \begin{pmatrix} \phi_1 \\ \phi_2 \end{pmatrix}, \tag{5.9}$$

with

$$\lambda = -\mathrm{i}\,\mu, \quad Q = \frac{1}{4}(mn)^{-\frac{1}{2}} \left(2\mathrm{i} + \left(\frac{m}{n}\right)\left(\frac{n}{m}\right)_x\right).$$

It is remarked that (5.9) is the isospectral problem of the (defocusing) mKdV equation

$$Q_\tau + \frac{1}{4} Q_{yyy} - \frac{3}{2} Q^2 Q_y = 0. \tag{5.10}$$

Motivated by these results, we are led to establish the Liouville correspondence between the dNLS hierarchy and the (defocusing) mKdV hierarchy by utilizing the Liouville transformation

$$\begin{aligned} y &= \int^x \Delta(t, \xi) \mathrm{d}\xi, \quad \tau = t, \\ Q &= \frac{1}{4\Delta}\left(2\mathrm{i} + \left(\frac{m}{n}\right)\left(\frac{n}{m}\right)_x\right), \quad \Delta = (mn)^{\frac{1}{2}} = |m|. \end{aligned} \tag{5.11}$$

First of all, under the coordinate transformation (5.11), the Hamiltonian pair $\mathcal{K}$ and $\mathcal{J}$ (5.3) of the dNLS hierarchy will yield the Hamiltonian pair of the (defocusing) mKdV hierarchy. The following theorem is thus established to illustrate the preceding claim.

Theorem 12. *By the coordinate transformation* (5.11), *the Hamiltonian pair* $\mathcal{K}$ *and* $\mathcal{J}$ (5.3) *admitted by the dNLS equation* (5.5) *is related to the Hamiltonian pair*

$$\overline{\mathcal{K}} = -\frac{1}{8}\,\partial_y^3 + \frac{1}{2}\,\partial_y\, Q\,\partial_y^{-1}\, Q\,\partial_y \qquad \text{and} \qquad \overline{\mathcal{J}} = -\frac{1}{4}\,\partial_y. \tag{5.12}$$

Proof. In view of the transformation (5.11), one has

$$\partial_x = \Delta\partial_y, \tag{5.13}$$

and then further concludes that the operator identity:

$$\Delta^{-1}\left(\frac{n}{m}\right)^{-\beta}\left(2\,\mathrm{i} + \frac{1}{\beta}\,\partial_x\right)\left(\frac{n}{m}\right)^{\beta} = 4Q + \frac{1}{\beta}\,\partial_y \tag{5.14}$$

holds for arbitary nonzero constant β. Next, define

$$\begin{aligned}\mathbf{T} &= -\frac{1}{4}\Delta\left((\partial_y + 2Q + 2Q_y\,\partial_y^{-1})\,m^{-1}(-\partial_y + 2Q + 2Q_y\,\partial_y^{-1})\,n^{-1}\right),\\ \mathbf{T}^* &= -\frac{1}{4}\begin{pmatrix}(n/m)^{\frac{1}{2}}\,(2\partial_y^{-1}\,Q\,\partial_y - \partial_y)\\ (n/m)^{-\frac{1}{2}}\,(2\partial_y^{-1}\,Q\,\partial_y + \partial_y)\end{pmatrix}.\end{aligned} \tag{5.15}$$

Hence, the Hamiltonian operators $\overline{\mathcal{K}}$ and $\overline{\mathcal{J}}$ follow from the formulae $\overline{\mathcal{K}} = \Delta^{-1}\mathbf{T}\mathcal{J}\mathbf{T}^*$ and $\overline{\mathcal{J}} = \Delta^{-1}\mathbf{T}\mathcal{K}\mathbf{T}^*$, respectively, where the identities (5.13) and (5.14) are used. This completes the proof of the theorem. ■

It is worth noting that $\overline{\mathcal{K}}$ and $\overline{\mathcal{J}}$, as given in (5.12), are the compatible operators admitted by the hierarchy initiated with the (defocusing) mKdV equation (5.10), whose bi-Hamiltonian structure takes the following form

$$Q_\tau = \overline{\mathcal{K}}\,\frac{\delta\overline{\mathcal{H}}_1}{\delta Q} = \overline{\mathcal{J}}\,\frac{\delta\overline{\mathcal{H}}_2}{\delta Q},$$

with

$$\overline{\mathcal{H}}_1 = \int Q^2\,\mathrm{d}y, \qquad \overline{\mathcal{H}}_2 = -\frac{1}{2}\int (Q_y^2 + Q^4)\,\mathrm{d}y.$$

As for the (defocusing) mKdV hierarchy, each member in the positive direction takes the form

$$Q_\tau = \overline{F}_l = \overline{\mathcal{K}}\,\frac{\delta\overline{\mathcal{H}}_{l-1}}{\delta Q} = \overline{\mathcal{J}}\,\frac{\delta\overline{\mathcal{H}}_l}{\delta Q}, \qquad l = 1, 2, \ldots, \tag{5.16}$$

where, the (defocusing) mKdV equation (5.10) is the second member in (5.16) and the seed equation corresponding to $l = 1$ is

$$Q_\tau = \overline{F}_1 = -\frac{1}{2}\,Q_y = \overline{\mathcal{J}}\,\frac{\delta\overline{\mathcal{H}}_1}{\delta Q}.$$

However, in the negative direction, the l-th negative flow is

$$\overline{\mathcal{R}}^l\, Q_\tau = 0, \qquad l = 1, 2, \ldots, \qquad \text{where} \qquad \overline{\mathcal{R}} = \overline{\mathcal{K}}\,\overline{\mathcal{J}}^{-1} = \frac{1}{2}\partial_y^2 - 2\partial_y\, Q\, \partial_y^{-1}\, Q$$

is the recursion operator of the (defocusing) mKdV hierarchy.

Hereafter, for each positive integer l, we denote the l-th equation in the positive and negative directions of the dNLS hierarchy by $(\mathrm{dNLS})_l$ and $(\mathrm{dNLS})_{-l}$, respectively, while the l-th positive and negative flows in the (defocusing) mKdV hierarchy are denoted by $(\mathrm{mKdV})_l$ and $(\mathrm{mKdV})_{-l}$, respectively. With this notation, we are now in a position to establish the following theorem, illustrating the Liouville correspondence between the two hierarchies.

Theorem 13. *Under the transformation* (5.11)*, for each integer* $l \in \mathbb{Z}^+$*, the* $(\mathrm{dNLS})_{-l}$ *equation is mapped into the* $(\mathrm{mKdV})_l$ *equation, and the* $(\mathrm{dNLS})_l$ *equation is mapped into the* $(\mathrm{mKdV})_{-(l-1)}$ *equation.*

In order to prove Theorem 13, a relation identity with regard to the two recursion operators admitted by the dNLS and (defocusing) mKdV hierarchies is required.

Lemma 2. *Let* $\mathcal{R} = \mathcal{K}\,\mathcal{J}^{-1}$ *and* $\overline{\mathcal{R}} = \overline{\mathcal{K}}\,\overline{\mathcal{J}}^{-1}$ *be the recursion operators admitted by the dNLS and (defocusing) mKdV hierarchies, respectively. Define*

$$\mathbf{D} = \begin{pmatrix} m & 0 \\ 0 & n \end{pmatrix}. \tag{5.17}$$

Then, for each integer $l \geq 1$*,*

$$\overline{\mathcal{R}}^l\, \left(\partial_y \quad -\partial_y\,\right) \mathbf{D}^{-1}\, \mathcal{R}^l = -4\Delta^{-1}\,\mathbf{T} \tag{5.18}$$

under the transformation (5.11)*, where* $\mathbf{T}$ *is the matrix differential operator defined in* (5.15).

Proof. Note first, in the case of $l = 1$, equation (5.18) is equivalent to

$$\Delta^{-1}\,\mathbf{T}\,\mathcal{J} = -\frac{1}{4}\overline{\mathcal{R}}\,\left(\partial_y \quad -\partial_y\,\right)\mathbf{D}^{-1}\,\mathcal{K},$$

which can be directly verified by utilizing the formulae (5.13) and (5.14).

In addition, using the relations between $\overline{\mathcal{K}}$, $\overline{\mathcal{J}}$ and $\mathcal{K}$, $\mathcal{J}$ allows us to deduce that $\overline{\mathcal{R}}$ satisfies

$$\overline{\mathcal{R}}\,\Delta^{-1}\,\mathbf{T}\,\mathcal{K} = \Delta^{-1}\,\mathbf{T}\,\mathcal{J} \qquad \text{and then} \qquad \overline{\mathcal{R}}\,\Delta^{-1}\,\mathbf{T}\,\mathcal{R} = \Delta^{-1}\,\mathbf{T}. \tag{5.19}$$

Hence, for the general case, we assume that (5.18) holds for $l = k$. Then for $l = k + 1$,

$$-\frac{1}{4}\overline{\mathcal{R}}^{l+1}\left(\partial_y - \partial_y\right)\mathbf{D}^{-1}\mathcal{R}^{l+1} = -\frac{1}{4}\overline{\mathcal{R}}\,\overline{\mathcal{R}}^l\left(\partial_y - \partial_y\right)\mathbf{D}^{-1}\mathcal{R}^l\mathcal{R} = \overline{\mathcal{R}}\Delta^{-1}\,\mathbf{T}\mathcal{R} = \Delta^{-1}\mathbf{T},$$

which completes the induction step and thus establishes (5.18) for $l \geq 1$. We thus verify (5.18) holds in general, proving the lemma. ∎

Proof of Theorem 13. As the first step, we deduce from the relation between Q and m given in transformation (5.11) that

$$Q_\tau = \Delta^{-1}\mathbf{T}\begin{pmatrix} m \\ n \end{pmatrix}_t, \tag{5.20}$$

where $\mathbf{T}$ is the matrix differential operator given in (5.15).

Next, we consider the $(\text{dNLS})_{-l}$ equation for $l \geq 1$. Note first that the $(\text{dNLS})_{-1}$ equation corresponding to $l = 1$ in (5.7) can be written as

$$\begin{pmatrix} m \\ n \end{pmatrix}_t = \frac{1}{2}\mathcal{J}\begin{pmatrix} (n/m)^{\frac{1}{2}} \\ (n/m)^{-\frac{1}{2}} \end{pmatrix} = \begin{pmatrix} mQ \\ -nQ \end{pmatrix}.$$

Substituting it into (5.20) yields

$$Q_\tau = \Delta^{-1}\mathbf{T}\begin{pmatrix} mQ \\ -nQ \end{pmatrix} = -\frac{1}{2}Q_y,$$

which reveals that, under the transformation (5.11), the $(\text{dNLS})_{-1}$ equation is mapped into the $(\text{mKdV})_1$ equation.

Furthermore, it follows from (5.18) that, for each $l \geq 2$, equation (5.20) can be rewritten as

$$Q_\tau = -\frac{1}{4}\overline{\mathcal{R}}^l\,(\partial_y \quad -\partial_y\,)\,\mathbf{D}^{-1}\mathcal{R}^l\begin{pmatrix} m \\ n \end{pmatrix}_t,$$

with $\mathbf{D}$ defined in (5.17). Hence, in general, if $m(t,x)$ is a solution of $(\text{dNLS})_{-l}$ equation, and $n(t,x) = \overline{m}(t,x)$, so that (5.7) holds for each $l \geq 2$, then the corresponding $Q(\tau, y)$ satisfies

$$Q_\tau = -\frac{1}{4}\overline{\mathcal{R}}^l\,(\partial_y \quad -\partial_y\,)\,\mathbf{D}^{-1}\mathbf{F}_0 = -\frac{1}{4}\overline{\mathcal{R}}^l\,(\partial_y \quad -\partial_y\,)\begin{pmatrix} c \\ -c \end{pmatrix} = \overline{\mathcal{R}}^{l-1}\left(-\frac{1}{2}Q_y\right),$$

which means that $Q(\tau, y)$ solves the $(\text{mKdV})_l$ equation (5.16), completing the first part.

Finally, concerning the opposite direction, it is worth noting that

$$\Delta^{-1}\mathbf{T}\begin{pmatrix} m_x \\ n_x \end{pmatrix} = -(Q\Delta)_y + \frac{1}{4}\partial_y\left(\frac{m}{n}\left(\frac{n}{m}\right)_x\right) = 0.$$

Hence, if $m(t,x)$ is a solution of $(\text{dNLS})_l$ equation (5.6), and $n(t,x) = \overline{m}(t,x)$, i.e.,

$$\begin{pmatrix} m \\ n \end{pmatrix}_t = (\mathcal{K}\,\mathcal{J})^{l-1}\begin{pmatrix} m \\ n \end{pmatrix}_x, \qquad l = 1, 2, \ldots,$$

then in view of (5.20), the corresponding $Q(\tau, y)$ satisfies

$$Q_\tau = \Delta^{-1}\mathbf{T}\,(\mathcal{K}\,\mathcal{J})^{l-1}\begin{pmatrix} m \\ n \end{pmatrix}_x,$$

and then

$$\overline{\mathcal{R}}^{l-1}Q_\tau = \overline{\mathcal{R}}^{l-1}\Delta^{-1}\mathbf{T}\mathcal{R}^{l-1}\begin{pmatrix} m \\ n \end{pmatrix}_x = \Delta^{-1}\mathbf{T}\begin{pmatrix} m \\ n \end{pmatrix}_x = 0.$$

This allows us to draw the conclusion that $Q(\tau, y)$ is a solution of the $(\mathrm{mKdV})_{-(l-1)}$ equation for $l \geq 1$. We thus complete the proof of Theorem 13 in general. ∎

In what follows, we investigate the effect of the transformation (5.11) on the two hierarchies of the Hamiltonian functionals $\{\mathcal{H}_l\}$ and $\{\overline{\mathcal{H}}_l\}$ of the dNLS equation and the (defocusing) mKdV equation. With the two pairs of Hamiltonian operators $\mathcal{K}$ and $\mathcal{J}$ admitted by the dNLS equation and $\overline{\mathcal{K}}$ and $\overline{\mathcal{J}}$ admitted by the (defocusing) mKdV equation, $\{\mathcal{H}_l\}$ and $\{\overline{\mathcal{H}}_l\}$ are determined by the recursive formulae

$$\mathcal{K}\,\delta\mathcal{H}_l = \mathcal{J}\,\delta\mathcal{H}_{l+1},\ \delta\mathcal{H}_l = \left(\frac{\delta\mathcal{H}_l}{\delta m}, \frac{\delta\mathcal{H}_l}{\delta n}\right)^T, \ \overline{\mathcal{K}}\,\frac{\delta\overline{\mathcal{H}}_l}{\delta Q} = \overline{\mathcal{J}}\,\frac{\delta\overline{\mathcal{H}}_{l+1}}{\delta Q}, \qquad l \in \mathbb{Z}. \tag{5.21}$$

Lemma 3. *Let $\{\mathcal{H}_n\}$ and $\{\overline{\mathcal{H}}_n\}$ be the hierarchies of conserved functionals determined by the recursive formulae* (5.21)*. Then, for each $l \in \mathbb{Z}$, their respective variational derivatives satisfy the relation*

$$\Delta^{-1}\,\mathbf{T}\,\mathcal{J}\,\delta\mathcal{H}_l = \overline{\mathcal{J}}\,\frac{\delta\overline{\mathcal{H}}_{-l}}{\delta Q}. \tag{5.22}$$

Proof. We first prove (5.22) for $l \geq 0$ by induction on l. In the case of $l = 0$, equation (5.22) follows from the fact $\mathcal{J}\delta\mathcal{H}_0 = \left(cm,\ -cn\right)^T$ and $\Delta^{-1}\,\mathbf{T}\left(cm,\ -cn\right)^T = 0$. Assume now (5.22) holds for $l = k$ with $k \geq 1$, say

$$\Delta^{-1}\,\mathbf{T}\,\mathcal{J}\,\delta\mathcal{H}_k = \overline{\mathcal{J}}\,\frac{\delta\overline{\mathcal{H}}_{-k}}{\delta Q}.$$

Then, on the one hand, by the assumption and in view of (5.19)

$$\overline{\mathcal{R}}\,\Delta^{-1}\,\mathbf{T}\,\mathcal{K}\,\delta\mathcal{H}_k = \Delta^{-1}\,\mathbf{T}\,\mathcal{J}\,\delta\mathcal{H}_k = \overline{\mathcal{J}}\,\frac{\delta\overline{\mathcal{H}}_{-k}}{\delta Q} = \overline{\mathcal{K}}\,\frac{\delta\overline{\mathcal{H}}_{-(k+1)}}{\delta Q}, \tag{5.23}$$

while, on the other hand, by the recursive formula (5.21),

$$\overline{\mathcal{R}}\,\Delta^{-1}\,\mathbf{T}\,\mathcal{K}\,\delta\mathcal{H}_k = \overline{\mathcal{R}}\,\Delta^{-1}\,\mathbf{T}\,\mathcal{J}\,\delta\mathcal{H}_{k+1},$$

which, in comparison with (5.23) produces

$$\overline{\mathcal{J}}^{-1}\,\Delta^{-1}\,\mathbf{T}\,\mathcal{J}\,\delta\mathcal{H}_{k+1} = \frac{\delta\overline{\mathcal{H}}_{-(k+1)}}{\delta Q}.$$

Then

$$\Delta^{-1}\,\mathbf{T}\,\mathcal{J}\,\delta\mathcal{H}_{k+1} = \overline{\mathcal{J}}\,\frac{\delta\overline{\mathcal{H}}_{-(k+1)}}{\delta Q}$$

follows, establishing (5.22) for $l \geq 0$.

Next, in the case of $l = -1$, the fact $\overline{\mathcal{J}}\delta\overline{\mathcal{H}}_1/\delta Q = -Q_y/2$ and $\mathcal{J}\delta\mathcal{H}_{-1} = \mathbf{D}(Q,\ -Q)^T$ shows that (5.22) holds for $l = -1$. Finally, we prove (5.22) for all $l \leq -1$ by induction. Assume that it holds for $l = k$, then, for $l = k - 1$, according to the assumption and using the formula (5.19) again, we arrive at

$$\Delta^{-1}\,\mathbf{T}\,\mathcal{J}\,\delta\mathcal{H}_{k-1} = \overline{\mathcal{R}}\,\Delta^{-1}\,\mathbf{T}\,\mathcal{K}\,\delta\mathcal{H}_{k-1} = \overline{\mathcal{R}}\,\Delta^{-1}\,\mathbf{T}\,\mathcal{J}\,\delta\mathcal{H}_{k} = \overline{\mathcal{K}}\,\frac{\delta\overline{\mathcal{H}}_{-k}}{\delta Q},$$

which completes the induction step and verifies (5.22) holds for all $l \in \mathbb{Z}$, proving the lemma. ■

In addition, we deduce a formula which reveals the change of the variational derivative under the transformation (5.11).

Lemma 4. *Let* $m(t,x)$, $n(t,x) = \overline{m}(t,x)$ *and* $Q(\tau,y)$ *be related by the Liouville transformation* (5.11). *If the Hamiltonian functionals* $\mathcal{H}(m,n) = \overline{\mathcal{H}}(Q)$, *then*

$$\delta\mathcal{H}(m,n) = \big(\delta\mathcal{H}/\delta m,\ \delta\mathcal{H}/\delta n\big)^{T} = \mathbf{T}^{*}\,\frac{\delta\overline{\mathcal{H}}}{\delta Q},$$

where $\mathbf{T}^{*}$ *is the formal adjoint of* $\mathbf{T}$ *given in* (5.15).

Finally, under the hypothesis of Lemma 4, we define a functional

$$\mathcal{G}_k(Q) \equiv \mathcal{H}_l(m,n),$$

for some $k \in \mathbb{Z}$. Then, it follows from Lemma 3 and Lemma 4 that

$$\overline{\mathcal{K}}\,\frac{\delta\overline{\mathcal{H}}_{-(l+1)}}{\delta Q} = \Delta^{-1}\,\mathbf{T}\,\mathcal{J}\,\delta\mathcal{H}_l(m,n) = \Delta^{-1}\,\mathbf{T}\,\mathcal{J}\,\mathbf{T}^{*}\,\frac{\delta\mathcal{G}_k}{\delta Q} = \overline{\mathcal{K}}\,\frac{\delta\mathcal{G}_k}{\delta Q},$$

which immediately leads to

$$\frac{\delta\overline{\mathcal{H}}_{-(l+1)}}{\delta Q} = \frac{\delta\mathcal{G}_k}{\delta Q},$$

and then

$$\mathcal{H}_l(m,n) = \mathcal{G}_k(Q) = \overline{\mathcal{H}}_{-(l+1)}(Q)$$

follows. Consequently, we conclude that there also exists a one-to-one correspondence between the Hamiltonian functionals admitted by the dNLS and (defocusing) mKdV equations.

Theorem 14. *Under the transformation* (5.11), *for each integer* l, *the Hamiltonian conservation law* $\mathcal{H}_l(m,n)$ *of the dNLS equation is related to the Hamiltonian conservation law* $\overline{\mathcal{H}}_l(Q)$ *of the (defocusing) mKdV equation, according to the following identity*

$$\mathcal{H}_l(m,n) = \overline{\mathcal{H}}_{-(l+1)}(Q), \qquad l \in \mathbb{Z}.$$

Acknowledgements

Kang's research was supported by NSFC under Grant 11631007 and Grant 11871395, and Science Basic Research Program of Shaanxi (No. 2019JC-28). Liu's research was supported in part by NSFC under Grant 11722111 and Grant 11631007. Qu's research was supported by NSFC under Grant 11631007 and Grant 11971251.

References

[1] Ablowitz M J, Kaup D J, Newell A C and Segur H, The inverse scattering transform - Fourier analysis for nonlinear problems, *Stud. Appl. Math.* **53**, 249–315, 1974.

[2] Antonowicz M and Fordy A P, Coupled KdV equations with multi-Hamiltonian structures, *Physica D* **28**, 345–358, 1987.

[3] Antonowicz M and Fordy A P, Coupled Harry Dym equations with multi-Hamiltonian structures, *J. Phys. A: Math. Gen.* **21**, 269–275, 1988.

[4] Antonowicz M and Fordy A P, Factorisation of energy dependent Schrödinger operators: Miura maps and modified systems, *Commun. Math. Phys.* **124**, 465–486, 1989.

[5] Beals R, Sattinger D H and Szmigielski J, Acoustic scattering and the extended Korteweg-de Vries hierarchy, *Adv. Math.* **140**, 190–206, 1998.

[6] Beals R, Sattinger D H and Szmigielski J, Multipeakons and the classical moment problem, *Adv. Math.* **154**, 229–257, 2000.

[7] Bies P M, Gorka P and Reyes E, The dual modified Korteweg-de Vries-Fokas-Qiao equation: geometry and local analysis, *J. Math. Phys.* **53**, 073710, 2012.

[8] Camassa R and Holm D D, An integrable shallow water equation with peaked solitons, *Phys. Rev. Lett.* **71**, 1661–1664, 1993.

[9] Camassa R, Holm D D and Hyman J, A new integrable shallow water equation, *Adv. Appl. Mech.* **31**, 1–33, 1994.

[10] Cao C S, Holm D D and Titi E S, Traveling wave solutions for a class of one-dimensional nonlinear shallow water wave models, *J. Dyn. Diff. Eq.* **16**, 167–178, 2004.

[11] Caudrey P J, Dodd R K and Gibbon J D, A new hierarchy of Korteweg-de Vries equations, *Proc. Roy. Soc. London Ser. A* **351**, 407–422, 1976.

[12] Chen D Y, Li Y S and Zeng Y B, The transformation operator between recursion operators of Bäcklund transformations, *Sci. China Ser. A* **28**, 907–922, 1985.

[13] Chen M, Liu S Q and Zhang Y J, A two-component generalization of the Camassa-Holm equation and its solutions, *Lett. Math. Phys.* **75**, 1–15, 2006.

[14] Chou K S and Qu C Z, Integrable equations arising from motions of plane curves I, *Physica D* **162**, 9–33, 2002.

[15] Clarkson P A, Fokas A S and Ablowitz M J, Hodograph transformations of linearizable partial differential equations, *SIAM J. Appl. Math.* **49**, 1188–1209, 1989.

[16] Constantin A and Ivanov R I, On an integrable two-component Camassa-Holm shallow water system, *Phys. Lett. A* **372**, 7129–7132, 2008.

[17] Constantin A and Lannes D, The hydrodynamical relevance of the Camassa-Holm and Degasperis-Procesi equations, *Arch. Rational Mech. Anal.* **192**, 165–186, 2009.

[18] Degasperis A, Holm D D and Hone A N W, A new integrable equation with peakon solutions, *Theor. Math. Phys.* **133**, 1463–1474, 2002.

[19] Degasperis A and Procesi M, Asymptotic integrability, *Symmetry and perturbation theory (Rome, 1998)*, World Sci. Publ. New York, 1999.

[20] Dorfman I, *Dirac Structures and Integrability of Nonlinear Evolution Equations*, John Wiley, New York, 1993.

[21] Fokas A S, On a class of physically important integrable equations, *Physica D* **87**, 145–150, 1995.

[22] Fokas A S and Fuchssteiner B, Bäcklund transformations for hereditary symmetries, *Nonlinear Anal.* **5**, 423–432, 1981.

[23] Fokas A S and Himonas A, Well-posedness of an integrable generalization of the nonlinear Schrödinger equation on the circle, *Lett. Math. Phys.* **96**, 169–189, 2011.

[24] Fokas A S, Olver P J and Rosenau P, A plethora of integrable bi-Hamiltonian equations, *Algebraic aspects of integrable systems*, (Progr. Nonlinear Differential Equations Appl.) **26**, 93-101, Birkhäuser, Boston, 1997.

[25] Fordy A P and Gibbons J, Some remarkable nonlinear transformations, *Phys. Lett. A* **75**, 325, 1980.

[26] Fuchssteiner B, Some tricks from the symmetry-toolbox for nonlinear equations: generalizations of the Camassa-Holm equation, *Physica D* **95**, 229–243, 1996.

[27] Fuchssteiner B and Fokas A S, Symplectic structures, their Bäcklund transformations and hereditary symmetries, *Physica D* **4**, 47–66, 1981/1982.

[28] Fuchssteiner B and Oevel W, The bi-Hamiltonian structure of some nonlinear fifth and seventh order differential equations and recursion formulas for their symmetries and conserved covariants, *J. Math. Phys.* **23**, 358–363, 1982.

[29] Geng X G and Xue B, An extension of integrable peakon equations with cubic nonlinearity, *Nonlinearity* **22** 1847–1856, 2009.

[30] Gui G L, Liu Y, Olver P J and Qu C Z, Wave-breaking and peakons for a modified Camassa-Holm equation, *Commun. Math. Phys.* **319**, 731–759, 2013.

[31] Hone A N W and Wang J P, Prolongation algebras and Hamiltonian operators for peakon equations, *Inverse Problems* **19**, 129–145, 2003.

[32] Hone A N W and Wang J P, Integrable peakon equations with cubic nonlinearity, *J. Phys. A: Math. Theor.* **41**, 372002, 2008.

[33] Ito M, Symmetries and conservation laws of a coupled nonlinear wave equation, *Phys. Lett. A* **91**, 335–338, 1982.

[34] Ivanov R I and Lyons T, Integrable models for shallow water with energy dependent spectral problems, *J. Nonlinear Math. Phys.* **19**, 72–88, 2012.

[35] Johnson R S, Camassa-Holm, Korteweg-de Vries and related models for water waves, *J. Fluid Mech.* **455**, 63–82, 2002.

[36] Kang J, Liu X C, Olver P J and Qu C Z, Liouville correspondence between the modified KdV hierarchy and its dual integrable hierarchy, *J. Nonlinear Sci.* **26**, 141–170, 2016.

[37] Kang J, Liu X C, Olver P J and Qu C Z, Liouville correspondences between integrable hierarchies, *SIGMA Symmetry Integrability Geom. Methods Appl.* **13**, 035, 26pp, 2017.

[38] Kang J, Liu X C, Olver P J and Qu C Z, Liouville correspondence between multicomponent integrable hierarchy, *Theor. Math. Phys.* **204**, 843–873, 2020.

[39] Kaup D J, A higher-order water-wave equation and the method for solving it, *Prog. Theor. Phys.* **54**, 396–408, 1975.

[40] Kaup D J, On the inverse scattering problem for cubic eigenvalue problems of the class $\psi_{xxx} + 6Q\psi_x + 6R\psi = \lambda\psi$, *Stud. Appl. Math.* **62**, 189–216, 1980.

[41] Kundu A, Landau-Lifshitz and higher-order nonlinear systems gauge generated from nonlinear Schrödinger-type equations, *J. Math. Phys.* **25**, 3433–3438, 1984.

[42] Kupershmidt B A, A super Korteweg-de Vries equation: an integrable system, *Phys. Lett. A* **102**, 213–215, 1984.

[43] Lax P D, Integrals of nonlinear equations of evolution and solitary waves, *Commun. Pure Appl. Math.* **21**, 467–490, 1968.

[44] Lenells J, The correspondence between KdV and Camassa-Holm, *Int. Math. Res. Not.* **71**, 3797–3811, 2004.

[45] Lenells J, Traveling wave solutions of the Camassa-Holm and Korteweg-de Vries equations, *J. Nonlinear Math. Phys.* **11**, 508–520, 2004.

[46] Lenells J, Dressing for a novel integrable generalization of the nonlinear Schrödinger equation, *J. Nonlinear Sci.* **20**, 709–722, 2010.

[47] Lenells J and Fokas A S, On a novel integrable generalization of the nonlinear Schrödinger equation, *Nonlinearity* **22**, 11–27, 2009.

[48] Lenells J and Fokas A S, An integrable generalization of the nonlinear Schrödinger equation on the half-line and solitons, *Inverse Problems* **25**, 115006, 32pp, 2009.

[49] Li H M and Chai W, A new Liouville transformation for the Geng-Xue system, *Commun. Nonlinear Sci. Numer. Simulat.* **49**, 93–101, 2017.

[50] Li N H and Liu Q P, On bi-hamiltonian structure of two-component Novikov equation, *Phys. Lett. A* **377**, 257–261, 2013.

[51] Li N H and Niu X X, A reciprocal transformation for the Geng-Xue equation, *J. Math. Phys.* **55**, 053505, 2014.

[52] Li Y A, Olver P J and Rosenau P, Non-analytic solutions of nonlinear wave models, *Nonlinear Theory of Generalized Functions*, Chapman & Hall/CRC, New York, 1999.

[53] Liu X C, Liu Y and Qu C Z, Orbital stability of the train of peakons for an integrable modified Camassa-Holm equation, *Adv. Math.* **255**, 1–37, 2014.

[54] Liu Y, Olver P J, Qu C Z and Zhang S H, On the blow-up of solutions to the integrable modified Camassa-Holm equation, *Anal. Appl.* **12**, 355–368, 2014

[55] Lundmark H and Szmigielski J, An inverse spectral problem related to the Geng-Xue two-component peakon equation, *Mem. Amer. Math. Soc.* **244**, 1155, 87pp, 2016.

[56] Lundmark H and Szmigielski J, Dynamics of interlacing peakons (and shock-peakons) in the Geng-Xue equation, *J. Integrable Syst.* **2**, xyw014, 65pp, 2017.

[57] Magri F, A simple model of the integrable Hamiltonian equation, *J. Math. Phys.* **19**, 1156–1162, 1978.

[58] Matsuno Y, Bäcklund transformation and smooth multisoliton solutions for a modified Camassa-Holm equation with cubic nonlinearity *J. Math. Phys.* **54**, 051504, 2013.

[59] Matveev V B and Salle M A, *Darboux transformations and solitons*, Springer-Verlag, Berlin, 1991.

[60] McKean H P, The Liouville correspondence between the Korteweg-de Vries and the Camassa-Holm hierarchies, *Commun. Pure Appl. Math.* **56** 998–1015, 2003.

[61] Milson R, Liouville transformation and exactly solvable Schrödinger equations, *Int. J. Theor. Phys.* **37**, 1735–1752, 1998.

[62] Miura R, Korteweg-de Vries equation and generalizations. I. A remarkable explicit nonlinear transformation, *J. Math. Phys.* **9**, 1202–1204, 1968.

[63] Novikov V, Generalizations of the Camassa-Holm equation, *J. Phys. A: Math. Theor.* **42**, 342002, 2009.

[64] Olver F W J, *Asymptotics and Special Functions*, Academic Press, New York, 1974.

[65] Olver P J, Evolution equations possessing infinitely many symmetries, *J. Math. Phys.* **18**, 1212–1215, 1977.

[66] Olver P J, Applications of Lie Groups to Differential Equations (Second edition), *Graduate Texts in Mathematics* **107**, Springer-Verlag, New York, 1993

[67] Olver P J and Rosenau P, Tri-Hamiltonian duality between solitons and solitary-wave solutions having compact support, *Phys. Rev. E* **53**, 1900–1906, 1996.

[68] Qiao Z, A new integrable equation with cuspons and W/M-shape-peaks solitons, *J. Math. Phys.* **47**, 112701, 9pp, 2006.

[69] Qu C Z, Liu X C and Liu Y, Stability of peakons for an integrable modified Camassa-Holm equation with cubic nonlinearity, *Commun. Math. Phys.* **322**, 967–997, 2013.

[70] Rogers C and Schief W K, *Bäcklund and Darboux Transformations*, Cambridge University Press, Cambridge, 2002.

[71] Sawada K and Kotera T, A method for finding N-soliton solutions of the K.d.V. equation and K.d.V.-like equation, *Prog. Theor. Phys.* **51**, 1355–1367, 1974.

[72] Schiff J, Zero curvature formulations of dual hierarchies, *J. Math. Phys.* **37**, 1928–1938, 1996.

[73] Schiff J, The Camassa-Holm equation: a loop group approach, *Physica D* **121**, 24–43, 1998.

[74] Wadati M and Sogo K, Gauge transformations in soliton theory, *J. Phys. Soc. Japan* **53**, 394–398, 1983.

[75] Xu J and Fan E G, Long-time asymptotics for the Fokas-Lenells equation with decaying initial value problem: without solitons, *J. Differential Equations* **259**, 1098–1148, 2015.

A5. Darboux transformations for supersymmetric integrable systems: A brief review

Q P Liu[a] *and Lingling Xue*[b]

[a]*Department of Mathematics, China University of Mining and Technology, Beijing 100083, P.R. China*

[b]*Department of Mathematics, Ningbo University, Zhejiang 315211, P.R. China*

Abstract

This chapter concerns Darboux transformations for supersymmetric integrable systems. We concentrate on the most typical supersymmetric integrable system, namely the supersymmetric Korteweg-de Vries equation, and demonstrate how its proper Darboux transformation may be constructed systematically. As applications, both simple solutions and discrete systems of the supersymmetric Korteweg-de Vries equation are generated. A list of supersymmetric integrable systems is also given.

1 Introduction

The notation of symmetry has long been playing fundamental roles in physics. Supersymmetry, first introduced in quantum field theory in the early seventies of last century as symmetry between bosons and fermions, was expected to unify the space-time symmetries and internal symmetries. Apart from the particle physics, supersymmetry has a role in other research fields such as statistical physics. The mathematics behind supersymmetry is the super-mathematics, which roughly speaking is the mathematics involving both commuting and anti-commuting quantities, or graded algebraic structures. It is interesting to note that some parts of super-mathematics were studied much earlier than the development of supersymmetry in physics. We refer to Muller-Kirsten and Wiedemann [60], Rogers [68], Wegner [80] and Ferrara et al. [16], *etc.* and the references therein for more on supersymmetry and its applications in mathematics and physics.

Soliton theory or the modern theory of integrable systems is originated from the study of the Korteweg-de Vries (KdV) equation by Kruskal and his collaborators in the sixties of the last century. The theory has been studied and developed extensively during the last fifty years and it turns out that a large number of equations both continuous and discrete, which are of fundamental important in physics, are integrable systems.

In the late seventies, interaction between supersymmetry and integrable systems gave birth to the theory of supersymmetric integrable systems, in particular the supersymmetric Liouville equation and supersymmetric sine-Gordon equation were proposed and studied. Roughly speaking, a super system is a system of equations which involves both commutating (bosonic) variables and anti-commuting (fermionic) variables. We take the celebrated KdV equation

$$u_t = u_{xxx} + 6uu_x \tag{1.1}$$

as an example, where the field variable $u = u(x,t)$ is a real-valued (scalar) function of x and t. Here both dependent variable u and independent variables x, t are commuting variables. In 1984, Kupershmidt [34] introduced the following coupled system

$$u_t = u_{xxx} + 6uu_x + 12\rho\rho_{xx}, \tag{1.2a}$$

$$\rho_t = 4\rho_{xxx} + 6u\rho_x + 3u_x\rho, \tag{1.2b}$$

where $u = u(x,t)$ and $\rho = \rho(x,t)$ are commuting (bosonic) and anti-commuting (fermionic) variables, respectively. The derivatives of the fermionic variables may be calculated in a usual way and the only thing to remember is the sign rule: *a minus sign appears when the order of two quantities of fermionic type changes.* For example, $\rho^{(k)}\rho^{(\ell)} = -\rho^{(\ell)}\rho^{(k)}$ where $\rho^{(k)} = \frac{\partial^k \rho}{\partial x^k}$. Kupershmidt proved that this system, as the KdV equation itself, is integrable. Indeed, it has a Lax representation, infinity many conservation laws and bi-Hamiltonian structure, thus it is a typical integrable system.

One year later, in their study of supersymmetric KP hierarchy, a somewhat different super KdV equation was proposed by Manin and Radul [49] and it reads as

$$u_t = u_{xxx} + 6uu_x - 3\rho\rho_{xx}, \tag{1.3a}$$

$$\rho_t = \rho_{xxx} + 3(u\rho)_x, \tag{1.3b}$$

which is very similar to Kupershmidt's version. This system is also integrable, and it is a bi-Hamiltonian system, possesses Lax pair, and enjoys many other integrable properties.

Due to the appearance of the fermionic variable ρ, both (1.2) and (1.3) are super systems. In addition, when the fermionic variable ρ vanishes, both systems reduce to the KdV equation, thus they are super extensions of the KdV equation (or super KdV equation). But what is the difference?

It is known that the system (1.3) may be reformulated in terms of superspace formalism. Indeed, for the space variable x, let θ be a fermionic variable so that the (x,θ) serves as a superspace coordinate. Then any field variable F is supposed to be a function of (x,θ) and temporal variable t. An important ingredient in this formalism is the super derivative $\mathcal{D}$ defined by $\mathcal{D} = \partial_\theta + \theta\partial_x$. Given a super field $F = F(x,\theta,t)$, bosonic or fermionic, we may expand it with respect to θ and obtain $F = f_1(x,t) + \theta f_2(x,t)$, where f_1 and f_2 are known as the components of F. Then a direct calculation

$$\begin{aligned}(\mathcal{D}^2F) &= (\mathcal{D}(\mathcal{D}F)) = (\mathcal{D}(\partial_\theta + \theta\partial_x)(f_1 + \theta f_2)) = (\mathcal{D}(f_2 + \theta f_{1,x}))\\ &= (\partial_\theta + \theta\partial_x)(f_2 + \theta f_{1,x}) = f_{1,x} + \theta f_{2,x} = F_x\end{aligned}$$

shows that $\mathcal{D}^2 = \partial_x$, namely $\mathcal{D}$ is a square root of ∂_x.

Now let α be a fermionic super field defined by $\alpha = \rho + \theta u$, where $\rho = \rho(x,t), u = u(x,t)$. Then the super KdV Equation (1.3) of Manin and Radul may be rewritten as

$$\alpha_t = \alpha_{xxx} + 3(\alpha(\mathcal{D}\alpha))_x. \tag{1.4}$$

The root for this rewriting is that the system (1.3) possesses a space supersymmetry as explained by Mathieu [52]. Indeed, a superspace translation is defined by $x \to x + \theta\varepsilon, \theta \to \theta + \varepsilon$, where ε is a fermionic constant. This translation induces for a fermionic super field $\alpha = \alpha(x,\theta,t) = \rho(x,t) + \theta u(x,t)$ the following transformation

$$\rho \mapsto \rho + \varepsilon u, \quad u \mapsto u + \varepsilon \rho_x$$

and it is easy to check that, for the system (1.3), $(u + \varepsilon\rho_x, \rho + \varepsilon u)$ is another solution provided (u, ρ) solves the system. However, Kupershmidt's system (1.2) does not have such property of invariance.

Thus, there exist two types of super extensions for a given integrable system, one possesses a (space) supersymmetry and the other does not. In other words, if a super system is not invariant under the space supersymmetry, it is referred as a super system and those systems with space supersymmetry are named as supersymmetry systems. To emphasize, *supersymmetry requires more than a mere coupling of a bosonic field to a fermionic field.*

This article is meant to be a brief review on the supersymmetric integrable systems. As is well known, Darboux transformations have been playing important roles in the theory of integrable systems [21, 55, 69] and we mainly concern with the Darboux transformations for the supersymmetric integrable systems. Such study began with the consideration of the supersymmetric KdV equation in [40], and up until now Darboux transformations for various supersymmetric integrable systems have been constructed. We will concentrate on the supersymmetric KdV equation and show how its Darboux transformation may be obtained naturally. Then, as an interesting application, we will show that some semi-discrete and fully discrete super integrable systems may be obtained. Also, we will provide a list of supersymmetric integrable systems and point out the relevant references related to them.

2 Darboux transformations

By taking a reduction of the supersymmetric KP hierarchy, Manin and Radul [49] obtained the following supersymmetric KdV system

$$v_t = v_{xxx} + 6vv_x + 3(\alpha(\mathcal{D}v))_x, \tag{2.1a}$$
$$\alpha_t = \alpha_{xxx} + 3\left(\alpha(\mathcal{D}\alpha)\right)_x + 6\left(\alpha v\right)_x, \tag{2.1b}$$

where $v = v(x,\theta,t)$, $\alpha = \alpha(x,\theta,t)$ are bosonic and fermionic variables, respectively. For this system, we may consider two further reductions: the KdV Equation (1.1) with $\alpha = 0$, and the supersymmetric KdV Equation (1.4) with $v = 0$.

According to [40, 44], for the spectral problem of (2.1)

$$(\partial^2 + \alpha\mathcal{D} + v)\psi = \lambda^2\psi \tag{2.2}$$

we have

Proposition 1. Let ψ_0 be a fermionic solution of (2.2) at $\lambda = \lambda_0$. Then for any solution ψ of (2.2),

$$\widehat{\psi} = (\mathcal{D} + \delta)\psi, \quad \delta = -\frac{\mathcal{D}\psi_0}{\psi_0},$$
$$\widehat{\alpha} = -\alpha - 2\delta_x, \quad \widehat{v} = v + (\mathcal{D}\alpha) + 2\delta(\alpha + \delta_x)$$

solve $(\partial^2 + \widehat{\alpha}\mathcal{D} + \widehat{v})\widehat{\psi} = \lambda^2\widehat{\psi}$.

It is noted that the Darboux operator $\mathcal{D} + \delta$, a natural extension of Darboux's original result, is an odd operator. We also note that δ satisfies

$$(\mathcal{D}\delta)_x - v + \alpha\delta - (\mathcal{D}\delta)^2 + \lambda_0^2 = 0,$$

which may be taken as the related Bäcklund transformation. While there are two reductions for (2.1), the above Darboux transformation allows neither of them. To solve this problem, we need to consider a Darboux transformation of even type. Indeed, we have:

Proposition 2. Assume that the fermionic variable Λ and bosonic variable a are defined by

$$(\mathcal{D}\Lambda)_x = -(\mathcal{D}\alpha) - 2\mathcal{D}(a\Lambda) - (\mathcal{D}\Lambda)^2 + \alpha\Lambda + \kappa, \tag{2.3a}$$
$$a_{xx} = -(v + a^2)_x + \Lambda(\mathcal{D}v) - (\alpha + 2\Lambda_x)(\mathcal{D}a) - 2a\Lambda(\alpha + \Lambda_x), \tag{2.3b}$$

where κ is a constant. Then

$$\widetilde{\psi} = (\partial - \Lambda\mathcal{D} - a)\psi, \tag{2.4a}$$
$$\widetilde{\alpha} = \alpha + 2\Lambda_x, \tag{2.4b}$$
$$\widetilde{v} = v + 2\Lambda(\alpha + \Lambda_x) + 2a_x. \tag{2.4c}$$

keep Equation (2.2) invariant.

It is observed that the system of Riccati type, namely (2.3), may be solved. To see it, first let $(\partial^2 + \alpha\mathcal{D} + v)\psi_k = \lambda_k^2\psi_k (k = 0,\ 1)$ with ψ_0 bosonic and ψ_1 fermionic. From the condition

$$a\psi_0 + \Lambda(\mathcal{D}\psi_0) = \psi_{0x}, \quad a\psi_1 + \Lambda(\mathcal{D}\psi_1) = \psi_{1x}, \tag{2.5}$$

we have

$$a = \frac{\psi_{0x}(\mathcal{D}\psi_1) + (\mathcal{D}\psi_0)\psi_{1x}}{\psi_0(\mathcal{D}\psi_1) + (\mathcal{D}\psi_0)\psi_1}, \tag{2.6a}$$
$$\Lambda = \frac{\psi_0\psi_{1x} - \psi_{0x}\psi_1}{\psi_0(\mathcal{D}\psi_1) - (\mathcal{D}\psi_0)\psi_1}, \tag{2.6b}$$

for $\psi_0(\mathcal{D}\psi_1) \neq 0$. Then (2.3b) is satisfied and (2.3a) gives $\kappa = \lambda_1^2 - \lambda_0^2$. It is noted here that this even Darboux transformation may be regarded as a 2-fold Darboux transformation of Proposition 1.

Now let us consider the reductions of Darboux transformation presented by Proposition 2.

On one hand, for the case of $\alpha = 0$, we set $\Lambda = \psi_1 = 0$ and $\lambda_1 = \lambda_0$. From the first equation of (2.5) we have $a = (\ln \psi_0)_x$, hence (2.4a,2.4c) leads to the standard Darboux transformation for KdV equation.

On the other hand, for the case of $v = 0$, we let $\psi_0 = 0$, $\lambda_0 = \lambda_1 = p_1$, and $a = -p_1$. The second equation of (2.5) yields

$$\Lambda = \frac{\psi_{1x} + p_1\psi_1}{(\mathcal{D}\psi_1)}.$$

In summary we have the following Darboux transformation for the supersymmetric KdV Equation (1.4) [81]:

Proposition 3. Let ψ_1 be a fermionic solution of

$$(\partial^2 + \alpha\mathcal{D})\psi = \lambda^2\psi \tag{2.7}$$

at $\lambda = p_1$. Then

$$\widetilde{\psi} = (\partial - \Lambda\mathcal{D} + p_1)\psi, \tag{2.8}$$

$$\widetilde{\alpha} = \alpha + 2\Lambda_x, \quad \Lambda = \frac{\psi_{1x} + p_1\psi_1}{(\mathcal{D}\psi_1)}, \tag{2.9}$$

satisfy

$$\widetilde{\psi}_{xx} + \widetilde{\alpha}(\mathcal{D}\widetilde{\psi}) = \lambda^2\widetilde{\psi}.$$

We now show that the Darboux transformation leads to the Bäcklund transformation for (1.4). Introducing β and $\widetilde{\beta}$ such that $\alpha = 2\beta_x$, $\widetilde{\alpha} = 2\widetilde{\beta}_x$, the first equation of (2.9) leads

$$\Lambda = \widetilde{\beta} - \beta, \tag{2.10}$$

where the constant of integration is set to zero. Moreover, the second equation of (2.9) implies

$$\Lambda_x + \alpha - 2p_1\Lambda + \Lambda\,(\mathcal{D}\Lambda) = 0,$$

namely

$$\left(\widetilde{\beta} + \beta\right)_x - 2p_1\left(\widetilde{\beta} - \beta\right) + (\widetilde{\beta} - \beta)\mathcal{D}\left(\widetilde{\beta} - \beta\right) = 0, \tag{2.11}$$

which is the (spatial-part) Bäcklund transformation for the supersymmetric KdV Equation (1.4).

Next as a simple application we take above Bäcklund transformation and calculate the 1-soliton solution for the supersymmetric KdV Equation (1.4). It is easy to have the potential form of (1.4)

$$\beta_t = \beta_{xxx} + 6\beta_x(\mathcal{D}\beta)_x, \tag{2.12}$$

which, up to introducing $\beta(x,t,\theta) = \xi(x,t) + \theta w(x,t)$, may be reformulated in components as

$$\xi_t = \xi_{xxx} + 6\xi_x w_x, \quad w_t = w_{xxx} + 6w_x^2 + 6\xi_{xx}\xi_x. \tag{2.13}$$

Let us take the trivial solution $\widetilde{\beta} = 0$. Then from (2.11) we have

$$\beta_x + 2p_1\beta + \beta\,(\mathcal{D}\beta) = 0,$$

which in components yields

$$w_x = -2p_1 w - w^2, \quad \xi_x = -2p_1\xi - \xi w.$$

Solving the above system, we obtain

$$w = \frac{-2p_1}{1+e^{2p_1x+d}}, \quad \xi = \frac{\eta}{1+e^{2p_1x+d}},$$

where d (bosonic) and η (fermionic) are independent of x and θ to be determined. Next by using (2.13), we have

$$\beta = \frac{\mu - 2p_1\theta}{1+e^{2p_1x+(2p_1)^3t+x_0}}, \quad \mu \text{ is a fermionic constant,}$$

thus we obtain the 1-soliton solution for (2.12)

$$\beta = (\mathcal{D}\log f), \quad f = 1 + e^{-2p_1x-(2p_1)^3t+\theta\mu-x_0}.$$

The multisoliton solutions were constructed in [9, 10, 11, 42].

3 Discrete systems

Apart from solution construction, Darboux and Bäcklund transformations also play important roles in integrable discretizations for nonlinear systems. In this section, we employ the results of the last section and build both integrable semi-discrete and fully discrete systems for the supersymmetric KdV Equation (1.4).

3.1 Semi-discrete system

By the following identifications

$$\beta \equiv \beta_n, \quad \widetilde{\beta} \equiv \beta_{n+1}, \quad \psi \equiv \psi_n, \quad \widetilde{\psi} \equiv \psi_{n+1},$$

we can interpret (2.11), i.e.,

$$(\beta_{n+1}+\beta_n)_x - 2p_1\,(\beta_{n+1}-\beta_n) + (\beta_{n+1}-\beta_n)\mathcal{D}\,(\beta_{n+1}-\beta_n) = 0, \tag{3.1}$$

as a differential-difference equation. In the next section, we will show that, after a continuum limit, the system (3.1) leads to the potential supersymmetric KdV equation, thus it constitutes a discretization of the potential supersymmetric KdV equation.

We now show that Equation (3.1) admits a Lax pair. To see it, let us rewrite the Darboux transformation (2.7, 2.8) in matrix form. Introducing the vector $\Psi_n = (\psi_n, \psi_{n,x}, (\mathcal{D}\psi_n), (\mathcal{D}\psi_n)_x)^{\mathrm{T}}$, with the help of (3.1) one may rewrite (2.7) and (2.8) as the following systems

$$\Psi_{n,x} = L_n \Psi_n, \quad L_n = \begin{pmatrix} 0 & 1 & 0 & 0 \\ \lambda^2 & 0 & -2\beta_{n,x} & 0 \\ 0 & 0 & 0 & 1 \\ 0 & 2\beta_{n,x} & \lambda^2 - 2(\mathcal{D}\beta_n)_x & 0 \end{pmatrix}, \tag{3.2}$$

$$\Psi_{n+1} = W_n \Psi_n, \quad W_n = \begin{pmatrix} p_1 & 1 & -\Lambda & 0 \\ \lambda^2 & p_1 & -(p_1+\Delta)\Lambda & -\Lambda \\ 0 & \Lambda & \Delta & 1 \\ \lambda^2\Lambda & (p_1+\Delta)\Lambda & \lambda^2 - p_1^2 + \Delta^2 & \Delta \end{pmatrix}, \tag{3.3}$$

where $\Delta = p_1 - (\mathcal{D}\Lambda)$, and $\Lambda = \beta_{n+1} - \beta_n$. The compatibility of the two linear systems (3.2, 3.3) is

$$W_{n,x} + W_n L_n - L_{n+1} W_n = 0, \tag{3.4}$$

which leads to (3.1). Therefore, this semi-discrete system, having (3.2, 3.3) as a Lax pair, is integrable.

3.2 Fully discrete system

To construct a possible difference-difference system, we introduce one more discrete variable m and define

$$\beta_n \equiv \beta_{n,m}, \quad \psi_n \equiv \psi_{n,m}.$$

To avoid the differential terms in W_n, we write out the super fields in terms of their components, that is,

$$\beta_{n,m} = \xi_{n,m} + \theta w_{n,m}, \quad \psi_{n,m} = \varphi_{n,m} + \theta f_{n,m}.$$

Then introducing $\chi_{n,m} = ((\varphi_{n,m}), (\varphi_{n,m})_x, (f_{n,m}), (f_{n,m})_x)^{\mathrm{T}}$, from (3.3) we have

$$\chi_{n+1,m} = \mathcal{W}_{n,m}\chi_{n,m}, \; \mathcal{W}_{n,m} = \begin{pmatrix} p_1 & 1 & -\eta & 0 \\ \lambda^2 & p_1 & -(p_1+g)\eta & -\eta \\ 0 & \eta & g & 1 \\ \lambda^2\eta & (p_1+g)\eta & \lambda^2 - p_1^2 + g^2 & g \end{pmatrix} \tag{3.5}$$

with $\eta = \xi_{n+1,m} - \xi_{n,m}$, $g = p_1 - (w_{n+1,m} - w_{n,m})$.

Next, to find a difference-difference system, we consider another Darboux transformation

$$\chi_{n,m+1} = \mathcal{V}_{n,m}\chi_{n,m}, \tag{3.6}$$

where the matrix $\mathcal{V}_{n,m}$ is the matrix $\mathcal{W}_{n,m}$ of (3.5) with p_1, $\xi_{n+1,m}$ and $w_{n+1,m}$ replaced by p_2, $\xi_{n,m+1}$ and $w_{n,m+1}$, respectively.

Now the compatibility condition of (3.5) and (3.6), namely

$$\mathcal{W}_{n,m+1}\mathcal{V}_{n,m} = \mathcal{V}_{n+1,m}\mathcal{W}_{n,m},$$

yields an integrable difference-difference system

$$\xi_{n+1,m+1} = \xi_{n,m} + \frac{(p_1+p_2)\left(\xi_{n+1,m} - \xi_{n,m+1}\right)}{p_2 - p_1 + w_{n+1,m} - w_{n,m+1}}, \tag{3.7a}$$

$$\begin{aligned} w_{n+1,m+1} &= w_{n,m} + \frac{(p_1+p_2)(w_{n+1,m} - w_{n,m+1})}{p_2 - p_1 + w_{n+1,m} - w_{n,m+1}} \\ &- \frac{2(p_2-p_1) + w_{n+1,m} - w_{n,m+1}}{p_2 - p_1 + w_{n+1,m} - w_{n,m+1}} \left(\xi_{n+1,m+1} - \xi_{n,m}\right)\left(\xi_{n,m+1} - \xi_{n,m}\right). \end{aligned} \tag{3.7b}$$

It is noted that if $\xi_{n,m} = \mu w_{n,m}$, (3.7) reduces to the well-known lattice potential KdV equation

$$w_{n+1,m+1} = w_{n,m} + \frac{(p_1+p_2)(w_{n+1,m}-w_{n,m+1})}{p_2-p_1+w_{n+1,m}-w_{n,m+1}},$$

which is consistent around a cube or possesses the CAC property [1]. Hence the discrete 1-soliton solution of lattice potential KdV equation [24] gives rise to 1-supersoliton solution of (3.7)

$$\xi_{n,m} = \mu w_{n,m}, \quad w_{n,m} = \frac{2p_3}{1 + c\left(\frac{p_1-p_3}{p_1+p_3}\right)^n \left(\frac{p_2-p_3}{p_2+p_3}\right)^m}, \tag{3.8}$$

where c is a bosonic constant. The Hirota bilinear form and multisoliton solutions of (3.7) were constructed in [10]. It was also shown in [81] that this system (3.7) has the CAC property as well but does not have the tetrahedron property.

In the next section, we will show that, after a double continuum limit, the system (3.7) leads to the potential supersymmetric KdV equation, thus it constitutes a discretization of the potential supersymmetric KdV equation.

Obviously, (3.7) can be defined on a lattice (see Figure 1(a)). Let us give values of $(w_{n,m}, \xi_{n,m})$, $(w_{n+1,m}, \xi_{n+1,m})$, and $(w_{n,m+1}, \xi_{n,m+1})$, depicted by black dots, then $(w_{n+1,m+1}, \xi_{n+1,m+1})$, denoted by an orange dot, may be calculated by (3.7). Therefore, the initial value problems for this system may be posed by assigning the values either on a staircase or on a corner (black dots as shown in Figure 1(b) or Figure 1(c)).

4 Continuum limits of the discrete systems

To gain a better understanding of the discrete systems constructed in the previous section, following [24, 61] we analyse their continuum limits. In this way, our claim, the differential-difference system (3.1) and the difference-difference system (3.7) as discrete versions of the potential supersymmetric KdV equation, will be justified.

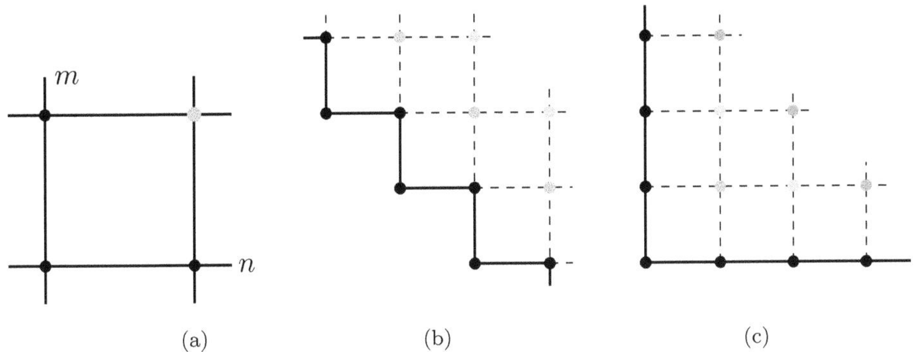

Figure 1. *The system* (3.7) *may be defined on a lattice as shown in (a); the initial values can be given in a staircase (b) or on a corner (c).*

4.1 The semi-continuous limits of the difference-difference system

In this part, we consider two different continuum limits of (3.7), namely the straight continuum limit and the skew continuum limit. In the first case, the limit will be taken for one of the discrete variables, and in the second the limit will be performed for a combination of both discrete variables.

4.1.1 Straight continuum limit

The difference-difference system (3.7) may be regarded as a discrete analogue of the differential-difference system (3.1). To see it, let us define

$$\xi_{n,m} \equiv \xi_n(x), \quad w_{n,m} \equiv w_n(x), \quad x = \frac{m}{p_2}.$$

For $\frac{1}{p_2}$ small, we have the following Taylor series expansions

$$\xi_{k,m+1} = \xi_k\left(x + \tfrac{1}{p_2}\right) = \xi_k + \tfrac{1}{p_2}\xi_{k,x} + \mathcal{O}\left(\tfrac{1}{p_2^2}\right),$$
$$w_{k,m+1} = w_k\left(x + \tfrac{1}{p_2}\right) = w_k + \tfrac{1}{p_2}w_{k,x} + \mathcal{O}\left(\tfrac{1}{p_2^2}\right),$$

where k can take the value n or $n+1$. Substituting the above expansions into (3.7), the leading terms yield

$$(\xi_{n+1} + \xi_n)_x = 2p_1(\xi_{n+1} - \xi_n) - (\xi_{n+1} - \xi_n)(w_{n+1} - w_n), \tag{4.1a}$$
$$(w_{n+1} + w_n)_x = 2p_1(w_{n+1} - w_n) - (w_{n+1} - w_n)^2 - 2(\xi_{n+1} - \xi_n)\xi_{n,x}, \tag{4.1b}$$

namely the component form of (3.1).

4.1.2 Skew continuum limit

In this case, we introduce a new discrete variable N and a continuous variable τ such that

$$N = n + m, \quad \tau = \epsilon m, \quad p_2 = p_1 + \epsilon. \tag{4.2}$$

Then

$$\begin{aligned} \xi_{n,m} \equiv& \xi_N(\tau), & \xi_{n+1,m} \equiv& \xi_{N+1}(\tau), & \xi_{n,m+1} \equiv& \xi_{N+1}(\tau+\epsilon), \\ w_{n,m} \equiv& w_N(\tau), & w_{n+1,m} \equiv& w_{N+1}(\tau), & w_{n,m+1} \equiv& w_{N+1}(\tau+\epsilon), \\ \chi_{n,m} \equiv& \chi_N(\tau), & \chi_{n+1,m} \equiv& \chi_{N+1}(\tau), & \chi_{n,m+1} \equiv& \chi_{N+1}(\tau+\epsilon). \end{aligned} \tag{4.3}$$

The skew continuum limit may be done directly on the level of the difference-difference system (3.7). Alternatively we consider the continuum limit of the corresponding Lax pair (3.5) and (3.6):

$$\chi_{n+1,m} = \mathcal{W}_{n,m}\chi_{n,m}, \quad \chi_{n,m+1} = \mathcal{V}_{n,m}\chi_{n,m}, \tag{4.4}$$

where the matrices $\mathcal{W}$ and $\mathcal{V}$ are defined in Section 3, then a differential-difference system together with its Lax pair may be derived. Indeed, using (4.2, 4.3) the first equation of the Lax pair (4.4) reads as

$$\chi_{N+1} = \mathcal{W}_N\,\chi_N, \quad \mathcal{W}_{_N} = \begin{pmatrix} p_1 & 1 & -\zeta & 0 \\ \lambda^2 & p_1 & -(p_1+h)\zeta & -\zeta \\ 0 & \zeta & h & 1 \\ \lambda^2\zeta & (p_1+h)\zeta & \lambda^2 - p_1^2 + h^2 & h \end{pmatrix}, \tag{4.5}$$

where $\zeta \equiv \xi_{N+1} - \xi_N$, $h \equiv p_1 - (w_{N+1} - w_N)$. Expanding the second equation of the Lax pair (4.4) in ϵ, we obtain

$$\chi_{N+1} + \epsilon\chi_{N+1,\tau} + \cdots = (\mathcal{W}_N + \epsilon\mathcal{T}_N + \cdots)\chi_N,$$

where

$$\mathcal{T}_N = \begin{pmatrix} 1 & 0 & -\xi_{N+1,\tau} & 0 \\ 0 & 1 & t_{23} & -\xi_{N+1,\tau} \\ 0 & \xi_{N+1,\tau} & 1 - w_{N+1,\tau} & 0 \\ \lambda^2\xi_{N+1,\tau} & -t_{23} & -2(p_1-h) - 2hw_{N+1,\tau} & 1 - w_{N+1,\tau} \end{pmatrix}$$

with

$$t_{23} \equiv -2\zeta + \zeta w_{N+1,\tau} - (p_1+h)\xi_{N+1,\tau}.$$

Next with the help of (4.5) we obtain as the coefficient of the leading term of order $\mathcal{O}(\epsilon)$ the following equation

$$\chi_{N+1,\tau} = \mathcal{T}_N\chi_N. \tag{4.6}$$

The consistency condition of (4.5) and (4.6)

$$\mathcal{W}_{N,\tau}\mathcal{W}_{N-1} + \mathcal{W}_N\mathcal{T}_{N-1} - \mathcal{T}_N\mathcal{W}_{N-1} = 0$$

leads to a differential-difference system

$$\xi_{N,\tau} = \frac{\xi_{N+1} - \xi_{N-1}}{w_{N+1} - w_{N-1} - 2p_1}, \tag{4.7a}$$

$$w_{N,\tau} = 1 + \frac{2p_1 + (w_{N+1} - w_{N-1} - 4p_1)(\xi_{N+1} - \xi_N)\xi_{N,\tau}}{w_{N+1} - w_{N-1} - 2p_1}. \tag{4.7b}$$

It is noted that under $\xi_n = \mu w_N$ (μ is a fermionic constant) the coupled system (4.7) may be reduced to

$$w_{N,\tau} = \frac{w_{N+1} - w_{N-1}}{w_{N+1} - w_{N-1} - 2p_1},$$

which is a differential-difference equation related to the Kac-van Moerbeke equation [29] (see [24] also). Thus, the system (4.7) is a super extension of the Kac-van Moerbeke equation. Moreover, by introducing $e^a = \frac{p_1+p_3}{p_1-p_3}$, (3.8) in the continuum limit leads to one-supersoliton solution of (4.7)

$$\xi_N = \mu w_N, \quad w_N = \frac{2p_1 \tanh \frac{a}{2}}{1 + e^{-aN - b\tau - \tau_0}}, \qquad b = -\frac{1}{p_1} \sinh a.$$

Further study of the system (4.7) may be found in [10].

4.2 Full continuum limit

The continuum limits of (3.7), namely (4.1) (or (3.1)) and (4.7) are still discrete systems, thus further continuum limits may be taken for them. In the following, we will show that both of them may lead to the potential supersymmetric KdV Equation (2.12).

To carry out the continuous limit of the differential-difference system (3.1) we define a new continuous variable τ as

$$\beta_n(x, \theta) \equiv \beta(x, \theta, \tau), \quad \tau = \frac{n}{p_1}.$$

Then

$$\beta_{n+1}(x, \theta) \equiv \beta\left(x, \theta, \tau + \frac{1}{p_1}\right)$$

can be expanded in $\frac{1}{p_1}$, and defining a new independent variable t in terms of τ and x such that

$$\partial_\tau = \partial_x + \frac{1}{12p_1^2}\partial_t,$$

in the continuous limit up to terms of order $\frac{1}{p_1}$ we obtain the potential supersymmetric KdV Equation (2.12).

For the differential-difference system (4.7), defining

$$\xi_N(\tau) \equiv \xi(s, \tau), \quad w_N(\tau) \equiv w(s, \tau), \quad s = \frac{N}{p_1},$$

expanding

$$\xi_{N\pm1}(\tau) \equiv \xi\left(s \pm \frac{1}{p_1}, \tau\right), \quad w_{N\pm1}(\tau) \equiv w\left(s \pm \frac{1}{p_1}, \tau\right)$$

in $\frac{1}{p_1}$, and redefining the independent variables from s, τ to x, t such that

$$\partial_s = \partial_x, \quad \partial_\tau = -\tfrac{1}{p_1^2}\partial_x - \tfrac{1}{6p_1^4}\partial_t,$$

we get (2.13), i.e., the components of potential supersymmetric KdV equation.

5 List of supersymmetric integrable equations

In this section, we provide a list of supersymmetric integrable systems. The list is guided by our interest and is not meant to be complete in any sense.

1. *Supersymmetric sine-Gordon equation*

$$\mathcal{D}_t\mathcal{D}_x\Phi = \sin\Phi. \tag{5.1}$$

This is one of the earliest studied supersymmetric integrable models [12, 15, 19, 26, 79]. This equation was bilinearised in [20] and its Darboux transformation was obtained in [71] (see also [83]).

2. *Supersymmetric Lund-Regge-Pohlmeyer equation*

$$\begin{aligned}\mathcal{D}_1\mathcal{D}_2R + im(e^R + e^{-R}\cos\Phi) &= 0,\\ \mathcal{D}_1\mathcal{D}_2\Phi + ime^{-R}\sin\Phi + \frac{2\cos\frac{\Phi}{2}}{\sin^3\frac{\Phi}{2}}(\mathcal{D}_1H)(\mathcal{D}_2H) &= 0,\\ \mathcal{D}_1\mathcal{D}_2H - \frac{1}{\sin\Phi}[(\mathcal{D}_1H)(\mathcal{D}_2\Phi) + (\mathcal{D}_1\Phi)(\mathcal{D}_2H)] &= 0,\end{aligned}$$

where $\mathcal{D}_k\,(k = 1, 2)$ are super derivatives, R, H and Φ are the super fields. This system first appeared in [13] (see (5.17) and (5.26) therein).

3. *Supersymmetric KdV equation*

$$\alpha_t = \alpha_{xxx} + 3(\alpha(\mathcal{D}\alpha))_x.$$

As mentioned above, this equation was obtained as a reduction of the supersymmetric KP hierarchy [49] and has been studied extensively. In particular, Mathieu, by considering a fermionic extension of the KdV equation with a supersymmetry, singled it out from a one-parameter family of systems. He further considered its Hamiltonian structure, infinitely many conserved quantities, modification and deformation [52]. The bi-Hamiltonian structure was worked out independently almost simultaneously by two groups, Oevel and Popowicz [63], Figueroa-O'Farrill, Mas and Ramos [17]. We also mention that Painlevé analysis and Hirota method were developed for it, see [25, 53] and [9, 11, 18, 42, 56].

4. *Supersymmetric Liouville equation*

$$\mathcal{D}_t\mathcal{D}_x\Phi = e^\Phi, \tag{5.2}$$

which was discussed in [12], where a Lax pair and a Bäcklund transformation were constructed for it.

5. *Supersymmetric principal $SU(N)$ Chiral model*

$$\mathcal{D}_2(\mathcal{D}_1\hat{g}\hat{g}^+) - \mathcal{D}_1(\mathcal{D}_2\hat{g}\hat{g}^+) = 0,$$

where $\hat{g} = \hat{g}(\xi, \eta, \theta_1, \theta_2) \in SU(N)$, $\hat{g}\hat{g}^+ = \hat{g}^+\hat{g} = I$. $D_1 = \partial_{\theta_2} - i\theta_2\partial_\xi, D_2 = -\partial_{\theta_1} + i\theta_1\partial_\eta$. This system was proposed by Mikhailov [57]. For more on it, we refer to the papers by Harnad and Jacques [23] and by Haider and Hassan [22] and the references therein.

6. *Supersymmetric MKdV equation*

$$\Psi_t + \Psi_{xxx} - 3\Psi(\mathcal{D}\Psi_x)\mathcal{D}\Psi - 3(\mathcal{D}\Psi)^2\Psi_x = 0. \tag{5.3}$$

This equation, related to the supersymmetric KdV Equation (1.4) via a Miura-type transformation, first appeared in [52, 84]. Its proper Hirota bilinear form was constructed in [43] and its Darboux transformation was worked out in [83]. For a study from the nonlocal symmetry viewpoint, we refer to the monograph by Krasil'shchik and Kersten [32].

7. *Supersymmetric Schwarzian KdV equation*

$$\Phi_t = \frac{1}{4}\left(\Phi_{xxx} - \frac{3\Phi_{yy}(\mathcal{D}\Phi_y)}{(\mathcal{D}\Phi)}\right),$$

which, obtained by Mathieu [54], relates to the supersymmetric MKdV equation (5.3) via a super version of Cole-Hopf transformation.

8. *Supersymmetric classical Boussinesq or two-boson system*

$$\begin{aligned} u_t &= -u_{xx} + 2uu_x + 2(\mathcal{D}\alpha)_x, \\ \alpha_t &= \alpha_{xx} + 2(u\alpha)_x, \end{aligned}$$

where $u = u(x, \theta, t)$ and $\alpha = \alpha(x, \theta, t)$ are bosonic and fermionic, respectively. The system was proposed by Brenelli and Das [4] in 1994 (see also (7.34) of [32]). Interestingly, this system is a bi-Hamiltonian system with two local Hamiltonian operators [5, 6]. The system was bilinearised [48, 14] and its Darboux transformation was constructed [62].

9. *OSP(3,2) KdV system*

$$\begin{aligned} \tau_t &= -\frac{1}{8}(\tau_{xx} - 6(\mathcal{D}a)_x + 3(\mathcal{D}\tau)\tau + 12a\tau)_x, \\ a_t &= -\frac{1}{8}\left((4a_{xx} + 3\tau(\mathcal{D}a) + 6a^2)_x + 3\tau(\mathcal{D}a)_x\right). \end{aligned}$$

This system was proposed and studied by Morosi and Pizzocchero [58].

10. *Supersymmetric Sawada-Kotera equation*

$$\phi_t + \phi_{xxxxx} + 5\phi_{xxx}(\mathcal{D}\phi) + 5\phi_{xx}(\mathcal{D}\phi_x) + 5\phi_x(\mathcal{D}\phi)^2 = 0. \tag{5.4}$$

This equation was constructed in [72] by a deep reduction of a general third order Lax operator $L = \partial_x^3 + \Psi\mathcal{D}^3 + U\partial + \Phi\mathcal{D} + V$. Subsequently, Popowicz showed that it is a bi-Hamiltonian system [66]. Very recently, its Darboux transformation was built [51].

11. *Supersymmetric Burgers equation*

$$\Phi_t + \Phi(\mathcal{D}\Phi)_x + \Phi_{xx} = 0$$

which was singled out by Hlavatý by Painlevé analysis [25]; but it seems that the equation is not be integrable.

12. *Supersymmetric Harry Dym systems*

The Harry Dym equation $u_t = u^3 u_{xxx}$ has several supersymmetric extensions. The first one reads as

$$U_t = \frac{1}{4}U^3 U_{xxx} + \frac{3}{8}U^2(\mathcal{D}U)(\mathcal{D}U)_{xx},$$

which was proposed by Brunelli, Das and Popowicz [8]. It is observed that above equation may be taken as a reduction of the supersymmetric system contained in [41].

In addition, two more supersymmetric Harry Dym equations were found in [75] and they are

$$U_t = \frac{1}{4}U^3 U_{xxx} + \frac{3}{4}U^2(\mathcal{D}U)(\mathcal{D}U)_{xx},$$

and

$$U_t = \frac{1}{4}U^3 U_{xxx} + \frac{3}{8}U^2 U_{xx}U_x - \frac{3}{8}U^2(\mathcal{D}U)_{xx}(\mathcal{D}U) - \frac{3}{16}(\mathcal{D}U)_x(\mathcal{D}U)U_x U.$$

13. *Supersymmetric Hunter-Saxton equation*

$$W_t = -\frac{3}{2}W(\mathcal{D}^{-1}W) - W_x(\mathcal{D}^{-3}W) - \frac{1}{2}(\mathcal{D}W)(\partial_x^{-1}W) \tag{5.5}$$

first appeared in [8] and was later rediscovered in [38]. This equation, a bi-Hamiltonian system with two local Hamiltonian structures, was termed as the even supersymmetric Hunter-Saxton equation in [46]. There is one more supersymmetric Hunter-Saxton equation or the odd supersymmetric Hunter-Saxton equation, see equation (35) of [46].

14. *Supersymmetric nonlinear Schrödinger equations*

Roelofs and Kersten [70], within theory of coverings, proposed two such systems, namely

- Type A

$$\phi_t = \phi_{xx} - 2\phi(\mathcal{D}\phi)(\mathcal{D}\psi) + 2\psi\phi_x\phi,$$
$$\psi_t = -\psi_{xx} + 2\psi(\mathcal{D}\psi)(\mathcal{D}\phi) - 2\phi\psi_x\psi,$$

- Type B

$$\phi_t = \phi_{xx} - 4\phi(\mathcal{D}\phi)(\mathcal{D}\psi) + a\psi\phi_x\phi,$$
$$\psi_t = -\psi_{xx} + 4\psi(\mathcal{D}\psi)(\mathcal{D}\phi) - a\phi\psi_x\psi,$$

For further study of the type A equation, we refer to [4] and [83], and the last paper contains a Darboux transformation for this system. It seems that the type B equation deserves a more in-depth investigation.

15. *Supersymmetric Kawamoto equation*

$$
\begin{aligned}
V_t = & (V^5 V_{xxxx})_x + \frac{5}{2} V^4 V_{xxx} V_{xx} + \frac{15}{4} V^3 V_{xxx} V_x^2 - \frac{5}{2} V^4 (\mathcal{D} V_{xxxx})(\mathcal{D} V) \\
& - 5V^4 (\mathcal{D} V_{xxx})(\mathcal{D} V_x) - \frac{15}{2} V^3 (\mathcal{D} V_{xxx})(\mathcal{D} V) V_x - \frac{15}{2} V^3 (\mathcal{D} V_{xx})(\mathcal{D} V_x) V_x \\
& - \frac{15}{4} V^3 (\mathcal{D} V_{xx})(\mathcal{D} V) V_{xx} - \frac{15}{8} V^2 (\mathcal{D} V_{xx})(\mathcal{D} V) V_x^2
\end{aligned}
$$

appeared in [46] in the context of the supersymmetric reciprocal transformations and is related to the supersymmetric Sawada-Kotera Equation (5.4).

16. *Supersymmetric 5th order KdV equation*

$$
\Phi_t = \left[\Phi_{4x} + 10\Phi_{2x}(\mathcal{D}\Phi) + 5\Phi_x(\mathcal{D}\Phi_x) + 15\Phi(\mathcal{D}\Phi)^2\right]_x - 15\Phi(\mathcal{D}\Phi)(\mathcal{D}\Phi_x). \quad (5.6)
$$

This equation, contained in [73], is not the next flow of the hierarchy of the supersymmetric KdV equation. A recursion operator and a Hamiltonian structure were obtained [73]. Very recently, a Lax pair was constructed for it [30].

17. *Supersymmetric 3rd order Burgers equation*

$$
\Psi_t = \Psi_{xxx} + b\Psi_x(\mathcal{D}\Psi_x) + b\Psi(\mathcal{D}\Psi_{xx}) + \frac{1}{3} b^2 \Psi(\mathcal{D}\Psi_x)(\mathcal{D}\Psi).
$$

This equation, having a 5th order symmetry, was found by means of symmetry approach in [73]. It is surprising that not much is known for this simple-looking equation.

18. *Supersymmetric 5th order MKdV equation*

$$
\begin{aligned}
\Phi_t = & \Phi_{5x} + 5\Phi_{3x}(\mathcal{D}\Phi_x) - 5\Phi_x(\mathcal{D}\Phi_{3x}) - 10\Phi_{3x}(\mathcal{D}\Phi)^2 - 15\Phi_{3x}\Phi_x\Phi \\
& - 15\Phi_{2x}(\mathcal{D}\Phi_x)(\mathcal{D}\Phi) - 10\Phi_x(\mathcal{D}\Phi_{2x})(\mathcal{D}\Phi) - 10\Phi_x(\mathcal{D}\Phi_x)^2 \\
& - 15\Phi(\mathcal{D}\Phi_{2x})(\mathcal{D}\Phi_x) - 15\Phi_{2x}\Phi_x(\mathcal{D}\Phi) + 15\Phi_x(\mathcal{D}\Phi_x)(\mathcal{D}\Phi)^2 \\
& - 15\Phi(\mathcal{D}\Phi_x)^2(\mathcal{D}\Phi) + 15\Phi_x(\mathcal{D}\Phi)^4 + 15\Phi(\mathcal{D}\Phi_x)(\mathcal{D}\Phi)^3,
\end{aligned}
$$

was found by Tian and Wang [77] in their classification study of supersymmetric integrable systems. It may be related to (5.6) via a Miura type transformation.

19. $N = 2$ *supersymmetric KdV equations*

$$
\phi_t = -\phi_{xxx} + 3(\phi\mathcal{D}_1\mathcal{D}_2\phi)_x + \frac{1}{2}(a-1)(\mathcal{D}_1\mathcal{D}_2\phi^2)_x + 3a\phi^2\phi_x
$$

where $\phi = \phi(t, x, \theta_1, \theta_2)$. This family of equations was introduced by Laberge and Mathieu [35, 36]. It is interesting to note that this family of equations is integrable for three values of a, namely $a = -2, 1, 4$ or SKdV$_{-2}$, SKdV$_1$, SKdV$_4$ (see [3, 35, 36, 64]). Hirota's bilinear method was applied so that both SKdV$_1$ and SKdV$_4$ are bilinearised [86]. Recently, we constructed the Darboux transformation for the SKdV$_{-2}$ equation [50].

20. *$N = 2$ supersymmetric Burgers equation*

$$\Phi_t = (\mathcal{D}_1\mathcal{D}_2\Phi)_x + \Phi\Phi_x$$

appeared in [31, 67]. This equation is actually the first flow of the SKdV$_4$ hierarchy (see also [87]).

21. *$N = 2$ supersymmetric Harry Dym equations*

Three such systems are known. The first two read as

$$\begin{aligned} W_t = \frac{1}{8}\big(&2W^3W_{xxx} - 6(\mathcal{D}_1\mathcal{D}_2W_x)(\mathcal{D}_1\mathcal{D}_2W)W^2 - 3(\mathcal{D}_2W_{xx})(\mathcal{D}_2W)W^2 \\ &-3(\mathcal{D}_1W_{xx})(\mathcal{D}_1W)W^2\big), \end{aligned}$$

and

$$\begin{aligned} W_t = \frac{1}{8}\big(&2W^3W_{xxx} - 3(\mathcal{D}_2W_{xx})(\mathcal{D}_2W)W^2 - 3(\mathcal{D}_1W_{xx})(\mathcal{D}_1W)W^2 \\ &+3(\mathcal{D}_2W)(\mathcal{D}_1W)(\mathcal{D}_1\mathcal{D}_2W_x)W\big). \end{aligned}$$

Both of them were obtained by Brunelli, Das and Popowicz [8] and were shown to be related with the $N = 2$ supersymmetric KdV equations [74].

A third $N = 2$ supersymmetric Harry Dym equation, proposed in [76], is given by

$$\begin{aligned} W_t = &-W^3W_{xxx} + \frac{3}{2}(\mathcal{D}_1\mathcal{D}_2W_x)(\mathcal{D}_1\mathcal{D}_2W)W^2 + \frac{3}{2}(\mathcal{D}_2W_{xx})(\mathcal{D}_2W)W^2 \\ &+\frac{3}{2}(\mathcal{D}_1W_{xx})(\mathcal{D}_1W)W^2 - \frac{3}{4}(\mathcal{D}_2W)(\mathcal{D}_1W)(\mathcal{D}_1\mathcal{D}_2W_x)W. \end{aligned}$$

Thus, there are three $N = 2$ supersymmetric Harry Dym equations related to the three $N = 2$ supersymmetric KdV equations, respectively.

22. *$N = 2$ supersymmetric nonlinear Schrödinger equation*

$$\begin{aligned} F_t &= F_{xx} - G^2F^3 + 2F(\mathcal{D}_1\mathcal{D}_2(GF)), \\ G_t &= G_{xx} + G^3F^2 - 2G(\mathcal{D}_1\mathcal{D}_2(GF)), \end{aligned}$$

which appeared in [65].

23. *$N = 2$ supersymmetric Boussinesq equation*

$$\begin{aligned} J_t &= 2T_x - \delta J_x + 4\alpha J J_x, \\ T_t &= -2J_{xxx} + \delta T_x - 20\alpha\partial(\bar{\mathcal{D}}J\mathcal{D}J) + 8\alpha J_x \delta J_x + 16\alpha^2 J^2 J_x \\ &\quad - 12\bar{\mathcal{D}}J\mathcal{D}T - 12\alpha\mathcal{D}J\bar{\mathcal{D}}T - 12\alpha J_x T - 4\alpha J T_x, \end{aligned}$$

where $\delta = [\mathcal{D}, \bar{\mathcal{D}}], \mathcal{D} = \partial_\theta - \frac{1}{2}\bar{\theta}\partial, \bar{\mathcal{D}} = \partial_{\bar{\theta}} - \frac{1}{2}\theta\partial$. For further properties of this equation, we refer to [2, 28, 85].

24. *$N = 2$ supersymmetric Camassa-Holm and Hunter-Saxton equations*

The family of the equations

$$M_t = -(MU)_x + \frac{1}{2}[(\mathcal{D}_1 M)(\mathcal{D}_1 U) + (\mathcal{D}_2 M)(\mathcal{D}_2 U)],$$

where $M = (i\mathcal{D}_1\mathcal{D}_2 - \gamma)U$ is a combination of the $N = 2$ supersymmetric Camassa-Holm equation and Hunter-Saxton equation. These equations were constructed by Popowicz [67] and studied by Lenells and Lechtenfeld [39].

25. *Supersymmetric KP hierarchy*

To define it, let (x, θ) be the superspace coordinate and $(t_1, \tau_1, t_2, \tau_2, ...)$ be the super time variables. The corresponding vector fields are given by

$$\partial_n = \frac{\partial}{\partial t_n}, \ D_n = \frac{\partial}{\partial \tau_n} - \sum_{m=1}^{\infty} \tau_m \frac{\partial}{\partial t_{n+m-1}}.$$

Then the supersymmetric KP hierarchy is represented as

$$\partial_n L = -[(L^{2n})_+, L], \ \ D_n L = -[L^{2n-1})_+, L] + 2L^{2n},$$

where $L = \mathcal{D} + \sum_{i=0}^{\infty} u_i \mathcal{D}^{-i}$. This hierarchy was introduced by Manin and Radul [49] and has been studied by Ueno and Yamada [78] and Mulase [59]. From this hierarchy, the following supersymmetric KP system may be obtained

$$\begin{aligned} 2u_{3,t_4} &= -3u_{3,xx} + 2v_{5,x}, \\ 3u_{4,t_4} &= -3u_{4,xx} + 6u_3 u_{3,x} - 4u_3 v_5 + 2v_{6x}, \\ v_{5,t_4} + 2u_{3,t_6} &= v_{5,xx} - 2u_{3,xxx} - 6u_3(\mathcal{D}u_{3,x}) - 6(u_3 u_4)_x - 2\mathcal{D}(u_3 v_5), \\ v_{6,t_4} + 2u_{4,t_6} &= v_{6,xx} + 2u_3(\mathcal{D}v_6) - 2u_{4,xxx} - 6u_3(\mathcal{D}u_{4,x}) - 6u_4 u_{4,x} + 2(\mathcal{D}u_4)v_5. \end{aligned}$$

26. *Supersymmetric Toda systems*

$$\mathcal{D}_t \mathcal{D}_x \Phi_i = -\exp\left(\sum_{j=0}^{r} K_{ji}\Phi_j\right),$$

where $K = (K_{ij})_{r\times r}$ is the Cartan matrix corresponding to certain Lie superalgebra (see [27, 37] and the references therein).

Acknowledgements

This work is supported by the National Natural Science Foundation of China (NNSFC) (Grant Nos. 11501312, 11775121, 1193101 and 11871471), Zhejiang Provincial Natural Science Foundation of China (Grant No. LQ15A010002), the Yue Qi Outstanding Scholar Project, China University of Mining & Technology, Beijing (Grant No. 00-800015Z1177), the K. C. Wong Magna Fund in Ningbo University, the Fundamental Research Funds for the Central Universities.

References

[1] Adler V E, Bobenko A I and Suris Yu B, Classification of integrable equations on quad-graphs. The consistency approach, *Commun. Math. Phys.*, **233**, 513-543, 2003.

[2] Belluci S, Ivanov E, Krivonos S and Pichugin A, $N = 2$ super Boussinesq hierarchy: Lax pairs and conservation laws, *Phys. Lett. B*, **312**, 467-470, 1993.

[3] Bourque S and Mathieu P, The Painlevé analysis for $N = 2$ super KdV equations, *J. Math. Phys.*, **42**, 3517-3529, 2001.

[4] Brunelli J C and Das A, The supersymmetric two boson hierarchies, *Phys. Lett. B*, **337**, 303-307, 1994.

[5] Brunelli J C and Das A, Properties of nonlocal charges in the supersymmetric two boson hierarchy, *Phys. Lett. B*, **354**, 307-314, 1994.

[6] Brunelli J C and Das A, Supersymmetric two-boson equation, its reductions and the nonstandard supersymmetric KP hierarchy, *Int. J. Mod. Phys. Lett. A*, **10**, 4563-4599, 1995.

[7] Brunelli J C and Das A, Tests of integrability of the supersymmetric nonlinear Schrödinger equation, *J. Math. Phys.*, **36**, 268-280, 1995.

[8] Brunelli J C, Das A and Popowicz Z, Supersymmetric extensions of the Harry Dym hierarchy, *J. Math. Phys.*, **44**, 4756-4767, 2004.

[9] Carstea A S, Extension of the bilinear formalism to supersymmetric KdV-type equations, *Nonlinearity*, **13**, 1645-1656, 2000.

[10] Carstea A S, Constructing soliton solutions and superbilinear form of lattice supersymmetric KdV equation, *J. Phys. A: Math. Theor.*, **48**, 285201, 2015.

[11] Carstea A S, Ramani A and Grammaticos B, Constructing the soliton solutions for the $N = 1$ supersymmetric KdV hierarchy, *Nonlinearity*, **14**, 1419-1423, 2001.

[12] Chaichian M and Kulish P P, On the method of inverse scattering problem and Bäcklund transformations for supersymmetric equations, *Phys. Lett. B*, **78**, 413-416, 1978.

[13] D'Auria R and Sciuto S, Group theoretical construction of two-dimensional supersymmetric models, *Nucl. Phys. B*, **171**, 189-208, 1980.

[14] Fan E, New bilinear Bäcklund transformation and Lax pair for the supersymmetric two-Boson equation, *Stud. Appl. Math.*, **127**, 284-301, 2011.

[15] Ferrara S, Girardello L and Sciuto S, An infinite set of conservation laws of the supersymmetric sine-Gordon theory, *Phys. Lett. B*, **76**, 303-306, 1978.

[16] Ferrara S, Fioresi R and Varadarajan V S, *Supersymmetry in Mathematics and Physics: UCLA Los Angeles, USA 2010*, Springer-Verlag, Berlin, Heidelberg, 2011.

[17] Figueroa-O'Farrill J M, Mas J and Ramos E, Bi-Hamiltonian structure of the supersymmetric SKdV hierarchy, *Rev. Math. Phys.*, **3**, 479, 1991.

[18] Gao X N, Lou S Y and Tang X Y, Bosonization, singularity analysis, nonlocal symmetry reductions and exact solutions of supersymmetric KdV equation, *JHEP*, **2013**, 29, 2013.

[19] Girardello L and Sciuto S, Inverse scattering-like problem for supersymmetric models, *Phys. Lett. B*, **77**, 267-269, 1978.

[20] Grammaticos B, Ramanai A and Carstea A S, Bilinearization and soliton solutions of the $N = 1$ supersymmetric sine-Gordon equation, *J. Phys. A: Math. Gen.*, **34**, 4881-4886, 2001.

[21] Gu C H, Hu H S and Zhou Z X, *Darboux Transformations in Integrable Systems: Theory and their Applications to Geometry*, Springer, Netherlands, 2005.

[22] Haider B and Hassan M, Quasideterminant multisoliton solutions of a supersymmetric chiral field model in two dimensions, *J. Phys. A: Math. Theor.*, **43**, 035204, 2010.

[23] Harnad J and Jacques M, The supersymmetric soliton correlation matrix, *J. Geom. Phys.*, **2**, 1-15, 1985.

[24] Hietarinta J, Joshi N and Nijhoff F W, *Discrete Systems and Integrability*, Cambridge University Press, Cambridge, 2016.

[25] Hlavatý L, The Painlevé analysis of fermionic extensions of KdV and Burgers equations, *Phys. Lett. A*, **137**, 173-178, 1989.

[26] Hruby J, On the supersymmetric sine-Gordon model and a two-dimensional bag, *Nucl. Phys. B*, **131**, 275-284, 1977.

[27] Inami T and Kanno H, Lie superalgebraic approach to super Toda lattice and generalized super KdV equations, *Commun. Math. Phys.*, **136**, 519-542, 1991.

[28] Ivanov E and Krivonos S, Superfield realizations of $N = 2$ super-W_3, *Phys. Lett. B*, **291**, 63-70, 1992.

[29] Kac M and van Moerbeke P, On an explicitly soluble system on nonlinear differential equations related to certain Toda lattices, *Adv. Math.*, **16**, 160-169, 1975.

[30] Kiselev A V and Krutov A O, On the (non)removability of spectral parameters in Z_2-graded zero-curvature representations and its applications, *Acta Appl. Math.*, **160**, 129-167, 2019.

[31] Kiselev A V and Wolf T, Supersymmetric representations and integrable fermionic extensions of the Burgers and Boussinesq equations, *SIGMA*, **2**, 30 (19pp), 2006.

[32] Krasil'shchik I S and Kersten P H M, *Symmetries and Recursion Operators for Classical and Supersymmetric Differential Equations*, Springer, Dordrecht, 2000.

[33] Krivonos S, Pashev A and Popowicz Z, Lax pairs for $N = 2, 3$ supersymmetric KdV equations and their extensions, *Mod. Phys. Lett. A*, **13**, 1435-1443, 1998.

[34] Kupershmidt B A, A super Korteweg-de Vries equation: an integrable system, *Phys. Lett. A*, **104**, 213-215, 1984.

[35] Laberge C -A and Mathieu P, $N = 2$ superconformal algebra and integrable $O(2)$ fermionic extensions of the Korteweg-de Vries equation, *Phys. Lett. B*, **215**, 718-722, 1988.

[36] Laberge C -A and Mathieu P, A new $N = 2$ supersymmetric Korteweg-de Vries equation, *J. Math. Phys.*, **32**, 923-927, 1991.

[37] Leites D A, Saveliev M V and Serganova V V, Embedding of Lie superalgebra OSp(1/2) and the associated nonlinear supersymmetric equations. *Preprint*, IHEP85-81, 1985.

[38] Lenells J, A bi-Hamiltonian supersymmetric geodesic equation, *Lett. Math. Phys.*, **85**, 55-63, 2008.

[39] Lenells J and Lechtenfeld O, On the N=2 supersymmetric Camassa-Holm and Hunter-Saxton equations, *J. Math. Phys.*, **50**, 012704 (17pp), 2009.

[40] Liu Q P, Darboux transformations for supersymmetric Korteweg-de Vries equations, *Lett. Math. Phys.*, **35**, 115-122, 1995.

[41] Liu Q P, Supersymmetric Harry Dym type equations, *J. Phys. A: Math. Gen.*, **28**, L245-L248, 1995.

[42] Liu Q P and Hu X B, Bilinearization of $N = 1$ supersymmetric Korteweg-de Vries equation revisited, *J. Phys. A: Math. Gen.*, **38**, 6371-6378, 2005.

[43] Liu Q P, Hu X B and Zhang M X, Supersymmetric modified Korteweg-de Vries equation: bilinear approach, *Nonlinearity*, **18**, 1597-1603, 2005.

[44] Liu Q P and Manas M, Crum transformation and Wronskian type solutions for the supersymmetric KdV equation, *Phys. Lett. B*, **396**, 133-140, 1997.

[45] Liu Q P, Popowicz Z and Tian K, Supersymmetric reciprocal transformation and its applications, *J. Math. Phys.*, **51**, 093511 (24pp), 2010.

[46] Liu Q P, Popowicz Z and Tian K, The even and odd supersymmetric Hunter - Saxton and Liouville equations, *Phys. Lett. A*, **375**, 25-29, 2010.

[47] Liu Q P and Xie Y F, Nonlinear superposition formula for $N = 1$ supersymmetric KdV equation, *Phys. Lett. A*, **325**, 139-143, 2004.

[48] Liu Q P and Yang X-Y, Supersymmetric two-boson equation: Bilinearization and solutions, *Phys. Lett. A*, **351**, 131-135, 2006.

[49] Manin Yu I and Radul A O, A supersymmetric extension of the Kadomtsev-Petviashvili hierarchy, *Commun. Math. Phys.*, **98**, 65-77, 1985.

[50] Mao H and Liu Q P, Bäcklund-Darboux transformations and discretizations of $N = 2$ $a = -2$ supersymmetric KdV equation, *Phys. Lett. A*, **382,** 253-258, 2018.

[51] Mao H, Liu Q P and Xue L, Supersymmetric Sawada-Kotera equation: Bäcklund-Darboux transformations and applications, *J. Nonlinear Math. Phys.*, **25**, 375-386, 2018.

[52] Mathieu P, Supersymmetric extension of Korteweg-de Vries equation, *J. Math. Phys.*, **29**, 2499-2506, 1988.

[53] Mathieu P, The Painlevé property for fermionic extensions of the Korteweg-de Vries equation, *Phys. Lett. A*, **128**, 169-171, 1988.

[54] Mathieu P, in *Integrable and Superintegrable Systems*, Kupershmidt B A (Ed), World Scientific, Singapore, 1990.

[55] Matveev V B and Salle M A, *Darboux Transformations and Solitons*, Springer-Verlag, Berlin, Heidelberg, 1991.

[56] McArthur I and Yung C M, Hirota bilinear form for the super-KdV hierarchy, *Mod. Phys. Lett. A*, **8**, 1739-1745, 1993.

[57] Mikhailov A V, Integrability of supersymmetrical generalizations of classical chiral models in two-dimensional space time, *JETP Lett.*, **28**, 512-516, 1978.

[58] Morosi C and Pizzocchero L, $OSP(3,2)$ and $GL(3,3)$ supersymmetric KdV hierarchies, *Phys. Lett. A*, **185**, 241-259, 1994.

[59] Mulase M, Solvability of the super KP equation and a generalization of the Birkhoff decomposition, *Invent. Math.*, **92**, 1-46, 1988.

[60] Muller-Kirsten H J W and Wiedemann A, *Introduction to Supersymmetry*, World Scientific, Singapore, 2010.

[61] Nijhoff F W and Capel H W, The discrete Korteweg-de Vries equation *Acta Appl. Math.*, **39**, 133-158, 1995.

[62] Niu X, Liu Q P and Xue L, Darboux transformations for supersymmetric two-boson equation, *preprint*, 2017.

[63] Oevel W and Popowicz Z, The bi-Hamiltonian structure of fully supersymmetric Korteweg-de Vries systems, *Commun. Math. Phys.*, **139**, 441-460, 1991.

[64] Popowicz Z, The Lax formulation of the "new" $N = 2$ SUSY KdV equation, *Phys. Lett. A*, **174**, 411-415, 1993.

[65] Popowicz Z, The extended supersymmetrization of the nonlinear Schrödinger equation, *Phys. Lett. A*, **194**, 375-379, 1994.

[66] Popowicz Z, Odd Hamiltonian structure for supersymmetric Sawada-Kotera equation, *Phys. Lett. A*, **373**, 3315-3323, 2009.

[67] Popowicz Z, A 2-component or $N = 2$ supersymmetric Camassa-Holm equation, *Phys. Lett. A*, **354**, 110-114, 2006.

[68] Rogers A, *Supermanifolds: Theory and Applications*, World Scientific, Singapore, 2007.

[69] Rogers C and Schief W, *Bäcklund and Darboux Transformations: Geometry and Modern Applications in Soliton Theory*, Cambridge University Press, Cambridge, 2002.

[70] Roelofs G H M and Kersten P H M, Supersymmetric extensions of the nonlinear Schrödinger equation: symmetries and coverings, *J. Math. Phys.*, **33**, 2185-2206, 1992.

[71] Siddiq M, Hassan M and Saleem U, On Darboux transformation of the supersymmetric sine-Gordon equation, *J. Phys. A: Math. Gen.*, **39**, 7313-7318, 2006.

[72] Tian K and Liu Q P, A supersymmetric Sawada-Kotera equation, *Phys. Lett. A*, **373**, 1807-1810, 2009.

[73] Tian K and Liu Q P, Supersymmetric fifth order evolution equations. in *Nonlinear and Modern Mathematical Physics*, AIP conference proceedings, **1212**, 81-88, 2010.

[74] Tian K and Liu Q P, The transformation between $N = 2$ supersymmetric Korteweg-de Vries and Harry Dym equations, *J. Math. Phys.*, **53**, 053503, 2012.

[75] Tian K and Liu Q P, Two new supersymmetric equations of Harry Dym type and their supersymmetric reciprocal transformations, *Phys. Lett. A*, **376**, 2334-2340, 2012.

[76] Tian K, Popowicz Z and Liu Q P, A non-standard Lax formulation of the Harry Dym hierarchy and its supersymmetric extension, *J. Phys. A: Math. Theor.*, **45**, 122001 (8pp), 2012.

[77] Tian K and Wang J P, Symbolic representation and classification of $N = 1$ supersymmetric evolutionary equations, *Stud. Appl. Math.*, **138**, 467-498, 2017.

[78] Ueno K and Yamada H, Supersymmetric extension of the Kadomtsev-Petviashvili hierarchy and the universal super Grassmann manifold, *Advanced Studies in Pure Math.*, **16**, 373-426, 1988.

[79] Vecchia P D and Ferrara S, Classical solutions in two-dimensional supersymmetric field theories, *Nucl. Phys. B*, **130**, 93-104, 1977.

[80] Wegner F, *Supermathematics and its Applications in Statistical Physics*, Springer, Heidelberg, 2016.

[81] Xue L, Levi D and Liu Q P, Supersymmetric KdV equation: Darboux transformation and discrete systems, *J. Phys. A: Math. Theor.*, **46**, 502001 (11pp) 2013.

[82] Xue L and Liu Q P, On Bäcklund transformation for supersymmetric two-boson equation, *Phys. Lett. A*, **377**, 828-832, 2013.

[83] Xue L and Liu Q P, A supersymmetric AKNS problem and its Darboux-Backlund transformations and discrete systems, *Stud. Appl. Math.*, **135**, 35-62, 2015.

[84] Yamanaka I and Sasaki R, Super Virasoro algebra and solvable supersymmetric quantum field theories, *Prog. Theor. Phys.*, **79**, 1167-1184, 1988.

[85] Yung C M, The $N = 2$ supersymmetric Boussinesq hierarchies, *Phys. Lett. B*, **309**, 75-84, 1993.

[86] Zhang M X, Liu Q P, Shen Y L and Wu K, Bilinear approach to $N = 2$ supersymmetric KdV equations, *Sci. China Ser. A*, **52**, 1973-1981, 2009.

[87] Zuo D, Supersymmetric Euler equations associated to the $N \leq 3$ Neveu-Schwarz algebra, *J. Math. Phys.*, **60**, 123505 (11p), 2019.

A6. Nonlocal symmetries of nonlinear integrable systems

S Y Lou

School of Physical Science and Technology, Ningbo University, Ningbo, 315211, P.R. China

Abstract

In additional to local symmetries, there are infinitely many nonlocal symmetries for nonlinear integrable systems. The nonlocal symmetries may be potential symmetries, residual symmetries, square eigenfunction symmetries, infinitesimal Darboux and Bäcklund transformations and so on. Some types of nonlocal symmetries may be localized to find related group invariant solutions and finite Darboux transformations. In (1+1)-dimensions, the local symmetries correspond to the positive integrable hierarchies, while the nonlocal symmetries are related to the negative integrable hierarchies. Usually, there are more negative hierarchies than positive hierarchies. For instance, there is no even order positive Korteweg-de Vries (KdV) hierarchy, while both the even order and the odd order KdV hierarchies do exist. In (2+1)-dimensions, the nonlocal potential symmetries are responsible for the positive integrable hierarchy, while the square eigenfunction related to nonlocal symmetries are responsible for the negative hierarchies. Nonlocal symmetries can be used to nonlinearize linear Lax pairs to find new nonlinear integrable hierarchies via applying the nonlinearization approach, introducing self-consistent sources, and using fermionic nonlocal symmetries. It is also shown that the reciprocal transformations must be companied to study the properties and exact solutions of the Camassa-Holm type systems.

1 Introduction

Symmetry study is fundamental to find or establish universal models like the standard model [8] in particle physics. There are some types of symmetry methods to solve complicated nonlinear physical problems. The symmetry approach is more attractive in the study of integrable systems because of the existence of infinitely many local and nonlocal symmetries [1, 29]. Local symmetries are widely used to get symmetry invariant solutions, to reduce the dimensions of partial differential equations and to find new integrable systems. Recently, it is found that nonlocal symmetries are also very useful to find exact novel types of exact solutions, integrable models, and the relations among different types of integrable properties.

In Section 2 of this chapter, we review four types of methods to find nonlocal symmetries including the eigenvalue problem method for recursion operators, the truncated Painlevé approach, the infinitesimal forms of the finite transformations (like the Darboux transformations and Bäcklund transformations) and the formal series symmetry method for high dimensional integrable systems. In Section 3, the nonlocal symmetries are used to find some types of integrable systems via the nonlinearization procedure for the linear Lax pairs. The most generalization of symmetry constraints can be applied in some different ways including to find general group invariants of the original models, to be considered as a general mixing of the members of positive and negative hierarchies, to find first and second types of integrable models with self-consistent sources and so on. In Section 4, two types of integrable models with bosonic and fermionic sources are discussed. In Section 5, the nonlocal symmetries are used to find Darboux transformations and the group invariant solutions by using localization approach. The localization procedure of nonlocal symmetries for the Camassa-Holm type integrable systems are also discussed. The last section offers a short summary and some discussions.

2 Nonlocal symmetries in integrable systems

2.1 Nonlocal symmetries from eigenvalue problems of recursion operators

In (1+1)-dimensional cases, an integrable system possesses a recursion operator [29], ϕ, which maps one symmetry, σ_0, to a set of infinitely many symmetries, $\sigma_n = \phi^n\sigma_0$, $n = 0,\ 1,\ 2,\ \ldots,\ \infty$. For instance, for the Kortweg-de Vries (KdV) equation

$$u_t = u_{xxx} + 6uu_x, \tag{2.1}$$

the recursion operator possesses the form

$$\phi = \partial_x^2 + 4u + 2u_x\partial_x^{-1}, \quad \partial_x\partial_x^{-1} = \partial_x^{-1}\partial_x = 1. \tag{2.2}$$

Thus, we have a positive integrable KdV hierarchy

$$u_{t_{2n+1}} = \phi^n u_x \equiv K_{2n+1}, \quad n = 0,\ 1,\ 2,\ \ldots,\ \infty, \tag{2.3}$$

where K_{2n+1} are all local symmetries of the KdV equation.

A symmetry (σ) of the KdV equation is a solution of its linearized equation

$$\sigma_t = \sigma_{xxx} + 6\sigma u_x + 6u\sigma_x. \tag{2.4}$$

It is known that the eigenvalue problem of the recursion operator

$$\phi_\lambda\sigma \equiv (\phi - 4\lambda)\sigma = 0 \tag{2.5}$$

and the symmetry Equation (2.4) constitute a Lax pair of the KdV equation. In fact, ϕ_λ defined in (2.5) is still a recursion operator. The eigenvalue problem (2.5) can be formally solved by means of the usual Lax pair and the kernels of ϕ_λ [18]. The final result reads

$$\sigma = \phi_\lambda^{-1}0 = \partial_x\psi^2\partial_x^{-1}\psi^{-2}\partial_x^{-1}\psi^{-2}\partial_x^{-1}\psi^2 0 = c_1(\psi^2)_x + c_2(\psi\psi_1)_x + c_3(\psi_1^2)_x, \tag{2.6}$$

where $\psi_1 \equiv \psi\partial_x^{-1}\psi^{-2}$ and ψ is the spectral function of the Lax pair

$$\psi_{xx} + u\psi = \lambda\psi, \tag{2.7a}$$

$$\psi_t = 4\psi_{xxx} + 6u\psi_x + 3u_x\psi. \tag{2.7b}$$

In fact, both ψ and ψ_1 are solutions of the Lax pair (2.7). Because $(\psi^2)_x$, $(\psi\psi_1)_x$ and $(\psi_1^2)_x$ are all kernels of the recursion operator ϕ_λ, one can directly verify that they are all nonlocal symmetries and $(\psi^2)_x$ (or equivalently $(\psi_1^2)_x$) is just the known square eigenfunction symmetry of the KdV equation. Thus, we have two sets of independent nonlocal symmetries or two negative integrable hierarchies

$$u_{t_{-2n-1}} = \phi_\lambda^{-n}(\psi^2)_x \equiv K_{-2n-1}, \quad n = 0,\ 1,\ 2,\ \ldots,\ \infty, \tag{2.8a}$$
$$u_{t_{-2n-2}} = \phi_\lambda^{-n}(\psi\psi_1)_x \equiv K_{-2n-2}, \quad n = 0,\ 1,\ 2,\ \ldots,\ \infty. \tag{2.8b}$$

It should be mentioned that all the models in the positive KdV hierarchy (2.3) and the negative KdV hierarchy (2.8a) are odd order systems. However, all the models in the negative hierarchy (2.8b) are even order equations.

2.2 Nonlocal symmetries from truncated Painlevé expansions

For the Painlevé integrable systems [5], the nonlocal symmetries can be simply obtained by using the truncated Painlevé expansions.

For the KdV Equation (2.1), the truncated Painlevé expansion possesses the form

$$u = \frac{u_0}{f^2} + \frac{u_1}{f} + u_2. \tag{2.9}$$

Substituting (2.9) into (2.1), we have

$$\begin{aligned} &u_{2t} - u_{2xxx} - 6u_2u_{2x} + (u_{1t} - u_{1xxx} - 6u_1u_{2x} - 6u_2u_{1x})f^{-1} + \ldots \\ &+[2u_0(6u_2f_x + f_{xxx} - f_t) - 6f_x(u_1f_x)_x + 6u_1^2f_x + 6(u_{0x}f_x - u_0u_1)_x]f^{-3} \\ &+6[u_1f_x^3 - 3f_x(u_0f_x)_x + 3u_0u_1f_x - u_0u_{0x}]f^{-4} + 12u_0f_x(u_0 + 2f_x^2)f^{-5} = 0. \end{aligned} \tag{2.10}$$

Vanishing the coefficients of f^{-n}, $n = 0,\ 1,\ \ldots,\ 5$ for Eq. (2.10), we have

$$u_0 = -2f_x^2,\ u_1 = 2f_{xx},\ u_2 = \lambda - \frac{1}{2}\frac{f_{xxx}}{f_x} + \frac{1}{4}\frac{f_{xx}^2}{f_x^2}, \tag{2.11}$$

while f is a solution of the Schwarz KdV equation

$$f_t = 6\lambda f_x + f_{xxx} - \frac{3}{2}f_{xx}^2f_x^{-1}. \tag{2.12}$$

From the coefficients of f^0 and f^{-1} in (2.10), we know that u_2 is also a solution of the KdV Equation (2.1) and u_1 is a symmetry related to the solution u_2. Thus,

$$\sigma = f_{xx} \tag{2.13}$$

with $u = \lambda - \frac{1}{2}f_{xxx}f_x^{-1} + \frac{1}{4}f_{xx}^2f_x^{-2}$ is a nonlocal symmetry (residual symmetry [7, 13]) of the KdV equation where λ is just the spectral parameter. The residual symmetry f_{xx} is related to the square eigenfunction symmetry by $f_x \sim \psi^2$.

2.3 Nonlocal symmetries from Darboux transformations

Darboux transformations (DT) and Bäcklund transformations (BT) exactly transform one solution to another, while a symmetry approximately transforms one solution to another. Thus, one can naturally find some kinds of nonlocal symmetries from DT [24] and BT [25] by taking some proper constants as infinitesimal parameters.

For the Kadomtsev-Petviashvili (KP) equation

$$(u_t - 6uu_x - u_{xxx})_x - 3\gamma^2 u_{yy} = 0 \tag{2.14}$$

with the Lax pair

$$\gamma\psi_y = \psi_{xx} + u\psi, \tag{2.15a}$$
$$\psi_t = 4\psi_{xxx} + 6u\psi_x + 3[u_x + \gamma(\partial_x^{-1} u_y)]\psi, \tag{2.15b}$$

and the dual Lax pair

$$\gamma\psi_y^* = -\psi_{xx}^* - u\psi^*, \tag{2.16a}$$
$$\psi_t^* = 4\psi_{xxx}^* + 6u\psi_x^* + 3[u_x - \gamma(\partial_x^{-1} u_y)]\psi^*, \tag{2.16b}$$

the DT theorem reads:
Theorem 1 [27, 23]. If u is a solution of the KP Equation (2.14), and ψ is a spectral function of the Lax pair (2.15), then $\bar{u} = u + 2(\ln\psi)_{xx}$ is also a solution of the KP equation.

In [23], we had also proved that the infinitesimal form of the DT Theorem 1 is just the square eigenfunction symmetry theorem.
Theorem 2 [23]. If ψ and ψ^* are the solutions of the Lax pair (2.15) and the dual Lax pair (2.16), then

$$\sigma = (\psi\psi^*)_x \tag{2.17}$$

is a nonlocal symmetry of the KP Equation (2.14).

The square eigenfunction symmetry (2.17) is also equivalent to the residual symmetry f_{xx} with f being a solution of the Schwarz KP equation. Though there is no similar recursion operator for the KP equation, the whole set of nonlocal symmetries for the KP Equation (2.14) can also be obtained by means of the Lax and dual Lax operators [19]. The more general nonlocal symmetries of the KP equation have been investigated for the bilinear KP equation [9].

2.4 Nonlocal symmetries and formal series symmetries

In (2+1)-dimensions, the symmetries related to the positive hierarchies are also nonlocal. Though there are no proper recursion operators to get infinitely many symmetries related to (2+1)-dimensional positive hierarchies, one can really find a whole set of nonlocal symmetries for many integrable systems including the KP equation [15], Toda equation [16], Nizhnik-Novikov-Veselov equation [17] and so on by means of the formal series symmetry approach.

For instance, for the KP Equation (2.14), the positive hierarchy can be written as

$$u_{t_n} = \frac{1}{2(n-1)!3^n} L^n y^{n-1} \equiv K_n, \; n = 1,\, 2,\, \ldots,\, \infty. \tag{2.18}$$
$$L \equiv \partial_x^3 + 6\partial_x u + 3\gamma^2 \partial_x^{-1}\partial_y^2 - \partial_t.$$

Except for $K_1 = u_x$ and $K_2 = 2\gamma^2 u_y$, all K_n for $n > 2$ are nonlocal. Here are the simplest two special cases,

$$K_3 = \gamma^2(3\gamma^2\partial_x^{-1}u_{yy} + \phi u_x), \tag{2.19a}$$
$$K_4 = 4\gamma^4(\gamma^2\partial_x^{-2}u_{yyy} + \phi u_y), \tag{2.19b}$$

where ϕ is just the recursion operator of the KdV equation defined by (2.2). From Eq. (2.19b), one can find an interesting fact that the fourth equation of the KP hierarchy (2.18) is a generalization of the breaking soliton equation [2].

3 Nonlocal symmetries and nonlinearizations

The nonlinearization of the linear Lax pair was firstly introduced and applied to find algebro-geometric solutions and integrable models by Cao [3]. Cheng and Li called the method as the symmetry constrained method. In fact, the method can be used to find various new integrable systems.

In this subsection, we just write down a special type of symmetry constraints on the KP Equation (2.14). Let ψ_m, $m = 1,\, 2,\, \ldots,\, M$ are solutions of the Lax pair (2.15) and ψ_n^*, $n = 1,\, 2,\, \ldots,\, N$ are solutions of the dual Lax pair (2.16), then $(\psi_m\psi_n^*)_x$ are all nonlocal symmetries of the KP Equation (2.14) for arbitrary m and n. Thus, applying the symmetry constrained condition

$$u_z = \sum_{m=1}^{M}\sum_{n=1}^{N} a_{mn}(\psi_m\psi_n^*)_x \tag{3.1}$$

to the space parts of the Lax pair (2.15a) and the dual Lax pair (2.16a), we get the following (2+1)-dimensional integrable (M+N)-component coupled Ablowitz-Kaup-Newell-Segue (AKNS) systems [24]

$$\gamma\psi_{iy} = \psi_{ixx} + \psi_i \sum_{m=1}^{M}\sum_{n=1}^{N} a_{mn}\partial_z^{-1}(\psi_m\psi_n^*)_x, \; i = 1,\, 2,\, \ldots,\, M, \tag{3.2a}$$

$$-\gamma\psi_{jy}^* = \psi_{jxx}^* + \psi_j^* \sum_{m=1}^{M}\sum_{n=1}^{N} a_{mn}\partial_z^{-1}(\psi_m\psi_n^*)_x, \; j = 1,\, 2,\, \ldots,\, N. \tag{3.2b}$$

The model (3.2) contains various well-known integrable situations. When $M = N = 1$, $z = x$, $\gamma = \mathrm{i} = \sqrt{-1}$, $y \to t$, the model is just the usual AKNS system with many special reductions such as the nonlinear Schrödinger (NLS) equations and two-place nonlocal NLS reductions [21, 22]. When $M = N = 2$, $z = x$, $\gamma = \mathrm{i} = \sqrt{-1}$, $y \to t$, the model (3.2) becomes a generalized integrable two-component AKNS system with many interesting reductions including the generalized Manakov system [26] and two-place nonlocal Manakov systems, two-place and four place nonlocal NLS equations [21, 22]. When $z = y \to t$, the model is the generalized multi-component Yajima-Oikawa (YO) equations. For general case ($y \to t,\ z \to y$), the model (3.2) is a multi-component generalized asymmetric AKNS systems including the asymmetric Davey-Stewartson (ADS) system, the long-wave short-wave interaction model, the Maccari system and so on.

Applying the symmetry constrained condition (3.1) to the time parts of the Lax pair (2.15b) and the dual Lax pair (2.16b), we get the following (2+1)-dimensional integrable (M+N)-component higher order AKNS systems ($i = 1,\ 2,\ \ldots,\ M,\ j = 1,\ 2,\ \ldots,\ N$)

$$\psi_{it} = -4\psi_{ixxx} + 6\sum_{m=1}^{M}\sum_{n=1}^{N} a_{mn}[\psi_{ix}\partial_z^{-1}(\psi_m\psi_n^*)_x + \psi_i\partial_z^{-1}(\psi_{mx}\psi_n^*)_x], \tag{3.3a}$$

$$\psi_{jt}^* = -4\psi_{jxxx}^* + 6\sum_{m=1}^{M}\sum_{n=1}^{N} a_{mn}[\psi_{jx}^*\partial_z^{-1}(\psi_m\psi_n^*)_x + \psi_j^*\partial_z^{-1}(\psi_m\psi_{nx}^*)_x]. \tag{3.3b}$$

The (M+N)-component higher order AKNS system (3.3) includes also many interesting models such as the generalizations of the multi-component asymmetric NNV (ANNV) systems, the multi-component modified ANNV systems, the multi-component higher order long-wave sort wave interaction models and the multi-component higher order YO systems.

4 Nonlocal symmetries and self-consistent sources

4.1 Integrable systems with self-consistent sources

It is natural that the symmetry constraint condition (3.1) can be extended to

$$u_z + \sum_{p=1}^{P} a_p K_p = \sum_{m=1}^{M}\sum_{n=1}^{N} a_{mn}(\psi_m\psi_n^*)_x \tag{4.1}$$

with K_p being defined by (2.18) and

$$\gamma\psi_{my} = \psi_{mxx} + u\psi_m, \tag{4.2a}$$

$$\gamma\psi_{ny}^* = -\psi_{nxx}^* - u\psi_n^*, \tag{4.2b}$$

$$\psi_{mt} = 4\psi_{mxxx} + 6u\psi_{mx} + 3[u_x + \gamma(\partial_x^{-1}u_y)]\psi_m. \tag{4.3a}$$

$$\psi_{nt}^* = 4\psi_{nxxx}^* + 6u\psi_{nx}^* + 3[u_x - \gamma(\partial_x^{-1}u_y)]\psi_n^* \tag{4.3b}$$

The integrability of the system (2.14), (4.1), (4.2) and (4.3) is trivial because it is only a symmetry reduction of the KP Equation (2.14). If $z = t$, then the system

(4.1), (4.2) and (4.3) is integrable because it is a combination of the integrable positive KP hierarchy and negative KP hierarchy.

It is interesting that the system (4.1) and (4.2) is still integrable which is a generalization of the integrable KP equation ($a_3 \neq 0$) with a first type of self-consistent sources. The models (4.1) and (4.2) include some physically meaningful models such as the DS III models [4], the Manakov systems, the YO systems [30] and the Melnikov systems [28]. When $z = t$, the system (4.1) and (4.3) is also integrable which can be considered as a generalization of the integrable KP equation ($a_3 \neq 0$) with a second type of self-consistent sources.

For $z \neq t$, the integrability of the (3+1)-dimensional system (4.1) and (4.3) is open for any parameters though it is an interesting problem.

4.2 Super-integrable systems as integrable models with self-consistent fermionic sources

Because the Lax pairs are linear for the spectral functions, we can extend the spectral functions as fermion fields. For the KdV Equation (2.1), we rewrite the Lax pair (2.7) in fermionic form

$$\xi_{xx} + u\xi = \lambda\xi, \tag{4.4a}$$
$$\xi_t = 4\xi_{xxx} + 6u\xi_x + 3u_x\xi, \tag{4.4b}$$

where ξ is a fermion field. It is easy to verify that $\xi\xi_{xx}$ is a nonlocal symmetry of the KdV Equation (2.1). Thus, a nontrivial integrable KdV equation with a self-consistent fermion source becomes

$$u_t = u_{xxx} + 6uu_x + \xi\xi_{xx}, \tag{4.5a}$$
$$\xi_t = 4\xi_{xxx} + 6u\xi_x + 3u_x\xi, \tag{4.5b}$$

which is just the so-called super-KdV (or Kuper-KdV) equation proposed by Kupershmidt [11].

The most general integrable KdV equations with fermionic and bosonic sources possess the form

$$u_t = u_{xxx} + 6uu_x + \sum_{p=1}^{P} a_p\xi_p\xi_{pxx} + \sum_{m=1}^{M}\sum_{n=1}^{N} a_{mn}(\psi_m\phi_n)_x, \tag{4.6a}$$
$$\xi_{pt} = 4\xi_{pxxx} + 6u\xi_{px} + 3u_x\xi_p,\ p = 1,\ 2,\ \ldots,\ P, \tag{4.6b}$$
$$\psi_{mt} = 4\psi_{mxxx} + 6u\psi_{mx} + 3u_x\psi_m,\ m = 1,\ 2,\ \ldots,\ M, \tag{4.6c}$$
$$\phi_{nt} = 4\phi_{nxxx} + 6u\phi_{nx} + 3u_x\phi_n,\ n = 1,\ 2,\ \ldots,\ N, \tag{4.6d}$$

where ξ_p are fermion fields and ψ_m and ϕ_n are boson fields. The model (4.6) is Lax integrable and possesses bi-Hamiltonian.

5 Localizations of nonlocal symmetries

5.1 Darboux transformations and symmetry reductions related to nonlocal symmetries

In addition to integrable systems from nonlocal symmetry constraints, one can also find many kinds of exact solutions especially the interaction solutions among different types of nonlinear waves such as the solitary waves, periodic waves, Painlevé waves and so on by using the localization procedure of nonlocal symmetries.

For the KdV equation, one can readily prove that the nonlocal symmetry

$$V = 2\sum_{i=1}^{n} c_i f_{ixx}\partial_u, \tag{5.1}$$

in its vector form can be localized to

$$\begin{aligned} V &= 2\sum_{i=1}^{n} c_i h_i \partial_u - \sum_{i=1}^{n}\left[c_i f_i^2 + \sum_{j\neq i}^{n} \frac{c_j}{4}\frac{(g_i h_j - h_i g_j)^2}{g_i g_j(\lambda_i-\lambda_j)^2}\right]\partial_{f_i} \\ &\quad - \sum_{i=1}^{n}\left[2c_i g_i f_i + \sum_{j\neq i}^{n} c_j \frac{g_i h_j - g_j h_i}{\lambda_i - \lambda_j}\right]\partial_{g_i} \\ &\quad - \sum_{i=1}^{n}\left\{2c_i(h_i f_i + g_i^2) + \sum_{j\neq i}^{n}\frac{c_j}{2}\left[4g_i g_j + \frac{g_i^2 h_j^2 - g_j^2 h_i^2}{g_i g_j(\lambda_i - \lambda_j)}\right]\right\}\partial_{h_i} \end{aligned} \tag{5.2}$$

where

$$u = \lambda_i - \frac{1}{2}\frac{f_{ixxx}}{f_{ix}} + \frac{1}{4}\frac{f_{ixx}^2}{f_{ix}^{-2}}, \quad g_i = f_{ix}, \quad h_i = g_{ix}, \quad i = 1,\ 2,\ \ldots,\ n. \tag{5.3}$$

In other words, the symmetry (5.1) is nonlocal for the single KdV Equation (2.1); however, it is local for the extended system (2.1) and (5.3).

The finite transformation of the symmetry (5.2) can be obtained by the standard Lie approach. The result can be summarized in the following theorem:

Theorem 3 [13]. If $\{u,\ f_i,\ g_i,\ h_i\}$ is a solution of the prolonged KdV system (2.1) and (5.3), so is $\{\bar{u} = u + 2(\ln\Delta)_{xx},\ \bar{f_i} = -\Delta_i/\Delta,\ \bar{g}_i = \bar{f}_{ix},\ \bar{h}_i = \bar{g}_{ix}\}$ with $\Delta = \det(M)$, $M_{ji} = \epsilon c_i w_{ji}$, $j \neq i$, $M_{ii} = \epsilon c_i f_i - 1$ and $\Delta_k = \det(M_k)$, $M_{k,ji} = \epsilon c_i w_{ji}$, $i \neq j,\ k$, $M_{k,jk} = w_{jk}$, $j \neq k$, $M_{k,ii} = \epsilon c_i f_i - 1$, $i \neq k$, $M_{k,kk} = f_k$.

For the square eigenfunction symmetries of the KdV Equation (2.1), there are equivalent localization results [12].

Combining the local and nonlocal symmetries and using the localization procedure for the nonlocal symmetries, one can find some interesting interaction solutions among different types of nonlinear waves. For simplicity, we consider only one nonlocal symmetry combined with local Lie point symmetries for the KdV equation in the vector form,

$$V = X\partial_x + T\partial_t + U\partial_u + P\partial_\psi + Q\partial_{\psi_1} + F\partial_f, \tag{5.4}$$

where $X = c_1(x-12\lambda t)+c_4$, $T = 3c_1t+c_2$, $U = 4c_3\psi\psi_1-2c_1(u-\lambda)$, $P = c_3f\psi+c_5\psi$, $Q = c_3(\psi^3 + f\psi_1) + (c_5 - c_1)\psi_1$, $F = c_3f^2 + (c_1 + 2c_5)f + c_6$, ψ is the spectral function of the Lax pair, $\psi_1 = \psi_x$, $f_x = \psi^2$ and $c_1, c_2, \ldots, c_6$ are arbitrary constants. Using the standard Lie point symmetry method for the symmetry (5.4), many interaction solutions among solitons, Painlevé II waves (including rational waves, special Bessel and/or Airy function waves) and cnoidal periodic waves have been explicitly obtained [10]. Here, we only write down a special interaction solution between a soliton and a cnoidal wave for the field u:

$$u = -2U^2 + \frac{C_2}{2k_2}U - \frac{C_1 - \omega_2}{6k_2} + 2k_2U_X \tanh(\xi) - 2U^2 \tanh(\xi)^2 \tag{5.5}$$

where the cnoidal wave U is determined by the elliptic integral

$$\int \frac{\mathrm{d}U}{\sqrt{4k_2U^4 - C_2U^3 + C_1U^2 - (k_2\omega_1 - k_1\omega_2)U}} = \pm\frac{X - X_0}{k_2^{3/2}}, \tag{5.6}$$

$\xi = k_1x+\omega_1t+W(X)$, $X = k_2x+\omega_2t$, $W(X)$ is related to U by $W_X = (U-k_1)/k_2$ and k_1, k_2, ω_1, ω_2, C_1, C_2 and X_0 are all arbitrary constants. It may be more convenient to find the interaction solution (5.5) by using the so-called consistent Riccati expansion (CRE) method [20].

5.2 Nonlocal symmetries and reciprocal links

Recently, the Camassa-Holm (CH) type and the Harry-Dym (HD) type equations have attracted much of the attention of mathematicians. For the CH equation,

$$m_t + 2mv_x + um_x = 0, \ m = u - u_{xx}, \tag{5.7}$$

the square eigenfunction symmetry for the field u possesses the form,

$$\sigma^u = (\psi^2)_x, \tag{5.8}$$

where ψ is a spectral function of the Lax pair

$$\psi_{xx} = \frac{1}{4}(1 + 2\lambda m)\psi, \tag{5.9a}$$

$$\psi_t = \frac{1}{2}(\lambda^{-1} - u)\psi_x + \frac{1}{2}u_x\psi. \tag{5.9b}$$

Similar to the KdV case, in order to find the finite transformation or symmetry reductions related to the nonlocal symmetry (5.8), we have to localize the nonlocal symmetry by extending the CH equation to a prolonged system. The final localization result of (5.8) possesses the form

$$\begin{aligned} V \ = \ & \lambda\psi^2\partial_x + (2\phi + \lambda u_1\psi)\psi\partial_u + \left(2\phi^2 + \frac{1}{2}\psi^2 + \lambda u\psi^2\right)\partial_{u_1} + 4m\lambda\phi\psi\partial_m \\ & +(f + \lambda\psi\phi)\psi\partial_\psi + f^2\partial_f + \left(f\phi + \frac{1}{4}\lambda\psi^3\right)\partial_\phi \end{aligned} \tag{5.10}$$

in its vector form, where

$$u_1 = u_x,\ \phi = \psi_x,\ f_x = -\frac{1}{2}\lambda^2 m\psi^2,\ f_t = \frac{\psi^2}{4} - \phi^2 + \frac{\lambda^2}{2}um\psi^2. \tag{5.11}$$

Different from the KdV case, to localize the nonlocal square eigenfunction symmetry (5.8), the reciprocal transformation must be companied because of the appearance of the first term on the left of (5.10).

The finite transformation of (5.10) is just the DT of the CH equation. Here, we just write down the Bäcklund transformation theorem for the field u:

Theorem 4. If $\{u,\ m,\ u_1,\ \psi,\ \psi_1, f\}$ is a solution of the prolonged systems (5.7), (5.9) and (5.11), then $\bar{u}$ given by

$$\bar{u} = \left.\frac{u(1-\epsilon f + \epsilon\psi\phi)^2}{(1-\epsilon f)^2}\right|_{x\to x+\frac{\epsilon\psi^2}{\epsilon f-1}} \tag{5.12}$$

is a solution of the CH equation where ϵ is an arbitrary constant.

6 Summary and discussions

In summary, nonlocal symmetries can be obtained by using various methods including to find kernels or the eigenfunctions of recursion operators [18], to get the residual symmetries from the truncated Painlevé expansions [7], to find infinitesimal forms of finite transformations such as the Darboux transformations [24] and the Bäcklund transformations [9], to use formal series symmetry approaches [16], to find potential or pseudo-potential symmetries [6] and so on.

The nonlocal symmetries are usually related to integrable negative hierarchies. Many negative hierarchies have not yet been studied because of their complexity and the nonlocality. For the KdV equation, the even order negative hierarchy (2.8b) may be more important because there is no even order positive KdV hierarchy. The model (2.8b) with $n = 0$ possesses an alternative variant form

$$v_t = z^2 + w^2 + v_0,\ z_x = \sin(v),\ w_x = \cos(v) \tag{6.1}$$

with arbitrary constant v_0. Cancelling the field v in (6.1), the model can be expressed in a symmetric elegant form for the fields z and w,

$$w_{xt} = -(z^2 + w^2 + c)z_x,\ z_{xt} = (z^2 + w^2 + c)w_x. \tag{6.2}$$

The similar phenomena exist for other integrable systems. For instance, for the Sawada-Kotera hierarchies, there is no positive $6n+p$ order equations for $p = 0,\ 2,\ 3$ and 4; however, there are all negative $-6n - p$ equations for all $p = 0,\ 1,\ \ldots,\ 5$ [14].

In (2+1)-dimensional cases, the symmetries related to positive hierarchies are also nonlocal. For the KP hierarchy (2.18), the odd order equations ($u_{t_{2n+1}} = K_{2n+1}$) are the generalizations of the usual KdV hierarchy (2.3) while the even order equations ($u_{t_{2n+2}} = K_{2n+2}$) are the generalizations of the KdV type breaking soliton hierarchy ($u_{t_{2n+2}} = \phi^n u_y$).

By using the constraints related to nonlocal symmetries, one can also obtain many types of integrable systems via some different ways such as the nonlinearizations of linear Lax pairs and the uses of self-consistent sources. Especially, the super- (or Kuper-) integrable systems can be considered as the integrable models with self-consistent fermionic sources. Applying inner parameter related symmetry (u_z, $z \neq x$, y, t) constraints (say (4.1)) on (n+1)-dimensional integrable systems, one can still obtain integrable models in the same dimensions. Furthermore, the second type of models with self-consistent sources may lead to models in higher dimensions. A (3+1)-dimensional model (4.1) and (4.3) is proposed in this review paper. The integrability of the model (4.1) and (4.3) is still open except for some special cases with $z = x$, y and t.

For the CH, HD, Degasperis-Procesi type, and Novikov type equations, the localization procedures show us that the reciprocal transformations should be companied when one tries to study the properties and exact solutions of these types of systems.

The nonlocal symmetries can be successfully used to find many kinds of exact solutions such as the algebro-geometric solutions (via the nonlinearization approach) [3], the solutions which can be obtained from DT (via the Lie's first theorem and the localization of nonlocal symmetries) [12] and the interaction solutions among different types of nonlinear waves (via the symmetry reduction method and localization approach) [25].

Acknowledgements

The author is grateful to Professors Q. P. Liu, X. B. Hu, D. J. Zhang and K. Tian for their helpful discussions. The work was sponsored by the National Natural Science Foundation of China (Nos.11975131,11435005) and K. C. Wong Magna Fund in Ningbo University.

References

[1] Bluman G W and Kumei S, *Symmetries and differential equations*, Springer, Berlin, 1989.

[2] Calogero F and Degasperis A, Nonlinear evolution equation solvable by the inverse spectral transformation, *Nuovo Cimento B*, **31**, 201–242, 1976.

[3] Cao C W, Nonlinearization of the Lax system for AKNS hierarchy, *Sci. China A*, **33**, 528-536, 1990.

[4] Chvartatskyi O I and Sydorenko Yu M, A new bidirectional generalization of (2+1)-dimensional matrix k-constrained KP hierarchy, *arXiv: 1303.6510v2 [nlin.SI]*, 2013.

[5] Conte R (Ed), *The Painlevé Property One Century Later*, Springer, New York, 1999.

[6] Galas F, New non-local symmetries with pseudopotentials, *J. Phys. A: Math. Gen.*, **25**, L981-L986, 1992.

[7] Gao X N, Lou S Y and Tang X Y, Bosonization, singularity analysis, nonlocal symmetry reductions and exact solutions of supersymmetric KdV equation, *JHEP*, **05**, 029, 2013.

[8] Higgs P W, Broken symmetries and the masses of gauge bosons, *Phys. Rev. Lett.*, **13**, 508-509, 1964.

[9] Hu X B, Lou S Y and Qian X M, Nonlocal Symmetries for Bilinear Equations and Their Applications, *Stud. Appl. Math.*, **122**, 305-324, 2009.

[10] Hu X R, Lou S Y and Chen Y, Explicit solutions from eigenfunction symmetry of the Korteweg-de Vries equation, *Phys. Rev. E*, **85**, 056607, 2012.

[11] Kupershmidt B A, A super Korteweg-de Vries equation: An integrable system, *Phys. Lett. A*, **102**, 213-215, 1984.

[12] Li Y Q, Chen J C, Chen Y and Lou S Y, Darboux Transformations via Lie Point Symmetries: KdV Equation, *Chin. Phys. Lett.*, **31**, 010201, 2014.

[13] Liu S J, Tang X Y and Lou S Y, Multiple Darboux-Backlund transformations via truncated Painlevé expansion and Lie point symmetry approach, *Chin. Phys. B*, **27**, 060201, 2018.

[14] Lou S Y, Twelve sets of the symmetries of the Caudry-Dodd-Gibbon-Sawada-Kortera equation, *Phys. Lett. A*, **175**, 23-26, 1993.

[15] Lou S Y, Symmetries of the Kadomtsev-Petviashvili equation, *J. Phys. A: Math. Gen.*, **26**, 4387-4394, 1993.

[16] Lou S Y, Generalized symmetries and w_∞ algebras in three dimensional Toda field theory, *Phys. Rev. Lett.*, **71**, 4099-4102, 1993.

[17] Lou S Y, Symmetry algebras of the potential Nizhnik-Novikov-Veselov model, *J. Math. Phys.*, **35**, 1755-1762, 1994.

[18] Lou S Y, Eigenvectors of the recursion operator and symmetry structure for the coupled KdV hierarchies, *J. Nonl. Math. Phys.*, **1**, 401-413, 1994.

[19] Lou S Y, Negative Kadomtsev-Petviashvili hierarchy, *Phys. Scr.*, **57**, 481-485, 1998.

[20] Lou S Y, Consistent Riccati Expansion for Integrable Systems, *Stud. Appl. Math.* , **134**, 372-402, 2015.

[21] Lou S Y, Alice-Bob systems, $\hat{P}$–$\hat{T}$–$\hat{C}$ symmetry invariant and symmetry breaking soliton solutions, *J. Math. Phys.*, **59** , 083507, 2018.

[22] Lou S Y, Multi-place physics and multi-place nonlocal systems, *Commun. Theor. Phys.*, **72**, 057001, 2020.

[23] Lou S Y and Hu X B, Non-local symmetries via Darboux transformations, *J. Phys. A: Math. Gen.*, **30**, L95-L100, 1997.

[24] Lou S Y and Hu X B, Infinitely many Lax pairs and symmetry constraints of the KP equation, *J. Math. Phys.*, **38**, 6401–6427, 1997.

[25] Lou S Y, Hu X R and Chen Y, Nonlocal symmetries related to Bäcklund transformation and their applications, *J. Phys. A: Math. Theor.* **45**, 155209, 2012.

[26] Manakov S V, On the theory of two-dimensional stationary self-focusing of electromagnetic waves, *Sov. Phys. JETP*, **38**, 248-253, 1974.

[27] Matveev V B and Salle M A, *Darboux Transformations and Solitons*, Springer, Berlin, 1990.

[28] Mel'nikov V K, A direct method for deriving a multi-soliton solution for the problem of interaction of waves on the x,y plane, *Commun. Math. Phys.*, **112**, 639-652, 1987.

[29] Olver P J, *Applications of Lie groups to differential equations*, Springer, New York, 1993.

[30] Yajima N and Oikawa M, Formation and interaction of Sonic-Langmuir solitons: Inverse scattering method, *Prog. Theor. Phys.*, **56**, 1719–1739, 1977.

A7. High-order soliton matrix for an extended nonlinear Schrödinger equation

Huijuan Zhou[a] *and Yong Chen*[a b c]

[a]*School of Mathematical Sciences, Shanghai Key Laboratory of PMMP, Shanghai Key Laboratory of Trustworthy Computing, East China Normal University, Shanghai, 200062, P.R. China*

[b]*College of Mathematics and Systems Science, Shandong University of Science and Technology, Qingdao, 266590, P.R. China*

[c]*Department of Physics, Zhejiang Normal University, Jinhua, 321004, China*

Abstract

The extended nonlinear Schrödinger (ENLS) equation with third-order term and fourth-order term which describe the wave propagation in the optical fibers is more accurate than the NLS equation. A study of high-order soliton matrix is presented for an ENLS equation in the framework of the Riemann-Hilbert problem (RHP). Through a standard dressing procedure and the generalized Darboux transformation (gDT), soliton matrix for simple zeros and elementary high-order zeros in the RHP for the ENLS equation are constructed. Then the N-soliton solutions and high-order soliton solutions for the ENLS equation can be determined. Moreover, collision dynamics along with the asymptotic behavior for the two-solitons and long-time asymptotic estimations for the high-order one-soliton are concretely analyzed. For the given spectral parameters, we can control the propagation direction, velocity, width and other physical quantities of solitons by adjusting the free parameters of ENLS equations.

1 Introduction

The phenomenon of the solitary wave, which was discovered by the famous British scientist John Scott Russell in 1834. He thought that this kind of wave should be a stable solution of fluid motion and named it "solitary wave", but he had not confirmed the existence of solitary wave in theory. In 1895, Diederik Korteweg and his student Gustav de Vries pointed out that such waves could be approximated as long waves with small amplitude, and thus established the KdV equation. The existence of solitary waves is explained theoretically by the KdV equation. In 1955, Enrico Fermi, John Pasta and Stanislaw Ulam published *Studies of Nonlinear Problems*, so that the study of solitary waves is active again. Ten years later, Martin Kruskal and Norman Zabusky, two American mathematicians, studied the whole process of the interaction between two waves of the KdV equation in detail through numerical calculation using advanced computers. The result indicated that the solitary waves have the property of elastic collision, which is similar to the colliding property of particles. Therefore, Kruskal and Zabusky named them "solitons" [47]. From then on, the research on solitons began to flourish.

The inverse scattering transform (IST) method was discovered by Gardner, Greene, Kruskal and Miura (GGKM) [17] in 1967 as a method to solve the initial value problem that decay sufficiently rapidly at infinity for the KdV equations. In 1968, Lax gave Lax pair of KdV equations [23] and pointed out that this IST was general and could solve the initial value problem of multiple equations, and established the general framework of the solving theory of the IST. Zakharov and Shabat [50] promoted IST by using Lax's thought, and gave the solution of the higher-order KdV equation and cubic Schrödinger equation in 1972, which is the first time to give an example to prove the generality of the IST. In 1973, the initial value problem for the sine-Gordon equation was solved by the IST [3]. In the same year, Ablowitz, Newell, Manakov and Shabat et al. studied the long-time behavior of KdV equation and NLS equation according to the IST [2, 26, 30]. In 1975-1976, the continuation method (W-E method) of nonlinear PDE, with only two independent variables, was proposed by Wahlquist and Estabrook [14, 37], and an important application of which was to obtain Lax pairs of the equation with the help of Lie algebra, providing a necessary condition for solving the equation with IST. However, obtainment of the solution by the W-E method is more complicated. IST is one of the important discoveries in the field of mathematical physics in the 20th century.

The IST method was originally solved by using the Gel'Fand-Levitan-Marchenko (GLM) integral equation, although GLM equation can be used to obtain the solution of the equation, the solution process is very complex. The RHP, derived by two famous mathematicians Riemann and Hilbert, was first introduced by Riemann in his doctoral thesis in 1851, and then generalized by Hilbert to a more formal form in 1900, and presented at the international congress of mathematicians in Paris. In 1976, Zakharov and Manakov used precise steps to give a long-time asymptotic formula for the solution of NLS equation that explicitly depends on the initial value [51]. In the 1980s, Jimbo et al. [22] applied the IST to the long-time properties of a quantum solvable model. In essence, these methods in [51, 22] requiring a priori judgment on the asymptotic form of the solution have implied the ideas of classical RHP. RHP as a more general method than the IST began to be applied to integrable systems since the 1980s. For example, GLM theory is equivalent to the RHP for second-order spectral problems, while since there is no GLM theory for the high-order spectral problems, inverse scattering problems need to be transformed into RHP. Most importantly, the exact long-time asymptotic property of the solution can be obtained through RHP.

Inspired by the work of Zakharov and Manakov, Its [21] developed the isomonodromy method and converted the long-time behavior of the initial value problems for the NLS equation into a small neighborhood of the local RHP in 1981, providing a set of practical and strict approaches for analyzing the long-time behavior of integrable equations. However, this method still cannot get rid of the prior judgment on the asymptotic form of the solution. In 1989, Zhou [52] studied the connection between the Riemann-Hilbert factorization on self-intersecting contours and a class of singular integral equations with a pair of decomposing algebras, providing an effective way to treat the IST problem of first-order systems. Deift and Zhou [10] perfected Its's methods and proposed nonlinear steepest descent method in 1993. The RHP corresponding to the initial value problem of mKdV equation was studied directly by using the nonlinear steepest descent method, and the long-time behavior of the exact solution of mKdV was obtained. In 1997 and 2003, using the nonlinear steepest descent method, long-time behavior of the initial value problem of small dispersion KdV equation [12] and the solutions of NLS equation for weighted Sobolev space initial data [13] were studied successively by Deift and Zhou. In 2002, Vartanian [35] studied the long-time behavior of solutions of NLS equations with finite dense initial values. Tovbis et al. [33] studied the asymptotic properties of the first term of the solution to the semi-classical limit initial value problem of NLS equation in 2006. In 2009, Monvel, Its and Kotlyarov [27] studied the long-time properties of solutions of focusing NLS equation under periodic boundary conditions on a half-line. In 2010, Yang and other collaborators used RHP to study the initial value problems of nonlinear integrable system long-time asymptotic behavior of soliton solution [45, 32]. In 2011, Deift and Park [11] studied the solution of the focused NLS equation with Robin boundary conditions at the origin on a half line long-time behavior. Fokas [15, 24, 25] published three papers in 2012 on solving the RHP in integrable systems. Since 2013, Fan's group began to study the RHP [39, 40], and they have studied the long-time asymptotic behavior of Fokas-Lenells equation under zero boundary conditions based on the nonlinear steepest descent method. The initial boundary value problems of Sasa-Satsuma equations and three wave equations with more complex spectral problems were studied by using the Fokas method. In the last five years, many papers have been published on solving the initial boundary value problem and long-time behavior of integrable equations under the RHP framework. For example, according to the Deift-Zhou nonlinear steepest descent method, Biondini [7] studied the long-time asymptotics for the focusing NLS equation with nonzero boundary conditions at infinity and asymptotic stage of modulational instability. Miller et al. [9] investigated rogue waves of infinite order and the Painléve-III hierarchy by using the nonlinear steepest descent method. Bilman [8] gave the large-order asymptotics for multiple-pole solitons of the focusing NLS equation.

It is well known that the classical method of IST shows that the poles of the scattering coefficients (or zeros of the RHP) can produce soliton solutions. The soliton solutions are usually derived by using one of the several well-known techniques, such as the dressing method or the RHP approach. However, in most literatures only soliton solutions from simple poles are considered. It is usually assumed that a multiple-pole solution can be obtained in a straightforward way by coalescing several distinct poles [28] which describe multi-soliton solutions. Indeed, the soliton dressing matrix corresponding to a multi-soliton solution is a rational matrix function which has distinct simple poles, while the coalescing procedure must produce multiple poles. Soliton solutions corresponding to multiple poles, i.e., the high-order solitons, have been investigated in the literatures [31, 56]. As described in [16], this high-order soliton can be used to describe the weak bound state of a soliton solution and may appear in the study of line-propagation solitons with nearly the same speed and amplitude height. High-order soliton solutions of some equations, such as sine-Gordon, Schrödinger, Kadomtsev-Petviashvili I, N-wave system, derivative NLS and Landau-Lifshitz equations have been studied in the following literatures [1, 6, 18, 32, 34, 36].

Recently, we have also done some research related to the RHP and high-order soliton, in [44] we studied the high-order soliton matrix for Sasa-Satsuma equation in the framework of the RHP. It is noted that pairs of zeros are simultaneously tackled in the situation of the higher-order zeros, which is different from other NLS-type equations. Moreover, collision dynamics along with the asymptotic behavior for the two solitons were analyzed, and long-time asymptotic estimations for the higher-order soliton solution were concretely calculated. In this case, two double-humped solitons with nearly equal velocities and amplitudes can be observed. In the same year, we also studied the generalized NLS equation by IST [53]. In this chapter, the high-order rogue wave of generalized NLS equation with nonzero boundary was given based on the robust IST method. This method is more convenient than before because we do not have to take a limit. A study of high-order solitons in three nonlocal NLS equations including the PT-symmetric, reverse-time, and reverse-space-time was presented in 2018 [43]. General high-order solitons in three different equations were derived from the same Riemann-Hilbert solutions of the AKNS hierarchy, except for the difference in the corresponding symmetry relations on the "perturbed" scattering data. Dynamics of general high-order solitons in these equations were further analyzed. It was shown that the high-order fundamental-soliton is moving on different trajectories in nearly equal velocities, and they can be nonsingular or repeatedly collapsing, depending on the choices of the parameters. It was also shown that the high-order multi-solitons could have more complicated wave structures and behaviors which are different from higher-order fundamental solitons.

There is also a wide range of literature concerning the behavior of solitons and their interactions in various integrable systems such as soliton scattering, breather solutions, and soliton bound states have been published recently [38, 41, 42, 54]. As is known to all, some integrable nonlinear PDEs in mathematical physics have rich mathematical structures and extensive physics applications [29, 46]. In particular, it is always possible to find explicit solutions to these equations, such as they often have multi-soliton solutions. Among these integral PDEs, the NLS equation

$$iu_t + u_{xx} + 2|u|^2u = 0, \tag{1.1}$$

has been considered as the most important mathematical model. Eq.(1.1) can be used to describe wave evolution in scientific fields such as water waves [5, 48], plasma physics [49], condensed matter physics, fluids, arterial mechanics and fiber optics [19, 20]. However, several phenomena observed in the experiment cannot be explained by NLS equation, as the short soliton pulses get shorter, some additional effects become important. The NLS-type equations with high-order terms have important effects in fiber optics, Heisenberg spin chain and ocean waves. In order to describe the dynamics of a one-dimensional continuum anisotropic Heisenberg ferromagnetic spin chain with the octuple-dipole interaction or the alpha helical protein with higher-order excitation and interaction under the continuum approximation, an

ENLS equation with higher-order odd (third-order) and even (fourth-order) terms has been studied [4]. The ENLS equation is as follows:

$$\begin{aligned} & iu_t + \tfrac{1}{2}u_{xx} + |u|^2u - i\alpha\left(u_{xxx} + 6u_x|u|^2\right) \\ & +\gamma\left(u_{xxxx} + 6u_x^2u^* + 4u\left|u_x\right|^2 + 8u_{xx}|u|^2 + 2u_{xx}^*u^2 + 6u|u|^4\right) = 0. \end{aligned} \tag{1.2}$$

Here, t is the propagation variable and x is the retarded time in the moving frame, with the function $u(x,t)$ being the envelope of the wave field. The notation is standard in the theory of nonlinear waves. Sometimes x and t are interchanged in optics and water wave theory. All coefficients in this equation are fixed except for the α and γ. The coefficients α and γ are two real parameters which control independently the values of third-order dispersion u_{xxx} and that of fourth-order dispersion u_{xxxx}. When coefficients α and γ are equal to zero, the remaining part is the standard normalized NLS equation. If $\alpha \neq 0, \gamma = 0$, the equation is integrable and is known as the Hirota equation. Furthermore, when $\alpha = 0, \gamma \neq 0$ the equation is also integrable and known as the Lakshmanan-Porsezian-Daniel (LPD) equation. Ankiewicz and Akhmediev [4] have indicated the integrability and derived the soliton solutions of (1.2) by DT, which motivates us to search and analyze more exact solutions. To the best of our knowledge, the high-order solitons of the ENLS equation have never been reported.

The main subject of the present chapter is to research the high-order solitons of the ENLS equation in the framework of the RHP. Through a standard dressing procedure, we can find the soliton matrix for the nonregular RHP with simple zeros. Then combined with gDT, soliton matrix for elementary high-order zeros in the RHP for the ENLS equation are constructed. Moreover, the influence of free parameter (α, γ) in soliton solutions of ENLS equation on soliton propagation, collision dynamics along with the asymptotic behavior for the two solitons and long-time asymptotic estimations for the high-order one soliton are concretely analyzed. The propagation direction, velocity, width and other physical quantities of solitons can be modulated by adjusting the free parameters of ENLS equation. Our work may be helpful to observe the light pulse waves in optical fibers and guide optical experiments.

This chapter is organized as follows. In Section 2, the inverse scattering theory is established for the 2×2 spectral problems, and the corresponding matrix RHP is formulated. In Section 3, the N-soliton formula for ENLS equation is derived by considering the simple zeros in the RHP. In Section 4, the high-order soliton matrix and the generalized DT is constructed and the explicit high-order N-soliton formula is obtained, which corresponds to the elementary high-order zeros in the RHP. The final section is devoted to conclusion and discussion.

2 Inverse scattering theory for ENLS equation

In this section, we consider the scattering and inverse scattering problem for ENLS equation.

2.1 Scattering theory of the spectral problem

Considering the spectral problem of the ENLS Equation (1.2):

$$Y_x = LY, \tag{2.1}$$

$$Y_t = BY, \tag{2.2}$$

with 2×2 matrices L and B in the forms of:

$$\begin{aligned}
L &= -i\zeta\Lambda + Q,\\
B &= (-i\zeta^2 + 4i\alpha\zeta^3 + 8i\gamma\zeta^4)\Lambda + V_1,\\
V_1 &= (\zeta - 4\alpha\zeta^2)Q + (\frac{1}{2} - 2\alpha\zeta)V - \alpha K + \gamma V_p.
\end{aligned}$$

Where

$$\begin{aligned}
\Lambda &= \begin{pmatrix} 1 & 0 \\ 0 & -1 \end{pmatrix},\\
Q &= \begin{pmatrix} 0 & u \\ -u^* & 0 \end{pmatrix},\\
V &= \begin{pmatrix} \mathrm{i}|u|^2 & iu_x \\ iu_x^* & -\mathrm{i}|u|^2 \end{pmatrix},\\
K &= \begin{pmatrix} uu_x^* - u^*u_x & -\left(2|u|^2u + u_{xx}\right) \\ 2|u|^2u^* + u_{xx}^* & -\left(uu_x^* - u^*u_x\right) \end{pmatrix},\\
V_p &= \begin{pmatrix} \mathrm{i}A_p(x,t) & B_p(x,t) \\ -B_p^*(x,t) & -\mathrm{i}A_p(x,t) \end{pmatrix},
\end{aligned} \tag{2.3}$$

with

$$\begin{aligned}
A_p(x,t) &= 3|u|^4 - |u_x|^2 + uu_{xx}^* + u^*u_{xx} - 2\mathrm{i}\zeta\left(u^*u_x - uu_x^*\right) - 4\zeta^2|u|^2,\\
B_p(x,t) &= 6\mathrm{i}|u|^2u_x + \mathrm{i}u_{xxx} + 2\zeta u_{xx} + 4\zeta|u|^2u - 4\mathrm{i}\zeta^2u_x - 8\zeta^3u.
\end{aligned}$$

Here, ζ is a spectral parameter, $Y(x,t,\zeta)$ is a vector function, and the superscript "*" represents complex conjugation. The spatial linear operator (2.1) and the temporal linear operator (2.2) are the Lax pair of the ENLS Equation (1.2). Supposing $u(x) = u(x,0) \to 0$ sufficiently fast as $x \to \pm\infty$. For a prescribed initial condition $u(x,0)$, we seek the solution $u(x,t)$ at any later time t. That is, we solve an initial value problem for the ENLS equation.

Notation

$$E_1 = e^{-i\zeta\Lambda x - (i\zeta^2 - 4i\alpha\zeta^3 - 8i\gamma\zeta^4)\Lambda t}, \tag{2.4}$$

$$J = YE_1^{-1}, \tag{2.5}$$

so that the new matrix function J is (x,t)-independent at infinity. Inserting (2.5) into (2.1)-(2.2), we find that the Lax pair (2.1)-(2.2) becomes

$$J_x = -i\zeta[\Lambda, J] + QJ, \tag{2.6}$$

$$J_t = -(i\zeta^2 - 4i\alpha\zeta^3 - 8i\gamma\zeta^4)[\Lambda, J] + V_1 J, \tag{2.7}$$

where $[\Lambda, J] = \Lambda J - J\Lambda$ is the commutator. Notice that both matrices Q and V_1 are anti-Hermitian, i.e.,

$$Q^\dagger = -Q, \quad V_1^\dagger = -V_1, \tag{2.8}$$

where the superscript "†" represents the Hermitian of a matrix. In addition, their traces are both equal to zero, i.e., $trQ = trV_1 = 0$.

Now we let time t be fixed and a dummy variable, and thus it will be suppressed in our notation. In the scattering problem, we first introduce matrix Jost solutions $J_\pm(x, \zeta)$ of (2.6) with the following asymptotic at large distances:

$$J_\pm(x, \zeta) \to I, \quad x \to \pm\infty. \tag{2.9}$$

Here, I is the 2×2 unit matrix. Now we will delineate Jost solutions $J_\pm(x, \zeta)$ analytical properties first.

Introducing the notation $E = e^{-i\zeta\Lambda x}$, $\Phi \equiv J_- E$ and $\Psi \equiv J_+ E$. (Φ, Ψ) satisfy the scattering Equation (2.1), i.e.,

$$Y_x + i\zeta\Lambda Y = QY. \tag{2.10}$$

When we treat the QY term as an inhomogeneous term, E is the solution to the homogeneous equation $Y_x + i\zeta\Lambda Y = 0$, then using the method of variation of parameters as well as the boundary conditions (2.9), we can turn (2.10) into the following Volterra integral equations:

$$J_\pm(x, \zeta) = I + \int_{\pm\infty}^{x} e^{-i\zeta\Lambda(x-y)} Q(y) J_\pm(y, \zeta) e^{-i\zeta\Lambda(y-x)} dy, \tag{2.11}$$

Thus $J_\pm(x, \zeta)$ allow analytical continuations off the real axis $\zeta \in R$ as long as the integrals on the right sides of the above Volterra equations converge. Due to the structure (2.3) of the potential Q, we can easily get the following proposition.

Proposition 1. *The first column of J_- and the second column of J_+ can be analytically continued to the upper half-plane $\zeta \in \mathbb{C}_+$, while the second column of J_- and the first column of J_+ can be analytically continued to the lower half-plane $\mathbb{C}_-$.*

Proof. The integral Equation (2.11) for the first column of J_- , say $(\varphi_1, \varphi_2)^T$, is

$$\varphi_1 = 1 + \int_{-\infty}^{x} u(y)\varphi_2(y, \zeta) dy, \tag{2.12}$$

$$\varphi_2 = -\int_{-\infty}^{x} u^*(y)\varphi_1(y, \zeta) e^{2i\zeta(x-y)} dy. \tag{2.13}$$

When $\zeta \in \mathbb{C}_+$, since $e^{2i\zeta(x-y)}$ in (2.13) is bounded, and $u(x)$ decays to zero sufficiently fast at large distances, both integrals in the above two equations converge. Thus the Jost solution $(\varphi_1, \varphi_2)^T$ can be analytically extended to $\mathbb{C}_+$. The analytic properties of the other Jost solutions J_+ can be obtained similarly. ■

From Abel's identity, we find that $|J(x,\zeta)|$ is a constant for all x. Then using the boundary conditions (2.9), we see that

$$|J_{\pm}(x,\zeta)| = 1, \tag{2.14}$$

for all (x,ζ). Since $\Phi(x,\zeta)$ and $\Psi(x,\zeta)$ are both solutions of the linear Equation (2.1), they are linearly related by a scattering matrix $S(\zeta) = (s_{ij})_{2\times 2}$:

$$\Phi(x,\zeta) = \Psi(x,\zeta)S(\zeta), \quad \zeta \in \mathbb{R}. \tag{2.15}$$

$\mathbb{R}$ is the set of real numbers.

Because we need to use scattering matrix $S(\zeta)$ to reconstruct the potential $u(x,t)$, now we need to delineate the analytical properties of $S(\zeta)$. If we express (Φ,Ψ) as a collection of columns

$$\Phi = (\phi_1^+, \phi_2^-), \quad \Psi = (\psi_1^-, \psi_2^+). \tag{2.16}$$

Where the superscripts "$\pm$" indicate the half-plane of analyticity for the underlying quantities. Since

$$S = \Psi^{-1}\Phi = \begin{pmatrix} \hat{\psi}_1^{\,+} \\ \hat{\psi}_2^{\,-} \end{pmatrix} (\phi_1^+, \phi_2^-), \tag{2.17}$$

$$S^{-1} = \Phi^{-1}\Psi = \begin{pmatrix} \hat{\phi}_1^{\,-} \\ \hat{\psi}_2^{\,+} \end{pmatrix} (\psi_1^-, \psi_2^+). \tag{2.18}$$

We see immediately that scattering matrices S and S^{-1} have the following analyticity structures:

$$S = \begin{pmatrix} s_{11}^+, s_{12} \\ s_{21}, s_{22}^- \end{pmatrix}, \quad S^{-1} = \begin{pmatrix} \hat{s}_{11}^-, \hat{s}_{12} \\ \hat{s}_{21}, \hat{s}_{22}^+ \end{pmatrix}. \tag{2.19}$$

Elements without superscripts indicate that such elements do not allow analytical extensions to $\mathbb{C}_{\pm}$ in general. Because S is 2×2 matrix with unit determinant, then we can get

$$\hat{s}_{11} = s_{22}, \quad \hat{s}_{22} = s_{11}, \quad \hat{s}_{12} = -s_{12}, \quad \hat{s}_{21} = -s_{21}. \tag{2.20}$$

Hence analytic properties of S^{-1} can be directly read off from analytic properties of S.

In order to construct the RHP, we define the Jost solutions

$$P^+ = (\phi_1, \psi_2)e^{i\zeta\Lambda x} = J_- H_1 + J_+ H_2 \tag{2.21}$$

are analytic in $\zeta \in \mathbb{C}_+$, here

$$H_1 \equiv diag(1,0), \quad H_2 \equiv diag(0,1). \tag{2.22}$$

In addition, from the Volterra integral Equations (2.11), we see that the large ζ asymptotics of these analytical functions are

$$P^{+}(x,\zeta) \to I, \zeta \in \mathbb{C}_{+} \to \infty, \tag{2.23}$$

If we express Φ^{-1} and Ψ^{-1} as a collection of rows

$$\Phi^{-1} = \begin{pmatrix} \hat{\phi}_1 \\ \hat{\phi}_2 \end{pmatrix}, \qquad \Psi^{-1} = \begin{pmatrix} \hat{\psi}_1 \\ \hat{\psi}_2 \end{pmatrix}. \tag{2.24}$$

Then by techniques similar to those used above, we can show that the adjoint Jost solutions

$$P^{-} = e^{-i\zeta\Lambda x} \begin{pmatrix} \hat{\phi}_1 \\ \hat{\psi}_2 \end{pmatrix} = H_1 J_{-}^{-1} + H_2 J_{+}^{-1} \tag{2.25}$$

are analytic in $\zeta \in \mathbb{C}_{-}$. In addition,

$$P^{-}(x,\zeta) \to I, \zeta \in \mathbb{C}_{-} \to \infty. \tag{2.26}$$

The anti-Hermitian property (2.8) of the potential matrix Q gives rise to involution properties in the scattering matrix as well as in the Jost solutions. Indeed, by taking the Hermitian of the scattering Equation (2.6) and utilizing the anti-Hermitian property of the potential matrix $Q^{\dagger} = -Q$, we get

$$J_{\pm}^{\dagger}(\zeta^{*}) = J_{\pm}^{-1}(\zeta). \tag{2.27}$$

From this involution property as well as the definitions (2.21) and (2.25) for $P^{\pm}$, we see that the analytic solutions $P^{\pm}$ satisfy the involution property as well:

$$(P^{+})^{\dagger}(\zeta^{*}) = P^{-}(\zeta). \tag{2.28}$$

In addition, in view of the scattering relation (2.15) between J_{+} and J_{-}, we see that S also satisfies the involution property:

$$S^{\dagger}(\zeta^{*}) = S^{-1}(\zeta). \tag{2.29}$$

2.2 Matrix Riemann-Hilbert problem

Hence we have constructed two matrices functions $P^{\pm}(x,\zeta)$ which are analytic for ζ in $\mathbb{C}_{\pm}$, respectively. On the real line, using (2.15), (2.21) and (2.25), we easily get

$$P^{-}(x,\zeta)P^{+}(x,\zeta) = G(x,\zeta), \quad \zeta \in \mathbb{R}, \tag{2.30}$$

where

$$G = E(H_1 + H_2 S)(H_1 + S^{-1}H_2)E^{-1} = E \begin{pmatrix} 1 & \hat{s}_{12} \\ s_{21} & 1 \end{pmatrix} E^{-1}. \tag{2.31}$$

Equation (2.30) forms a matrix RHP. The normalization condition for this RHP can be obtained from (2.23) and (2.26) as

$$P^{\pm}(x,\zeta) \rightarrow I, \quad \zeta \rightarrow \infty, \tag{2.32}$$

which is the canonical normalization condition.

Recalling the definitions (2.21) and (2.25) of $P^{\pm}$ as well as the scattering relation (2.15), we see that

$$|P^{+}| = \widehat{s}_{22} = s_{11}, \quad |P^{-}| = s_{22} = \widehat{s}_{11}. \tag{2.33}$$

We consider the solution of the regular RHP first, i.e., : $|P^{\pm}| \neq 0$, i.e. :$\widehat{s}_{22} = s_{11} \neq 0$ and $s_{22} = \widehat{s}_{11} \neq 0$ in their respective planes of analyticity. Under the canonical normalization condition (2.32), the solution to this regular RHP is unique [45]. This unique solution to the regular matrix RHP (2.30) defines explicit expressions. But its formal solution can be given in terms of a Fredholm integral equation.

To use this Plemelj-Sokhotski formula on the regular RHP (2.30), we first rewrite (2.30) as

$$\left(P^{+}\right)^{-1}(\zeta) - P^{-}(\zeta) = \widehat{G}(\zeta)\left(P^{+}\right)^{-1}(\zeta), \quad \zeta \in \mathbb{R},$$

where

$$\widehat{G} = I - G = -E\begin{pmatrix} 0 & \hat{s}_{12} \\ s_{21} & 0 \end{pmatrix} E^{-1}.$$

$\left(P^{+}\right)^{-1}(\zeta)$ is analytic in $\mathbb{C}_{+}$, and $P^{-}(\zeta)$ is analytic in $\mathbb{C}_{-}$. Applying the Plemelj-Sokhotski formula and utilizing the canonical boundary conditions (2.32), the solution to the regular RHP (2.30) is provided by the following integral equation:

$$\left(P^{+}\right)^{-1}(\zeta) = I + \frac{1}{2\pi i}\int_{-\infty}^{\infty} \frac{\widehat{G}(\xi)\left(P^{+}\right)^{-1}(\xi)}{\xi - \zeta} d\xi, \quad \zeta \in \mathbb{C}_{+}.$$

In the more general case, the RHP (2.30) is not regular, i.e., $|P^{+}(\zeta)|$ and $|P^{-}(\zeta)|$ can be zero at certain discrete locations $\zeta_k \in \mathbb{C}_{+}$ and $\bar{\zeta}_k \in \mathbb{C}_{-}, 1 \leq k \leq N$, where N is the number of these zeros. In view of (2.33), we see that $\left(\zeta_k, \bar{\zeta}_k\right)$ are zeros of the scattering coefficients $\hat{s}_{22}(\zeta)$ and $s_{22}(\zeta)$. Due to the involution property (2.29), we have the involution relation

$$\bar{\zeta}_k = \zeta_k^{*}. \tag{2.34}$$

For simplicity, we assume that all zeros $\left\{\left(\zeta_k, \bar{\zeta}_k\right), k = 1, \ldots, N\right\}$ are simple zeros of $(\hat{s}_{22}, s_{22})$ which is the generic case. In this case, both $\ker(P^{+}(\zeta_k))$ and $\ker(P^{-}(\bar{\zeta}_k))$ are spanned by one-dimensional column vector $|v_k\rangle$ and row vector $\langle v_k|$, respectively.

$$P^{+}(\zeta_k)|v_k\rangle = 0, \quad \langle v_k| P^{-}\left(\bar{\zeta}_k\right) = 0, \quad 1 \leq k \leq N. \tag{2.35}$$

Taking the Hermitian of the first equation in (2.35) and utilizing the involution properties (2.28) and (2.34), we see that eigenvectors $(|v_k\rangle, \langle v_k|)$ satisfy the involution property $\langle v_k| = |v_k\rangle^\dagger$, vectors $|v_k\rangle$ and $\langle v_k|$ are x dependent, our starting point is (2.35) for $|v_k\rangle$ and $\langle v_k|$. Taking the x derivative to the $|v_k\rangle$ equation and recalling that P^+ satisfies the scattering Equation (2.6), we get

$$|v_k(x)\rangle = e^{-i\zeta_k \Lambda x} |v_{k0}\rangle, \tag{2.36}$$

where $|v_{k0}\rangle = |v_k(x)\rangle|_{x=0}$.

Following similar calculations for $\bar{v}_k$, we readily get

$$\langle v_k(x)| = \langle \bar{v}_{k0}| e^{i\zeta_k \Lambda x}.$$

These two equations give the simple x dependence of vectors $|v_k(x)\rangle$ and $\langle \bar{v}_k(x)|$. The zeros $\{(\zeta_k, \bar{\zeta}_k)\}$ of $|P^\pm(\zeta)|$ as well as vectors $|v_k\rangle, \langle v_k|$ in the kernels of $P^+(\zeta_k)$ and $P^-(\bar{\zeta}_k)$ constitute the discrete scattering data which is also needed to solve the general RHP (2.30).

Now we construct a matrix function which could remove all the zeros of this RHP. For this purpose, we will introduce the rational matrix function:

$$\Gamma_j = I + \frac{\bar{\zeta}_j - \zeta_j}{\zeta - \bar{\zeta}_j} \frac{|v_j\rangle \langle v_j|}{\langle v_j|v_j\rangle},$$

and its inverse matrix

$$\Gamma_j^{-1} = I + \frac{\zeta_j - \bar{\zeta}_j}{\zeta - \zeta_j} \frac{|v_j\rangle \langle v_j|}{\langle v_j|v_j\rangle},$$

where

$$|v_i\rangle \in \mathrm{Ker}\left(P^+\Gamma_1^{-1} \cdots \Gamma_{i-1}^{-1}(\zeta_i)\right), \quad \langle v_j| = |v_j\rangle^\dagger.$$

Therefore, if one is introducing the matrix function:

$$\Gamma = \Gamma_N \Gamma_{N-1} \cdots \Gamma_1,$$

$$\Gamma(\zeta) = I + \sum_{j,k=1}^{N} \frac{|v_j\rangle \left(M^{-1}\right)_{jk} \langle v_k|}{\zeta - \bar{\zeta}_k},$$

$$\Gamma^{-1}(\zeta) = I - \sum_{j,k=1}^{N} \frac{|v_j\rangle \left(M^{-1}\right)_{jk} \langle v_k|}{\zeta - \zeta_j}.$$

M is an $N \times N$ matrix with its (j,k) th element given *by*

$$M_{jk} = \frac{\langle v_j|v_k\rangle}{\bar{\zeta}_j - \zeta_k}, \quad 1 \le j,k \le N, \tag{2.37}$$

then $\Gamma(x,\zeta)$ cancels all the zeros of $P_{\pm}$, and the analytic solutions can be represented as

$$P^{+}(\zeta)=\widehat{P}^{+}(\zeta)\Gamma(\zeta),$$
$$P^{-}(\zeta)=\Gamma^{-1}(\zeta)\widehat{P}^{-}(\zeta).$$

Here, $\widehat{P}^{\pm}(\zeta)$ are meromorphic 2×2 matrix functions in $\mathbb{C}_{+}$ and $\mathbb{C}_{-}$, respectively, with finite number of poles and specified residues. Therefore, all the zeros of RHP have been eliminated and we can formulate a regular RHP

$$\widehat{P}^{-}(\zeta)\widehat{P}^{+}(\zeta)=\Gamma(\zeta)G(\zeta)\Gamma^{-1}(\zeta),\quad \zeta\in\mathbb{R},$$

with boundary condition: $\widehat{P}^{\pm}(\zeta)=P^{\pm}(\zeta)\Gamma^{-1}\to I$ as $\zeta\to\infty$. As a result, when $\zeta\to\infty$ we have $P^{+}(\zeta)=\Gamma$.

2.3 Solution of the Riemann-Hilbert problem

In this subsection, we discuss how to solve the matrix RHP (2.30) in the complex ζ plane. This inverse problem can be solved by expanding $P^{\pm}$ at large ζ as

$$P^{\pm}(x,\zeta)=I+\zeta^{-1}P_{1}^{\pm}(x)+O(\zeta^{-2}),\quad \zeta\to\infty, \tag{2.38}$$

and inserting (2.38) into (2.6), then by comparing terms of the same power in ζ^{-1}, ζ^{0}. We found that

$$diag(P_{1}^{+})_{x}=diag(QP_{1}^{+}). \tag{2.39}$$

$$Q=i[\wedge,P_{1}^{+}]=-i[\wedge,P_{1}^{-}]. \tag{2.40}$$

Hence the solution u can be reconstructed by

$$u=2i(P_{1}^{+})_{12}=-2i(P_{1}^{-})_{12}. \tag{2.41}$$

This completes the inverse scattering process. How to solve the matrix RHP (2.30) will be discussed in the next subsection.

2.4 Time evolution of scattering data

In this subsection, we determine the time evolution of the scattering data. First we determine the time evolution of the scattering matrices S and S^{-1}. Our starting point is the definition (2.15) for the scattering matrix, which can be rewritten as

$$J_{-}E=J_{+}ES,\quad \zeta\in\mathbb{R}.$$

Since $J_{\pm}$ satisfies the temporal Equation (2.7) of the Lax pair, then multiplying (2.7) by the time-independent diagonal matrix $E=e^{-i\zeta\Lambda x}$, we see that $J_{-}E$, i.e., $J_{+}ES$, satisfies the same temporal Equation (2.7) as well. Thus, by inserting $J_{+}ES$ into (2.7), taking the limit $x\to+\infty$, and recalling the boundary condition (2.9) for J_{+} as well as the fact that $V\to 0$ as $x\to\pm\infty$, we get

$$S_{t}=-(i\zeta^{2}-4i\alpha\zeta^{3}-8i\gamma\zeta^{4})[\Lambda,S].$$

Similarly, by inserting $J_- E S^{-1}$ into (2.7), taking the limit $x \to -\infty$, and recalling the asymptotics (2.9) for J_-, we get

$$\left(S^{-1}\right)_t = -(i\zeta^2 - 4i\alpha\zeta^3 - 8i\gamma\zeta^4)\left[\Lambda, S^{-1}\right].$$

From these two equations, we get

$$\frac{\partial \hat{s}_{22}}{\partial t} = \frac{\partial s_{22}}{\partial t} = 0, \tag{2.42}$$

and

$$\frac{\partial \hat{s}_{12}}{\partial t} = -(i\zeta^2 - 4i\alpha\zeta^3 - 8i\gamma\zeta^4)\hat{s}_{12}, \quad \frac{\partial s_{21}}{\partial t} = (i\zeta^2 - 4i\alpha\zeta^3 - 8i\gamma\zeta^4)s_{21}. \tag{2.43}$$

The two equations in (2.42) show that $\hat{s}_{22}$ and s_{22} are time independent. Recall that ζ_k and $\bar{\zeta}_k$ are zeros of $|P^{\pm}(\zeta)|$, i.e., they are zeros of $\hat{s}_{22}(\zeta)$ and $s_{22}(\zeta)$ in view of (2.33). Thus ζ_k and $\bar{\zeta}_k$ are also time independent. The two equations in (2.43) give the time evolution for the scattering data $\hat{s}_{12}$ and s_{21}, which is

$$\hat{s}_{12}(t;\zeta) = \hat{s}_{12}(0;\zeta)e^{-(i\zeta^2 - 4i\alpha\zeta^3 - 8i\gamma\zeta^4)t}, \quad s_{21}(t;\zeta) = s_{21}(0;\zeta)e^{(i\zeta^2 - 4i\alpha\zeta^3 - 8i\gamma\zeta^4)t}.$$

Next we determine the time dependence of the scattering data $|v_k\rangle$ and $\langle v_j|$. We start with (2.35) for $|v_k\rangle$ and $\langle v_j|$. Taking the time derivative to the $|v_k\rangle$ equation and recalling that P^+ satisfies the temporal Equation (2.35), we get

$$P^+(\zeta_k; x, t)\left(\frac{\partial |v_k\rangle}{\partial t} + (i\zeta^2 - 4i\alpha\zeta^3 - 8i\gamma\zeta^4)\Lambda |v_k\rangle\right) = 0,$$

thus

$$\frac{\partial |v_k\rangle}{\partial t} + (i\zeta^2 - 4i\alpha\zeta^3 - 8i\gamma\zeta^4)|v_k\rangle = 0.$$

Combining it with the spatial dependence (2.36), we get the temporal and spatial dependence for the vector $|v_k\rangle$ as

$$|v_k\rangle(x,t) = e^{-i\zeta\Lambda x - (i\zeta^2 - 4i\alpha\zeta^3 - 8i\gamma\zeta^4)\Lambda t}|v_{k0}\rangle, \tag{2.44}$$

where $|v_{k0}\rangle$ is a constant. Similar calculations for $\langle v_k|$ give

$$\langle v_k|(x,t) = \langle v_{k0}| e^{i\bar{\zeta}_k x + (i\bar{\zeta}^2 - 4i\alpha\bar{\zeta}^3 - 8i\gamma\bar{\zeta}^4)t}.$$

We see that the scattering data needed to solve this non-regular RHP is

$$\left\{s_{21}(\xi), \hat{s}_{12}(\xi), \xi \in \mathbb{R}; \quad \zeta_k, \bar{\zeta}_k, |v_j\rangle, \langle v_k|, 1 \le k \le N\right\}. \tag{2.45}$$

This is called minimal scattering data. From this scattering data at any later time, we can solve the non-regular RHP (2.30) with zeros (2.35), and thus reconstruct the solution $u(x,t)$ at any later time from the formula (2.41). So far, the IST process for ENLS Equation (1.2) has been completed.

3 N-soliton solutions

It is well known that when scattering data $\hat{s}_{12} = s_{21} = 0$, the soliton solutions correspond to the reflectionless potential. Then jump matrix $G = I$, $\widehat{G} = 0$. Due to $P^{+}(\zeta) = \Gamma, \zeta \to \infty$. Recall to (2.41), we can get

$$u(x,t) = 2i(\sum_{j,k=1}^{N} |v_j\rangle (M^{-1})_{jk} \langle v_k|)_{12}. \tag{3.1}$$

Here vectors $|v_j\rangle$ are given by (2.44), $\langle v_k| = |v_k\rangle^{\dagger}$, and matrix M is given by (2.37). Without loss of generality, we let $|v_{k0}\rangle = (c_k, 1)^T$. In addition, we introduce the notation

$$\theta_k = -i\zeta_k x - (i\zeta_k^2 - 4i\alpha\zeta_k^3 - 8i\gamma\zeta_k^4)t. \tag{3.2}$$

Then the above solution u can be written out explicitly as

$$u(x,t) = 2i \sum_{j,k=1}^{N} c_j e^{\theta_j - \theta_k^*} (M^{-1})_{jk}, \tag{3.3}$$

where the elements of the $N \times N$ matrix M are given by

$$M_{jk} = \frac{1}{\zeta_j^* - \zeta_k} [e^{-(\theta_k + \theta_j^*)} + c_j^* c_k e^{\theta_k + \theta_j^*}]. \tag{3.4}$$

Notice that M^{-1} can be expressed as the transpose of $M's$ cofactor matrix divided by $|M|$. Also recall that the determinant of a matrix can be expressed as the sum of its elements along a row or column multiplying their corresponding cofactor. Hence the solution (3.3) can be rewritten as

$$u(x,t) = -2i\frac{|F|}{|M|}, \tag{3.5}$$

where F is the following $(N+1) \times (N+1)$ matrix:

$$\begin{pmatrix} 0 & e^{-\theta_1^*} & \dots & e^{-\theta_N^*} \\ c_1 e^{\theta_1} & M_{11} & \dots & M_{N1} \\ \cdot & \cdot & \cdot & \cdot \\ \cdot & \cdot & \cdot & \cdot \\ \cdot & \cdot & \cdot & \cdot \\ c_N e^{\theta_N} & M_{1N} & \dots & M_{NN} \end{pmatrix}. \tag{3.6}$$

Let $N = 1$, $\zeta 1 = \xi + i\eta$, $c1 = 1$ the solution (3.5) is

$$\begin{aligned} u &= \frac{2\, i c_1\, (\zeta_1^* - \zeta_1) \mathrm{e}^{-\theta_1^* + \theta_1}}{(|c_1|)^2 \mathrm{e}^{\theta_1^* + \theta_1} + \mathrm{e}^{-\theta_1^* - \theta_1}} \\ &= 2\eta\, \mathrm{sech}(2\,\eta\, A)\, \mathrm{e}^{2\,i((8\,\eta^4\gamma - 48\,\eta^2\gamma\,\xi^2 + 8\,\xi^4\gamma - 12\,\alpha\,\eta^2\xi + 4\,\xi^3\alpha + \eta^2 - \xi^2)t - x\xi)}, \end{aligned} \tag{3.7}$$

where

$$A = ((32\,\eta^2\gamma\,\xi - 32\,\gamma\,\xi^3 + 4\,\alpha\,\eta^2 - 12\,\alpha\,\xi^2 + 2\,\xi)t + x).$$

This solution is a solitary wave. Its amplitude function $|u|$ has the shape of a hyperbolic secant with peak amplitude 2η, and its velocity is $-(32\,\eta^2\gamma\,\xi - 32\,\gamma\,\xi^3 + 4\,\alpha\,\eta^2 - 12\,\alpha\,\xi^2 + 2\,\xi)$. The phase of this solution depends linearly on both space x and time t. The spatial gradient of the phase is proportional to the speed of the wave. This solution is called a single-soliton solution of the ENLS Equation (1.2).

Solve the equation $|u|^2 = b$, $0 < b < 4\eta^2$, we can get

$$x_1 = \frac{-128\gamma\eta^3 t\xi + 128\gamma\eta t\xi^3 - 16\alpha\eta^3 t + 48\alpha\eta t\xi^2 - 8\eta t\xi + ln(\frac{8\eta^2+4\sqrt{4\eta^4-b\eta^2}-b}{b})}{4\eta},$$

$$x_2 = \frac{-128\gamma\eta^3 t\xi + 128\gamma\eta t\xi^3 - 16\alpha\eta^3 t + 48\alpha\eta t\xi^2 - 8\eta t\xi + ln(\frac{8\eta^2-4\sqrt{4\eta^4-b\eta^2}-b}{b})}{4\eta}. \tag{3.8}$$

Notice d is the width of the wave and

$$d = \frac{ln\frac{8\eta^2-4\sqrt{-\eta^2(-4\eta^2+b)}-b}{b} - ln\frac{8\eta^2+4\sqrt{-\eta^2(-4\eta^2+b)}-b}{b})}{4\eta}. \tag{3.9}$$

This means that the wave width is only related to the imaginary part of the spectral parameter and is not affected by the coefficients α and γ. The dispersion term and the non-linear term in the higher order term of the ENLS equation play a good balance, which makes the system energy conservation.

Further, we can derive the center trajectory of the single-soliton solution

$$x = -(32\gamma\xi(\eta^2 - \xi^2) + 4\alpha(\eta^2 - (\sqrt{3}\xi)^2) + 2\xi)t. \tag{3.10}$$

The angle between the center trajectory and the t-axis is $arctan(-32\eta^2\gamma\xi + 32\gamma\xi^3 - 4\alpha\eta^2 + 12\alpha\xi^2 - 2\xi)$. Generally speaking, the third-order and fourth-order coefficients (α, γ) affect both the velocity of the soliton and the slope of the central trajectory. But when the spectral parameter has only an imaginary part ($\xi = 0$) or $|\eta| = |\xi|$, the fourth-order coefficient γ no longer affects the above quantities. When $|\eta| = |\sqrt{3}\xi|$, the third-order coefficient α no longer affects the above quantities. Without loss of generality, it can be divided into the following cases:

case 1: $\eta = \xi = 1$. The velocity of the soliton and the slope of the central trajectory are equal to $8\alpha - 2$; take three special cases: $\alpha = 1/8, \alpha = 1/4, \alpha = 1/2$.

case 2: $\xi = 0, \eta = 1$. The velocity of the soliton and the slope of the central trajectory are equal to $-4\alpha\eta^2$; take three special cases: $\alpha = -1$, $\alpha = 0$, $\alpha = 1$.

case 3: $\eta = 1, \xi = \frac{1}{\sqrt{3}}$. The velocity of the soliton and the slope of the central trajectory are equal to $-(\frac{64\sqrt{3}}{9}\gamma + \frac{2\sqrt{3}}{3})$; take three special cases: $\gamma = -3/64, \gamma = -3/32, \gamma = -3/16$.

case 4: $\eta = 1, \xi = \frac{1}{2}$. The velocity of the soliton and the slope of the central

trajectory are equal to $-(12\gamma+\alpha+1)$, fix $\alpha=-1$; take three special cases: $\gamma=-1, \gamma=0, \gamma=1$.
We can control the propagation direction and speed of solitons by adjusting the parameters. The images of the central trajectory under different parameters is shown in Figure 1.

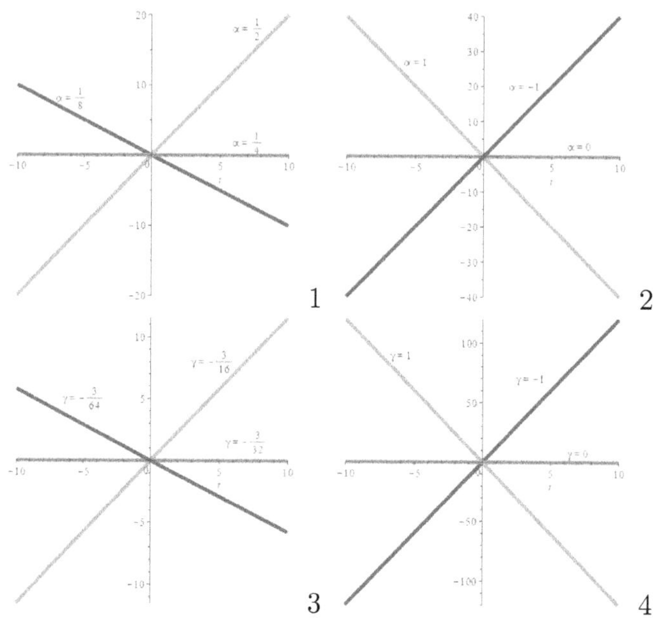

Figure 1. Central trajectory for the single-soliton solution of cases 1, 2, 3 and 4.

When $N=2$, the two-soliton solutions of ENLS equation can be written out explicitly as follows:

$$u=\frac{h_1\,\mathrm{e}^{\Theta_1'+\Theta_1}+h_2\,\mathrm{e}^{\Theta_2'-\Theta_2}+h_3\,\mathrm{e}^{\Theta_1-\Theta_1'}+h_4\,\mathrm{e}^{\Theta_2'+\Theta_2}}{d_1\,\mathrm{e}^{\Theta_1'-\Theta_2}+d_2\,\mathrm{e}^{\Theta_1'+\Theta_2}+d_3\,\mathrm{e}^{-\Theta_1'+\Theta_2}+d_4\,\mathrm{e}^{\Theta_1'-\Theta_1}+d_5\,\mathrm{e}^{-\Theta_1'-\Theta_2}+d_6\,\mathrm{e}^{-\Theta_2'+\Theta_1}}, \tag{3.11}$$

where

$$\begin{aligned}
d_1 &= (\zeta_1^*-\zeta_2^*)(\zeta_2-\zeta_1),\\
d_2 &= |c_2|^2(\zeta_1^*-\zeta_2)(\zeta_2^*-\zeta_1),\\
d_3 &= |{c_1}^2{c_2}^2|(\zeta_2^*-\zeta_1^*)(\zeta_1-\zeta_2),\\
d_4 &= c_1{c_2}^*(\zeta_1^*-\zeta_1)(\zeta_2-{\zeta_2}^*),\\
d_5 &= |c_1|^2(\zeta_2^*-\zeta_1)(\zeta_1^*-\zeta_2),\\
d_6 &= c_1^*c_2(\zeta_2^*-\zeta_2)(\zeta_1-\zeta_1^*),\\
h_1 &= 2ic_2\,(\zeta_2-\zeta_1^*)(\zeta_2^*-\zeta_1^*)(\zeta_2^*-\zeta_2),\\
h_2 &= -2ic_1\,(\zeta_1-\zeta_1^*)(\zeta_2^*-\zeta_1^*)(\zeta_2^*-\zeta_1),\\
h_3 &= -2ic_2|c_1|^2(\zeta_2^*-\zeta_2)(\zeta_1-\zeta_2)(\zeta_2^*-\zeta_1),\\
h_4 &= 2ic_1\,|c_2|^2(\zeta_1-\zeta_2)(\zeta_2-\zeta_1^*)(\zeta_1-\zeta_1^*),\\
\Theta_1 &= \theta_2-\theta_2^*,\quad \Theta_1'=-\theta_1-\theta_1^*,\quad \Theta_2=\theta_2^*+\theta_2,\quad \Theta_2'=\theta_1-\theta_1^*.
\end{aligned}$$

Let $\zeta_1 = \xi_1 + \eta_1$, $\zeta_2 = \xi_2 + \eta_2$.

Starting with a simple case, when the spectral parameters ζ_1, ζ_2 are pure imaginary numbers, i.e., $\xi_1 = \xi_2 = 0$

$$Re(\theta_1) = \eta_1(4\,t\alpha\,{\eta_1}^2 + x), \qquad Re(\theta_2) = \eta_2(4\,t\alpha\,{\eta_2}^2 + x). \tag{3.12}$$

When $\alpha = 0$, the two constituent solitons have equal velocities, thus they will stay together and form a bound state. In a frame moving at this speed, this bound state will be spatially localized, and its amplitude function $|u(x,t)|$ will oscillate periodically with time. Let $\zeta_1 = 0.7i, \zeta_2 = 0.4i, c_1 = c_2 = 1$, such a bound state is illustrated in Figure 2 with **Case A** and **Case D**. It can be seen that the "width" of this solution changes periodically with time, thus this solution is called a "breather" in literature. When $\alpha \neq 0$, two solitons do not form bound states, but the attraction ability between solitons will change with the change of parameters, see the figures of **Case A** and **Case D** at below.

In the following figures we notice:

Case A: $\alpha = 0$, $\gamma = 1$. In this case the ENLS equation will be decayed to LPD equation, and $u(x,t)$ is the soliton solution of LPD equation.

Case B: $\alpha = 1$, $\gamma = 1$. This is the ENLS equation.

Case C: $\alpha = 1$, $\gamma = 0$. In this case the ENLS equation will be decayed to Hirota equation, and $u(x,t)$ is the soliton solution of the Hirota equation.

Case D: $\alpha = 0$, $\gamma = 0$. In this case the ENLS equation will be decayed to NLS equation, and $u(x,t)$ is the soliton solution of the NLS equation.

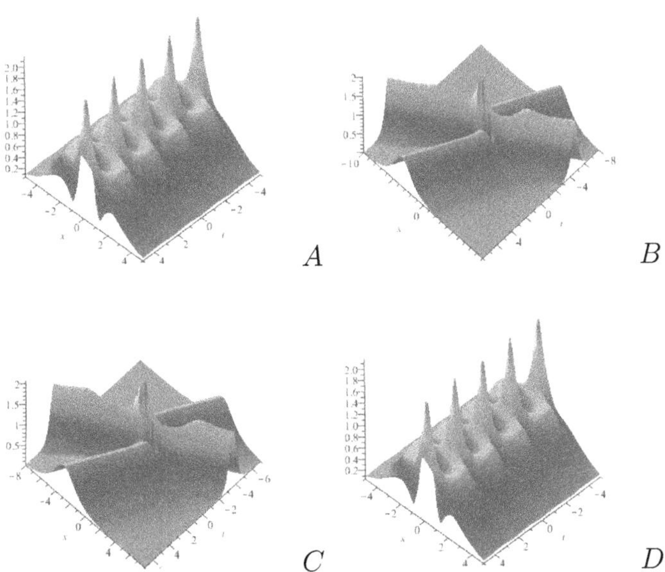

Figure 2. 3-D plot for the 2-soliton solution evolution of cases A, B, C and D. $\zeta 1 = 0.7i, \zeta 2 = 0.4i$.

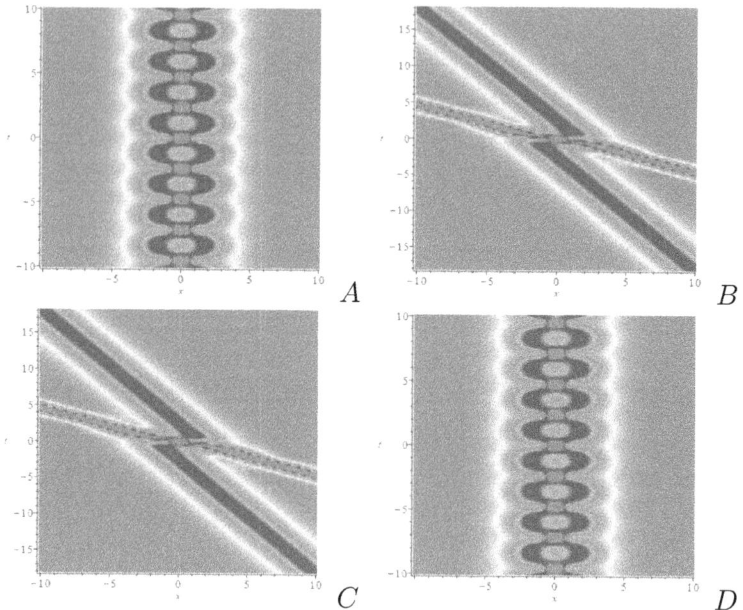

Figure 3. The plot for the 2-soliton solution evolution of cases A, B, C and D. $\zeta 1 = 0.7i, \zeta 2 = 0.4i$.

Except for the case where both spectral parameters are purely imaginary, let's consider the more complex case.
Let $\zeta_1 = 0.1 + 0.7i, \zeta_2 = -0.1 + 0.4i, c_1 = c_2 = 1$. We see from Figure 4 that as $t \to -\infty$, the solution consists of two single solitons which are far apart and moving toward each other. When they collide, they interact strongly. But when $t \to \infty$, these solitons re-emerge out of interactions without any change of shape and velocity, and there is no energy radiation emitted to the far field. Thus the interaction of these solitons is elastic (see Figure 6). This elastic interaction is a remarkable property which signals that the ENLS equation is integrable. There is still some trace of the interaction, however. Indeed, after the interaction, each soliton acquires a position shift and a phase shift (see Figure 5). The position of each soliton is always shifted forward (toward the direction of propagation), as if the soliton accelerates during interactions.

Figure 4 is typical of all two-soliton solutions (3.11) except $\xi_1 = \xi_2 = 0$, we can easily find that (α, γ) will change the velocity phase of the soliton figure. We analyze the asymptotic states of the solution (3.11) as $t \to \pm\infty$ and (α, γ) is non-negative. Without loss of generality, Let $\zeta_k = \xi_k + i\eta_k$ and assume that $|\xi_1| > |\xi_2|$. This means that at $t = -\infty$, soliton-1 is on the right side of soliton-2 and moves slower. Note also that $\eta_k > 0$ and $\eta_2 > \eta_1$, since $\zeta_k \in \mathbb{C}_+$. In the moving frame with velocity

-$(32\,{\eta_1}^2\gamma\,\xi_1 - 32\,\gamma\,{\xi_1}^3 + 4\,\alpha\,{\eta_1}^2 - 12\,\alpha\,{\xi_1}^2 + 2\,\xi_1)$, so (α, γ) will influence the velocity.

$$
\begin{aligned}
&Re(\theta_1) = \eta_1\left(32\,t\gamma\,\xi_1\,{\eta_1}^2 - 32\,t\gamma\,{\xi_1}^3 + 4\,t\alpha\,{\eta_1}^2 - 12\,t\alpha\,{\xi_1}^2 + 2t\xi_1 + x\right) = O(1),\\
&Re(\theta_2) = \eta_2\left(32\,t\gamma\,\xi_2\,{\eta_1}^2 - 32\,t\gamma\,{\xi_1}^3 + 4\,t\alpha\,{\eta_1}^2 - 12\,t\alpha\,{\xi_1}^2 + 2t\xi_1 + x\right)\\
&+\eta_2((32\,t\gamma\,\xi_2\,{\eta_2}^2 - 32\,t\gamma\,\xi_2\,{\eta_1}^2) + (32\,\gamma\,({\xi_1}^3 - {\xi_2}^3) + 4\,\alpha\,({\eta_2}^2 - {\eta_1}^2)\\
&+12\,\alpha\,({\xi_1}^2 - {\xi_2}^2) + 2\,(\xi_2 - 2\,\xi_1)))t.
\end{aligned}
\tag{3.13}
$$

When $t \to -\infty, \mathrm{Re}\,(\theta_2) \to +\infty$. When $t \to +\infty, \mathrm{Re}\,(\theta_2) \to -\infty$. In this case, simple calculations show that the asymptotic state of the solution (3.5) is

$$
u(x,t) \to \begin{cases} 2i\,(\zeta_1^* - \zeta_1)\,\dfrac{c_1^- e^{\theta_1 - \theta_1^*}}{e^{-\left(\theta_1 + \theta_1^*\right)} + |c_1^-|^2 e^{\theta_1 + \theta_1^*}}, & t \to -\infty,\\[2ex] 2i\,(\zeta_1^* - \zeta_1)\,\dfrac{c_1^+ e^{\theta_1 - \theta_1^*}}{e^{-\left(\theta_1 + \theta_1^*\right)} + |c_1^+|^2 e^{\theta_1 + \theta_1^*}}, & t \to +\infty, \end{cases}
\tag{3.14}
$$

where $c_1^- = \frac{c_1(\zeta_1 - \zeta_2)}{(\zeta_1 - \zeta_2^*)}$, $c_1^+ = \frac{c_1(\zeta_1 - \zeta_2^*)}{(\zeta_1 - \zeta_2)}$. Comparing this expression with (3.11), we see that this asymptotic solution is a single-soliton solution with peak amplitude $2\eta_1$ and velocity $32\,{\eta_1}^2\gamma\,\xi_1 - 32\,\gamma\,{\xi_1}^3 + 4\,\alpha\,{\eta_1}^2 - 12\,\alpha\,{\xi_1}^2 + 2\,\xi_1$. This indicates that if we fix the parameters, this soliton does not change its shape and velocity after collision.

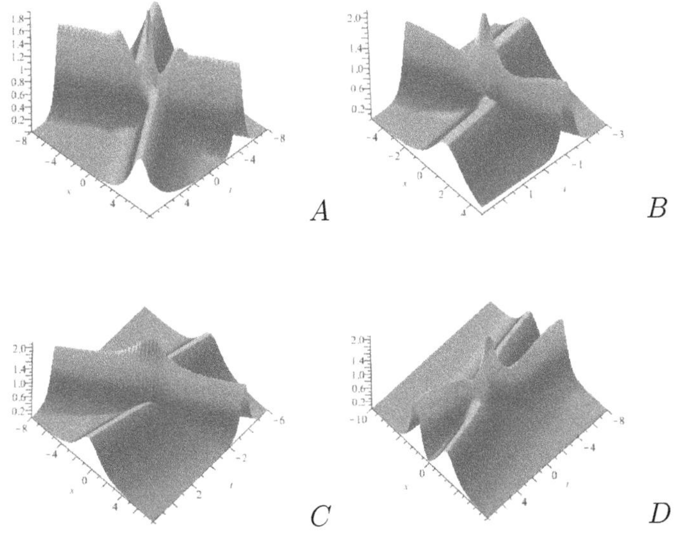

Figure 4. 3-D plot for the 2-soliton solution evolution of cases A, B, C and D. $\zeta 1 = 0.1 + 0.7i, \zeta 2 = -0.1 + 0.4i$.

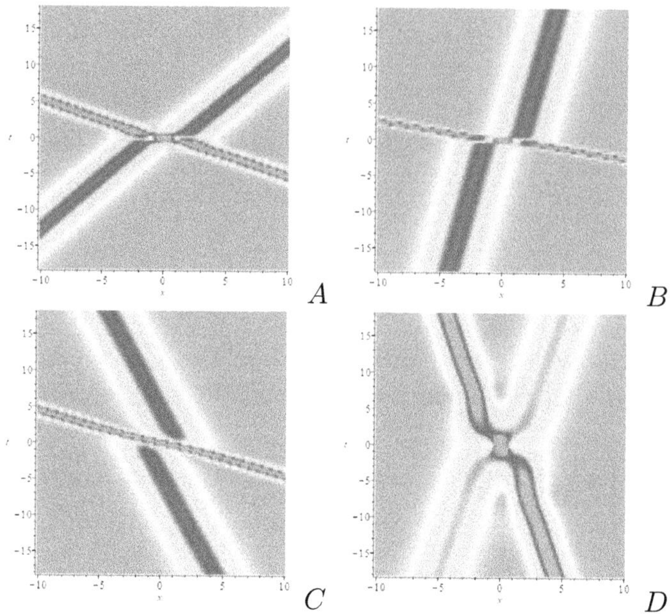

Figure 5. The plot for the 2-soliton solution evolution of cases A, B, C and D. $\zeta 1 = 0.1 + 0.7i, \zeta 2 = -0.1 + 0.4i$.

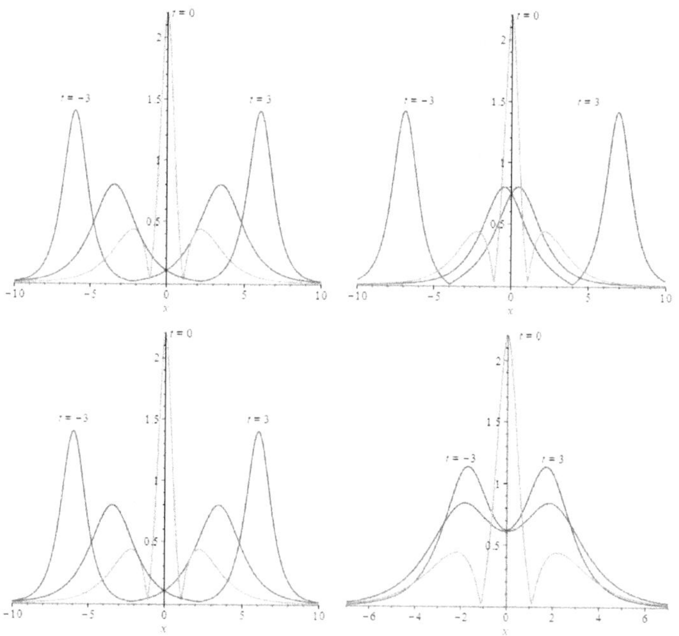

Figure 6. The plot for the 2-soliton solution evolution of cases A, B, C and D. $\zeta 1 = 0.1 + 0.7i, \zeta 2 = -0.1 + 0.4i$.

4 Soliton matrix for high-order zeros

In this case, following the discussion of simple zeros, we consider the high-order zeros in RHP of ENLS equation. First of all, we let functions $P^+(\zeta)$ and $P^-(\zeta)$ from the above RHP have only one pair of zero of order n, i.e., $\zeta_1, \bar{\zeta}_1$.

$$|P^+(\zeta)| = (\zeta - \zeta_1)^n \varphi(\zeta), \quad |P^-(\zeta)| = (\zeta - \bar{\zeta}_1)^n \bar{\varphi}(\zeta), \tag{4.1}$$

where $|\varphi(\zeta_1)| \neq 0$ and $|\bar{\varphi}(\bar{\zeta}_1)| \neq 0$.

Following the idea proposed in [31], we first consider the elementary zero case under the assumption that the geometric multiplicity of ζ_1 and $\bar{\zeta}_1$ has the same number. Hence, one needs to construct the dressing matrix $\Gamma(\zeta)$ whose determinant is $\frac{(\zeta-\zeta_1)^n}{(\zeta-\bar{\zeta}_1)^n}$. As a special case, we first consider the elementary zeros which have geometric multiplicity 1. In this case, Γ is constituted of n elementary dressing factors, i.e., $\Gamma = \chi_n \chi_{n-1} \cdots \chi_1$, where

$$\chi_i(\zeta) = \mathbb{I} + \frac{\bar{\zeta}_1 - \zeta_1}{\zeta - \bar{\zeta}_1} \mathbb{P}_i, \quad \mathbb{P}_i = \frac{|v_i\rangle\langle\bar{v}_i|}{\langle\bar{v}_i|v_i\rangle}, \quad |v_i\rangle \in \mathrm{Ker}(P_+ \chi_1^{-1} \cdots \chi_{i-1}^{-1}(\zeta_1)).$$

In addition, if we let $\hat{P}^+(\zeta) = P^+(\zeta)\chi_1^{-1}(\zeta)$ and $\hat{P}^-(\zeta) = \chi_1(\zeta)P^-(\zeta)$, then it is proved that matrices $\hat{P}^+(\zeta)$ and $\hat{P}^-(\zeta)$ are still holomorphic in the respective half-plane of $\mathbb{C}$. Moreover, ζ_1 and $\bar{\zeta}_1$ are still a pair of zeros of $|\hat{P}^+(\zeta)|$ and $|\hat{P}^-(\zeta)|$, respectively. Thus, $\Gamma(\zeta)^{-1}$ cancels all the high-order zeros for $|P^+(\zeta)|$. Moreover, it is necessary to reformulate the dressing factor into summation of fractions, then we derive the soliton matrix $\Gamma(\zeta)$ and its inverse for a pair of an elementary high-order zero. The results can be formulated in the following lemma.

Lemma 1. *Consider a pair of an elementary high-order zero of order n : $\{\zeta_1\}$ in $\mathbb{C}_+$ and $\{\bar{\zeta}_1\}$ in $\mathbb{C}_-$. Then the corresponding soliton matrix and its inverse can be cast in the following form*

$$\Gamma^{-1}(\zeta) = I + (|p_1\rangle, \cdots, |p_n\rangle)\, \mathcal{D}(\zeta) \begin{pmatrix} \langle q_n| \\ \vdots \\ \langle q_1| \end{pmatrix}, \quad \Gamma(\zeta) = I + (|\bar{q}_n\rangle, \cdots, |\bar{q}_1\rangle)\, \bar{\mathcal{D}}(\zeta) \begin{pmatrix} \langle \bar{p}_1| \\ \vdots \\ \langle \bar{p}_n| \end{pmatrix}, \tag{4.2}$$

where $\mathcal{D}(\zeta)$ and $\bar{\mathcal{D}}(\zeta)$ are $n \times n$ block matrices,

$$\mathcal{D}(\zeta) = \begin{pmatrix} (\zeta-\zeta_1)^{-1} & (\zeta-\zeta_1)^{-2} & \cdots & (\zeta-\zeta_1)^{-n} \\ 0 & \ddots & \ddots & \vdots \\ \vdots & \ddots & (\zeta-\zeta_1)^{-1} & (\zeta-\zeta_1)^{-2} \\ 0 & \cdots & 0 & (\zeta-\zeta_1)^{-1} \end{pmatrix},$$

$$\bar{\mathcal{D}}(\zeta) = \begin{pmatrix} (\zeta-\zeta_1)^{-1} & 0 & \cdots & 0 \\ (\zeta-\zeta_1)^{-2} & (\zeta-\zeta_1)^{-1} & \ddots & \vdots \\ \vdots & \ddots & \ddots & 0 \\ (\zeta-\zeta_1)^{-n} & \cdots & (\zeta-\zeta_1)^{-2} & (\zeta-\zeta_1)^{-1} \end{pmatrix}.$$

This lemma can be proved by induction as in [31]. Besides, we notice that in the expressions for $\Gamma^{-1}(\zeta)$ and $\Gamma(\zeta)$, only half of the vector parameters, i.e., $|p_1\rangle, \cdots, |p_n\rangle$ and $\langle\bar{p}_1|, \cdots, \langle\bar{p}_n|$ are independent. In fact, the rest of the vector parameters in (4.2) can be derived by calculating the poles of each order in the identity $\Gamma(\zeta)\Gamma^{-1}(\zeta) = I$ at $\zeta = \zeta_1$

$$\Gamma(\zeta_1)\begin{pmatrix} |p_1\rangle \\ \vdots \\ |p_n\rangle \end{pmatrix} = 0,$$

where

$$\Gamma(\zeta) = \begin{pmatrix} \Gamma(\zeta) & 0 & \cdots & 0 \\ \frac{d}{d\zeta}\Gamma(\zeta) & \Gamma(\zeta) & \ddots & \vdots \\ \vdots & \ddots & \ddots & 0 \\ \frac{1}{(n-1)!}\frac{d^{n-1}}{d\zeta^{n-1}}\Gamma(\zeta) & \cdots & \frac{d}{d\zeta}\Gamma(\zeta) & \Gamma(\zeta) \end{pmatrix}.$$

Hence, in terms of the independent vector parameters, results (4.2) can be formulated in a more compact form as in [31] and here we just avoid these overlapped parts. In the following, we derive this compact formula via the method of gDT [18]. We intend to investigate the relation between dressing matrices and DT for ENLS equation in the high-order zero case. The essence of DT is a kind of gauge transformation. Following the project proposed in [6], we can construct the gDT for ENLS equation as well.

The elementary form of DT has already been constructed in [44], then it is obvious to notice that: $G_1(\zeta_1+\epsilon)|v_1(\zeta_1+\epsilon)\rangle = 0$. Denoting $\left|\chi_1^{[0]}(\zeta_1)\right\rangle = |v_1(\zeta_1)\rangle$, and considering the following limitation:

$$\left|\chi_1^{[1]}(\zeta_1)\right\rangle \triangleq \lim_{\epsilon\to 0}\frac{G_1(\zeta_1+\epsilon)\left|\chi_1^{[0]}(\zeta_1+\epsilon)\right\rangle}{\epsilon} = \frac{d}{d\zeta}\left[G_1(\zeta)\left|\chi_1^{[0]}(\zeta)\right\rangle\right]_{\zeta=\zeta_1},$$

then $\left|\chi_1^{(1)}\right\rangle$ can be used to construct the next step DT, i.e.,

$$G_1^{[1]}(\zeta) = \left(I + \frac{\bar{\zeta}_1-\zeta_1}{\zeta-\bar{\zeta}_1}\mathbb{P}_1^{[1]}\right), \quad \mathbb{P}_1^{[1]} = \frac{\left|\chi_1^{[1]}\right\rangle\left\langle\chi_1^{[1]}\right|}{\left\langle\chi_1^{[1]}|\chi_1^{[1]}\right\rangle}.$$

Generally, continuing this process we obtain:

$$\left|\chi_1^{[N]}\right\rangle = \lim_{\epsilon\to 0}\frac{G_1^{[N-1]}\ldots G_1^{[1]}G_1^{[0]}(\zeta_1+\epsilon)\left|\chi_1^{[0]}(\zeta_1+\epsilon)\right\rangle}{\epsilon^N}.$$

The N-times generalized Darboux matrix can be represented as:

$$T_N(\zeta) = G_1^{[N-1]}\ldots G_1^{[1]}G_1^{[0]}(\zeta),$$

where

$$G_1^{[i]}(\zeta) = \left(I + \frac{\bar{\zeta}_i - \zeta_i}{\zeta - \bar{\zeta}_i}\mathbb{P}_1^{[i]}\right), \quad \mathbb{P}_1^{[i]} = \frac{\left|\chi_1^{[i]}\right\rangle\left\langle\chi_1^{[i]}\right|}{\left\langle\chi_1^{[i]}|\chi_1^{[i]}\right\rangle}.$$

In addition, the transformation between different potential matrices is:

$$Q^{(N)} = Q + i\left[\Lambda, \sum_{j=0}^{N-1}\left(\bar{\zeta}_1 - \zeta_1\right)\mathbb{P}_1^{[j]}\right].$$

In this expression, $P_1^{[i]}$ is rank 1 matrix, so $G_1^{[i]}(\zeta)$ can be also decomposed into the summation of simple fraction; that means the multiple product form of T_N can be directly simplified by the conclusion of Lemma 1. In other words, the above generalized Darboux matrix for ENLS equation can be given in the following theorem:

Theorem 1. *In the case of one pair of elementary high-order zero, the generalized Darboux matrix for ENLS equation can be represented as [44]:*

$$T_N = I - YM^{-1}\bar{\mathcal{D}}(\zeta)Y^{\dagger},$$

where $\bar{\mathcal{D}}(\zeta)$ is $N \times N$ block Toeplitz matrix which has been given before, Y is a $2 \times N$ matrix:

$$Y = \left(|v_1\rangle, \ldots, \frac{|v_1\rangle^{(N-1)}}{(N-1)!}\right),$$

$$|v_1\rangle^{(j)} = \lim_{\epsilon\to 0}\frac{d^j}{d\epsilon^j}\left|v_1\left(\zeta_1 + \epsilon\right)\right\rangle,$$

and M is $N \times N$ matrix:

$$M = \left(\ M^{[ij]}\ \right), \quad M^{[ij]} = \left(M_{l,m}^{[i,j]}\right)_{N\times N},$$

with

$$M_{l,m}^{[i,j]} = \lim_{\epsilon,\bar{\epsilon}\to 0}\frac{1}{(l-1)!(m-1)!}\frac{\partial^{m-1}}{\partial\epsilon^{m-1}}\frac{\partial^{l-1}}{\partial\bar{\epsilon}^{l-1}}\left[\frac{\langle y_i|y_j\rangle}{\zeta_j - \bar{\zeta}_i + \epsilon - \bar{\epsilon}}\right].$$

Theorem 1 can be proved via direct calculation as in [6].

Therefore, if $\Phi^{|N|} = T_N\Phi$, then $\Phi^{[N]}$ indeed solves spectral problem (2.1). Substituting T_N into the above relation and letting spectral ζ go to infinity, we have the relation:

$$Q^{[N]} = Q - i\left[\Lambda, \left(|v_1\rangle, \ldots, \frac{|v_1\rangle^{(N-1)}}{(N-1)!}\right)M^{-1}\begin{pmatrix}\langle v_1| \\ \vdots \\ \frac{\langle v_1|^{(N-1)}}{(N-1)!}\end{pmatrix}\right].$$

Moreover, the transformations between the potential functions are

$$Q_{j,l}^{[N]} = Q_{j,l}^{[0]} + 2i\frac{|A_{j,l}|}{|M|}, \quad A_{j,l} = \begin{bmatrix} M & Y[l]^{\dagger} \\ Y[j] & 0\end{bmatrix}, 1 \le j, l \le 2. \tag{4.3}$$

Here the subscript $_{j,l}$ denotes the j th row and l th column element of matrix A, and $Y[l]$ represents the j th row of matrix Y.

5 Dynamics of high-order solitons in the ENLS equation

For simple, we consider the second-order fundamental soliton, which corresponds to a single pair of purely imaginary eigenvalues, $\zeta_1 = i\eta_1 \in i\mathbb{R}_+$, and $\bar{\zeta}_1 = i\bar{\eta}_1 \in i\mathbb{R}_-$, where $\eta_1 > 0$ and $\bar{\eta}_1 = -\eta_1 < 0$. In this case, taking $v_{10}(\epsilon) = \left[1, e^{i\theta_{10}-\theta_{11}\epsilon}\right]^T$ and $\bar{v}_{10}(\bar{\epsilon}) = \left[1, e^{i\bar{\theta}_{10}-\bar{\theta}_{11}\bar{\epsilon}}\right]^T$, where $\theta_{10}, \theta_{11}, \bar{\theta}_{10}, \bar{\theta}_{11}$ are real constants. Substituting these expressions into high-order soliton formula (4.3) with $N = 2, Q^{[0]}_{1,2} = 0$. Then we obtain an analytic expression for the second-order fundamental soliton solution of (1.2)

$$u(x,t) = 2(\eta_1 - \bar{\eta}_1)\frac{t_{11}e^{2\eta_1 x+(2i\eta_1^2+8\alpha\eta_1^3+16i\gamma\eta_1^4)t+i\bar{\theta}_{10}} + t_{12}e^{2\bar{\eta}_1 x+(2i\bar{\eta}_1^2+8\alpha\bar{\eta}_1^3+16i\gamma\bar{\eta}_1^4)t-i\theta_{10}}}{4\cosh^2(w)+F}, \tag{5.1}$$

$$w = (\eta_1 - \bar{\eta}_1)x + [i(\bar{\eta}_1^2 - \eta_1^2) + 4\alpha(\bar{\eta}_1^3 - \eta_1^3) + 8i\gamma(\bar{\eta}_1^4 - \eta_1^4)]t - \frac{i}{2}(\theta_{10} + \bar{\theta}_{10}),$$
$$t_{11} = (\bar{\eta}_1 - \eta_1)((64\bar{\eta}_1^3\gamma - 24i\bar{\eta}_1^2\alpha + 4\bar{\eta}_1)t - 2ix) - 2i,$$
$$t_{12} = (\eta_1 - \bar{\eta}_1)((64\eta_1^3\gamma - 24i\eta_1^2\alpha + 4\eta_1)t - 2ix) - 2i,$$
$$F(x,t) = (t_{11} + 2i)(t_{12} + 2i).$$

Let $\bar{\eta}_1 = -\eta_1, \bar{\theta}_{10} = -\theta_{10}$, $u(x,t)$ can be written in the form of a traveling solitary wave:

$$u(x,t) = \psi(x,t)e^{i(2\eta_1^2+16\gamma\eta_1^4-\theta_{10})}, \tag{5.2}$$

$$\psi(x,t) = 4\eta_1\frac{(t_{11}e^{2\eta_1 x+8\alpha\eta_1^3 t} + t_{12}e^{-2\eta_1 x-8\alpha\eta_1^3 t})}{4\cosh^2(2\eta_1 x - 8\alpha\eta_1^3 t) + F}, \tag{5.3}$$

$$|\psi(x,t)|^2 = 16\eta_1^2\frac{(t_{11}^2 e^{4\eta_1 x+16\alpha\eta_1^3 t} + t_{12}^2 e^{-4\eta_1 x-16\alpha\eta_1^3 t} + 2t_{11}t_{12})}{(4\cosh^2(2\eta_1 x - 8\alpha\eta_1^3 t) + F)^2}. \tag{5.4}$$

The center trajectory Σ_+ and Σ_- for this solution can be approximately described by the following two curves:

$$\Sigma_+ : (\eta_1 - \bar{\eta}_1)x + 4\alpha(\bar{\eta}_1^3 - \eta_1^3)t + \frac{1}{2}ln|F| = 0,$$
$$\Sigma_- : (\eta_1 - \bar{\eta}_1)x + 4\alpha(\bar{\eta}_1^3 - \eta_1^3)t - \frac{1}{2}ln|F| = 0.$$

Moreover, regardless of the effect brought by the logarithmic part when $t \to \pm\infty$, two solitons separately move along each curve at nearly the same velocity, which is approximate to $4\alpha(\bar{\eta}_1^2 + \eta_1\bar{\eta}_1 + \eta_1^2)$.

Due to $\eta_1 - \bar{\eta}_1 > 0$, with simple calculation, it is found that $|u(x,t)|$ possesses the following asymptotic estimation:

$$|u(x,t)| \to 0, \quad x \to \pm\infty.$$

However, with the development of time, a simple asymptotic analysis with estimation on the leading-order terms shows that when soliton (5.1) is moving on Σ_+ or Σ_-, its amplitudes $|u|$ can approximately vary as

$$u(x,t) \sim \begin{cases} \dfrac{2|\eta_1-\bar{\eta}_1|\mathrm{e}^{(\eta_1+\bar{\eta}_1)x}}{\left|\mathrm{e}^{(4i(\bar{\eta}_1^2-\eta_1^2)+32i\gamma(\bar{\eta}_1^4-\eta_1^4))t-i(\arg[\mathcal{F}(x,t)]+2k\pi)+i(\theta_{10}+\bar{\theta}_{10})}+1\right|}, & t\sim+\infty, \\ \\ \dfrac{2|\eta_1-\bar{\eta}_1|\mathrm{e}^{-(\eta_1+\bar{\eta}_1)x}}{\left|\mathrm{e}^{-(4i(\bar{\eta}_1^2-\eta_1^2)+32i\gamma(\bar{\eta}_1^4-\eta_1^4))t-i(\arg[\mathcal{F}(x,t)]+2k\pi)-i(\theta_{10}+\bar{\theta}_{10})}+1\right|}, & t\sim-\infty, \end{cases} \tag{5.5}$$

$k\in\mathbb{Z}$.

Let $\eta_1=\frac{i}{2},\bar{\eta}_1=-\frac{i}{2},\theta_{10}=\bar{\theta}_{10}=\theta_{11}=\bar{\theta}_{11}=0$. Because of the effect brought by the logarithmic part, two solitons separately move along each curve with different velocity, direction and shape with the different values of (α,γ) (see Figure 7).

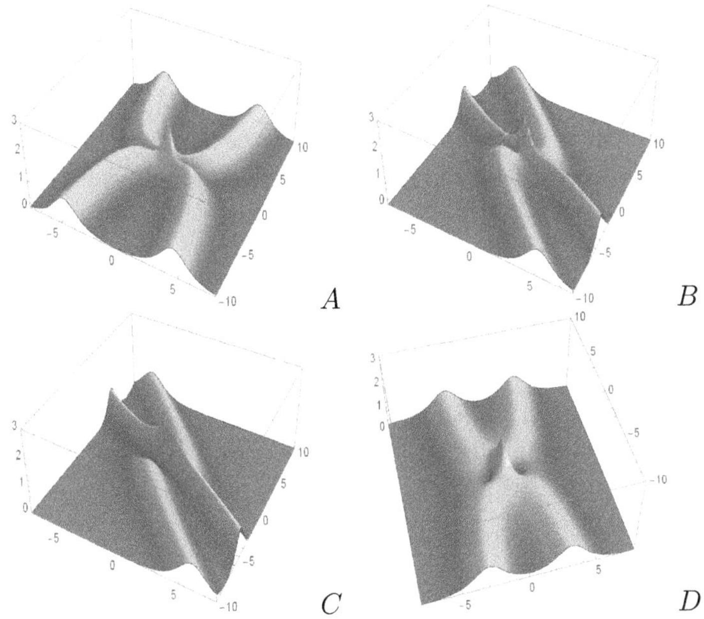

Figure 7. 3-D plot for the high-order solution evolution of cases A, B, C and D.

6 Conclusion

In the present chapter, we research the ENLS equation which can be used to describe the wave propagation in the optical fibers more accurately than the NLS equation. We find the soliton matrix corresponds to simple zeros for the ENLS equation in the framework of the RHP firstly. Then combined with gDT, soliton matrix for elementary high-order zeros in the RHP for the ENLS equation is constructed.

By analyzing the single-soliton solution, we found the wave width is only related to the imaginary part of the spectral parameter ζ and is not affected by the coefficients α or γ. However, the propagation direction and velocity of solitons can be controlled by adjusting the free parameters (α,γ). The special case is when the spectral parameter ζ has only an imaginary part ($\xi = 0$) or $|\eta| = |\xi|$, the fourth-order coefficient γ no longer affects the above quantities. When $|\eta| = |\sqrt{3}\xi|$, the third-order coefficients α no longer affect the above quantities.

Further, the asymptotic behavior for the two-solitons and the influence of free parameter (α, γ) in soliton solutions of ENLS equation on collision dynamics are studied. For the two-solitons, when the spectral parameters ζ_1, ζ_2 are pure imaginary numbers and $\alpha = 0$, the two constituent solitons will form a bound state and its amplitude function $|u(x,t)|$ will oscillate periodically with time. When $\alpha \neq 0$, two solitons do not form bound states, but the attraction ability between solitons will change with the change of parameters (α, γ).

Finally, for two-order one soliton, long-time asymptotic estimations for the high-order one-soliton are concretely analyzed. Obviously, because the effect brought by the logarithmic part of center trajectory Σ_+ and Σ_- , two solitons separately move along each curve with different velocity, direction and shape with the different values of (α, γ).

The above analysis uncovers the dynamic behaviors of solitons and high-order solitons clearly and can be used to guide optical experiments. In this paper, we research the high-order solitons of the ENLS equation with zero boundary by RHP. Inspired by [55], we will further explore the high-order soliton problem of integrable equations with non-zero boundary which is more complex but more pervasive in future study.

Acknowledgements

The project is supported by the National Natural Science Foundation of China (No.11675054), Shanghai Collaborative Innovation Center of Trustworthy Software for Internet of Things (No.ZF1213), and Science and Technology Commission of Shanghai Municipality (No.18dz2271000).

References

[1] Ablowitz M J, Chakravarty S, Trubatch A D and Villarroel J, A novel class of solutions of the non-stationary Schrödinger and the Kadomtsev-Petviashvili I equations, *Phys. Lett. A.* **267**, 132-146, 2000.

[2] Ablowitz M J and Newell A C, The decay of the continuous spectrum for solutions of the KdV equations, *J. Math. Phys.* **14**, 1277-1284, 1973.

[3] Ablowitz M J, Kaup D J, Newell A C and Segur H, Method for solving the sine-Gordon equation, *Phys. Rev. Lett.* **30**, 1262-1264, 1973.

[4] Ankiewicz A and Akhmediev N, Higher-order integrable evolution equation and its soliton solutions, *Phys. Lett. A.* **378**, 358, 2014.

[5] Benneyand D J and Newell A C, The propagation of nonlinear wave envelopes, *J. Math. Phys.* **46**, 133-139, 1967.

[6] Bian D F, Guo B L and Ling L M, High-order soliton solution of Landau-Lifshitz equation, *Stud. Appl. Math.* **134**, 181-214, 2015.

[7] Biondini G and Mantzavinos D, Long-time asymptotics for the focusing nonlinear Schrödinger equation with nonzero boundary conditions at infinity and asymptotic stage of modulational instability, *Commun. Pure Appl. Math.* **70**, 2300-2365, 2017.

[8] Bilman D, Buckingham R and Wang D S, Large-order asymptotics for multiple-pole solitons of the focusing nonlinear Schrödinger equation II: Far-field behavior, arXiv:1911.04327v1, 2019.

[9] Bilman D, Ling L M and Miller D P, Extreme superposition: Rogue waves of infinite order and the Painléve-III hierarchy, arXiv:1806.00545v1, 2018.

[10] Deift P and Zhou X, A steepest descent method for oscillatory Riemann-Hilbert problems asymptotics for the mKdV equation, *Ann. of Math.* **137**, 295-368, 1993.

[11] Deift P and Park J, Long-time asymptotics for solutions of the NLS equation with a delta potential and even initial data, *Lett. Math. Phys.* **96**, 143-156, 2011.

[12] Deift P, Venakides S and Zhou X, New results in small dispersion KdV by an extension of the steepest descent method for Riemann-Hilbert problems, *Int. Math. Res. Notices* **6**, 285-299, 1997.

[13] Deift P and Zhou X, Long-time asymptotics for solutions of the NLS equation with initial data in a weighted Sobolev space, *Commun. Pure Appl. Math.* **56**, 1029-1077, 2003.

[14] Estabrook F B and Wahlquist H D, Prolongation structures of non-linear evolutions equations II. The non-linear Schrödinger equation. *J. Math. Phys.* **17**, 1293-1297, 1976.

[15] Fokas A S and Lenells J, The unified method: I non-linearizable problems on the half-line, *J. Phys. A: Math. Theor.* **45**, 195201, 2012.

[16] Gagnon L and Stiévenart N, N soliton interaction in optical fibers: the multiple pole case, *Opt. Lett.* **19**, 619-621, 1994.

[17] Gardner C S, Greene I M, Kruskal M D and Miura R M, Method for solving the Korteweg-de Vries equation, *Phys. Rev. Lett.* **19**, 1095-1097, 1967.

[18] Guo B L, Ling L M and Liu Q P, High order solutions and generalized Darboux transformations of derivative nonlinear Schrödinger equations, *Stud. Appl. Math.* **130**, 317-344, 2012.

[19] Hasegawa A and Tappert F, Transmission of stationary nonlinear optical pulses in dispersive dielectric fibres I, Anomalous dispersion, *Appl. Phys. Lett.* **23**, 142-144, 1973.

[20] Hasegawa A and Tappert F, Transmission of stationary nonlinear optical pulses in dispersive dielectric fibres II, Normal dispersion, *Appl. Phys. Lett.* **23**, 171-172, 1973.

[21] Its A R, Asymptotics of solutions of the nonlinear Schrödinger equation and isomonodromic deformations of systems of linear differential equations, *Sov. Math. Dokl.* **24**, 452-456, 1981.

[22] Jimbo M, Miwa T, Mori Y and Sato M, Density matrix an impenetrable bose gas and the Painleve transcendent, *Physica D* **1**, 80-139, 1980.

[23] Lax P D, Integrals of nonlinear equations of evolution and solitary waves, *Commun. Pure Appl. Math.* **21**, 467-490, 1968.

[24] Lenells J and Fokas A S, The unified method: II. NLS on the half-line t-periodic boundary conditions, *J. Phys. A: Math. Theor.* **45**, 195202, 2012.

[25] Lenells J and Fokas A S, The unified method: III. Nontinearizable problem on the interval, *J. Phys. A: Math. Theor.* **45**, 195203, 2012.

[26] Manakov S V, Nonlinear Fraunhofer diffraction, *Zh. Eksp. Teor. Fiz.* **65**, 1392-1398, 1973.

[27] Monvel A B, Its A and Kotlyarov V, Long-time asymptotics for the focusing NLS equation with time-periodic boundary condition on half-line, *Commun. Math. Phys.* **290**, 479-522, 2009.

[28] Novikov S, Manakov S V, Pitaevskii L P and Zakharov V E, *Theory of Solitons*, Plenum, New York, 1984.

[29] Peng W Q, Tian S F, Wang X B, Zhang T T and Fang Y, Riemann-Hilbert method and multi-soliton solutions for three-component coupled nonlinear Schrödinger equations, *J. Geom. Phys.* **146**, 103508, 2019.

[30] Shabat A B, The Korteweg de Vries equation. *Sov. Math. Dokl.* **14**, 1266-1269, 1973.

[31] Shchesnovich V S and Yang J K, Higher-order solitons in the N-wave system. *Stud. Appl. Math.* **110**, 297-332, 2003.

[32] Shchesnovich V S and Yang J K, General soliton matrices in the Riemann-Hilbert problem for integrable nonlinear equations. *J. Math. Phys.* **44**, 4604-4639, 2003.

[33] Tovbis A, Venakides S and Zhou X, On long time behavior of semiclassical (zero dispersion) limit of the focusing NLS equation: pure radiaton case, *Commun. Pure Appl. Math.* **59**, 1379-1432, 2006.

[34] Tsuru H and Wadati M, The multiple pole solutions of the sine-Gordon equation, *J. Phys. Soc. Jpn.* **53**, 2908-2921, 1984.

[35] Vartanian A H, Long-time asymptotics of solutions to the Cauchy problem for the defocusing nonlinear Schrödinger equation with finite-density initial data, *Math. Phys. Anal. Geom.* **5**, 319-413, 2002.

[36] Villarroel J and Ablowitz M J, On the discrete spectrum of the non-stationary Schrödinger equation and multipole lumps of the Kadomtsev-Petviashvili I equation, *Comm. Math. Phys.* **207**, 1-42, 1999.

[37] Wahlquist H D and Estabrook F B, Prolongation structures of non-linear evolutions equations I. The KdV equation. *J. Math. Phys.* **16**, 1-7, 1975.

[38] Wang H T and Wen X Y, Soliton elastic interactions and dynamical analysis of a reduced integrable nonlinear Schrödinger system on a triangular-lattice ribbon, *Nonlinear Dyn.* **100**, 1571-1587, 2020.

[39] Xu J and Fan E G, The unified transformation method for the Sasa-Satsuma equation on the half-line, *Proc. R. Soc. A.* **469**, 20130068, 2013.

[40] Xu J and Fan E G, Long-time asymptotics for the Fokas-Lenells equation with decaying initial value problem: Without solitons, *J. Differ. Equations.* **259**, 1098-1148, 2015.

[41] Xu T and He G, The coupled derivative nonlinear Schrödinger equation: conservation laws, modulation instability and semirational solutions, *Nonlinear Dyn.* **100**, 2823-2837, 2020.

[42] Xu T and He G, Higher-order interactional solutions and rogue wave pairs for the coupled Lakshmanan-Porsezian-Daniel equations, *Nonlinear Dyn.* **98**, 1731-1744, 2019.

[43] Yang B and Chen Y, Dynamics of high-order solitons in the nonlocal nonlinear Schrödinger equations, *Nonlinear Dyn.* **94**, 489-502, 2018.

[44] Yang B and Chen Y, High-order soliton matrices for Sasa-Satsuma equation via local Riemann-Hilbert problem, *Nonlinear Anal-Real.* **45**, 918-941, 2019.

[45] Yang J K, *Nonlinear Waves in Integrable and Non integrable Systems*, SIAM, Philadelphia, 2010.

[46] Yue Y F, Huang L L and Chen Y, Modulation instability, rogue waves and spectral analysis for the sixth-order nonlinear Schrödinger equation, *Commun. Nonlinear Sci. Numer. Simulat.* **89**, 105284, 2020.

[47] Zabusky N J and Kruskal M D, Interaction of solitons in a collisionless plasma and the recurrence of initial states, *Phys. Rev. Lett.* **15**, 240-243, 1965.

[48] Zakharov V E, Stability of periodic waves of finite amplitude on the surface of a deep fluid, *Sov. Phys. J. Appl. Mech. Tech.* **4**, 190-194, 1968.

[49] Zakharov V E, Collapse of langmuir waves, *Sov. Phys. J. Appl. Mech. Tech.* **35**, 908-914, 1972.

[50] Zakharov V E and Shabat A B, Exact theory of two-dimensional self-focusing and one-dimensional self-focusing and one-dimensional self-modulation of waves in nonlinear media, *Sov. Phys. JETP.* **34**, 62-69, 1972.

[51] Zakharov V E and Manakov S V, Asymptotic behavior of nonlinear wave systems integrated by the inverse scattering method, *Sov. Phys. JETP.* **44**, 106-122, 1976.

[52] Zhou X, The Riemann-Hilbert problem and inverse scattering, *SIAM J. Math. Anal.* **20**, 966-986, 1989.

[53] Zhang X E and Chen Y, Inverse scattering transformation for generalized non-blinear Schrödinger equation, *Appl. Math. Lett.* **98**, 306-313, 2019.

[54] Zhang Y S, Tao X and Xu S, The bound-state soliton solutions of the complex modified KdV equation, *Inverse probl.* **36**, 065003, 2020.

[55] Zhao Y and Fan E G, Inverse scattering transformation for the Fokas-Lenells equation with nonzero boundary conditions, arXiv: 1912.12400 [nlin.SI], 2019.

[56] Zhang Y S, Rao J and Cheng Y, Riemann-Hilbert method for the Wadati-Konno-Ichikawa equation: N simple poles and one higher-order pole, *Physica D* **399**, 173-185, 2019.

A8. Darboux transformation for integrable systems with symmetries

Zi-Xiang Zhou

School of Mathematical Sciences, Fudan University
Shanghai 200433, P.R. China
zxzhou@fudan.edu.cn

Abstract

The Darboux transformation method is useful in getting explicit solutions of integrable systems. When the system has no symmetries (i.e., all the entries of the coefficients of the Lax pair are independent), the construction of Darboux transformation is well known, which was given by Neugebauer and Meinel for 2×2 systems, then by Gu, Sattinger and Zurkowski for general systems. However, in practice, there should be some symmetries. To keep all these symmetries, the construction is much more difficult. In this chapter, we shall first review the general construction of Darboux transformation for systems without symmetries. Then the Darboux transformation for systems with unitary symmetry is discussed. The equivalent construction given by Zakharov and Mikhailov is also reviewed. By analyzing the spectrum, soliton solutions and breather solutions of MKdV equation are obtained. Finally, the Darboux transformation for a two-dimensional affine Toda equation, which has more complicated symmetries, is constructed.

1 Introduction

For the equation

$$-\phi_{xx} - u(x)\phi = \lambda\phi, \tag{1.1}$$

which is now called the one-dimensional stationary Schrödinger equation, Darboux introduced a transformation for both u and ψ in 1882 (see [5]). The transformation is as follows. Let λ_0 be a constant, and f be a solution of (1.1) with $\lambda = \lambda_0$, then

$$u'(x) = u(x) + 2(\ln f)_{xx}, \quad \phi'(x, \lambda) = \phi_x(x, \lambda) - \frac{f_x}{f}\phi(x, \lambda) \tag{1.2}$$

satisfy

$$-\phi'_{xx} - u'\phi' = \lambda\phi'. \tag{1.3}$$

This transformation $(u, \phi) \to (u', \phi')$ is now called a Darboux transformation.

Owing to the Miura transformation (see [18]), it was known in the 1960s that the KdV equation

$$u_t + 6uu_x + u_{xxx} = 0 \tag{1.4}$$

has a Lax pair

$$\begin{aligned} & -\phi_{xx} - u\phi = \lambda\phi, \\ & \phi_t = -4\phi_{xxx} - 6u\phi_x - 3u_x\phi. \end{aligned} \tag{1.5}$$

The first equation of (1.5) is the same as (1.1). Hence the Darboux transformation (1.2) was used to get new solutions of the KdV equation from a known one.

After the action of Darboux transformation, the solution ϕ' of the Lax pair for the derived u' is also known. Therefore, the Darboux transformations can be done successively by a purely algebraic way as

$$(u,\phi) \xrightarrow{\lambda_1, f_1} (u',\phi') \xrightarrow{\lambda_2, f_2'} (u'',\phi'') \xrightarrow{\lambda_3, f_3''} (u''',\phi''') \longrightarrow \cdots.$$

By using Miura transformation, a Darboux transformation for MKdV equation, whose Lax pair is in matrix form, could also be constructed (see [28]). Then the Darboux transformation, which is in matrix form, was applied to a lot of 1+1 dimensional and higher dimensional integrable systems after that.

We say that a Lax pair has no symmetries if the entries of the coefficient matrices of the Lax pair are independent. In this case, the general construction of Darboux transformations was given by Neugebauer and Meinel for 2×2 systems, then by Gu, Sattinger and Zurkowski for general systems. In these constructions, the Darboux transformation depends on the free choice of spectral parameters and the solutions of the Lax pairs. However, in practice, there should be some symmetries in the Lax pair, i.e., the entries of the coefficients of the Lax pairs are dependent. To keep all these symmetries, the spectral parameters and the solutions of the Lax pairs cannot be chosen arbitrarily. They are related to each other according to the symmetries of the Lax pair. If the symmetries are complicated, the construction is much more difficult. In this chapter, we start with general construction of Darboux transformations and focus on how to make the Darboux transformation keep all the symmetries of the Lax pairs.

In Section 2, the general construction of Darboux transformation for systems without symmetries is reviewed. Then in Section 3, the Darboux transformation for systems with unitary symmetry is discussed. The equivalent construction given by Zakharov and Mikhailov is also reviewed. The Darboux transformation for the nonlinear Schrödinger equation is listed as an example.

To illustrate the distribution of the spectral parameters, the soliton solution and breather solution of the MKdV equation are discussed in Section 4. The minimal number of real spectral parameters in constructing Darboux transformation is 2 and that of complex spectral parameters is 4. This leads to a Darboux transformation of degree 1 for getting soliton solutions and a Darboux transformation of degree 2 for getting breather solutions.

The two-dimensional affine Toda equations have much more complicated symmetries. In Section 5, the two-dimensional $D_{l+1}^{(2)}$ Toda equation is discussed as an example. The Lax pair has a unitary symmetry, a reality symmetry and a cyclic symmetry simultaneously. The Darboux transformation keeping all these symmetries is constructed by proper choice of spectral parameters and solutions of the Lax pairs.

2 Darboux transformation for general 1+1 dimensional Lax pairs

Consider the general $n \times n$ over-determined linear system

$$\Phi_x = U(x, t, \lambda)\Phi, \quad \Phi_t = V(x, t, \lambda)\Phi \tag{2.1}$$

in 1+1 dimensions with variables (x, t), where $U(x, t, \lambda), V(x, t, \lambda)$ are $n \times n$ matrices whose entries are functions of (x, t, λ) and are polynomial of λ, λ is a spectral parameter. Hereafter, we shall often write $U(\lambda)$ or simply U instead of $U(x, t, \lambda)$.

The linear system (2.1) has the integrability condition

$$U_t - V_x + [U, V] = 0, \tag{2.2}$$

which should be independent of the spectral parameter λ. (2.2) is a system of partial differential equations that we want to study, and (2.1) is called its the Lax pair. Here $[U, V] = UV - VU$.

Example. When

$$U = \begin{pmatrix} \lambda & u \\ -\bar{u} & -\lambda \end{pmatrix}, \quad V = \begin{pmatrix} -2\mathrm{i}\lambda^2 - \mathrm{i}|u|^2 & -2\mathrm{i}u\lambda - \mathrm{i}u_x \\ 2\mathrm{i}\bar{u}\lambda - \mathrm{i}\bar{u}_x & 2\mathrm{i}\lambda^2 + \mathrm{i}|u|^2 \end{pmatrix}, \tag{2.3}$$

(2.2) leads to the nonlinear Schrödinger equation

$$\mathrm{i}u_t = u_{xx} + 2|u|^2 u. \tag{2.4}$$

A Darboux transformation for the Lax pair (2.1) is given by a matrix $D(x, t, \lambda)$, which is called a Darboux matrix, such that for any solution Φ of the Lax pair (2.1), $\Phi' = D\Phi$ satisfies a Lax pair

$$\Phi'_x = U'(x, t, \lambda)\Phi', \quad \Phi'_t = V'(x, t, \lambda)\Phi' \tag{2.5}$$

where U', V' have the same form as U, V. Here "have the same form" should be specified for concrete Lax pair. If there is no symmetry in it, D is a Darboux matrix simply means that $U'(x, t, \lambda)$, $V'(x, t, \lambda)$ are polynomials of λ.

The integrability condition

$$U'_t - V'_x + [U', V'] = 0 \tag{2.6}$$

of (2.5) means that the Darboux transformation transforms a solution of the nonlinear Equation (2.2) to a solution of (2.6). Written explicitly, we have

$$\begin{aligned} U' &= DUD^{-1} + D_x D^{-1}, \\ V' &= DVD^{-1} + D_t D^{-1}. \end{aligned} \tag{2.7}$$

Since the positions of U and V are symmetric, hereafter we only consider U. Similar conclusion is true for V.

Darboux matrix of degree n is a Darboux matrix which is a polynomial of λ of degree n.

The simplest Darboux matrix is a Darboux matrix of degree 1. In this section, we suppose there is no symmetry, which means that all the entries of U are independent, so are the entries of V. The general construction is as follows (see [9]), which was first given by Neugerbauer and Meinel for $n=2$ and by Gu, Sattinger and Zurkowski for general n (see [8, 19, 25]).

Theorem 1. *Let $\lambda_1, \cdots, \lambda_n$ be n complex constants. Let h_j be a column solution of the Lax pair (2.1) with $\lambda = \lambda_j$ $(j = 1, \cdots, n)$. Denote $\Lambda = \operatorname{diag}(\lambda_1, \cdots, \lambda_n)$, $H = (h_1, \cdots, h_n)$, which are $n \times n$ matrices. When $\det H \neq 0$, let $S = H\Lambda H^{-1}$. Then $D = \lambda I - S$ is a Darboux matrix of degree 1 for the Lax pair (2.1).*

Proof. Obviously, if the first equation of (2.7) holds for certain polynomial $U'(\lambda)$ of λ with matrix coefficient, the degree of $U'(\lambda)$ is the same as that of $U(\lambda)$. Suppose $U(\lambda) = \sum_{k=0}^{m} U_k \lambda^{m-k}$, $U'(\lambda) = \sum_{k=0}^{m} U'_k \lambda^{m-k}$, then the first equation of (2.7) is

$$\Big(\sum_{k=0}^{m} U'_k \lambda^{m-k}\Big)(\lambda I - S) = (\lambda I - S)\Big(\sum_{k=0}^{m} U_k \lambda^{m-k}\Big) - S_x. \tag{2.8}$$

Comparing the coefficient of λ^k, we have

$$U'_0 = U_0, \quad U'_{k+1} - U'_k S = U_{k+1} - SU_k - S_x \delta_{km} \quad (k = 0, 1, \cdots, m) \tag{2.9}$$

where $U_{m+1} = U'_{m+1} = 0$. $\{U'_k\}_{k=1,\cdots,m}$ can be solved recursively from (2.9) with $k = 0, 1, \cdots, m-1$ so that

$$U'_k = U_k + \Big[\sum_{j=0}^{k-1} U_{k-1-j} S^j, S\Big] \quad (k = 0, 1, \cdots, m). \tag{2.10}$$

The equation with $k = 0$ in (2.9) becomes

$$S_x + \Big[S, \sum_{k=0}^{m} U_k S^{m-k}\Big] = 0. \tag{2.11}$$

Hence $D(\lambda)$ is a Darboux matrix if and only if (2.11) holds.

Now $h_{j,x} = U(\lambda_j) h_j$, so

$$H_x H^{-1} = \sum_{k=0}^{m} U_k H \Lambda^{m-k} H^{-1} = \sum_{k=0}^{m} U_k S^{m-k}. \tag{2.12}$$

Hence

$$S_x = H_x \Lambda H^{-1} - H\Lambda H^{-1} H_x H^{-1} = \Big[\sum_{k=0}^{m} U_k S^{m-k}, S\Big], \tag{2.13}$$

which coincides with (2.11). Therefore, $D(\lambda)$ is a Darboux matrix. The theorem is proved.

In this construction, $\lambda_1, \cdots, \lambda_n$ are usually chosen not to be all the same. Otherwise, the derived Darboux transformation will be trivial.

Note that the construction of D in Theorem 1 is equivalent to the conditions $D(x, t, \lambda_j)h_j = 0$ $(j = 1, 2, \cdots, n)$. This is an important idea given by [19].

Likewise, we can construct Darboux matrix of degree r, that is, a Darboux matrix of form

$$D(x,t,\lambda) = \lambda^r I + D_1(x,t)\lambda^{r-1} + \cdots + D_{r-1}(x,t)\lambda + D_r(x,t), \tag{2.14}$$

by the following theorem (see [9]).

Theorem 2. *Given nr complex numbers $\lambda_1, \cdots, \lambda_{nr}$ and the corresponding column solution h_j of the Lax pair (2.1) with $\lambda = \lambda_j$ $(j = 1, \cdots, nr)$, a Darboux matrix of degree r can be uniquely determined by $D(x, t, \lambda_j)h_j = 0$ $(j = 1, \cdots, nr)$, provided that* $\det(\lambda_k^{j-1} h_k)_{1\le j\le r, 1\le k\le nr} \neq 0$.

Proof. By (2.14), $D(x, t, \lambda_j)h_j = 0$ $(j = 1, \cdots, nr)$ are equivalent to

$$\lambda_j^r h_j + D_1(x,t)\lambda_j^{r-1} h_j + \cdots + D_{r-1}(x,t)\lambda_j h_j + D_r(x,t)h_j = 0. \tag{2.15}$$

This gives a linear system

$$(D_r, D_{r-1}, \cdots, D_2, D_1) \begin{pmatrix} h_1 & h_2 & \cdots & h_{nr} \\ \lambda_1 h_1 & \lambda_2 h_2 & \cdots & \lambda_{nr} h_{nr} \\ \vdots & \vdots & \ddots & \vdots \\ \lambda_1^{r-1} h_1 & \lambda_2^{r-1} h_2 & \cdots & \lambda_{nr}^{r-1} h_{nr} \end{pmatrix} \tag{2.16}$$
$$= -(\lambda_1^r h_1, \lambda_2^r h_2, \cdots, \lambda_{nr}^r h_{nr})$$

for matrices $D_1, \cdots, D_r$. It is uniquely solvable since the coefficient matrix is invertible.

Now we prove that $D(\lambda)$ is a Darboux matrix. Suppose $U(\lambda) = \sum_{k=0}^{m} U_k \lambda^{m-k}$. We want to determine $U'(\lambda) = \sum_{k=0}^{m} U'_k \lambda^{m-k}$ so that the first equation of (2.7) holds. Let

$$\Delta(\lambda) = U'(\lambda)D(\lambda) - D(\lambda)U(\lambda) - D(\lambda)_x, \tag{2.17}$$

then $\Delta(\lambda)$ is a polynomial of λ of degree $m+r$ with matrix coefficients. Setting the coefficients of λ^{m+r}, $\cdots$, λ^r in $\Delta(\lambda)$ to be zero determines U_0', U_1', $\cdots$, U_m'. Then $\Delta(\lambda)$ is a polynomial of λ of degree at most $r-1$ with matrix coefficients, and satisfies $\Delta(\lambda_j)h_j = U'(\lambda_j)D(\lambda_j)h_j - (D(\lambda_j)h_j)_x = 0$ for $j = 1, \cdots, nr$. Suppose $\Delta(\lambda) = \sum_{k=0}^{r-1} \Delta_k \lambda^{r-1-k}$, then we obtain a linear system

$$(\Delta_{r-1}, \cdots, \Delta_1, \Delta_0) \begin{pmatrix} h_1 & h_2 & \cdots & h_{nr} \\ \lambda_1 h_1 & \lambda_2 h_2 & \cdots & \lambda_{nr} h_{nr} \\ \vdots & \vdots & \ddots & \vdots \\ \lambda_1^{r-1} h_1 & \lambda_2^{r-1} h_2 & \cdots & \lambda_{nr}^{r-1} h_{nr} \end{pmatrix} = 0 \tag{2.18}$$

for Δ_0, Δ_1, $\cdots$, Δ_{r-1}. Since the coefficient matrix is invertible, $\Delta = 0$ holds identically. The theorem is proved.

Hereafter we say that a Darboux matrix exists generically if

$$\det(\lambda_k^{j-1} h_k)_{1\le j\le r, 1\le k\le nr} \neq 0 \tag{2.19}$$

holds.

Moreover, it can be proved that, generically, a Darboux matrix of degree n can be decomposed into n Darboux matrices of degree 1 (see [9]).

Example. For the $n \times n$ AKNS system (see [1, 2]), $U(x,t,\lambda) = \lambda J + P(x,t)$, $V(x,t,\lambda) = \sum_{j=0}^{m} V_j \lambda^{m-j}$. Here J is a constant diagonal matrix with distinct diagonal entries, P is off-diagonal, V_j ($j = 0, \cdots, m$) are determined by the integrability condition (2.2) up to some integral constants.

After the action of the Darboux matrix D in (2.14), (2.7) leads to

$$\begin{aligned} &(\lambda J + P')(\lambda^r I + D_1\lambda^{r-1} + \cdots + D_r) \\ &= (\lambda^r I + D_1 \lambda^{r-1} + \cdots + D_r)(\lambda J + P) + (D_{1,x}\lambda^{r-1} + \cdots + D_{r,x}). \end{aligned} \tag{2.20}$$

The coefficient of λ^r gives the transformation of P as

$$P' = P - [J, D_1]. \tag{2.21}$$

3 Darboux transformation for Lax pairs with unitary symmetry

Let K be an invertible real symmetric matrix. We say that the Lax pair (2.1) has the unitary symmetry with respect to K if the coefficients satisfy

$$U(x,t,-\bar{\lambda})^* = -KU(x,t,\lambda)K^{-1}, \quad V(x,t,-\bar{\lambda})^* = -KV(x,t,\lambda)K^{-1}, \tag{3.1}$$

where $*$ refers to the Hermitian conjugation of a matrix.

If $K = \mathrm{diag}(\underbrace{1, \cdots, 1}_{p}, \underbrace{-1, \cdots, -1}_{n-p})$, and λ is purely imaginary, the matrices $U(x,t,\lambda)$ and $V(x,t,\lambda)$ are in the Lie algebra $u(p, n-p)$. Especially, if $p = n$, then $U(x,t,\lambda)$ and $V(x,t,\lambda)$ are in the Lie algebra $u(n)$.

To keep the symmetries (3.1), neither the spectral parameters $\lambda_1, \cdots, \lambda_n$ nor the solutions $h_1, \cdots, h_n$ of the Lax pair can be chosen arbitrarily. In fact, they should satisfy the following conditions. Let μ be a complex constant which is not purely imaginary, $\lambda_j = \mu$ or $-\bar{\mu}$. If $\lambda_k = -\bar{\lambda}_j$, then

$$
\begin{aligned}
(h_j^* K h_k)_x &= h_j^* U(\lambda_j)^* K h_k + h_j^* K U(\lambda_k) h_k \\
&= -h_j^* K U(\lambda_k) h_k + h_j^* K U(\lambda_k) h_k = 0.
\end{aligned} \tag{3.2}
$$

This implies that $h_j^* K h_k = 0$ holds identically if it holds at one point. Hence we can always want that $h_j^* K h_k = 0$ holds identically whenever $\lambda_k = -\bar{\lambda}_j$. If so, the Darboux transformation constructed in the last section keeps the unitary symmetry. That is, U', V' satisfy

$$
U'(x,t,-\bar{\lambda})^* = -KU'(x,t,\lambda)K^{-1}, \quad V'(x,t,-\bar{\lambda})^* = -KV'(x,t,\lambda)K^{-1}. \tag{3.3}
$$

This is a consequence of the following Theorem 3. A direct proof can also be seen in [9].

In general, the Darboux transformation only exists locally (where $\det H \neq 0$ when degree is 1) for the system without symmetry or with this general unitary symmetry. However, the Darboux transformation exists globally for the system with $u(n)$ symmetry.

Without loss of generality, suppose $\lambda_1 = \cdots = \lambda_s = \mu$, $\lambda_{s+1} = \cdots = \lambda_n = -\bar{\mu}$, $\Lambda = \mathrm{diag}(\lambda_1, \cdots, \lambda_n)$. Let h_j be a column solution of the Lax pair (2.1) with $\lambda = \lambda_j$ $(j = 1, \cdots, n)$ such that $h_j^* K h_k = 0$ whenever $\lambda_k = -\bar{\lambda}_j$. Let $H = (h_1, \cdots, h_n)$, $S = H\Lambda H^{-1}$. Then we have seen that the Darboux matrix $\lambda I - S$ keeps the unitary symmetry (3.3).

Denote $L = (h_1, \cdots, h_s)$, $Z = (h_{s+1}, \cdots, h_n)$, then $H = (L\ Z)$, $L^* K Z = 0$,

$$
H^{-1} = (H^* K H)^{-1} H^* K = \begin{pmatrix} (L^* K L)^{-1} L^* K \\ (Z^* K Z)^{-1} Z^* K \end{pmatrix}. \tag{3.4}
$$

Hence,

$$
I = H(H^* K H)^{-1} H^* K = L(L^* K L)^{-1} L^* K + Z(Z^* K Z)^{-1} Z^* K. \tag{3.5}
$$

Therefore,

$$
\begin{aligned}
\lambda I - S &= H(\lambda I - \Lambda) H^{-1} \\
&= (L\ Z) \begin{pmatrix} (\lambda - \mu) I_s & \\ & (\lambda + \bar{\mu}) I_{n-s} \end{pmatrix} \begin{pmatrix} (L^* K L)^{-1} L^* K & \\ & (Z^* K Z)^{-1} Z^* K \end{pmatrix} \\
&= (\lambda - \mu) L (L^* K L)^{-1} L^* K + (\lambda + \bar{\mu}) Z (Z^* K Z)^{-1} Z^* K.
\end{aligned} \tag{3.6}
$$

Substituting (3.5) into it, we get

$$\begin{aligned}\lambda I - S &= (\lambda-\mu)L(L^*KL)^{-1}L^*K + (\lambda+\bar{\mu})(I - L(L^*KL)^{-1}L^*K)\\ &= (\lambda+\bar{\mu})I - (\mu+\bar{\mu})L(L^*KL)^{-1}L^*K\\ &= (\lambda+\bar{\mu})\Big(I - \frac{L\Gamma^{-1}L^*K}{\lambda+\bar{\mu}}\Big),\end{aligned} \tag{3.7}$$

where

$$\Gamma = \frac{L^*KL}{\mu+\bar{\mu}}. \tag{3.8}$$

Note that $\lambda I - S$ does not depend on Z.

To generalize this result, we get the construction of Darboux transformation of degree r given by Zakharov and Mikhailov (see [4, 29]).

Theorem 3. *Suppose K is an invertible real symmetric matrix. Let s be an integer with $1 \le s \le n-1$. Let μ_j $(j = 1, \cdots, r)$ be r complex constants such that $\mu_1, \cdots, \mu_r, -\bar{\mu}_1, \cdots, -\bar{\mu}_r$ are distinct. (We shall call these μ_j's to be the spectrum for constructing a Darboux matrix.) Let H_j be an $n \times s$ matrix solution of rank s of the Lax pair (2.1) with $\lambda = \mu_j$ $(j = 1, \cdots, r)$. Let*

$$\Gamma_{ij} = \frac{H_i^*KH_j}{\bar{\mu}_i + \mu_j} \quad (1 \le i, j \le r). \tag{3.9}$$

When $\det(\Gamma_{ij})_{1\le i,j\le r} \ne 0$, *define*

$$D(\lambda) = \prod_{k=1}^{r}(\lambda+\bar{\mu}_k)\left(I - \sum_{i,j=1}^{r}\frac{H_i(\Gamma^{-1})_{ij}H_j^*K}{\lambda+\bar{\mu}_j}\right). \tag{3.10}$$

Then,

$$D(\lambda)^{-1} = \prod_{k=1}^{r}(\lambda+\bar{\mu}_k)^{-1}\left(I + \sum_{i,j=1}^{r}\frac{H_i(\Gamma^{-1})_{ij}H_j^*K}{\lambda-\mu_i}\right), \tag{3.11}$$

$$D(-\bar{\lambda})^*KD(\lambda) = \prod_{k=1}^{r}(-\lambda+\mu_k)(\lambda+\bar{\mu}_k)K, \tag{3.12}$$

and $D(\lambda)$ is a Darboux matrix keeping the unitary symmetry.

Proof. The proof is divided into three parts:
(i) Proof of (3.11).

$$\begin{aligned}
&\sum_{i,j=1}^{r}\frac{H_i(\Gamma^{-1})_{ij}H_j^*K}{\lambda-\mu_i}\sum_{a,b=1}^{r}\frac{H_a(\Gamma^{-1})_{ab}H_b^*K}{\lambda+\bar\mu_b}\\
&=\sum_{i,j,a,b=1}^{r}\frac{1}{\bar\mu_b+\mu_i}\Big(\frac{1}{\lambda-\mu_i}-\frac{1}{\lambda+\bar\mu_b}\Big)H_i(\Gamma^{-1})_{ij}(\bar\mu_j+\mu_a)\Gamma_{ja}(\Gamma^{-1})_{ab}H_b^*K\\
&=\sum_{i,j,a,b=1}^{r}\frac{\bar\mu_j}{\bar\mu_b+\mu_i}\Big(\frac{1}{\lambda-\mu_i}-\frac{1}{\lambda+\bar\mu_b}\Big)H_i(\Gamma^{-1})_{ij}\Gamma_{ja}(\Gamma^{-1})_{ab}H_b^*K\\
&\quad+\sum_{i,j,a,b=1}^{r}\frac{\mu_a}{\bar\mu_b+\mu_i}\Big(\frac{1}{\lambda-\mu_i}-\frac{1}{\lambda+\bar\mu_b}\Big)H_i(\Gamma^{-1})_{ij}\Gamma_{ja}(\Gamma^{-1})_{ab}H_b^*K\\
&=\sum_{i,j=1}^{r}\frac{\bar\mu_j}{\bar\mu_j+\mu_i}\Big(\frac{1}{\lambda-\mu_i}-\frac{1}{\lambda+\bar\mu_j}\Big)H_i(\Gamma^{-1})_{ij}H_j^*K\\
&\quad+\sum_{i,b=1}^{r}\frac{\mu_i}{\bar\mu_b+\mu_i}\Big(\frac{1}{\lambda-\mu_i}-\frac{1}{\lambda+\bar\mu_b}\Big)H_i(\Gamma^{-1})_{ib}H_b^*K\\
&=\sum_{i,j=1}^{r}\Big(\frac{1}{\lambda-\mu_i}-\frac{1}{\lambda+\bar\mu_j}\Big)H_i(\Gamma^{-1})_{ij}H_j^*K.
\end{aligned}$$

This leads to

$$\begin{aligned}
&\left(I+\sum_{i,j=1}^{r}\frac{H_i(\Gamma^{-1})_{ij}H_j^*K}{\lambda-\mu_i}\right)\left(I-\sum_{a,b=1}^{r}\frac{H_a(\Gamma^{-1})_{ab}H_b^*K}{\lambda+\bar\mu_b}\right)\\
&=I+\sum_{i,j=1}^{r}\frac{H_i(\Gamma^{-1})_{ij}H_j^*K}{\lambda-\mu_i}-\sum_{i,j=1}^{r}\frac{H_i(\Gamma^{-1})_{ij}H_j^*K}{\lambda+\bar\mu_j}\\
&\quad-\sum_{i,j=1}^{r}\frac{H_i(\Gamma^{-1})_{ij}H_j^*K}{\lambda-\mu_i}\sum_{a,b=1}^{r}\frac{H_a(\Gamma^{-1})_{ab}H_b^*K}{\lambda+\bar\mu_b}=I.
\end{aligned}\tag{3.13}$$

Hence (3.11) is true.

(ii) Proof of (3.12).

Using the fact $\Gamma_{ij}^*=\Gamma_{ji}$, we have $\displaystyle\sum_{k=1}^{r}\Gamma_{ki}(\Gamma^{*-1})_{kj}=\sum_{k=1}^{r}\Gamma_{ik}^*(\Gamma^{*-1})_{kj}=\delta_{ij}$, which leads to $(\Gamma^{*-1})_{ij}=(\Gamma^{-1})_{ji}$.

$$\begin{aligned}
&\prod_{k=1}^{r}(-\lambda+\mu_k)^{-1}(\lambda+\bar\mu_k)^{-1}D(-\bar\lambda)^*KD(\lambda)\\
&=\left(I-\sum_{i,j=1}^{r}\frac{KH_j(\Gamma^{-1})_{ji}H_i^*}{-\lambda+\mu_j}\right)K\left(I-\sum_{a,b=1}^{r}\frac{H_a(\Gamma^{-1})_{ab}H_b^*K}{\lambda+\bar\mu_b}\right)\\
&=K-\sum_{i,j=1}^{r}\frac{KH_j(\Gamma^{-1})_{ji}H_i^*K}{-\lambda+\mu_j}-\sum_{a,b=1}^{r}\frac{KH_a(\Gamma^{-1})_{ab}H_b^*K}{\lambda+\bar\mu_b}\\
&\quad+\sum_{i,j,a,b=1}^{r}\frac{1}{\bar\mu_b+\mu_j}\Big(\frac{1}{-\lambda+\mu_j}+\frac{1}{\lambda+\bar\mu_b}\Big)KH_j(\Gamma^{-1})_{ji}(\bar\mu_i+\mu_a)\Gamma_{ia}(\Gamma^{-1})_{ab}H_b^*K.
\end{aligned}$$

Using the similar computation as in (i), we can verify that it equals K. Hence (3.12) holds.

(iii) Proof of $D(\lambda)$ being a Darboux matrix.
By using the symmetry $U(-\bar{\lambda})^* = -KU(\lambda)K^{-1}$, (3.9) gives

$$\Gamma_{ij,x} = \frac{H_i^* U(\mu_i)^* K H_j}{\bar{\mu}_i + \mu_j} + \frac{H_i^* K U(\mu_j) H_j}{\bar{\mu}_i + \mu_j} = \frac{H_i^* K\big(U(\mu_j) - U(-\bar{\mu}_i)\big) H_j}{\bar{\mu}_i + \mu_j}. \tag{3.14}$$

Then (3.10) leads to

$$\begin{aligned}
&\prod_{k=1}^{r}(\lambda + \bar{\mu}_k)^{-1} D(\lambda)_x = -\Big(\sum_{i,j=1}^{r} \frac{H_i(\Gamma^{-1})_{ij} H_j^* K}{\lambda + \bar{\mu}_j}\Big)_x \\
&= -\sum_{i,j=1}^{r} \frac{U(\mu_i) H_i (\Gamma^{-1})_{ij} H_j^* K}{\lambda + \bar{\mu}_j} + \sum_{i,j=1}^{r} \frac{H_i(\Gamma^{-1})_{ij} H_j^* K U(-\bar{\mu}_j)}{\lambda + \bar{\mu}_j} \\
&\quad + \sum_{i,j,a,b=1}^{r} \frac{H_i(\Gamma^{-1})_{ia} H_a^* K\big(U(\mu_b) - U(-\bar{\mu}_a)\big) H_b (\Gamma^{-1})_{bj} H_j^* K}{(\bar{\mu}_a + \mu_b)(\lambda + \bar{\mu}_j)}.
\end{aligned} \tag{3.15}$$

Hence

$$\begin{aligned}
U'(\lambda) &= \big(D(\lambda)U(\lambda) + D(\lambda)_x\big) D(\lambda)^{-1} \\
&= \Bigg(U(\lambda) - \sum_{i,j=1}^{r} \frac{H_i(\Gamma^{-1})_{ij} H_j^* K\big(U(\lambda) - U(-\bar{\mu}_j)\big)}{\lambda + \bar{\mu}_j} - \sum_{i,j=1}^{r} \frac{U(\mu_i) H_i(\Gamma^{-1})_{ij} H_j^* K}{\lambda + \bar{\mu}_j} \\
&\quad + \sum_{i,j,a,b=1}^{r} \frac{H_i(\Gamma^{-1})_{ia} H_a^* K\big(U(\mu_b) - U(-\bar{\mu}_a)\big) H_b(\Gamma^{-1})_{bj} H_j^* K}{(\bar{\mu}_a + \mu_b)(\lambda + \bar{\mu}_j)}\Bigg) \cdot \\
&\quad \cdot \Bigg(I + \sum_{p,q=1}^{r} \frac{H_p(\Gamma^{-1})_{pq} H_q^* K}{\lambda - \mu_p}\Bigg).
\end{aligned} \tag{3.16}$$

Formally, $U'(\lambda)$ is a rational function of λ with matrix coefficients. However, we shall show that it is really a polynomial of λ with matrix coefficients. In order to do so, we shall compute $\operatorname*{Res}_{\lambda=\mu_c} U'(\lambda)$ and $\operatorname*{Res}_{\lambda=-\bar{\mu}_c} U'(\lambda)$. From (3.16),

$$\begin{aligned}
&\operatorname*{Res}_{\lambda=\mu_c} U'(\lambda) \\
&= \Bigg(U(\mu_c) - \sum_{i,j=1}^{r} \frac{H_i(\Gamma^{-1})_{ij} H_j^* K\big(U(\mu_c) - U(-\bar{\mu}_j)\big)}{\mu_c + \bar{\mu}_j} - \sum_{i,j=1}^{r} \frac{U(\mu_i) H_i(\Gamma^{-1})_{ij} H_j^* K}{\mu_c + \bar{\mu}_j} \\
&\quad + \sum_{i,j,a,b=1}^{r} \frac{H_i(\Gamma^{-1})_{ia} H_a^* K\big(U(\mu_b) - U(-\bar{\mu}_a)\big) H_b(\Gamma^{-1})_{bj} H_j^* K}{(\bar{\mu}_a + \mu_b)(\mu_c + \bar{\mu}_j)}\Bigg) \cdot \\
&\quad \cdot \sum_{q=1}^{r} H_c(\Gamma^{-1})_{cq} H_q^* K
\end{aligned}$$

$$
\begin{aligned}
&= U(\mu_c)\sum_{q=1}^{r} H_c(\Gamma^{-1})_{cq}H_q^*K \\
&- \sum_{i,j,q=1}^{r} \frac{H_i(\Gamma^{-1})_{ij}H_j^*K\big(U(\mu_c)-U(-\bar{\mu}_j)\big)H_c(\Gamma^{-1})_{cq}H_q^*K}{\mu_c+\bar{\mu}_j} \\
&- \sum_{i,j,q=1}^{r} U(\mu_i)H_i(\Gamma^{-1})_{ij}\Gamma_{jc}(\Gamma^{-1})_{cq}H_q^*K \\
&\quad + \sum_{i,j,a,b,q=1}^{r} \frac{H_i(\Gamma^{-1})_{ia}H_a^*K\big(U(\mu_b)-U(-\bar{\mu}_a)\big)H_b(\Gamma^{-1})_{bj}\Gamma_{jc}(\Gamma^{-1})_{cq}H_q^*K}{\bar{\mu}_a+\mu_b} \\
&= U(\mu_c)\sum_{q=1}^{r} H_c(\Gamma^{-1})_{cq}H_q^*K \\
&- \sum_{i,j,q=1}^{r} \frac{H_i(\Gamma^{-1})_{ij}H_j^*K\big(U(\mu_c)-U(-\bar{\mu}_j)\big)H_c(\Gamma^{-1})_{cq}H_q^*K}{\mu_c+\bar{\mu}_j} \\
&- \sum_{q=1}^{r} U(\mu_c)H_c(\Gamma^{-1})_{cq}H_q^*K \\
&\quad + \sum_{i,a,q=1}^{r} \frac{H_i(\Gamma^{-1})_{ia}H_a^*K\big(U(\mu_c)-U(-\bar{\mu}_a)\big)H_c(\Gamma^{-1})_{cq}H_q^*K}{\bar{\mu}_a+\mu_c} = 0.
\end{aligned}
$$

Similarly, we have $\operatorname*{Res}_{\lambda=-\bar{\mu}_c} U'(\lambda) = 0$. Hence $U'(\lambda)$ is a polynomial of λ with matrix coefficients, which means that $D(\lambda)$ is a Darboux matrix without considering the unitary symmetry.

Moreover, by (2.7),

$$
\begin{aligned}
&(U'(-\bar{\lambda}))^* = D(-\bar{\lambda})^{*-1}U(-\bar{\lambda})^*D(-\bar{\lambda})^* + D(-\bar{\lambda})^{*-1}D(-\bar{\lambda})^*_x \\
&= KD(\lambda)K^{-1}\cdot(-KU(\lambda)K^{-1})\cdot KD(\lambda)^{-1}K^{-1} + KD(\lambda)K^{-1}\cdot K(D(\lambda)^{-1})_xK^{-1} \\
&= -K\Big(D(\lambda)U(\lambda)D(\lambda)^{-1} + D(\lambda)_xD(\lambda)^{-1}\Big)K^{-1} = -KU'(\lambda)K^{-1}.
\end{aligned}
\tag{3.17}
$$

Therefore, $D(\lambda)$ keeps the unitary symmetry. The theorem is proved.

Example. Soliton solutions of nonlinear Schrödinger equation.

The Lax pair (2.1), (2.3) of the nonlinear Schrödinger equation has the $u(2)$ symmetry. The spectrum for Darboux transformation of degree 1 on the complex plane is shown in the following figure.

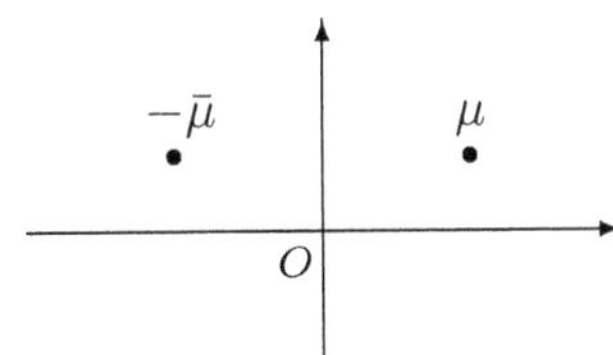

The Darboux transformation of degree 1 is constructed as follows. Let μ be a nonzero complex number, $\Lambda = \begin{pmatrix} \mu & \\ & -\bar{\mu} \end{pmatrix}$. Let $h = \begin{pmatrix} h_1 \\ h_2 \end{pmatrix}$ be a column solution of the Lax pair (2.1), (2.3) with $\lambda = \mu$, then $\begin{pmatrix} -\bar{h}_2 \\ \bar{h}_1 \end{pmatrix}$ is a solution of that Lax pair with $\lambda = -\bar{\mu}$. Let $H = (h_1, h_2)$, $S = H\Lambda H^{-1}$. Then $D(\lambda) = \lambda I - S$ is a Darboux matrix of degree 1. It is defined globally on $\mathbf{R}^2$. Written explicitly, we have

$$S = \frac{1}{1+|\sigma|^2} \begin{pmatrix} \mu - \bar{\mu}|\sigma|^2 & (\mu+\bar{\mu})\bar{\sigma} \\ (\mu+\bar{\mu})\sigma & -\bar{\mu}+\mu|\sigma|^2 \end{pmatrix} \tag{3.18}$$

where $\sigma = \dfrac{h_2}{h_1}$.

The Darboux transformation of higher degree can also be constructed according to Theorem 2. All soliton solutions are obtained from the zero seed solution in this way.

4 Darboux transformation for MKdV equation

4.1 Darboux transformation

The MKdV equation

$$u_t + 6u^2 u_x + u_{xxx} = 0 \tag{4.1}$$

has a Lax pair

$$\begin{aligned} \Phi_x &= U(x,t,\lambda)\Phi = (\lambda J + P)\Phi \equiv \begin{pmatrix} \lambda & u \\ -u & -\lambda \end{pmatrix} \Phi, \\ \Phi_t &= V(x,t,\lambda)\Phi = \begin{pmatrix} -4\lambda^3 - 2u^2\lambda & -4u\lambda^2 - 2u_x\lambda - 2u^3 - u_{xx} \\ 4u\lambda^2 - 2u_x\lambda + 2u^3 + u_{xx} & 4\lambda^3 + 2u^2\lambda \end{pmatrix} \Phi. \end{aligned} \tag{4.2}$$

It can be checked directly that

$$\begin{aligned} &U(-\lambda) = KU(\lambda)K^{-1}, \quad \overline{U(\bar{\lambda})} = U(\lambda), \quad \operatorname{tr} U(\lambda) = 0, \\ &V(-\lambda) = KV(\lambda)K^{-1}, \quad \overline{V(\bar{\lambda})} = V(\lambda), \quad \operatorname{tr} V(\lambda) = 0 \end{aligned} \tag{4.3}$$

where $K = \begin{pmatrix} 0 & -1 \\ 1 & 0 \end{pmatrix}$. This implies that the spectrum should have symmetries shown in the following figure ($\operatorname{tr} U(\lambda) = 0$ and $\operatorname{tr} V(\lambda) = 0$ do not give constraints on the spectrum).

$-\bar{\mu}$ μ O $-\mu$ $\bar{\mu}$

Suppose $U(\lambda)=\begin{pmatrix} a(\lambda) & b(\lambda) \\ c(\lambda) & -a(\lambda) \end{pmatrix}$ satisfies (4.3), then it can be checked directly that the entries of $U(\lambda)$ satisfy

$$\begin{aligned} &a(-\lambda)=-a(\lambda), \quad b(-\lambda)=-c(\lambda), \quad c(-\lambda)=-b(\lambda), \\ &\overline{a(\bar{\lambda})}=a(\lambda), \quad \overline{b(\bar{\lambda})}=b(\lambda), \quad \overline{c(\bar{\lambda})}=c(\lambda). \end{aligned} \tag{4.4}$$

Hence

$$U(-\bar{\lambda})^*=\begin{pmatrix} a(-\lambda) & c(-\lambda) \\ b(-\lambda) & -a(-\lambda) \end{pmatrix}=-\begin{pmatrix} a(\lambda) & b(\lambda) \\ c(\lambda) & -a(\lambda) \end{pmatrix}=-U(\lambda). \tag{4.5}$$

This means that the symmetries in (4.3) lead to the $u(2)$ symmetries $U(-\bar{\lambda})=-U(\lambda)$, $V(-\bar{\lambda})=-V(\lambda)$ in this 2×2 case.

From the symmetries (4.3), we have the symmetries of the solutions of the Lax pair.

Lemma 1. *Suppose Φ is a column solution of the Lax pair (2.1) with $\lambda=\mu$, then $K\Phi$ is a solution of (2.1) with $\lambda=-\mu$, and $\bar{\Phi}$ is a solution of (2.1) with $\lambda=\bar{\mu}$.*

4.2 Soliton solutions

Let $\mu\in\mathbf{R}$, then the spectrum for constructing Darboux matrix can be chosen as in the following figure where $\bar{\mu}$ and μ overlap.

$-\mu=-\bar{\mu}$ $\mu=\bar{\mu}$ O

Let h be a real column solution of the Lax pair (4.2) with $\lambda=\mu$. Write $h=\begin{pmatrix} h_1 \\ h_2 \end{pmatrix}$, then $Kh=\begin{pmatrix} -h_2 \\ h_1 \end{pmatrix}$. Hence we can take

$$\Lambda=\begin{pmatrix} \mu & \\ & -\mu \end{pmatrix}, \quad H=\begin{pmatrix} h_1 & -h_2 \\ h_2 & h_1 \end{pmatrix}. \tag{4.6}$$

The Darboux matrix is

$$D(\lambda) = \lambda I - H\Lambda H^{-1} = \begin{pmatrix} \lambda - \mu\dfrac{1-\sigma^2}{1+\sigma^2} & -\mu\dfrac{2\sigma}{1+\sigma^2} \\ -\mu\dfrac{2\sigma}{1+\sigma^2} & \lambda + \mu\dfrac{1-\sigma^2}{1+\sigma^2} \end{pmatrix} \tag{4.7}$$

where $\sigma = \dfrac{h_2}{h_1}$. After the action of $D(\lambda)$, (2.7) gives

$$u' = u + \frac{4\mu\sigma}{1+\sigma^2}, \tag{4.8}$$

which is a new solution of the MKdV equation.

Especially, take the seed solution $u = 0$ for the MKdV equation, then the Lax pair with $\lambda = \mu \neq 0$ has a column solution

$$\begin{pmatrix} h_1 \\ h_2 \end{pmatrix} = \begin{pmatrix} 0 \\ e^{-\mu x + 4\mu^3 t} \end{pmatrix} + e^{2\alpha} \begin{pmatrix} e^{\mu x - 4\mu^3 t} \\ 0 \end{pmatrix} \tag{4.9}$$

where α is a real constant. The Darboux matrix is

$$D(\lambda) = \lambda I - \frac{\mu}{\cosh v_1} \begin{pmatrix} \sinh v_1 & 1 \\ 1 & -\sinh v_1 \end{pmatrix}, \tag{4.10}$$

where $v_1 = 2\mu x - 8\mu^3 t + 2\alpha$. The corresponding solution is the single soliton solution

$$u' = 2\mu \operatorname{sech}(2\mu x - 8\mu^3 t + 2\alpha), \tag{4.11}$$

which is shown in the following figure. The parameters are $\alpha = 0$, $\mu = 1$. The range of x is $[-7, 7]$ and $t = -1, 0, 1$ for three figures, respectively.

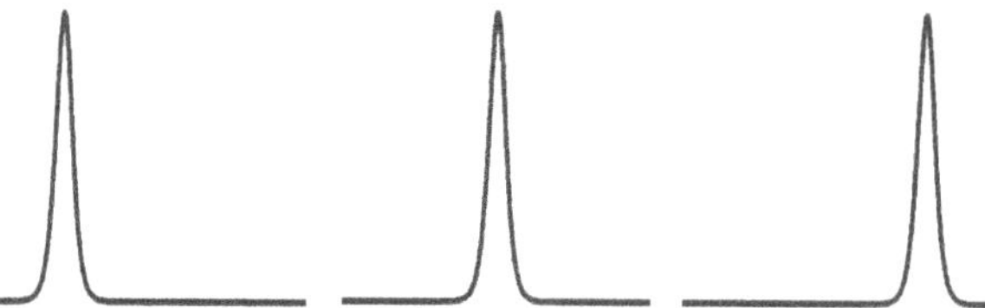

Multi-soliton solutions can be obtained by successive actions of Darboux transformations of degree 1. The following figure shows an example of the spectrum for a 3-soliton solution.

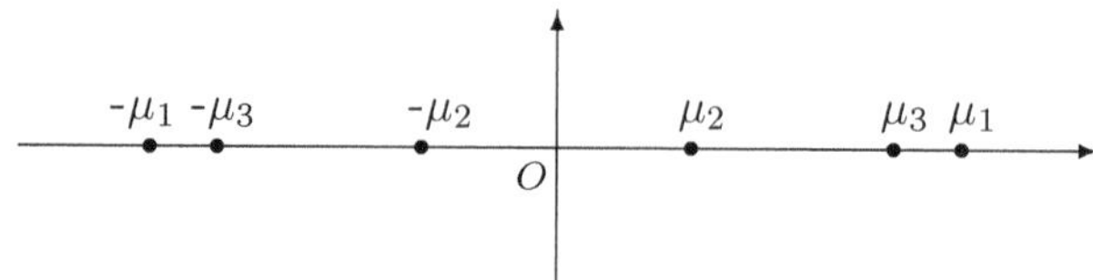

4.3 Breather solutions

If μ is not real, (4.3) implies that the spectrum should have the symmetries as shown in the following figure.

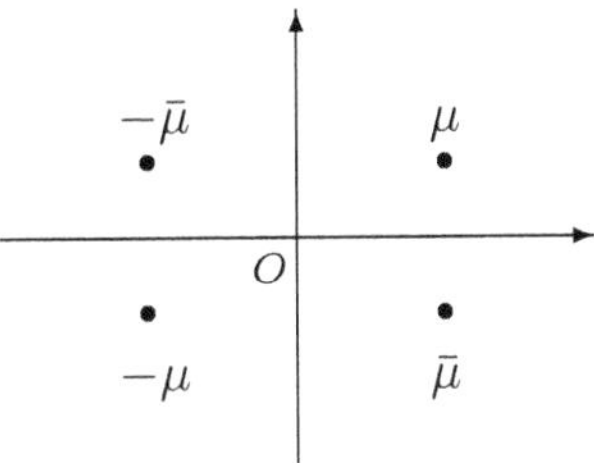

Since a Darboux matrix of degree 1 can only contain two spectral parameters, a Darboux matrix of degree 2 is necessary.

Let $h = \begin{pmatrix} h_1 \\ h_2 \end{pmatrix}$ be a solution of the Lax pair (4.2) with $\lambda = \mu_1 = \mu$, then by Lemma 1, $Kh = \begin{pmatrix} -h_2 \\ h_1 \end{pmatrix}$, $\bar{h} = \begin{pmatrix} \bar{h}_1 \\ \bar{h}_2 \end{pmatrix}$, $K\bar{h} = \begin{pmatrix} -\bar{h}_2 \\ \bar{h}_1 \end{pmatrix}$ are solutions of (4.2) with $\lambda = -\mu, \bar{\mu}, -\bar{\mu}$, respectively.

Let

$$\Lambda_1 = \begin{pmatrix} \mu & \\ & -\bar{\mu} \end{pmatrix}, \quad H_1 = \begin{pmatrix} h_1 & -\bar{h}_2 \\ h_2 & \bar{h}_1 \end{pmatrix}, \quad \Lambda_2 = \bar{\Lambda}_1, \quad H_2 = \bar{H}_1. \tag{4.12}$$

Λ_1 and H_1 have the same form as Λ and H for the nonlinear Schrödinger equation in the last section. Hence we can use (3.18) to get

$$S_1 = \frac{1}{1+|\sigma|^2} \begin{pmatrix} \mu - \bar{\mu}|\sigma|^2 & (\mu+\bar{\mu})\bar{\sigma} \\ (\mu+\bar{\mu})\sigma & -\bar{\mu}+\mu|\sigma|^2 \end{pmatrix}, \quad S_2 = \bar{S}_1 \tag{4.13}$$

where $\sigma = \dfrac{h_2}{h_1}$. After the action of $D_1(\lambda) = \lambda I - S_1$, the solution $\Phi(\lambda)$ of the Lax pair (4.2) is transformed to $D_1(\lambda)\Phi(\lambda)$, hence H_2 is transformed to $H_2' = H_2\Lambda_2 - S_1 H_2 = (S_2 - S_1)H_2$. The corresponding $S_2' = H_2'\Lambda_2 H_2'^{-1} = (S_2 - S_1)S_2(S_2 - S_1)^{-1}$. The Darboux matrix of degree 2 is

$$\begin{aligned} D''(\lambda) &= (\lambda I - S_2')(\lambda I - S_1) \\ &= \lambda^2 I - \lambda(S_2^2 - S_1^2)(S_2 - S_1)^{-1} + (S_2 - S_1)S_2(S_2 - S_1)^{-1}S_1. \end{aligned} \tag{4.14}$$

According to (2.21), after the action of this Darboux transformation, P is transformed to

$$P'' = P + [J, (S_2^2 - S_1^2)(S_2 - S_1)^{-1}]. \tag{4.15}$$

Hence the new solution of the MKdV Equation (4.1) is

$$
\begin{aligned}
u'' = P''_{12} &= u + 2((S_2^2 - S_1^2)(S_2 - S_1)^{-1})_{12} \\
&= u + \frac{4(\bar{\mu}^2 - \mu^2)(\bar{\mu}\bar{\sigma} - \mu\sigma - (\mu\bar{\sigma} - \bar{\mu}\sigma)|\sigma|^2)}{(\mu^2 + \bar{\mu}^2)(1 + \sigma^2)(1 + \bar{\sigma}^2) - 2|\mu|^2((1 + |\sigma|^2)^2 - (\sigma - \bar{\sigma})^2)}.
\end{aligned}
\tag{4.16}
$$

Now take the seed solution $u = 0$, then (4.9) gives $\sigma = \mathrm{e}^{-2\mu x + 8\mu^3 t - 2\alpha}$ where α is a constant. With $\alpha = 0$, the derived solution is

$$
u'' = \frac{4\mu_{\mathrm{R}}\mu_{\mathrm{I}}(\mu_{\mathrm{I}} \cosh p \cos\theta - \mu_{\mathrm{R}} \sinh p \sin\theta)}{\mu_{\mathrm{I}}^2 2\cosh^2 p + \mu_{\mathrm{R}}^2 \sin^2\theta}
\tag{4.17}
$$

where $p = 2\mu_{\mathrm{R}}(x - 4(\mu_{\mathrm{R}}^2 - 3\mu_{\mathrm{I}}^2)t)$, $\theta = 2\mu_{\mathrm{I}}(x - 4(3\mu_{\mathrm{R}}^2 - \mu_{\mathrm{I}}^2)t)$, the subscripts "R" and "I" refer to the real and imaginary part of a complex number, respectively.

The solution u'' is a breather solution (see [15, 24]), which was shown in the following figure. The parameters are $\alpha = 0$, $\mu = 1 + 0.5\mathrm{i}$. The range of x is $[-2, 6]$ and t is from 0.1 to 0.7 from the first figure to the 12th.

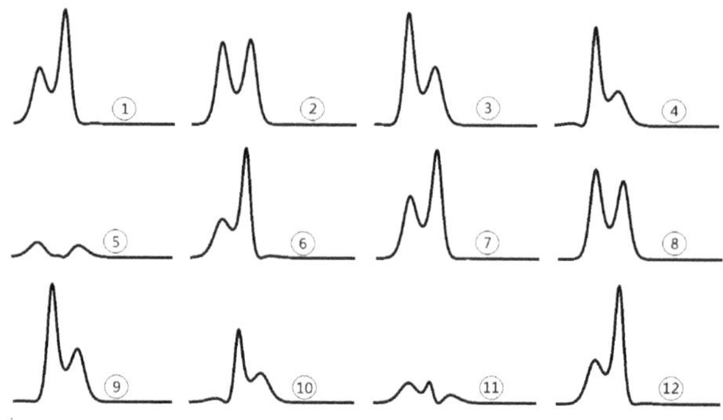

Multi-breather solutions can be obtained by successive actions of the above Darboux transformations. Here is an example of the spectrum for a 2-breather solution.

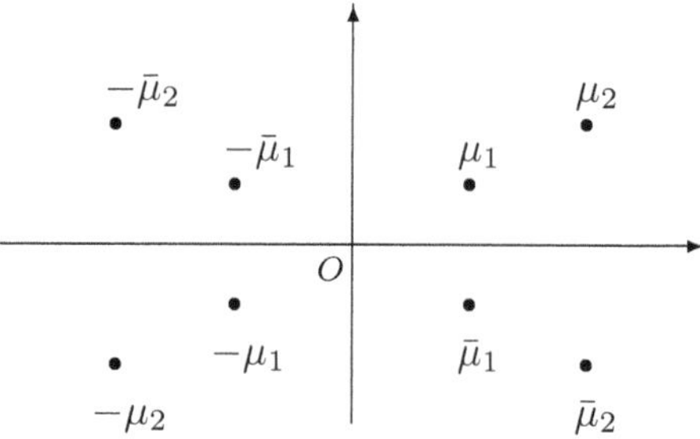

Soliton and breather can appear simultaneously. Here is an example of the spectrum for a 1-breather–1-soliton solution.

$-\bar{\mu}_1$ μ_1 $-\mu_2$ μ_2 O $-\mu_1$ $\bar{\mu}_1$

5 Darboux transformation for 2D $D_{l+1}^{(2)}$ Toda equation

In the construction of the breather solutions of the MKdV equation, we have seen that the whole spectrum should be included in the construction of Darboux transformation so that all the symmetries of the Lax pair can be kept. The symmetries of MKdV equation are not too complicated. In this section, we discuss a Lax pair with much more complicated symmetries, and take two-dimensional $D_{l+1}^{(2)}$ Toda equation as an example.

The Toda equations have profound mathematical structure and important applications (see [17]). There is quite a lot of work on the Darboux transformations for the two-dimensional affine Toda equations (see [13, 14, 20–23, 26, 30–32]) and their important applications in geometry and physics (see [3, 6, 7, 27, 33]).

The two-dimensional $D_{l+1}^{(2)}$ Toda equation

$$\begin{aligned} u_{j,xt} &= \mathrm{e}^{u_j - u_{j-1}} - \mathrm{e}^{u_{j+1} - u_j} \quad (2 \le j \le l-1), \\ u_{1,xt} &= \mathrm{e}^{u_1} - \mathrm{e}^{u_2 - u_1}, \quad u_{l,xt} = \mathrm{e}^{u_l - u_{l-1}} - \mathrm{e}^{-u_l} \end{aligned} \tag{5.1}$$

has a Lax pair

$$\begin{aligned} \Phi_x &= U(x,t,\lambda)\Phi = (\lambda J + P(x,t))\Phi, \\ \Phi_t &= V(x,t,\lambda)\Phi = \lambda^{-1} Q(x,t)\Phi, \end{aligned} \tag{5.2}$$

where J, P, Q are $(2l+2) \times (2l+2)$ matrices with

$$J = \begin{pmatrix} & 1 & & & & \\ & & 1 & & & \\ & & & \ddots & & \\ & & & & 1 & \\ & & & & & 1 \\ 1 & & & & & \end{pmatrix}, \tag{5.3}$$

$$P = \mathrm{diag}(u_{1,x}, \cdots, .u_{l,x}, 0, -u_{l,x}, -u_{1,x}, 0), \tag{5.4}$$

$$Q = \begin{pmatrix} & & & & & & & & e^{u_1} \\ e^{u_2-u_1} & & & & & & & & \\ & \ddots & & & & & & & \\ & & e^{u_l-u_{l-1}} & & & & & & \\ & & & e^{-u_l} & & & & & \\ & & & & e^{-u_l} & & & & \\ & & & & & e^{u_l-u_{l-1}} & & & \\ & & & & & & \ddots & & \\ & & & & & & & e^{u_2-u_1} & \\ & & & & & & & & e^{u_1} \end{pmatrix}. \tag{5.5}$$

Note that the integrability condition of (5.2) is

$$Q_x = [P, Q], \quad P_t + [J, Q] = 0. \tag{5.6}$$

In fact, (5.6) contains all the two-dimensional affine Toda equations (see [11, 12, 16]). When J, P, Q are given by (5.3)–(5.5), (5.6) is the two-dimensional $D_{l+1}^{(2)}$ Toda equation and the order of the matrices is $n = 2l + 2$.

Now the coefficients of the Lax pair satisfy the reality symmetry, cyclic symmetry and unitary symmetry

$$\begin{aligned} &\overline{U(\lambda)} = U(\bar{\lambda}), \quad \Omega U(\lambda)\Omega^{-1} = U(\omega\lambda), \quad U(-\bar{\lambda})^* = -KU(\lambda)K, \\ &\overline{V(\lambda)} = V(\bar{\lambda}), \quad \Omega V(\lambda)\Omega^{-1} = V(\omega\lambda), \quad V(-\bar{\lambda})^* = -KV(\lambda)K, \end{aligned} \tag{5.7}$$

where $\Omega = \operatorname{diag}(1, \omega^{-1}, \cdots, \omega^{-n+1})$, $\omega = e^{\frac{2\pi i}{n}}$ and

$$K = \begin{pmatrix} & & & & 1 & \\ & & & 1 & & \\ & & ⋰ & & & \\ & 1 & & & & \\ 1 & & & & & \\ & & & & & 1 \end{pmatrix}_{n\times n}. \tag{5.8}$$

K is symmetric and satisfies $K^2 = I$, $\Omega^* K = \omega^{n-2} K\Omega$.

(5.7) is equivalent to

$$\begin{aligned} &\bar{J} = J, \quad \Omega J \Omega^{-1} = \omega J, \quad KJK = J^T, \\ &\bar{P} = P, \quad \Omega P \Omega^{-1} = P, \quad KPK = -P^T, \\ &\bar{Q} = Q, \quad \Omega Q \Omega^{-1} = \omega^{-1} Q, \quad KQK = Q^T. \end{aligned} \tag{5.9}$$

Note that n is an even number. The symmetries of the Lax pair leads to the symmetries of its solutions.

Lemma 2. Suppose (5.9) holds. Let $\mu \in \mathbf{C}\backslash\{0\}$, $\Phi(x,t)$ be a solution of (5.2) with $\lambda = \mu$. Then

(i) $\bar{\Phi}(x,t)$ is a solution of (5.2) with $\lambda = \bar{\mu}$, and $\Omega\Phi(x,t)$ is a solution of (5.2) with $\lambda = \omega\mu$.

(ii) $\Omega^{\frac{n}{2}}\bar{\Phi}$ is a solution of (5.2) with $\lambda = -\bar{\mu}$, and $\Psi = K\Omega^{\frac{n}{2}}\Phi$ is a solution of the adjoint Lax pair $\Psi_x = -U(\mu)^T\Psi$, $\Psi_t = -V(\mu)^T\Psi$. Therefore, $(\Phi^T K\Omega^{\frac{n}{2}}\Phi)_x = 0$, $(\Phi^T K\Omega^{\frac{n}{2}}\Phi)_t = 0$.

Proof. It is only necessary to prove the x-part of (5.2).
The conclusion of (i) follows from

$$\begin{aligned} \bar{\Phi}_x &= \overline{U(\mu)}\bar{\Phi} = U(\bar{\mu})\Phi, \\ (\Omega\Phi)_x &= \Omega U(\lambda)\Phi = U(\omega\lambda)\Omega\Phi. \end{aligned} \tag{5.10}$$

The first two conclusions of (ii) follows from

$$(\Omega^{\frac{n}{2}}\bar{\Phi})_x = \overline{(\Omega^{-\frac{n}{2}}\Phi)_x} = \overline{U(\omega^{-\frac{n}{2}}\mu)\Omega^{-\frac{n}{2}}\Phi} = \overline{U(-\mu)\Omega^{-\frac{n}{2}}\Phi} = U(-\bar{\mu})\Omega^{\frac{n}{2}}\bar{\Phi}, \tag{5.11}$$

and

$$\Psi_x = K\Omega^{\frac{n}{2}}\Phi_x = K\Omega^{\frac{n}{2}}U(\mu)\Omega^{-\frac{n}{2}}K\Psi = KU(-\mu)K\Psi = -U(\mu)^T\Psi. \tag{5.12}$$

Therefore, we have

$$(\Psi^T\Phi)_x = -\Psi^T U(\mu)\Phi + \Psi^T U(\mu)\Phi = 0. \tag{5.13}$$

Then

$$(\Phi^T K\Omega^{\frac{n}{2}}\Phi)_x = 0. \tag{5.14}$$

Here we use the fact $\Omega^{\frac{n}{2}}K = K\Omega^{\frac{n}{2}}$ which follows from $\Omega^* K = \omega^{n-2}K\Omega$. The lemma is proved.

We use Theorem 3 to construct the Darboux transformation. The most important thing to do is to choose $\{\mu_j\}_{j=1,\cdots,r}$ properly, as we shall show below. If the Darboux matrix is constructed and satisfies

$$\begin{aligned} &\overline{D(\lambda)} = D(\bar{\lambda}), \quad \Omega D(\lambda)\Omega^{-1} = \gamma D(\omega\lambda), \\ &D(-\bar{\lambda})^* K D(\lambda) = C(\lambda)K \end{aligned} \tag{5.15}$$

where $\gamma = \pm 1$, $C(\lambda)$ is a scalar function, then the symmetries in (5.7) are kept. By (2.7), after the action of the Darboux transformation (2.14), the coefficients of the Lax pair are changed to $U' = \lambda J + P'$, $V = \dfrac{1}{\lambda}Q'$ where

$$P' = P - [J, D_1], \quad Q' = D_r Q D_r^{-1}. \tag{5.16}$$

According to Lemma 2, in order to let the Darboux transformation keep all the symmetries, $\omega^j\mu$ and $\overline{\omega^j\mu}$ $(j = 1, \cdots, N)$ should be in the spectrum if μ is. However, $\mu_j \neq -\bar{\mu}_i$ should hold for all $i, j = 1, \cdots, r$ in order to define Γ_{ij}. This leads to the following choice of μ_j's.

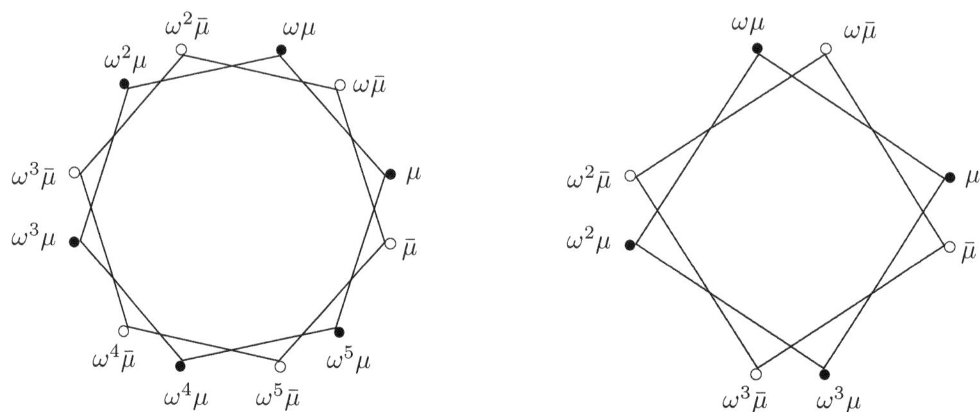

Let $\mu \in \mathbf{C}\backslash\{0\}$ with $\arg\mu \neq \dfrac{k\pi}{n}$ $(k \in \mathrm{Z})$, $\mu_j = \omega^{j-1}\mu$ $(j = 1 \cdots, n)$. Then $\mu_1, \cdots, \mu_n, \bar{\mu}_1, \cdots, \bar{\mu}_n$ are distinct. However, $-\bar{\mu}_j = \omega^{\frac{n}{2}}\bar{\mu}_j$ $(j = 1, \cdots, n)$. Hence we cannot put $\bar{\mu}_j$'s into the spectrum for constructing Darboux transformation as in Theorem 3. The spectrum is shown in the above left figure for $n = 6$ and right figure for $n = 4$. In the figures, the solid dots refer to μ_j's and small circles refer to $-\bar{\mu}_j$'s.

Let H be an $n \times \dfrac{n}{2}$ matrix solution of rank $\dfrac{n}{2}$ of the Lax pair (5.2) with $\lambda = \mu$ and satisfies $H^T K \Omega^{\frac{n}{2}} H = 0$ identically. (Due to (ii) of Lemma 2, this equality holds identically if it holds at one point.) Then $H_j = \Omega^{j-1} H$ $(j = 1, \cdots, n)$ are solutions of (5.2) with $\lambda = \mu_j$ $(j = 1, \cdots, n)$, respectively. The matrix $D(\lambda)$ is constructed by Theorem 3. Now we shall show that $D(\lambda)$ is a Darboux matrix which keeps all the symmetries in (5.7).

Theorem 4. *Let $\mu \in \mathbf{C}\backslash\{0\}$ with $\arg\mu \neq \dfrac{k\pi}{n}$ $(k \in \mathrm{Z})$, $\mu_j = \omega^{j-1}\mu$ $(j = 1 \cdots, n)$. Let H be an $n \times \dfrac{n}{2}$ matrix solution of rank $\dfrac{n}{2}$ of the Lax pair (5.2) with $\lambda = \mu$ and satisfies $H^T K \Omega^{\frac{n}{2}} H = 0$ identically. When $\det\Gamma \neq 0$, $D(\lambda)$ is a Darboux matrix which keeps all the symmetries in (5.7).*

To simplify the notations, the subscript can be an arbitrary integer so that for any vector $v = (v_1, \cdots, v_n)$, $v_j = v_k$ if $j - k \equiv 0 \mod n$. The same is true for matrices.

In the following proof, the unitary symmetry and the cyclic symmetry are verified directly. However, since $\bar{\mu}_j$'s are not included in the spectrum, it is not easy to verify the reality symmetry $\overline{U'(\lambda)} = U'(\bar{\lambda})$, $\overline{V'(\lambda)} = V'(\bar{\lambda})$ directly. Instead, let $\Delta(\lambda) = \overline{D(\bar{\lambda})} - D(\lambda)$, then the result follows from $\Delta(\mu_j)H_j = 0$, $\Delta(-\bar{\mu}_j)\Omega^{\frac{n}{2}}\bar{H}_j = 0$ $(j = 1, \cdots, n)$.

Proof. (i) **Unitary symmetry:** $D(-\bar{\lambda})^* K D(\lambda) = (\lambda^n - \mu^n)(\lambda^n - \bar{\mu}^n)K$.
This is a direct consequence of (3.12) in Theorem 3.

(ii) **Cyclic symmetry:** $D(\omega\lambda) = \Omega D(\lambda)\Omega^{-1}$.
By (3.9) and the relation $\Omega^* K = \omega^{n-2} K\Omega$,

$$\begin{aligned}\Gamma_{i+1,j-1} &= \frac{H_{i+1}^* K H_{j-1}}{\bar{\mu}_{i+1} + \mu_{j-1}} = \frac{H^*\Omega^{-i} K \Omega^{j-2} H}{\omega^{-i}\bar{\mu} + \omega^{j-2}\mu} \\ &= \frac{H^*\Omega^{-i+1}\Omega^* K \Omega^{-1}\Omega^{j-1} H}{\omega^{-1}(\omega^{-i+1}\bar{\mu} + \omega^{j-1}\mu)} = \omega^{-1}\Gamma_{ij}.\end{aligned} \tag{5.17}$$

That is, $\hat{J}\Gamma\hat{J} = \omega^{-1}\Gamma$ where $\hat{J} = J \otimes I_{\frac{n}{2}}$. Hence $\hat{J}\Gamma^{-1}\hat{J} = \omega^{-1}\Gamma^{-1}$, that is, $(\Gamma^{-1})_{i+1,j-1} = \omega^{-1}(\Gamma^{-1})_{ij}$. By (3.10),

$$\begin{aligned}\Omega D(\lambda)\Omega^{-1} &= \prod_{k=1}^{n}(\lambda + \bar{\mu}_k)\left(1 - \sum_{i,j=1}^{n} \frac{\Omega H_i (\Gamma^{-1})_{ij} H_j^* K \Omega^{-1}}{\lambda + \bar{\mu}_j}\right) \\ &= \prod_{k=1}^{n}(\omega\lambda + \bar{\mu}_{k-1})\left(1 - \sum_{i,j=1}^{n} \frac{\omega H_{i+1}(\Gamma^{-1})_{ij} H_{j-1}^* \Omega^{-1} K \Omega^{-1}}{\omega\lambda + \bar{\mu}_{j-1}}\right) \\ &= \prod_{k=1}^{n}(\omega\lambda + \bar{\mu}_k)\left(1 - \sum_{i,j=1}^{n} \frac{\omega H_{i+1}\cdot\omega(\Gamma^{-1})_{i+1,j-1}\cdot H_{j-1}^* \cdot \omega^{-2} K}{\omega\lambda + \bar{\mu}_{j-1}}\right) \\ &= \prod_{k=1}^{n}(\omega\lambda + \bar{\mu}_k)\left(1 - \sum_{i,j=1}^{n} \frac{H_i(\Gamma^{-1})_{ij} H_j^* K}{\omega\lambda + \bar{\mu}_j}\right) = D(\omega\lambda).\end{aligned} \tag{5.18}$$

(iii) **Reality symmetry:** $\overline{D(\bar{\lambda})} = D(\lambda)$.
According to (ii) of Lemma 2, if Φ is a solution of (5.2) with $\lambda = \mu$, then $\Omega^{\frac{n}{2}}\bar{\Phi}$ is a solution of (5.2) with $\lambda = -\bar{\mu}$. Using the facts $\mu_{j+\frac{n}{2}} = -\mu_j$ and $H_{j+\frac{n}{2}} = \Omega^{\frac{n}{2}} H_j$, (3.10) implies that for $k = 1, 2, \cdots, n$,

$$D(\mu_k)H_k = \prod_{s=1}^{n}(\mu_k + \bar{\mu}_s)\left(H_k - \sum_{i,j=1}^{n} \frac{H_i(\Gamma^{-1})_{ij} H_j^* K H_k}{\bar{\mu}_j + \mu_k}\right) = 0, \tag{5.19}$$

$$\begin{aligned}D(-\bar{\mu}_k)\Omega^{\frac{n}{2}}\bar{H}_k &= -\prod_{s\neq k}(\bar{\mu}_s - \bar{\mu}_k)\sum_{i=1}^{n} H_i(\Gamma^{-1})_{ik} H_k^* K \Omega^{\frac{n}{2}}\bar{H}_k \\ &= -\omega^{-2(k-1)}\prod_{s\neq k}(\bar{\mu}_s - \bar{\mu}_k)\sum_{i=1}^{n} H_i(\Gamma^{-1})_{ik}\overline{H^T K \Omega^{\frac{n}{2}} H} = 0,\end{aligned} \tag{5.20}$$

$$\overline{D(\bar{\mu}_k)}H_k = \overline{D(-\bar{\mu}_{k+\frac{n}{2}})\Omega^{\frac{n}{2}}\bar{H}_{k+\frac{n}{2}}} = 0, \tag{5.21}$$

$$\overline{D(-\mu_k)}\Omega^{\frac{n}{2}}\bar{H}_k = \overline{D(\mu_{k+\frac{n}{2}})H_{k+\frac{n}{2}}} = 0. \tag{5.22}$$

Let $\Delta(\lambda) = D(\lambda) - \overline{D(\bar{\lambda})}$ and write

$$\Delta(\lambda) = \sum_{j=0}^{n-1} \Delta_{n-1-j}\lambda^j, \tag{5.23}$$

then $\Delta(\mu_k)H_k = 0$, $\Delta(-\bar{\mu}_k)\Omega^{\frac{n}{2}}\bar{H}_k = 0$ $(k = 1, \cdots, n)$. Written in components, they are

$$\sum_{j=0}^{n-1} \Delta_{n-1-j}H_k\mu_k^j = 0, \quad \sum_{j=0}^{n-1} \Delta_{n-1-j}\Omega^{\frac{n}{2}}\bar{H}_k(-\bar{\mu}_k)^j = 0 \quad (k = 1, \cdots, n) \tag{5.24}$$

i.e.,

$$(\Delta_{n-1}, \Delta_{n-2}, \cdots, \Delta_0)\mathcal{M} = 0 \tag{5.25}$$

where $\mathcal{M} = (\mathcal{M}_{ij})_{1\le i\le n,\, 1\le j\le 2n}$ is an $n^2 \times n^2$ matrix with

$$\mathcal{M}_{ij} = \mu_j^{i-1}H_j, \quad \mathcal{M}_{i,n+j} = (-\bar{\mu}_j)^{i-1}\Omega^{\frac{n}{2}}\bar{H}_j, \quad i, j = 1, \cdots, n. \tag{5.26}$$

Now we shall prove that $\mathcal{M}$ is invertible. Define the $n^2 \times n^2$ matrix $\mathcal{N} = (\mathcal{N}_{ij})_{1\le i\le 2n,\, 1\le j\le n}$ with

$$\mathcal{N}_{ij} = (-\bar{\mu}_i)^{-j+1}H_i^*K, \quad \mathcal{N}_{n+i,j} = \mu_i^{-j+1}H_i^TK\Omega^{\frac{n}{2}}, \quad i, j = 1, \cdots, n. \tag{5.27}$$

Using the identity

$$\sum_{k=1}^{n} \left(\frac{\bar{\mu}_j}{\bar{\mu}_i}\right)^{k-1} = \sum_{k=1}^{n} \omega^{(k-1)(i-j)} = n\delta_{ij}, \tag{5.28}$$

we have, for $i, j = 1, \cdots, n$,

$$\begin{aligned}
\sum_{k=1}^{n} \mathcal{N}_{ik}\mathcal{M}_{kj} &= \sum_{k=1}^{n} (-\bar{\mu}_i)^{-k+1}\mu_j^{k-1}H_i^*KH_j \\
&= \frac{1 - \left(-\dfrac{\mu_j}{\bar{\mu}_i}\right)^n}{1 + \dfrac{\mu_j}{\bar{\mu}_i}} H_i^*KH_j = \bar{\mu}_i\left(1 - \left(\frac{\mu}{\bar{\mu}}\right)^n\right)\Gamma_{ij}, \\
\sum_{k=1}^{n} \mathcal{N}_{ik}\mathcal{M}_{k,n+j} &= \sum_{k=1}^{n} \left(\frac{\bar{\mu}_j}{\bar{\mu}_i}\right)^{k-1} H_i^*K\Omega^{\frac{n}{2}}\bar{H}_j = n\delta_{ij}\overline{H_i^TK\Omega^{\frac{n}{2}}H_j} \\
&= n\delta_{ij}\overline{H^T\Omega^{i-1}K\Omega^{\frac{n}{2}+j-1}H} = n\omega^{-2(i-1)}\delta_{ij}\overline{H^TK\Omega^{\frac{n}{2}+j-i}H} = 0, \\
\sum_{k=1}^{n} \mathcal{N}_{n+i,k}\mathcal{M}_{kj} &= \sum_{k=1}^{n} \left(\frac{\mu_j}{\mu_i}\right)^{k-1} H_i^TK\Omega^{\frac{n}{2}}H_j = 0, \\
\sum_{k=1}^{n} \mathcal{N}_{n+i,k}\mathcal{M}_{k,n+j} &= \mu_i^{-k+1}(-\bar{\mu}_j)^{k-1}H_i^TK\bar{H}_j = \mu_i\left(1 - \left(\frac{\bar{\mu}}{\mu}\right)^n\right)\bar{\Gamma}_{ij}.
\end{aligned} \tag{5.29}$$

Note that each Γ_{ij} is an $\frac{n}{2}\times\frac{n}{2}$ matrix ($i,j=1,\cdots,n$), we have

$$\det(\mathcal{N}\mathcal{M}) = |\mu|^{n^2}\left(1-\left(\frac{\mu}{\bar{\mu}}\right)^n\right)^{\frac{n^2}{2}}\left(1-\left(\frac{\bar{\mu}}{\mu}\right)^n\right)^{\frac{n^2}{2}}|\det\Gamma|^2. \tag{5.30}$$

Since $\arg(\mu)\notin\{\frac{k\pi}{n}\,|\,k\in \mathrm{Z}\}$, $\det\mathcal{M}\neq 0$ holds whenever $\det\Gamma\neq 0$. (5.25) implies $\Delta_j=0$ ($j=0,\cdots,n-1$). This proves the reality of $D(\lambda)$. The theorem is proved.

It is quite technical to simplify the expression of the derived solution. Here we only list the result. For details, please see [32]. (There was a mistake in notations there. The names $C_l^{(1)}$ and $D_{l+1}^{(2)}$ should be interchanged.)

Theorem 5. *Suppose* $(u_1,\cdots,u_l)$ *is a solution of the two-dimensional* $D_{l+1}^{(2)}$ *Toda Equation (5.1),* μ *is a nonzero complex number such that* $\arg(\mu)\neq\frac{k\pi}{2l+2}$ *for any integer* k*. Let* $H=\begin{pmatrix} h_1\\ \vdots\\ h_{2l+2}\end{pmatrix}$ *be a* $(2l+2)\times(l+1)$ *solution of rank* $l+1$ *of the Lax pair (2.1) with* $\lambda=\mu$ *such that* $H^TK\Omega^{l+1}K=0$ *holds identically. Let*

$$\eta_j = 1-h_j\left(\sum_{k=1}^{2l+2}(-1)^{k-j}\frac{\mu^{\{k-j\}}\bar{\mu}^{2l+2-\{k-j\}}}{\bar{\mu}^{2l+2}-\mu^{2l+2}}h^*_{2l+2-k}h_k\right)^{-1}h^*_{2l+2-j} \tag{5.31}$$

($j=1,2,\cdots,2l+2$) *where* $\{k\}\in\{0,1,\cdots,2l+1\}$ *is the remainder of* k *divided by* $2l+2$*. Then*

$$\widetilde{u}_j = u_j+\frac{1}{2}\ln\frac{\eta_j}{\eta_{2l+2-j}}\qquad (j=1,\cdots,l) \tag{5.32}$$

gives a new local solution of the two-dimensional $D_{l+1}^{(2)}$ *Toda Equation (5.1).*

Remark 1. Here we always suppose $\arg\mu\neq\frac{k\pi}{n}$ ($k\in\mathrm{Z}$). When $\arg\mu=\frac{k\pi}{n}$, the construction of Darboux matrix is more difficult, but the expression of the derived solutions is simpler. For details, please see [32].

6 Discussion

In this chapter, we have paid much attention on the construction of Darboux transformations for the 1+1 dimensional Lax pair with Lie algebraic symmetries. There are other types of symmetries for 1+1 dimensional as well as higher dimensional Lax pairs.

For the 1+1 dimensional integrable system, the general form of Darboux matrix of degree 1 is $D(x,t,\lambda)=R(x,t)\big(\lambda I-S(x,t)\big)$ where S is given in Theorem 1 and R

is an invertible matrix. Although R can be arbitrary when there is no symmetry, it is quite difficult to be determined if there is symmetry since there is no general way to do so like the construction of S. For example, R can always be chosen as I for the AKNS system, and cannot be a constant matrix for the Kaup-Newell system. The construction of R strongly depends on a specific system.

After constructing the Darboux matrices, the globality of the derived solution is an important problem. For the system with $u(n)$ symmetry, the globality of the derived solution is guaranteed automatically by the compactness of the Lie algebra $u(n)$. However, for noncompact Lie algebra, the derived solutions usually have singularities. For example, the Lax pair of the defocusing nonlinear Schrödinger equation has $u(1,1)$ symmetry. The solutions derived from zero seed solution will have singularities. Certain limiting processes can be used to get global solutions (e.g., see [10]).

All these related problems are not discussed in this chapter.

Acknowledgements

This work was supported by the National Natural Science Foundation of China (No. 11971114) and the Key Laboratory of Mathematics for Nonlinear Sciences of Ministry of Education of China.

References

[1] Ablowitz M J, Kaup D J, Newell A C and Segur H, Nonlinear evolution equations of physical significance, *Phys. Rev. Lett.* **31**, 125-127, 1973.

[2] Ablowitz M J and Segur H, *Solitons and the Inverse Scattering Transform*, SIAM, 1981.

[3] Bolton J and Woodward L M, The affine Toda equations and minimal surfaces, in: Fordy A P and Wood J C (Eds.), *Harmonic Maps and Integrable System*, Vieweg, 59–82, 1994.

[4] Cieśliński J L, Algebraic construction of the Darboux matrix revisited, *J. Phys. A* **42**, 404003, 2009.

[5] Darboux G, Sur une proposition relative aux équations linéaires, *Compts Rendus Hebdomadaires des Seances de l'Academie des Sciences, Paris* **94**, 1456–1460, 1882.

[6] Dorfmeister J F, Freyn W, Kobayashi S and Wang E X, Survey on real forms of the complex $A_2^{(2)}$-Toda equation and surface theory, *arXiv:1902.01558 [math.DG]*, 2019.

[7] Feng B F, Maruno K I and Ohta Y, On the τ-functions of the reduced Ostrovsky equation and the $A_2^{(2)}$ two-dimensional Toda system, *J. Phys. A* **45**, 355203, 2012.

[8] Gu C H, On the Darboux form of Bäcklund transformations, in: Song X C (Eds.), *Integrable System*, World Scientific, 162–168, 1989.

[9] Gu C H, Hu H S and Zhou Z X, *Darboux Transformations in Integrable Systems* Springer, 2005.

[10] Han J W, Yu J and He J S, Determinant representation of N-times Darboux transformations for the defocusing nonlinear Schrödinger equation, *Mod. Phys. Lett. B* **27**, 1350216, 2013.

[11] Kac V G, *Infinite Dimensional Lie Algebras*, 2nd edn., Cambridge University Press, 1990.

[12] Leznov A N and Saveliev M V, Theory of group representations and integration of nonlinear systems, *Physica D* **3**, 62–72, 1981.

[13] Leznov A N and Yuzbashjan E A, The general solution of 2-dimensional matrix Toda chain equations with fixed ends, *Lett. Math. Phys.* **35**, 345–349, 1995.

[14] Matveev V B, Darboux transformations and the explicit solutions of differential-difference and difference-difference evolution equations, *Lett. Math. Phys.* **3**, 217–222, 1979.

[15] Matveev V B and Salle M A, *Darboux Transformations and Solitons*, Springer, 1991.

[16] McIntosh I, Global solutions of the elliptic 2D periodic Toda lattice, *Nonlinearity* **7**, 85–108, 1994.

[17] Mikhailov A V, Integrability of a two-dimensional generalization of the Toda chain, *Soviet JETP Lett.* **30**, 414–418, 1979.

[18] Miura R M, Korteweg-de Vries equation and generalizations, I: A remarkable explicit nonlinear transformation, *J. Math. Phys.* **9**,1202–1204, 1968.

[19] Neugebauer G and Meinel R, General N-soliton solution of the AKNS class on arbitrary background, *Phys. Lett. A* **100**, 467–470, 1984.

[20] Nimmo J J C and Willox R, Darboux transformations for the two-dimensional Toda system, *Proc. Roy. Soc. London A* **453**, 2497–2525, 1997.

[21] Nirov Kh S and Razumov A V, Abelian Toda solitons revisited, *Rev. Math. Phys.* **20**, 1209–1248, 2008.

[22] Nirov Kh S and Razumov A V, More non-Abelian loop Toda solitons, *J. Phys. A* **42**, 285201, 2009.

[23] Razumov A V and Saveliev M V, Differential geometry of Toda systems, *Commun. Anal. Geom.* **2**, 461–511, 1994.

[24] Rogers C and Schief W K, *Bäcklund and Darboux Transformations, Geometry and Modern Applications in Soliton Theory*, Cambridge University Press, 2002.

[25] Sattinger D H and Zurkowski V D, Gauge theory of Bäcklund transformations II, *Physica D* **26**, 225–250, 1987.

[26] Saveliev M V, Bäcklund transformations for the generalized two-dimensional Toda lattice, *Phys. Lett. A* **122**, 312–316, 1987.

[27] Terng C L, Geometries and symmetries of soliton equations and integrable elliptic equations, in: Guest M, Miyaoka R and Ohnita Y (Eds.), *Surveys on Geometry and Integrable Systems, Advanced Studies in Pure Mathematics* **51**, 401–488, 2008.

[28] Wadati M, Sanuki H and Konno K, Relationships among inverse method, Bäcklund transformations and infinite number of conservation laws, *Prog. Theor. Phys.* **53**, 418–436, 1975.

[29] Zakharov V B and Mikhailov A V, On the integrability of classical spinor models in two-dimensional space-time, *Commun. Math. Phys.* **74**, 21–40, 1980.

[30] Zhou Z X, Darboux transformations and exact solutions of two dimensional $A_{2n}^{(2)}$ Toda equation, *J. Math. Phys.* **46**, 033515, 2005.

[31] Zhou Z X, Darboux transformations and exact solutions of two dimensional $C_l^{(1)}$ and $D_{l+1}^{(2)}$ Toda equations, *J. Phys. A* **39**, 5727–5737, 2006.

[32] Zhou Z X, Darboux transformations of lower degree for two dimensional $C_l^{(1)}$ and $D_{l+1}^{(2)}$ Toda equations, *Inv. Prob.*, **24**, 045016, 2008.

[33] Zhu Z and Caldi D G, Multi-soliton solutions of affine Toda models, *Nucl. Phys. B* **436**, 659–678, 1995.

A9. Frobenius manifolds and Orbit spaces of reflection groups and their extensions

Dafeng Zuo

School of Mathematical Science, University of Science and Technology of China, Hefei 230026, P. R. China

Abstract

We review recent developments about Frobenius manifold structures on the orbit spaces of reflection groups and their extensions and propose a conjecture for general cases.

1 Introduction

1.1 WDVV equations and Frobenius manifolds

In the beginning of 1990's, E.Witten, R.Dijkgraaf, E.Verlinde and H.Verlinde [8, 28] introduced one remarkable system of partial differential equations, that is WDVV equations of associativity (see below (1.2)–(1.5)), on two-dimensional topological field theory (2D TFT).

In the physical setting, the solutions of WDVV equations describe the moduli space of topological conformal field theory. From the point of mathematical view, some particular solutions of WDVV equations with certain good analytic properties are generating functions for the genus zero Gromov-Witten invariants of Käler, and more generally, of symplectic manifolds.

In order to understand a geometrical foundation of 2D TFT on the bases of WDVV equations, in 1993 B.Dubrovin extended Atiyah's axioms [3] of TFT for the two-dimensional case by the properties of the canonical modulo space of a TFT model. On this way he invented a nice geometrical object, that is, Frobenius manifold designed as a coordinate-free formulation of WDVV equations. It was shown that any model of 2D TFT was encoded by a Frobenius manifold and many constructions of TFT (integrable hierarchies for the partition function, their bi-Hamiltonian formulae and τ-functions, string equations, genus zero recursion relations for correlators) can be deduced from geometry of Frobenius manifolds. Let us first recall the definition of Frobenius manifold, see [9, 10] for details.

Definition 1. A *Frobenius algebra* is a pair $(A, < \ , \ >)$ where A is a commutative associative algebra with a unity e over a field $\mathcal{K}$ (in our case $\mathcal{K} = \mathbb{C}$) and $< \ , \ >$ is a $\mathcal{K}$-bilinear symmetric nondegenerate *invariant* form on A, i.e.,

$$< x \cdot y, z > = < x, y \cdot z >, \quad \forall\ x, y, z \in A.$$

Definition 2. A Frobenius structure of charge d on an n-dimensional manifold M is a structure of Frobenius algebra on the tangent spaces $T_tM = (A_t, < \ , \ >_t)$ depending (smoothly, analytically, etc.) on the point t. This structure satisfies the following axioms:

FM1. The metric $< \ , \ >_t$ on M is flat, and the unity vector field e is covariantly constant, i.e., $\nabla e = 0$. Here we denote ∇ the Levi-Civita connection for this flat metric.

FM2. Let c be the 3-tensor $c(x, y, z) :=< x \cdot y, z >$, $x, y, z \in T_tM$. Then the 4-tensor $(\nabla_w c)(x, y, z)$ is symmetric in $x, y, z, w \in T_tM$.

FM3. The existence on M of a vector field E, called the Euler vector field, which satisfies the conditions $\nabla\nabla E = 0$ and

$$[E, x \cdot y] - [E, x] \cdot y - x \cdot [E, y] = x \cdot y,$$

$$E < x, y > - < [E, x], y > - < x, [E, y] >= (2 - d) < x, y >$$

for any vector fields x, y on M.

A manifold M equipped with a Frobenius structure on it is called a Frobenius manifold.

Let us choose local flat coordinates $t^1, \cdots t^n$ for the invariant flat metric, then locally there exists a function $F(t^1, \cdots, t^n)$, called the *potential* of the Frobenius manifold, such that

$$< u \cdot v, w >= u^i v^j w^s \frac{\partial^3 F}{\partial t^i \partial t^j \partial t^s} \tag{1.1}$$

for any three vector fields $u = u^i \frac{\partial}{\partial t^i}$, $v = v^j \frac{\partial}{\partial t^j}$, $w = w^s \frac{\partial}{\partial t^s}$. Here and in what follows summations over repeated indices are assumed. By definition, we can also choose the coordinates t^1 such that $e = \frac{\partial}{\partial t^1}$. Then in the flat coordinates the components of of the flat metric $< \frac{\partial}{\partial t^i}, \frac{\partial}{\partial t^j} >$ can be expressed in the form

$$\frac{\partial^3 F}{\partial t^1 \partial t^i \partial t^j} = \eta_{ij}, \quad i, j = 1, \ldots, n. \tag{1.2}$$

The associativity of the Frobenius algebras is equivalent to the following overdetermined system of equations for the function F

$$\frac{\partial^3 F}{\partial t^i \partial t^j \partial t^\lambda} \eta^{\lambda\mu} \frac{\partial^3 F}{\partial t^\mu \partial t^k \partial t^m} = \frac{\partial^3 F}{\partial t^k \partial t^j \partial t^\lambda} \eta^{\lambda\mu} \frac{\partial^3 F}{\partial t^\mu \partial t^i \partial t^m} \tag{1.3}$$

for arbitrary indices i, j, k, m from 1 to n.

We assume that flat coordinates have been chosen so that the Euler vector field E has the form

$$E = \sum_{i=1}^{n} (\hat{d}_i t^i + r_i) \frac{\partial}{\partial t^i} \tag{1.4}$$

for some constants $\hat{d}_i, r_i$, $i = 1, \ldots, n$ which satisfy $\hat{d}_1 = 1, r_1 = 0$. From the axiom FM3, it follows that the potential F satisfies the quasi-homogeneity condition

$$\mathcal{L}_E F = (3 - d)F + \text{quadratic polynomial in t}. \tag{1.5}$$

The system (1.2)–(1.5) is called the *WDVV equations of associativity* which is equivalent to the above definition of Frobenius manifold in the chosen system of local coordinates.

Let us also recall an important geometrical structure on a Frobenius manifold M, the *intersection form* of M. This is a symmetric bilinear form $(\ ,\ )^*$ on T^*M defined by the formula

$$(w_1, w_2)^* = i_E(w_1 \cdot w_2), \tag{1.6}$$

here the product of two 1-forms w_1, w_2 at a point $t \in M$ is defined by using the algebra structure on T_tM and the isomorphism

$$T_tM \to T_t^*M \tag{1.7}$$

established by the invariant flat metric $<\ ,\ >$. In the flat coordinates $t^1, \cdots, t^n$ of the invariant metric, the intersection form can be represented by

$$(dt^i, dt^j)^* = \mathcal{L}_E F^{ij} = (d - 1 + \hat{d}_i + \hat{d}_j)F^{ij}, \tag{1.8}$$

where

$$F^{ij} = \eta^{ii'}\eta^{jj'}\frac{\partial^2 F}{\partial t^{i'}\partial t^{j'}} \tag{1.9}$$

and $F(t)$ is the potential of the Frobenius manifold. Denote by $\Sigma_0 \subset M$ the *discriminant* of M on which the intersection form degenerates, then an important property of the intersection form is that on $M \setminus \Sigma_0$ its inverse defines a new flat metric.

1.2 Frobenius mainfolds and orbit spaces

A clue to understanding of a rich differential-geometric structure of the orbit spaces can be found in the singularity theory due to V.I.Anorld and K.Saito, etc. [2, 18]. According to this, the complexified orbit space of an irreducible Coxeter group is bi-holomorphic equivalent to the universal unfolding of a simple singularity. Under this identification, the Coxeter group coincides with the monodromy group of vanishing cycles of the singularity. The discriminant of the Coxeter group (the set of irregular orbits) is identified with the bifurcation diagram of the singularity. The invariant Euclidean inner product coincides with the pairing on the cotangent bundle defined by the intersection form of vanishing cycles. The bi-holomorphic equivalence is given by the period mapping.

Additional differential-geometric structures on a universal unfolding of an isolated hypersurface singularity are determined by the Grothéndieck residues [11, 21]. For *ADE* cases the formulae for the residues were rediscovered by E.Witten R.Dijkgraaf, E.Verlinde and H.Verlinde. Moreover, they discussed 2D TFT from the point of view of the theory of singularities. B.Dubrovin [10, 11] gave an intrinsic formula for calculation of the Grothéndieck residues for arbitrary Coxeter group without using the construction of the correspondent universal unfolding and obtained a complete differential-geometric characterization of the space of orbits in terms of these structures. Furthermore, B.Dubrovin defined a monodromy group $W(M)$ for an arbitrary l-dimensional Frobenius manifold M which acts on an l-dimensional linear space. For instance,

Example 1. [$W(M)$=Coxeter group A_1] $l = 1$, $M = \mathbb{C}$, $t = t^1$,

$$F(t) = \frac{1}{6}t^3, \quad E = t\frac{\partial}{\partial t}, \quad e = \frac{\partial}{\partial t}, \quad \eta^{11} =< \frac{\partial}{\partial t}, \frac{\partial}{\partial t} >= 1.$$

Example 2. [$W(M)$=extended affine Weyl group $\widetilde{W}(A_1)$]
Quantum cohomology of $\mathbb{CP}^1$:

$$F(t^1, t^2) = \frac{1}{2}(t^1)^2t^2 + e^{t^2}, E = t^1\frac{\partial}{\partial t^1} + 2\frac{\partial}{\partial t^2}, e = \frac{\partial}{\partial t^1}.$$

It might be conjectured that for a Frobenius manifold with good analytic properties the monodromy group acts discretely in some domain of the space. The Frobenius manifold itself can be identified with the orbit space of the group in the sense to be specified for each class of monodromy groups. A natural question is how to construct the Frobenius manifold structure on the orbit space of some groups. Equivalently, *"Which kind of groups can be realized as monodromy groups of Frobenius manifolds?"*.

Nowdays the theory of Frobenius manifolds have established remarkable relationships between some rather distant mathematical theories, including the theory of Gromov-Witten invariants, singularity theory, differential geometry of the orbit spaces of reflection groups and of their extensions, the Hamiltonian theory of integrable hierarchies, and so on. In this chapter we only review some recent developments about the Frobenius manifold structures on the orbit spaces of reflection groups and their extensions.

2 Frobenius manifolds and reflection groups

Let W be a finite Coxeter group, i.e., a finite group of linear transformations of an l-dimensional Euclidean space V generated by reflections. The orbit space $\mathcal{M} = V/W$ has a natural structure of affine variety: the coordinate ring of $\mathcal{M}$ coincides with the ring of W-invariant polynomial functions on V. Due to Chevalley theorem, this is a polynomial ring with the generators $y^1, \cdots, y^l$ being invariant homogeneous polynomials. The basic invariant polynomials are not specified uniquely. But their degrees $d_1, \cdots, d_l$ are invariants of the group. The maximal degree h of the polynomials is called Coxeter number of the group W.

We denote by $(\ , \ )^*$ the metric on the cotangent bundle $T^*\mathcal{M}$ induced by the W-invariant Euclidean structure on V. There are two marked vector fields on $\mathcal{M}$: the Euler vector field

$$E := \sum_{i=1}^{l} d_i y^i \frac{\partial}{\partial y^i} \tag{2.1}$$

and the unity vector field

$$e := \frac{\partial}{\partial y^1} \tag{2.2}$$

corresponding to the polynomial of the *maximal* degree $\deg y^1 = h$. The vector field e is defined uniquely up to a constant factor. The *Saito metric* on $\mathcal{M}$ is defined as

$$\langle\ ,\ \rangle^* := \mathcal{L}_e(\ ,\ )^*.$$

This is a flat globally defined metric on $\mathcal{M}$ [22, 23]. Moreover B.Dubrovin obtained the following:

Theorem 1. *([10, 11]) There exists a unique (up to rescaling)* **polynomial** *Frobenius structure on the orbit space of a finite irreducible Coxeter group with the charges and dimension*

$$q_\alpha = 1 - \frac{d_\alpha}{h}, \quad d = 1 - \frac{2}{h},$$

the unity e, the Euler vector field $\frac{1}{h}E$, and the Saito invariant metric such that for any two invariant polynomials f, g the following formula holds

$$i_E(df \cdot dg) = (df, dg)^*. \tag{2.3}$$

Here i_E is the operator of inner derivative (contraction) along the vector field E.

Furthermore, B.Dubrovin [10, 11] presented the following conjecture,

> *"Any massive polynomial Frobenius manifold with positive invariant degrees is isomorphic to the orbit space of a finite Coxeter group",*

which was proved by C.Hertling [16].

In [4], M.Bertola noticed that on the orbit space of the Coxeter group B_l, there are l flat pencils of metrics having as common basepoint. Moreover by analogy with [1, 4, 11], he started from a superpotential

$$\lambda_{k,l}(p) = p^{-2(l-k)}\left(\sum_{a=1}^{l} p^{2(l-a)}y^a + p^{2l}\right) \tag{2.4}$$

to compute the corresponding potential F, generally which is not polynomial with respect to $y^1, \cdots, y^l$. Here

$$y^i = \sigma_i((x^1)^2, \cdots, (x^l)^2), \quad i = 1, \cdots, l, \tag{2.5}$$

where σ_i is the i-th elementary symmetric polynomial.

In [29], we used the analytic method to construct the Frobenius manifold on the orbit space of the Coxeter group B_l and D_l, which is different from M.Bertola's construction [4]. More precisely,

Theorem 2. *([29]) For any fixed integer $1 \leq k \leq l$, there exists a unique Frobenius structure of charge $d = 1 - \frac{1}{k}$ on the orbit space $\mathcal{M} \backslash \{t^l = 0\}$ of B_l (or D_l) polynomial in $t^1, t^2, \cdots, t^l, \frac{1}{t^l}$ such that*

1. *the unity vector field e coincides with $\frac{\partial}{\partial y^k} = \frac{\partial}{\partial t^k}$;*
2. *the Euler vector field has the form*

$$E = \sum_{\alpha=1}^{l} \tilde{ds}_\alpha t^\alpha \frac{\partial}{\partial t^\alpha}, \tag{2.6}$$

where $\tilde{d}_1, \ldots, \tilde{d}_l$ are defined as follows

$$\tilde{d}_j = \frac{j}{k}, \quad \tilde{d}_m = \frac{2k(l-m)+l}{2k(l-k)}; \tag{2.7}$$

3. *the invariant flat metric and the intersection form of the Frobenius structure coincide, respectively, with the metric $(\eta^{ij}(t))$ and $(g^{ij}(t))$ on $\mathcal{M} \setminus \{t^l = 0\}$.*

Example 3. ([29]) Let R be the root system of type B_3. The $W(B_3)$-invariant ring is generated by

$$\begin{aligned} y^1 &= (x^1)^2 + (x^2)^2 + (x^3)^2, \\ y^2 &= (x^1)^2\,(x^2)^2 + (x^1)^2\,(x^3)^2 + (x^2)^2\,(x^3)^2, \\ y^3 &= (x^1)^2\,(x^2)^2(x^3)^2. \end{aligned}$$

and the intersection form is given by $g^{ij}(y) = \sum_{s=3}^{l} \frac{\partial y^i}{\partial x^s}\frac{\partial y^j}{\partial x^s}$ and $\eta^{ij}(y) = \mathcal{L}_e g^{ij}(y)$.

Case 1. $k = 1$, *i.e.*, $e = \frac{\partial}{\partial y^1}$. The flat coordinates of $\eta^{ij}(y)$ are

$$t_1 = y^1, \quad t_2 = y^2\,(y^3)^{-\frac{1}{4}}, \quad t_3 = (y^3)^{\frac{1}{4}}$$

and the intersection form reads

$$\begin{aligned} &g^{11}(t) = 4t_1, \quad g^{12} = 5t_2, \quad g^{13} = 3t_3, \\ &g^{22} = 12t_3^2 + \frac{4t_2^3}{t_3^3}, \; g^{23} = 2t_1 - \frac{4t_2^2}{t_3^2}, \; g^{33} = \frac{4t_2}{t_3}. \end{aligned}$$

The potential has the form

$$F = \frac{1}{2}t_1t_2t_3 + \frac{1}{24}{t_1}^3 + \frac{1}{6}{t_3}^4 + \frac{1}{48}\frac{{t_2}^3}{t_3}$$

and the Euler vector field is given by

$$E = t_1\partial_1 + \frac{5}{4}t_2\partial_2 + \frac{3}{4}t_3\partial_3,$$

where $\partial_\alpha = \frac{\partial}{\partial t^\alpha}$.

Case 2. $k = 2$, *i.e.*, $e = \frac{\partial}{\partial y^2}$. The flat coordinates are

$$t_1 = y^1, \quad t_2 = y^2 - \frac{1}{8}(y^1)^2, \quad t_3 = \sqrt{y^3}$$

and the intersection form has the expression

$$g^{11}(t) = 4t_1, \quad g^{12} = 8t_2, \quad g^{22} = \frac{1}{4}{t_1}^3 + 12\,t_3^2,$$
$$g^{13} = 6t_3, \; g^{23} = \frac{5}{2}t_1t_3, \; g^{33} = {t_2}^2 + \frac{1}{8}{t_1}^2.$$

The potential has the form

$$F = \frac{1}{2}{t_3}^2t_2 + \frac{1}{16}t_1{t_2}^2 + \frac{1}{16}{t_1}^2{t_3}^2 + \frac{1}{7680}{t_1}^5$$

and the Euler vector field is given by

$$E = \frac{1}{2}t_1\partial_1 + t_2\partial_2 + \frac{3}{4}t_3\partial_3.$$

Case 3. $k = 3$, *i.e.*, $e = \frac{\partial}{\partial y^3}$. The flat coordinates are

$$t_1 = y^1, \quad t_2 = y^2 - \frac{1}{4}(y^1)^2, \quad t_3 = y^3 - \frac{1}{6}y^1y^2 + \frac{7}{216}(y^1)^3$$

and the intersection form has the expression

$$g^{11}(t) = 4t_1, \quad g^{12} = 8t_2, \quad g^{22} = \frac{1}{9}{t_1}^3 - 2\,t_1t_2 + 12\,t_3,$$
$$g^{13} = 12t_3, \; g^{23} = \frac{4}{9}{t_1}^2t_2 - \frac{4}{3}{t_2}^2, \; g^{33} = \frac{5}{9}t_1{t_2}^2 + \frac{1}{324}{t_1}^5.$$

The potential has the form

$$F = \frac{1}{24}t_1{t_3}^2 + \frac{1}{24}t_3{t_2}^2 - \frac{1}{432}t_1{t_2}^3 + \frac{1}{2592}{t_1}^3{t_2}^2 + \frac{1}{3265920}{t_1}^7$$

and the Euler vector field is given by

$$E = \frac{1}{3}t_1\partial_1 + \frac{2}{3}t_2\partial_2 + t_3\partial_3.$$

3 Frobenius manifolds and extended affine Weyl groups

3.1 Simple-pole extensions

Motivated by Example 2, B.Dubrovin and Y.Zhang [12] introduced certain extension of affine Weyl groups, called extended affine Weyl groups. More precisely, let R be an irreducible reduced root system defined in an l-dimensional Euclidean space V with the Euclidean inner product $(\ ,\ )$. We fix a basis of simple roots $\alpha_1, \ldots, \alpha_l$ and denote by $\alpha_j^\vee$, $j = 1, 2, \cdots, l$ the corresponding coroots. The Weyl group W is generated by the reflections

$$\mathbf{x} \mapsto \mathbf{x} - (\alpha_j^\vee, \mathbf{x})\alpha_j, \quad \forall \mathbf{x} \in V,\ j = 1, \ldots, l. \tag{3.1}$$

Recall that the *Cartan matrix* of the root system has integer entries $A_{ij} = \left(\alpha_i, \alpha_j^\vee\right)$ satisfying $A_{ii} = 2$, $A_{ij} \leq 0$ for $i \neq j$. The semi-direct product of W by the lattice of coroots yields the affine Weyl group W_a that acts on V by the affine transformations

$$\mathbf{x} \mapsto w(\mathbf{x}) + \sum_{j=1}^{l} m_j {\alpha^\vee}_j, \quad w \in W,\ m_j \in \mathbb{Z}. \tag{3.2}$$

We denote by $\omega_1, \ldots, \omega_l$ the fundamental weights defined by the relations

$$(\omega_i, \alpha_j^\vee) = \delta_{ij}, \quad i, j = 1, \ldots, l. \tag{3.3}$$

Note that the root system R is one of the types $A_l, B_l, C_l, D_l, E_6, E_7, E_8,\ F_4, G_2$. In what follows the Euclidean space V and the basis $\alpha_1, \ldots, \alpha_l$ of the simple roots will be defined as in Plate I-IX [7]. Let us fix a simple root α_k and define an extension of the affine Weyl group W_a in a similar way as was done in [12].

Definition 3. ([12]) The extended affine Weyl group $\widetilde{W} = \widetilde{W}^{(k)}(R)$ acts on the extended space

$$\widetilde{V} = V \oplus \mathbb{R},$$

and is generated by the transformations

$$x = (\mathbf{x}, x_{l+1}) \mapsto (w(\mathbf{x}) + \sum_{j=1}^{l} m_j {\alpha^\vee}_j,\ x_{l+1}), \quad w \in W,\ m_j \in \mathbb{Z}, \tag{3.4}$$

and

$$x = (\mathbf{x}, x_{l+1}) \mapsto (\mathbf{x} + \gamma\,\omega_k,\ x_{l+1} - \gamma). \tag{3.5}$$

Here $1 \leq k \leq l$, $\gamma = 1$ except for the cases when $R = B_l, k = l$ and $R = F_4, k = 3$ or $k = 4$, in these three cases $\gamma = 2$.

Coordinates $x_1, \ldots, x_l$ may be introduced on the space V via the expression

$$\mathbf{x} = x_1\alpha_1^\vee + \cdots + x_l\alpha_l^\vee. \tag{3.6}$$

Let $f = \det(A_{ij})$, the determinant of the Cartan matrix of the root system R.

Definition 4 ([12]). $\mathcal{A} = \mathcal{A}^{(k)}(R)$ is the ring of all $\widetilde{W}$-invariant Fourier polynomials of the form

$$\sum_{m_1,\ldots,m_{l+1} \in \mathbb{Z}} a_{m_1,\ldots,m_{l+1}} e^{2\pi i(m_1x_1+\cdots+m_lx_l+\frac{1}{f}m_{l+1}x_{l+1})}$$

bounded in the limit

$$\mathbf{x} = \mathbf{x}^0 - i\omega_k\tau, \quad x_{l+1} = x^0_{l+1} + i\tau, \quad \tau \to +\infty \tag{3.7}$$

for any $x^0 = (\mathbf{x}^0, x^0_{l+1})$.

We introduce a set of numbers

$$d_j = (\omega_j, \omega_k), \quad j = 1, \dots, l \tag{3.8}$$

and define the following Fourier polynomials [12]

$$\widetilde{y}_j(x) = e^{2\pi i d_j x_{l+1}} y_j(\mathbf{x}), \quad j = 1, \dots, l, \tag{3.9}$$

$$\widetilde{y}_{l+1}(x) = e^{\frac{2\pi i}{\gamma} x_{l+1}}. \tag{3.10}$$

Here $y_1(\mathbf{x}), \dots, y_l(\mathbf{x})$ are the basic W_a-invariant Fourier polynomials defined by

$$y_j(\mathbf{x}) = \frac{1}{n_j} \sum_{w \in W} e^{2\pi i(\omega_j, w(\mathbf{x}))}, \quad n_j = \#\{w \in W | e^{2\pi i(\omega_j, w(\mathbf{x}))} = e^{2\pi i(\omega_j, \mathbf{x})}\}. \tag{3.11}$$

Chevalley-type theorem was shown in [12] for the ring $\mathcal{A}$, i.e.,

Theorem 3. *([12]) For certain particular choices of the simple root α_k, it is isomorphic to the polynomial ring generated by $\widetilde{y}_1, \dots, \widetilde{y}_{l+1}$.*

We thus know that the orbit space of the extended affine Weyl group $\widetilde{W}$ *defined* as $\mathcal{M} = \operatorname{Spec} \mathcal{A}$ is an affine algebraic variety of dimension $l+1$.

Theorem 4. *([12]) On such an orbit space, there exists a Frobenius manifold structure whose potential is a polynomial of $t^1, \dots, t^{l+1}, e^{t^{l+1}}$. Here $t^1, \dots, t^{l+1}$ are the flat coordinates of the Frobenius manifold.*

For the root system of type A_l, there are in fact no restrictions on the choice of α_k. However, for the root systems of types $B_l, C_l, D_l, E_6, E_7, E_8, F_4, G_2$ there is only one choice for each system.

Slodowy [26] pointed out that the Chevalley-type theorem of [12] is a consequence of the results of Looijenga and Wirthmüller [19, 20, 27], and in fact it holds true for any choice of the base element α_k, or equivalently, for any fixed vertex of the Dynkin diagram. Hence:

Theorem 5. *([19, 20, 27, 26]) The ring $\mathcal{A}$ is isomorphic to the ring of polynomials of $\tilde{y}_1(x), \cdots, \tilde{y}_{l+1}(x)$.*

A natural question, as was raised in [12, 26], is *whether the geometric structures revealed in [12] also exist on the orbit spaces of the extended affine Weyl groups for an arbitrary choice of the root α_k.* In [13, 14], we gave an affirmative answer to this question for the root systems of types B_l, C_l and D_l (recall that for the root system of type A_l, the question was already answered affirmatively in [12]).

Let $\mathcal{M}$ be the orbit space of the extended affine Weyl group $\widetilde{W}^{(k)}(C_l)$ and $\widetilde{\mathcal{M}}$ the universal covering of $\mathcal{M} \setminus \{\widetilde{y}_{l+1} = 0\}$. There is an induced symmetric bilinear form on $T^*\widetilde{\mathcal{M}}$ defined by the matrix

$$g^{ij}(y) := \sum_{a,b=1}^{l+1} \frac{\partial y^i}{\partial x_a}\frac{\partial y^j}{\partial x_b}(dx_a, dx_b)^{\sim}, \tag{3.12}$$

in the coordinates $y^1 = \widetilde{y}_1, \ldots, y^l = \widetilde{y}_l,\ y^{l+1} = \log \widetilde{y}_{l+1} = 2\pi i x_{l+1}$. Denote the *discriminant* of the extended affine Weyl group $\widetilde{W}^{(k)}(C_l)$ by

$$\Sigma = \{(y^1, \ldots, y^l, e^{y^{l+1}}) \in \mathcal{M} \,|\, \det(g^{ij}(y)) = 0\}, \tag{3.13}$$

which is an algebraic subvariety of $\mathcal{M}$ and consists of the orbits of points $(\mathbf{x}, x_{l+1}) \in \widetilde{V}$ on the *mirrors* of the group with $(\beta, \mathbf{x}) \in \mathbb{Z}$ for some positive root β. Furthermore, we construct another symmetric bilinear form on $T^*\mathcal{M}$ by

$$\eta^{ij}(y) := \mathcal{L}_e g^{ij}(y). \tag{3.14}$$

Here the vector field e has the form

$$e = \sum_{j=k}^{l} c_j \frac{\partial}{\partial y^j}, \tag{3.15}$$

it depends on the choice of an integer m in the range $0 \le m \le l-k$. Namely, for a given m the coefficients $c_k, \ldots, c_l$ are defined by the generating function

$$\sum_{j=k}^{l} c_j u^{l-j} = (u+2)^m (u-2)^{l-k-m}.$$

The symmetric bilinear forms (η^{ij}) is non-degenerate on $\mathcal{M} \setminus \Sigma_1 \cup \Sigma_2$, where

$$\Sigma_1 = \{f_1 = 0\} \subset \mathcal{M}, \quad \Sigma_2 = \{f_2 = 0\} \subset \mathcal{M}$$

are the loci of zeros of the following $\widetilde{W}^{(k)}(C_l)$-invariant polynomials

$$f_1 = e^{2\pi i k\, x_{l+1}} \prod_{j=1}^{l} \cos^2 \pi(x_j - x_{j-1}), \quad f_2 = e^{2\pi i k\, x_{l+1}} \prod_{j=1}^{l} \sin^2 \pi(x_j - x_{j-1}) \tag{3.16}$$

with $x_0 = 0$, and it gives the flat metric of the Frobenius manifold structure that we are to construct. Denote

$$\mathcal{M}_{k,m}(C_l) := \begin{cases} \mathcal{M} \setminus \{\tilde{y}_{l+1} = 0\} \cup \Sigma_1 \cup \Sigma_2, & \text{when } 0 < m < l-k; \\ \mathcal{M} \setminus \{\tilde{y}_{l+1} = 0\} \cup \Sigma_1, & \text{when } m = 0; \\ \mathcal{M} \setminus \{\tilde{y}_{l+1} = 0\} \cup \Sigma_2, & \text{when } m = l-k. \end{cases}$$

Theorem 6. *([14]) For any fixed integer $0 \leq m \leq l-k$, there exists a unique Frobenius manifold structure of charge $d = 1$ on the orbit space $\mathcal{M}_{k,m}(C_l)$ of $\widetilde{W}^{(k)}(C_l)$ such that in the flat coordinates $t^1, \ldots, t^{l+1}$ of the metric (3.14) defined on certain covering of $\mathcal{M}_{k,m}(C_l)$ the Frobenius manifold structure is polynomial in $t^1, \cdots, t^{l+1}, \frac{1}{t^{l-m}}, \frac{1}{t^l}, e^{t^{l+1}}$. In these coordinates*

$$e = \frac{\partial}{\partial t^k} \tag{3.17}$$

and

$$E = \sum_{\alpha=1}^{l} \tilde{d}_\alpha t^\alpha \frac{\partial}{\partial t^\alpha} + \frac{1}{k}\frac{\partial}{\partial t^{l+1}}, \tag{3.18}$$

where $\tilde{d}_1, \ldots, \tilde{d}_l$ are defined in

$$\begin{aligned} \tilde{d}_j &= \deg t^j := \frac{j}{k}, \quad 1 \leq j \leq k, \\ \tilde{d}_s &= \deg t^s := \frac{2l-2m-2s+1}{2(l-m-k)}, \quad k+1 \leq s \leq l-m, \\ \tilde{d}_\alpha &= \deg t^\alpha := \frac{2l-2\alpha+1}{2m}, \quad l-m+1 \leq \alpha \leq l. \end{aligned} \tag{3.19}$$

We remark that for the root systems of type B_l and D_l,the resulting Frobenius manifolds are isomorphic to those obtained from the root system of type C_l.

For the case of A_l an alternative construction of the Frobenius manifold structure was given in [12]. This structure was given in terms of a LG superpotential construction. In particular, it was shown that the extended affine Weyl group $\widetilde{W}^{(k)}(A_l)$ describes the monodromy of roots of trigonometric polynomials - the superpotential - with a given bidegree being of the form

$$\lambda(\varphi) = e^{\mathbf{i}k\varphi} + a_1 e^{\mathbf{i}(k-1)\varphi} + \cdots + a_{l+1} e^{\mathbf{i}(k-l-1)\varphi}, \quad a_{l+1} \neq 0.$$

A natural question is "*Does there exist a similar construction for the root systems of type B_l, C_l and D_l?*"

Let us denote by $\mathfrak{M}_{k,m,n}$ the space of a particular class of cosine Laurent series or superpotentials of one variable with a given tri-degree $(2k, 2m, 2n)$ being of the form

$$\lambda(\varphi) = \left(\cos^2(\varphi) - 1\right)^{-m} \sum_{j=0}^{k+m+n} a_j \cos^{2(k+m-j)}(\varphi),$$

where all $a_j \in \mathbb{C}$, $m, n \in \mathbb{Z}_{\geq 0}$, $k \in \mathbb{N}$, and the coefficients $a_0, \ldots, a_{k+m+n}$ satisfy certain conditions. The space $\mathfrak{M}_{k,m,n}$ carries a natural structure of Frobenius manifold.

Theorem 7. *([14]) Let $\mathfrak{h} : \mathcal{M}_{k,m}(C_l) \to \mathfrak{M}_{k,m,n}$ be induced by the map*

$$(x_1, \cdots, x_{l+1}) \mapsto (\varphi_1, \cdots, \varphi_{l+1}) \tag{3.20}$$

a with

$$\varphi_1 = \pi x_1, \quad \varphi_j = \pi(x_j - x_{j-1}), \quad \varphi_{l+1} = \pi x_{l+1}, \quad j = 2, \cdots, l.$$

Then $\mathfrak{h}$ is a k-fold covering map, which is also a local isomorphism between the Frobenius manifolds $\mathcal{M}_{k,m}(C_l)$ and $\mathfrak{M}_{k,m,n}$, where $l = k + m + n$.

3.2 Multi-poles extensions

It is well known that there is a Frobenius manifold structure on the Hurwitz space $\mathfrak{M}_{0;0,0,0}$ ([10]), which is the space of the rational functions of the form

$$\lambda(p) = p + \frac{a}{p-b} + \frac{c}{p-d},$$

where a, b, c, d are arbitrary complex parameters. The monodromy group of this Frobenius manifold is a certain extension of affine Weyl group $W_a(A_2)$, which is different from those in [12, 14]. Motivated by this observation, in [30] we proposed a class of new extensions of affine Weyl groups denoted by $\widetilde{W}^{(k,k+1)}(R)$, where the new class of extended affine Weyl groups $\widetilde{W}^{(k,k+1)}(R)$ act on the extended space $V \oplus \mathbb{R}^2$ generated by the transformations

$$x = (\mathbf{x},\ x_{l+1}, x_{l+2}) \mapsto (w(\mathbf{x}) + \sum_{j=1}^{l} m_j \alpha_j^\vee,\ x_{l+1}, x_{l+2}), \quad w \in W, \quad m_j \in \mathbb{Z},$$

and

$$x = (\mathbf{x},\ x_{l+1}, x_{l+2}) \mapsto (\mathbf{x} + \gamma_k(R)\,\omega_k,\ x_{l+1} - \gamma_k(R),\ x_{l+2})$$

and

$$x = (\mathbf{x},\ x_{l+1}, x_{l+2}) \mapsto (\mathbf{x} + \gamma_{k+1}(R)\,\omega_{k+1},\ x_{l+1},\ x_{l+2} - \gamma_{k+1}(R)).$$

Here $1 \leq k \leq l-1$ and $\gamma_s(R) = 1$ for $1 \leq s \leq l$ except three cases $\gamma_l(B_l) = \gamma_3(F_4) = \gamma_4(F_4) = 2$. Up to now, we could not obtain any flat pencil of metrics and Frobenius manifold structures on the orbit spaces $\mathcal{M}^{(1,2)}(B_2)$, $\mathcal{M}^{(1,2)}(C_2)$ and $\mathcal{M}^{(1,2)}(G_2)$. But for the type A_l case, we have shown that:

Theorem 8. *([30]) For any fixed integer $1 \leq k < l$, there exists a unique Frobenius manifold structure of charge $d = 1$ on the orbit space $\mathcal{M}^{(k,k+1)}(A_l) \setminus \{\tilde{y}_{l+1} = 0\} \cup \{\tilde{y}_{l+2} = 0\}$ of $\widetilde{W}^{(k,k+1)}(A_l)$ such that the potential $F(t) = \widehat{F}(t) + \frac{1}{2}(t^{k+1})^2 \log(t^{k+1})$, where $\widehat{F}(t)$ is a weighted homogeneous polynomial in $t^1, t^2, \cdots, t^{l+2}, e^{t^{l+1}}, e^{t^{l+2}-t^{l+1}}$, satisfying*

(1). the unity vector field e coincides with $\dfrac{\partial}{\partial t^k} = \dfrac{\partial}{\partial y^k} + \dfrac{\partial}{\partial y^{k+1}}$;

(2). the Euler vector field has the form

$$E = \sum_{\alpha=1}^{l} d_\alpha t^\alpha \frac{\partial}{\partial t^\alpha} + \frac{1}{l-k}\frac{\partial}{\partial t^{l+1}} + \frac{l}{k(l-k)}\frac{\partial}{\partial t^{l+2}} \tag{3.21}$$

$$= \sum_{\alpha=1}^{l} d_\alpha y^\alpha \frac{\partial}{\partial y^\alpha} + \frac{1}{k}\frac{\partial}{\partial y^{l+1}} + \frac{1}{l-k}\frac{\partial}{\partial y^{l+2}},$$

where $d_1, \ldots, d_l$ are defined by

$$d_s = \frac{s}{k}, s = 1, \cdots, k; \quad d_j = \frac{l-j+1}{l-k}, j = k+1, \cdots, l.$$

Let $\mathbb{M}_{k,l-k+1,1}$ be the space of a particular class of LG superpotentials consisting of trigonometric-Laurent series of one variable with tri-degree $(k+1, l-k, 1)$, these being functions of the form

$$\lambda(\varphi) = (e^{\mathbf{i}\varphi} - a_{l+2})^{-1}(e^{\mathbf{i}(k+1)\varphi} + a_1 e^{\mathbf{i}k\varphi} + \cdots + a_{l+1}e^{\mathbf{i}(k-l)\varphi}), \quad a_{l+1}a_{l+2} \neq 0,$$

where $a_j \in \mathbb{C}$ for $j = 1, \cdots, l+2$. The space $\mathbb{M}_{k,l-k+1,1}$ carries a natural structure of Frobenius manifold. We have shown that:

Theorem 9. *([30]) Let $\mathfrak{f} : \mathcal{M}^{(k,k+1)}(A_l) \setminus \{\tilde{y}_{l+1} = 0\} \cup \{\tilde{y}_{l+2} = 0\} \to \mathbb{M}_{k,m+1,1}$ be induced by the map*

$$(x_1, \cdots, x_{l+2}) \mapsto (\varphi_1, \cdots, \varphi_{l+2}) \tag{3.22}$$

with

$$\begin{aligned} &\varphi_1 = 2\pi(\rho + x_1), \quad \varphi_j = 2\pi(\rho + x_j - x_{j+1}), \quad j = 2, \cdots, l, \\ &\varphi_{l+1} = 2\pi(\rho - x_l), \quad \varphi_{l+2} = 2\pi x_{l+1}, \end{aligned} \tag{3.23}$$

where $\rho = \frac{m+1}{l+1}x_{l+1} + \frac{m}{l+1}x_{l+2}$. Then $\mathfrak{f}$ is an m-fold covering map, which is also a local isomorphism between the Frobenius manifolds $\mathcal{M}^{(k,k+1)}(A_l) \setminus \{\tilde{y}_{l+1} = 0\} \cup \{\tilde{y}_{l+2} = 0\}$ and $\mathbb{M}_{k,m+1,1}$.

3.3 More general cases

In the above, we mainly discuss $\widetilde{W}^{(k)}(R)$ and $\widetilde{W}^{(k,k+1)}(A_l)$. Similarly, we could define general extended affine groups $\widetilde{W}^{(k_1,\cdots,k_r)}(R)$ for arbitrary irreducible reduced root system R. A challenging problem is to understand whether there are Frobenius manifold structures on the orbit space of $\widetilde{W}^{(k_1,\cdots,k_r)}(R)$. We have the following conjecture.

Conjecture 1. (1). There is no Frobenius manifold structure of charge $d = 1$ on the orbit space $\mathcal{M}^{(k_1,\cdots,k_r)}(R)$ of $\widetilde{W}^{(k_1,\cdots,k_r)}(R)$ for $R \neq A_l, B_l, C_l, D_l$ and $r > 2$.

(2). There are Frobenius manifold structures of charge $d = 1$ on the orbit space $\mathcal{M}^{(k_1,\cdots,k_r)}(R)$ of $\widetilde{W}^{(k_1,\cdots,k_r)}(R)$ for $R = A_l, B_l, C_l, D_l$.

Especially, for the case $\widetilde{W}^{(k,k+S)}(A_l)$ we have a delicate conjecture, where $k, S \in \mathbb{N}$ and $1 \le k < k+S \le l$. For the brevity, we recall some known facts about Weyl groups of type A_l, see [7] for details. Let $\mathbb{R}^{l+1}$ be a $(l+1)$-dimensional Euclidean space with Euclidean inner product $(\ ,\ )$ and an orthonormal basis $\epsilon_1, \cdots, \epsilon_{l+1}$. Let A_l be an irreducible reduced root system in the hyperplane $V = \left\{ \sum_{s=1}^{l+1} v_s \epsilon_s \in \mathbb{R}^{l+1} \middle| \sum_{s=1}^{l+1} v_s = 0 \right\}$. We fix a basis

$$\alpha_1 = \epsilon_1 - \epsilon_2, \cdots, \alpha_l = \epsilon_l - \epsilon_{l+1}$$

of simple roots. The corresponding coroots are $\alpha_j^\vee = \alpha_j$ for $j = 1, \cdots, l$. The Weyl group $W = W(A_l)$ is generated by the reflections

$$\mathbf{x} \mapsto \mathbf{x} - (\alpha_j^\vee, \mathbf{x})\alpha_j, \quad \forall \mathbf{x} \in V,\ j = 1, \ldots, l. \tag{3.24}$$

W acts on V by permutations of the coordinates $v_1, \cdots, v_{l+1}$. The basic W-invariant Fourier polynomials coincide with the elementary symmetric functions

$$y_j(\mathbf{x}) = \sigma_j(e^{2\pi i v_1}, \cdots, e^{2\pi i v_{l+1}}), \quad j = 1, \cdots, l. \tag{3.25}$$

Definition 5. Let $k, S \in \mathbb{N}$ and $1 \le k < k+S \le l$, we call $\widetilde{W} = \widetilde{W}^{(k,k+S)}(A_l)$ to be an extended affine Weyl group of type A if it acts on $\widetilde{V} = V \oplus \mathbb{R}^2$ generated by the transformations

$$x = (\mathbf{x},\ x_{l+1}, x_{l+2}) \mapsto (w(\mathbf{x}) + \textstyle\sum_{j=1}^{l} m_j \alpha_j^\vee,\ x_{l+1}, x_{l+2}), \quad w \in W, \quad m_j \in \mathbb{Z},$$

and

$$x = (\mathbf{x},\ x_{l+1}, x_{l+2}) \mapsto (\mathbf{x} + \omega_k,\ x_{l+1} - 1, x_{l+2}), \tag{3.26}$$

and

$$x = (\mathbf{x},\ x_{l+1}, x_{l+2}) \mapsto (\mathbf{x} + \omega_{k+1},\ x_{l+1}, x_{l+2} - 1). \tag{3.27}$$

Coordinates $x_1, \cdots, x_l$ may be introduced on the space V via the expression

$$\mathbf{x} = x_1 \alpha_1^\vee + \cdots + x_l \alpha_l^\vee. \tag{3.28}$$

That is to say,

$$v_1 = x_1, \quad v_i = x_j - x_{j-1}, \quad v_{l+1} = -x_l, \quad j = 2, \cdots, l. \tag{3.29}$$

Definition 6. $\mathcal{A} = \mathcal{A}^{(k,k+S)}(A_l)$ is the ring of all $\widetilde{W}$-invariant Fourier polynomials of $x_1, \cdots, x_l, \frac{1}{l+1}x_{l+1}, \frac{1}{l+1}x_{l+2}$ that are bounded in the following limit conditions

$$\mathbf{x} = \mathbf{x}^0 - i\omega_k\tau,\ x_{l+1} = x_{l+1}^0 + i\tau, x_{l+2} = x_{l+2}^0,\ \tau \to +\infty \tag{3.30}$$

and

$$\mathbf{x} = \mathbf{x}^0 - i\omega_{k+S}\tau,\ x_{l+1} = x_{l+1}^0,\ x_{l+2} = x_{l+2}^0 + i\tau,\ \tau \to +\infty \tag{3.31}$$

for any $x^0 = (\mathbf{x}^0, x_{l+1}^0, x_{l+2}^0)$.

We introduce a set of numbers

$$d_{j,k} := (\omega_j, \omega_k) = \begin{cases} \frac{j(l-k+1)}{l+1}, & j = 1, \cdots, k, \\ \frac{k(l-j+1)}{l+1}, & j = k+1, \cdots, l \end{cases} \tag{3.32}$$

and define the following Fourier polynomials

$$\begin{aligned} \widetilde{y}_j(x) &= e^{2\pi i (d_{j,k}\, x_{l+1} + d_{j,k+S}\, x_{l+2})} y_j(\mathbf{x}), \quad j = 1, \cdots, l, \\ \widetilde{y}_{l+1}(x) &= e^{2\pi i x_{l+1}}, \quad \widetilde{y}_{l+2}(x) = e^{2\pi i x_{l+2}}. \end{aligned} \tag{3.33}$$

By analogy with the proof in [30], we could show that

Theorem 10. *(Chevalley-type theorem) The ring $\mathcal{A}$ is isomorphic to the ring of polynomials of $\widetilde{y}_1(x), \cdots, \widetilde{y}_{l+2}(x)$.*

Let us denote $\mathcal{M}^{(k,k+S)} := \widetilde{V} \otimes_{\mathbb{R}} \mathbb{C}/\widetilde{W}$, called the *orbit space* of the extended Weyl group $\widetilde{W}$. We define an indefinite flat metric $(dx_i, dx_j)^{\sim}$ on $\widetilde{V}_{\mathbb{C}} = \widetilde{V} \otimes_{\mathbb{R}} \mathbb{C}$, where $\widetilde{V}$ is the orthogonal direct sum of V and $\mathbb{R}^2$. Here V is endowed with the W-invariant Euclidean metric

$$(dx_a, dx_b)^{\sim} = \frac{1}{4\pi^2}(\omega_a, \omega_b), \quad 1 \le a, b \le l \tag{3.34}$$

and $\mathbb{R}^2$ is endowed with the metric

$$(dx_{l+1}, dx_{l+1})^{\sim} = -\frac{\tau_{11}}{4\pi^2}, \quad (dx_{l+1}, dx_{l+2})^{\sim} = -\frac{\tau_{12}}{4\pi^2}, \quad (dx_{l+2}, dx_{l+2})^{\sim} = -\frac{\tau_{22}}{4\pi^2}, \tag{3.35}$$

where

$$\begin{pmatrix} \tau_{11} & \tau_{12} \\ \tau_{12} & \tau_{22} \end{pmatrix} = \begin{pmatrix} d_{k,k} & d_{k,k+S} \\ d_{k+S,k} & d_{k+S,k+S} \end{pmatrix}^{-1}. \tag{3.36}$$

The set of generators for the ring $\mathcal{A}$ are defined by (3.33). They form a system of global coordinates on $\mathcal{M}^{(k,k+S)}$. We now introduce a system of local coordinates on $\mathcal{M}^{(k,k+S)}$ as follows

$$y^1 = \tilde{y}_1, \ldots, y^l = \tilde{y}_l, \; y^{l+1} = \log \tilde{y}_{l+1} = 2\pi i x_{l+1}, \; y^{l+2} = \log \tilde{y}_{l+2} = 2\pi i x_{l+2}. \tag{3.37}$$

They live on the universal covering $\widetilde{\mathcal{M}}$ of $\mathcal{M}$, where $\mathcal{M} := \mathcal{M}^{(k,k+S)} \setminus \{\tilde{y}_{l+1} = 0\} \cup \{\tilde{y}_{l+2} = 0\}$. The projection

$$\mathrm{Pr} : \widetilde{V} \to \widetilde{\mathcal{M}}, \qquad (x_1, \cdots, x_{l+2}) \mapsto (y^1, \cdots, y^{l+2}) \tag{3.38}$$

induces a symmetric bilinear form on $T^*\widetilde{\mathcal{M}}$

$$(dy^i, dy^j)^{\sim} \equiv g^{ij}(y) := \sum_{a,b=1}^{l+2} \frac{\partial y^i}{\partial x_a} \frac{\partial y^j}{\partial x_b} (dx_a, dx_b)^{\sim}. \tag{3.39}$$

Conjecture 2. (i). The matrix entries $g^{ij}(y)$ of (3.39) are weighted homogeneous polynomials in $y^1,\cdots,y^l$, $e^{y^{l+1}}$, $e^{y^{l+2}}$ of the degree $\deg g^{ij}(y)=\deg y^i+\deg y^j$, here $\deg y^{l+1}=\deg y^{l+1}=0$.

(ii). Let us take a vector field

$$e=\sum_{a=0}^{S}\left(\begin{array}{c}S\\a\end{array}\right)\frac{\partial}{\partial y^{a+k}}$$

and define

$$\eta^{ij}(y)=\mathcal{L}_e g^{ij}(y),$$

then $g^{ij}(y)$ and $\eta^{ij}(y)$ form a flat pencil of metrics, where $\mathcal{L}_e$ is the Lie derivative.

Furthermore, we have:

Conjecture 3. Assume that $k,S\in\mathbb{N}$ and $1\le k<k+S\le l$, there exists a unique Frobenius manifold structure of charge $d=1$ on the orbit space $\mathcal{M}^{(k,k+S)}(A_l)\setminus\{t^{k+S}=0\}$ of $\widetilde{W}^{(k,k+S)}(A_l)$ such that the potential $F(t)=\tilde{F}(t)+\frac{1}{2}(t^{k+1})^2\log(t^{k+S})$, where $\tilde{F}(t)$ is a weighted homogeneous polynomial in $t^1,\cdots,t^l$, $(1-\delta^1_S-\delta^2_S)\,\frac{1}{t^{k+S}}$, $e^{t^{l+1}}$, $e^{t^{l+2}-t^{l+1}}$, satisfying

(i). the unity vector field is

$$e=\frac{\partial}{\partial t^k}=\sum_{a=0}^{S}\left(\begin{array}{c}S\\a\end{array}\right)\frac{\partial}{\partial y^{a+k}};$$

(ii). the Euler vector field has the form

$$E=\sum_{\alpha=1}^{l}d_\alpha t^\alpha\frac{\partial}{\partial t^\alpha}+\frac{1}{l-k-S+1}\frac{\partial}{\partial t^{l+1}}+\frac{l-S+1}{k(l-k-S+1)}\frac{\partial}{\partial t^{l+2}},$$

where $d_1,\ldots,d_l$ are defined as follows

$$d_j=\begin{cases}\frac{j}{k}, & j=1,\cdots,k\ ;\\ \frac{k+S-j}{S}, & j=k+1,\cdots,k+S-1\ ;\\ \frac{l-j+1}{l-k-S+1}, & j=k+S,\cdots,l\ .\end{cases}$$

Let us remark that for the case $S=1$, the conjecture is true, that is the above **Theorem 8**. To end this section, we illustrate some examples to verify the above Conjecture.

Example 4. $[A_3,k_1=1,k_2=3]$ Let R be the root system of type A_3 and $k_1=1$, $k_2=3$, i.e., $k=1$, $S=2$, then

$$y^1=e^{\frac{\pi i}{2}(3x_4+x_5)}\sum_{j=1}^{4}\xi_a,\quad y^2=e^{\pi i(x_4+x_5)}\sum_{1\le a<b\le 4}\xi_a\xi_b,$$
$$y^3=e^{\frac{\pi i}{2}(x_4+3x_5)}\sum_{1\le a<b<m\le 4}\xi_a\xi_b\xi_m,\quad y^4=2i\pi x_4,\quad y^5=2i\pi x_5,$$

where $\xi_j = e^{2\pi\, i x_j}$ for $j = 1, 2, 3$ and $\xi_4 = e^{-2\pi\, i(x_1+x_2+x_3)}$. The metric $(\ ,\ )^{\sim}$ has the form

$$((dx_i, dx_j)^{\sim}) = \frac{1}{4\pi^2}\begin{pmatrix} \frac{3}{4} & \frac{1}{2} & \frac{1}{4} & 0 & 0 \\ \frac{1}{2} & 1 & \frac{1}{2} & 0 & 0 \\ \frac{1}{4} & \frac{1}{2} & \frac{3}{4} & 0 & 0 \\ 0 & 0 & 0 & -\frac{3}{2} & \frac{1}{2} \\ 0 & 0 & 0 & \frac{1}{2} & -\frac{3}{2} \end{pmatrix}.$$

To write down the flat coordinates, we first introduce the following variables

$$z^1 = y^1,\ z^2 = -y^1 + y^3,\ z^3 = -y^1 + y^2 - y^3 + e^{y^4} + e^{y^5}, \quad z^4 = y^5,\ z^5 = y^4 + y^5,$$

then the flat coordinates are given by

$$t_1 = z^1 - 2e^{z^5 - z^4}, \quad t_2 = z^2 + 2e^{z^5-z^4} - e^{2z^4},$$
$$t_3 = \sqrt{z^3}, \quad t_4 = z^4, \quad t_5 = z^5$$

The potential has the expression

$$\begin{aligned} F = \quad & \frac{1}{2}{t_1}^2 t_5 + t_1 t_4 t_2 + \frac{1}{2}{t_2}^2 t_4 - t_1 {t_3}^2 - \frac{1}{2} t_2 {t_3}^2 - \frac{1}{24}{t_3}^4 \\ & + e^{t_5} - t_2 e^{t_4} + {t_3}^2 e^{t_4} + (t_2 + {t_3}^2) e^{t_5 - t_4} + \frac{1}{2}{t_2}^2 \log(t_3) \end{aligned}$$

and the unit vector field is

$$e = \partial_1$$

and the Euler vector field is given by

$$E = t_1\partial_1 + t_2\partial_2 + \frac{1}{2} t_3\partial_3 + \partial_4 + 2\partial_5,$$

where $\partial_i = \frac{\partial}{\partial t_i}$ and $t_i = t^i$.

Example 5. $[A_4, k_1 = 1, k_2 = 4]$ Let R be the root system of type A_3 and $k_1 = 1$, $k_2 = 4$, i.e., $k = 1$, $S = 3$. The potential has the expression

$$\begin{aligned} F = \quad & \frac{1}{2}{t_1}^2 t_6 + t_1 t_2 t_5 + \frac{1}{2}{t_2}^2 t_5 - \frac{1}{3} t_3 t_1 t_4 - \frac{1}{699840}{t_4}^6 \\ & - \frac{1}{216}{t_4}^2 {t_3}^2 + \frac{1}{648}{t_4}^3 t_2 - \frac{1}{6} t_2 t_3 t_4 - t_2 e^{t_5} + t_2 e^{t_6 - t_5} \\ & + \frac{1}{3} e^{t_6 - t_5} t_3 t_4 + \frac{1}{54} e^{t_6 - t_5} {t_4}^3 + e^{t_6} + \frac{1}{3} t_3 t_4 e^{t_5} - \frac{1}{54} e^{t_5} {t_4}^3 \\ & + \frac{1}{2}\frac{{t_3}^2 t_2}{t_4} - \frac{1}{12}\frac{{t_3}^4}{{t_4}^2} + \frac{1}{2}{t_2}^2 \log(t_4) \end{aligned}$$

and the unit vector field is

$$e = \partial_1$$

and the Euler vector field is given by

$$E = t_1\partial_1 + t_2\partial_2 + \frac{2}{3} t_3\partial_3 + \frac{1}{3} t_4\partial_4 + \partial_5 + 2\partial_6.$$

4 Jacobi groups and their extensions

Complex crystallographic Coxeter groups $\widetilde{W}(\tau)$ were introduced by J.Bernstein and O.Schwarzman, implicitly also by E.Looijenga. Bernstein and Schwarzman also found an analogue of the Chevalley theorem for $\widetilde{W}$. A natural question is whether there are Frobenius manifold structures on their orbit spaces. If it exists, the first step is to construct a $\widetilde{W}(\tau)$-invariant metric with some particular property. Unfortunately such metric does not exist.

In order to solve this problem, B.Dubrovin considered a certain extension of $\widetilde{W}(\tau)$, i.e., the Jacobi group $\mathbf{J}(\mathcal{G})$ which was proposed by M.Eichler and D.Zagier. Here $\mathcal{G}$ is a simple Lie algebra with Weyl group W. The corresponding orbit space is called the Jacobi form $J(\mathcal{G})$. B.Dubrovin in 1993 ([10]) studied the Frobenius manifold structure related to the Jacobi forms $J(A_1)$. Afterwards, M.Bertola ([5, 6]) generalized this construction and obtained a Frobenius manifold structure on $J(\mathcal{G})$ for $\mathcal{G} = A_l, B_l, G_2$. Especially, in the case of type G_2, M.Bertola obtained two different Frobenius manifold structures. The first structure comes from the embedding of $J(G_2)$ in $J(A_2)$, while the second was constructed by K.Saito. I.Satake ([24]) studied the Frobenius manifold structures on $J(\mathcal{G})$ for $\mathcal{G} = E_6, D_4$, and in [25] showed that there is a Frobenius manifold structure on the complex orbit space of the reflection group for an elliptic root system of codimension 1.

Very recently, Guilherme F.Almeida ([15]) introduced a new group denoted by $\mathfrak{J}(\tilde{A}_1)$ which is a combination of the extended affine $\widetilde{W}^{(1)}(A_1)$ and the Jacobi group of type A_1. Moreover, he also constructed Frobenius manifold structure on the orbit space of the extended affine Jacobi Group $\mathfrak{J}(\tilde{A}_1)$. For general cases, it deserves further study.

Acknowledgments

The author is grateful to Professor Youjin Zhang for helpful suggestions. This work is partially supported by NSFC (No.11671371, No.11871446, NO.12071451) and Wu Wen-Tsun Key Laboratory of Mathematics, USTC, CAS.

References

[1] S.Aoyama and Y.Kodama, Topological Landau-Ginzburg with a rational potential, and the dispersionless KP hierarchy, Comm. Math. Phys. 182 (1996), 185-219.

[2] V.I.Arnol'd, Normal forms of functions close to degenerate critical points. The Weyl groups A_k, D_k, E_k, and Lagrangian singularities, Functional Anal., 6 (1972), 3–25.

[3] M.F.Atiyah, Topological quantum field theories, Publ. Math. I.H.E.S., 68 (1988), 175.

[4] M.Bertola, Jacobi groups, Jacobi forms and their applications, Phd. dissertation. 1999 (SISSA).

[5] M.Bertola, Frobenius manifold structure on orbit space of Jacobi groups. I. Differential Geom. Appl. 13 (2000), no. 1, 19–41.

[6] M.Bertola, Frobenius manifold structure on orbit space of Jacobi groups. II. Differential Geom. Appl. 13 (2000), no. 3, 213–233.

[7] N. Bourbaki, Groupes et Algèbres de Lie, Chapitres 4, 5 et 6, Masson, Paris-New York-Barcelone-Milan-Mexico-Rio de Janeiro, 1981.

[8] R.Dijkgraaf, E.Verlinde, and H.Verlinde, Topological strings in $d < 1$, Nuclear Phys. B 352(1991), 59–86.

[9] B.Dubrovin, Integrable systems in topological field theory, Nucl. Phys. B 379 (1992), 627–689.

[10] B. Dubrovin, Geometry of 2D topological field theories, in: M. Francaviglia and S. Greco, eds, Integrable Sytems and Quantum Groups, Montecatini Terme, 1993, Lecture Notes in Math. 1620 (Springer, Berlin, 1996), 120–384.

[11] B. Dubrovin, Differential geometry of the space of orbits of a Coxeter group. Surveys in differential geometry: integral systems [integrable systems], 181–211, Surv. Differ. Geom., IV, Int. Press, Boston, MA, 1998.

[12] B. Dubrovin and Y.Zhang, Extended affine Weyl groups and Frobenius manifolds, Compositio Mathematica 111(1998), 167–219.

[13] B. Dubrovin, Y. Zhang and D. Zuo, Extended affine Weyl groups and Frobenius manifolds–II, Preprint arXiv:052365 (Unpublished).

[14] B.Dubrovin, I.A.B.Strachan, Y.Zhang and D.Zuo, Extended affine Weyl groups of BCD-type: their Frobenius manifolds and Landau–Ginzburg superpotentials, Adv. Math. 351 (2019), 897–946.

[15] Guilherme F. Almeida, Differential Geometry of Orbit space of Extended Affine Jacobi Group A_1, SIGMA 17 (2021), 022, 39 pages.

[16] C.Hertling, Frobenius manifolds and moduli spaces for singularities. Cambridge Tracts in Mathematics, 151 Cambridge University Press, Cambridge, 2002.

[17] J.E.Humphtryd, Introduction to Lie algebras and representation theory, Graduate texts in mathematics 9, Springer-Verlage, New York,Hidelberg, Berlin, 1972.

[18] E.Looijenga, A period mapping for certain semiuniversal deformations, Compos. Math. 30 (1975), 299–316.

[19] E. Looijenga, Root systems and elliptic curves, Invent. Math., 38 (1976), 17-32.

[20] E. Looijenga, Invariant theory of generalized root systems, Invent. Math., 61 (1980), 1-32.

[21] K.Saito, Extended affine root systems II (flat invariants), Publ. RIMS 26 (1990), 15–78.

[22] K.Saito, On a linear structure of the quotient variety by a finite reflexion group. Publ. Res. Inst. Math. Sci., 29(4)(1993), 535–579.

[23] K. Saito, T. Yano and J. Sekiguchi, On a certain generator system of the ring of invariants of a finite reflection group, Comm. Algebra 8(4) (1980), 373–408.

[24] I.Satake, Flat structure for the simple elliptic singularity of type E_6 and Jacobi form, Proc. Japan Acad. Ser. A Math. Sci. 69 (1993), no. 7, 247–251.

[25] I.Satake, Frobenius manifolds for elliptic root systems. Osaka J. Math. 47 (2010), no. 1, 301–330.

[26] P. Slodowy, A remark on a recent paper by B. Dubrovin and Y. Zhang, Preprint 1997 (Unpublished).

[27] K. Wirthmüller, Torus embeddings and deformations of simple space curves, Acta Math. 157 (1986), 159–241.

[28] E.Witten, On the structure of the topological phase of two-dimensional gravity, Nucl.Phys.B 340(1990), 281–332.

[29] D.Zuo, Frobenius manifolds associated to B_l and D_l, revisited, International Mathematics Research Notices, Vol. 2007, Article ID rnm020, 25 pages.

[30] D.Zuo, Frobenius manifolds and a new class of extended affine Weyl groups of A-type. Lett. Math. Phys. 110 (2020), no. 7, 1903–1940.

B1. On finite Toda type lattices and multipeakons of the Camassa-Holm type equations

Xiangke Chang

LSEC, ICMSEC, Academy of Mathematics and Systems Science, Chinese Academy of Sciences, P.O. Box 2719, Beijing 100190, P.R. China; and School of Mathematical Sciences, University of Chinese Academy of Sciences, Beijing 100049, P.R. China;
changxk@lsec.cc.ac.cn

Abstract

The celebrated Toda lattice was originally obtained as a simple model for describing a chain of particles with nearest neighbour exponential interaction and has been generalized in multiple ways. A class of nonlinear integrable partial differential equations (PDEs) admit some special weak solutions called "peakons", which are characterised by systems of ordinary differential equations (ODEs), namely peakon lattices. It is shown that Toda and peakon type lattices can be regarded as isospectral deformations in opposite directions related to certain orthogonal functions. We will take the ordinary Toda lattice and the Camassa-Holm peakons, the Kac-van Moerbeke lattice and a two-component modified Camassa-Holm interlacing peakons, the B-Toda lattice and the Novikov peakons, the C-Toda lattice and the Degasperis-Procesi peakons, the Frobenius-Stickelberger-Thiele lattice and the modified Camassa-Holm peakons as examples to illustrate this point of view.

1 Background

Since the initial value problem of the Korteweg-de Vries (KdV) equation was successfully solved by the inverse scattering transformation in 1967, integrable systems have been extensively studied by mathematicians and physicists. Besides the KdV equation, the nonlinear Schrödinger equation, the Toda lattice and the Camassa-Holm (CH) equation have been typical examples that arouse in the course of the development on integrable systems. The present chapter focusses on the Toda type lattices and their connections to the Camassa-Holm (CH) type equations with their multipeakons.

1.1 On Toda lattice and its generalizations

The Toda lattice

$$\dot{x}_k = y_k, \qquad \dot{y}_k = e^{x_{k-1}-x_k} - e^{x_k-x_{k+1}}$$

was originally introduced by Toda [83] as a simple model for describing a chain of particles with nearest neighbor exponential interaction. Here x_k is the displacement of the k-th unit mass from equilibrium, and y_k is the corresponding momentum. Later Flaschka [51] found a nonlinear transformation (now called Flaschka's transformation) of the infinite Toda lattice flow so that the lattice can be written as an isospectral flow of an infinite tridiagonal matrix–a Jacobi matrix. Thus, effectively, he found a Lax pair for the flow. Moreover, he showed that the flow can be linearized by using a discrete version of KdV inverse scattering theory. The periodic Toda lattice was solved by Date and Tanaka [44] and by Kac and van Moerbeke [61] in terms of hyperelliptic integrals associated with an algebraic curve; Dubrovin, Matveev, and Novikov [48] expressed the solution in terms of the theta function of the curve. The case of a finite tied lattice ($x_0 = -\infty, x_{n+1} = +\infty$) was investigated by Moser [69], who pointed out the connection of the isospectral problem with Stieltjes' work on continued fractions and the spectral theory of Jacobi matrices. Due to the intimate relationship between Jacobi matrix and orthogonal polynomials (OPs), the ordinary OPs may be regarded as wave functions of the Lax pair of the Toda lattice undergoing a one-parameter deformation of the spectral measure [12, 38, 47, 75].

Because of the intriguing properties of the Toda lattice, there is a multitude of Toda-like integrable lattices known to date. Some remarkable systems include the Lotka-Volterra (LV) lattice (sometimes also called the Kac-van Moerbeke lattice or the Langmuir lattice) [62], the Ablowitz-Ladik lattice [1, 2], the Narita-Itoh-Bogoyavlensky lattice [24–26, 59, 71], the relativistic Toda lattice [78], and the full Kostant-Toda lattice [50, 64], etc. The links between the theory of orthogonal polynomials (OPs) and the integrable systems of Toda type have been extensively investigated and used by both integrable systems and special functions communities since the early 1990s. As a second example, the LV lattice [38,62] can be obtained as a one-parameter deformation of the measure associated to symmetric OPs. Furthermore, by considering a time deformation of Laurent biorthogonal polynomials, one may get the relativistic Toda lattice [63]. Lax pairs for the Ablowitz-Ladik system and also the Schur flow are associated with one-parameter deformation for OPs on the unit circle [70, 72]. An integrable lattice connected to the isospectral evolution of the polynomials of type R_I was also presented in [84]. For more examples, one may refer [4, 10, 11, 13–15, 30, 31], etc.

In some sense, the work of Adler and van Moerbeke was fundamental to the connection between integrable systems of Toda type and OPs. They showed how the Gauss-Borel factorization problem appears in the theory of the 2D Toda hierarchy and what they called the discrete KP hierarchy in [3, 5–9] etc. These papers clearly established—using a group-theoretical setup—the reasons why orthogonal polynomials and integrability of nonlinear equations of Toda type were just different manifestations of the same underlying structure.

1.2 Peakons

It is essential to search for simple mathematical models for effectively describing nonlinear phenomena in nature, such as, the breakdown of regularity. The KdV equation is a simple mathematical model for gravity waves in water, but it fails to model such phenomena. Whitham, in his book [85], emphasized the need for a water wave model exhibiting a soliton interaction, the existence of peaked waves, and, at the same time, allowing for breaking waves. In 1993, Camassa and Holm [27] derived such a shallow water wave model by executing an asymptotic expansion of the Hamiltonian for Euler's equations in the shallow water regime. Nowadays the resulting equation is known as the Camassa-Holm (CH) equation and reads

$$u_t - u_{xxt} + 2\kappa u_x + 3uu_x - 2u_x u_{xx} - uu_{xxx} = 0,$$

where u represents the fluid velocity (or equivalently the height of the water's free surface above a flat bottom) and κ is a nonnegative constant.

This equation has a number of remarkable properties [27]. For example, it leads to a meaningful model of wave-breaking [39]. In addition, it is an integrable system with a bi-Hamiltonian structure that possesses infinitely many conserved quantities and a Lax pair. Actually, it firstly appeared in the work of Fuchssteiner and Fokas [54] as an abstract bi-Hamiltonian equation with infinitely many conservation laws. But it attracted no special attention until its rediscovery by Camassa and Holm.

As a shallow water wave equation, it is natural to expect solutions in the form of the travelling wave

$$u(x,t) = U(x - ct).$$

It was shown by Camassa and Holm [28] that, when $\kappa > 0$, the solution is smooth but implicit. In the limit of $\kappa \to 0$, one gets the travelling wave solution

$$u(x,t) = ce^{|x-ct|} + \mathcal{O}(\kappa \log \kappa).$$

In fact, when $\kappa = 0$, the CH equation admits the peakon solutions (simply called peakons) as its solitary wave solutions with peaks. In the case of n peaks, the solution, somewhat unexpectedly, can be obtained as a simple superposition of individual peaks

$$u(x,t) = \sum_{j=1}^{n} m_j(t) e^{-|x - x_j(t)|}, \tag{1.1}$$

where $m_j(t)$ and $x_j(t)$ are constrained to satisfy an appropriate ODE system. Clearly, the peakons are smooth solutions except at the peak positions $x = x_j(t)$ where the x derivative of u is discontinuous, forcing us to interpret peakons in a suitable weak sense.

Observe that the CH equation for $\kappa = 0$ may be written as

$$m_t + (um)_x + mu_x = 0, \qquad m = u - u_{xx}, \tag{1.2}$$

by introducing the momentum variable m. Hence, for u given by (1.1), m is a sum of delta distributions

$$m(x,t) = 2\sum_{j=1}^{n} m_j(t)\delta(x - x_j(t)).$$

By using distributional calculus, it is known that the first equation of (1.2) is satisfied in a weak sense if the positions $(x_1, \ldots, x_n)$ and momenta $(m_1, \ldots, m_n)$ of the peakons obey the following system of $2n$ ODEs:

$$\dot{x}_k = \sum_{j=1}^{n} m_j e^{-|x_j - x_k|}, \qquad \dot{m}_k = \sum_{j=1}^{n} \operatorname{sgn}(x_k - x_j) m_j e^{-|x_j - x_k|}.$$

The mathematics of peakons has attracted a great deal of attention since they were discovered in the CH equation. It would be nearly impossible to mention all relevant contributions, so we select only those which are directly relevant to the present chapter.

1. The dynamics of the CH peakons can be described by an ODE system, which was explicitly solved by the inverse spectral method in [16, 17] by Beals, Sattinger and Szmigielski. The closed form can be expressed in terms of Hankel determinants and orthogonal polynomials.

2. The CH peakons are orbitally stable, in the sense that a solution that initially is close to a peakon solution is also close to some peakon solution at a later time. This was proved by Constantin and Strauss [41, 42].

3. There exists a bijection between a certain Krein type string problem, which in turn is equivalent to the CH peakon problem, and the Jacobi spectral problem, the isospectral flow of which gives the celebrated Toda lattice. In a well-defined sense, the CH peakon ODEs could be regarded as a negative flow of the finite Toda lattice [18, 77].

4. The peakon solutions seem to capture main attributes of solutions of the CH equation: the breakdown of regularity which can be interpreted as collisions of peakons, and the nature of long-time asymptotics which can be loosely described as peakons becoming free particles in the asymptotic region [17, 68].

Not surprisingly, the initial success of the CH equation also prompted researchers to look for other equations exhibiting similar properties, especially peakon solutions, giving rise to numerous new equations that have been proposed and studied over the last two decades. Among those equations that attracted considerable attention are

1. the Degasperis–Procesi (DP) equation [46]

$$m_t + m_x u + 3m u_x = 0, \qquad m = u - u_{xx},$$

2. the Novikov equation [57, 58, 73]

$$m_t + (m_x u + 3m u_x)u = 0, \qquad m = u - u_{xx},$$

3. the modified CH (mCH) equation [52, 53, 74, 76]

$$m_t + [m(u^2 - u_x^2)]_x = 0, \qquad m = u - u_{xx}$$

4. multicomponent generalizations of CH [37,55,79], for example, the two-component mCH equation [79]

$$\begin{aligned} m_t + [(u - u_x)(v + v_x)m]_x &= 0, \\ n_t + [(u - u_x)(v + v_x)n]_x &= 0, \\ m = u - u_{xx}, \qquad n &= v - v_{xx}. \end{aligned}$$

   For our convenience, we shall call it 2-mCH equation (it is also called SQQ equation in some references).

1.3 Motivation

The finite Toda lattice introduced by Moser [69] can be solved by inverse spectral method related to Jacobi matrix. It was shown by Beals, Sattinger and Szmigielski [17] that the spectral problem of the CH peakon lattice is related to a finite discrete string problem, which in turn can also be solved by use of inverse spectral method involving Stieltjes' theorem on continued fractions. The explicit formulae of the solutions can both be expressed in terms of Hankel determinants with moments of discrete measures and associated orthogonal polynomials. In a follow-up work [18], Beals, Sattinger and Szmigielski indicated that there exists a bijective map from a discrete string problem with positive weights to normalized Jacobi matrices, which allows the pure peakon flow of the CH equation to be realized as an isospectral Jacobi flow as well. This gives a unified picture for the Toda and the CH peakon flows. Indeed, this also implies that the CH peakon and Toda lattices can be viewed as opposite flows related to the ordinary OPs. We remark that a mapping of the flow of n peakons into an isospectral flow of an $n \times n$ Jacobi matrix was also indicated in [77], where the starting point is r-matrix.

The work of Beals et al. opened the door to establishing the relationships among peakon, Toda lattices and the corresponding OPs. In subsequent years, many integrable systems supporting peakon solutions have been discovered. In parallel, there also exist various types of Toda lattices and accompanying OPs. A natural question then is to ask if there exists similar relationships. However, no more work has been done until the paper [34], where it was shown that the 2-mCH interlacing multipeakon lattice and the finite Kac-van Moerbeke (KvM) lattice can be regarded as opposite flows associated with the symmetric OPs (sOPs).

Hankel determinants appear in the expressions of multipeakons of the CH equation [16]. If one substitutes a multipeakon solution into the CH peakon ODEs and computes all necessary derivatives of the corresponding determinants, the CH peakon ODEs are nothing but certain determinant identities; this result was proven in [29]. In other words, the CH peakon ODEs can be regarded as an isospectral flow on the manifold cut out by determinant identities. Moreover, the CH equation has a reciprocal link to the first negative flow in the KdV hierarchy [53] which in turn belongs to a larger class of flows, called the AKP type [56, 60], whose closed form solutions in terms of the τ-function are all determinantal solutions. It therefore makes sense to expect similar determinantal description for the CH peakons. Similar argument applies for the 2-mCH equation.

By contrast, the DP equation is connected with a negative flow in the Kaup-Kupershmidt (KK) hierarchy [45], while the Novikov equation is related to a negative flow in the Sawada-Kotera (SK) hierarchy via reciprocal transformations [58]. The SK hierarchy belongs to BKP type, while KK hierarchy belongs to CKP type. (Note that the letters "A,B,C" refer to different types of infinite dimensional Lie groups and the corresponding Lie algebras which are associated with respective hierarchies [43, 60]. The transformation groups of AKP, BKP, CKP type equations are $GL(\infty)$, $O(\infty)$ and $Sp(\infty)$, respectively.)

Due to much more complicated structures than the AKP type equations, the solutions of the BKP type equations are expressed not as determinants but as Pfaffians [56]. Therefore, it is natural to guess that Pfaffians might be appropriate objects to describe multipeakons of the Novikov equation. This result was reported in [33], where the peakon solutions of the Novikov equation is rewritten in terms of Pfaffians and the Novikov peakon dynamical system is connected with the finite Toda lattice of BKP type (B-Toda lattice). As for the DP peakon case, it was dealt with in [31], where the main object is certain bimoment determinants with respect to the Cauchy kernel leading to a Toda lattice of CKP type (C-Toda lattice) as an opposite flow to the DP peakon lattice.

It is interesting that new types of polynomials may be deduced during the research of integrable systems. In fact the Cauchy Bi-OPs (CBOPs) are such an example and are deduced in the search of explicit solutions of DP multipeakons. Here we mention that they can be useful in the investigation of various problems in random matrix theory [19, 20, 22, 23]. This implies that the DP peakon lattice and the C-Toda lattice may be viewed as opposite flows related to the CBOPs. In order to figure out the corresponding OPs related to the Novikov peakon lattice and the B-Toda lattice and also with the aim to studying the average of characteristic polynomials of the Bures random matrix ensemble, the concept of partial-skew-orthogonal polynomials (PSOPs) was proposed [30, 33]. Therefore, the Novikov peakon lattice and the B-Toda lattice might be seen as opposite flows related to PSOPs [33].

Moreover, the mCH equation is an intriguing modification of the CH equation. Its multipeakon problem was analyzed in [36] by use of inverse spectral method, which leads to an explicit formula in terms of the so-called Cauchy-Stieltjes-Vandermond (CSV) determinants. Later, it was shown that the mCH peakons is associated to a negative flow related to the Frobenius-Stickelberger-Thiele (FST) polynomials. However, an isospectral flow of the FST polynomials in the positive direction was not derived until in the work [35], where their connection was also studied.

2 Main results

In this section, we first summarise the core of the main results in the table below. The details are presented in the subsections.

Table 1. Known relevance between Toda type lattices and peakon lattices

Toda lattice $e^{tx}\mu(x;0)$	Peakon lattice $e^{\frac{t}{x}}\mu(x;0)$	Lax pair	"τ-structure"	Moments $c_{i,j}$
A-Toda	CH	OPs	$\det((c_{i,j}))$	$\int x^{i+j}\mu(x)dx$
KvM	2-mCH	sOPs	$\det((c_{i,j}))$	$\int x^{i+j}\mu(x)dx$
B-Toda	Novikov	PSOPs	$Pf((c_{i,j}))$	$\iint \frac{x-y}{x+y}x^iy^j\mu(x)\mu(y)dxdy$
C-Toda	DP	CBOPs	$\det((c_{i,j}))$	$\iint \frac{x^iy^j}{x+y}\mu(x)\mu(y)dxdy$
FST	mCH	FSTPs	$\det(CSV)$	$\int \frac{x^{i+j}}{\prod_{k=1}^n(x+e_k)}\mu(x)dx$

2.1 Finite A-Toda, CH peakon lattices and ordinary OPs

Under Flaschka's variables, the finite A-Toda lattice reads

$$\dot{u}_k = u_k(b_k - b_{k-1}), \qquad k = 1, \dots, n-1, \tag{2.1a}$$

$$\dot{b}_k = u_{k+1} - u_k, \qquad k = 0, \dots, n-1, \tag{2.1b}$$

with the boundary condition $u_0 = u_n = 0$. It can be equivalently written as the Lax equation

$$\dot{L} = [L, B]$$

with

$$L = \begin{pmatrix} b_0 & 1 & & & \\ u_1 & b_1 & 1 & & \\ & \ddots & \ddots & \ddots & \\ & & u_{n-2} & b_{n-2} & 1 \\ & & & u_{n-1} & b_{n-1} \end{pmatrix}, \qquad B = \begin{pmatrix} 0 & 0 & & & \\ u_1 & 0 & 0 & & \\ & \ddots & \ddots & \ddots & \\ & & u_{n-2} & 0 & 0 \\ & & & u_{n-1} & 0 \end{pmatrix}.$$

It is not hard to see that the ordinary OPs appear as wave functions of the Lax pair of the finite A-Toda lattice. Indeed, the finite A-Toda lattice can arise as a one-parameter deformation of the measure of the ordinary OPs.

For positive initial m_k, the multipeakon problem was solved in [17] by use of inverse spectral method (see [49] for the indefinite case). The relation between the finite A-Toda and the CH peakon lattices was discovered in [18], where a mapping is established via the corresponding spectral problems. Here, in anticipation of a more unified frame, we give a slightly different presentation from that in [18]. In summary, the A-Toda and CH peakon lattices can be seen as opposite nonlinear flows related to ordinary OPs.

Theorem 1. *Given*

$$A_i(0) = \sum_{p=1}^{n} \xi_p^i(0)c_p(0), \qquad \textit{with} \qquad 0 < \xi_1(0) < \xi_2(0) < \cdots < \xi_n(0), \qquad c_p(0) > 0,$$

let $\tau_k^{(l)}(0)$ be defined as Hankel determinants[1]

$$\tau_k^{(l)}(0)=\begin{vmatrix} A_l(0) & A_{l+1}(0) & \cdots & A_{l+k-1}(0)\\ A_{l+1}(0) & A_{l+2}(0) & \cdots & A_{l+k}(0)\\ \vdots & \vdots & \ddots & \vdots\\ A_{l+k-1}(0) & A_{l+k}(0) & \cdots & A_{l+2k-2}(0)\end{vmatrix}$$

for any nonnegative integer k, and $\tau_0^{(l)}(0)=1$, $\tau_k^{(l)}(0)=0$ for $k<0$.

1. *Introduce the variables $\{x_k(0),m_k(0)\}_{k=1}^n$ defined by*

$$x_{k'}(0)=\ln\left(\frac{2\tau_k^{(0)}(0)}{\tau_{k-1}^{(2)}(0)}\right),\qquad m_{k'}(0)=\frac{\tau_k^{(0)}(0)\tau_{k-1}^{(2)}(0)}{\tau_k^{(1)}(0)\tau_{k-1}^{(1)}(0)},$$

where $k'=n+1-k$. If $\{\xi_p(t),c_p(t)\}_{p=1}^n$ evolve as

$$\dot{\xi}_p=0,\qquad \dot{c}_p=\frac{c_p}{\xi_p},$$

then $\{x_k(t),m_k(t)\}_{k=1}^n$ satisfy the CH peakon ODEs:

$$\dot{x}_k=\sum_{j=1}^n m_je^{-|x_j-x_k|},\qquad \dot{m}_k=\sum_{j=1}^n \operatorname{sgn}(x_k-x_j)m_je^{-|x_j-x_k|}.$$

2. *Introduce the variables $\left\{\{u_k(0)\}_{k=1}^{n-1},\{b_k(0)\}_{k=0}^{n-1}\right\}$ defined by*

$$u_k(0)=\frac{\tau_{k+1}^{(1)}(0)\tau_{k-1}^{(1)}(0)}{(\tau_k^{(1)}(0))^2},\qquad b_k=\frac{\tau_{k+1}^{(1)}(0)\tau_{k-1}^{(2)}(0)}{\tau_k^{(2)}(0)\tau_k^{(1)}(0)}+\frac{\tau_{k+1}^{(2)}(0)\tau_k^{(1)}(0)}{\tau_k^{(2)}(0)\tau_{k+1}^{(1)}(0)}.$$

If $\{\xi_p(t),c_p(t)\}_{p=1}^n$ evolve as

$$\dot{\xi}_p=0,\qquad \dot{c}_p=\xi_pc_p,$$

then $\left\{\{u_k(0)\}_{k=1}^{n-1},\{b_k(0)\}_{k=0}^{n-1}\right\}$ satisfy the finite A-Toda lattice (2.1) with $u_0=u_n=0$.

3. *There exists a mapping from $\{x_k(0),m_k(0)\}_{k=1}^n$ to $\left\{\{u_k(0)\}_{k=1}^{n-1},\{b_k(0)\}_{k=0}^{n-1}\right\}$ according to*

$$u_k(0)=\frac{e^{x_{k'-1}(0)-x_{k'}(0)}}{m_{k'-1}(0)m_{k'}(0)\left(1-e^{x_{k'-1}(0)-x_{k'}(0)}\right)^2},$$

$$b_k(0)=\frac{1}{m_{k'-1}(0)(1+e^{-x_{k'-1}(0)})}\left(\frac{1+e^{-x_{k'}(0)}}{1-e^{x_{k'-1}(0)-x_{k'}(0)}}+\frac{1+e^{-x_{k'-1}(0)}}{1-e^{x_{k'-2}(0)-x_{k'-1}(0)}}\right)$$

with the convention

$$x_0(0)=-\infty,\quad x_{n+1}(0)=+\infty.$$

[1]The positivity for such determinants with $1\le k\le n$ is guaranteed (see, e.g., [17, 29]).

2.2 Finite KvM, 2-mCH interlacing peakon lattices and symmetric OPs

The finite KvM

$$\dot{d}_k = \frac{1}{2} d_k (d_{k+1}^2 - d_{k-1}^2), \quad k = 1, \cdots 2K-1,$$

with the boundary condition $d_0 = d_{2K} = 0$ was derived as a discretization of the Korteweg–de Vries equation in the original work of Kac and van Moerbeke [62]. It admits the Lax representation

$$\dot{J} = \frac{1}{2}[B, J]$$

with

$$J = \begin{pmatrix} 0 & d_1 & 0 & \cdots & 0 \\ d_1 & 0 & d_2 & 0 & \\ 0 & d_2 & 0 & & \\ \vdots & & & \ddots & d_{2K-1} \\ 0 & & & d_{2K-1} & 0 \end{pmatrix},$$

$$B = \begin{pmatrix} 0 & 0 & d_1 d_2 & 0 & \cdots & 0 \\ 0 & 0 & 0 & d_2 d_3 & \cdots & 0 \\ -d_1 d_2 & 0 & 0 & \ddots & \ddots & 0 \\ \vdots & \ddots & \ddots & \ddots & 0 & d_{2K-2} d_{2K-1} \\ & & \ddots & 0 & 0 & 0 \\ 0 & & & -d_{2K-2} d_{2K-1} & 0 & 0 \end{pmatrix}.$$

Due to the intimate connection between the A-Toda lattice and the KvM lattice, it is natural to ask which peakon flow is related to the finite KvM lattice. Unexpectedly, the finite KvM lattice was shown to project to the 2-mCH interlacing peakon flow [32, 34], whose solutions both admit the Hankel determinant structure.

The 2-mCH was proposed by Song, Qu and Qiao [79] as a two-component integrable extension of the mCH Equation (2.15):

$$\begin{aligned} &m_t + [(u - u_x)(v + v_x)m]_x = 0, \\ &n_t + [(u - u_x)(v + v_x)n]_x = 0, \\ &\quad m = u - u_{xx}, \qquad n = v - v_{xx}, \end{aligned} \tag{2.2}$$

This system of equations is known to possess a Lax formulation, infinitely many conservation laws, as well as a bi-Hamiltonian structure [82]. It is elementary to verify that Equations (2.2) are compatibility conditions resulting from the Lax pair

$$\frac{\partial}{\partial x}\begin{pmatrix} \Psi_1 \\ \Psi_2 \end{pmatrix} = \frac{1}{2} U \begin{pmatrix} \Psi_1 \\ \Psi_2 \end{pmatrix}, \qquad \frac{\partial}{\partial t}\begin{pmatrix} \Psi_1 \\ \Psi_2 \end{pmatrix} = \frac{1}{2} V \begin{pmatrix} \Psi_1 \\ \Psi_2 \end{pmatrix}, \tag{2.3}$$

where

$$U = \begin{pmatrix} -1 & \lambda m \\ -\lambda n & 1 \end{pmatrix},$$

$$V = \begin{pmatrix} 4\lambda^{-2} + Q & -2\lambda^{-1}(u - u_x) - \lambda m Q \\ 2\lambda^{-1}(v + v_x) + \lambda n Q & -Q \end{pmatrix},$$

with $Q = (u - u_x)(v + v_x)$. The interlacing peakon ansatz means that u and v are given by the form

$$u = \sum_{k=1}^{K} m_{2k-1}(t) e^{-|x - x_{2k-1}(t)|}, \qquad v = \sum_{k=1}^{K} n_{2k}(t) e^{-|x - x_{2k}(t)|},$$

with $x_1 < x_2 < \cdots < x_{2K}$. Clearly, in this case m and n are discrete measures, albeit with disjoint support:

$$m(x,t) = 2\sum_{k=1}^{K} m_{2k-1}(t)\delta(x - x_{2k-1}(t)), \qquad n = 2\sum_{k=1}^{K} n_{2k}\delta(x - x_{2k}(t)).$$

Using elementary distribution calculus, one is led to the interlacing peakon ODEs

$$\dot{x}_{2k-1} = \left(2\sum_{j=1}^{k-1} m_{2j-1}e^{x_{2j-1}} + m_{2k-1}e^{x_{2k-1}}\right)\left(2\sum_{j=k}^{K} n_{2j}e^{-x_{2j}}\right), \tag{2.4a}$$

$$\dot{x}_{2k} = \left(2\sum_{j=1}^{k} m_{2j-1}e^{x_{2j-1}}\right)\left(n_{2k}e^{-x_{2k}} + 2\sum_{j=k+1}^{K} n_{2j}e^{-x_{2j}}\right), \tag{2.4b}$$

$$\dot{m}_{2k-1} = 0, \qquad \dot{n}_{2k} = 0. \tag{2.4c}$$

For positive constants m_{2k-1}, n_{2k}, this ODE system can be solved by use of inverse spectral method so that the explicit formulae of the interlacing multipeakons of the 2-mCH equation were obtained [34], where the relevance of the 2-mCH interlacing peakon ODE system and the finite KvM lattice was also given.

Theorem 2. *Given*

$$A_i(0) = \sum_{p=1}^{K} \lambda_p^{2i} c_p(0),$$

with

$$0 < \lambda_1(0)^2 < \lambda_2(0)^2 < \cdots < \lambda_K(0)^2, \qquad c_p(0) > 0,$$

let $\tau_k^{(l)}(0)$ *be defined as Hankel determinants*

$$\tau_k^{(l)}(0) = \begin{vmatrix} A_l(0) & A_{l+1}(0) & \cdots & A_{l+k-1}(0) \\ A_{l+1}(0) & A_{l+2}(0) & \cdots & A_{l+k}(0) \\ \vdots & \vdots & \ddots & \vdots \\ A_{l+k-}(0) & A_{l+k}(0) & \cdots & A_{l+2k-2}(0) \end{vmatrix}$$

for any nonnegative integer k, *and* $\tau_0^{(l)}(0) = 1$, $\tau_k^{(l)}(0) = 0$ *for* $k < 0$.

1. *Introduce the variables* $\{x_k(0), m_k(0)\}_{k=1}^K$ *defined by*

$$x_{2k'-1} = \ln\left(\frac{1}{m_{2k'-1}} \cdot \frac{(\tau_k^{(0)})^2}{\tau_k^{(1)}\tau_{k-1}^{(1)}}\right), \quad x_{2k'} = \ln\left(n_{2k'} \cdot \frac{\tau_k^{(0)}\tau_{k-1}^{(0)}}{(\tau_{k-1}^{(1)})^2}\right)$$

where $k' = K + 1 - k$.
If $\{\lambda_p(t), c_p(t)\}_{p=1}^n$ *evolve as*

$$\dot{\lambda}_p = 0, \qquad \dot{c}_p = \frac{c_p}{\lambda_p^2},$$

then $\{x_k(t), m_k(0)\}_{k=1}^n$ *satisfy the 2-mCH interlacing peakon lattice.*

2. *Introduce the variables* $\{d_k(0)\}_{k=1}^{2K-1}$ *defined by*

$$d_{2k-1} = \sqrt{\frac{\tau_k^{(1)}\tau_{k-1}^{(0)}}{\tau_k^{(0)}\tau_{k-1}^{(1)}}}, \qquad d_{2k} = \sqrt{\frac{\tau_{k-1}^{(1)}\tau_{k+1}^{(0)}}{\tau_k^{(1)}\tau_k^{(0)}}}.$$

If $\{\lambda_p(t), c_p(t)\}_{p=1}^n$ *evolve as*

$$\dot{\lambda}_p = 0, \qquad \dot{c}_p = \lambda_p^2 c_p,$$

then $\{d_k(0)\}_{k=1}^{2K-1}$ *satisfy the finite KvM lattice.*

3. *There exists a mapping from* $\{x_k(0), m_k(0)\}_{k=1}^n$ *to* $\{d_k(0)\}_{k=1}^{2K-1}$ *according to*

$$d_{2k-1} = \sqrt{\frac{1}{g_{k'}h_{k'}}}, \quad d_{2k} = \sqrt{\frac{1}{g_{k'}h_{k'-1}}},$$
$$g_k = m_{2k-1}e^{x_{2k-1}}, \qquad h_k = n_{2k}e^{-x_{2k}}.$$

2.3 Novikov peakon, finite B-Toda lattices and PSOPs

The Novikov equation

$$m_t + m_x u^2 + 3muu_x = 0, \qquad m = u - u_{xx} \tag{2.5}$$

is an integrable system with cubic nonlinearity, which was derived by V. Novikov [73] in a symmetry classification of nonlocal partial differential equations and first published in the paper by Hone and Wang [58]. It possesses a matrix Lax pair and admits a bi-Hamiltonian structure. There also exist infinitely many conservation quantities.

The Lax pair of the Novikov equation [58] reads:

$$D_x\Phi = U\Phi, \qquad D_t\Phi = V\Phi \tag{2.6}$$

with

$$U = \begin{pmatrix} 0 & \lambda m & 1 \\ 0 & 0 & \lambda m \\ 1 & 0 & 0 \end{pmatrix}, \quad V = \begin{pmatrix} -uu_x & \lambda^{-1}u_x - \lambda u^2 m & u_x^2 \\ \lambda^{-1}u & -\lambda^{-2} & -\lambda^{-1}u_x - \lambda u^2 m \\ -u^2 & \lambda^{-1}u & uu_x \end{pmatrix}.$$

In other words, the compatibility condition

$$(D_xD_t - D_tD_x)\Phi = 0$$

implies the zero curvature condition

$$U_t - V_x + [U, V] = 0,$$

which is exactly the Novikov equation.

The Novikov Equation (2.5) admits the multipeakon solution of the form

$$u = \sum_{k=1}^{n} m_k(t)e^{-|x-x_k(t)|}$$

in some weak sense if the positions and momenta satisfy the following ODE system:

$$\dot{x}_k = u(x_k)^2, \qquad \dot{m}_k = -m_k u(x_k)\langle u_x\rangle(x_k), \qquad 1 \le k \le n. \tag{2.7}$$

Again, $\langle f\rangle(a)$ denotes the average of left and right limits at the point a.

For the positive momenta $m_k(0)$, the ODE system (2.7) was solved by Hone, Lundmark and Szmigielski in [57] who used the inverse spectral method to explicitly construct the peakon solutions.

The finite B-Toda lattice reads

$$\dot{u}_k = u_k(b_k - b_{k-1}), \qquad k = 1, \ldots, n-1, \tag{2.8a}$$

$$\dot{b}_k = u_{k+1}(b_{k+1} + b_k) - u_k(b_k + b_{k-1}), \quad k = 0, \ldots, n-1, \tag{2.8b}$$

with the boundary condition $u_0 = u_n = 0$. The Lax representation of the finite B-Toda lattice was constructed in [30]. In fact, an appropriate deformation of the so-called partial-skew orthogonal polynomials (PSOPs) may lead to the finite B-Toda lattice, that is, the finite B-Toda lattice can be equivalently written as

$$\dot{L} = [B, L],$$

where

$$L = L_1^{-1}L_2, \qquad B = L_1^{-1}B_2,$$

$$L_1 = \begin{pmatrix} 1 & & & \\ -u_1 & 1 & & \\ & \ddots & \ddots & \\ & & -u_{n-1} & 1 \end{pmatrix}, \quad B_2 = \begin{pmatrix} 0 & & & \\ \mathcal{A}_1 & 0 & & \\ & \ddots & \ddots & \\ & & \mathcal{A}_{n-1} & 0 \end{pmatrix},$$

$$L_2 = \begin{pmatrix} \mathcal{B}_0 & 1 & & & & \\ \mathcal{C}_1 & \mathcal{B}_1 & 1 & & & \\ \mathcal{D}_2 & \mathcal{C}_2 & \mathcal{B}_2 & 1 & & \\ & \ddots & \ddots & \ddots & \ddots & \\ & & \mathcal{D}_{n-2} & \mathcal{C}_{n-2} & \mathcal{B}_{n-2} & 1 \\ & & & \mathcal{D}_{n-1} & \mathcal{C}_{n-1} & \mathcal{B}_{n-1} \end{pmatrix}$$

with

$$\mathcal{A}_k = -u_k(b_k + b_{k-1}), \quad \mathcal{B}_k = b_k - u_k, \quad \mathcal{C}_k = u_k(b_k - u_{k+1}), \quad \mathcal{D}_k = (u_k)^2 u_{k-1}.$$

The correspondence of the Novikov peakon and finite B-Toda lattices was investigated in the recent work [33], where the B-Toda lattice and Novikov peakon lattice are viewed as opposite flows to each other.

Theorem 3. *Given Pfaffian entries*

$$Pf(i,j)|_{t=0} = \sum_{p=1}^{n}\sum_{q=1}^{n} \frac{\zeta_p(0) - \zeta_q(0)}{\zeta_p(0) + \zeta_q(0)} (\zeta_p(0))^i (\zeta_q(0))^j c_p(0) c_q(0),$$

$$Pf(d_0, i)|_{t=0} = \sum_{p=1}^{n} c_p(0)(\zeta_p(0))^i,$$

with

$$0 < \zeta_1(0) < \zeta_2(0) < \cdots < \zeta_n(0), \qquad c_p(0) > 0,$$

let $\tau_k^{(l)}(0)$ *be defined as Pfaffians*[2]

$$\tau_{2k}^{(l)}(0) = Pf(l, l+1, \cdots, l+2k-1)|_{t=0}, \quad \tau_{2k-1}^{(l)}(0) = Pf(d_0, l, l+1, \cdots, l+2k-2)|_{t=0}$$

for any nonnegative integers k, *and* $\tau_0^{(l)}(0) = 1$, $\tau_k^{(l)}(0) = 0$ *for* $k < 0$. *Define*

$$W_k^{(l)}(0) = \tau_k^{(l+1)}(0)\tau_k^{(l)}(0) - \tau_{k-1}^{(l+1)}(0)\tau_{k+1}^{(l)}(0).$$

1. *Introduce the variables* $\{x_k(0), m_k(0)\}_{k=1}^n$ *defined by*

$$x_{k'}(0) = \frac{1}{2}\ln\frac{W_k^{(-1)}(0)}{W_{k-1}^{(0)}(0)}, \qquad m_{k'}(0) = \frac{\sqrt{W_k^{(-1)}(0)W_{k-1}^{(0)}(0)}}{\tau_k^{(0)}(0)\tau_{k-1}^{(0)}(0),}$$

where $k' = n + 1 - k$. *If* $\{\zeta_p(t), c_p(t)\}_{j=1}^n$ *evolve as*

$$\dot{\zeta}_p = 0, \qquad \dot{c}_p = \frac{c_p}{\zeta_p},$$

then $\{x_k(t), m_k(t)\}_{k=1}^n$ *satisfy the Novikov peakon ODEs:*

$$\dot{x}_k = \left(\sum_{j=1}^{n} m_j e^{-|x_j - x_k|}\right)^2,$$

$$\dot{m}_k = \left(\sum_{j=1}^{n} m_j e^{-|x_j - x_k|}\right)\left(\sum_{j=1}^{n} \operatorname{sgn}(x_k - x_j) m_j e^{-|x_j - x_k|}\right).$$

[2]The positivity of $\tau_k^{(l)}$ and $W_k^{(l)}$ for $1 \le k \le n$ is ensured (see, e.g., [33, 57]).

2. *Introduce the variables* $\{\{u_k(0)\}_{k=1}^{n-1},\{b_k(0)\}_{k=0}^{n-1}\}$ *defined by*

$$u_k(0)=\frac{\tau_{k+1}^{(0)}(0)\tau_{k-1}^{(0)}(0)}{(\tau_k^{(0)}(0))^2},$$

$$b_k(0)=\frac{\tau_{k+1}^{(0)}(0)\tau_{k-1}^{(0)}(0)}{(\tau_k^{(0)}(0))^2}+\frac{\tau_{k+2}^{(0)}(0)\tau_k^{(0)}(0)}{(\tau_{k+1}^{(0)}(0))^2}$$

$$+\frac{\left(\tau_{k+1}^{(0)}(0)\right)^2W_{k-1}^{(0)}(0)}{\left(\tau_k^{(0)}(0)\right)^2W_k^{(0)}(0)}+\frac{\left(\tau_k^{(0)}(0)\right)^2W_{k+1}^{(0)}(0)}{\left(\tau_{k+1}^{(0)}(0)\right)^2W_k^{(0)}(0)}.$$

If $\{\zeta_p(t),c_p(t)\}_{j=1}^n$ *evolve as*

$$\dot\zeta_p=0,\qquad \dot c_p=\zeta_p c_p,$$

then $\{\{u_k(0)\}_{k=1}^{n-1},\{b_k(0)\}_{k=0}^{n-1}\}$ *satisfy the finite B-Toda lattice* (2.8) *with* $u_0=u_n=0$.

3. *There exists a mapping from* $\{x_k(0),m_k(0)\}_{k=1}^n$ *to* $\left\{\{u_k(0)\}_{k=1}^{n-1},\{b_k(0)\}_{k=0}^{n-1}\right\}$ *according to*

$$u_k(0)=\frac{1}{2m_{k'}(0)m_{k'-1}(0)\cosh x_{k'}(0)\cosh x_{k'-1}(0)(\tanh x_{k'}(0)-\tanh x_{k'-1}(0))},$$

$$b_k(0)=u_k(0)\left(1+\frac{m_{k'}(0)e^{x_{k'-1}(0)}}{m_{k'-1}(0)e^{x_{k'}(0)}}\right)+u_{k+1}(0)\left(1+\frac{m_{k'-2}(0)e^{x_{k'-1}(0)}}{m_{k'-1}(0)e^{x_{k'-2}(0)}}\right)$$

with the convention

$$x_0(0)=-\infty,\quad x_{n+1}(0)=+\infty.$$

2.4 DP peakon, finite C-Toda lattices and CBOPs

The Degasperis–Procesi (DP) equation

$$m_t+(um)_x+2u_xm=0,\qquad m=u-u_{xx} \tag{2.9}$$

was found by Degasperis and Procesi [46] to pass the necessary (but not sufficient) test of asymptotic integrability, and later shown by Degasperis, Holm, and Hone [45] to be integrable in the sense of Lax, and has a bi-Hamiltonian structure as well as infinitely many conservation laws.

Like the CH equation, the DP equation may also be regarded as a model for the propagation of shallow water waves, and a rigorous explanation was given in [40]. It also admits peakon solutions [66, 67]. In [80, 81], collisions of DP peakons was considered. Furthermore, the study on the DP peakon problem induces new questions regarding Nikishin systems [21] studied in approximation theory, and random two-matrix models [20, 23]. Despite its superficial similarity to the CH equation, the DP equation has in addition shock solutions (see [65] for the onset of shocks in the form of shockpeakons). Indeed, there has been considerable interest in the DP equation.

When the multipeakon ansatz

$$u(x,t)=\sum_{j=1}^{n} m_j(t)e^{-|x-x_j(t)|}$$

is taken into account, it follows from (2.9) that m can be regarded as a discrete measure

$$m(x,t)=2\sum_{k=1}^{n} m_k(t)\delta(x-x_k(t)).$$

By using distributional calculus, it is known that the first equation of (2.9) is satisfied in a weak if the positions $(x_1,\ldots,x_n)$ and momenta $(m_1,\ldots,m_n)$ of the peakons obey the following system of $2n$ ODEs [45, 67]:

$$\dot{x}_k=u(x_k)=\sum_{j=1}^{n} m_j e^{-|x_j-x_k|}, \tag{2.10a}$$

$$\dot{m}_k=-2\langle u_x\rangle(x_k)=2\sum_{j=1}^{n}\operatorname{sgn}(x_k-x_j)m_j e^{-|x_j-x_k|}. \tag{2.10b}$$

Recall that the DP equation admits the Lax pair

$$(\partial_x-\partial_x^3)\psi=zm\psi, \tag{2.11a}$$

$$\psi_t=[z^{-1}(1-\partial_x^2)+u_x-u\partial_x]\psi. \tag{2.11b}$$

Due to Lax integrability in the peakon sector, Lundmark and Szmigielski [66, 67] studied the discrete cubic string problem and employed inverse spectral method to give an explicit construction of DP multipeakons in the case of positive initial m_k.

The finite C-Toda lattice [31],

$$\dot{u}_k=u_k(b_k-b_{k-1}), \qquad k=1,\ldots,n-1, \tag{2.12a}$$

$$\dot{b}_k=2(\sqrt{u_{k+1}b_{k+1}b_k}-\sqrt{u_kb_kb_{k-1}}), \quad k=0,\ldots,n-1, \tag{2.12b}$$

can be regarded as the isospectral deformation of CBOPs. It admits the Lax representation:

$$\dot{L}=[B,L],$$

where

$$L=L_1^{-1}L_2, \qquad B=L_1^{-1}B_2$$

$$L_1=\begin{pmatrix}1&&&\\ \mathcal{A}_1&1&&\\ &\ddots&\ddots&\\ &&\mathcal{A}_{n-1}&1\end{pmatrix},\quad L_2=\begin{pmatrix}\mathcal{B}_0&1&&&&\\ \mathcal{C}_1&\mathcal{B}_1&1&&&\\ \mathcal{D}_2&\mathcal{C}_2&\mathcal{B}_2&1&&\\ &\ddots&\ddots&\ddots&\ddots&\\ &&\mathcal{D}_{n-2}&\mathcal{C}_{n-2}&\mathcal{B}_{n-2}&1\\ &&&\mathcal{D}_{n-1}&\mathcal{C}_{n-1}&\mathcal{B}_{n-1}\end{pmatrix},$$

$$B_2=\begin{pmatrix}0&&&\\ 2(\mathcal{B}_0-\mathcal{A}_0)\mathcal{A}_1&0&&\\ &\ddots&\ddots&\\ &&2(\mathcal{B}_{n-2}-\mathcal{A}_{n-2})\mathcal{A}_{n-1}&0\end{pmatrix},$$

with

$$\mathcal{A}_k=-\sqrt{\frac{b_ku_k}{b_{k-1}}},\quad \mathcal{B}_k=\frac{1}{2}b_k-\sqrt{\frac{b_ku_k}{b_{k-1}}},\quad \mathcal{C}_k=-u_k+\frac{1}{2}\sqrt{b_ku_kb_{k-1}},\quad \mathcal{D}_k=u_{k-1}\sqrt{\frac{b_ku_k}{b_{k-1}}}.$$

The relation between nonlinear variables of DP peakon ODEs and the finite C-Toda lattice is interpreted by the following theorem.

Theorem 4. *Given*

$$\beta_i(0)=\sum_{p=1}^n\xi_p(0)^ic_p(0),\qquad J_{i,j}(0)=J_{j,i}(0)=\sum_{p=1}^n\sum_{q=1}^n\frac{\xi_p(0)^i\xi_q(0)^j}{\xi_p(0)+\xi_q(0)}c_p(0)c_q(0),$$

with

$$0<\xi_1(0)<\xi_2(0)<\cdots<\xi_n(0),\qquad c_p(0)>0,$$

let $\tau_k^{(i,j)}(0)$ and $\sigma_k^{(i,j)}(0)$ be defined as[3]

$$\tau_k^{(i,j)}(0)=\begin{vmatrix}J_{i,j}(0)&J_{i,j+1}(0)&\cdots&J_{i,j+k-1}(0)\\ J_{i+1,j}(0)&J_{i+1,j+1}(0)&\cdots&J_{i+1,j+k-1}(0)\\ \vdots&\vdots&\ddots&\vdots\\ J_{i+k-1,j}(0)&J_{i+k-1,j+1}(0)&\cdots&J_{i+k-1,j+k-1}(0)\end{vmatrix}=\tau_k^{(j,i)}(0)$$

for any nonnegative integer k, and $\tau_0^{(i,j)}(0)=1$, $\tau_k^{(i,j)}(0)=0$ for $k<0$, and

$$\sigma_k^{(i,j)}(0)=\begin{vmatrix}\beta_i(0)&J_{i,j}(0)&J_{i,j+1}(0)&\cdots&J_{i,j+k-2}(0)\\ \beta_{i+1}(0)&J_{i+1,j}(0)&J_{i+1,j+1}(0)&\cdots&J_{i+1,j+k-2}(0)\\ \vdots&\vdots&\vdots&\ddots&\vdots\\ \beta_{i+k-1}(0)&J_{i+k-1,j}(0)&J_{i+k-1,j+1}(0)&\cdots&J_{i+k-1,j+k-2}(0)\end{vmatrix}\tag{2.13}$$

with the convention $\sigma_1^{(i,j)}(0)=\beta_i(0)$ and $\sigma_k^{(i,j)}(0)=0$ for $k<1$.

[3]The positivity of such determinants for $1\leq k\leq n$ is guaranteed (see, e.g., [21,67]).

1. *Introduce the variables* $\{x_k(0), m_k(0)\}_{k=1}^n$ *defined by*

$$x_{k'}(0) = \frac{1}{2}\log\frac{2\tau_k^{(0,1)}(0)}{\tau_{k-1}^{(1,2)}(0)}, \qquad m_{k'}(0) = \frac{\tau_k^{(0,1)}(0)\tau_{k-1}^{(1,2)}(0)}{\tau_k^{(1,1)}(0)\tau_{k-1}^{(1,1)}(0)},$$

where $k' = n+1-k$. *If* $\{\xi_p(t), c_p(t)\}_{p=1}^n$ *evolve as*

$$\dot{\xi}_p = 0, \qquad \dot{c}_p = \frac{c_p}{\xi_p},$$

then $\{x_k(t), m_k(t)\}_{k=1}^n$ *satisfy the DP peakon ODEs* (2.10).

2. *Introduce the variables* $\left\{\{u_k(0)\}_{k=1}^{n-1}, \{b_k(0)\}_{k=0}^{n-1}\right\}$ *defined by*

$$u_k(0) = \frac{\tau_{k+1}^{(1,1)}(0)\tau_{k-1}^{(1,1)}(0)}{(\tau_k^{(1,1)}(0))^2}, \qquad b_k = \frac{(\sigma_{k+1}^{(1,1)}(0))^2}{\tau_{k+1}^{(1,1)}(0)\tau_k^{(1,1)}(0)}.$$

If $\{\xi_p(t), c_p(t)\}_{p=1}^n$ *evolve as*

$$\dot{\xi}_p = 0, \qquad \dot{c}_p = \xi_p c_p,$$

then $\left\{\{u_k(0)\}_{k=1}^{n-1}, \{b_k(0)\}_{k=0}^{n-1}\right\}$ *satisfy the finite C-Toda lattice* (2.12) *with* $u_0 = u_n = 0$.

3. *There exists a mapping from* $\{x_k(0), m_k(0)\}_{k=1}^n$ *to* $\left\{\{u_k(0)\}_{k=1}^{n-1}, \{b_k(0)\}_{k=0}^{n-1}\right\}$ *according to*

$$u_k(0) = \frac{e^{2x_{k'-1}(0)-2x_{k'}(0)}}{m_{k'-1}(0)m_{k'}(0)\left(1-e^{x_{k'-1}(0)-x_{k'}(0)}\right)^4},$$
$$b_k(0) = \frac{2(e^{x_{k'}(0)-x_{k'-1}(0)} - e^{x_{k'-2}(0)-x_{k'-1}(0)})^2}{m_{k'-1}(0)(1-e^{x_{k'-2}(0)-x_{k'-1}(0)})^2(e^{x_{k'}(0)-x_{k'-1}(0)}-1)^2}$$

with the convention

$$x_0(0) = -\infty, \quad x_{n+1}(0) = +\infty.$$

2.5 Finite FST, mCH peakon lattices and FST polynomials

The finite FST lattice was obtained by imposing an isospectral deformation on the measure of the FST polynomials in [35]. In fact, by considering the compatibility of the overdetermined system

$$L_{[2K]}\Psi_{[2K]} = zE_{[2K]}\Psi_{[2K]}, \qquad \dot{\Psi}_{[2K]} = F_{[2K]}R_{[2K]}\Psi_{[2K]},$$

where

$$L_{[2K]} = \begin{pmatrix} d_1 & -e_1 & & & \\ 1 & d_2 & -e_2 & & \\ & \ddots & \ddots & \ddots & \\ & & 1 & d_{2K-1} & -e_{2K-1} \\ & & & 1 & d_{2K} \end{pmatrix}, \quad E_{[2K]} = \begin{pmatrix} 0 & 1 & & & \\ & 0 & 1 & & \\ & & \ddots & \ddots & \\ & & & 0 & 1 \\ & & & & 0 \end{pmatrix},$$

$$R_{[2K]} = \begin{pmatrix} 0 & d_1 \sum_{j=1}^{K} d_{2j} + e_1 & & & \\ 0 & \sum_{j=2}^{K} d_{2j} & -d_1 \sum_{j=2}^{K} d_{2j} & & \\ & \ddots & \ddots & \ddots & \\ & & -1 & \sum_{j=0}^{K-2} d_{2j+1} & d_{2K} \sum_{j=0}^{K-1} d_{2j+1} + e_{2K-1} \\ & & & 0 & 0 \end{pmatrix},$$

with the nonzero entries being given by

$$\begin{aligned}
&L_{i,i} = d_i, \quad L_{i+1,i} = 1, \quad L_{i,i+1} = -e_i, \quad E_{i,i+1} = 1, \\
&R_{2i,2i} = \sum_{j=i+1}^{K} d_{2j}, \quad R_{2i,2i+1} = -\left(\sum_{j=i+1}^{K} d_{2j}\right)\left(\sum_{j=0}^{i-1} d_{2j+1}\right), \\
&R_{2i-1,2i-1} = \sum_{j=0}^{i-2} d_{2j+1}, \quad R_{2i-1,2i-2} = -1, \\
&R_{2i-1,2i} = \left(\sum_{j=0}^{i-1} d_{2j+1}\right)\left(\sum_{j=i}^{K} d_{2j}\right) + e_{2i-1}
\end{aligned}$$

one is led to the finite FST lattice

$$\dot{d}_{2k-1} = d_{2k-1}\left(\sum_{j=k}^{K} d_{2j}\right)\left(2\sum_{j=0}^{k-2} d_{2j+1} + d_{2k-1}\right) - e_{2k-2}\left(\sum_{j=0}^{k-2} d_{2j+1}\right) + e_{2k-1}\left(\sum_{j=0}^{k-1} d_{2j+1}\right), \tag{2.14a}$$

$$\dot{d}_{2k} = -d_{2k}\left(\sum_{j=0}^{k-1} d_{2j+1}\right)\left(d_{2k} + 2\sum_{j=k+1}^{K} d_{2j}\right) - e_{2k-1}\left(\sum_{j=k}^{K} d_{2j}\right) + e_{2k}\left(\sum_{j=k+1}^{K} d_{2j}\right), \tag{2.14b}$$

for $k = 1, 2, \ldots, K$, where e_k are some known distinct constants.

The mCH equation

$$m_t + \left((u^2 - u_x^2)m\right)_x = 0, \qquad m = u - u_{xx} \tag{2.15}$$

is an intriguing modification of the CH equation. As the CH equation and the Equation (2.15) can be obtained from the general method of tri-Hamiltonian duality applied to the bi-Hamiltonian representation of the KdV equation and the modified KdV equation respectively, we refer to Equation (2.15) as the mCH equation here. We note that the history of the mCH equation is long and convoluted [52,53,74,76]; therefore, it is sometimes called FORQ equation.

The mCH equation can be obtained by the compatibility condition of the overdetermined system (2.3) with the constraints $u = v, m = n$. It also admits multipeakon solution in the form of

$$u(x,t) = \sum_{j=1}^{n} m_j(t) e^{-|x-x_j(t)|}.$$

Due to the Lax integrability, it has been shown in [36] that, assuming the peakon ansatz as well as the ordering condition $x_1 < x_2 < \cdots < x_n$, (2.15) can be viewed as a distribution equation provided the ODE system

$$\dot{m}_j = 0, \qquad \dot{x}_j = 2 \sum_{\substack{1 \le k \le n, \\ k \ne j}} m_j m_k e^{-|x_j - x_k|} + 4 \sum_{1 \le i < j < k \le n} m_i m_k e^{-|x_i - x_k|} \tag{2.16}$$

holds.

In the case that all m_k are positive and distinct, an inverse spectral method [36] has been formulated to solve the mCH peakon ODEs (2.16) and hence (2.15).

A correspondence between the mCH peakon lattice (2.16) and the finite FST lattice (2.14) was given in [35] by establishing a connection between the boundary value problem associated to mCH peakons and a finite family of FST polynomials.

Theorem 5. *Given positive and distinct constants $e_k, 1 \le k \le 2K$, let*

$$\beta_j(0) = \sum_{i=1}^{K} \zeta_i(0)^j b_i(0), \qquad V_k(0) = \sum_{i=1}^{K} \frac{b_i(0)}{\zeta_i(0) + e_k},$$

with

$$0 < \zeta_1(0) < \zeta_2(0) < \cdots < \zeta_K(0), \qquad b_i(0) > 0,$$

For any positive integer k, index p, and l such that $0 \le l \le k$, define $\tau_k^{(l,p)}(0)$ as

$$\tau_k^{(l,p)}(0) = \det \begin{pmatrix} e_1^p V(e_1) & e_1^{p+1} V(e_1) & \cdots & e_1^{p+l-1} V(e_1) & 1 & e_1 & \cdots & e_1^{k-l-1} \\ e_2^p V(e_2) & e_2^{p+1} V(e_2) & \cdots & e_2^{p+l-1} V(e_2) & 1 & e_2 & \cdots & e_2^{k-l-1} \\ \vdots & \vdots & \ddots & \vdots & \vdots & \vdots & \ddots & \vdots \\ e_k^p V(e_k) & e_k^{p+1} V(e_k) & \cdots & e_k^{p+l-1} V(e_k) & 1 & e_k & \cdots & e_k^{k-l-1} \end{pmatrix},$$

as well as $\tau_0^{(l,p)}(0) = 1$, $\tau_k^{(l,p)}(0) = 0$ for $k < 0$ or $l > k$.

1. Let the variables $\{x_k(0), m_k(0)\}_{k=1}^{2K}$ be defined by

$$x_{k'}(0) = \ln \frac{(-1)^{\lfloor \frac{k}{2} \rfloor} \mathbf{e}_{[1,k]} \tau_k^{(\lfloor \frac{k+1}{2} \rfloor, 0)}(0) \tau_{k-1}^{(\lfloor \frac{k}{2} \rfloor, 0)}(0)}{m_{k'} \tau_k^{(\lfloor \frac{k}{2} \rfloor, 1)}(0) \tau_{k-1}^{(\lfloor \frac{k-1}{2} \rfloor, 1)}(0)}, \qquad m_{k'}(0) = \frac{1}{\sqrt{e_k}},$$

where $k' = 2K + 1 - k$, $\mathbf{e}_{[1,k]} = \prod_{i=1}^{k} e_i$. If $\{\zeta_i(t), b_i(t)\}_{i=1}^{K}$ evolve as

$$\dot{\zeta}_i = 0, \qquad \dot{b}_i = \frac{2b_i}{\zeta_i},$$

then $\{x_k(t), m_k(t)\}_{k=1}^{2K}$ satisfy the mCH peakon ODEs (2.16) with $n = 2K$.

2. *Let the variables* $\{d_k(0)\}_{k=1}^{2K}$ *be defined by*

$$d_{2p+1}(0) = (-1)^p \frac{\tau_{2p+1}^{(p,0)}(0)}{\tau_{2p+1}^{(p+1,0)}(0)} - (-1)^{p-1} \frac{\tau_{2p-1}^{(p-1,0)}(0)}{\tau_{2p-1}^{(p,0)}(0)},$$

$$d_{2p+2}(0) = (-1)^{p+1} \frac{\tau_{2p+2}^{(p+2,0)}(0)}{\tau_{2p+2}^{(p+1,0)}(0)} - (-1)^{p} \frac{\tau_{2p}^{(p+1,0)}(0)}{\tau_{2p}^{(p,0)}(0)}.$$

If $\{\zeta_i(t), b_i(t)\}_{i=1}^{K}$ *evolve as*

$$\dot{\zeta}_i = 0, \qquad \dot{b}_i = \zeta_i b_i,$$

then $\{d_k(t)\}_{k=1}^{2K}$ *satisfy the finite FST lattice* (2.14).

3. *The initial data of the mCH peakon problem* $\{x_k(0), m_k(0)\}_{k=1}^{2K}$ *is mapped to the initial data of the FST lattice* $\{d_k(0)\}_{k=1}^{2K}$ *as follows*

$$d_{2p}(0) = -\left(\frac{1}{h_{(2p)'}(0)} + \frac{1}{h_{(2p-1)'}(0)}\right) \frac{h_{1'}(0) \prod_{i=1}^{p-1} g_{(2i)'}(0) h_{(2i+1)'}(0)}{g_{1'}(0) \prod_{i=1}^{p-1} h_{(2i)'}(0) g_{(2i+1)'}(0)},$$

$$d_{2p+1}(0) = \left(\frac{1}{h_{(2p+1)'}(0)} + \frac{1}{h_{(2p)'}(0)}\right) \frac{\prod_{i=1}^{p} g_{(2i-1)'}(0) h_{(2i)'}(0)}{\prod_{i=1}^{p} h_{(2i-1)'}(0) g_{(2i)'}(0)},$$

where $g_j(0) = m_j(0)e^{-x_j(0)}$, $h_j(0) = m_j(0)e^{x_j(0)}$.

3 Conclusion and discussion

We briefly reviewed the developments pertaining to the relation between Toda lattices and peakon solutions and emphasized their isospectral structures in terms of orthogonal functions. We illustrated the emerging picture on examples of the A-Toda and CH peakon, KvM and 2-mCH interlacing peakon, B-Toda and Novikov peakon, C-Toda and DP peakon, FST and mCH peakon lattices, respectively. In our opinion, this direction remains to be developed.

Acknowledgements

We thank Jacek Szmigielski for his useful comments. This work was supported in part by the National Natural Science Foundation of China (#11688101, 11731014, 11701550) and the Youth Innovation Promotion Association CAS.

References

[1] M. J. Ablowitz and L. J. F. Nonlinear differential-difference equations. *J. Math. Phys.*, 16:598–603, 1975.

[2] M. J. Ablowitz and L. J. F. Nonlinear differential-dfference equations and fourier analysis. *J. Math. Phys.*, 17:1011–1018, 1976.

[3] M. Adler, E. Horozov, and P. van Moerbeke. Group factorization, moment matrices and Toda lattices. *Int. Math. Res. Notices*, 12:556–572, 1997.

[4] M. Adler, E. Horozov, and P. van Moerbeke. The Pfaff lattice and skew-orthogonal polynomials. *Int. Math. Res. Notices*, 1999(11):569–588, 1999.

[5] M. Adler and P. van Moerbeke. Matrix integrals, Toda symmetries, Virasoro constraints and orthogonal polynomials. *Duke Math. J.*, 80, 1995.

[6] M. Adler and P. van Moerbeke. String-orthogonal polynomials, string equations, and 2-Toda symmetries. *Commun. Pure Appl. Math.*, 50:241–290, 1997.

[7] M. Adler and P. van Moerbeke. Generalized orthogonal polynomials, discrete KP and Riemann–Hilbert problems. *Commun. Math. Phys.*, 207(3):589–620, 1999.

[8] M. Adler and P. van Moerbeke. The spectrum of coupled random matrices. *Ann. of Math.*, 149:921–976, 1999.

[9] M. Adler and P. van Moerbeke. Vertex operator solutions to the discrete KP hierarchy. *Commun. Math. Phys.*, 203:185–210, 1999.

[10] C. Álvarez Fernández and M. Mañas. Orthogonal Laurent polynomials on the unit circle, extended CMV ordering and 2D Toda type integrable hierarchies. *Adv. Math.*, 240:132–193, 2013.

[11] C. Álvarez Fernández, U. F. Prieto, and M. Mañas. Multiple orthogonal polynomials of mixed type: Gauss–Borel factorization and the multi-component 2D Toda hierarchy. *Adv. Math.*, 227:1451–1525, 2011.

[12] A. Aptekarev, A. Branquinho, and F. Marcellán. Toda-type differential equations for the recurrence coefficients of orthogonal polynomials and Freud transformation. *J. Comput. Appl. Math.*, 78(1):139–160, 1997.

[13] A. Aptekarev, M. Derevyagin, H. Miki, and W. Van Assche. Multidimensional Toda lattices: continuous and discrete time. *SIGMA*, 12:054, 2016.

[14] G. Ariznabarreta and M. Mañas. Matrix orthogonal Laurent polynomials on the unit circle and Toda type integrable systems. *Adv. Math.*, 264:396–463, 2014.

[15] G. Ariznabarreta and M. Mañas. Multivariate orthogonal polynomials and integrable systems. *Adv. Math.*, 302:628–739, 2016.

[16] R. Beals, D. H. Sattinger, and J. Szmigielski. Multi-peakons and a theorem of Stieltjes. *Inverse Problems*, 15:L1–L4, 1999.

[17] R. Beals, D. H. Sattinger, and J. Szmigielski. Multipeakons and the classical moment problem. *Adv. Math.*, 154(2):229–257, 2000.

[18] R. Beals, D. H. Sattinger, and J. Szmigielski. Peakons, strings, and the finite Toda lattice. *Commun. Pure Appl. Math.*, 54(1):91–106, 2001.

[19] M. Bertola and T. Bothner. Universality conjecture and results for a model of several coupled positive-definite matrices. *Commun. Math. Phys.*, 337(3):1077–1141, 2015.

[20] M. Bertola, M. Gekhtman, and J. Szmigielski. The Cauchy two-matrix model. *Commun. Math. Phys.*, 287(3):983–1014, 2009.

[21] M. Bertola, M. Gekhtman, and J. Szmigielski. Cauchy biorthogonal polynomials. *J. Approx. Theory*, 162(4):832–867, 2010.

[22] M. Bertola, M. Gekhtman, and J. Szmigielski. Strong asymptotics for Cauchy biorthogonal polynomials with application to the Cauchy two-matrix model. *J. Math. Phys.*, 54(4):043517, 2013.

[23] M. Bertola, M. Gekhtman, and J. Szmigielski. Cauchy-Laguerre two-matrix model and the Meijer-G random point field. *Commun. Math. Phys.*, 326(1):111–144, 2014.

[24] O. Bogoyavlensky. Integrable discretizations of the KdV equation. *Phys. Lett. A*, 134:34–38, 1988.

[25] O. Bogoyavlensky. Integrable dynamical systems associated with the KdV equation. *Math. USSR Izv.*, 31:435–454, 1988.

[26] O. Bogoyavlensky. Some constructions of integrable dynamical systems. *Math. USSR Izv.*, 31:47–75, 1988.

[27] R. Camassa and D. Holm. An integrable shallow water equation with peaked solitons. *Phys. Rev. Lett.*, 71:1661–1664, 1993.

[28] R. Camassa, D. D. Holm, and J. M. Hyman. A new integrable shallow water equation. *Adv. Appl. Mech.*, 31:1–33, 1994.

[29] X. Chang, X. Chen, and X. Hu. A generalized nonisospectral Camassa-Holm equation and its multipeakon solutions. *Adv. Math.*, 263:154–177, 2014.

[30] X. Chang, Y. He, X. Hu, and S. Li. Partial-skew-orthogonal polynomials and related integrable lattices with Pfaffian tau-functions. *Commun. Math. Phys.*, 364(3):1069–1119, 2018.

[31] X. Chang, X. Hu, and S. Li. Degasperis-Procesi peakon dynamical system and finite Toda lattice of CKP type. *Nonlinearity*, 31:4746–4775, 2018.

[32] X. Chang, X. Hu, and S. Li. Moment modification, multipeakons, and nonisospectral generalizations. *J. Differential Equations*, 265:3858–3887, 2018.

[33] X. Chang, X. Hu, S. Li, and J. Zhao. An application of Pfaffians to multipeakons of the Novikov equation and the finite Toda lattice of BKP type. *Adv. Math.*, 338:1077–1118, 2018.

[34] X. Chang, X. Hu, and J. Szmigielski. Multipeakons of a two-component modified Camassa-Holm equation and the relation with the finite Kac-van Moerbeke lattice. *Adv. Math.*, 299:1–35, 2016.

[35] X. Chang, X. Hu, J. Szmigielski, and A. Zhedanov. Isospectral flows related to Frobenius-Stickelberger-Thiele polynomials. *Commun. Math. Phys.*, 377:387–419, 2020.

[36] X. Chang and J. Szmigielski. Lax integrability and the peakon problem for the modified Camassa-Holm equation. *Commun. Math. Phys.*, 358(1):295–341, 2018.

[37] M. Chen, S. Liu, and Y. Zhang. A two-component generalization of the Camassa-Holm equation and its solutions. *Lett. Math. Phys.*, 75(1):1–15, 2006.

[38] M. Chu. Linear algebra algorithms as dynamical systems. *Acta Numer.*, 17:1–86, 2008.

[39] A. Constantin and J. Escher. Wave breaking for nonlinear nonlocal shallow water equations. *Acta Math.*, 181(2):229–243, 1998.

[40] A. Constantin and D. Lannes. The hydrodynamical relevance of the Camassa-Holm and Degasperis-Procesi equations. *Arch. Ration. Mech. Anal.*, 192(1):165–186, 2009.

[41] A. Constantin and W. A. Strauss. Stability of peakons. *Commun. Pure App. Math.*, 53:603–610, 2000.

[42] A. Constantin and W. A. Strauss. Stability of the Camassa-Holm solitons. *J. Nonlinear Sci.*, 12:415–422, 2002.

[43] E. Date, M. Jimbo, M. Kashiwara, and T. Miwa. Transformation groups for soliton equations. VI. KP hierarchies of orthogonal and symplectic type. *J. Phys. Soc. Japan*, 50(11):3813–3818, 1981.

[44] E. Date and S. Tanaka. Analogue of inverse scattering theory for the discrete Hill's equation and exact solutions for the periodic Toda lattice. *Prog. Theor. Phys.*, 55(2):457–465, 1976.

[45] A. Degasperis, D. Holm, and A. Hone. A new integrable equation with peakon solutions. *Theor. Math. Phys.*, 133(2):1463–1474, 2002.

[46] A. Degasperis and M. Procesi. Asymptotic integrability. In A. Degasperis and G. Gaeta, editors, *Symmetry and perturbation theory (Rome, 1998)*, pages 23–37. World Scientific Publishing, River Edge, NJ, 1999.

[47] P. Deift. *Orthogonal polynomials and random matrices: a Riemann-Hilbert approach*, volume 3. Courant Lecture Notes, Vol. 3, New York University, 2000.

[48] B. A. Dubrovin, V. B. Matveev, and S. P. Novikov. Non-linear equations of Korteweg-de Vries type, finite-zone linear operators, and Abelian varieties. *Russian Math. surveys*, 31(1):59, 1976.

[49] J. Eckhardt and A. Kostenko. An isospectral problem for global conservative multi-peakon solutions of the Camassa-Holm equation. *Commun. Math. Phys.*, 329(3):893–918, 2014.

[50] N. Ercolani, H. Flaschka, and S. Singer. The Geometry of the Full Kostant-Toda lattice. In *Integrable Systems, Vol. 115 of Progress in Mathematics*, pages 181–226. Birkhäuser, 1993.

[51] H. Flaschka. The Toda lattice. I. Existence of integrals. *Phys. Rev. B*, 9(4):1924, 1974.

[52] A. Fokas. The Korteweg-de Vries equation and beyond. *Acta Appl. Math.*, 39(1-3):295–305, 1995.

[53] B. Fuchssteiner. Some tricks from the symmetry-toolbox for nonlinear equations: generalizations of the Camassa-Holm equation. *Physica D: Nonlinear Phenomena*, 95(3):229–243, 1996.

[54] B. Fuchssteiner and A. Fokas. Symplectic structures, their Bäcklund transformations and hereditary symmetries. *Phys.D*, 4(1):47–66, 1981.

[55] X. Geng and B. Xue. An extension of integrable peakon equations with cubic nonlinearity. *Nonlinearity*, 22(8):1847–1856, 2009.

[56] R. Hirota. *The direct method in soliton theory.* Cambridge University Press, Cambridge, 2004. Translated by Nagai, A., Nimmo, J. and Gilson, C.

[57] A. Hone, H. Lundmark, and J. Szmigielski. Explicit multipeakon solutions of Novikov's cubically nonlinear integrable Camassa-Holm type equation. *Dyn. Partial Differ. Equ.*, 6(3):253–289, 2009.

[58] A. Hone and J. Wang. Integrable peakon equations with cubic nonlinearity. *J. Phys. A*, 41(37):372002, 2008.

[59] Y. Itoh. Integrals of a Lotka-Volterra system of odd number of variables. *Theoret. Phys.*, 78:507–510, 1987.

[60] M. Jimbo and T. Miwa. Solitons and infinite dimensional Lie algebras. *Publ. RIMS*, 19(3):943–1001, 1983.

[61] M. Kac and P. van Moerbeke. A complete solution of the periodic Toda problem. *Proc. Natl. Acad. Sci.*, 72(8):2879, 1975.

[62] M. Kac and P. van Moerbeke. On an explicitly soluble system of nonlinear differential equations related to certain Toda lattices. *Adv. Math.*, 16(2):160–169, 1975.

[63] S. Kharchev, A. Mironov, and A. Zhedanov. Faces of relativistic Toda chain. *Int. J. Mod. Phys. A*, 12(15):2675–2724, 1997.

[64] B. M. Kostant. The solution to a generalized Toda lattice and representation theory. *Adv. Math.*, 34:195–338, 1979.

[65] H. Lundmark. Formation and dynamics of shock waves in the Degasperis–Procesi equation. *J. Nonlinear Sci.*, 17(3):169–198, 2007.

[66] H. Lundmark and J. Szmigielski. Multi-peakon solutions of the Degasperis-Procesi equation. *Inverse Problems*, 19(6):1241–1245, 2003.

[67] H. Lundmark and J. Szmigielski. Degasperis-Procesi peakons and the discrete cubic string. *Int. Math. Res. Pap.*, 2005(2):53–116, 2005.

[68] H. P. McKean. Breakdown of the Camassa-Holm equation. *Comm. Pure Appl. Math.*, 57(3):416–418, 2004.

[69] J. Moser. Finitely many mass points on the line under the influence of an exponential potential—an integrable system. In *Dynamical Systems, Theory and Applications, Lecture Notes in Phys. Vol. 38, J. Moser ed.*, pages 467–497. Springer,Berlin, 1975.

[70] A. Mukaihira and Y. Nakamura. Schur flow for orthogonal polynomials on the unit circle and its integrable discretization. *J. Comput. Appl. Math.*, 139(1):75–94, 2002.

[71] K. Narita. Soliton solution to extended Volterra equation. *J. Phys. Soc. Japan*, 51:1682–1685, 1982.

[72] I. Nenciu. Lax pairs for the Ablowitz-Ladik system via orthogonal polynomials on the unit circle. *Int. Math. Res. Not.*, 11:647–686, 2005.

[73] V. Novikov. Generalisations of the Camassa-Holm equation. *J. Phys. A*, 42(34):342002, 2009.

[74] P. Olver and P. Rosenau. Tri-Hamiltonian duality between solitons and solitary-wave solutions having compact support. *Phys. Rev. E*, 53(2):1900, 1996.

[75] F. Peherstorfer, V. P. Spiridonov, and A. S. Zhedanov. Toda chain, Stieltjes function, and orthogonal polynomials. *Theor. Math. Phys.*, 151(1):505–528, 2007.

[76] Z. Qiao. A new integrable equation with cuspons and W/M-shape-peaks solitons. *J. Math. Phys.*, 47(11):112701, 2006.

[77] O. Ragnisco and M. M. Bruschi. Peakons, r-matrix and Toda lattice. *Physica A*, 228:150–159, 1996.

[78] S. N. M. Ruijsenaars. Relativistic Toda systems. *Commun. Math. Phys.*, 133:217–247, 1990.

[79] J. Song, C. Qu, and Z. Qiao. A new integrable two-component system with cubic nonlinearity. *J. Math. Phys.*, 52(1):013503, 2011.

[80] J. Szmigielski and L. Zhou. Colliding peakons and the formation of shocks in the Degasperis-Procesi equation. *Proceedings of the Royal Society A: Mathematical, Physical and Engineering Science*, 469(2158), 2013.

[81] J. Szmigielski and L. Zhou. Peakon-antipeakon interaction in the Degasperis-Procesi equation. *Contemporary Mathematics*, 593:83–107, 2013.

[82] K. Tian and Q. Liu. Tri-Hamiltonian duality between the Wadati-Konno-Ichikawa hierarchy and the Song-Qu-Qiao hierarchy. *J. Math. Phys.*, 54(4):043513, 2013.

[83] M. Toda. Vibration of a chain with nonlinear interaction. *J. Phys. Soc. Japan*, 22(2):431–436, 1967.

[84] L. Vinet and A. Zhedanov. An integrable chain and bi-orthogonal polynomials. *Lett. Math. Phys.*, 46(3):233–245, 1998.

[85] G. Whitham. *Linear and Nonlinear Waves*. John Wiley & Sons, New York, 1974.

B2. Long-time asymptotics for the generalized coupled derivative nonlinear Schrödinger equation

Mingming Chen, Xianguo Geng[1], *Kedong Wang and Bo Xue*

[a] *School of Mathematics and Statistics, Zhengzhou University, 100 Kexue Road, Zhengzhou, Henan 450001, P.R. China*

Abstract

Based on spectral analysis of 3×3 matrix spectral problems, the initial value problem for the generalized coupled derivative nonlinear Schrödinger equation is transformed into a 3×3 matrix oscillatory Riemann-Hilbert problem. The long-time asymptotics of the initial value problem for the generalized coupled derivative nonlinear Schrödinger equation in the situation of solitonless is obtained by using the nonlinear steepest descent method.

1 Introduction

The principal subject of this chapter concerns the long-time asymptotic behavior for the initial value problem of the generalized coupled derivative nonlinear Schrödinger (GCDNLS) equation

$$\begin{cases} iu_t = u_{xx} + 2iuq^*q_x + (4-3\ell_0)i|u|^2u_x + (2-3\ell_0)i(|q|^2u_x + u^2u_x^* + uqq_x^*) \\ \qquad -\frac{3}{2}\ell_0(1-\frac{3}{2}\ell_0)u(|u|^2+|q|^2)^2, \\ iq_t = q_{xx} + 2iu^*qu_x + (4-3\ell_0)i|q|^2q_x + (2-3\ell_0)i(|u|^2q_x + q^2q_x^* + uqu_x^*) \\ \qquad -\frac{3}{2}\ell_0(1-\frac{3}{2}\ell_0)q(|u|^2+|q|^2)^2, \\ u(x,0) = u_0(x), \quad q(x,0) = q_0(x), \end{cases} \tag{1.1}$$

by resorting to the nonlinear steepest descent method [12], where $u(x,t)$ and $q(x,t)$ are two potentials, and ℓ_0 is a constant. The initial value functions $u_0(x)$ and $q_0(x)$ lie in the Schwartz space $\mathscr{S}(\mathbb{R}) = \{f(x)\in C^\infty(\mathbb{R}) : \sup_{x\in\mathbb{R}}|x^\alpha\partial^\beta f(x)| < \infty, \forall\alpha,\beta\in\mathbb{N}\}$. Moreover, $u_0(x)$ and $q_0(x)$ are assumed to be generic so that $\det a(\lambda)$ defined in the following context is nonzero in D_- (Figure 1). If we choose the proper ℓ_0, GCDNLS Equation (1.1) reduces to three interesting cases [20, 31–33]:

I. The coupled derivative nonlinear Schrödinger equation ($\ell_0 = 0$)

$$\begin{aligned} iu_t &= u_{xx} + 2i(|u|^2u + |q|^2u)_x, \\ iq_t &= q_{xx} + 2i(|q|^2q + |u|^2q)_x. \end{aligned} \tag{1.2}$$

[1]Corresponding author. *E-mail address*: xggeng@zzu.edu.cn

II. The coupled Chen-Lee-Liu derivative nonlinear Schrödinger equation ($\ell_0 = \frac{2}{3}$)

$$\begin{aligned} iu_t &= u_{xx} + 2i(|u|^2 u_x + uq^* q_x) = 0, \\ iq_t &= q_{xx} + 2i(|q|^2 q_x + qu^* u_x) = 0. \end{aligned} \tag{1.3}$$

III. The coupled Gerdjikov-Ivanov (GI) equation ($\ell_0 = \frac{4}{3}$)

$$\begin{aligned} iu_t &= u_{xx} + 2iuq^* q_x - 2i(|q|^2 u_x + u^2 u_x^* + uqq_x^*) + 2u(|u|^2 + |q|^2)^2, \\ iq_t &= q_{xx} + 2iu^* qu_x - 2i(|u|^2 q_x + q^2 q_x^* + uqu_x^*) + 2q(|u|^2 + |q|^2)^2. \end{aligned} \tag{1.4}$$

The coupled derivative nonlinear Schrödinger equation is one of the most important integrable models in the mathematical physics [31, 32], which can be used to describe the propagation of nonlinear Alfvén waves in magnetized plasmas with right and left circular polarizations [10]. This equation has been widely studied by using various methods. For example, the Darboux transformation for the coupled derivative nonlinear Schrödinger equation has been constructed, from which N-soliton solutions are obtained in [20, 27] with the help of iterating the Darboux transformation [28, 29]. The existence and stability of standing waves for the coupled derivative nonlinear Schrödinger equation was studied in [15]. The coupled derivative nonlinear Schrödinger equation was discussed in the framework of the Riemann-Hilbert problem and a compact N-soliton solution formula is found [14]. Based on the theory of algebraic curves, the explicit theta function representations of solutions for the Kaup-Newell coupled derivative nonlinear Schrödinger system are obtained [22] by resorting to the Baker-Akhiezer functions and the meromorphic functionss [21, 36]. In Ref. [32], the properties of the coupled Chen-Lee-Liu derivative nonlinear Schrödinger equation was studied with the help of the Lax pair, the conservation laws and the gauge transformation. The Fokas unified method is used to analyze the initial-boundary value problem of two-component Gerdjikov-Ivanov equation on the half-line. It is shown that the solution of the initial-boundary problem can be expressed in terms of the solution of a 3×3 Riemann-Hilbert problem. The Dirichlet to Neumann map is obtained through the global relation [39].

The nonlinear steepest descent method (also called Deift-Zhou method) was first introduced by Deift and Zhou [12], which provides a powerful tool to obtain the long-time asymptotic behavior of solutions of integrable nonlinear partial differential equation. Subsequently, various integrable equations associated with 2×2 matrix spectral problems have been studied [2, 3, 7, 9, 13, 16, 23, 24, 26, 34, 37]. However, there are rarely literatures about the integrable equations associated with 3×3 matrix spectral problems [5, 6, 18, 19, 30]. To our knowledge, there have been no results about the long-time asymptotic behavior for the initial value problem of the GCDNLS Equation (1.1).

The main result of this chapter is expressed as follows:

Theorem 1.1. *Let $(u(x,t), q(x,t))$ be the solution for the initial value problem for GCDNLS Equation* (1.1) *with the initial values u_0, $q_0 \in \mathscr{S}(\mathbb{R})$. Then, for $x < 0$ and $\lambda_0 > C$, the leading asymptotics of $(u(x,t), q(x,t))$ has the form*

$$(u(x,t), q(x,t)) = \frac{1}{\sqrt{16\pi t\lambda_0^2}} 2^{-2i\nu} e^{Z} \nu\Gamma(-i\nu)[(\delta_A^0)^2 + (\delta_B^0)^2]\gamma(-\lambda_0) Y_0 + O\left(\frac{\log t}{t\lambda_0^2}\right),$$

(1.5)

where C is a fixed constant, $\Gamma(\cdot)$ is the Gamma function and the vector function $\gamma(\lambda)$ is defined in (2.26), *and*

$$\begin{aligned}
&\delta_A^0 = (8t\lambda_0^4)^{\frac{i\nu}{2}}\exp\{-\frac{1}{2}it\lambda_0^4 + \chi_+(-\lambda_0) + \chi_-(-\lambda_0) + \hat{\chi}_+(-\lambda_0) + \hat{\chi}_-(-\lambda_0)\},\\
&\delta_B^0 = (8t\lambda_0^4)^{\frac{i\nu}{2}}\exp\{-\frac{1}{2}it\lambda_0^4 + \chi_+(\lambda_0) + \chi_-(\lambda_0) + \hat{\chi}_+(\lambda_0) + \hat{\chi}_-(\lambda_0)\},\\
&\tilde{\delta}_A^0(\lambda) = (8t\lambda^4)^{\frac{i\nu}{2}}\exp\{-\frac{1}{2}it\lambda^4 + \chi_+(-\lambda) + \chi_-(-\lambda) + \hat{\chi}_+(-\lambda) + \hat{\chi}_-(-\lambda)\},\\
&\tilde{\delta}_B^0(\lambda) = (8t\lambda^4)^{\frac{i\nu}{2}}\exp\{-\frac{1}{2}it\lambda^4 + \chi_+(\lambda) + \chi_-(\lambda) + \hat{\chi}_+(\lambda) + \hat{\chi}_-(\lambda)\},\\
&\lambda_0 = \sqrt{\frac{-x}{2t}}, \quad \nu = -\frac{1}{2\pi}\log(1 - |\gamma(\lambda_0)|^2),\\
&\tilde{\nu}(\lambda) = -\frac{1}{2\pi}\log(1 - |\gamma(\lambda)|^2),\\
&Z = \frac{3\pi i}{4} + \frac{\pi\nu}{2} + (\frac{i}{2} + \frac{3}{4}i\ell_0)\int_{-\infty}^{\lambda_0}\left[2 + \left(\frac{\tilde{\delta}_A^0(s)}{\tilde{\delta}_B^0(s)}\right)^2 + \left(\frac{\tilde{\delta}_B^0(s)}{\tilde{\delta}_A^0(s)}\right)^2\right]\frac{\tilde{\nu}(s)}{s}\,\mathrm{d}s,\\
&Y_0 = I - (i + \frac{3}{4}i\ell_0)\int_{-\infty}^{\lambda_0}\left[2 + \left(\frac{\tilde{\delta}_A^0(s)}{\tilde{\delta}_B^0(s)}\right)^2 + \left(\frac{\tilde{\delta}_B^0(s)}{\tilde{\delta}_A^0(s)}\right)^2\right]\frac{\tilde{\nu}(s)}{2s(1 - e^{-2\pi\tilde{\nu}(s)})}\gamma^\dagger(s)\gamma(s)\,\mathrm{d}s,\\
&\chi_\pm(\lambda_l) = \frac{1}{2\pi i}\int_{\pm\lambda_0}^{0}\log\left(\frac{1 - |\gamma(\xi)|^2}{1 - |\gamma(\lambda_0)|^2}\right)\frac{\mathrm{d}\xi}{\xi - \lambda_l},\\
&\hat{\chi}_\pm(\lambda_l) = \frac{i}{2\pi}\int_0^{\pm\infty}\log\left(1 + |\gamma(i\xi)|^2\right)\mathrm{d}\log(i\xi - \lambda_l),\\
&l \in \{1, 2\}, \quad \lambda_1 = \lambda_0, \quad \lambda_2 = -\lambda_0.
\end{aligned}$$

This chapter is organized as follows. In Section 2, we start with the 3×3 matrix Lax pair of the GCDNLS equation and write it in the 2×2 block form. We introduce the Jost solutions from the Volterra integral equations, from which a corresponding Riemann-Hilbert (RH) problem is formulated by the Jost solutions and the scattering matrix. It is proved that the solution of the GCDNLS equation can be expressed by the solution of this RH problem. In Section 3, a key step is to transform the known RH problem into an equivalent RH problem on an augmented contour. Then a model RH problem is obtained by truncating the contour with strict error estimation, which can be solved explicitly. Finally, the long-time asymptotic behavior for the initial value problem of the GCDNLS equation is obtained.

2 Riemann-Hilbert problem

In this section, we will analyze the inverse scattering problem of (1.1) and derive the corresponding RH problem, so that the solution of GCDNLS equation can be expressed. We first consider the 3×3 matrix spectral problems associated with

GCDNLS Equation (1.1)

$$\psi_x = (-\frac{1}{2}i\lambda^2\sigma + \tilde{U})\psi, \tag{2.1a}$$

$$\psi_t = (-\frac{1}{2}i\lambda^4\sigma + \tilde{V})\psi, \tag{2.1b}$$

where λ is a spectral parameter, $\sigma = \mathrm{diag}\,(-1,1,1)$, and

$$\tilde{U} = \begin{pmatrix} w & \lambda u & \lambda q \\ \lambda u^* & -\frac{1}{2}w & 0 \\ \lambda q^* & 0 & -\frac{1}{2}w \end{pmatrix}, \quad \tilde{V} = \begin{pmatrix} \tilde{V}_{11} & \tilde{V}_{12} & \tilde{V}_{13} \\ \tilde{V}_{21} & \tilde{V}_{22} & \tilde{V}_{23} \\ \tilde{V}_{31} & \tilde{V}_{32} & \tilde{V}_{33} \end{pmatrix}, \tag{2.2a}$$

$$\begin{aligned}
&\tilde{V}_{11} = i(|u|^2+|q|^2)\lambda^2 + \tfrac{3}{2}\ell_0[u_x u^* - u u_x^* + q_x q^* - q q_x^* + (3-3\ell_0)i(|u|^2+|q|^2)^2],\\
&\tilde{V}_{12} = u\lambda^3 - i\lambda[u_x + (2-\tfrac{3}{2}\ell_0)iu(|u|^2+|q|^2)],\\
&\tilde{V}_{13} = q\lambda^3 - i\lambda[q_x + (2-\tfrac{3}{2}\ell_0)iq(|u|^2+|q|^2)],\\
&\tilde{V}_{21} = u^*\lambda^3 - i\lambda[-u_x^* + (2-\tfrac{3}{2}\ell_0)iu^*(|u|^2+|q|^2)], \quad \tilde{V}_{22} = -i\lambda^2|u|^2, \quad \tilde{V}_{23} = -i\lambda^2 u^* q,\\
&\tilde{V}_{31} = q^*\lambda^3 - i\lambda[-q_x^* + (2-\tfrac{3}{2}\ell_0)iq^*(|u|^2+|q|^2)], \quad \tilde{V}_{32} = -i\lambda^2 u q^*, \quad \tilde{V}_{33} = -i\lambda^2|q|^2,
\end{aligned}$$

with $w = \frac{3}{2}i\ell_0(|u|^2+|q|^2)$, ℓ_0 is a constant, and u and q are two potentials. A direct calculation shows that the zero-curvature equation, $\tilde{U}_t - \tilde{V}_x + [\tilde{U}, \tilde{V}] = 0$, leads to the GCDNLS Equation (1.1).

In the following, we write the 3×3 matrix $A = (a_{ij})$ as a block form

$$A = \begin{pmatrix} A_{11} & A_{12} \\ A_{21} & A_{22} \end{pmatrix}, \tag{2.3}$$

where A_{11} is scalar, A_{12} is a two-dimensional row vector, A_{21} is a two-dimensional column vector, and A_{22} is a 2×2 matrix. For example, let $p(x,t) = (u(x,t), q(x,t))$ and we can rewrite $\tilde{U}$ of (2.2a) as the block form

$$\tilde{U} = \begin{pmatrix} w & \lambda p \\ \lambda p^\dagger & -\frac{1}{2}wI_{2\times 2} \end{pmatrix}.$$

where the superscript "$\dagger$" denotes the Hermitian conjugate.

Now we introduce a block matrix

$$D(x,t) = \begin{pmatrix} e^{\int_{(-\infty,0)}^{(x,t)} y_1(x',t')} & 0 \\ 0 & Y(x,t) \end{pmatrix}, \tag{2.4}$$

where

$$y_1(x,t) = (ipp^\dagger + \frac{3}{2}i\ell_0 pp^\dagger)\mathrm{d}x + \frac{3}{2}\ell_0\left((3i\ell_0 - 3i)|p|^4 + pp_x^\dagger - p_x p^\dagger\right)\mathrm{d}t, \tag{2.5}$$

and $Y(x,t)$ satisfies

$$\begin{cases} Y_x = (-i - \frac{3}{4}i\ell_0)p^\dagger p Y, \\ Y_t = \left((3i - 3i\ell_0)|p|^2 p^\dagger p - p_x^\dagger p + p^\dagger p_x\right)Y \end{cases} \tag{2.6}$$

with the condition $Y(x,t) \to I$ as $x \to -\infty, t = 0$. Then, (2.6) is equivalent to the equation

$$\begin{aligned} Y(x,t) = I + \int_{(-\infty,0)}^{(x,t)} & (-i - \frac{3}{4} i\ell_0) p^\dagger p Y(x',t') \mathrm{d}x' \\ & + \Big((3i - 3i\ell_0)|p|^2 p^\dagger p - p_x^\dagger p + p^\dagger p_x \Big) Y(x',t') \mathrm{d}t'. \end{aligned} \tag{2.7}$$

Define the new eigenfunction $\mu = D^{-1}\psi e^{\frac{1}{2}i\lambda^2\sigma x + \frac{1}{2}i\lambda^4\sigma t}$, $e^\sigma = \mathrm{diag}\,\big(e^{-1}, e, e\big)$. Then

$$\mu = I + O(\frac{1}{\lambda}), \qquad \lambda \to \infty. \tag{2.8}$$

The spectral problems (2.1a) and (2.1b) can be written as

$$\mu_x = -\frac{1}{2} i\lambda^2 [\sigma, \mu] + U\mu, \tag{2.9a}$$

$$\mu_t = -\frac{1}{2} i\lambda^4 [\sigma, \mu] + V\mu, \tag{2.9b}$$

where $[\cdot,\cdot]$ is the commutator, $[\sigma,\mu] = \sigma\mu - \mu\sigma$, and

$$U = D^{-1}\tilde{U}D - D^{-1}D_x, \qquad V = D^{-1}\tilde{V}D - D^{-1}D_t. \tag{2.10}$$

Then U has the block form

$$U = \begin{pmatrix} ipp^\dagger & \lambda e^{-\int_{(-\infty,0)}^{(x,t)} y_1} pY \\ \lambda Y^{-1} p^\dagger e^{\int_{(-\infty,0)}^{(x,t)} y_1} & iY^{-1}p^\dagger pY \end{pmatrix}. \tag{2.11}$$

The matrix Jost solutions $\mu_\pm$ of (2.9a) are defined by the following Volterra integral equation

$$\mu_\pm(\lambda; x, t) = I + \int_{\pm\infty}^{x} e^{\frac{1}{2}i\lambda^2\hat{\sigma}(\xi - x)} U(\xi,t)\mu_\pm(\lambda;\xi,t)\,\mathrm{d}\xi, \tag{2.12}$$

with the asymptotic conditions $\mu_\pm \to I$ as $x \to \pm\infty$, where $\hat{\sigma}$ acts on a 3×3 matrix X by $\hat{\sigma}X = [\sigma, X]$ and $e^{\hat{\sigma}}X = e^\sigma X e^{-\sigma}$. Write $\mu_\pm$ into two columns, that is $\mu_\pm = (\mu_{\pm L}, \mu_{\pm R})$, where $\mu_{\pm L}$ represent the first column of $\mu_\pm$, and $\mu_{\pm R}$ denote the last two columns. Moreover, as shown in Figure 1, μ_{-R} and μ_{+L} are continuous in $D_+ \cup \hat{\Gamma}$ and analytic in D_+, μ_{+L} and μ_{-R} are continuous in $D_- \cup \hat{\Gamma}$ and analytic in D_-, where

$$\begin{aligned} &\hat{\Gamma} = \{\lambda : \mathrm{Im}(\lambda^2) = 0\}, \\ &D_+ = \{\lambda \in \mathbb{C} |\, \arg\lambda \in (\pi/2, \pi) \cup (-\pi/2, 0)\}, \\ &D_- = \{\lambda \in \mathbb{C} |\, \arg\lambda \in (0, \pi/2) \cup (-\pi, -\pi/2)\}. \end{aligned}$$

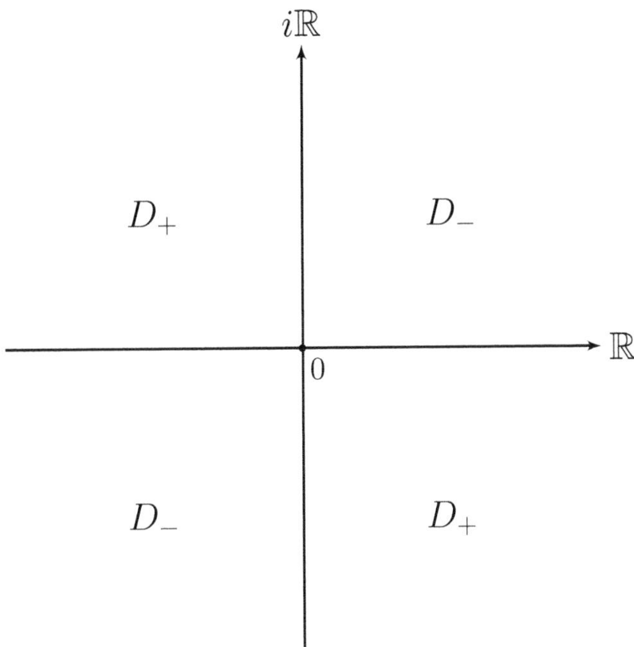

Figure 1: The domain D

Since $D\mu_\pm e^{-\frac{1}{2}i\lambda^2\sigma x-\frac{1}{2}i\lambda^4\sigma t}$ satisfy the same Equations (2.1a) and (2.1b), they are linearly dependent. Then there exists a scattering matrix $s(\lambda)$ such that

$$\mu_- = \mu_+ e^{-\frac{1}{2}i\lambda^2\hat{\sigma}x-\frac{1}{2}i\lambda^4\hat{\sigma}t}s(\lambda), \quad \lambda\in\hat{\Gamma}. \tag{2.13}$$

Then matrix U in (2.2a) is traceless implies that $\det\mu_\pm$ are independent of x. Particularly evaluating $\det\mu_\pm$ at $x=\pm\infty$, we have $\det\mu_\pm = 1$. Furthermore, from (2.13) we obtain that $\det s(\lambda)=1$. Notice that matrix U has two important symmetry properties

$$\sigma U^\dagger(\lambda^*)\sigma = -U(\lambda), \quad \sigma U(-\lambda)\sigma = U(\lambda), \tag{2.14}$$

According to (2.14), a direct calculation shows that the Jost solutions $\mu_\pm$ and the scattering matrix $s(\lambda)$ satisfy the symmetry properties

$$\sigma\mu_\pm^\dagger(\lambda^*)\sigma = \mu_\pm^{-1}(\lambda), \quad \sigma\mu_\pm(-\lambda)\sigma = \mu_\pm(\lambda); \tag{2.15}$$

$$\sigma s^\dagger(\lambda^*)\sigma = s^{-1}(\lambda), \quad \sigma s(-\lambda)\sigma = s(\lambda). \tag{2.16}$$

It follows from $\det s(\lambda)=1$ and (2.16) that

$$s_{11}(\lambda) = \det[s_{22}^\dagger(\lambda^*)], \quad s_{21}(\lambda) = \mathrm{adj}[s_{22}^\dagger(\lambda^*)]s_{12}^\dagger(\lambda^*), \tag{2.17}$$

where $\mathrm{adj}A$ denote the adjoint of matrix A. For convenience, the scattering matrix $s(\lambda)$ can be written as the block form

$$s(\lambda) = \begin{pmatrix} \det[a^\dagger(\lambda^*)] & b(\lambda) \\ \mathrm{adj}[a^\dagger(\lambda^*)]b^\dagger(\lambda^*) & a(\lambda) \end{pmatrix}, \tag{2.18}$$

with

$$a(\lambda) = a(-\lambda), \quad b(\lambda) = -b(-\lambda). \tag{2.19}$$

When $t = 0$, the evaluation of (2.13) as $x \to +\infty$ gives that

$$s(\lambda) = \lim_{x \to +\infty} e^{\frac{1}{2}i\lambda^2\hat{\sigma}x}\mu_-(\lambda; x, 0) = I + \int_{-\infty}^{+\infty} e^{\frac{1}{2}i\lambda^2\hat{\sigma}\xi}U(\xi, 0)\mu_-(\lambda; \xi, 0)\, d\xi, \tag{2.20}$$

which implies

$$\begin{aligned} a(\lambda) = I_{2\times 2} + \int_{-\infty}^{+\infty} & [\lambda Y^{-1}(x, 0)p^\dagger(\xi, 0)e^{(i+\frac{3}{2}i\ell_0)\int_{-\infty}^{x} p(x',0)p^\dagger(x',0)\,\mathrm{d}x'}\mu_{-12}(\lambda; \xi, 0) \\ & + iY^{-1}(x, 0)p^\dagger(x, 0)p(x, 0)Y(x, 0)\mu_{-22}(\lambda; \xi, 0)]\,\mathrm{d}\xi, \end{aligned} \tag{2.21}$$

$$\begin{aligned} b(\lambda) = \int_{-\infty}^{+\infty} & e^{-i\lambda^2\xi}[ip(\xi, 0)p^\dagger(\xi, 0)\mu_{-12}(\lambda; \xi, 0) \\ & + \lambda e^{(-i-\frac{3}{2}i\ell_0)\int_{-\infty}^{x} p(x',0)p^\dagger(x',0)\,\mathrm{d}x'}p(\xi, 0)Y(x, 0)\mu_{-22}(\lambda; \xi, 0)]\,\mathrm{d}\xi. \end{aligned} \tag{2.22}$$

We define sectionally analytic function

$$M(\lambda; x, t) = \begin{cases} \left(\frac{\mu_{-L}(\lambda)}{\det[a^\dagger(\lambda^*)]}, \mu_{+R}(\lambda)\right), & \lambda \in D_+, \\ \left(\mu_{+L}(\lambda), \mu_{-R}(\lambda)a^{-1}(\lambda)\right), & \lambda \in D_-. \end{cases} \tag{2.23}$$

From (2.13) and (2.23), we can infer that the matrix $M(\lambda; x, t)$ is analytic for $\lambda \in \mathbb{C}\backslash\hat{\Gamma}$ and solves an oscillatory RH problem

$$\begin{cases} M_+(\lambda; x, t) = M_-(\lambda; x, t)J(\lambda; x, t), & \lambda \in \hat{\Gamma}, \\ M(\lambda; x, t) \to I, & \lambda \to \infty, \end{cases} \tag{2.24}$$

where $M_+(\lambda; x, t)$ and $M_-(\lambda; x, t)$ denote the limiting values as λ approaches the contour $\hat{\Gamma}$ from the left and the right along the contour, and

$$J(\lambda; x, t) = \begin{pmatrix} 1 - \gamma(\lambda)\gamma^\dagger(\lambda^*) & -e^{-2it\theta}\gamma(\lambda) \\ e^{2it\theta}\gamma^\dagger(\lambda^*) & I \end{pmatrix}, \tag{2.25}$$

$$\theta(\lambda; x, t) = -\frac{x}{2t}\lambda^2 - \frac{1}{2}\lambda^4, \quad \gamma(\lambda) = b(\lambda)a^{-1}(\lambda). \tag{2.26}$$

Here we assume that $a(\lambda)$ is reversible, and $\gamma(\lambda)$ lies in Schwartz space and satisfies

$$\gamma(\lambda) = -\gamma(-\lambda), \quad \sup_{\lambda \in \hat{\Gamma}} |\gamma(\lambda)| < 1. \tag{2.27}$$

It is easy to see from (2.27) that the jump matrix $J(\lambda; x, t)$ is positive definite. Hence, using the Vanishing Lemma [1], we find that the solution of RH problem (2.24) is existent and unique. We then have the next theorem:

Theorem 2.1. *The solution for the initial value problem of GCDNLS Equation (1.1) is given by*

$$p(x,t)=(u(x,t),q(x,t))=e^{(i+\frac{3}{2}i\ell_0)\int_{-\infty}^{x}p(x')p^\dagger(x')\,\mathrm{d}x'}\tilde{p}(x,t)Y^{-1}(x,t), \tag{2.28}$$

where $\tilde{p}(x,t)$ is a two-dimensional row vector given by

$$\tilde{p}(x,t)=-i\lim_{\lambda\to\infty}(\lambda M(\lambda;x,t))_{12}. \tag{2.29}$$

Proof. For the large λ, substituting asymptotic expansion of $M(\lambda;x,t)$ into (2.9a) and comparing the coefficients, we obtain (2.29). ■

3 Long-time asymptotic behavior

In this section, we reduce the oscillatory RH problem (2.24) to a model RH problem with constant coefficients, which can be solved explicitly in terms of the parabolic cylinder functions. Finally, we obtain the long-time asymptotic behavior for the initial value problem of GCDNLS Equation (1.1) by the Deift-Zhou method.

We first make the following basic notations. For a matrix M, we define $\|M(\cdot)\|_p=\||M(\cdot)|\|_p$, where $|M|=(\mathrm{tr}M^\dagger M)^{\frac{1}{2}}$ denotes the norm of matrix M. For two quantities A and B, if there exists a constant $C>0$ such that $|A|\leqslant CB$, we denote it as $A\lesssim B$. If C depends on the parameter α, then we say that $A\lesssim_\alpha B$. For any oriented contour Σ, we define the left side to be positive and the right side to be negative.

3.1 First transformation: reoriented contour

Notice that the jump matrix $J(\lambda;x,t)$ has the upper/lower triangular factorization

$$J=\begin{pmatrix}1 & -e^{-2it\theta}\gamma(\lambda)\\ 0 & I\end{pmatrix}\begin{pmatrix}1 & 0\\ e^{2it\theta}\gamma^\dagger(\lambda^*) & I\end{pmatrix},\quad \forall\lambda\in(-\infty,-\lambda_0)\cup(\lambda_0,+\infty),$$

and the lower/upper triangular factorization

$$J=\begin{pmatrix}1 & 0\\ \frac{e^{2it\theta}\gamma^\dagger(\lambda^*)}{1-\gamma(\lambda)\gamma^\dagger(\lambda^*)} & I\end{pmatrix}\begin{pmatrix}1-\gamma(\lambda)\gamma^\dagger(\lambda^*) & 0\\ 0 & (I-\gamma^\dagger(\lambda^*)\gamma(\lambda))^{-1}\end{pmatrix}\begin{pmatrix}1 & \frac{-e^{-2it\theta}\gamma(\lambda)}{1-\gamma(\lambda)\gamma^\dagger(\lambda^*)}\\ 0 & I\end{pmatrix},$$

$\forall\lambda\in(-\lambda_0,\lambda_0)\cup(-i\infty,+i\infty)$. A crucial observation is that the complex plane can be decomposed according to the signature of $\mathrm{Re}(it\theta)$ in Figure 2.

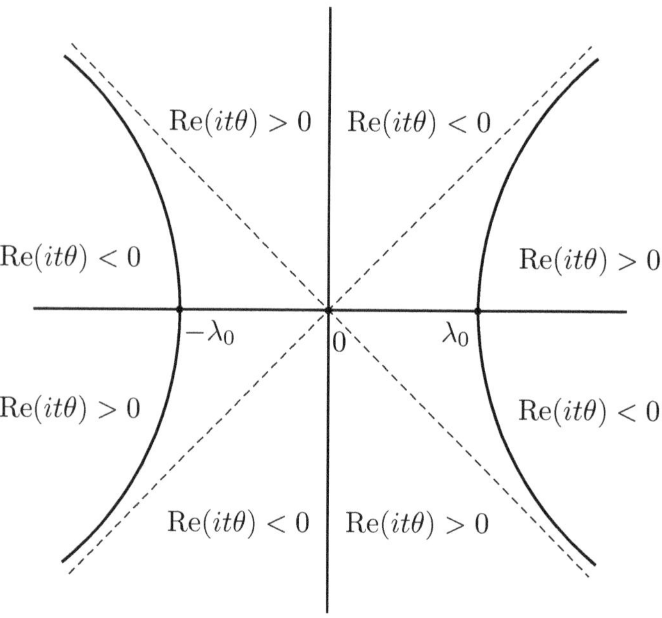

Figure 2: Signature graph of $\mathrm{Re}(it\theta)$

In order to make analytic and decay properties of the two factorizations consistent, we introduce a 2×2 matrix-valued function $\delta(\lambda)$, which satisfies the following RH problem

$$\begin{cases} \delta_+(\lambda) = (I - \gamma^\dagger(\lambda^*)\gamma(\lambda))\delta_-(\lambda), & \lambda \in (-\lambda_0, \lambda_0) \cup (-i\infty, +i\infty), \\ \delta(\lambda) \to I, & \lambda \to \infty. \end{cases} \tag{3.1}$$

The jump matrix $I - \gamma^\dagger(\lambda^*)\gamma(\lambda)$ is positive definite by (2.27), hence it follows from Vanishing Lemma that the RH problem (3.1) has a unique solution $\delta(\lambda)$. By (3.1), we have the scalar RH problem

$$\begin{cases} \det \delta_+(\lambda) = (\det \delta_-(\lambda))(1 - \gamma(\lambda)\gamma^\dagger(\lambda^*)), & \lambda \in (-\lambda_0, \lambda_0) \cup (-i\infty, +i\infty), \\ \det \delta(\lambda) \to 1, & \lambda \to \infty. \end{cases} \tag{3.2}$$

The solution of the above scalar RH problem (3.2) can be expressed by Plemelj formula [1]

$$\det \delta(\lambda) = \left(\left(\frac{\lambda - \lambda_0}{\lambda} \right) \left(\frac{\lambda + \lambda_0}{\lambda} \right) \right)^{-i\nu} e^{\chi_+(\lambda)} e^{\chi_-(\lambda)} e^{\hat{\chi}_+(\lambda)} e^{\hat{\chi}_-(\lambda)}, \tag{3.3}$$

where

$$\nu = -\frac{1}{2\pi} \log(1 - |\gamma(\lambda_0)|^2),$$
$$\chi_\pm(\lambda) = \frac{1}{2\pi i} \int_{\pm\lambda_0}^{0} \log \left(\frac{1 - |\gamma(\xi)|^2}{1 - |\gamma(\lambda_0)|^2} \right) \frac{d\xi}{\xi - \lambda},$$
$$\hat{\chi}_\pm(\lambda) = \frac{1}{2\pi i} \int_{i0}^{\pm i\infty} \log \left(1 - \gamma(\xi)\gamma^\dagger(\xi^*) \right) \frac{d\xi}{\xi - \lambda}.$$

By uniqueness, we have

$$\delta(\lambda)=(\delta^{\dagger}(\lambda^*))^{-1}, \quad \delta(\lambda)=\delta(-\lambda). \tag{3.4}$$

Inserting (3.4) in (3.1), we find

$$|\delta_+(\lambda)|^2=\begin{cases}2-\gamma(\lambda)\gamma^{\dagger}(\lambda^*), & \lambda\in(-\lambda_0,\lambda_0)\cup(-i\infty,+i\infty),\\ 2, & \lambda\in(-\infty,-\lambda_0)\cup(\lambda_0,+\infty),\end{cases}$$
$$|\delta_-(\lambda)|^2=\begin{cases}2+\frac{\gamma(\lambda)\gamma^{\dagger}(\lambda^*)}{1-\gamma(\lambda)\gamma^{\dagger}(\lambda^*)}, & \lambda\in(-\lambda_0,\lambda_0)\cup(-i\infty,+i\infty),\\ 2, & \lambda\in(-\infty,-\lambda_0)\cup(\lambda_0,+\infty).\end{cases}$$
$$|\det\delta_+(\lambda)|^2=\begin{cases}1-\gamma(\lambda)\gamma^{\dagger}(\lambda^*), & \lambda\in(-\lambda_0,\lambda_0)\cup(-i\infty,+i\infty),\\ 1, & \lambda\in(-\infty,-\lambda_0)\cup(\lambda_0,+\infty),\end{cases}$$
$$|\det\delta_-(\lambda)|^2=\begin{cases}\frac{1}{1-\gamma(\lambda)\gamma^{\dagger}(\lambda^*)}, & \lambda\in(-\lambda_0,\lambda_0)\cup(-i\infty,+i\infty),\\ 1, & \lambda\in(-\infty,-\lambda_0)\cup(\lambda_0,+\infty).\end{cases}$$

Therefore, by the maximum principle, we arrive at

$$|\delta(\lambda)|\leqslant \text{const}<\infty, \quad |\det\delta(\lambda)|\leqslant\text{const}<\infty, \tag{3.5}$$

for all $\lambda\in\mathbb{C}$. In addition, by the relation (3.4) this implies that we also have

$$|\delta^{-1}(\lambda)|\leqslant\text{const}<\infty, \quad |(\det\delta(\lambda))^{-1}|\leqslant\text{const}<\infty. \tag{3.6}$$

Set the vector spectral function

$$\rho(\lambda)=\begin{cases}\gamma(\lambda), & \lambda\in(-\infty,-\lambda_0]\cup[\lambda_0,+\infty),\\ -\dfrac{\gamma(\lambda)}{1-\gamma(\lambda)\gamma^{\dagger}(\lambda^*)}, & \lambda\in(-\lambda_0,\lambda_0)\cup(-i\infty,+i\infty),\end{cases} \tag{3.7}$$

and define

$$M^{\Delta}(\lambda;x,t)=M(\lambda;x,t)\Delta^{-1}(\lambda), \tag{3.8}$$

where

$$\Delta(\lambda)=\begin{pmatrix}\det\delta(\lambda) & 0\\ 0 & \delta^{-1}(\lambda)\end{pmatrix}.$$

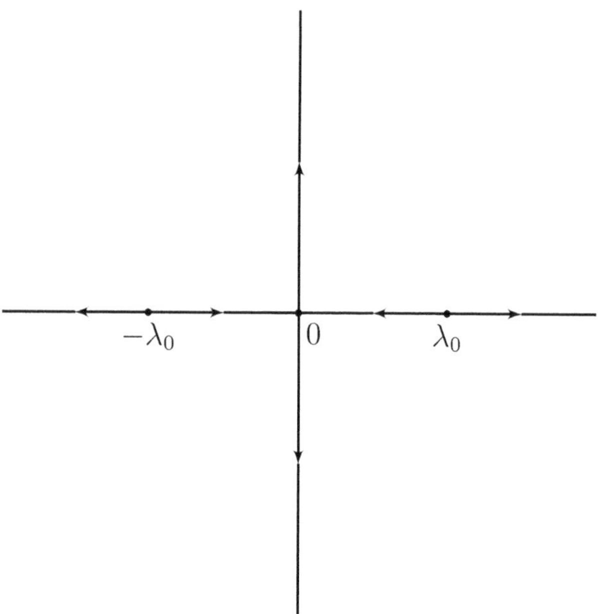

Figure 3: The reoriented contour

We reverse the direction for $\lambda \in (-\infty, -\lambda_0) \cup (\lambda_0, +\infty)$ as shown in Figure 3. Then, a direct calculation shows that RH problem (2.24) is transformed into a new RH problem

$$\begin{cases} M_+^{\Delta}(\lambda; x, t) = M_-^{\Delta}(\lambda; x, t) J^{\Delta}(\lambda; x, t), & \lambda \in \hat{\Gamma}, \\ M^{\Delta}(\lambda; x, t) \to I, & \lambda \to \infty, \end{cases} \tag{3.9}$$

where the jump matrix $J^{\Delta}(\lambda; x, t)$ has a lower/upper triangular decomposition

$$J^{\Delta}(\lambda; x, t) = (b_-)^{-1} b_+ = \begin{pmatrix} 1 & 0 \\ -\frac{e^{2it\theta}\delta_-^{-1}(\lambda)\rho^{\dagger}(\lambda^*)}{\det \delta_-(\lambda)} & I \end{pmatrix} \begin{pmatrix} 1 & e^{-2it\theta}\rho(\lambda)\delta_+(\lambda)[\det \delta_+(\lambda)] \\ 0 & I \end{pmatrix}. \tag{3.10}$$

3.2 Second transformation: augmented contour

The main purpose of this subsection is to reformulate the RH problem (3.9) as an equivalent RH problem on the augmented contour Σ in Figure 4. Set

$$\Sigma = \hat{\Gamma} \cup L \cup L^*,$$

where $L = L^0 \cup L^1$, and

$$L^0 = \{\lambda = h\lambda_0 e^{\frac{3\pi i}{4}} \,|\, -\infty < h < +\infty\};$$

$$L^1 = \{\lambda = \lambda_0 + h\lambda_0 e^{-\frac{3\pi i}{4}} \,|\, -\infty < h \leqslant \frac{\sqrt{2}}{2}\} \cup \{\lambda = -\lambda_0 + h\lambda_0 e^{\frac{\pi i}{4}} \,|\, -\infty < h \leqslant \frac{\sqrt{2}}{2}\}.$$

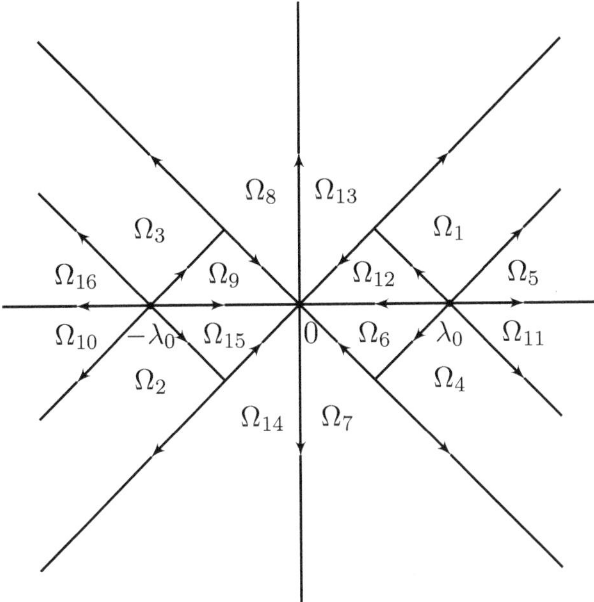

Figure 4: The oriented jump contour Σ

In addition, we define $L_\epsilon = L_\epsilon^0 \cup L_\epsilon^1$, where

$$L_\epsilon^0 = \{\lambda = h\lambda_0 e^{\frac{3\pi i}{4}} | h \in [-\frac{\sqrt{2}}{2}, -\epsilon) \cup (\epsilon, \frac{\sqrt{2}}{2}]\};$$

$$L_\epsilon^1 = \{\lambda = \lambda_0 + h\lambda_0 e^{-\frac{3\pi i}{4}} | \epsilon < h \leqslant \frac{\sqrt{2}}{2}\} \cup \{\lambda = -\lambda_0 + h\lambda_0 e^{\frac{\pi i}{4}} | \epsilon < h \leqslant \frac{\sqrt{2}}{2}\}.$$

Theorem 3.1. *There exists a decomposition of the vector-valued function $\rho(\lambda)$ in* (3.7)

$$\rho(\lambda) = f_1(\lambda) + f_2(\lambda) + R(\lambda), \quad \lambda \in \hat{\Gamma},$$

where $f_1(\lambda)$ is analytic on $\hat{\Gamma}$, $R(\lambda)$ is a piecewise-rational function and $f_2(\lambda)$ has an analytic continuation to L. What's more, in the case $\lambda_0 > C$, for an arbitrary positive integer l, they satisfy the following estimates

$$|e^{-2it\theta(\lambda)} f_1(\lambda)| \lesssim \frac{1}{(1+|\lambda|^2)(t\lambda_0^2)^l}, \quad \lambda \in \hat{\Gamma}, \tag{3.11}$$

$$|e^{-2it\theta(\lambda)} f_2(\lambda)| \lesssim \frac{1}{(1+|\lambda|^2)(t\lambda_0^2)^l}, \quad \lambda \in L, \tag{3.12}$$

$$|e^{-2it\theta(\lambda)} R(\lambda)| \lesssim e^{-2\epsilon^2\lambda_0^4 t}, \quad \lambda \in L_\epsilon, \tag{3.13}$$

where $\epsilon \in \mathbb{R}$ is sufficiently small. Taking the Hermitian conjugate

$$\rho^\dagger(\lambda^*) = R^\dagger(\lambda^*) + f_1^\dagger(\lambda^*) + f_2^\dagger(\lambda^*),$$

leads to the same estimates for $e^{2it\theta(\lambda)} f_1^\dagger(\lambda^*)$, $e^{2it\theta(\lambda)} f_2^\dagger(\lambda^*)$ and $e^{2it\theta(\lambda)} R^\dagger(\lambda^*)$ on $\hat{\Gamma} \cup L^*$.

Proof. The process of proof follows from proposition 4.2 in [23]. ■

Observe from (3.10) that, for $b_{\pm}$, we then have the further decomposed

$$\begin{aligned} b_+ &= b_+^o b_+^a = (I_{3\times3}+\omega_+^o)(I_{3\times3}+\omega_+^a) \\ &= \begin{pmatrix} 1 & e^{-2it\theta}[\det\delta_+(\lambda)]f_1(\lambda)\delta_+(\lambda) \\ 0 & I_{2\times2} \end{pmatrix}\begin{pmatrix} 1 & e^{-2it\theta}[\det\delta_+(\lambda)][f_2(\lambda)+R(\lambda)]\delta_+(\lambda) \\ 0 & I_{2\times2} \end{pmatrix}, \\ b_- &= b_-^o b_-^a = (I_{3\times3}-\omega_-^o)(I_{3\times3}-\omega_-^a) \\ &= \begin{pmatrix} 1 & 0 \\ \dfrac{e^{2it\theta}\delta_-^{-1}(\lambda)f_1^\dagger(\lambda^*)}{\det\delta_-(\lambda)} & I_{2\times2} \end{pmatrix}\begin{pmatrix} 1 & 0 \\ \dfrac{e^{2it\theta}\delta_-^{-1}(\lambda)[f_2^\dagger(\lambda^*)+R^\dagger(\lambda^*)]}{\det\delta_-(\lambda)} & I_{2\times2} \end{pmatrix}. \end{aligned}$$

Set

$$M^\sharp(\lambda;x,t) = \begin{cases} M^\Delta(\lambda;x,t), & \lambda\in\Omega_1\cup\Omega_2\cup\Omega_3\cup\Omega_4, \\ M^\Delta(\lambda;x,t)(b_-^a)^{-1}, & \lambda\in\Omega_{11}\cup\Omega_{12}\cup\Omega_{13}\cup\Omega_{14}\cup\Omega_{15}\cup\Omega_{16}, \\ M^\Delta(\lambda;x,t)(b_+^a)^{-1}, & \lambda\in\Omega_5\cup\Omega_6\cup\Omega_7\cup\Omega_8\cup\Omega_9\cup\Omega_{10}. \end{cases} \tag{3.14}$$

From RH problem (3.9) and formulae (3.14), we find by a simple computation that (3.9) is equivalent to the following RH problem

$$\begin{cases} M_+^\sharp(\lambda;x,t) = M_-^\sharp(\lambda;x,t)J^\sharp(\lambda;x,t), & \lambda\in\Sigma \\ M^\sharp(\lambda;x,t)\to I, & \lambda\to\infty, \end{cases} \tag{3.15}$$

where

$$J^\sharp(\lambda;x,t) = (b_-^\sharp)^{-1}b_+^\sharp = \begin{cases} I^{-1}b_+^a, & \lambda\in L, \\ (b_-^a)^{-1}I, & \lambda\in L^*, \\ (b_-^o)^{-1}b_+^o, & \lambda\in\hat\Gamma. \end{cases} \tag{3.16}$$

The canonical normalization condition of $M^\sharp(\lambda;x,t)$ can be derived, such as, $(b_+^a)^{-1}$ converges to I as $\lambda\to\infty$ in $\Omega_5\cup\Omega_7\cup\Omega_8\cup\Omega_{10}$. We consider first the domain Ω_5 for fixed x,t. By the boundedness of $\delta(\lambda)$ and $\det\delta(\lambda)$ in (3.5) and the definition of $R(\lambda)$, we have

$$|e^{-2it\theta}[\det\delta(\lambda)][f_2(\lambda)+R(\lambda)]\delta(\lambda)| \lesssim |e^{-2it\theta}f_2(\lambda)| + |e^{-2it\theta}R(\lambda)|, \quad \forall\lambda\in\Omega_5,$$

$$|e^{-2it\theta}R(\lambda)| \lesssim \frac{|\sum_{j=0}^m \mu_j(\lambda-\lambda_0)^j|}{|(\lambda+i)^{m+5}|} \lesssim \frac{1}{|\lambda+i|^5}, \quad \forall\lambda\in\Omega_5.$$

Using the convergence of $e^{-2it\theta}f_2(\lambda)$ in Theorem 3.1, we finally obtain that $M^\sharp(\lambda;x,t)$ $\to I$ as $\lambda\to\infty$ in Ω_5, and other cases are similar.

The above RH problem (3.15) can be solved as follows (see [12], P.322 and [4]). Set Cauchy operators $C_\pm$ on Σ by

$$(C_\pm f)(\lambda) = \int_\Sigma \frac{f(\xi)}{\xi-\lambda_\pm}\frac{d\xi}{2\pi i}, \quad \lambda\in\Sigma,\ f\in\mathscr{L}^2(\Sigma), \tag{3.17}$$

where $C_+ f$ ($C_- f$) denotes the left (right) boundary value for the oriented contour Σ in Figure 4. Define $C_{\omega^\sharp} : \mathscr{L}^2(\Sigma) + \mathscr{L}^\infty(\Sigma) \to \mathscr{L}^2(\Sigma)$ by

$$C_{\omega^\sharp} f = C_+\left(f\omega_-^\sharp\right) + C_-\left(f\omega_+^\sharp\right) \tag{3.18}$$

for a 3×3 matrix-valued function f. Let $\mu^\sharp(\lambda; x, t) \in \mathscr{L}^2(\Sigma) + \mathscr{L}^\infty(\Sigma)$ be the solution of the basic inverse equation

$$\mu^\sharp = I + C_{\omega^\sharp}\mu^\sharp,$$

where $\omega^\sharp = \omega_+^\sharp + \omega_-^\sharp$, $\omega_\pm^\sharp = \pm(b_\pm^\sharp - I)$. Then

$$M^\sharp(\lambda; x, t) = I + \int_\Sigma \frac{\mu^\sharp(\xi; x, t)\omega^\sharp(\xi; x, t)}{\xi - \lambda}\frac{d\xi}{2\pi i}, \quad \lambda \in \mathbb{C}\backslash\Sigma, \tag{3.19}$$

is the solution of the RH problem (3.15).

Theorem 3.2. *The solution $p(x,t)$ of the GCDNLS Equation* (1.1) *can be expressed by*

$$\tilde{p}(x,t) = \frac{1}{2\pi}\left(\int_\Sigma \left((1 - C_{\omega^\sharp})^{-1} I\right)(\xi)\omega^\sharp(\xi)\, d\xi\right)_{12}. \tag{3.20}$$

Proof. From Formula (2.29) and Definitions (3.8), (3.14) and (3.19), we learn that

$$\begin{aligned}
\tilde{p}(x,t) &= \lim_{\lambda\to\infty} -i\left(\lambda M^\sharp(\lambda; x.t)\right)_{12} \\
&= \frac{1}{2\pi}\left(\int_\Sigma \mu^\sharp(\xi; x, t)\omega^\sharp(\xi)\, d\xi\right)_{12} \\
&= \frac{1}{2\pi}\left(\int_\Sigma ((1 - C_{\omega^\sharp})^{-1} I)(\xi)\omega^\sharp(\xi)\, d\xi\right)_{12}.
\end{aligned} \tag{3.21}$$

■

3.3 Third transformation: truncated contour

In this subsection, we shall reduce the RH problem (3.15) on the contour Σ to a RH problem on the truncated contour Σ', where $\Sigma' = \Sigma\backslash(\hat{\Gamma} \cup L_\epsilon \cup L_\epsilon^*)$ as shown in Figure 5. After a complicated analysis, we estimate the errors between these two RH problems. Set $\omega^e = \omega^a + \omega^b + \omega^c$, we then have the following:

$$\begin{cases}
\omega^a = \omega^\sharp|_{\hat{\Gamma}} \text{ is supported on } \hat{\Gamma} \text{ and is composed of terms of type } f_1(\lambda)\, and\, f_1^\dagger(\lambda^*). \\
\omega^b = \omega^\sharp|_{L\cup L^*} \text{ is supported on } L \cup L^* \text{ and is composed of terms of type } f_2(\lambda)\, and\, f_2^\dagger(\lambda^*). \\
\omega^c = \omega^\sharp|_{L_\epsilon \cup L_\epsilon^*} \text{ is supported on } L_\epsilon \cup L_\epsilon^* \text{ and is composed of terms of type } R(\lambda)\, and\, R^\dagger(\lambda^*).
\end{cases}$$

We define $\omega^\sharp = \omega^e + \omega'$, and observe that ω'=0 on $\Sigma\backslash\Sigma'$. Hence, ω' is supported on Σ' with contribution to $\omega^\sharp$ from rational terms $R(\lambda)$ and $R^\dagger(\lambda^*)$.

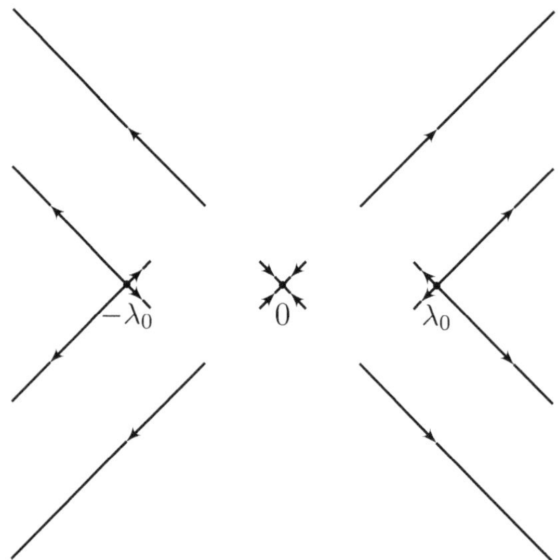

Figure 5: The truncated contour Σ'

Lemma 3.1. *For $\lambda_0 > C$, we have*

$$\|\omega^a\|_{\mathscr{L}^1(\hat{\Gamma})\cap\mathscr{L}^2(\hat{\Gamma})\cap\mathscr{L}^\infty(\hat{\Gamma})} \lesssim (t\lambda_0^2)^{-l}, \tag{3.22}$$

$$\|\omega^b\|_{\mathscr{L}^1(L\cup L^*)\cap\mathscr{L}^2(L\cup L^*)\cap\mathscr{L}^\infty(L\cup L^*)} \lesssim (t\lambda_0^2)^{-l}, \tag{3.23}$$

$$\|\omega^c\|_{\mathscr{L}^1(L_\epsilon\cup L_\epsilon^*)\cap\mathscr{L}^2(L_\epsilon\cup L_\epsilon^*)\cap\mathscr{L}^\infty(L_\epsilon\cup L_\epsilon^*)} \lesssim e^{-2\epsilon^2\lambda_0^4 t}, \tag{3.24}$$

$$\|\omega'\|_{\mathscr{L}^2(\Sigma)} \lesssim (t\lambda_0^2)^{-\frac{1}{4}}, \quad \|\omega'\|_{\mathscr{L}^1(\Sigma)} \lesssim (t\lambda_0^2)^{-\frac{1}{2}}. \tag{3.25}$$

Proof. By Theorem 3.1, the estimates (3.22), (3.23), and (3.24) are given. From the definition of $R(\lambda)$ on the contour $\{\lambda = \lambda_0 + \lambda_0 h e^{-\frac{3\pi i}{4}} | -\infty < h < \frac{\sqrt{2}}{2}\}$, we have

$$|R(\lambda)| \lesssim (1 + |\lambda|^5)^{-1}.$$

Noting that $\mathrm{Re}(i\theta) \geqslant \lambda_0^4 h^2$ on this contour, with the boundedness of $\delta(\lambda)$ and $\det\delta(\lambda)$ (3.5), we see that

$$|e^{-2it\theta}[\det\delta(\lambda)]R(\lambda)\delta(\lambda)| \lesssim e^{-2t\lambda_0^4 h^2}(1 + |\lambda|^5)^{-1}.$$

Then the estimate (3.25) holds by a direct computation. ■

Proposition 3.1. *As $t \to \infty$, for $\lambda_0 > C$, $(1 - C_{\omega'})^{-1} : \mathscr{L}^2(\Sigma) \to \mathscr{L}^2(\Sigma)$ exists and is uniformly bounded:*

$$\|(1 - C_{\omega'})^{-1}\|_{\mathscr{L}^2(\Sigma)} \lesssim 1.$$

Furthermore, $\|(1 - C_{\omega^\sharp})^{-1}\|_{\mathscr{L}^2(\Sigma)} \lesssim 1$.

Proof. The result follows from proposition 2.23 and corollary 2.25 in [12]. ■

Theorem 3.3. *As $t \to \infty$, then*

$$\int_{\Sigma} ((1-C_{\omega^\sharp})^{-1}I)(\xi)\omega^\sharp(\xi)\,\mathrm{d}\xi = \int_{\Sigma} ((1-C_{\omega'})^{-1}I)(\xi)\omega'(\xi)\,\mathrm{d}\xi + O((t\lambda_0^2)^{-l}). \quad (3.26)$$

Proof. Using the second resolvent identity, we find that

$$\begin{aligned}((1-C_{\omega^\sharp})^{-1}I)\omega^\sharp &= ((1-C_{\omega'})^{-1}I)\,\omega' + \omega^e + ((1-C_{\omega'})^{-1}(C_{\omega^e}I))\,\omega^\sharp \\ &\quad + ((1-C_{\omega'})^{-1}(C_{\omega'}I))\omega^e + ((1-C_{\omega'})^{-1}C_{\omega^e}(1-C_{\omega^\sharp})^{-1})\,(C_{\omega^\sharp}I)\omega^\sharp,\end{aligned} \quad (3.27)$$

then from Lemma 3.1 and Proposition 3.1, we have the following results

$$\begin{aligned}&\|\omega^e\|_{\mathscr{L}^1(\Sigma)} \leqslant \|\omega^a\|_{\mathscr{L}^1(\hat{\Gamma})} + \|\omega^b\|_{\mathscr{L}^1(L\cup L^*)} + \|\omega^c\|_{\mathscr{L}^1(L_\epsilon\cup L_\epsilon^*)} \lesssim (t\lambda_0^2)^{-l}, \\ &\|\,((1-C_{\omega'})^{-1}(C_{\omega^e}I))\,\omega^\sharp\|_{\mathscr{L}^1(\Sigma)} \leqslant \|(1-C_{\omega'})^{-1}\|_{\mathscr{L}^2(\Sigma)}\|C_{\omega^e}I\|_{\mathscr{L}^2(\Sigma)}\|\omega^\sharp\|_{\mathscr{L}^2(\Sigma)} \\ &\qquad\qquad \lesssim \|\omega^e\|_{\mathscr{L}^2(\Sigma)}\|\omega^\sharp\|_{\mathscr{L}^2(\Sigma)} \lesssim (t\lambda_0^2)^{-l-\frac{1}{4}}, \\ &\|\,((1-C_{\omega'})^{-1}(C_{\omega'}I))\,\omega^e\|_{\mathscr{L}^1(\Sigma)} \leqslant \|(1-C_{\omega'})^{-1}\|_{\mathscr{L}^2(\Sigma)}\|C_{\omega'}I\|_{\mathscr{L}^2(\Sigma)}\|\omega^e\|_{\mathscr{L}^2(\Sigma)} \\ &\qquad\qquad \lesssim \|\omega'\|_{\mathscr{L}^2(\Sigma)}\|\omega^e\|_{\mathscr{L}^2(\Sigma)} \lesssim (t\lambda_0^2)^{-l-\frac{1}{4}}, \\ &\|\,((1-C_{\omega'})^{-1}C_{\omega^e}(1-C_{\omega^\sharp})^{-1})\,(C_{\omega^\sharp}I)\omega^\sharp\|_{\mathscr{L}^1(\Sigma)} \\ &\leqslant \|(1-C_{\omega'})^{-1}\|_{\mathscr{L}^2(\Sigma)}\|(1-C_{\omega^\sharp})^{-1}\|_{\mathscr{L}^2(\Sigma)}\|C_{\omega^e}\|_{\mathscr{L}^2(\Sigma)}\|C_{\omega^\sharp}I\|_{\mathscr{L}^2(\Sigma)}\|\omega^\sharp\|_{\mathscr{L}^2(\Sigma)} \\ &\lesssim \|\omega^e\|_{\mathscr{L}^\infty(\Sigma)}\|\omega^\sharp\|^2_{\mathscr{L}^2(\Sigma)} \lesssim (t\lambda_0^2)^{-l-\frac{1}{2}}.\end{aligned}$$

Inserting the above estimates into (3.27), we obtain (3.26). ■

Note. Noticing that for $\lambda \in \Sigma\backslash\Sigma'$, $\omega'(\lambda) = 0$, we set $C_{\omega'}|_{\mathscr{L}^2(\Sigma')}$ denote the restriction of $C_{\omega'}$ to $\mathscr{L}^2(\Sigma')$. For simplicity, we write $C_{\omega'}|_{\mathscr{L}^2(\Sigma')}$ as $C_{\omega'}$. Thus we have

$$\int_{\Sigma} ((1-C_{\omega'})^{-1}I)(\xi)\omega'(\xi)\,\mathrm{d}\xi = \int_{\Sigma'} ((1-C_{\omega'})^{-1}I)(\xi)\omega'(\xi)\,\mathrm{d}\xi.$$

Lemma 3.2. *As $t \to \infty$, then*

$$\tilde{p}(x,t) = \frac{1}{2\pi}\left(\int_{\Sigma'} ((1-C_{\omega'})^{-1}I)(\xi)\omega'(\xi)\,\mathrm{d}\xi\right)_{12} + O((t\lambda_0^2)^{-l}), \quad t \to \infty. \quad (3.28)$$

Proof. Combining with (3.20) and (3.26), one can derive the asymptotic estimate directly. ■

Next, we construct the RH problem on Σ'. Set $L' = L\backslash L_\epsilon$ and $\mu' = (1-C_{\omega'})^{-1}I$. Again utilizing the theory in [4], we learn that

$$M'(\lambda; x,t) = I + \int_{\Sigma'} \frac{\mu'(\xi;x,t)\omega'(\xi;x,t)}{\xi-\lambda}\frac{\mathrm{d}\xi}{2\pi i}$$

solves the RH problem

$$\begin{cases} M'_+(\lambda;x,t) = M'_-(\lambda;x,t)J'(\lambda;x,t), & \lambda \in \Sigma', \\ M'(\lambda;x,t) \to I, & \lambda \to \infty, \end{cases} \tag{3.29}$$

where

$$\begin{aligned} &J' = (b'_-)^{-1}b'_+ = (I - \omega'_-)^{-1}(I + \omega'_+), \\ &\omega' = \omega'_+ + \omega'_-, \\ &b'_+ = \begin{pmatrix} 1 & e^{-2it\theta}[\det\delta(\lambda)]R(\lambda)\delta(\lambda) \\ 0 & I_{2\times 2} \end{pmatrix}, \quad b'_- = I_{3\times 3}, \quad \text{on } L', \\ &b'_+ = I_{3\times 3}, \quad b'_- = \begin{pmatrix} 1 & 0 \\ \dfrac{e^{2it\theta}\delta^{-1}(\lambda)R^\dagger(\lambda^*)}{\det\delta(\lambda)} & I_{2\times 2} \end{pmatrix}, \quad \text{on } (L')^*. \end{aligned}$$

3.4 Fourth transformation: three disjoint crosses

In this subsection, we will further simplify the RH problem (3.29) on the truncated contour Σ'. To this end, we separate Σ' into three disjoint crosses defined by

$$\begin{aligned} \Sigma'_A =& \{\lambda = -\lambda_0 + h\lambda_0 e^{\frac{\pi i}{4}} \mid -\infty < h \leqslant \epsilon\} \\ &\cup \{\lambda = -\lambda_0 + h\lambda_0 e^{-\frac{\pi i}{4}} \mid -\infty < h \leqslant \epsilon\}, \\ \Sigma'_B =& \{\lambda = \lambda_0 + h\lambda_0 e^{-\frac{3\pi i}{4}} \mid -\infty < h \leqslant \epsilon\} \\ &\cup \{\lambda = \lambda_0 + h\lambda_0 e^{\frac{3\pi i}{4}} \mid -\infty < h \leqslant \epsilon\}, \\ \Sigma'_C =& \Sigma' - \Sigma'_A - \Sigma'_B. \end{aligned}$$

Set

$$\omega'_\pm = \omega'_{A\pm} + \omega'_{B\pm} + \omega'_{C\pm},$$

where

$$\begin{cases} \omega'_{A\pm}(\lambda) = 0, & \lambda \in \Sigma' \backslash \Sigma'_A; \\ \omega'_{B\pm}(\lambda) = 0, & \lambda \in \Sigma' \backslash \Sigma'_B; \\ \omega'_{C\pm}(\lambda) = 0, & \lambda \in \Sigma' \backslash \Sigma'_C. \end{cases}$$

Lemma 3.3. *Define the operators $C_{\omega'_A}$, $C_{\omega'_B}$ and $C_{\omega'_C}$: $\mathscr{L}^2(\Sigma') + \mathscr{L}^\infty(\Sigma') \to \mathscr{L}^2(\Sigma')$ just as (3.18). Thus, $C_{\omega'} = C_{\omega'_A} + C_{\omega'_B} + C_{\omega'_C}$, and for $\alpha \neq \beta \in \{A, B, C\}$, we have*

$$\|C_{\omega'_\alpha} C_{\omega'_\beta}\|_{\mathscr{L}^2(\Sigma')} \lesssim_{\lambda_0} (t\lambda_0^2)^{-\frac{1}{2}}, \quad \|C_{\omega'_\alpha} C_{\omega'_\beta}\|_{\mathscr{L}^\infty(\Sigma') \to \mathscr{L}^2(\Sigma')} \lesssim_{\lambda_0} (t\lambda_0^2)^{-\frac{3}{4}}, \quad t \to \infty. \tag{3.30}$$

Proof. Similarly to Lemma 3.5 in [12], we can obtain the estimate. ■

Theorem 3.4. *As $t \to \infty$, then*

$$\begin{aligned}\tilde{p}(x,t) =&\frac{1}{2\pi}\left(\int_{\Sigma'}(\omega'_A+\omega'_B+\omega'_C+\sum_{\alpha,\beta\in\{A,B,C\}}C_{\omega'_\alpha}(1-C_{\omega'_\alpha})^{-1}\omega'_\beta)\,\mathrm{d}\xi\right)_{12}+O\left(\frac{1}{t\lambda_0^2}\right)\\ =&\frac{1}{2\pi}\left(\int_{\Sigma'_A}(1-C_{\omega'_A})^{-1}I(\xi)\omega'_A(\xi)\,\mathrm{d}\xi+\int_{\Sigma'_B}(1-C_{\omega'_B})^{-1}I(\xi)\omega'_B(\xi)\,\mathrm{d}\xi\right.\\ &\left.+\int_{\Sigma'_C}(1-C_{\omega'_C})^{-1}I(\xi)\omega'_C(\xi)\,\mathrm{d}\xi\right)_{12}+O\left(\frac{1}{t\lambda_0^2}\right).\end{aligned} \tag{3.31}$$

Proof. From identity

$$(1-C_{\omega'_A}-C_{\omega'_B}-C_{\omega'_C})D_{\Sigma'}=1-E_{\Sigma'}, \tag{3.32}$$

where

$$\begin{aligned}D_{\Sigma'}&=1+C_{\omega'_A}(1-C_{\omega'_A})^{-1}+C_{\omega'_B}(1-C_{\omega'_B})^{-1}+C_{\omega'_C}(1-C_{\omega'_C})^{-1},\\ E_{\Sigma'}&=\sum_{\alpha,\beta\in\{A,B,C\}}(1-\delta_{\alpha\beta})C_{\omega'_\alpha}C_{\omega'_\beta}(1-C_{\omega'_\beta})^{-1},\end{aligned}$$

and $\delta_{\alpha\beta}$ denote the Kronecker delta, one can derive that

$$(1-C_{\omega'_A}-C_{\omega'_B}-C_{\omega'_C})^{-1}=D_{\Sigma'}+D_{\Sigma'}(1-E_{\Sigma'})^{-1}E_{\Sigma'}. \tag{3.33}$$

Noticing that $\|C_{\omega'_\alpha}\|_{\mathscr{L}^2(\Sigma')}\lesssim\|\omega'_\alpha\|_{\mathscr{L}^\infty(\Sigma')}\lesssim 1$, by triangular inequality we have

$$\begin{aligned}\|D_{\Sigma'}\|_{\mathscr{L}^2(\Sigma')}&\lesssim 1,\\ \|E_{\Sigma'}I\|_{\mathscr{L}^2(\Sigma')}\leqslant&\|\sum_{\alpha,\beta\in\{A,B,C\}}(1-\delta_{\alpha\beta})C_{\omega'_\alpha}C_{\omega'_\beta}I\|_{\mathscr{L}^2(\Sigma')}\\ &+\|\sum_{\alpha,\beta\in\{A,B,C\}}(1-\delta_{\alpha\beta})C_{\omega'_\alpha}C_{\omega'_\beta}(1-C_{\omega'_\beta})^{-1}C_{\omega'_\beta}I\|_{\mathscr{L}^2(\Sigma')}\\ \lesssim_{\lambda_0}&\frac{1}{(t\lambda_0^2)^{3/4}}.\end{aligned}$$

From the Cauchy-Schwarz inequality and Lemma 3.1, we have

$$\|E_{\Sigma'}I(\omega'_A+\omega'_B+\omega'_C)\|_{\mathscr{L}^1(\Sigma')}\leqslant\|E_{\Sigma'}I\|_{\mathscr{L}^2(\Sigma')}\|\omega'\|_{\mathscr{L}^2(\Sigma')}\lesssim_{\lambda_0}(t\lambda_0^2)^{-1},$$

and

$$D_{\Sigma'}I(\omega'_A+\omega'_B+\omega'_C)=\omega'_A+\omega'_B+\omega'_C+\sum_{\alpha,\beta\in\{A,B,C\}}C_{\omega'_\alpha}(1-C_{\omega'_\alpha})^{-1}I\omega'_\beta,$$

To estimate $C_{\omega'_\alpha}(1-C_{\omega'_\alpha})^{-1}I\omega'_\beta$, where $\alpha\neq\beta$ and $\alpha,\beta\in\{A,B,C\}$, we first consider the case of $\alpha=A$ and $\beta=B$,

$$\begin{aligned}&\left|\int_{\Sigma'}\left(C_{\omega'_A}(1-C_{\omega'_A})^{-1}I\right)(\xi)\omega'_B(\xi)\,\mathrm{d}\xi\right|\\ \leqslant&\left|\int_{\Sigma'}\left(C_{\omega'_A}I\right)(\xi)\omega'_B(\xi)\,\mathrm{d}\xi\right|+\left|\int_{\Sigma'}\left(C_{\omega'_A}(1-C_{\omega'_A})^{-1}C_{\omega'_A}I\right)(\xi)\omega'_B(\xi)\,\mathrm{d}\xi\right|.\end{aligned}$$

Observe that $\text{dist}(\Sigma'_A, \Sigma'_B) \geqslant c\lambda_0, c > 0$, evaluating the first and the second terms in the above expression yields

$$\left|\int_{\Sigma'} \left(C_{\omega'_A} I\right)(\xi)\omega'_B(\xi)\,\mathrm{d}\xi\right| = \left|\int_{\Sigma'_B} \left(\int_{\Sigma'_A} \frac{\omega'_A(\eta)}{\eta-\xi}\frac{\mathrm{d}\eta}{2\pi i}\right)\omega'_B(\xi)\,\mathrm{d}\xi\right|$$
$$\lesssim \frac{1}{\lambda_0}\|\omega'_A\|_{\mathscr{L}^1(\Sigma'_A)}\|\omega'_B\|_{\mathscr{L}^1(\Sigma'_B)} \lesssim_{\lambda_0} (t\lambda_0^2)^{-1},$$

and

$$\left|\int_{\Sigma'} \left(C_{\omega'_A}(1-C_{\omega'_A})^{-1}C_{\omega'_A} I\right)(\xi)\omega'_B(\xi)\,\mathrm{d}\xi\right|$$
$$= \left|\int_{\Sigma'_B} \left(\int_{\Sigma'_A} \frac{\left((1-C_{\omega'_A})^{-1}C_{\omega'_A} I\right)(\eta)\omega'_A(\eta)}{\eta-\xi}\frac{\mathrm{d}\eta}{2\pi i}\right)\omega'_B(\xi)\,\mathrm{d}\xi\right|$$
$$\lesssim \frac{1}{\lambda_0}\|(1-C_{\omega'_A})^{-1}C_{\omega'_A} I\|_{\mathscr{L}^2(\Sigma'_A)}\|\omega'_A\|_{\mathscr{L}^2(\Sigma'_A)}\|\omega'_B\|_{\mathscr{L}^1(\Sigma'_B)}$$
$$\lesssim \frac{1}{\lambda_0}\|\omega'_A\|^2_{\mathscr{L}^2(\Sigma'_A)}\|\omega'_B\|_{\mathscr{L}^1(\Sigma'_B)} \lesssim_{\lambda_0} (t\lambda_0^2)^{-1}.$$

The other cases can be similarly proved. Combining with Lemma 3.2, this proof is completed.

■

3.5 Rescaling and further reduction of RH problems

From (3.31), we find that $\tilde{p}(x,t)$ is composed of the integrals on the contours Σ'_A, Σ'_B and Σ'_C, respectively. In this subsection, we first extend the above contours and define three scaling operators, then we have the RH problems on the crosses. Extend the contours Σ'_A, Σ'_B and Σ'_C to the following contours

$$\hat{\Sigma}'_A = \{-\lambda_0 + h\lambda_0 e^{\pm\frac{\pi i}{4}} | h \in \mathbb{R}\},$$
$$\hat{\Sigma}'_B = \{\lambda_0 + h\lambda_0 e^{\pm\frac{\pi i}{4}} | h \in \mathbb{R}\},$$
$$\hat{\Sigma}'_C = \{h\lambda_0 e^{\pm\frac{\pi i}{4}} | h \in \mathbb{R}\},$$

and define $\hat{\omega}'_{A\pm}$, $\hat{\omega}'_{B\pm}$ and $\hat{\omega}'_{C\pm}$ on $\hat{\Sigma}'_A$, $\hat{\Sigma}'_B$ and $\hat{\Sigma}'_C$, respectively, by

$$\hat{\omega}'_{A\pm} = \begin{cases} \omega'_{A\pm}, & \lambda \in \Sigma'_A, \\ 0, & \lambda \in \hat{\Sigma}'_A \backslash \Sigma'_A, \end{cases} \quad \hat{\omega}'_{B\pm} = \begin{cases} \omega'_{B\pm}, & \lambda \in \Sigma'_B, \\ 0, & \lambda \in \hat{\Sigma}'_B \backslash \Sigma'_B, \end{cases} \quad \hat{\omega}'_{C\pm} = \begin{cases} \omega'_{C\pm}, & \lambda \in \Sigma'_C, \\ 0, & \lambda \in \hat{\Sigma}'_C \backslash \Sigma'_C. \end{cases}$$

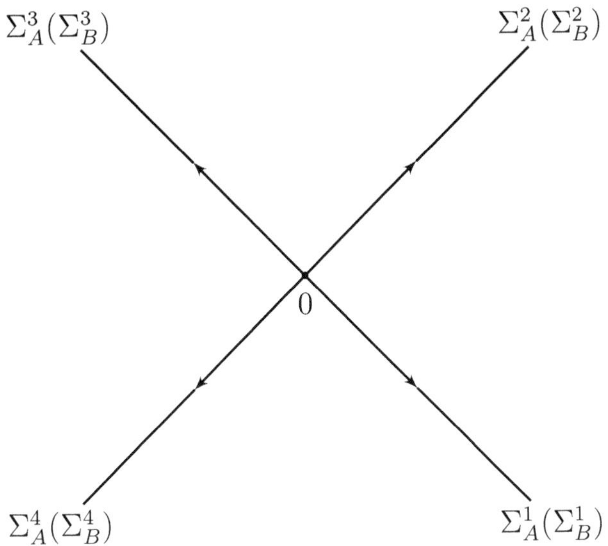

Figure 6: The oriented contour Σ_A or Σ_B

Let Σ_A, Σ_B in Figure 6 and Σ_C denote the contours $\{\lambda = h\lambda_0 e^{\pm\frac{\pi i}{4}} | h \in \mathbb{R}\}$ oriented outward as in Σ'_A, $\hat{\Sigma}'_A$, outward as in Σ'_B, $\hat{\Sigma}'_B$, and outward/inward as in Σ'_C, $\hat{\Sigma}'_C$. We introduce the scaling operators

$$\begin{aligned} N_A :& \mathscr{L}^2(\hat{\Sigma}'_A) \to \mathscr{L}^2(\Sigma_A) \\ & f(\lambda) \mapsto (N_A f)(\lambda) = f\left(-\lambda_0 + \frac{\lambda}{\sqrt{8t\lambda_0^2}}\right); \end{aligned} \tag{3.34}$$

$$\begin{aligned} N_B :& \mathscr{L}^2(\hat{\Sigma}'_B) \to \mathscr{L}^2(\Sigma_B) \\ & f(\lambda) \mapsto (N_B f)(\lambda) = f\left(\lambda_0 + \frac{\lambda}{\sqrt{8t\lambda_0^2}}\right); \end{aligned} \tag{3.35}$$

$$\begin{aligned} N_C :& \mathscr{L}^2(\hat{\Sigma}'_C) \to \mathscr{L}^2(\Sigma_C) \\ & f(\lambda) \mapsto (N_C f)(\lambda) = f\left(\frac{\lambda}{\sqrt{4t\lambda_0^2}}\right). \end{aligned} \tag{3.36}$$

When the scaling operators act on the exponential term and $\det\delta(\lambda)$, we have

$$N_k(e^{-it\theta}\det\delta(\lambda)) = \delta_k^0\delta_k^1(\lambda), \quad k \in \{A, B, C\}, \tag{3.37}$$

where δ_k^0 is independent of λ,

$$\delta_A^0 = (8t\lambda_0^4)^{\frac{i\nu}{2}} \exp\{-\frac{1}{2}it\lambda_0^4 + \chi_+(-\lambda_0) + \chi_-(-\lambda_0) + \hat{\chi}_+(-\lambda_0) + \hat{\chi}_-(-\lambda_0)\}, \tag{3.38}$$

$$\delta_B^0 = (8t\lambda_0^4)^{\frac{i\nu}{2}} \exp\{-\frac{1}{2}it\lambda_0^4 + \chi_+(\lambda_0) + \chi_-(\lambda_0) + \hat{\chi}_+(\lambda_0) + \hat{\chi}_-(\lambda_0)\}, \tag{3.39}$$

$$\delta_C^0 = \exp\{\chi_+(0) + \chi_-(0) + \hat{\chi}_+(0) + \hat{\chi}_-(0)\}, \tag{3.40}$$

$$\chi_\pm(\lambda_l) = \frac{1}{2\pi i}\int_{\pm\lambda_0}^{0} \log\left(\frac{1-|\gamma(\xi)|^2}{1-|\gamma(\lambda_0)|^2}\right)\frac{\mathrm{d}\xi}{\xi-\lambda_l},$$

$$\hat{\chi}_\pm(\lambda_l) = \frac{i}{2\pi}\int_0^{\pm\infty} \log\left(1+|\gamma(i\xi)|^2\right)\,\mathrm{d}\log(i\xi-\lambda_l),$$

$$\chi_\pm(0) = \frac{1}{2\pi i}\int_{\pm\lambda_0}^{0} \log\left(\frac{1-|\gamma(\xi)|^2}{1-|\gamma(\lambda_0)|^2}\right)\mathrm{d}\log|\xi|,$$

$$\hat{\chi}_\pm(0) = -\frac{1}{2\pi i}\int_0^{\pm\infty} \log\left(1+|\gamma(i\xi)|^2\right)\frac{\mathrm{d}\xi}{\xi},$$

$$l \in \{1,2\}, \quad \lambda_1 = \lambda_0, \quad \lambda_2 = -\lambda_0.$$

Let

$$L_A = \{\sqrt{8t\lambda_0^2}h\lambda_0 e^{\frac{\pi i}{4}} \,|\, -\infty < h < \epsilon\},$$

$$L_B = \{\sqrt{8t\lambda_0^2}h\lambda_0 e^{-\frac{3\pi i}{4}} \,|\, -\infty < h < \epsilon\},$$

$$L_C = \{\sqrt{4t\lambda_0^2}h\lambda_0 e^{-\frac{\pi i}{4}} \,|\, -\infty < h < +\infty\}.$$

Lemma 3.4. *For $\lambda \in L_A^*, L_B^*$ and L_C^*, then*

$$|N_A(R^\dagger(\lambda^*))(\delta_A^1)^{-2} - R^\dagger(-\lambda_0\pm)(-2\lambda)^{2i\nu}e^{-\frac{1}{2}i\lambda^2}| \lesssim_{\lambda_0} \frac{\log t}{\sqrt{t\lambda_0^2}}, \quad \lambda \in L_A^*, \tag{3.41}$$

$$|N_B(R^\dagger(\lambda^*))(\delta_B^1)^{-2} - R^\dagger(\lambda_0\pm)(2\lambda)^{2i\nu}e^{-\frac{1}{2}i\lambda^2}| \lesssim_{\lambda_0} \frac{\log t}{\sqrt{t\lambda_0^2}}, \quad \lambda \in L_B^*, \tag{3.42}$$

$$|N_C(R^\dagger(\lambda^*))(\delta_C^1)^{-2} - R^\dagger(0\pm)(-\lambda_0^2)^{2i\nu}e^{\frac{1}{2}i\lambda^2}| \lesssim_{\lambda_0} \frac{\log t}{\sqrt{t\lambda_0^2}}, \quad \lambda \in L_C^*, \tag{3.43}$$

where

$$R(-\lambda_0-) = \lim_{\mathrm{Re}\lambda<-\lambda_0} R(\lambda) = \gamma(-\lambda_0), \quad R(-\lambda_0+) = \lim_{\mathrm{Re}\lambda>-\lambda_0} R(\lambda) = -\frac{\gamma(-\lambda_0)}{1-|\gamma(-\lambda_0)|^2},$$

$$R(\lambda_0-) = \lim_{\mathrm{Re}\lambda<\lambda_0} R(\lambda) = -\frac{\gamma(\lambda_0)}{1-|\gamma(\lambda_0)|^2}, \quad R(\lambda_0+) = \lim_{\mathrm{Re}\lambda>\lambda_0} R(\lambda) = \gamma(\lambda_0).$$

Proof. See [12] and [23], the process of proof and results are similar. ■

Note: When $\lambda \in L_A$, L_B and L_C, we have the estimates

$$|N_A(R(\lambda))(\delta_A^1)^2 - R(-\lambda_0\pm)(-2\lambda)^{-2i\nu}e^{\frac{1}{2}i\lambda^2}| \lesssim_{\lambda_0} \frac{\log t}{\sqrt{t\lambda_0^2}}, \quad \lambda \in L_A,$$

$$|N_B(R(\lambda))(\delta_B^1)^2 - R(\lambda_0\pm)(2\lambda)^{-2i\nu}e^{\frac{1}{2}i\lambda^2}| \lesssim_{\lambda_0} \frac{\log t}{\sqrt{t\lambda_0^2}}, \quad \lambda \in L_B,$$

$$|N_C(R(\lambda))(\delta_C^1)^2 - R(0\pm)(-\lambda_0^2)^{-2i\nu}e^{-\frac{1}{2}i\lambda^2}| \lesssim_{\lambda_0} \frac{\log t}{\sqrt{t\lambda_0^2}}, \quad \lambda \in L_C.$$

Set

$$\omega_A = (\delta_A^0)^{\sigma}N_A\hat{\omega}'_A(\delta_A^0)^{-\sigma}, \quad \omega_B = (\delta_B^0)^{\sigma}N_B\hat{\omega}'_B(\delta_B^0)^{-\sigma}, \quad \omega_C = (\delta_C^0)^{\sigma}N_C\hat{\omega}'_C(\delta_C^0)^{-\sigma}. \tag{3.44}$$

Define $\widetilde{\Delta}_A$, $\widetilde{\Delta}_B$ and $\widetilde{\Delta}_C$ when they act on function f

$$\widetilde{\Delta}_A f = f(\delta_A^0)^{-\sigma}, \quad \widetilde{\Delta}_B f = f(\delta_B^0)^{-\sigma}, \quad \widetilde{\Delta}_C f = f(\delta_C^0)^{-\sigma}. \tag{3.45}$$

A simple calculation shows that

$$\begin{aligned} C_{\hat{\omega}'_A} &= N_A^{-1}(\widetilde{\Delta}_A)^{-1}C_{\omega_A}\widetilde{\Delta}_A N_A, \\ C_{\hat{\omega}'_B} &= N_B^{-1}(\widetilde{\Delta}_B)^{-1}C_{\omega_B}\widetilde{\Delta}_B N_B, \\ C_{\hat{\omega}'_C} &= N_C^{-1}(\widetilde{\Delta}_C)^{-1}C_{\omega_C}\widetilde{\Delta}_C N_C, \end{aligned} \tag{3.46}$$

where $C_{\omega_A}(C_{\omega_B}, C_{\omega_C}) : \mathscr{L}^2(\Sigma_A)(\mathscr{L}^2(\Sigma_B), \mathscr{L}^2(\Sigma_C)) \to \mathscr{L}^2(\Sigma_A)(\mathscr{L}^2(\Sigma_B), \mathscr{L}^2(\Sigma_C))$ are all bounded. So we have

$$\omega_A = \omega_{A+} = \begin{pmatrix} 0 & (\delta_A^0)^{-2}(N_A s_1)(\lambda) \\ 0 & 0 \end{pmatrix} \tag{3.47}$$

on L_A and

$$\omega_A = \omega_{A-} = \begin{pmatrix} 0 & 0 \\ (\delta_A^0)^{2}(N_A s_2)(\lambda) & 0 \end{pmatrix} \tag{3.48}$$

on L_A^*, where

$$s_1 = e^{-2it\theta}[\det\delta(\lambda)]R(\lambda)\delta(\lambda), \quad s_2 = -\frac{e^{2it\theta}\delta^{-1}(\lambda)R^{\dagger}(\lambda^*)}{\det\delta(\lambda)}. \tag{3.49}$$

Lemma 3.5. *As $t \to \infty$ and $\lambda \in L_A$, for an arbitrary positive integer l,*

$$|(N_A\tilde{\delta})(\lambda)| \lesssim t^{-l}, \tag{3.50}$$

where

$$\tilde{\delta}(\lambda) = e^{-2it\theta}[R(\lambda)\delta(\lambda) - (\det\delta(\lambda))R(\lambda)].$$

Proof. Because $\delta(\lambda)$ and $\det \delta(\lambda)$ satisfy the RH Problems (3.1) and (3.2), we infer that $\tilde{\delta}(\lambda)$ satisfies

$$\begin{cases} \tilde{\delta}_+(\lambda) = \tilde{\delta}_-(\lambda)(1-|\gamma(\lambda)|^2) + e^{-2it\theta}T(\lambda), & \lambda \in (-\lambda_0, 0), \\ \tilde{\delta}(\lambda) \to 0, & \lambda \to \infty, \end{cases} \tag{3.51}$$

where $T(\lambda) = R(\lambda)(|\gamma|^2 I - \gamma^\dagger\gamma)\delta_-(\lambda)$. This RH problem has a solution as below

$$\tilde{\delta}(\lambda) = X(\lambda)\int_{-\lambda_0}^{0} \frac{e^{-2it\theta(\xi)}T(\xi)}{X_+(\xi)(\xi-\lambda)}\frac{d\xi}{2\pi i},$$

$$X(\lambda) = \exp\{\frac{1}{2\pi i}\int_{-\lambda_0}^{0} \frac{\log(1-|\gamma(\xi)|^2)}{\xi-\lambda}d\xi\}.$$

We deduce that $\rho|\gamma|^2 - \rho\gamma^\dagger\gamma = 0$ because of the definition of $\rho(\lambda)$ in (3.7). Therefore, we have

$$\begin{aligned} R(|\gamma|^2 I - \gamma^\dagger\gamma) &= (R-\rho)|\gamma|^2 - (R-\rho)\gamma^\dagger\gamma \\ &= (f_1+f_2)\begin{pmatrix} 0 & 1 \\ -1 & 0 \end{pmatrix}(\gamma^\dagger\gamma)^T\begin{pmatrix} 0 & 1 \\ -1 & 0 \end{pmatrix}. \end{aligned}$$

Define L_t in Figure 7 by

$$\begin{aligned} L_t = &\left\{\lambda = h\lambda_0 e^{\frac{3\pi i}{4}} : 0 \leqslant h \leqslant \frac{\sqrt{2}}{2}(1-\frac{1}{t})\right\} \\ &\cup \left\{\lambda = \frac{\lambda_0}{t} - \lambda_0 + h\lambda_0 e^{\frac{\pi i}{4}} : 0 \leqslant h \leqslant \frac{\sqrt{2}}{2}(1-\frac{1}{t})\right\}. \end{aligned}$$

Similar to Theorem 3.1, $T(\lambda)$ has a decomposition: $T(\lambda) = T_1(\lambda) + T_2(\lambda)$, where $T_2(\lambda)$ has an analytic continuation to L_t. Moreover, for $l \geqslant 2$, $T_1(\lambda)$ and $T_2(\lambda)$ satisfy

$$|e^{-2it\theta}T_1(\lambda)| \lesssim \frac{1}{(1+|\lambda|^2)(t\lambda_0^2)^l}, \quad \lambda \in \mathbb{R}, \tag{3.52}$$

$$|e^{-2it\theta}T_2(\lambda)| \lesssim \frac{1}{(1+|\lambda|^2)(t\lambda_0^2)^l}, \quad \lambda \in L_t. \tag{3.53}$$

When $\lambda \in L_A$, a direct calculation shows that

$$\begin{aligned} (N_A\tilde{\delta})(\lambda) =& X(\frac{\lambda}{\sqrt{8t\lambda_0^2}} - \lambda_0)\int_{-\lambda_0}^{\frac{\lambda_0}{t}-\lambda_0} \frac{e^{-2it\theta(\xi)}T(\xi)}{X_+(\xi)(\xi+\lambda_0-\frac{\lambda}{\sqrt{8t\lambda_0^2}})}\frac{d\xi}{2\pi i} \\ &+ X(\frac{\lambda}{\sqrt{8t\lambda_0^2}} - \lambda_0)\int_{\frac{\lambda_0}{t}-\lambda_0}^{0} \frac{e^{-2it\theta(\xi)}T_1(\xi)}{X_+(\xi)(\xi+\lambda_0-\frac{\lambda}{\sqrt{8t\lambda_0^2}})}\frac{d\xi}{2\pi i} \\ &+ X(\frac{\lambda}{\sqrt{8t\lambda_0^2}} - \lambda_0)\int_{\frac{\lambda_0}{t}-\lambda_0}^{0} \frac{e^{-2it\theta(\xi)}T_2(\xi)}{X_+(\xi)(\xi+\lambda_0-\frac{\lambda}{\sqrt{8t\lambda_0^2}})}\frac{d\xi}{2\pi i} \\ =& I_1 + I_2 + I_3, \end{aligned}$$

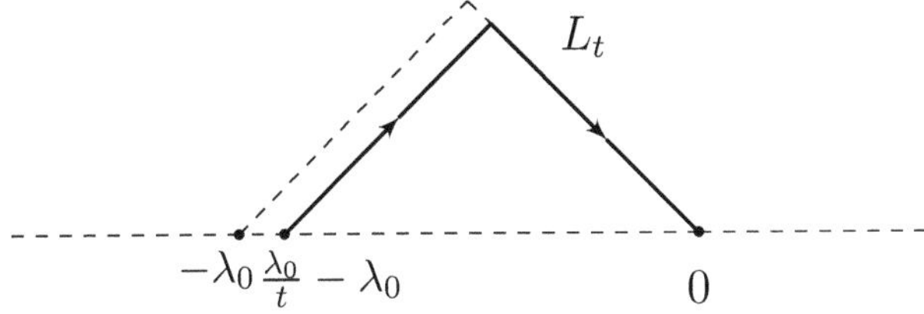

Figure 7: The contour L_t

and

$$|I_1| \lesssim \int_{-\lambda_0}^{\frac{\lambda_0}{t}-\lambda_0} \frac{|e^{-2it\theta}T(\xi)|}{|\xi+\lambda_0-\frac{\lambda}{\sqrt{8t\lambda_0^2}}|} d\xi \lesssim t^{-l-1},$$

$$|I_2| \lesssim \int_{\frac{\lambda_0}{t}-\lambda_0}^{0} \frac{|e^{-2it\theta}T_1(\xi)|}{|\xi+\lambda_0-\frac{\lambda}{\sqrt{8t\lambda_0^2}}|} d\xi \lesssim t^{-l}\frac{\sqrt{2}t}{\lambda_0}(\lambda_0-\frac{\lambda_0}{t}) \lesssim t^{-l+1}.$$

As a consequence of Cauchy's Theorem, we can evaluate I_3 along the contour L_t instead of the interval $(\frac{\lambda_0}{t}-\lambda_0, 0)$ and obtain $|I_3| \lesssim t^{-l+1}$. ■

Note: Similarly, when $\lambda \in L_A^*$ and $t \to \infty$, we have

$$|(N_A\hat{\delta})(\lambda)| \lesssim t^{-l}, \tag{3.54}$$

where $\hat{\delta}(\lambda) = e^{2it\theta}[\delta^{-1}(\lambda) - (\det \delta(\lambda))^{-1}I]R^\dagger(\lambda^*)$.

Let $J^{A^0} = (I - \omega_{A^0-})^{-1}(I + \omega_{A^0+})$, where

$$\omega_{A^0} = \omega_{A^0+} = \begin{cases} \begin{pmatrix} 0 & -(-2\lambda)^{-2i\nu}e^{\frac{1}{2}i\lambda^2}\dfrac{\gamma(-\lambda_0)}{1-|\gamma(-\lambda_0)|^2} \\ 0 & 0 \end{pmatrix}, & \lambda \in \Sigma_A^2, \\ \begin{pmatrix} 0 & (-2\lambda)^{-2i\nu}e^{\frac{1}{2}i\lambda^2}\gamma(-\lambda_0) \\ 0 & 0 \end{pmatrix}, & \lambda \in \Sigma_A^4, \end{cases} \tag{3.55}$$

$$\omega_{A^0} = \omega_{A^0-} = \begin{cases} \begin{pmatrix} 0 & 0 \\ (-2\lambda)^{2i\nu}e^{-\frac{1}{2}i\lambda^2}\dfrac{\gamma^\dagger(-\lambda_0)}{1-|\gamma(-\lambda_0)|^2} & 0 \end{pmatrix}, & \lambda \in \Sigma_A^1, \\ \begin{pmatrix} 0 & 0 \\ -(-2\lambda)^{2i\nu}e^{-\frac{1}{2}i\lambda^2}\gamma^\dagger(-\lambda_0) & 0 \end{pmatrix}, & \lambda \in \Sigma_A^3. \end{cases} \tag{3.56}$$

It follows from Lemma 3.4 that

$$\|\omega_A - \omega_{A^0}\|_{\mathscr{L}^1(\Sigma_A)\cap\mathscr{L}^2(\Sigma_A)\cap\mathscr{L}^\infty(\Sigma_A)} \lesssim_{\lambda_0} \frac{\log t}{\sqrt{t\lambda_0^2}}. \tag{3.57}$$

When $\lambda \in \Sigma_B$ and $\lambda \in \Sigma_C$, we have similar consequences. Let $J^{B^0} = (I - \omega_{B^0-})^{-1}(I + \omega_{B^0+})$, where

$$\omega_{B^0} = \omega_{B^0+} = \begin{cases} \begin{pmatrix} 0 & (2\lambda)^{-2i\nu} e^{\frac{1}{2}i\lambda^2}\gamma(\lambda_0) \\ 0 & 0 \end{pmatrix}, & \lambda \in \Sigma_B^2, \\ \begin{pmatrix} 0 & -(2\lambda)^{-2i\nu} e^{\frac{1}{2}i\lambda^2}\dfrac{\gamma(\lambda_0)}{1-|\gamma(\lambda_0)|^2} \\ 0 & 0 \end{pmatrix}, & \lambda \in \Sigma_B^4, \end{cases} \tag{3.58}$$

$$\omega_{B^0} = \omega_{B^0-} = \begin{cases} \begin{pmatrix} 0 & 0 \\ -(2\lambda)^{2i\nu} e^{-\frac{1}{2}i\lambda^2}\gamma^\dagger(\lambda_0) & 0 \end{pmatrix}, & \lambda \in \Sigma_B^1, \\ \begin{pmatrix} 0 & 0 \\ (2\lambda)^{2i\nu} e^{-\frac{1}{2}i\lambda^2}\dfrac{\gamma^\dagger(\lambda_0)}{1-|\gamma(\lambda_0)|^2} & 0 \end{pmatrix}, & \lambda \in \Sigma_B^3. \end{cases} \tag{3.59}$$

For $\lambda \in \Sigma_C$, using the symmetry property of $\gamma(\lambda)$ in (2.27) yields

$$\gamma(-0) = -\gamma(0), \quad \gamma(-i0) = -\gamma(i0), \tag{3.60}$$

which means that $\gamma(0) = 0$ and $\gamma(i0) = 0$. So $\omega_{C^0+} = \omega_{C^0-} = 0$ and the jump matrix $J^{C^0} = (I - \omega_{C^0-})^{-1}(I + \omega_{C^0+}) = I$ for $\lambda \in \Sigma_C$.

Theorem 3.5. *As $t \to \infty$, then*

$$\begin{aligned} \tilde{p}(x,t) =& \frac{1}{2\pi\sqrt{8t\lambda_0^2}}(\delta_A^0)^2 \left(\int_{\Sigma_A} (1 - C_{\omega_{A^0}})^{-1} I(\xi)\omega_{A^0}(\xi)\,\mathrm{d}\xi\right)_{12} \\ &+ \frac{1}{2\pi\sqrt{8t\lambda_0^2}}(\delta_B^0)^2 \left(\int_{\Sigma_B} (1 - C_{\omega_{B^0}})^{-1} I(\xi)\omega_{B^0}(\xi)\,\mathrm{d}\xi\right)_{12} \\ &+ \frac{1}{2\pi\sqrt{4t\lambda_0^2}}(\delta_C^0)^2 \left(\int_{\Sigma_C} (1 - C_{\omega_{C^0}})^{-1} I(\xi)\omega_{C^0}(\xi)\,\mathrm{d}\xi\right)_{12} + O\left(\frac{\log t}{t\lambda_0^2}\right). \end{aligned} \tag{3.61}$$

Proof. Notice that

$$\begin{aligned} &\left((1 - C_{\omega_A})^{-1} I\right)\omega_A - \left((1 - C_{\omega_{A^0}})^{-1} I\right)\omega_{A^0} \\ =& \left((1 - C_{\omega_A})^{-1} I\right)(\omega_A - \omega_{A^0}) + (1 - C_{\omega_A})^{-1}(C_{\omega_A} - C_{\omega_{A^0}})(1 - C_{\omega_{A^0}})^{-1} I\omega_{A^0} \\ =& (\omega_A - \omega_{A^0}) + \left((1 - C_{\omega_A})^{-1} C_{\omega_A} I\right)(\omega_A - \omega_{A^0}) + (1 - C_{\omega_A})^{-1}(C_{\omega_A} - C_{\omega_{A^0}}) I\omega_{A^0} \\ &+ (1 - C_{\omega_A})^{-1}(C_{\omega_A} - C_{\omega_{A^0}})(1 - C_{\omega_{A^0}})^{-1} C_{\omega_{A^0}} I\omega_{A^0}. \end{aligned}$$

Utilizing the triangle inequality and the boundedness in (3.57), we have

$$\begin{aligned} &\int_{\Sigma_A} \left((1 - C_{\omega_A})^{-1} I\right)(\xi)\omega_A(\xi)\,\mathrm{d}\xi \\ =& \int_{\Sigma_A} \left((1 - C_{\omega_{A^0}})^{-1} I\right)(\xi)\omega_{A^0}(\xi)\,\mathrm{d}\xi + O\left(\frac{\log t}{\sqrt{t\lambda_0^2}}\right). \end{aligned}$$

According to (3.46), we obtain that

$$\begin{aligned}
&\frac{1}{2\pi}\left(\int_{\Sigma'_A}\left((1-C_{\omega'_A})^{-1}I\right)(\xi)\omega'_A(\xi)\,\mathrm{d}\xi\right)_{12}\\
=&\frac{1}{2\pi}\left(\int_{\hat{\Sigma}'_A}\left(N_A^{-1}(\widetilde{\Delta}_A)^{-1}(1-C_{\omega_A})^{-1}\widetilde{\Delta}_A N_A I\right)(\xi)\hat{\omega}'_A(\xi)\,\mathrm{d}\xi\right)_{12}\\
=&\frac{1}{2\pi\sqrt{8t\lambda_0^2}}\left((\delta_A^0)^{-\sigma}\int_{\Sigma_A}\left((1-C_{\omega_A})^{-1}I\right)(\xi)\omega_A(\xi)\,\mathrm{d}\xi(\delta_A^0)^{\sigma}\right)_{12}\\
=&\frac{1}{2\pi\sqrt{8t\lambda_0^2}}(\delta_A^0)^2\left(\int_{\Sigma_A}\left((1-C_{\omega_{A^0}})^{-1}I\right)(\xi)\omega_{A^0}(\xi)\,\mathrm{d}\xi\right)_{12}+O\left(\frac{\log t}{t\lambda_0^2}\right).
\end{aligned}$$

The other two cases are treated similarly. Together with (3.31), one can obtain (3.61). ■

For $\lambda\in\mathbb{C}\backslash\Sigma_A$, set

$$M^{A^0}(\lambda;x,t)=I+\int_{\Sigma_A}\frac{\left((1-C_{\omega_{A^0}})^{-1}I\right)(\xi)\omega_{A^0}(\xi)}{\xi-\lambda}\frac{\mathrm{d}\xi}{2\pi i}. \tag{3.62}$$

Then $M^{A^0}(\lambda;x,t)$ is the solution of the RH problem

$$\begin{cases}M_+^{A^0}(\lambda;x,t)=M_-^{A^0}(\lambda;x,t)J^{A^0}(\lambda;x,t), & \lambda\in\Sigma_A,\\ M^{A^0}(\lambda;x,t)\to I, & \lambda\to\infty.\end{cases} \tag{3.63}$$

In particular

$$M^{A^0}(\lambda)=I+\frac{M_1^{A^0}}{\lambda}+O(\lambda^{-2}),\quad \lambda\to\infty. \tag{3.64}$$

which, together with (3.62) and (3.64), yields

$$M_1^{A^0}=-\int_{\Sigma_A}\left((1-C_{\omega_{A^0}})^{-1}I\right)(\xi)\omega_{A^0}(\xi)\frac{d\xi}{2\pi i}. \tag{3.65}$$

There are analogous RH problems on Σ_B and Σ_C, respectively. On Σ_B,

$$\begin{cases}M_+^{B^0}(\lambda;x,t)=M_-^{B^0}(\lambda;x,t)J^{B^0}(\lambda;x,t), & \lambda\in\Sigma_B,\\ M^{B^0}(\lambda;x,t)\to I, & \lambda\to\infty,\end{cases} \tag{3.66}$$

where $J^{B^0}(\lambda;x,t)$ is defined in (3.58) and (3.59). Moreover, we have

$$M^{B^0}(\lambda)=I+\frac{M_1^{B^0}}{\lambda}+O(\lambda^{-2}),\quad \lambda\to\infty. \tag{3.67}$$

Then, the jump matrix $J^{C^0}(\lambda;x,t)=I$ on Σ_C because of (3.60). The coefficient matrix $M_1^{C^0}$ of λ^{-1} in the expansion of $M^{C^0}(\lambda)$ satisfies $(M_1^{C^0})_{12}=0$.

From the expressions (3.55), (3.56), (3.58) and (3.59), we have the symmetry relation

$$J^{A^0}(\lambda) = \sigma(J^{B^0})(-\lambda)\sigma.$$

By the uniqueness of the RH problem,

$$M^{A^0}(\lambda) = \sigma(M^{B^0})(-\lambda)\sigma.$$

Combining with the expansion (3.64) and (3.67), one can verify that

$$\begin{aligned} M_1^{A^0} &= -\sigma(M_1^{B^0})\sigma, \\ (M_1^{A^0})_{12} &= (M_1^{B^0})_{12}. \end{aligned}$$

Therefore we obtain from (3.61) and (3.65) that

$$\begin{aligned} \tilde{p}(x,t) =& \frac{1}{2\pi\sqrt{8t\lambda_0^2}}(-2\pi i)\left((\delta_A^0)^2 M_1^{A^0} + (\delta_B^0)^2 M_1^{B^0} + (\delta_C^0)^2 M_1^{C^0}\right)_{12} + O\left(\frac{\log t}{t\lambda_0^2}\right) \\ =& -\frac{i}{\sqrt{8t\lambda_0^2}}\left((\delta_A^0)^2 + (\delta_B^0)^2\right)\left(M_1^{A^0}\right)_{12} + O\left(\frac{\log t}{t\lambda_0^2}\right). \end{aligned} \tag{3.68}$$

3.6 Solving the model problem

In this subsection, we shall solve $(M_1^{A^0})_{12}$ in (3.68) explicitly. Let

$$\Psi(\lambda) = M^{A^0}(\lambda)(-2\lambda)^{i\nu\sigma}e^{-\frac{1}{4}i\lambda^2\sigma}. \tag{3.69}$$

Then, it follows from (3.63) that

$$\Psi_+(\lambda) = \Psi_-(\lambda)v(-\lambda_0), \quad v = e^{\frac{1}{4}i\lambda^2\hat{\sigma}}(-2\lambda)^{-i\nu\hat{\sigma}}J^{A^0}(\lambda). \tag{3.70}$$

The jump matrix is independent of λ along each ray of four rays $\Sigma_A^1, \Sigma_A^2, \Sigma_A^3, \Sigma_A^4$, we have

$$\frac{\mathrm{d}\Psi_+(\lambda)}{\mathrm{d}\lambda} = \frac{\mathrm{d}\Psi_-(\lambda)}{\mathrm{d}\lambda}v(-\lambda_0). \tag{3.71}$$

From (3.70) and (3.71), we arrive at

$$\frac{\mathrm{d}\Psi_+(\lambda)}{\mathrm{d}\lambda} + \frac{1}{2}i\lambda\sigma\Psi_+(\lambda) = \left(\frac{\mathrm{d}\Psi_-(\lambda)}{\mathrm{d}\lambda} + \frac{1}{2}i\lambda\sigma\Psi_-(\lambda)\right)v(-\lambda_0). \tag{3.72}$$

which implies that $(\mathrm{d}\Psi/\mathrm{d}\lambda + i\lambda\sigma\Psi)\Psi^{-1}$ has no jump discontinuity along each of the four rays. Directly from the relation between $\Psi(\lambda)$ and $M^{A^0}(\lambda)$, we have

$$\begin{aligned} \left(\frac{\mathrm{d}\Psi(\lambda)}{\mathrm{d}\lambda} + \frac{1}{2}i\lambda\sigma\Psi(\lambda)\right)\Psi^{-1}(\lambda) =& \frac{\mathrm{d}M^{A^0}(\lambda)}{\mathrm{d}\lambda}(M^{A^0}(\lambda))^{-1} - \frac{1}{2}i\lambda M^{A^0}(\lambda)\sigma(M^{A^0}(\lambda))^{-1} \\ &+ \frac{i\nu}{\lambda}M^{A^0}(\lambda)\sigma(M^{A^0}(\lambda))^{-1} + \frac{1}{2}i\lambda\sigma \\ =& O(\lambda^{-1}) + \frac{1}{2}i[\sigma, M_1^{A^0}]. \end{aligned}$$

By Liouville's Theorem, we arrive at

$$\frac{d\Psi(\lambda)}{d\lambda} + \frac{1}{2}i\lambda\sigma\Psi(\lambda) = \beta\Psi(\lambda), \tag{3.73}$$

where

$$\beta = i[\sigma, M_1^{A^0}] = \begin{pmatrix} 0 & \beta_{12} \\ \beta_{21} & 0 \end{pmatrix}.$$

Particularly,

$$(M_1^{A^0})_{12} = i\beta_{12}. \tag{3.74}$$

Set

$$\Psi(\lambda) = \begin{pmatrix} \Psi_{11}(\lambda) & \Psi_{12}(\lambda) \\ \Psi_{21}(\lambda) & \Psi_{22}(\lambda) \end{pmatrix}.$$

By (3.73) and its differential we have

$$\begin{aligned}
&\frac{d^2\Psi_{11}(\lambda)}{d\lambda^2} + \left(-\frac{i}{2} + \frac{\lambda^2}{4} - \beta_{12}\beta_{21}\right)\Psi_{11}(\lambda) = 0,\\
&\beta_{12}\Psi_{21}(\lambda) = \frac{d\Psi_{11}(\lambda)}{d\lambda} - \frac{i}{2}\lambda\Psi_{11}(\lambda),\\
&\frac{d^2\beta_{12}\Psi_{22}(\lambda)}{d\lambda^2} + \left(\frac{i}{2} + \frac{\lambda^2}{4} - \beta_{12}\beta_{21}\right)\beta_{12}\Psi_{22}(\lambda) = 0,\\
&\Psi_{12}(\lambda) = \frac{1}{\beta_{12}\beta_{21}}\left(\frac{d\beta_{12}\Psi_{22}(\lambda)}{d\lambda} + \frac{1}{2}i\lambda\beta_{12}\Psi_{22}(\lambda)\right).
\end{aligned}$$

Here we introduce Weber's equation

$$\frac{d^2g(\zeta)}{d\zeta^2} + \left(a + \frac{1}{2} - \frac{\zeta^2}{4}\right)g(\zeta) = 0$$

which has the solution

$$g(\zeta) = c_1D_a(\zeta) + c_2D_a(-\zeta),$$

where c_1, c_2 are constants, and $D_a(\cdot)$ denotes the standard parabolic-cylinder function which satisfies

$$\frac{dD_a(\zeta)}{d\zeta} + \frac{\zeta}{2}D_a(\zeta) - aD_{a-1}(\zeta) = 0, \tag{3.75}$$

$$D_a(\pm\zeta) = \frac{\Gamma(a+1)e^{\frac{i\pi a}{2}}}{\sqrt{2\pi}}D_{-a-1}(\pm i\zeta) + \frac{\Gamma(a+1)e^{-\frac{i\pi a}{2}}}{\sqrt{2\pi}}D_{-a-1}(\mp i\zeta), \tag{3.76}$$

where $\Gamma(\cdot)$ is the Gamma function. As $\zeta \to \infty$, it follows from [35] that

$$D_a(\zeta) = \begin{cases} \zeta^a e^{-\frac{\zeta^2}{4}}(1 + O(\zeta^{-2})), & |\arg\zeta| < \frac{3\pi}{4},\\ \zeta^a e^{-\frac{\zeta^2}{4}}(1 + O(\zeta^{-2})) - \frac{\sqrt{2\pi}}{\Gamma(-a)}e^{a\pi i + \frac{\zeta^2}{4}}\zeta^{-a-1}(1 + O(\zeta^{-2})), & \frac{\pi}{4} < \arg\zeta < \frac{5\pi}{4},\\ \zeta^a e^{-\frac{\zeta^2}{4}}(1 + O(\zeta^{-2})) - \frac{\sqrt{2\pi}}{\Gamma(-a)}e^{-a\pi i + \frac{\zeta^2}{4}}\zeta^{-a-1}(1 + O(\zeta^{-2})), & -\frac{5\pi}{4} < \arg\zeta < -\frac{\pi}{4}. \end{cases} \tag{3.77}$$

Set $ia = \beta_{12}\beta_{21}$,

$$\Psi_{11}(\lambda) = c_1 D_a\left(e^{-\frac{\pi i}{4}}\lambda\right) + c_2 D_a\left(e^{\frac{3\pi i}{4}}\lambda\right), \tag{3.78}$$

$$\beta_{12}\Psi_{22}(\lambda) = c_3 D_{-a}\left(e^{\frac{\pi i}{4}}\lambda\right) + c_4 D_{-a}\left(e^{-\frac{3\pi i}{4}}\lambda\right). \tag{3.79}$$

For $\arg\lambda \in (-\pi, -\frac{3\pi}{4}) \cup (\frac{3\pi}{4}, \pi)$ and $\lambda \to \infty$, then we have

$$\Psi_{11}(\lambda)(-2\lambda)^{i\nu}e^{-\frac{i\lambda^2}{4}} \to 1, \quad \Psi_{22}(\lambda)(-2\lambda)^{-i\nu}e^{\frac{i\lambda^2}{4}} \to I,$$

so that

$$\Psi_{11}(\lambda) = 2^{-i\nu}e^{\frac{\pi\nu}{4}}D_a\left(e^{\frac{3\pi i}{4}}\lambda\right), \quad \nu = \beta_{12}\beta_{21},$$

$$\beta_{12}\Psi_{22}(\lambda) = \beta_{12}2^{i\nu}e^{\frac{\pi\nu}{4}}D_{-a}\left(e^{-\frac{3\pi i}{4}}\lambda\right),$$

and further,

$$\beta_{12}\Psi_{21}(\lambda) = 2^{-i\nu}e^{\frac{\pi\nu}{4}}e^{\frac{3\pi i}{4}}aD_{a-1}\left(e^{\frac{3\pi i}{4}}\lambda\right),$$

$$\Psi_{12}(\lambda) = \beta_{12}2^{i\nu}e^{-\frac{\pi i}{4}}e^{\frac{\pi\nu}{4}}D_{-a-1}\left(e^{-\frac{3\pi i}{4}}\lambda\right).$$

For $\arg\lambda \in (-\frac{3\pi}{4}, -\frac{\pi}{4})$ and $\lambda \to \infty$, in a similar way we find that

$$\Psi_{11}(\lambda)(-2\lambda)^{i\nu}e^{-\frac{i\lambda^2}{4}} \to 1, \quad \Psi_{22}(\lambda)(-2\lambda)^{-i\nu}e^{\frac{i\lambda^2}{4}} \to I,$$

so that

$$\Psi_{11}(\lambda) = 2^{-i\nu}e^{\frac{\pi\nu}{4}}D_a\left(e^{\frac{3\pi i}{4}}\lambda\right),$$

$$\beta_{12}\Psi_{22}(\lambda) = \beta_{12}2^{i\nu}e^{-\frac{3\pi\nu}{4}}D_{-a}\left(e^{\frac{\pi i}{4}}\lambda\right),$$

and further,

$$\beta_{12}\Psi_{21}(\lambda) = 2^{-i\nu}e^{\frac{\pi\nu}{4}}e^{\frac{3\pi i}{4}}aD_{a-1}\left(e^{\frac{3\pi i}{4}}\lambda\right),$$

$$\Psi_{12}(\lambda) = \beta_{12}2^{i\nu}e^{\frac{3\pi i}{4}}e^{-\frac{3\pi\nu}{4}}D_{-a-1}\left(e^{\frac{\pi i}{4}}\lambda\right).$$

Along $\arg\lambda = -\frac{3\pi}{4}$, we arrive at

$$\Psi_+(\lambda) = \Psi_-(\lambda)\begin{pmatrix} 1 & \gamma(-\lambda_0) \\ 0 & I \end{pmatrix}. \tag{3.80}$$

From (3.80), we can obtain the following equality

$$\beta_{12}2^{i\nu}e^{\frac{\pi(3i-3\nu)}{4}}D_{-a-1}(e^{\frac{\pi i}{4}}\lambda) = 2^{-i\nu}e^{\frac{\pi\nu}{4}}D_a(e^{\frac{3\pi i}{4}}\lambda)\gamma(-\lambda_0) + \beta_{12}2^{i\nu}e^{\frac{\pi(\nu-i)}{4}}D_{-a-1}(e^{-\frac{3\pi i}{4}}\lambda). \tag{3.81}$$

We obtain by (3.76) that

$$D_a(e^{\frac{3\pi i}{4}}\lambda) = \frac{\Gamma(a+1)e^{\frac{\pi\nu}{2}}}{\sqrt{2\pi}} D_{-a-1}(e^{-\frac{3\pi i}{4}}\lambda) + \frac{\Gamma(a+1)e^{-\frac{\pi\nu}{2}}}{\sqrt{2\pi}} D_{-a-1}(e^{\frac{\pi i}{4}}\lambda). \quad (3.82)$$

Substituting (3.82) into (3.81) and separating the coefficients of the two independent functions yield

$$\beta_{12} = \frac{2^{-2i\nu}e^{-\frac{3\pi i}{4}}e^{\frac{\pi\nu}{2}}\Gamma(a+1)}{\sqrt{2\pi}}\gamma(-\lambda_0) = \frac{2^{-2i\nu}e^{\frac{3\pi i}{4}}e^{\frac{\pi\nu}{2}}\nu\Gamma(-i\nu)}{\sqrt{2\pi}}\gamma(-\lambda_0). \quad (3.83)$$

Based on (2.28), we can obtain

$$pp^\dagger = \tilde{p}\tilde{p}^\dagger, \quad p^\dagger p = Y\tilde{p}^\dagger\tilde{p}Y^{-1}. \quad (3.84)$$

Particularly, we denote

$$\tilde{\nu}(\lambda) = -\frac{1}{2\pi}\log(1-|\gamma(\lambda)|^2), \quad (3.85)$$

$$\tilde{\delta}_B^0(\lambda) = (8t\lambda^4)^{\frac{i\nu}{2}}\exp\{-\frac{1}{2}it\lambda^4 + \chi_+(\lambda) + \chi_-(\lambda) + \hat{\chi}_+(\lambda) + \hat{\chi}_-(\lambda)\}, \quad (3.86)$$

$$\tilde{\delta}_A^0(\lambda) = (8t\lambda^4)^{\frac{i\nu}{2}}\exp\{-\frac{1}{2}it\lambda^4 + \chi_+(-\lambda) + \chi_-(-\lambda) + \hat{\chi}_+(-\lambda) + \hat{\chi}_-(-\lambda)\}, \quad (3.87)$$

and further,

$$\begin{aligned}\int_{-\infty}^{x} pp^\dagger\,\mathrm{d}x' &= \int_{-\infty}^{x} |\tilde{p}|^2\,\mathrm{d}x' \\ &= \frac{1}{2t}\int_{-\infty}^{x}[2 + \left(\frac{\delta_A^0}{\delta_B^0}\right)^2 + \left(\frac{\delta_B^0}{\delta_A^0}\right)^2]\frac{-\nu}{4\lambda_0^2}\,\mathrm{d}x' + O\left(\frac{\log t}{\sqrt{t\lambda_0^2}}\right) \\ &= \int_{-\infty}^{\lambda_0}[2 + \left(\frac{\tilde{\delta}_A^0(s)}{\tilde{\delta}_B^0(s)}\right)^2 + \left(\frac{\tilde{\delta}_B^0(s)}{\tilde{\delta}_A^0(s)}\right)^2]\frac{\tilde{\nu}(s)}{2s}\,\mathrm{d}s + O\left(\frac{\log t}{\sqrt{t\lambda_0^2}}\right).\end{aligned} \quad (3.88)$$

It follows from (2.6) and (3.84) that

$$\begin{aligned}Y^{-1} =& I - (i + \frac{3}{4}i\ell_0)\int_{-\infty}^{\lambda_0}\left[2 + \left(\frac{\tilde{\delta}_A^0(s)}{\tilde{\delta}_B^0(s)}\right)^2 + \left(\frac{\tilde{\delta}_B^0(s)}{\tilde{\delta}_A^0(s)}\right)^2\right] \\ &\times \frac{\tilde{\nu}(s)}{2s(1-e^{-2\pi\tilde{\nu}(s)})}\gamma^\dagger(s)\gamma(s)\,\mathrm{d}s + O\left(\frac{\log t}{\sqrt{t\lambda_0^2}}\right).\end{aligned} \quad (3.89)$$

Finally, we obtain (1.5) from (2.28), (3.68), (3.74), (3.83), (3.88) and (3.89).

Acknowledgements

This work is supported by National Natural Science Foundation of China (Grant Nos. 11871440, 11931017 and 11971442).

References

[1] Ablowitz M J and Fokas A S, *Complex variables: introduction and applications*, Cambridge University Press, Cambridge, 2003.

[2] Arruda L K and Lenells J, Long-time asymptotics for the derivative nonlinear Schrödinger equation on the half-line, *Nonlinearity.* **30**, 4141–4172, 2017.

[3] Andreiev K, Egorova I, Lange T L and Teschl G, Rarefaction waves of the Korteweg-de Vries equation via nonlinear steepest descent, *J. Differential Equations.* **261**, 5371–5410, 2016.

[4] Beals R and Coifman R R, Scattering and inverse scattering for first order systems, *Comm. Pure Appl. Math.* **37**, 39–90, 1984.

[5] Boutet de Monvel A and Shepelsky D, A Riemann-Hilbert approach for the Degasperis-Procesi equation, *Nonlinearity.* **26**, 2081–2107, 2013.

[6] Boutet de Monvel A, Lenells J and Shepelsky D, Long-time asymptotics for the Degasperis-Procesi equation on the half-line, *Ann. Inst. Fourier.* **69**, 171–230, 2019.

[7] Boutet de Monvel A, Kostenko A, Shepelsky D and Teschl G, Long-time asymptotics for the Camassa-Holm equation, *SIAM J. Math. Anal.* **41**, 1559–1588, 2009.

[8] Chen H H, Lee Y C and Liu C S, Integrability of nonlinear Hamiltonian systems by inverse scattering method, *Phys. Scr.* **20**, 490–492, 1979.

[9] Cheng P J, Venakides S and Zhou X, Long-time asymptotics for the pure radiation solution of the sine-Gordon equation, *Comm. Partial Differential Equations.* **24**, 1195–1262, 1999.

[10] Chan H N, Malomed B A, Chow K W and Ding E, Rogue waves for a system of coupled derivative nonlinear Schrödinger equations, *Phys. Rev. E.* **93**, 012217, 2016.

[11] Daniel M and Veerakumar V, Propagation of electromagnetic soliton in antiferromagnetic medium, *Phys. Lett. A.* **302**, 77–86, 2002.

[12] Deift P and Zhou X, A steepest descent method for oscillatory Riemann-Hilbert problems. Asymptotics for the MKdV equation, *Ann. Math.* **137**, 295–368, 1993.

[13] Deift P, Its A R and Zhou X, *Long-time asymptotics for integrable nonlinear wave equations, in "Important Developments" in "Soliton Theory"*, Springer, Berlin, 1993.

[14] Guo B L and Ling L M, Riemann-Hilbert approach and N-soliton formula for coupled derivative Schrödinger equation, *J. Math. Phys.* **53**, 073506, 2012.

[15] Guo B L, Huang D W and Wang Z Q, Existence and stability of standing waves for coupled derivative Schrödinger equations, *J. Math. Phys.* **56**, 103510, 2015.

[16] Grunert K and Teschl G, Long-time asymptotics for the Korteweg-de Vries equation via nonlinear steepest descent, *Math. Phys. Anal. Geom.* **12**, 287–324, 2009.

[17] Gerdjikov V S and Ivanov M I, The quadratic bundle of general form and the nonlinear evolution equations: hierarchies of Hamiltonian structures, *Bulg. J. Phys.* **10**, 130–43, 1983.

[18] Geng X G and Liu H, The nonlinear steepest descent method to long-time asymptotics of the coupled nonlinear Schrödinger equation, *J. Nonlinear Sci.* **28**, 739–763, 2018.

[19] Geng X G, Chen M M and Wang K D, Long-time asymptotics of the coupled modified Korteweg-de Vries equation, *J. Geom. Phys.* **142**, 151–167, 2019.

[20] Geng X G, Li R M and Xue B, A vector general nonlinear Schrödinger equation with $(m+n)$ components, *J. Nonlinear Sci.* **30**, 991–1013, 2020.

[21] Geng X G, Zhai Y Y and Dai H H, Algebro-geometric solutions of the coupled modified Korteweg-de Vries hierarchy, *Adv. Math.* **263**, 123–153, 2014.

[22] Geng X G, Li Z, Xue B and Guan L, Explicit quasi-periodic solutions of the Kaup-Newell hierarchy, *J. Math. Anal. Appl.* **425**, 1097–1112, 2015.

[23] Kitaev A V and Vartanian A H, Leading-order temporal asymptotics of the modified nonlinear Schrödinger equation: solitonless sector, *Inverse Problems.* **13**, 1311–1339, 1997.

[24] Kitaev A V and Vartanian A H, Asymptotics of solutions to the modified nonlinear Schrödinger equation: solution on a nonvanishing continuous background, *SIAM J. Math. Anal.* **30**, 787–832, 1999.

[25] Kaup D J and Newell A C, An exact solution for a derivative nonlinear Schrödinger equation, *J. Math. Phys.* **19**, 798–801, 1978.

[26] Lenells J, The nonlinear steepest descent method: asymptotics for initial-boundary value problems, *SIAM J. Math. Anal.* **48**, 2076–2118, 2016.

[27] Ling L M and Liu Q P, Darboux transformation for a two-component derivative nonlinear Schrödinger equation, *J. Phys. A.* **43**, 434023, 2010.

[28] Li R M and Geng X G, On a vector long wave-short wave-type model, *Stud. Appl. Math.* **144**, 164–184, 2020.

[29] Li R M and Geng X G, Rogue periodic waves of the sine-Gordon equation, *Appl. Math. Lett.* **102**, 106147, 2020.

[30] Liu H, Geng X G and Xue B, The Deift-Zhou steepest descent method to long-time asymptotics for the Sasa-Satsuma equation, *J. Differential Equations.* **265**, 5984–6008, 2018.

[31] Morris H C and Dodd R K, The two component derivative nonlinear Schrödinger equation, *Phys. Scr.* **20**, 505–508, 1979.

[32] Tsuchida T and Wadati M, New integrable systems of derivative nonlinear Schrödinger equations with multiple components, *Phys. Lett. A.* **257**, 53–64, 1999.

[33] Tsuchida T and Wadati M, Complete integrability of derivative nonlinear Schrödinger-type equations, *Inverse Problems.* **15**, 1363-1373, 1999.

[34] Vartanian A H, Higher order asymptotics of the modified non-linear Schrödinger equation, *Comm. Partial Differential Equations.* **25**, 1043–1098, 2000.

[35] Whittaker E T and Watson G N, *A Course of Modern Analysis*, 4th, Cambridge University Press, Cambridge, 1927.

[36] Wei J, Geng X G and Zeng X, The Riemann theta function solutions for the hierarchy of Bogoyavlensky lattices, *Trans. Amer. Math. Soc.* **371**, 1483–1507, 2019.

[37] Xu J and Fan E G, Long-time asymptotics for the Fokas-Lenells equation with decaying initial value problem: without solitons, *J. Differential Equations.* **259**, 1098–1148, 2015.

[38] Xu T, Tian B, Zhang C, Meng X H and Lv X, Alfvén solitons in the coupled derivative nonlinear Schrödinger system with symbolic computation, *J. Phys. A.* **42**, 415201, 2009.

[39] Zhu Q Z, Fan E G and Xu J, Initial-boundary value problem for two-component Gerdjikov-Ivanov equation with 3×3 Lax pair on half-line, *Commun. Theor. Phys.* **68**, 425–438, 2017.

B3. Bilinearization of nonlinear integrable evolution equations: Recursion operator approach

Xing-Biao Hu[a] *and Guo-Fu Yu*[b]

[a]*LSEC, ICMSEC, Academy of Mathematics and Systems Science, Chinese Academy of Sciences, P.O.Box 2719, Beijing 100190, P.R. China*
Department of Mathematical Sciences, University of the Chinese Academy of Sciences, Beijing, P.R. China

[b]*Department of Mathematics, Shanghai Jiao Tong University, Shanghai, 200240, P.R. China*

Abstract

In this chapter, by using recursion operators of integrable nonlinear evolution equations, we will construct so-called unified bilinear forms for several well-known or less-known integrable hierarchies of equations. Examples include unified bilinear forms for AKNS hierarchy, NLS hierarchy, Boussinesq hierarchy, Levi hierarchy, and Yajima-Oikawa hierarchy.

1 Introduction

Since the KdV equation was solved by the inverse scattering transform in 1967 [8], it has been found that many soliton equations can be solved exactly by various methods. Among those methods, Hirota's method is a powerful way to get explicit solutions. A crucial point in Hirota's method is that equations are expressed with bilinear forms through suitable dependent variable transformations. In many cases, compared with the original nonlinear forms, bilinear equations are much easier to deal with. Therefore it is very important in the study of soliton theory to consider whether a nonlinear equation can be transformed into bilinear form and deduce more new integrable bilinear equations. In the 1980s Sato School gave out bilinear KP, BKP hierarchies of equations and a variety of (1+1)-reductions together with their corresponding τ function solutions by Kac-Moody Lie-algebraic representation. This is a great progress and more details can be found in [6, 17, 27, 33, 34]. Bilinear equations of KdV, MKdV hierarchies were also derived by Matsuno in [22–24]. In [10] bilinear KdV, MKdV and classical Boussinesq hierarchies were obtained with the help of recursion operators. Deriving bilinear forms with recursion operators has many advantages, such as less computation and simpler unified explicit expressions. A unified and clear form will bring us many benefits. For example, in [10] from these simple unified forms, the corresponding Bäcklund transformation and nonlinear superposition formulae for KdV and MKdV hierarchy were easily obtained. In [21], rational solutions of classical Boussinesq hierarchy were obtained from its unified bilinear form.

In this chapter, we will use recursion operator method to further deduce the corresponding bilinear form of AKNS hierarchy, NLS hierarchy, Boussinesq hierarchy, Levi hierarchy, and Yajima-Oikawa hierarchy, respectively. Before going into the details, let us first list several examples of unified bilinear forms for some known soliton hierarchies.

Example 1.1: The unified bilinear form for the KdV hierarchy [10, 11] is

$$(D_xD_{t_{2n+1}}+\frac{2}{3}D_{t_{2n-1}}D_x^3-\frac{1}{3}D_{t_3}D_{t_{2n-1}})f\cdot f=0, \qquad n=1,2,\cdots,t_1=x. \tag{1.1}$$

Example 1.2: The bilinear mKdV hierarchy [10] is

$$D_x^2 f\cdot g=0 \tag{1.2}$$

$$(4D_{t_{2n+1}}+D_x^2D_{t_{2n-1}})f\cdot g=0, \qquad n=1,2,\cdots,t_1=x. \tag{1.3}$$

Example 1.3: The classical Boussinesq hierarchy [10] is

$$(D_xD_{t_{n+1}}-\frac{1}{2}D_{t_2}D_{t_n}-\frac{i}{4}D_x^2D_{t_n})f\cdot g=0 \tag{1.4}$$

$$(iD_{t_{n+1}}+\frac{1}{2}D_xD_{t_n})f\cdot g=0, \qquad n=1,2,\cdots,t_1=x. \tag{1.5}$$

Example 1.4: The bilinear Sawada-Kotera hierarchy [13] is

$$(D_1D_5-D_1^6)f\cdot f=0 \tag{1.6}$$

$$(9D_1^7D_m+5D_mD_7-35D_1D_{m+6}+21D_1^2D_5D_m)f\cdot f=0, \tag{1.7}$$

where m is an odd integer, $m\neq 3k, m,k\in Z_+$.

Example 1.5: The bilinear Ito hierarchy [35] is

$$D_3(D_3+D_1^3)f\cdot f=0 \tag{1.8}$$

$$(D_1D_5-\frac{1}{9}D_1^6-\frac{10}{9}D_1^3D_3)f\cdot f=0 \tag{1.9}$$

$$\begin{aligned}(105D_1D_{2m+5}&+D_1^7D_{2m-1}-70D_1D_3^2D_{2m-1}\\&+21D_1^2D_5D_{2m-1}+15D_7D_{2m-1})f\cdot f=0,\end{aligned} \tag{1.10}$$

where m is a positive integer.

Example 1.6: The bilinear Ramani hierarchy [39] is

$$(D_1^6-5D_1^3D_3-5D_3^2)f\cdot f=0, \tag{1.11}$$

$$(D_1^8+7D_1^5D_3-35D_1^2D_3^2+90D_1D_7)f\cdot f=0, \tag{1.12}$$

$$(D_1^{10}+15D_1^7D_3+75D_1^4D_3^2-100D_1D_3^3+900D_1D_9)f\cdot f=0, \tag{1.13}$$

$$\begin{aligned}(D_1D_{m+10}&-\frac{1}{11}D_{11}D_m-\frac{1}{9}D_9D_mD_1^2-\frac{1}{7}D_1D_3D_7D_m-\frac{2}{81}D_1^2D_3^3D_m\\&+\frac{1}{135}D_1^5D_3^2D_m+\frac{2}{945}D_1^8D_3D_m+\frac{1}{22275}D_1^{11}D_m)f\cdot f=0,\end{aligned} \tag{1.14}$$

where m is an odd integer, $m\neq 5k, m,k\in Z_+$.

Example 1.7: The bilinear Lotka-Volterra hierarchy [15] is

$$[D_{t_1}\sinh(\frac{1}{2}D_n)+\cosh(\frac{3}{2}D_n)-\cosh(\frac{1}{2}D_n)]f_n\cdot f_n=0, \tag{1.15}$$

$$\begin{aligned}&[D_{t_k}\sinh(\frac{1}{2}D_n)+\frac{1}{2}D_{t_{k-1}}\sinh(\frac{3}{2}D_n)+\frac{1}{2}D_{t_{k-1}}\sinh(\frac{1}{2}D_n)\\&\quad-\frac{1}{2}D_{t_1}D_{t_{k-1}}\cosh(\frac{1}{2}D_n)]f_n\cdot f_n=0.\end{aligned} \tag{1.16}$$

Example 1.8: The bilinear negative Volterra hierarchy [31] is

$$D_1e^{D_n}f_n\cdot f_n=f_n^2-f_{n+1}f_{n-1}, \tag{1.17}$$

$$(2D_{j+1}e^{D_n}+D_jD_1)f_n\cdot f_n=0. \tag{1.18}$$

The chapter is organized as follows. In Section 2, we present a unified bilinear AKNS hierarchy with the help of its recursion operator. The bilinear NLS hierarchy with a unified structure is deduced from its recursion operator in Section 3. Next, in Section 4 we give a unified form of bilinear Boussinesq hierarchy via its recursion operator. Section 5 is devoted to deriving bilinear Levi hierarchy, while bilinear Yajima-Oikawa hierarchy is presented in Section 6 by the recursion operator method. Finally conclusion and discussions are given in Section 7.

2 Bilinear form of AKNS hierarchy

From the spectral problem

$$\Psi_x=\begin{pmatrix}-i\lambda & q(x,t)\\ r(x,t) & i\lambda\end{pmatrix}\Psi,\qquad \Psi=\begin{pmatrix}\psi_1\\ \psi_2\end{pmatrix} \tag{2.1}$$

we can obtain the following AKNS hierarchy [1, 2, 25]

$$\begin{pmatrix}q_t\\ r_t\end{pmatrix}=L^n\begin{pmatrix}q_x\\ r_x\end{pmatrix},\qquad (n\geq 1) \tag{2.2}$$

where

$$L=\frac{1}{2i}\begin{pmatrix}-\partial_x+2q\partial_x^{-1}r & 2q\partial_x^{-1}q\\ -2r\partial_x^{-1}r & \partial_x-2r\partial_x^{-1}q\end{pmatrix}. \tag{2.3}$$

Introducing infinite time variables $x=t_1,t_2,t_3,\cdots$ and regarding q,r as functions of $t=t(t_1,t_2,t_3\cdots)$, we have the following equivalent equation

$$\begin{pmatrix}q_{t_{n+1}}\\ r_{t_{n+1}}\end{pmatrix}=\frac{1}{2i}\begin{pmatrix}-\partial_x+2q\partial_x^{-1}r & 2q\partial_x^{-1}q\\ -2r\partial_x^{-1}r & \partial_x-2r\partial_x^{-1}q\end{pmatrix}\begin{pmatrix}q_{t_n}\\ r_{t_n}\end{pmatrix}\qquad (n\geq 1) \tag{2.4}$$

or equivalently

$$q_{t_{n+1}}=-\frac{1}{2i}q_{xt_n}-iq\partial_x^{-1}(qr)_{t_n}, \tag{2.5}$$

$$r_{t_{n+1}}=\frac{1}{2i}r_{xt_n}+ir\partial_x^{-1}(qr)_{t_n}. \tag{2.6}$$

By dependent variable transformation

$$q = \frac{\sigma}{\tau}, \qquad r = \frac{\rho}{\tau}, \tag{2.7}$$

nonlinear Equations (2.5)-(2.6) can be changed into

$$(D_{t_{n+1}} - \frac{i}{2} D_{t_1} D_{t_n}) \sigma \cdot \tau = 0, \tag{2.8}$$

$$(D_{t_{n+1}} + \frac{i}{2} D_{t_1} D_{t_n}) \rho \cdot \tau = 0, \tag{2.9}$$

$$D_{t_1}^2 \tau \cdot \tau = -2\sigma\rho. \tag{2.10}$$

In fact, (2.5) can be written as

$$\begin{aligned}
\frac{D_{t_{n+1}}\sigma \cdot \tau}{\tau^2} &= -\frac{1}{2i}\frac{(\sigma_x\tau - \sigma\tau_x)_{t_n}\tau^2 - 2\tau\tau_{t_n}(\sigma_x\tau - \sigma\tau_x)}{\tau^4} - i\frac{\sigma}{\tau}\partial_x^{-1}\left(\frac{\sigma\rho}{\tau^2}\right)_{t_n} \\
&= -\frac{1}{2i}\frac{(\sigma_{xt_n}\tau + \sigma_x\tau_{t_n} - \sigma_{t_n}\tau_x - \sigma\tau_{xt_n})\tau^2 - 2\tau\tau_{t_n}(\sigma_x\tau - \sigma\tau_x)}{\tau^4} \\
&\qquad + \frac{i\sigma}{2\tau}\partial_x^{-1}\left(\frac{D_{t_1}^2\tau \cdot \tau}{\tau^2}\right)_{t_n} \\
&= -\frac{1}{2i}\frac{\sigma_{xt_n}\tau - \sigma_x\tau_{t_n} - \sigma_{t_n}\tau_x + \sigma\tau_{xt_n}}{\tau^2} \\
&= -\frac{1}{2i}\frac{D_x D_{t_n}\sigma \cdot \tau}{\tau^2},
\end{aligned}$$

so (2.8) is obtained. We can deduce (2.9) from (2.6) similarly. It should be pointed out that the bilinear form of the AKNS hierarchy has been given in [25], but the recursion operator is not explicitly used in the derivation process.

3 Bilinear form of NLS hierarchy

The nonlinear Schrödinger (NLS) hierarchy can be deduced from the reduction $r = \pm q^*$ in spectral problem (2.1). In the following we only consider the case $r = -q^*$. In the case we can have the following NS hierarchy

$$\begin{pmatrix} q \\ -q^* \end{pmatrix}_t = \bar{L}^n \begin{pmatrix} q_x \\ -q_x^* \end{pmatrix}, \qquad n \geq 1 \tag{3.1}$$

where

$$\bar{L} = \frac{1}{2i}\begin{pmatrix} -\partial_x - 2q\partial_x^{-1}q^* & 2q\partial_x^{-1}q \\ -2q^*\partial_x^{-1}q^* & \partial_x + 2q^*\partial_x^{-1}q \end{pmatrix}. \tag{3.2}$$

As to the hierarchy (3.1) we can obtain similar bilinear form as AKNS hierarchy. In the following we consider one sub-hierarchy of (3.1)

$$\begin{pmatrix} q \\ -q^* \end{pmatrix}_t = \bar{L}^{2n+1} \begin{pmatrix} q_x \\ -q_x^* \end{pmatrix} \tag{3.3}$$

Introducing infinite time variables $t_2, t_4, \cdots$ and regarding q, q^* as function of $t = (x, t_2, t_4, \cdots)$, then we have the equivalent equations

$$\begin{pmatrix} q_{t_2} \\ -q^*_{t_2} \end{pmatrix} = \bar{L} \begin{pmatrix} q_x \\ -q^*_x \end{pmatrix}, \tag{3.4}$$

$$\begin{pmatrix} q_{t_{2(n+1)}} \\ -q^*_{t_{2(n+1)}} \end{pmatrix} = M \begin{pmatrix} q_{t_{2n}} \\ -q^*_{t_{2n}} \end{pmatrix}, \qquad n \geq 1, \tag{3.5}$$

where we have taken the notation

$$M = \bar{L}^2 = -\frac{1}{4} \begin{pmatrix} A_{11} & A_{12} \\ A_{21} & A_{22} \end{pmatrix}, \tag{3.6}$$

and

$$\begin{aligned} A_{11} &= \partial_x^2 + 2\partial_x q \partial_x^{-1} q^* + 2q\partial_x^{-1} q^* \partial_x, & (3.7)\\ A_{12} &= -2\partial_x q \partial_x^{-1} q + 2q\partial_x^{-1} q \partial_x, & (3.8)\\ A_{21} &= 2q^* \partial_x^{-1} q + 2q\partial_x^{-1} q^*, & (3.9)\\ A_{22} &= \partial^2 + 2\partial_x q^* \partial_x^{-1} q + 2q^* \partial_x^{-1} q \partial_x. & (3.10) \end{aligned}$$

(3.4) and (3.5) can be written as

$$q_{t_2} = \frac{1}{2} i q_{xx} + i q^2 q^*, \tag{3.11}$$

$$q_{t_{2(n+1)}} = -\frac{1}{4} q_{xxt_{2n}} - \frac{1}{2} \partial_x q \partial_x^{-1} (qq^*)_{t_{2n}} - \frac{1}{2} q \partial_x^{-1} (q_{xt_{2n}} q^* - q q^*_{xt_{2n}}), \tag{3.12}$$

and the corresponding conjugate equations. Let $q = \frac{G}{F}$, F is a real function, then (3.11)-(3.12) can be transformed into the following bilinear form

$$D_x^2 F \cdot F = 2GG^*, \tag{3.13}$$

$$(D_{t_2} - \frac{i}{2} D_x^2) G \cdot F = 0, \tag{3.14}$$

$$(D_{t_{2(n+1)}} + \frac{1}{8} D_{t_{2n}} D_x^2 + \frac{1}{4i} D_{t_2} D_{t_{2n}}) G \cdot F = 0. \tag{3.15}$$

If we set $\phi = \ln \frac{G}{F}$, $\rho = \ln(GF)$, then from (3.13) we have

$$2qq^* = 2\frac{GG^*}{F^2} = \frac{D_x^2 F \cdot F}{F^2} = 2(\ln F)_{xx} = (\rho - \phi)_{xx}, \tag{3.16}$$

so we have

$$\rho_{xx} = \phi_{xx} + 2qq^*. \tag{3.17}$$

By use of (A1) and (A2), (3.14) can be transformed into

$$\phi_{t_2} = \frac{i}{2}(\rho_{xx} + \phi_x^2) = \frac{i}{2}(\phi_{xx} + 2qq^* + \phi_x^2). \tag{3.18}$$

Together with $\phi = \ln q$, Eq.(3.11) can be obtained. Similarly by use of (A1), (A3), (A4), (3.15) can be changed into

$$
\begin{aligned}
0 &= \phi_{t_{2(n+1)}} + \frac{1}{8}(\phi_{xxt_{2n}} + 2\phi_x\rho_{xt_{2n}} + \phi_{t_{2n}}\phi_x^2) + \frac{1}{4i}(\rho_{t_2t_{2n}} + \phi_{t_2}\phi_{t_{2n}}) \\
&= \phi_{t_{2(n+1)}} + \frac{1}{8}\{\phi_{xxt_{2n}} + 2\phi_x[\phi_{xt_{2n}} \\
&\quad +2\partial_x^{-1}(qq^*)_{t_{2n}}] + \phi_{t_{2n}}(\phi_{xx} + 2qq^*) + \phi_{t_{2n}}\phi_x^2\} \\
&\quad +\frac{1}{4i}[\phi_{t_2t_{2n}} + \partial_x^{-2}(2qq^*)_{t_2t_{2n}}] + \frac{1}{4i}\phi_{t_2}\phi_{t_{2n}} \\
&= \phi_{t_{2(n+1)}} + \frac{1}{8}(\phi_{xx} + \phi_x^2 + 2qq^*)_{t_{2n}} - \frac{1}{4}(qq^*)_{t_{2n}} + \frac{1}{2}\phi_x\partial_x^{-1}(qq^*)_{t_{2n}} \\
&\quad +\frac{1}{8}\phi_{t_{2n}}(\phi_{xx} + 2qq^* + \phi_x^2) + \frac{1}{4i}\phi_{t_2}\phi_{t_{2n}} + \frac{1}{4}\partial_x^{-1}(q_xq^* - qq_x^*)_{t_{2n}} + \frac{1}{4i}\phi_{t_2}\phi_{t_{2n}} \\
&= \phi_{t_{2(n+1)}} + \frac{1}{2i}\phi_{t_2t_{2n}} - \frac{1}{4}(qq^*)_{t_{2n}} + \frac{1}{2}\phi_x\partial_x^{-1}(qq^*)_{t_{2n}} \\
&\quad +\frac{1}{2i}\phi_{t_{2n}}\phi_{t_2} + \frac{1}{4}\partial_x^{-1}(q_xq^* - qq_x^*)_{t_{2n}} \\
&= \frac{q_{t_{2(n+1)}}}{q} + \frac{1}{2i}(\frac{q_{t_2}}{q})_{t_{2n}} - \frac{1}{4}(qq^*)_{t_{2n}} + \frac{1}{2}\frac{q_x}{q}\partial_x^{-1}(qq^*)_{t_{2n}} + \frac{1}{2i}\frac{q_{t_2}q_{t_n}}{q^2} \\
&\quad +\frac{1}{4}\partial_x^{-1}(q_xq^* - qq_x^*)_{t_{2n}} \\
&= \frac{q_{t_{2(n+1)}}}{q} + \frac{1}{2i}\frac{1}{q}(\frac{i}{2}q_{xx} + iq^2q^*)_{t_{2n}} - \frac{1}{4}(qq^*)_{t_{2n}} + \frac{1}{2}\frac{q_x}{q}\partial_x^{-1}(qq^*)_{t_{2n}} \\
&\quad +\frac{1}{4}\partial_x^{-1}(q_xq^* - qq_x^*)_{t_{2n}},
\end{aligned}
$$

which leads to Eq. (3.12). It should be remarked here that the bilinearization of the NLS hierarchy (3.4)-(3.5) was also discussed in [23], but here we give much simpler unified bilinear form of the NLS hierarchy.

4 Bilinear form of Boussinesq hierarchy

The well-known Boussinesq equation is

$$U_{tt} + (\frac{1}{3}U_{xx} + U^2)_{xx} = 0. \tag{4.1}$$

If we set $U = 2(\ln\tau)_{xx}$, (4.1) can be transformed into the bilinear form

$$(D_x^4 + 3D_t^2)\tau \cdot \tau = 0. \tag{4.2}$$

If we set $U_t = H_x$, (4.1) can be written as

$$\begin{pmatrix} U \\ H \end{pmatrix}_t = \begin{pmatrix} H_x \\ -(\frac{1}{3}U_{xx} + U^2)_x \end{pmatrix} \tag{4.3}$$

It is known that one recursion operator for (4.3) is [7, 36]

$$M = \begin{pmatrix} M_{11} & M_{12} \\ M_{21} & M_{22} \end{pmatrix} \tag{4.4}$$

where

$$M_{11} = 9H + 6H_x\partial_x^{-1}, M_{12} = 4\partial_x^2 + 6U + 3U_x\partial_x^{-1},$$
$$M_{21} = -\frac{4}{3}\partial_x^4 - 10U\partial_x^2 - 15U_x\partial_x - (9U_{xx} + 12U^2) - (2U_{xx} + 12UU_x)\partial_x^{-1},$$
$$M_{22} = 9H + 3H_x\partial_x^{-1}.$$

Then we have the following Boussinesq hierarchy

$$\begin{pmatrix} U \\ H \end{pmatrix}_t = M^n \begin{pmatrix} H_x \\ -(\frac{1}{3}U_{xx} + U^2)_x \end{pmatrix}, \qquad n \geq 0. \tag{4.5}$$

Introducing infinite time variables $x = t_1, t_2, t_4, t_5, \cdots, t_{3n+1}, t_{3n+2}, \cdots$ and viewing U, H as function of $t = (x, t_2, t_4, t_5, \cdots)$, then we have the following equivalent form of (4.5)

$$\begin{pmatrix} U \\ H \end{pmatrix}_{t_2} = \begin{pmatrix} H_x \\ -(\frac{1}{3}U_{xx} + U^2)_x \end{pmatrix} \tag{4.6}$$

$$\begin{pmatrix} U \\ H \end{pmatrix}_{t_{m+3}} = M \begin{pmatrix} U_{t_m} \\ H_{t_m} \end{pmatrix}, \qquad m \neq 3k, \quad k, m \in \mathbf{Z}^+ \tag{4.7}$$

Specially when $m = 1$, Eq. (4.7) becomes

$$U_{t_4} = 4H_{xxx} + 12(UH)_x \tag{4.8}$$

$$H_{t_4} = -\frac{4}{3}U_{5x} - 12UU_{xxx} - 24U_xU_{xx} - 24U^2U_x + 12HH_x \tag{4.9}$$

In the following discussion, we assume that U and H decay rapidly to zero as $|x| \to \infty$ and introduce new function W which satisfies the relation

$$W_x = U. \tag{4.10}$$

Then from (4.6) we have

$$H = W_{t_2}. \tag{4.11}$$

Next we shall show that under the relation (4.11), Eqs. (4.6) and (4.7) become

$$U_{t_2t_2} = -(\frac{U_{xx}}{3} + U^2)_{xx} \tag{4.12}$$

$$W_{t_{m+3}} = 3W_{xt_2t_m} + \frac{1}{4}\partial_x^{-1}W_{t_4t_m} + 6W_{t_2}W_{t_m} + 3U\partial_x^{-1}W_{t_2t_m}, \tag{4.13}$$

respectively, where $m \neq 3k$ and $k, m \in \mathbf{Z}^+$. Under the relation (4.10) and (4.11), Eq. (4.12) is equivalent to (4.6), so we only need to prove (4.13) is equivalent to (4.7). We first consider the case $m = 1$ in (4.13), or the equivalence of (4.8)-(4.9) and (4.13). Suppose (4.8)-(4.9) hold, integrate (4.8) and by use of (4.11), we have

$$W_{t_4} = 4W_{t_2xx} + 12UW_{t_2}, \tag{4.14}$$

which is just the case $m = 1$ in (4.13). On the other side, by use of (4.14) and (4.11) we can obtain (4.8). Differentiating Eq.(4.14) with respect to t_2, we have

$$\begin{aligned} H_{t_4} &= W_{t_2 t_4} = 4W_{t_2 t_2 xx} + 12UW_{t_2 t_2} + 12U_{t_2} W_{t_2} \\ &= -4(\frac{1}{3}U_{xx} + U^2)_{xxx} + 12HH_x - 12U(\frac{1}{3}U_{xx} + U^2)_x \\ &= -\frac{4}{3}U_{5x} - 12UU_{xxx} - 24U_x U_{xx} - 24U^2 U_x + 12HH_x, \end{aligned}$$

which is just Eq. (4.9). By now we have proved the equivalence between (4.8)-(4.9) and (4.13) when $m = 1$. For general case, if (4.7) holds, we have

$$\begin{aligned} U_{t_{m+3}} &= 9HU_{t_m} + 6H_x \partial_x^{-1} U_{t_m} + 4H_{xxt_m} + 6UH_{t_m} + 3U_x \partial_x^{-1} H_{t_m} \\ &= 9W_{t_2} U_{t_m} + 6U_{t_2} \partial_x^{-1} U_{t_m} + 4W_{xxt_2 t_m} + 6UW_{t_2 t_m} + 3U_x \partial_x^{-1} W_{t_2 t_m} \\ &= 3W_{xxt_2 t_m} + 6W_{t_2} U_{t_m} + 3UW_{t_2 t_m} + 6U_{t_2} \partial_x^{-1} U_{t_m} \\ &\qquad + 3U_x \partial_x^{-1} W_{t_2 t_m} + \frac{1}{4} W_{t_4 t_m}. \end{aligned}$$

Integrate once with respect to x, then Eq. (4.13) is obtained. Conversely, if (4.13) holds, we have

$$U_{t_{m+3}} = 9HU_{t_m} + 6H_x \partial_x^{-1} U_{t_m} + 4H_{xxt_m} + 6UH_{t_m} + 3U_x \partial_x^{-1} H_{t_m}.$$

Differentiating Eq. (4.13) with variable t_2 and by use of (4.10)-(4.12) and (4.9), we have

$$\begin{aligned} & H_{t_{m+3}} = W_{t_2 t_{m+3}} \\ & = 3W_{xt_2 t_2 t_m} + \frac{1}{4}\partial_x^{-1} W_{t_2 t_4 t_m} \\ & + 6W_{t_2 t_2} W_{t_m} + 6W_{t_2} W_{t_2 t_m} + 3U_{t_2} \partial_x^{-1} W_{t_2 t_m} + 3U\partial_x^{-1} W_{t_2 t_2 t_m} \\ & = 3\left[-(\frac{U_{xx}}{3} + U^2)\right]_{xxt_m} + \frac{1}{4}\partial_x^{-1}\left[-\frac{4}{3}U_{5x} - 12UU_{3x}\right. \\ & \left. - 24U_x U_{xx} - 24U^2 U_x + 12HH_x\right]_{t_m} \\ & - 6W_{t_m}(\frac{U_{xx}}{3} + U^2)_x + 6HH_{t_m} + 3H_x \partial_x^{-1} H_{t_m} - 3U\partial_x^{-1}(\frac{U_{xx}}{3} + U^2)_{xt_m} \\ & = \left[-\frac{4}{3}\partial_x^4 - 10U\partial_x^2 - 15U_x \partial_x - (9U_{xx} + 12U^2)\right. \\ & \left. - 2(U_{xxx} + 12UU_x)\partial_x^{-1}\right] U_{t_m} + (9H + 3H_x \partial_x^{-1}) H_{t_m}, \end{aligned}$$

so Eq. (4.7) holds. By now we have given the equivalence between (4.7) and (4.13). Let $U = 2(\ln \tau)_{xx}$ and by use of (A5)-(A7), we can deduce the following bilinear equations from (4.12) and (4.13)

$$(D_1^4 + 3D_2^2)\tau \cdot \tau = 0, \tag{4.15}$$

$$(D_1 D_{m+3} - 3D_1^2 D_2 D_m - \frac{1}{4}D_4 D_m)\tau \cdot \tau = 0, \quad m \neq 3k, \quad m, k \in \mathbf{Z}^+ \tag{4.16}$$

where we have denoted $D_k D_m = D_{t_k} D_{t_m}$ for simplicity. Equations (4.15)-(4.16) are the bilinear form of the Boussinesq hierarchy.

5 Bilinear form of Levi hierarchy

The Levi hierarchy is [3, 18–20]

$$\begin{pmatrix} u_t \\ v_t \end{pmatrix} = L^{*n} \begin{pmatrix} u_x \\ v_x \end{pmatrix}, \qquad (n \geq 1) \tag{5.1}$$

where

$$L^* = \begin{pmatrix} \partial_x + v - \partial_x u \partial_x^{-1} & u + \partial_x u \partial_x^{-1} \\ -v - \partial_x v \partial_x^{-1} & -\partial_x - u + \partial_x v \partial_x^{-1} \end{pmatrix}.$$

is a recursion operator. In particular, when $n = 1$, (5.1) becomes

$$u_t = u_{xx} + 2(uv)_x - 2uu_x, \tag{5.2}$$

$$v_t = -v_{xx} - 2(uv)_x + 2vv_x. \tag{5.3}$$

By introducing an infinite number of time variables $x = t_1, t_2, t_3, \cdots$ and regarding u, v as functions of $t = t(t_1, t_2, t_3 \cdots)$, we have an equivalent form of (5.1)

$$\begin{pmatrix} u_{t_{n+1}} \\ v_{t_{n+1}} \end{pmatrix} = L \begin{pmatrix} u_{t_n} \\ v_{t_n} \end{pmatrix}, \qquad (n \geq 1) \tag{5.4}$$

i.e.

$$u_{t_{n+1}} = u_{xt_n} + (uv)_{t_n} - (u\partial_x^{-1} u_{t_n})_x + (u\partial_x^{-1} v_{t_n})_x, \tag{5.5}$$

$$v_{t_{n+1}} = -v_{xt_n} - (uv)_{t_n} - (v\partial_x^{-1} u_{t_n})_x + (v\partial_x^{-1} v_{t_n})_x. \tag{5.6}$$

Through the dependent variable transform

$$u = -2\frac{BG^*}{FA}, v = 2\frac{AG}{FB},$$

Equations (5.5) and (5.6) are transformed into the bilinear form

$$D_x F \cdot A = -2BG^* \tag{5.7}$$

$$D_x F \cdot B = 2AG, \tag{5.8}$$

$$D_x^2 F \cdot F = 8GG^*, \tag{5.9}$$

$$(D_{t_{n+1}} + \frac{1}{2} D_x D_{t_n}) F \cdot A = D_{t_n} B \cdot G^*, \tag{5.10}$$

$$(D_{t_{n+1}} - \frac{1}{2} D_x D_{t_n}) F \cdot B = D_{t_n} A \cdot G. \tag{5.11}$$

In fact, we have, by using (5.7) and (5.8),

$$u = \frac{D_x F \cdot A}{FA} = (\ln \frac{F}{A})_x, v = \frac{D_x F \cdot B}{FB} = (\ln \frac{F}{B})_x.$$

Thus (5.5) becomes

$$
\begin{aligned}
\frac{D_{t_{n+1}}F\cdot A}{FA} &= (\ln\frac{F}{A})_{t_{n+1}} = (\ln\frac{F}{A})_{xt_n} + \partial_x^{-1}\left[-4\frac{BG^*}{FA}\frac{AG}{FB}\right]_{t_n} \\
&\quad -\frac{D_xF\cdot A}{FA}(\ln\frac{F}{A})_{t_n} + \frac{D_xF\cdot A}{FA}(\ln\frac{F}{B})_{t_n} \\
&= \left(\frac{D_xF\cdot A}{FA}\right)_{t_n} - 4\partial_x^{-1}\left[\frac{GG^*}{F^2}\right]_{t_n} - \frac{D_xF\cdot A}{FA}\frac{D_{t_n}F\cdot A}{FA} \\
&\quad + \frac{D_xF\cdot A}{FA}\frac{D_{t_n}F\cdot B}{FB} \\
&= \left(\frac{F_x}{F} - \frac{A_x}{A}\right)_{t_n} - \frac{1}{2}\partial_x^{-1}\left[\frac{D_x^2F\cdot F}{F^2}\right]_{t_n} \\
&\quad -\frac{F_xF_{t_n}A^2 - F_xA_{t_n}FA - F_{t_n}A_xFA + F^2A_xA_{t_n}}{F^2A^2} \\
&\quad +\frac{F_xF_{t_n}AB - F_xB_{t_n}FA - F_{t_n}A_xFB + F^2A_xB_{t_n}}{F^2AB} \\
&= \left(\frac{F_x}{F} - \frac{A_x}{A}\right)_{t_n} - \left(\frac{F_x}{F}\right)_{t_n} - \frac{-F_xA_{t_n}FA + F^2A_xA_{t_n}}{F^2A^2} \\
&\quad + \frac{-F_xB_{t_n}FA + F^2A_xB_{t_n}}{F^2AB} \\
&= -\frac{A_{xt_n}}{A} + \frac{F_xA_{t_n}}{FA} - \frac{F_xB_{t_n}}{FB} + \frac{A_xB_{t_n}}{AB} \\
&= \frac{-FA_{xt_n} + F_xA_{t_n}}{FA} - \frac{B_{t_n}D_xF\cdot A}{FAB} \\
&= \frac{-FA_{xt_n} + F_xA_{t_n}}{FA} + 2\frac{B_{t_n}BG^*}{FAB} \\
&= \frac{-\frac{1}{2}D_xD_{t_n}F\cdot A + \frac{1}{2}(D_xF\cdot A)_{t_n} + 2B_{t_n}G^*}{FA} \\
&= \frac{-\frac{1}{2}D_xD_{t_n}F\cdot A - (BG^*)_{t_n} + 2B_{t_n}G^*}{FA} \\
&= \frac{-\frac{1}{2}D_xD_{t_n}F\cdot A + D_{t_n}B\cdot G^*}{FA}
\end{aligned}
$$

which implies that (5.10) holds. Similarly we can get (5.11) from (5.6).

Remark. We have from (5.7), (5.8) and (5.9) that

$$
\begin{aligned}
0 &= \frac{1}{2}\frac{D_x^2 F\cdot F}{F^2} - 4\frac{GA}{FA}\frac{G^*B}{FB}\\
&= \frac{1}{2}\frac{D_x^2 F\cdot F}{F^2} + \frac{(D_xF\cdot A)(D_xF\cdot B)}{F^2AB}\\
&= \frac{1}{2}\frac{D_x^2 F\cdot A}{FA} + \frac{1}{2}\frac{(D_xF\cdot A)_x}{FA} - \frac{B_xD_xF\cdot A}{FAB}\\
&= \frac{1}{2}\frac{D_x^2 F\cdot A}{FA} - \frac{(BG^*)_x}{FA} + 2\frac{B_xBG^*}{FAB}\\
&= \frac{1}{2}\frac{D_x^2 F\cdot A}{FA} + \frac{D_xB\cdot G^*}{FA}
\end{aligned}
$$

which implies that (5.10) with $n=1$ can be rewritten as

$$(D_{t_2} + D_x^2)F\cdot A = 0. \tag{5.12}$$

Similarly (5.11) with $n=1$ can be rewritten as

$$(D_{t_2} - D_x^2)F\cdot B = 0. \tag{5.13}$$

So (5.7), (5.8), (5.9), (5.12) and (5.13) are another bilinear form for (5.2) which was obtained in [26].

6 Bilinear form of Yajima-Oikawa hierarchy

The Yajima-Oikawa hierarchy reads [4]

$$\begin{pmatrix} q_t \\ 2u_t \\ r_t \end{pmatrix} = \Phi^n \begin{pmatrix} q \\ 0 \\ -r \end{pmatrix}, \qquad (n\geq 1) \tag{6.1}$$

$$\begin{pmatrix} q_t \\ 2u_t \\ r_t \end{pmatrix} = \Phi^n \begin{pmatrix} q_x \\ 2u_x \\ r_x \end{pmatrix}, \qquad (n\geq 1) \tag{6.2}$$

where

$$\Phi = \begin{pmatrix} A_{11} & A_{12} & A_{13} \\ A_{21} & A_{22} & A_{23} \\ A_{31} & A_{32} & A_{33} \end{pmatrix}$$

is a recursion operator, and

$$
\begin{aligned}
&A_{11} = \partial_x^2 + 2u + \frac{3}{2}q\partial_x^{-1}r, A_{12} = \frac{3}{4}q + \frac{1}{2}q_x\partial_x^{-1},\\
&A_{13} = \frac{3}{2}q\partial_x^{-1}q, A_{21} = \frac{3}{2}r\partial_x + \frac{1}{2}r_x,\\
&A_{22} = \frac{1}{4}\partial_x^2 + 2u + u_x\partial_x^{-1}, A_{23} = -\frac{3}{2}q\partial_x - \frac{1}{2}q_x,\\
&A_{31} = -\frac{3}{2}r\partial_x^{-1}r, A_{32} = \frac{3}{4}r + \frac{1}{2}r_x\partial_x^{-1},\\
&A_{33} = \partial_x^2 + 2u - \frac{3}{2}r\partial_x^{-1}q.
\end{aligned}
$$

By introducing an infinite number of time variables $x = t_1, t_2, t_3, \cdots$ and regarding q, u, r as functions of $t = t(t_1, t_2, t_3 \cdots)$, we have an equivalent form of (6.1) and (6.2)

$$\begin{pmatrix} q_{t_2} \\ 2u_{t_2} \\ r_{t_2} \end{pmatrix} = \Phi \begin{pmatrix} q \\ 0 \\ -r \end{pmatrix} = \begin{pmatrix} q_{xx} + 2uq \\ 2(qr)_x \\ -r_{xx} - 2ur \end{pmatrix}, \qquad (n \geq 1) \tag{6.3}$$

$$\begin{pmatrix} q_{t_{n+2}} \\ 2u_{t_{n+2}} \\ r_{t_{n+2}} \end{pmatrix} = \Phi \begin{pmatrix} q_{t_n} \\ 2u_{t_n} \\ r_{t_n} \end{pmatrix}, \qquad (n \geq 1) \tag{6.4}$$

Equation (6.3) is just the Yajima-Oikawa equation, and Eq. (6.4) can be rewritten as

$$q_{t_{n+2}} = q_{xxt_n} + 2uq_{t_n} + \frac{3}{2}qu_{t_n} + q_x\partial_x^{-1}u_{t_n} + \frac{3}{2}q\partial_x^{-1}(qr)_{t_n}, \tag{6.5}$$

$$\begin{aligned} 2u_{t_{n+2}} &= \frac{3}{2}rq_{xt_n} + \frac{1}{2}r_xq_{t_n} + \frac{1}{2}u_{xxt_n} + 4uu_{t_n} \\ &\quad +2u_x\partial_x^{-1}u_{t_n} - \frac{3}{2}qr_{xt_n} - \frac{1}{2}q_xr_{t_n}, \end{aligned} \tag{6.6}$$

$$r_{t_{n+2}} = r_{xxt_n} + 2ur_{t_n} + \frac{3}{2}ru_{t_n} + r_x\partial_x^{-1}u_{t_n} + \frac{3}{2}r\partial_x^{-1}(qr)_{t_n} \tag{6.7}$$

Through the dependent variable transforms

$$q = \frac{\rho}{\tau}, r = \frac{\sigma}{\tau}, u = (\ln \tau)_{xx},$$

the Yajima-Oikawa hierarchy is transformed into the following bilinear forms

$$(D_2 - D_1^2)\rho \cdot \tau = 0, \tag{6.8}$$

$$(D_2 + D_1^2)\sigma \cdot \tau = 0, \tag{6.9}$$

$$D_1D_2\tau \cdot \tau = 2\rho\sigma, \tag{6.10}$$

$$(4D_{n+2} - D_1^2D_n - 3D_2D_n)\rho \cdot \tau = 0, \tag{6.11}$$

$$(4D_{n+2} - D_1^2D_n + 3D_2D_n)\sigma \cdot \tau = 0, \tag{6.12}$$

$$(\frac{1}{6}D_1^3D_n - D_1D_{n+2} + \frac{1}{3}D_3D_n)\tau \cdot \tau = D_n\sigma \cdot \rho, \tag{6.13}$$

where we have denoted $D_1D_2 \equiv D_{t_1}D_{t_2}$, etc. It is not hard to verify that (6.8), (6.9) and (6.10) are the bilinear forms of (6.3) [5]. Now we verify that (6.11), (6.12) and (6.13) are the bilinear forms of (6.5), (6.7) and (6.6), respectively. In fact, we set $\phi = \ln \frac{\rho}{\tau}, \psi = \ln(\rho\tau)$. Then we have from (6.8) and (6.10)

$$\phi_{t_2} = \psi_{xx} + \phi_x^2 \tag{6.14}$$

$$qr = \frac{\rho\sigma}{\tau^2} = \frac{1}{2}\frac{D_1D_2\tau \cdot \tau}{\tau^2} = \frac{1}{2}(\psi - \phi)_{t_1t_2} \tag{6.15}$$

from which it follows that

$$\psi_{t_2} = \psi_{xx} + \phi_x^2 + 2\partial_x^{-1}(qr). \tag{6.16}$$

Further, from (6.11) we have by use of (A1)-(A4)

$$\begin{aligned}
0 &= 4\phi_{t_{n+2}} - (\phi_{xxt_n} + 2\phi_x\psi_{xt_n} + \phi_{t_n}\psi_{xx} + \phi_{t_n}\phi_x^2) \\
&\quad -3(\psi_{t_2t_n} + \phi_{t_2}\phi_{t_n}) \\
&= 4\phi_{t_{n+2}} - (\phi_{xxt_n} + 2\phi_x\psi_{xt_n} + \phi_{t_n}\psi_{xx} + \phi_{t_n}\phi_x^2) \\
&\quad -3(\psi_{xx} + \phi_x^2 + 2\partial_x^{-1}(qr))_{t_n} - 3(\psi_{xx} + \phi_x^2)\phi_{t_n} \\
&= 4\phi_{t_{n+2}} - \phi_{xxt_n} - 4\phi_{t_n}\phi_x^2 - 6\phi_x\psi_{xt_n} - 6\partial_x^{-1}(qr)_{t_n} \\
&\quad -4(\psi_{xx} + 2u)\phi_{t_n} - 2\phi_x\partial_x^{-1}(\psi_{xx} + 2u)_{t_n} - 3(\psi_{xx} + 2u)_{t_n} \\
&= 4\phi_{t_{n+2}} - 4\phi_{xxt_n} - 4\phi_{t_n}\phi_x^2 - 8\phi_x\psi_{xt_n} - 4\psi_{xx}\phi_{t_n} \\
&\quad -6\partial_x^{-1}(qr)_{t_n} - 8u\phi_{t_n} - 4\phi_x\partial_x^{-1}u_{t_n} - 6u_{t_n}
\end{aligned}$$

which implies that (6.5) holds by noticing $q = e^{\phi}$. Similarly we can deduce (6.7) from (6.9), (6.10) and (6.12). Next, from (6.13) we obtain,

$$w_{t_3} = \frac{3}{4}(q_x r - q r_x) + \frac{1}{4}w_{xxx} + \frac{2}{3}u^2 \tag{6.17}$$

and

$$\begin{aligned}
&\frac{1}{3}(6uw_{t_n} + uu_{xt_n}) - 2w_{t_{n+2}} + \frac{2}{3}\partial_x^{-1}w_{t_nt_3} \\
&= \frac{\sigma_{t_n}\rho - \sigma\rho_{t_n}}{\tau^2} \\
&= \frac{\sigma_{t_n}\tau - \sigma\tau_{t_n}}{\tau^2}\frac{\rho}{\tau} - \frac{\sigma}{\tau}\frac{\rho_{t_n}\tau - \rho\tau_{t_n}}{\tau^2} \\
&= qr_{t_n} - q_{t_n}r
\end{aligned} \tag{6.18}$$

where $u = w_x = (\ln\tau)_{xx}$. Differentiating (6.18) with respect to x and using (6.17), we immediately obtain (6.6).

Remark. Cheng and Zhang [5] have obtained the bilinear equations for the k-constrained KP hierarchy. As a special case, the bilinear equations for the Yajima-Oikawa hierarchy are

$$e^{\sum_{i=1}^{\infty} y_iD_i}\sigma\cdot\rho = \sum_{j=0}^{\infty} p_j(-2y)p_{j+2}(\tilde{D})e^{\sum_{i=1}^{\infty} y_iD_i}\tau\cdot\tau$$

$$e^{\sum_{i=1}^{\infty} y_iD_i}\tau\cdot\rho = \sum_{j=0}^{\infty} p_j(-2y)p_{j+2}(\tilde{D})e^{\sum_{i=1}^{\infty} y_iD_i}\rho\cdot\tau$$

$$e^{\sum_{i=1}^{\infty} y_iD_i}\sigma\cdot\tau = \sum_{j=0}^{\infty} p_j(-2y)p_{j+2}(\tilde{D})e^{\sum_{i=1}^{\infty} y_iD_i}\tau\cdot\sigma$$

where $y = (y_1, y_2, \cdots)$, $\tilde{D} = (D_1, \frac{1}{2}D_2, \frac{1}{3}D_3, \cdots)$ and $p_j(t), j = 1, 2, \cdots$ are the Schur polynomials defined by

$$e^{\sum_{n=1}^{\infty}\lambda^n t_n} = \sum_{j=0}^{\infty} p_j(t)\lambda^j.$$

However, the bilinear Eqs. (6.8)-(6.12) given by us possess unified and explicit structures in the form.

7 Conclusion and discussions

In this chapter, we have shown unified bilinear forms for several integrable hierarchies of equations including KdV, Sawada-Koterra, AKNS, Boussinesq, NLS, Ito, Ramani, Levi and Yajima-Oikawa hierarchies. As is seen, resulted bilinear equations possess unified and explicit structures in the form. Such a simple structure would be easier to treat, and lead much convenience in calculation when the whole hierarchy of equations is considered. For example, in [12, 14–16, 39] we have derived bilinear Bäcklund transformations for unified bilinear forms of the Boussinesq, Sawada-Koterra, Ito, Ramani, Lotka-Volterra and negative Lotka-Volterra hierarchies. In Appendix B, we will give a list of bilinear Bäcklund transformations for all these mentioned hierarchies. Next, we give another example of unified bilinear form application. It is known [9] that under certain conditions the classical Boussinesq equation can be transformed into the NLS equation. We can now easily generalize this result to the classical Boussinesq hierarchy (1.4) and (1.5) and the NLS hierarchy (3.13), (3.14) and (3.15). In fact, set $\phi = f/g, \rho = 2(\ln g)_x$ and assume g to be real. We have from (1.4) and (1.5),

$$\phi_{t_2} = \frac{i}{2}\phi_{xx} + \frac{i}{2}\phi\rho_x \tag{7.1}$$

$$\phi_{xt_2} + \phi\rho_{t_2} = \frac{i}{2}\phi_{xxx} + \frac{3}{2}i\phi_x\rho_x. \tag{7.2}$$

Differentiating (7.1) with respect to x and subtracting (7.2) from it, we get

$$\phi\rho_{t_2} - i\phi_x\rho_x + \frac{i}{2}\phi\rho_{xx} = 0. \tag{7.3}$$

Notice that ρ is real. Multiplying (7.3) by ϕ^* and subtracting the complex conjugate of it, we get

$$i\rho_{xx}\phi\phi^* - i\rho_x(\phi\phi^*)_x = 0,$$

which gives

$$\rho_x = c_0\phi\phi^*, \tag{7.4}$$

where c_0 is an integration constant. We choose $c_0 > 0$ without loss of generality, and set $F = g, G = \sqrt{\frac{c_0}{2}}f$. In this case, (7.4) becomes (3.13). Further, we have from (1.4) and (1.5) that

$$(D_{t_2} - \frac{i}{2}D_x^2)G \cdot F = 0 \tag{7.5}$$

$$(iD_{t_{2(n+1)}} + \frac{1}{2}D_xD_{t_{2n+1}})G \cdot F = 0 \tag{7.6}$$

$$(D_xD_{t_{2n+1}} - \frac{1}{2}D_{t_2}D_{t_{2n}} - \frac{i}{4}D_x^2D_{t_{2n}})G \cdot F = 0, \tag{7.7}$$

which imply that (3.13), (3.14) and (3.15) hold, i.e., the classical Boussinesq hierarchy can be transformed into the NLS hierarchy.

Next, it is noted that in the literature there are some studies on so-called generalized soliton solutions, see, e.g., [28–30, 32, 37, 38]. As an illustrative example, let us see the generalized solition solutions for the bilinear KdV equation $D_x(D_t + D_x^3)f \cdot f = 0$:

$$
\begin{aligned}
f &= 1 + \sum_{n=1}^{\infty} f_n \epsilon^n \\
&= 1 + \sum_{n=1}^{\infty} \frac{\epsilon^n}{n!} \int_\Gamma \int_\Gamma \cdots \int_\Gamma \det(\Delta_n) \exp\left[\sum_{j=1}^{n} (p_j x - p_j^3 t)\right] \prod_{j=1}^{n} d\tau(p_j)
\end{aligned}
$$

where

$$(\Delta_n)_{jk} = \frac{2p_j}{p_j + p_k}, \qquad j, k = 1, 2, \cdots, n.$$

In particular,

$$f_1 = \int_\Gamma \exp(px - p^3 t) d\tau(p).$$

Here $\int_\Gamma d\tau(p)$ denotes the contour integral along the contour Γ which lies in the left half of the complex p plane (that is, $Re\ p < 0$) and which goes from $p = -i\infty$ to $p = i\infty$.

Along this line and with our unified bilinear form for the KdV hierarchy (1.1), we have the following generalized soliton solutions to (1.1):

$$f = 1 + \sum_{n=1}^{\infty} \frac{\epsilon^n}{n!} \int_\Gamma \int_\Gamma \cdots \int_\Gamma \det(\Delta_n) \exp\left[\sum_{j=1}^{n} (p_j x - p_j^3 t_3 + p_j^5 t_5 - \cdots)\right] \prod_{j=1}^{n} d\tau(p_j) \tag{7.8}$$

where

$$(\Delta_n)_{jk} = \frac{2p_j}{p_j + p_k}, \qquad j, k = 1, 2, \cdots, n.$$

Finally, it is remarked that unified bilinear forms may provide us with a convenient way to seek Virasoro symmetries, say, for the bilinear KdV hierarchy, directly.

Acknowledgements

This work was partially supported by the National Natural Science Foundation of China (Grant No. 11871336, 11931017, 12071447).

A Bilinear identities

The following bilinear identities hold for arbitrary functions a, b:

$$\frac{D_z a \cdot b}{ab} = \phi_z \tag{A1}$$

$$\frac{(D_x^2 a \cdot b)}{ab} = \rho_{xx} + \phi_x^2 \tag{A2}$$

$$\frac{(D_y D_z a \cdot b)}{ab} = \rho_{yz} + \phi_y \phi_z \tag{A3}$$

$$\frac{D_x^2 D_y a \cdot b}{ab} = \phi_{xxy} + 2\phi_x \rho_{xy} + \phi_y \rho_{xx} + \phi_y \phi_x^2 \tag{A4}$$

where $\phi = \ln \frac{a}{b}, \quad \rho = \ln(ab)$.

$$\frac{D_x D_t a \cdot a}{a^2} = w_z \tag{A5}$$

$$\frac{(D_y D_z a \cdot a)}{a^2} = \partial_x^{-1} w_{yz} \tag{A6}$$

$$\frac{D_x^2 D_y D_z a \cdot a}{a^2} = 2w_y w_z + u\partial_x^{-1} w_{yz} + u_{yz}. \tag{A7}$$

where $u = 2(\ln a)_{xx}, \quad w_x = u$.

B Bilinear Bäcklund transformations for the Boussinesq, Sawada-Koterra, Ito, Ramani, Lotka-Volterra and negative Volterra hierarchies

For the sake of completeness, we list bilinear Bäcklund transformations for the Boussinesq, Sawada-Koterra, Ito, Ramani, Lotka-Volterra and negative Lotka-Volterra hierarchies in the following:

Example B.1 A bilinear Bäcklund transformation for the Boussinesq hierarchy (4.15) and (4.16) is [14]

$$(D_1^2 + D_2 + \lambda)\tau \cdot \tau' = 0, \tag{B1}$$

$$(D_1^3 - 3\lambda D_1 - 3D_1 D_2 + \mu)\tau \cdot \tau' = 0, \tag{B2}$$

$$(12D_{m+3} + D_1^3 D_m - 3D_1 D_2 D_m - 3\lambda D_1 D_m - 2\mu D_m)\tau \cdot \tau' = 0, \tag{B3}$$

where λ and μ are two arbitrary constants and $m \neq 3k, \quad m, k \in \mathbf{Z}^+$.

Example B.2 A bilinear Bäcklund transformation for the Sawada-Kotera hierarchy (1.6) and (1.7) is [12]

$$(D_1^3-\lambda)f\cdot f'=0, \tag{B4}$$
$$(D_5+\frac{15}{2}\lambda D_1^2+\frac{3}{2}D_1^5)f\cdot f'=0, \tag{B5}$$
$$(D_7+\frac{3}{4}D_1^7+\frac{21}{4}\lambda D_1^4+21\lambda^2 D_1)f\cdot f'=0, \tag{B6}$$
$$(D_{m+6}+\frac{3}{5}D_1^6D_m-\frac{3}{5}D_1D_5D_m+6\lambda D_1^3D_m+15\lambda^2D_m)f\cdot f'=0, \tag{B7}$$

where λ is an arbitrary constant and m is an odd integer, $m\neq 3k, m,k\in Z_+$.

Example B.3 A bilinear Bäcklund transformation for the Ito hierarchy (1.8), (1.9) and (1.10) is [39]

$$D_1D_3f\cdot f'=0, \tag{B8}$$
$$(D_3+D_1^3)f\cdot f'=0, \tag{B9}$$
$$(D_5+\frac{1}{6}D_1^5-\frac{5}{6}D_1^2D_3)f\cdot f'=0, \tag{B10}$$
$$(D_{2m+5}-\frac{1}{45}D_1^6D_{2m-1}+\frac{1}{5}D_1D_5D_{2m-1}+\frac{2}{9}D_1^3D_3D_{2m-1})f\cdot f'=0, \tag{B11}$$

where m is a positive integer.

Example B.4 A bilinear Bäcklund transformation for the Ramani hierarchy (1.11), (1.12), (1.13) and (1.14) is [39]

$$(D_1^3-D_3)f\cdot f'=0, \tag{B12}$$
$$(D_1^5+5D_1^2D_3)f\cdot f'=0, \tag{B13}$$
$$(180D_7+\frac{1}{4}D_1^7-\frac{245}{4}D_1^4D_3+70D_1D_3^3)f\cdot f'=0, \tag{B14}$$
$$(\frac{5}{2}D_1^9+\frac{165}{2}D_1^6D_3+300D_1^3D_3^2+200D_3^3-1800D_9)f\cdot f'=0, \tag{B15}$$
$$\begin{aligned}(\frac{1139}{475200}D_1^{11}&+\frac{983}{20160}D_1^8D_3-\frac{2}{5}D_1^5D_3^2+\frac{47}{108}D_1^2D_3^3\\&-\frac{43}{6}D_1^2D_9+\frac{20}{11}D_{11}+\frac{27}{7}D_1D_3D_7)f\cdot f'=0,\end{aligned} \tag{B16}$$
$$\begin{aligned}(D_{m+10}&+\frac{1}{2025}D_1^{10}D_m-\frac{2}{945}D_1^7D_3D_m+\frac{1}{27}D_1^4D_3^2D_m\\&-\frac{4}{81}D_1D_3^3D_m+\frac{1}{7}D_3D_7D_m+\frac{1}{9}D_m)f\cdot f'=0,\end{aligned} \tag{B17}$$

where m is an odd integer, $m\neq 5k, m,k\in Z_+$.

Example B.5 A bilinear Bäcklund transformation for the Lotka-Volterra hierarchy (1.15) and (1.16) is [15]

$$(e^{D_n} - \lambda - \mu e^{-D_n}) f_n \cdot g_n = 0, \tag{B18}$$

$$(D_{t_1} - \lambda e^{-D_n} - \gamma) f_n \cdot g_n = 0, \tag{B19}$$

$$(D_{t_k} + \frac{\lambda}{2} D_{t_{k-1}} e^{-D_n} - \frac{\lambda}{2\mu} D_{t_{k-1}} e^{D_n} - \frac{\lambda^2}{2\mu} D_{t_{k-1}}) f_n \cdot g_n = 0, \tag{B20}$$

where λ, μ and γ are arbitrary constants.

Example B.6 A bilinear Bäcklund transformation for the negative Volterra hierarchy (1.17) and (1.18) is [16]

$$(e^{\frac{1}{2}D_n} - \lambda e^{\frac{3}{2}D_n} - \mu e^{-\frac{1}{2}D_n}) f_n \cdot g_n = 0, \tag{B21}$$

$$(D_1 + \lambda e^{D_n} + \gamma) f_n \cdot g_n = 0, \tag{B22}$$

$$\begin{aligned}(\frac{\lambda}{\mu} D_{j+1} e^{\frac{3}{2}D_n} - 2\lambda D_j e^{\frac{1}{2}D_n} + \frac{1}{\mu} D_{j+1} e^{\frac{1}{2}D_n} \\ + D_{j+1} e^{-\frac{1}{2}D_n} + \theta e^{\frac{1}{2}D_n}) f_n \cdot g_n = 0,\end{aligned} \tag{B23}$$

where λ, μ, γ and θ are arbitrary constants.

References

[1] Ablowitz M J, Kaup D J, Newell A C and Segur H, The inverse scattering transform-Fourier analysis for nonlinear problems, *Stud. Appl. Math.* **53**, 249-315, 1974.

[2] Ablowitz M J and Segur H, *Solitons and the Inverse Scattering Transform*, SIAM Philadelphia, 1985.

[3] Antonowicz M and Fordy A P, Factorisation of energy dependent Schrödinger operators: Miura maps and modified systems, *Comm. Math. Phys.* **124**, no. 3, 465-486, 1989.

[4] Cheng Y, Constraints of the Kadomtsev-Petviashvili hierarchy, *J. Math. Phys.* **33**, no. 11, 3774-3782, 1992.

[5] Cheng Y and Zhang Y J, Bilinear equations for the constrained KP hierarchy, *Inverse Problems* **10**, no. 2, L11-L17, 1994.

[6] Date E, Kashiwara M, Jimbo M and Miwa T, *Nonlinear Integrable Systems-Classical Theory and Quantum Theory*, Proc. RIMS Symp. Miwa T and Jimbo M (eds.), Singapore: World Scientific, 39-119, 1983

[7] Fokas A S and Anderson R L, On the use of isospectral eigenvalue problems for obtaining hereditary symmetries for Hamiltonian systems, *J. Math. Phys.* **23**, 1066-1073, 1982.

[8] Gardner C S, Greene J M, Kruskal M D and Miura R M, Method for solving the Korteweg-de Vries equation, *Phys. Rev. Lett.* **19**, 1095-1097, 1967.

[9] Hirota R, Classical Boussinesq equation is a reduction of the modified KP equation, *J. Phys. Soc. Japan* **54**, no. 7, 2409-2415, 1985.

[10] Hu X B, *Integrable systems and related problems*, Doctoral Dissertation, Computing Center of Chinese Academy of Sciences, 1990.

[11] Hu X B, Nonlinear superposition formula of the KdV equation with a source, *J. Phys. A* **24**, no. 24, 5489-5497, 1991.

[12] Hu X B and Bullough R, A Bäcklund transformation and nonlinear superposition formula of the Caudrey-Dodd-Gibbon-Kotera-Sawada hierarchy, *J. Phys. Soc. Japan* **67**, no. 3, 772-777, 1998.

[13] Hu X B and Li Y, Some results on the Caudrey-Dodd-Gibbon-Kotera-Sawada equation, *J. Phys. A* **24**, no. 14, 3205-3212, 1991.

[14] Hu X B, Li Y and Liu Q M, Nonlinear superposition formula of the Boussinesq hierarchy, *Acta Math. Appl. Sinica (English Ser.)* **9**, no. 1, 17-27, 1993.

[15] Hu X B and Springael J, A Bäcklund transformation and nonlinear superposition formula for the Lotka-Volterra hierarchy, *ANZIAM J.* **44**, 121-128, 2002.

[16] Hu X B and Xue W M, A bilinear Bäcklund transformation and nonlinear superposition formula for the negative Volterra hierarchy, *J. Phys. Soc. Japan* **72**, 3075-3078, 2003.

[17] Jimbo M and Miwa T, Solitons and infinite dimensional Lie algebras, *Publ. RIMS, Kyoto Univ.* **19**, 943-1001, 1983.

[18] Kupershmidt B A, Mathematics of dispersive water waves, *Comm. Math. Phys.* **99**, no. 1, 51-73, 1985.

[19] Levi D, Nonlinear differential-difference equations as Bäcklund transformations, *J. Phys. A* **14**, no. 5, 1083-1098, 1981.

[20] Levi D, Neugebauer G and Meinel R, A new nonlinear Schrödinger equation, its hierarchy and N-soliton solutions, *Phys. Lett. A* **102**, no. 1-2, 1-6, 1984.

[21] Liu Q M, Hu X B and Li Y, Rational solutions of classical Boussinesq hierarchy, *J. Phs. A. Math. Gen.* **23**, 585-591, 1990.

[22] Matsuno Y, Bilinearization of nonlinear evolution equations, *J. Phys. Soc. Jpn.* **48**, 2138-2143, 1980.

[23] Matsuno Y, Bilinearization of nonlinear evolution equations II. Higher-order modified Korteweg-de Vries equations, *J. Phys. Soc. Jpn.* **49**, 787-794, 1980.

[24] Matsuno Y, *Bilinear Transformation Method*, New York: Academic, 1984.

[25] Newell A C, *Solitons in Mathematics and Physics*, SIAM Philadelphia, 1985.

[26] Nimmo J J C, Wronskians and Multi-component KP hierarchies, in *Nonlinear evolution equations: integrability and spectral methods*, ed. A. Degasperis, A.P. Fordy and M. Lakshmanan, Manchester University Press, 139-148, 1990

[27] Ohta Y, Satsuma J, Takahashi D and Tokihiro T, An Elementary Introduction to Sato Theory, *Prog. Theor. Phys.* Suppl. No. 94, 210, 1988.

[28] Oishi S, A method of constructing generalized soliton solutions for certain bilinear soliton equations, *J. Phys. Soc. Japan* **47**, no. 4, 1341-1346, 1979.

[29] Oishi S, Relationship between Hirota's method and the inverse spectral method—the Korteweg-de Vries equation's case, *J. Phys. Soc. Japan* **47**, no. 3, 1037-1038, 1979.

[30] Oishi S, A method of analysing soliton equations by bilinearization, *J. Phys. Soc. Japan* **48** (1980), no. 2, 639-646, 1980.

[31] Pritula G M and Vekslerchik V E, Negative Volterra flows, *J. Phys. A* **36**, 213-226, 2003.

[32] Rosales R, Exact solutions of some nonlinear evolution equations, *Stud. Appl. Math.* **59**, no. 2, 117-151, 1978.

[33] Sato M (notes taken by Masatoshi Noumi) Soliton Equations and the Universal Grassmann Manifold (in Japanese), Sophia Kokyuroku in Mathematics 18, Department of Mathematics, Sophia University, 134 pages, 1984.

[34] Sato M and Sato Y, Soliton equations as dynamical systems on an infinite dimensional Grassmann manifold, *Proc. US-Japan Seminar, Nonlinear Partial Differential Equations in Applied Scinece*, Fujita H, Lax P D and Strang G (eds.) Kinokuniya: North-Holland, 1982.

[35] Springael J, Hu X B and Loris I, Bilinear characterization of higher order Ito-equations, *J. Phys. Soc. Japan* **65**, 1222-1226, 1996.

[36] Weiss J, The Painlevé Property and Bäcklund transformations for the sequence of Boussinesq equations, *J. Math. Phys.* **26**, 258-269, 1985.

[37] Yamagata H, On c-solitons, *Math. Japon.* **22**, no. 2, 193-217, 1977.

[38] Yamagata H, On blizzard solutions as 'superpositions' of c solitons with real speeds, *Math. Japon* **23**, no. 4, 439-455, 1978/79.

[39] Yu G F and Duan Q H, An implementation for the algorithm of the Hirota bilinear Bäcklund transformation of integrable hierarchies, *J. Phys. A: Math. Theor.* **43**, 395202, 2010.

B4. Rogue wave patterns and modulational instability in nonlinear Schrödinger hierarchy

Liming Ling [a] *and Li-Chen Zhao* [b]

[a] *School of Mathematics, South China University of Technology, Guangzhou, China, 510641*

[b] *School of Physics, Northwest University, Xi'an, 710069, China*
Shaanxi Key Laboratory for Theoretical Physics Frontiers, Xi'an, 710069, China

Abstract

In this chapter, we firstly solve the Lax pair with the plane wave background, which could be utilized to obtain the modulational instability analysis by combining the squared eigenfunction method. In this way, we can understand that the modulational instability analysis can be directly derived through the solutions of the Lax pair. Then we construct the exact solutions for the scalar and vector system, which involves the breathers solutions and rogue wave solutions. Some of the results on the exact solutions are given in the previous literature. But the analysis for the dynamics on the solutions is given firstly, which is hopeful to extend the future studies on the localized wave of dynamic behavior.

1 Introduction

The nonlinear Schrödinger (NLS) equation plays a crucial role in the nonlinear science, which is considered as the universal model for the nonlinear waves [59]. Rogue wave is a striking type of nonlinear waves, which comes from nowhere without trace [3]. The rational solutions of NLS can describe well the rogue wave dynamics in many nonlinear systems. Recently, there are lots of studies on the modulational instability and rogue waves for this model [2, 3, 4, 6, 27]. Besides the theoretic work, there have been many experimental observations of rogue waves in the nonlinear optics, water tanks and other NLS models describing physical systems [7, 15, 16, 40].

The construction of exact solutions for the integrable NLS equations has a long history [1, 5, 41, 54, 60]. It is hard to give a systematic review on the relative results. So we just list some of them to review the procedure on the topic of rogue waves. The NLS equations are important models for the integrable system, which has the Lax pair. The exact solutions can be constructed with the aid of the Lax pair. We are just concerned with the localized waves solution, which can be used to describe many different nonlinear wave dynamics in physical systems. An effective technique to construct the localized wave solutions is utilizing the Darboux transformation method [13, 23, 30, 32, 38, 45, 56, 73], which is a pure algebraic method and also can be considered as the simple method of formal inverse scattering transformation without non-reflection coefficients [11, 12, 14]. For the construction of rogue wave solutions, the Darboux transformation shows its advantages. For the other methods, we refer the readers to the literature [33, 58]. By the generalized Darboux transformation technique [30], the rogue waves for the scalar NLS equations can be constructed widely [13, 26, 30, 32, 45, 63, 73]. The extension to vector NLS systems is highly nontrivial, due to nonlinear coupling effects and relative wave vector between different components. Many different rogue wave patterns have been shown [8, 10, 29, 43, 50, 69, 74, 77, 82, 83, 84].

Modulational instability (MI) refers to the Benjamin-Feir instability [62, 72], which mainly describes the growth of weak perturbations on certain backgrounds. It can be used to explain the formation of rogue waves qualitatively [8]. The linear stability analysis can merely be used to describe weak perturbations, once the amplitude of rogue wave tends to large, which will not be suitable for the description of rogue waves, since the nonlinear system can not be interpreted by the linear theory completely. But on the other hand, the exact rogue waves or Akhmediev breathers describe the whole procedure of special perturbations. Then it is possible to establish the quantitative relations between nonlinear waves and MI character [46, 50, 78], which deepen the understanding of MI in the nonlinear system, and provide possible ways to excite several nonlinear waves controllably [24, 25, 51, 78, 81].

In this work, we rewrite our previous work in a uniform form, and present some novel results for breather dynamics and high-order rogue wave decomposition. Firstly, in Section 2, we give the generalized Darboux transformation method and the integrable hierarchy. The corresponding solutions of the Lax pair with the plane wave solutions are solved in the uniform form, in which the fundamental solutions are constructed by the linear algebra and the limit technique for the branch points. At the branch points, the rogue waves can be constructed by the rational functions. At the non-real non-branch points, the breather solutions can be constructed by the exponential functions. On the other hand, by the squared eigenfunction method, the solutions of linearized equations also can be constructed from the solutions of Lax pair, which can used to yield the solutions of the spectral problem of linearized operator. Here we will use them to yield the MI character, and find the MI and rogue waves or Akhmediev breathers possess the same origin through the solutions of Lax pair. These results interpret that the existence conditions of rogue wave or Akhmediev breather is equivalent to the dispersion relation for MI [46, 50].

For the stories of exact breather solutions, we analyze the breather solutions of scalar NLS equation, in which are the Ahkmediev breather, Kuznetusov-Ma breather and the non-zero velocity breathers. For the non-zero velocity breathers, we give the definition of oscillator breather by defining the width and oscillating period of breather. The envelope of oscillating breathers also can be obtained through the expression of solutions approximately. By the detailed analysis of single breathers, we can construct the multi-breathers. As for the rogue wave solutions, we propose a method to decompose the high-order rogue waves into lower-order rogue waves by choosing the large parameters. The description on the exact solutions for the scalar and vector system are given in Section 3 and 4. The conclusion and discussions are included in the final section.

2 Vector NLS equations and Hirota equations

The vector nonlinear Schrödinger (vNLS) equations and vector Hirota equations read in a general form:

$$\mathrm{i}\mathbf{q}_t + \gamma\left(\frac{1}{2}\mathbf{q}_{xx} + \mathbf{q}\mathbf{q}^\dagger\Lambda\mathbf{q}\right) + \mathrm{i}\,\delta\left(\mathbf{q}_{xxx} + 3\mathbf{q}_x\mathbf{q}^\dagger\Lambda\mathbf{q} + 3\mathbf{q}\mathbf{q}^\dagger\Lambda\mathbf{q}_x\right) = 0, \tag{2.1}$$

where $\Lambda = \mathrm{diag}\,(s_1, s_2, \cdots, s_n)$,, $\mathbf{q} = (q_1, q_2, \cdots, q_n)^{\mathrm{T}}$ and $s_i = \pm 1$, where $\gamma, \delta \in \mathbb{R}$, $\gamma \neq 0$. If $\delta = 0$, $\gamma \neq 0$, Equations (2.1) are the vector NLS equations [8, 10, 29, 43, 50, 69, 74, 77, 82, 83, 84]. If $\delta\gamma \neq 0$, the equations (2.1) are the vector Hirota equations [52, 53]. The parameters s_is regulate the focusing, defocusing or the mixed case [20, 21, 22, 48, 49, 55, 57]. If all $s_i = 1$, this is the focusing case. If all $s_i = -1$, this is the defocusing case. For the other cases, this corresponds to the mixed case. Equations (2.1) admit the following Lax pair

$$\begin{aligned}\Phi_x =& \mathbf{U}\Phi, \qquad \mathbf{U} = \mathrm{i}\left(\lambda\sigma_3 + \mathbf{Q}\right), \quad \sigma_3 = \mathrm{diag}\left(1, -\mathbb{I}_n\right), \quad \mathbf{Q} = \begin{bmatrix} 0 & \mathbf{q}^\dagger\Lambda \\ \mathbf{q} & 0 \end{bmatrix} \\ \Phi_t =& \mathbf{V}\Phi, \qquad \mathbf{V} = \mathrm{i}(\gamma + 4\delta\lambda)\left[\lambda^2\sigma_3 + \lambda\mathbf{Q} - \frac{\sigma_3}{2}(\mathbf{Q}^2 + \mathrm{i}\mathbf{Q}_x)\right] \\ & \qquad\qquad - \delta\left(\mathrm{i}\mathbf{Q}_{xx} + 2\mathrm{i}\mathbf{Q}^3 + \mathbf{Q}_x\mathbf{Q} - \mathbf{Q}\mathbf{Q}_x\right),\end{aligned} \tag{2.2}$$

here $\lambda \in \mathbb{C}$ is a spectral parameter, which is the mixture of second and third flow for the vector NLS hierarchy by the AKNS scheme. The compatibility conditions of Lax pair (2.2):

$$\mathbf{U}_t - \mathbf{V}_x + [\mathbf{U}, \mathbf{V}] = 0, \qquad [\mathbf{U}, \mathbf{V}] = \mathbf{U}\mathbf{V} - \mathbf{V}\mathbf{U}, \tag{2.3}$$

yields the vector Hirota Equation (2.1).

Proposition 1. *If the wave function $\Phi(\lambda; x, t)$ with $\Phi(\lambda; 0, 0) = \mathbb{I}_{n+1}$, then it possesses the symmetric condition*

$$\mathbf{U}(\lambda; x, t) = -\Omega\mathbf{U}^\dagger(\lambda^*; x, t)\Omega, \qquad \mathbf{V}(\lambda; x, t) = -\Omega\mathbf{V}^\dagger(\lambda^*; x, t)\Omega, \tag{2.4}$$

is equivalent to

$$\Phi(\lambda; x, t)\Omega\Phi^\dagger(\lambda^*; x, t)\Omega = \mathbb{I}_{n+1}, \tag{2.5}$$

where $\Omega = \mathrm{diag}\,(1, \Lambda)$.

Proof. By the symmetric relationship for the coefficient matrices $\mathbf{U}$ and $\mathbf{V}$: (2.4), if $\Phi(\lambda; x, t)$ satisfies the Lax pair (2.2), it follows that the matrix function $\Omega\Phi^\dagger(\lambda^*; x, t)\Omega$ satisfies the adjoint Lax pair:

$$\begin{aligned}-\Omega\Phi_x^\dagger(\lambda^*; x, t)\Omega &= \Omega\Phi^\dagger(\lambda^*; x, t)\Omega U(\lambda; x, t), \\ -\Omega\Phi_t^\dagger(\lambda^*; x, t)\Omega &= \Omega\Phi^\dagger(\lambda^*; x, t)\Omega V(\lambda; x, t).\end{aligned} \tag{2.6}$$

On the other hand, the matrix function $\Phi^{-1}(\lambda; x, t)$ also satisfies the adjoint Lax pair. By the uniqueness and existence of differential equations and $\Phi(\lambda; 0, 0) = \mathbb{I}$, we arrive at the condition on the wave function (2.5). The converse part of this proposition is obvious. ■

Now we proceed to find the elementary Darboux matrix. Consider the linear fractional transformation:

$$\mathbf{T}_1(\lambda;x,t)=\mathbb{I}-\frac{\lambda_1-\lambda_1^*}{\lambda-\lambda_1^*}\mathbf{P}_1(x,t),\qquad \mathbf{P}_1^2=\mathbf{P}_1,\qquad \mathbf{P}_1=\mathbf{P}_1^\dagger. \tag{2.7}$$

By the proposition 1, we require the condition $\mathbf{T}_1(\lambda;x,t)\Omega\mathbf{T}_1^\dagger(\lambda^*;x,t)\Omega=\mathbb{I}_{n+1}$. Together with the holomorphic property of $\mathbf{T}_1(\lambda;x,t)\Phi(\lambda;x,t)\mathbf{T}_1^{-1}(\lambda;0,0)$, we obtain the elementary Darboux transformation is

$$\mathbf{T}_1(\lambda;x,t)=\mathbb{I}_{n+1}-\frac{\lambda_1-\lambda_1^*}{\lambda-\lambda_1^*}\frac{\phi_1\phi_1^\dagger\Omega}{\phi_1^\dagger\Omega\phi_1} \tag{2.8}$$

where $\phi_1=\Phi(\lambda_1;x,t)\mathbf{v}_1$, $\mathbf{v}_1$ is a constant vector. In generally, we have the following generalized Darboux transformation:

Theorem 1. *Suppose we have a smooth solution $q_i(x,t)\in\mathbf{L}^\infty(\mathbb{R}^2)\cup\mathbf{C}^\infty(\mathbb{R}^2)$, and the matrix solution $\Phi(\lambda;x,t)$ is analytic in the whole complex plane $\mathbb{C}$, the Darboux transformation*

$$\mathbf{T}_N(\lambda;x,t)=\mathbb{I}+\mathbf{Y}_N\mathbf{M}^{-1}\mathbf{D}\mathbf{Y}_N^\dagger\Omega,\qquad \mathbf{M}=\mathbf{X}^\dagger\mathbf{S}\mathbf{X}, \tag{2.9}$$

where

$$\begin{gathered}
\mathbf{Y}_N=\left[\phi_1^{[0]},\phi_1^{[1]},\cdots,\phi_1^{[r_1-1]},\phi_2^{[0]},\phi_2^{[1]},\cdots,\phi_2^{[r_2-1]},\cdots,\phi_k^{[0]},\phi_k^{[1]},\cdots,\phi_k^{[r_k-1]}\right],\\
\mathbf{D}=\begin{bmatrix}\mathbf{D}_1&0&\cdots&0\\0&\mathbf{D}_2&\cdots&0\\\vdots&\vdots&\ddots&\vdots\\0&0&\cdots&\mathbf{D}_k\end{bmatrix},\qquad
\mathbf{D}_i=\begin{bmatrix}\frac{1}{\lambda-\lambda_i^*}&0&\cdots&0\\\frac{1}{(\lambda-\lambda_i^*)^2}&\frac{1}{\lambda-\lambda_i^*}&\cdots&0\\\vdots&\vdots&\ddots&\vdots\\\frac{1}{(\lambda-\lambda_i^*)^{r_i-1}}&\frac{1}{(\lambda-\lambda_i^*)^{r_i-1}}&\cdots&\frac{1}{\lambda-\lambda_i^*}\end{bmatrix},\\
\mathbf{X}=\begin{bmatrix}\mathbf{X}_1&0&\cdots&0\\0&\mathbf{X}_2&\cdots&0\\\vdots&\vdots&\vdots&\vdots\\0&0&\cdots&\mathbf{X}_k\end{bmatrix},\qquad
\mathbf{X}_i=\begin{bmatrix}\phi_i^{[0]}&\phi_i^{[1]}&\cdots&\phi_i^{[r_i-1]}\\0&\phi_i^{[0]}&\cdots&\phi_i^{[r_i-2]}\\\vdots&\vdots&\ddots&\vdots\\0&0&\cdots&\phi_i^{[0]}\end{bmatrix},\\
\mathbf{S}=\begin{bmatrix}\mathbf{S}_{11}&\mathbf{S}_{12}&\cdots&\mathbf{S}_{1k}\\\mathbf{S}_{21}&\mathbf{S}_{22}&\cdots&\mathbf{S}_{2k}\\\vdots&\vdots&\ddots&\vdots\\\mathbf{S}_{k1}&\mathbf{S}_{k2}&\cdots&\mathbf{S}_{kk}\end{bmatrix},
\end{gathered} \tag{2.10}$$

and $\phi_i^{[k]}=\frac{1}{k!}\left(\frac{\mathrm{d}}{\mathrm{d}\lambda}\right)^k\Phi(\lambda;x,t)\mathbf{v}_i|_{\lambda=\lambda_i}$,

$$\mathbf{S}_{i,j}=\begin{bmatrix}\binom{0}{0}\frac{\Omega}{\lambda_i^*-\lambda_j}&\binom{1}{0}\frac{\Omega}{(\lambda_i^*-\lambda_j)^2}&\cdots&\binom{r_j}{0}\frac{\Omega}{(\lambda_i^*-\lambda_j)^{r_j}}\\\binom{1}{1}\frac{(-1)\Omega}{(\lambda_i^*-\lambda_j)^2}&\binom{2}{1}\frac{(-1)\Omega}{(\lambda_i^*-\lambda_j)^3}&\cdots&\binom{r_j}{1}\frac{(-1)\Omega}{(\lambda_i^*-\lambda_j)^{r_j+1}}\\\vdots&\vdots&\ddots&\vdots\\\binom{r_i-1}{r_i-1}\frac{(-1)^{r_i-1}\Omega}{(\lambda_i^*-\lambda_j)^{r_i}}&\binom{r_i}{r_i-1}\frac{(-1)^{r_i-1}\Omega}{\lambda_i^*-\lambda_j)^{r_i+1}}&\cdots&\binom{r_i+r_j-2}{r_i-1}\frac{(-1)^{r_i-1}\Omega}{(\lambda_i^*-\lambda_j)^{r_i+r_j-1}}\end{bmatrix},$$

converts the Lax pair (2.2) *into the new one by replacing the potential functions*

$$\mathbf{q}^{[n]} = \mathbf{q} + 2\mathbf{Y}_{N,2}\mathbf{M}^{-1}\mathbf{Y}_{N,1}^{\dagger}, \tag{2.11}$$

which is the Bäcklund transformation, where the subscript $\mathbf{Y}_{N,1}$ *denotes the first row vector of* $\mathbf{Y}_N$*, the subscript* $\mathbf{Y}_{N,2}$ *denotes the second up to* $n+1$*-th row vector of* $\mathbf{Y}_N$.

Based on the above Theorem 1, several distinct localized waves can be constructed by the Bäcklund transformation, which involves the rogue waves, breathers and their high-order ones. To this end, we firstly need to solve the Lax pair with the background seed solutions of the vector NLS equations or vector Hirota equations [13, 23, 30, 32, 38, 45, 56, 73]. In this work, we mainly consider the plane wave seed solutions.

2.1 Solutions of Lax pair with the plane wave solution

Now we consider the plane wave solution for Equation (2.1). It is readily found that the plane wave solution for Equation (2.1) is:

$$\mathbf{q}(x,t) = \left[a_1 e^{i\theta_i}, a_2 e^{i\theta_i}, \cdots, a_n e^{i\theta_i}\right]^{\mathrm{T}}, \tag{2.12}$$

where $\theta_i = b_i x + c_i t$ and

$$c_i = b_i^2\left(\delta b_i - \frac{\gamma}{2}\right) + (\gamma - 3\delta b_i)\left(\sum_{i=1}^{n} s_i a_i^2\right) - 3\delta\sum_{i=1}^{n} s_i b_i a_i^2,$$

$a_i, b_i \in \mathbb{R}$.

Then we show the details how to solve the Lax pair (2.2) with the plane wave solution (2.12). Inserting the plane wave solution (2.12) into Lax pair (2.2), then the Lax pair can be converted into the following form:

$$\begin{aligned}
\Phi_{0,x} =& \mathbf{U}_0\Phi_0, \qquad \Phi_{0,t} = \mathbf{V}_0\Phi_0, \qquad \mathbf{U}_0 = \mathbf{K}_n\mathbf{C}\mathbf{K}_n^{-1} \\
\mathbf{V}_0 =& \mathbf{K}_n\left[-\delta\mathbf{C}^3 - i\left(\frac{\gamma}{2} + 3\delta\lambda\right)\mathbf{C}^2 + \left[3\delta\left(\lambda^2 - \sum_{i=1}^{n} s_i a_i^2\right) + \gamma\lambda\right]\mathbf{C}\right. \\
&\left. + 3i\delta\left(\sum_{i=1}^{n} s_i b_i a_i^2 - \lambda\sum_{i=1}^{n} s_i a_i^2 - \lambda^3\right) + i\gamma\left(\sum_{i=1}^{n} s_i a_i^2 + \frac{\lambda^2}{2}\right)\right]\mathbf{K}_n^{-1}
\end{aligned} \tag{2.13}$$

where $\mathbf{K}_n = \operatorname{diag}\left(1, e^{i\theta_1}, \cdots, e^{i\theta_n}\right)$ and

$$\mathbf{C} = i\begin{bmatrix} \lambda & s_1 a_1 & \cdots & s_n a_n \\ a_1 & -(\lambda + b_1) & \cdots & 0 \\ \vdots & \vdots & \ddots & \vdots \\ a_n & 0 & \cdots & -(\lambda + b_n) \end{bmatrix}. \tag{2.14}$$

Further, by the diagonalization of matrix $\mathbf{C}$, we have

$$\mathbf{C} = \mathbf{L}\mathbf{H}\mathbf{L}^{-1}, \tag{2.15}$$

which implies the fundamental solutions of Lax pair:

$$\Phi_0 = \mathbf{K}_n \mathbf{L} \exp\left(\mathbf{H}x + \mathbf{G}t\right) \tag{2.16}$$

where

$$\begin{aligned}\mathbf{G} = -\delta\mathbf{H}^3 - \mathrm{i}\left(\frac{\gamma}{2} + 3\delta\lambda\right)\mathbf{H}^2 + \left[3\delta\left(\lambda^2 - \sum_{i=1}^{n} s_i a_i^2\right) + \gamma\lambda\right]\mathbf{H} \\ + 3\mathrm{i}\delta\left(\sum_{i=1}^{n} s_i b_i a_i^2 - \lambda\sum_{i=1}^{n} s_i a_i^2 - \lambda^3\right) + \mathrm{i}\gamma\left(\sum_{i=1}^{n} s_i a_i^2 + \frac{\lambda^2}{2}\right).\end{aligned} \tag{2.17}$$

Because the matrix $\mathbf{C}$ involves the free parameters a_is and b_is, we classify the matrices $\mathbf{L}$ and $\mathbf{H}$ into the four distinct cases:

1. When $a_i > 0$ and $b_i \neq b_j$ with $i \neq j$, the matrices $\mathbf{H}$ and $\mathbf{L}$ are given by:

$$\begin{aligned}&\mathbf{H} = \mathrm{i}\,\mathrm{diag}\left(\xi_1 - \lambda, \xi_2 - \lambda, \cdots, \xi_{n+1} - \lambda\right), \\ &\mathbf{L} = \begin{bmatrix} 1 & 1 & \cdots & 1 \\ \frac{a_1}{\xi_1+b_1} & \frac{a_1}{\xi_2+b_1} & \cdots & \frac{a_1}{\xi_{n+1}+b_1} \\ \vdots & \vdots & \ddots & \vdots \\ \frac{a_n}{\xi_1+b_n} & \frac{a_n}{\xi_2+b_n} & \cdots & \frac{a_n}{\xi_{n+1}+b_n} \end{bmatrix}\end{aligned} \tag{2.18}$$

and the ξ_is are the roots of the characteristic equation

$$\prod_{i=1}^{n}(\xi + b_i)\left[2\lambda - \xi + \sum_{i=1}^{n}\frac{s_i a_i^2}{\xi + b_i}\right] = 0. \tag{2.19}$$

2. For the degenerate case, if $a_1 = a_2 = \cdots = a_k = 0$ and $b_i \neq b_j$ with $i \neq j$, then we have

$$\begin{aligned}&\mathbf{H} = \mathrm{i}\,\mathrm{diag}\left(\lambda + b_1, \lambda + b_2, \cdots, \lambda + b_k, \xi_1 - \lambda, \cdots, \xi_{n+1-k} - \lambda\right), \\ &\mathbf{L} = \begin{bmatrix} 0 & \cdots & 0 & 1 & 1 & \cdots & 1 \\ 1 & \cdots & 0 & 0 & 0 & \cdots & 0 \\ \vdots & \ddots & \vdots & \vdots & \vdots & \ddots & \vdots \\ 0 & \cdots & 1 & 0 & 0 & \cdots & 0 \\ 0 & \cdots & 0 & \frac{a_{k+1}}{\xi_1+b_{k+1}} & \frac{a_{k+1}}{\xi_2+b_{k+1}} & \cdots & \frac{a_{k+1}}{\xi_{n+1-k}+b_{k+1}} \\ \vdots & \ddots & \vdots & \vdots & \vdots & \ddots & \vdots \\ 0 & \cdots & 0 & \frac{a_n}{\xi_1+b_n} & \frac{a_n}{\xi_2+b_n} & \cdots & \frac{a_n}{\xi_{n+1-k}+b_n} \end{bmatrix}\end{aligned} \tag{2.20}$$

and the ξ_is are the characteristic equation

$$\prod_{i=k+1}^{n}(\xi + b_i)\left[2\lambda - \xi + \sum_{i=k+1}^{n}\frac{s_i a_i^2}{\xi + b_i}\right] = 0. \tag{2.21}$$

3. In this case, we consider the degenerate property from the parameters b_is. Here we just consider a simple case with $b_1 = b_2 = \cdots = b_k = b_{k+1}$. The general case, for instance $b_1 = b_2 = \cdots = b_{k+1}$ and $b_{k+l} = b_{k+l+1} = \cdots = b_{k+l+s}$ for $l \geq 2$ and $s \geq 1$, can be considered in a similar procedure. If $a_i > 0$, $b_1 = b_2 = \cdots = b_k = b_{k+1}$, and $b_i \neq b_j$ with $i \neq j$ for $i, j \geq k+1$, then we have

$$\mathbf{H} = \mathrm{i}\,\mathrm{diag}\left(\lambda + b_1, \lambda + b_2, \cdots, \lambda + b_k, \xi_1 - \lambda, \cdots, \xi_{n+1-k} - \lambda\right),$$

$$\mathbf{L} = \begin{bmatrix} 0 & \cdots & 0 & 1 & 1 & \cdots & 1 \\ s_2 a_2 & \cdots & s_k a_k & \frac{a_1}{\xi_1+b_1} & \frac{a_1}{\xi_2+b_1} & \cdots & \frac{a_1}{\xi_{n+1-k}+b_1} \\ -s_1 a_1 & \cdots & 0 & \frac{a_2}{\xi_1+b_2} & \frac{a_2}{\xi_2+b_2} & \cdots & \frac{a_2}{\xi_{n+1-k}+b_2} \\ \vdots & \ddots & \vdots & \vdots & \vdots & \ddots & \vdots \\ 0 & \cdots & 0 & \frac{a_{k-1}}{\xi_1+b_{k-1}} & \frac{a_{k-1}}{\xi_2+b_{k-1}} & \cdots & \frac{a_{k-1}}{\xi_{n+1-k}+b_{k-1}} \\ 0 & \cdots & -s_1 a_1 & \frac{a_k}{\xi_1+b_k} & \frac{a_k}{\xi_2+b_k} & \cdots & \frac{a_k}{\xi_{n+1-k}+b_k} \\ 0 & \cdots & 0 & \frac{a_{k+1}}{\xi_1+b_{k+1}} & \frac{a_{k+1}}{\xi_2+b_{k+1}} & \cdots & \frac{a_{k+1}}{\xi_{n+1-k}+b_{k+1}} \\ \vdots & \ddots & \vdots & \vdots & \vdots & \ddots & \vdots \\ 0 & \cdots & 0 & \frac{a_n}{\xi_1+b_n} & \frac{a_n}{\xi_2+b_n} & \cdots & \frac{a_n}{\xi_{n+1-k}+b_n} \end{bmatrix} \tag{2.22}$$

and the ξ_is are the roots of characteristic equation

$$\prod_{i=k+1}^{n} (\xi + b_i) \left[2\lambda - \xi + \sum_{i=k+1}^{n} \frac{s_i a_i^2}{\xi + b_i}\right] = 0. \tag{2.23}$$

4. Cases (2) and (3) can be mixed together to construct a general degenerated case, the explicit forms of matrices **H** and **L** are similar to the above two cases.

For the characteristic Equation (2.19), there exist the multiple roots or the branch points, which correspond to the rational solutions. The condition of multiple roots of characteristic Equation (2.19) is determined by the following equation

$$1 + \sum_{i=1}^{n} \frac{s_i a_i^2}{(\xi + b_i)^2} = 0. \tag{2.24}$$

Under this case, the matrix **C** cannot be diagonalized. Thus we have to look for the other methods to find the fundamental solutions for the Lax pair.

There are two ways to look for the fundamental solutions at the branch points. One of them is utilizing the limitation technique. Suppose we have a root ξ_1 with multiplicity m of Equation (2.24), then we have

$$\lambda_1 = \frac{1}{2}\left(\xi_1 - \sum_{i=1}^{n} \frac{s_i a_i^2}{\xi_1 + b_i}\right). \tag{2.25}$$

Then we substitute the expansions $\lambda(\epsilon)=\lambda_1+\epsilon^m$ into the characteristic Equation (2.19), which yields the expansions of ξ with the form

$$\xi(\epsilon)=\xi_1+\sum_{i=1}^{\infty}\xi_1^{[i]}\epsilon^i. \tag{2.26}$$

By the symmetric property, it is easy to find that $(\lambda(\epsilon),\xi(\epsilon))\to(\lambda(\mathrm{e}^{2k\pi\mathrm{i}/m}\epsilon),\xi(\mathrm{e}^{2k\pi\mathrm{i}/m}\epsilon))$ is invariant under the transformation in the neighborhood of (λ_1,ξ_1), so there exist m pairs $(\lambda(\mathrm{e}^{2k\pi\mathrm{i}/m}\epsilon),\xi(\mathrm{e}^{2k\pi\mathrm{i}/m}\epsilon))$, $k=0,1,2,\cdots,m-1$, which solves the characteristic Equation (2.19). We would like to use the following linear combinations

$$\begin{aligned}\Phi_s(\epsilon)&=\frac{1}{m\epsilon^s}\mathbf{K}_n\sum_{k=0}^{m-1}\mathrm{e}^{-\frac{2ks\pi\mathrm{i}}{m}}\mathbf{L}_1(\mathrm{e}^{\frac{2k\pi\mathrm{i}}{m}}\epsilon)\exp\left(\mathrm{i}(\xi(\mathrm{e}^{\frac{2k\pi\mathrm{i}}{m}}\epsilon)-\lambda(\mathrm{e}^{\frac{2k\pi\mathrm{i}}{m}}\epsilon))x+\eta(\mathrm{e}^{\frac{2k\pi\mathrm{i}}{m}}\epsilon)t\right)\\&=\mathbf{K}_n\sum_{l=0}^{\infty}\Psi_1^{[lm+s]}\epsilon^{lm}=\mathbf{K}_n\sum_{l=0}^{\infty}\Psi_1^{[lm+s]}(\lambda-\lambda_k)^l,\end{aligned} \tag{2.27}$$

where

$$\begin{aligned}\Psi_1(\epsilon)&=\mathbf{L}_1(\epsilon)\exp\left[\mathrm{i}(\xi(\epsilon)-\lambda(\epsilon))x+\eta(\epsilon)t\right]=\sum_{l=0}^{\infty}\Psi_1^{[l]}\epsilon^l,\\ \mathbf{L}_1(\epsilon)&=\left[1,\frac{a_1}{\xi(\epsilon)+b_1},\cdots,\frac{a_1}{\xi(\epsilon)+b_1}\right]^{\mathrm{T}},\\ \eta(\epsilon)&=\mathrm{i}\delta[\xi(\epsilon)-\lambda(\epsilon)]^3+\mathrm{i}\left(\frac{\gamma}{2}+3\delta\lambda(\epsilon)\right)[\xi(\epsilon)-\lambda(\epsilon)]^2\\&\quad+3\mathrm{i}\delta\left(\sum_{i=1}^{n}s_ib_ia_i^2-\lambda(\epsilon)\sum_{i=1}^{n}s_ia_i^2-\lambda(\epsilon)^3\right)+\mathrm{i}\gamma\left(\sum_{i=1}^{n}s_ia_i^2+\frac{\lambda(\epsilon)^2}{2}\right)\\&\quad+\mathrm{i}\left[3\delta\left(\lambda(\epsilon)^2-\sum_{i=1}^{n}s_ia_i^2\right)+\gamma\lambda(\epsilon)\right](\xi(\epsilon)-\lambda(\epsilon)),\end{aligned} \tag{2.28}$$

and the coefficient terms $\Psi_1^{[l]}$ are the rational polynomials with respect to x and t. With the help of $\Phi_s(\epsilon)$, m vector solutions $\mathbf{K}_n\Psi_1^{[s]}$, $s=0,1,\cdots,m-1$ are the solutions for the Lax pair at the points of multiplicity m. In the procedure of constructing the solutions of integrable equations, the terms of $\lambda(\epsilon)$ in exponential functions can be thrown away for the construction of rational solutions since they are invariant under the transformation $\epsilon\to\mathrm{e}^{2k\pi\mathrm{i}/m}\epsilon$.

Another method is to utilize the Jordan matrix [30, 45]. With this approach, we are able to iterate the Darboux transformation once. For the construction of high-order rogue waves, we still need to utilize the expansions of $\Phi(\lambda;x,t)$ in the neighborhood of branch points [30]. Thus we do not show the details here.

2.2 Linear stability analysis for the plane waves

In this subsection, we utilize the squared eigenfunction method [22, 36, 37] to construct the solutions of linearized vector Hirota equations, and further perform linear stability on plane waves based on Lax-pair. To this purpose, we depart from the stationary zero curvature equation:

$$\Psi_x = [\mathbf{U}, \Psi], \qquad \Psi_t = [\mathbf{V}, \Psi], \qquad \Psi = \begin{bmatrix} \Psi_{11} & \Psi_{12} \\ \Psi_{21} & \Psi_{22} \end{bmatrix}, \tag{2.29}$$

where the matrices Ψ_{11}, Ψ_{12}, Ψ_{21} and Ψ_{22} have the shape 1×1, $1 \times n$, $n \times 1$ and $n \times n$, respectively. Then we write the corresponding component form:

$$\begin{aligned}
\Psi_{11,x} =& \mathrm{i}\left(\mathbf{q}^\dagger \Lambda \Psi_{21} - \Psi_{12}\mathbf{q}\right), \\
\Psi_{12,x} =& \mathrm{i}\left(2\lambda\Psi_{12} + \mathbf{q}^\dagger \Lambda \Psi_{22} - \Psi_{11}\mathbf{q}^\dagger \Lambda\right), \\
\Psi_{21,x} =& \mathrm{i}\left(\mathbf{q}\Psi_{11} - \Psi_{22}\mathbf{q} - 2\lambda\Psi_{21}\right), \\
\Psi_{22,x} =& \mathrm{i}\left(\mathbf{q}\Psi_{12} - \Psi_{21}\mathbf{q}^\dagger \Lambda\right).
\end{aligned} \tag{2.30}$$

Further, taking the second order derivative of Ψ_{12} and Ψ_{21} with respect to x, together with the above Equations (2.30) we obtain that

$$\begin{aligned}
\Psi_{12,xx} =& -2\left[\left(\lambda\mathbf{q}^\dagger\Lambda - \frac{\mathrm{i}}{2}\mathbf{q}_x^\dagger\Lambda\right)\Psi_{22} - \Psi_{11}\left(\lambda\mathbf{q}^\dagger\Lambda - \frac{\mathrm{i}}{2}\mathbf{q}_x^\dagger\Lambda\right) + 2\lambda^2\Psi_{12}\right] \\
& - \mathbf{q}^\dagger\Lambda\left(\mathbf{q}\Psi_{12} - \Psi_{21}\mathbf{q}^\dagger\Lambda\right) + \left(\mathbf{q}^\dagger\Lambda\Psi_{21} - \Psi_{12}\mathbf{q}\right)\mathbf{q}^\dagger\Lambda \\
\Psi_{21,xx} =& 2\left[(\lambda\mathbf{q} + \frac{\mathrm{i}}{2}\mathbf{q}_x)\Psi_{11} - \Psi_{22}(\lambda\mathbf{q} + \frac{\mathrm{i}}{2}\mathbf{q}_x) - 2\lambda^2\Psi_{21}\right] \\
& - \left(\mathbf{q}^\dagger\Lambda\Psi_{21} - \Psi_{12}\mathbf{q}\right) + \left(\mathbf{q}\Psi_{12} - \Psi_{21}\mathbf{q}^\dagger\Lambda\right)\mathbf{q}
\end{aligned} \tag{2.31}$$

Taking the derivative of Ψ_{12} and Ψ_{21} with respect to x once more and combining Equation (2.30), we have

$$\begin{aligned}
\Psi_{12,xxx} =& -\left[4\mathrm{i}\lambda^3 - 2\mathrm{i}\mathbf{q}^\dagger\Lambda\mathbf{q}\lambda - \mathbf{q}_x^\dagger\Lambda\mathbf{q} + \mathbf{q}^\dagger\Lambda\mathbf{q}_x\right]\Psi_{12} \\
& - \Psi_{12}\left[4\mathrm{i}\lambda^3\mathbb{I}_n - 2\mathrm{i}\mathbf{q}^\dagger\Lambda\mathbf{q}\lambda + \mathbf{q}_x\mathbf{q}^\dagger\Lambda - \mathbf{q}\mathbf{q}_x^\dagger\Lambda\right] \\
& - \left[4\mathrm{i}\lambda^2\mathbf{q}^\dagger + 2\mathbf{q}_x^\dagger\Lambda\lambda - \mathrm{i}\mathbf{q}_{xx}^\dagger\Lambda - 2\mathrm{i}\mathbf{q}^\dagger\Lambda\mathbf{q}\mathbf{q}^\dagger\Lambda\right]\Psi_{22} \\
& + \Psi_{11}\left[4\mathrm{i}\lambda^2\mathbf{q}^\dagger + 2\mathbf{q}_x^\dagger\Lambda\lambda - \mathrm{i}\mathbf{q}_{xx}^\dagger\Lambda - 2\mathrm{i}\mathbf{q}^\dagger\Lambda\mathbf{q}\mathbf{q}^\dagger\Lambda\right] - 3\mathbf{q}^\dagger\Lambda\mathbf{q}\Psi_{12,x} \\
& + 3\left(\mathbf{q}^\dagger\Lambda\Psi_{21} - \Psi_{12}\mathbf{q}\right)\mathbf{q}_x^\dagger\Lambda - 3\Psi_{12,x}\mathbf{q}\mathbf{q}^\dagger\Lambda - 3\mathbf{q}_x^\dagger\Lambda\left(\mathbf{q}\Psi_{12} - \Psi_{21}\mathbf{q}^\dagger\Lambda\right) \\
\Psi_{21,xxx} =& -\left[4\mathrm{i}\lambda^2\mathbf{q} - 2\mathbf{q}_x\lambda - \mathrm{i}\mathbf{q}_{xx} - 2\mathrm{i}\mathbf{q}\mathbf{q}^\dagger\Lambda\mathbf{q}\right]\Psi_{11} \\
& + \Psi_{22}\left[4\mathrm{i}\lambda^2\mathbf{q} - 2\mathbf{q}_x\lambda - \mathrm{i}\mathbf{q}_{xx} - 2\mathrm{i}\mathbf{q}\mathbf{q}^\dagger\Lambda\mathbf{q}\right] \\
& + \Psi_{21}\left[4\mathrm{i}\lambda^3 - 2\mathrm{i}\mathbf{q}^\dagger\Lambda\mathbf{q}\lambda - \mathbf{q}_x^\dagger\Lambda\mathbf{q} + \mathbf{q}^\dagger\Lambda\mathbf{q}_x\right] \\
& + \left[4\mathrm{i}\lambda^3\mathbb{I}_n - 2\mathrm{i}\mathbf{q}^\dagger\Lambda\mathbf{q}\lambda + \mathbf{q}_x\mathbf{q}^\dagger\Lambda - \mathbf{q}\mathbf{q}_x^\dagger\Lambda\right]\Psi_{21} - 3\mathbf{q}\mathbf{q}^\dagger\Lambda\Psi_{21,x} \\
& - 3\mathbf{q}_x\left(\mathbf{q}^\dagger\Lambda\Psi_{21} - \Psi_{12}\mathbf{q}\right) - 3\Psi_{21,x}\mathbf{q}^\dagger\Lambda\mathbf{q} + 3\left(\mathbf{q}\Psi_{12} - \Psi_{21}\mathbf{q}^\dagger\Lambda\right)\mathbf{q}_x
\end{aligned} \tag{2.32}$$

For the $t-$part of stationary zero-curvature equations, we write the corresponding component form:

$$\begin{aligned}
\Psi_{11,t} &= V_{12}\Psi_{21} - \Psi_{12}V_{21},\\
\Psi_{12,t} &= V_{12}\Psi_{22} - \Psi_{11}V_{12} + V_{11}\Psi_{12} - \Psi_{12}V_{22},\\
\Psi_{21,t} &= V_{21}\Psi_{11} - \Psi_{22}V_{21} + V_{22}\Psi_{21} - \Psi_{21}V_{11},\\
\Psi_{22,t} &= V_{21}\Psi_{12} + V_{22}\Psi_{22} - \Psi_{21}V_{12} - \Psi_{22}V_{22}
\end{aligned} \tag{2.33}$$

where

$$\begin{aligned}
V_{11} =& \delta\left[4\mathrm{i}\left(\lambda^3 - \frac{1}{2}\mathbf{q}^\dagger\Lambda\mathbf{q}\lambda\right) - \mathbf{q}_x^\dagger\Lambda\mathbf{q} + \mathbf{q}^\dagger\Lambda\mathbf{q}_x\right] + \gamma\mathrm{i}\left(\lambda^2 - \frac{1}{2}\mathbf{q}^\dagger\Lambda\mathbf{q}\right),\\
V_{12} =& \delta\left[4\mathrm{i}\left(\lambda^2\mathbf{q}^\dagger\Lambda - \frac{\mathrm{i}}{2}\mathbf{q}_x^\dagger\Lambda\lambda\right) - \mathrm{i}\mathbf{q}_{xx}^\dagger\Lambda - 2\mathrm{i}\mathbf{q}^\dagger\Lambda\mathbf{q}\mathbf{q}^\dagger\Lambda\right] + \gamma\mathrm{i}\left(\lambda\mathbf{q}^\dagger\Lambda - \frac{\mathrm{i}}{2}\mathbf{q}_x^\dagger\Lambda\right),\\
V_{21} =& \delta\left[4\mathrm{i}\left(\lambda^2\mathbf{q} + \frac{\mathrm{i}}{2}\mathbf{q}_x\lambda\right) - \mathrm{i}\mathbf{q}_{xx} - 2\mathrm{i}\mathbf{q}\mathbf{q}^\dagger\Lambda\mathbf{q}\right] + \gamma\mathrm{i}\left(\lambda\mathbf{q} + \frac{\mathrm{i}}{2}\mathbf{q}_x\right),\\
V_{22} =& \delta\left[-4\mathrm{i}\left(\lambda^3\mathbb{I}_n - \frac{1}{2}\mathbf{q}\mathbf{q}^\dagger\Lambda\lambda\right) - \mathbf{q}_x\mathbf{q}^\dagger\Lambda + \mathbf{q}\mathbf{q}_x^\dagger\Lambda\right] + \gamma\mathrm{i}\left(\lambda^2\mathbb{I}_n + \frac{1}{2}\mathbf{q}\mathbf{q}^\dagger\Lambda\right).
\end{aligned} \tag{2.34}$$

Combining with Equations (2.31), (2.32) and (2.33), we will obtain the linearized equations:

$$\begin{aligned}
\mathrm{i}\Psi_{21,t} =& -\gamma\left[\frac{1}{2}\Psi_{21,xx} + \left(\Psi_{21}\mathbf{q}^\dagger\Lambda\mathbf{q} - \mathbf{q}\Psi_{12}\mathbf{q} + \mathbf{q}\mathbf{q}^\dagger\Lambda\Psi_{21}\right)\right]\\
&+ \delta\mathrm{i}\left[-\Psi_{21,xxx} - 3\mathbf{q}\mathbf{q}^\dagger\Lambda\Psi_{21,x} - 3\mathbf{q}_x\left(\mathbf{q}^\dagger\Lambda\Psi_{21} - \Psi_{12}\mathbf{q}\right)\right.\\
&\left.-3\Psi_{21,x}\mathbf{q}^\dagger\Lambda\mathbf{q} + 3\left(\mathbf{q}\Psi_{12} - \Psi_{21}\mathbf{q}^\dagger\Lambda\right)\mathbf{q}_x\right],\\
\mathrm{i}\Psi_{12,t} =& \gamma\left[\frac{1}{2}\Psi_{12,xx} + \left(\Psi_{12}\mathbf{q}\mathbf{q}^\dagger\Lambda - \mathbf{q}^\dagger\Lambda\Psi_{21}\mathbf{q}^\dagger\Lambda + \mathbf{q}^\dagger\Lambda\mathbf{q}\Psi_{12}\right)\right]\\
&+ \delta\mathrm{i}\left[-\Psi_{12,xxx} - 3\mathbf{q}^\dagger\Lambda\mathbf{q}\Psi_{12,x} + 3\left(\mathbf{q}^\dagger\Lambda\Psi_{21} - \Psi_{12}\mathbf{q}\right)\mathbf{q}_x^\dagger\Lambda\right.\\
&\left.-3\Psi_{12,x}\mathbf{q}\mathbf{q}^\dagger\Lambda - 3\mathbf{q}_x^\dagger\Lambda\left(\mathbf{q}\Psi_{12} - \Psi_{21}\mathbf{q}^\dagger\Lambda\right)\right].
\end{aligned} \tag{2.35}$$

Moreover, the symmetry relationship between Ψ_{12} and Ψ_{21}: $\Psi_{12} = -\Psi_{21}^\dagger\Lambda$ guarantees the above linearized equation to satisfy the linearized multi-component Hirota equations.

We construct the solutions of Ψ by the wavefunction of Lax pair. Suppose we have a vector solution $\phi_i(\lambda)$ for Lax pair, by the symmetric proposition we know that $\phi_j^\dagger(\lambda^*)\Omega$ satisfies the adjoint Lax pair, which implies that the matrix function $\Psi = \phi_i(\lambda)\phi_j^\dagger(\lambda^*)\Omega$ solves the stationary zero-curvature Equations (2.29), where $\phi_i(\lambda) = [\phi_{i,1}(\lambda), \phi_{i,2}^{\mathrm{T}}(\lambda)]^{\mathrm{T}}$. Similarly, we know that $\Psi = \phi_j(\lambda^*)\phi_i^\dagger(\lambda)\Omega$ also solves Equations (2.35). Since Equations (2.35) are unrelated with the spectral parameters

λ. Thus the solutions

$$\phi_{i,2}(\lambda)\phi_{j,1}^*(\lambda^*) - \phi_{j,2}(\lambda^*)\phi_{i,1}^*(\lambda) \tag{2.36}$$

solves the linearized multi-component Hirota equations automatically.

In what follows, we consider how to use the above procedures to give the linear stability analysis for the multi-component Hirota equation with the plane wave solution (2.12). The process is distinctive from the linearized process for MI analysis [8, 24, 25, 51, 62, 72, 78, 81]. Suppose we consider the perturbation form:

$$\Psi_{21} = \begin{bmatrix} \left(g_1 e^{i\xi(x+\zeta t)} + f_1^* e^{-i\xi(x+\zeta^* t)}\right) e^{i(b_1 x + c_1 t)} \\ \left(g_2 e^{i\xi(x+\zeta t)} + f_2^* e^{-i\xi(x+\zeta^* t)}\right) e^{i(b_2 x + c_2 t)} \\ \vdots \\ \left(g_n e^{i\xi(x+\zeta t)} + f_n^* e^{-i\xi(x+\zeta^* t)}\right) e^{i(b_n x + c_n t)} \end{bmatrix}, \tag{2.37}$$

and $\phi_i(\lambda) = \mathbf{K}_n \mathbf{L}_i(\lambda) e^{i[\xi_i(\lambda)-\lambda]x + \eta_i(\lambda)t}$ and $\phi_j(\lambda) = \mathbf{K}_n \mathbf{L}_j(\lambda) e^{i[\xi_j(\lambda)-\lambda]x + \eta_j(\lambda)t}$ provides a solution

$$g_k = \frac{a_i}{\xi_i(\lambda) + b_k}, \qquad f_k = -\frac{a_i}{\xi_j(\lambda) + b_k} \tag{2.38}$$

where $k = 1, 2, \cdots, n$ and

$$\begin{aligned} &\xi = \xi_i(\lambda) - \xi_j(\lambda) \in \mathbb{R}, \\ &\zeta = \frac{\gamma}{2}(\xi_i(\lambda) + \xi_j(\lambda)) + \delta\left(\xi_i(\lambda)^2 + \xi_j(\lambda)^2 + \xi_i(\lambda)\xi_j(\lambda) - 3\sum_{i=1}^{n} s_i a_i^2\right) \end{aligned} \tag{2.39}$$

the parameter ξ determines the perturbation frequency and the parameter ζ determines the gain index. As for the fixed ξ, the parameter $\xi_i(\lambda)$ can be determined by the following equations

$$1 + \sum_{k=1}^{n} \frac{s_k a_k^2}{(\xi_i(\lambda) + b_k)((\xi_i(\lambda) - \xi + b_k))} = 0. \tag{2.40}$$

Especially, if $\gamma = 1$ and $\delta = 0$ which corresponds to the multi-component NLS equations, the gain index $\zeta = \xi_i(\lambda) - \frac{\xi}{2}$ will solve the equations

$$1 + \sum_{k=1}^{n} \frac{s_k a_k^2}{(\zeta + b_k)^2 - \frac{\xi^2}{4}} = 0. \tag{2.41}$$

In this way, we establish the relations between MI and nonlinear wave solutions based on Lax-pair analysis. These results interpret that the existence conditions of rogue wave or Akhmediev breather is equivalent to the dispersion relation for MI [46, 50]. In the general case, if $\zeta \in \mathbb{R}$ with fixed ξ, the background is so called modulational stable, in which the W-shaped solitons can be constructed [24, 52, 80, 81]. Otherwise, if $\zeta \notin \mathbb{R}$ with fixed ξ, the background is so called modulational unstable, in which the rogue waves can be constructed. Since the perturbation form (2.37) just possesses the mixture of two conjugated plane waves, we can just interpret the solutions with two linear independent vector solutions. For the more complicated case, we need to use the other perturbation form together with the Bäcklund transformation with more than two independent linear solutions.

For the Hirota equation, $n = 1$ and $\gamma\delta \neq 0$, the the perturbation frequency and gain index are given by $\xi = \sqrt{a_1^2 + (\lambda + \frac{b_1}{2})^2} \in \mathbb{R}$ and

$$\zeta = \left[4\left(\lambda - \frac{1}{4}b_1\right)^2 + \frac{3}{4}b_1^2 - 2a_1^2\right]\delta + \gamma\left(\lambda - \frac{b_1}{2}\right) \tag{2.42}$$

respectively. By the range of $\xi \in \mathbb{R}$, we know that $\lambda \in \mathbb{R} \cup [-\frac{b_1}{2} - a_1\mathrm{i}, -\frac{b_1}{2} + a_1\mathrm{i}]$. If $\lambda \in \mathbb{R}$, it follows that $\zeta \in \mathbb{R}$. While for $\lambda \in [-\frac{b_1}{2} - a_1\mathrm{i}, -\frac{b_1}{2} + a_1\mathrm{i}]$, the gain index ζ will be a non-real complex number as $b_1 \neq \frac{\gamma}{6\delta}$ that corresponds to the modulational unstable background, ζ is still a real number as $b_1 = \frac{\gamma}{6\delta}$ that corresponds to modulational stable background [52]. In what follows, we will show the different dynamics for the modulation stable or unstable background.

3 Breathers, rogue waves and W-shape solitons

Throughout the Bäcklund transformation, lots of exact solutions can be constructed by fixing the parameters. Departing from the zero seed solution, the solitons can be constructed. By the plane wave seed solution, the breathers, rogue waves and rational traveling wave can be constructed. Firstly, we still consider the scalar NLS equation.

3.1 Breather and rogue waves for the scalar NLS equation

To construct the exact solutions for the scalar NLS equation, we start from the seed solution $q^{[0]} = \mathrm{e}^{\mathrm{i}t}$ with $n = 1$. Then the breathers and rogue waves can be obtained by the Bäcklund transformation (2.11). So we need to find the fundamental solution for the Lax pair, which is constructed in the last section. Here we rewrite it with a compact form:

$$\Phi_0(\lambda; x, t) = \mathbf{K}_1 \begin{bmatrix} \frac{\mathrm{i}\lambda\sin(\xi(x+\lambda t))}{\xi} + \cos(\xi(x+\lambda t)) & \frac{\mathrm{i}\sin(\xi(x+\lambda t))}{\xi} \\ \frac{\mathrm{i}\sin(\xi(x+\lambda t))}{\xi} & -\frac{\mathrm{i}\lambda\sin(\xi(x+\lambda t))}{\xi} + \cos(\xi(x+\lambda t)) \end{bmatrix}, \tag{3.1}$$

which is an analytic solution in the whole complex plane $\mathbb{C}$, where $\mathbf{K}_1 = \mathrm{diag}\left(\mathrm{e}^{-\mathrm{i}t/2}, \mathrm{e}^{\mathrm{i}t/2}\right)$ and $\xi^2 = 1 + \lambda^2$. By the Bäcklund transformation and introducing $\xi_1 = \cosh(\alpha_1 + \mathrm{i}\beta_1)$, $\lambda_1 = \sinh(\alpha_1 + \mathrm{i}\beta_1)$, then the general breather solution can be represented as the compact form:

$$q^{[1]} = \left[\frac{\cosh(\alpha_1)\cosh(A_1 + 2\mathrm{i}\beta_1) + \sin(\beta_1)\cos(A_2 + 2\mathrm{i}\alpha_1)}{\cosh(\alpha_1)\cosh(A_1) - \sin(\beta_1)\cos(A_2)}\right]\mathrm{e}^{\mathrm{i}t} \tag{3.2}$$

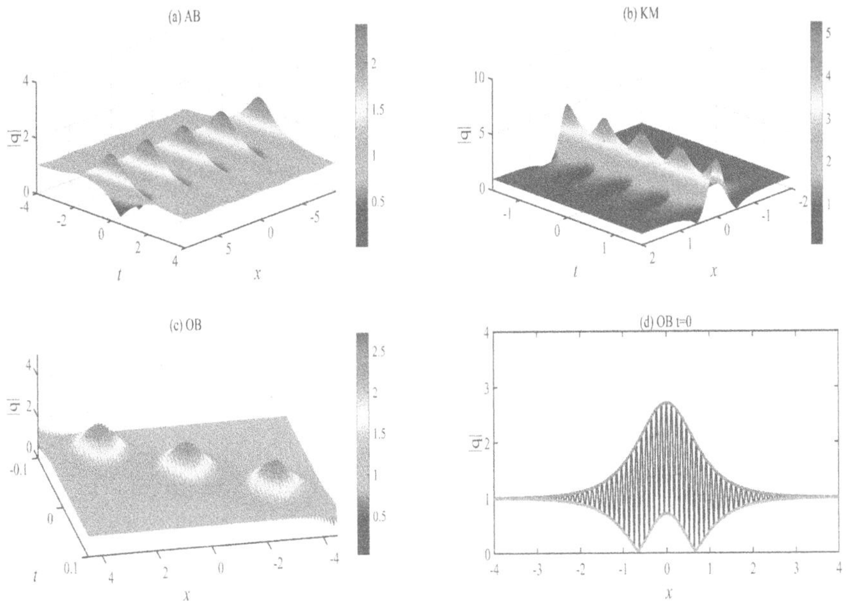

Figure 1. (a) Akhmediev breather: $\alpha = 0$, $\beta = \frac{\pi}{4}$; (b) Kuznetsov-Ma breather: $\alpha = \ln(4)$, $\beta = \frac{\pi}{2}$; (c) Oscillating breather $\alpha = 4$, $\beta = \frac{\pi}{100}$; (d) The profile of oscillating breather at $t = 0$.

where

$$
\begin{aligned}
A_1 &= 2\sin(\beta_1)\left[\sinh(\alpha_1)x + \cos(\beta_1)\cosh(2\alpha_1)t\right], \\
A_2 &= 2\cosh(\alpha_1)\left[\cos(\beta_1)x + \sinh(\alpha_1)\cos(2\beta_1)t\right].
\end{aligned}
\tag{3.3}
$$

3.1.1 Types of breathers

Now we proceed to analyze the dynamics for the breather solution (3.2). There are three types of breather dynamics by choosing the distinct parameters α_1 and β_1.

1) If $\alpha_1 = 0$, $\beta_1 \neq \frac{\pi}{2}$, the breather solution is spatial periodic and temporal localized, i.e., so-called Akhmediev breather (Fig. 1 a), whose spatial period is $\frac{\pi}{\cos(\beta_1)}$. It is readily seen that the period is greater than π. On the other hand, by the squared eigenfunction method, we know that the background is modulational stable under the perturbations (whose period is less than π), which is completely consistent with the dynamics of Ahkmediev breather. If we understand the whole procedure by combining the Darboux transformation and squared eigenfunction method, then we can know the reason they are completely consistent since both Darboux transformation and squared eigenfunction are constructed by the solutions of Lax pair. The maximal points of $|q(x,t)|$ are located on $(x,t) = (k\pi/\cos(\beta_1), 0)$, $k \in \mathbb{Z}$, with the value $1 + 2\sin(\beta_1)$. It is easy to find that the maximal value of the Ahkmediev breathers is less than 3, which is the peak value of the first-order rogue wave.

2) If $\alpha_1 \neq 0$, $\beta_1 = \frac{\pi}{2}$, the breather is temporal periodic and spatial localized, i.e., the Kuznetsov-Ma solution (Fig. 1 b). The temporal period is $\frac{2\pi}{\sinh(2\alpha_1)}$. The peak of $|q(x,t)|$ is located on $(x,t) = (0, 2k\pi/\sinh(2\alpha_1))$, $k \in \mathbb{Z}$, with the value $1 + 2\cosh(\alpha_1)$. It is easy to find that if α_1 is large enough, the period is small. Then we can see the dynamics of breathers is close to the solitons, which can be considered as the soliton added to the plane wave [79]. Conversely, if the parameter α_1 is close to 0, it will exhibit the clear structure with the rogue waves on the periodic unit.

3) If $\alpha_1 \neq 0$, $\beta_1 \neq \frac{\pi}{2}$, the breather has the velocity $v = -\cos(\beta_1)\cosh(2\alpha_1)/\sinh(\alpha_1)$, i.e., the general breather. The peak of $|q(x,t)|$ is located on

$$(t,x) = \left(\frac{-2k\pi\tanh(\alpha_1)}{\cosh(2\alpha_1)+\cos(2\beta_1)}, \frac{2k\pi\cosh(2\alpha_1)\cos(\beta_1)}{\cosh(\alpha_1)(\cosh(2\alpha_1)+\cos(2\beta_1))}\right),$$

$k \in \mathbb{Z}$, with the value $h = 1+2\cosh(\alpha_1)\sin(\beta_1)$. The distance between two adjacent peaks is

$$\frac{2\pi\sqrt{\cosh^2(2\alpha_1)\cos^2(\beta_1)+\sinh^2(\alpha_1)}}{\cosh(\alpha_1)(\cosh(2\alpha_1)+\cos(2\beta_1))}. \tag{3.4}$$

In this case, there is an interesting type of breather, which is oscillating during the wave packet and is called the oscillating breather (Fig. 1 c,d). The oscillating breathers are enveloped by the upper and lower bound functions, where the upper bound function

$$q_{up} = \left[\frac{\cosh(\alpha_1)\cosh(A_1+2\mathrm{i}\beta_1)+\sin(\beta_1)\cosh(2\alpha_1)}{\cosh(\alpha_1)\cosh(A_1)-\sin(\beta_1)}\right]\mathrm{e}^{\mathrm{i}t} \tag{3.5}$$

and the lower bound function

$$q_{lower} = \left[\frac{\cosh(\alpha_1)\cosh(A_1+2\mathrm{i}\beta_1)-\sin(\beta_1)\cosh(2\alpha_1)}{\cosh(\alpha_1)\cosh(A_1)+\sin(\beta_1)}\right]\mathrm{e}^{it}. \tag{3.6}$$

To introduce the oscillating breather, we first give the following definitions: the width of breather is

$$w_{100} = \frac{1}{2\sinh(\alpha_1)\sin(\beta_1)}\ln\left(\frac{100|\cosh(2\alpha_1)+\exp(2\mathrm{i}\beta_1)|}{2\cosh^2(\alpha_1)}\right) \tag{3.7}$$

where w_{100} stands for the width of one percent height of q_{up}, and the oscillating period is

$$op = \frac{\pi}{\cosh(\alpha_1)|\cos(\beta_1)|}. \tag{3.8}$$

If $w_{100}/op \geq 8$, the breathers exhibit the evident oscillating behavior, which is the condition of oscillating breather. Furthermore, if the height of breather $h > 2$, the lower bound function of breather will exhibit the W-shape. As for the other cases, they are ordinary breathers.

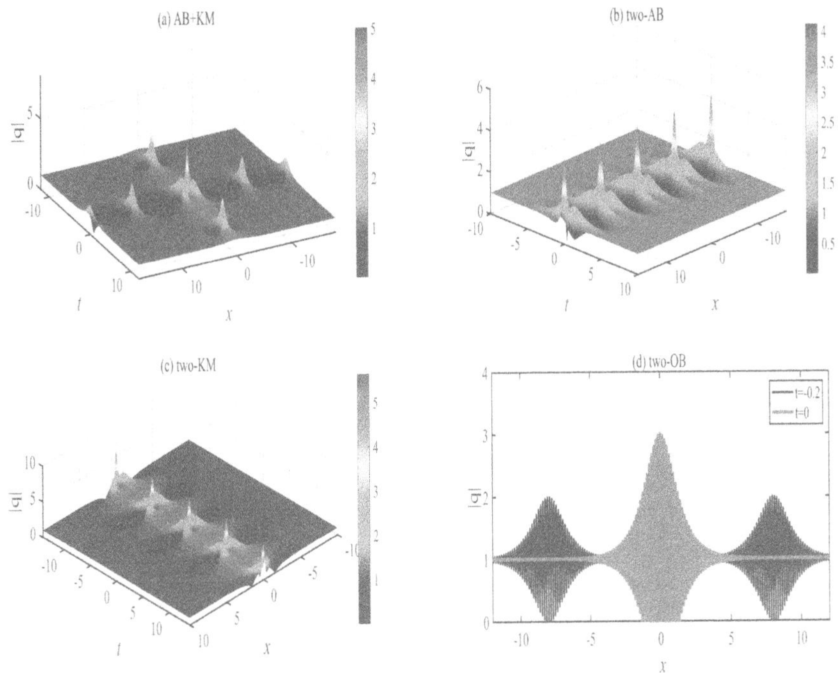

Figure 2. (a) The breathers Kuznetsov-Ma and Akhmediev breather: $k_1 = \frac{5+12\mathrm{i}}{13}$, $k_2 = \frac{3\mathrm{i}}{2}$; (b) Two Akhmediev breather: $k_1 = \exp(\mathrm{i}\arctan(12/5))$, $k_2 = \exp(\mathrm{i}\arccos(10/13))$, $T = \frac{\pi}{\cos(\arctan(12/5))}$; (c) Two Kuznetsov-Ma breather: $k_1 = \frac{3\mathrm{i}}{2}$, $k_2 = \exp(\frac{\mathrm{i}}{2}\mathrm{arcsinh}(2\sinh(2\ln(3/2))))$, the period along t direction is $\frac{2\pi}{\sinh(2\ln(3/2))}$; (d) The profile for two oscillating breathers: $k_1 = 40 + \mathrm{i}$, $k_2 = -40 + \mathrm{i}$.

3.1.2 Two-breather

Furthermore, we consider the two-breather solutions, whose expressions can be obtained through the formula of Bäcklund transformation (2.11):

$$q^{[2]} = \left[\frac{\det(\mathbf{M}+\mathbf{H})}{\det(\mathbf{M})}\right]\mathrm{e}^{\mathrm{i}t} \tag{3.9}$$

where $\phi_i = \mathbf{K}_1^{-1}\Phi_0(\lambda_i; x, t)\mathbf{v}_i$,

$$\mathbf{M} = \begin{bmatrix} \frac{\phi_1^\dagger\phi_1}{\lambda_1-\lambda_1^*} & \frac{\phi_1^\dagger\phi_2}{\lambda_2-\lambda_1^*} \\ \frac{\phi_2^\dagger\phi_1}{\lambda_1-\lambda_2^*} & \frac{\phi_2^\dagger\phi_2}{\lambda_2-\lambda_2^*} \end{bmatrix}, \qquad \mathbf{H} = \mathrm{i}\begin{bmatrix} \phi_{1,1}^* \\ \phi_{2,1}^* \end{bmatrix}\begin{bmatrix} \phi_{1,2} & \phi_{2,2} \end{bmatrix}, \tag{3.10}$$

the vectors are choosing as $\mathbf{v}_i = [1, \mathrm{i}]^{\mathrm{T}}$ such that the module of solutions attaining the maximal value at the origin, $\lambda_i = \frac{1}{2}(k_i - \frac{1}{k_i})$, $\xi_i = \frac{1}{2}(k_i + \frac{1}{k_i})$ and $\phi_{i,k}$ stands for the k-th component of ϕ_i. Here we just give the distinct parameter setting to construct the colorful dynamics. We consider the following special types:

1) The nonlinear superposition with the Kuznetsov-Ma breather and Akhmediev breather will form a cross-type solution (Fig. 2 a). At the origin, two-breather solution exhibits the dynamics of second-order rogue waves. If we adjust the parameter setting, the other patterns will be formed at the intersection point.

2) The two-breathers between two Akhmediev breathers. The first case, we consider the breathers with period T and $\frac{n}{m}T$, which will form a breather with period mnT. And each of them shows the similar dynamics with the shape of second-order rogue waves (Fig. 2 b). Then if two breathers have the period with T and kT, here k is an irrational number, they will show that the quasi-periodic dynamics.

3) The two-breathers between two Kuznetsov-Ma breathers. This is similar to the case of Akhmediev breathers. The first case, we consider the breathers with period T and $\frac{n}{m}T$; they form a Kuznetsov-Ma breathers with the period mnT. During the period, it shows the dynamics is also of a similar structure as the second-order rogue wave (Fig. 2 c). Secondly, we can also consider the breathers with period T and kT, here k is an irrational number, which exhibits the quasi-periodic dynamics behavior in the t direction.

4) We will show two oscillating breathers with distinct velocity, which shows the interaction dynamics is elastic (Fig. 2 d). The rigorous proof for the N-breathers to the complex short pulse equation was given in [44]. For the NLS equation, it can be done similarly.

If two breathers possess the same parameters, the high order breathers will be formed. In other words, they can be considered as the limitation for the two breathers. Through the formulas of breathers, we can obtain the multi-breathers. The detailed analysis for the dynamics of these multi-breathers can be performed by the further analysis in a similar manner as in the literature [44].

3.1.3 Rogue waves

In what follows, we will turn to analyze the rogue waves and high order rogue waves solutions. Firstly, we give the general formulas for the rogue waves:

$$q^{[n]} = \left[\frac{\det(H)}{\det(M)}\right] \mathrm{e}^{it} \tag{3.11}$$

where

$$\begin{gathered} M = F^{\dagger}BF + G^{\dagger}BG, \qquad H = M - 2(F_1^{\dagger} + G_1^{\dagger})(F_1 - G_1), \\ F = \begin{bmatrix} f_1 & f_2 & \cdots & f_n \\ 0 & f_1 & \cdots & f_{n-1} \\ \vdots & \vdots & \ddots & \vdots \\ 0 & 0 & \cdots & f_1 \end{bmatrix}, \qquad G = \begin{bmatrix} g_1 & g_2 & \cdots & g_n \\ 0 & g_1 & \cdots & g_{n-1} \\ \vdots & \vdots & \ddots & \vdots \\ 0 & 0 & \cdots & g_1 \end{bmatrix} \end{gathered} \tag{3.12}$$

and $B = \left(\binom{i+j-2}{i-1}\right)_{1\le i,j\le n}$ is a Pascal matrix, f_is and g_is can be given by the following Taylor expansions:

$$\begin{aligned}(\mathrm{i}\lambda-1)\frac{\sin(\xi(x+\lambda t))}{\xi}+c\cos(\xi(x+\lambda t)) &= \sum_{i=1}^{\infty} f_i\left[\frac{\mathrm{i}}{2}(\lambda-\mathrm{i})\right]^{i-1},\\ \cos(\xi(x+\lambda t))+c\,(\mathrm{i}\lambda+1)\frac{\sin(\xi(x+\lambda t))}{\xi} &= \sum_{i=1}^{\infty} g_i\left[\frac{\mathrm{i}}{2}(\lambda-\mathrm{i})\right]^{i-1}\\ c &= \sum_{i=1}^{n-1} c_i\left[\frac{\mathrm{i}}{2}(\lambda-\mathrm{i})\right]^{i-1}\end{aligned} \tag{3.13}$$

F_1 and G_1 denotes the first row of matrix F and G, respectively. Through the above formula (3.11), we know the expression of the first- and second-order fundamental rogue waves are

$$\begin{aligned} q_{1rw} &= \left[-1+\frac{4(1+2\mathrm{i}t)}{1+4(x^2+t^2)}\right]\mathrm{e}^{\mathrm{i}t}, \qquad q_{2rw}=(1+R_{2rw})\,\mathrm{e}^{\mathrm{i}t},\\ R_{2rw} &= \frac{R_1-24\mathrm{i}t\left(16t^4+32t^2x^2+16x^4+8t^2-24x^2-15\right)}{64t^6+(192x^2+432)\,t^4+(192x^4-288x^2+396)\,t^2+R_2},\\ R_1 &= -960t^4+\left(-1152x^2-864\right)t^2-192x^4-288x^2+36,\\ R_2 &= 64x^6+48x^4+108x^2+9.\end{aligned} \tag{3.14}$$

Then we give the classifications on the rogue waves. The first crucial type is the fundamental high order rogue waves, which shows the extreme superposition on the rogue waves. At the origin, we have $q_n(0,0) = 2n+1$ [2, 13, 67], which can be deduced from the determinant formula by setting $x = t = 0$. It is natural to ask about the dynamics when $n \to \infty$. It is proved that the leading dynamics for the near field of this extreme superposition will generate new solutions of NLS equation with zero background, which is related with the Painlevé-III equations [13].

The other types of rogue waves are the separated rogue waves, which will form the distinct spatial-temporal pattern, such as the triangle shape, the $2n+1$-gon type, the cyclic type, and so on. In what follows, we will show how to use the asymptotic method to analyze the separated rogue waves. The separated rogue waves are originated from the large parameters of the solutions. Under the large parameters, the rogue waves can be decomposed into the linear combinations of fundamental rogue waves. This was argued firstly in [39], we here demonstrate the decomposition method for the argument.

We consider the following two simple examples:

1) The fourth-order triangle shape rogue waves. By choosing the parameter

$c = \frac{\mathrm{i}}{2}(\lambda - \mathrm{i})a^3\mathrm{e}^{3\mathrm{i}\theta}$, we have the following $f_{1\leq i\leq 4}$ and $g_{1\leq i\leq 4}$

$$
\begin{aligned}
f_1 &= -2\chi, \qquad f_2 = 4\mathrm{i}t + \frac{4}{3}\chi^3 + 2\chi + a^3\mathrm{e}^{3\mathrm{i}\theta},\\
f_3 &= -8\mathrm{i}\chi^2 t - \frac{8}{3}\chi^3 - \frac{4\chi^5}{15} - 4\mathrm{i}t - 2a^3\mathrm{e}^{3i\theta}\chi^2,\\
f_4 &= -16\chi t^2 + 16\mathrm{i}\chi^2 t + \frac{8}{3}\mathrm{i}\chi^4 t + \frac{4}{5}\chi^5 + \frac{8\chi^7}{315} + \frac{4}{3}\chi^3\\
&\quad + a^3\mathrm{e}^{3i\theta}\left(8\mathrm{i}\chi t + 2\chi^2 + \frac{2}{3}\chi^4\right),\\
g_1 &= 1, \qquad g_2 = -2\chi^2, \qquad g_3 = 8\mathrm{i}\chi t + 2\chi^2 + \frac{2}{3}\chi^4 + 2a^3\mathrm{e}^{3i\theta}\chi,\\
g_4 &= 8t^2 - 8\mathrm{i}\chi t - \frac{16}{3}\mathrm{i}\chi^3 t - \frac{4}{3}\chi^4 - \frac{4\chi^6}{45} - 2a^3\mathrm{e}^{3i\theta}\left(2\mathrm{i}t + \frac{2}{3}\chi^3\right),
\end{aligned}
\tag{3.15}
$$

and $\chi = x + it$. Inserting the above expressions (3.15) and $n = 4$ into the formula (3.11), we will obtain the fourth-order triangle shape rogue waves if the parameter a is large enough (Fig. 3 b).

Proposition 2. *The fourth-order rogue waves can be decomposed into the following form:*

$$
q_4 = \left[1 - \sum_{i=1}^{10} \frac{4(1 + 2\mathrm{i}(t - t_i))}{1 + 4[(x - x_i)^2 + (t - t_i)^2]} + O\left(\frac{1}{a}\right)\right]\mathrm{e}^{\mathrm{i}t} \tag{3.16}
$$

where $x_i = a\mathrm{Re}(z_i\mathrm{e}^{\mathrm{i}\theta})$ *and* $t_i = a\mathrm{Im}(z_i\mathrm{e}^{\mathrm{i}\theta})$, $z_i s$ *are the roots for the following characteristic equation:*

$$
(512z^9 - 2880z^6 - 4725)z = 0. \tag{3.17}
$$

Proof. We look for the center of lower order rogue waves by the proper scale. On the other hand, we know that a is large enough, so the highest polynomial will play the main role in the dynamical behavior together with the large parameter a. To balance them, we need to use the ansatz $\xi = x - a\,\mathrm{Re}(\mathrm{e}^{\mathrm{i}\theta}z)$ and $\eta = t - a\,\mathrm{Im}(\mathrm{e}^{\mathrm{i}\theta}z)$ under the neighborhood of $(\xi, \eta) = (0, 0)$. In other words, we can plug the replacement $x = \xi + a\,\mathrm{Re}(\mathrm{e}^{\mathrm{i}\theta}z)$ and $t = \eta + a\,\mathrm{Im}(\mathrm{e}^{\mathrm{i}\theta}z)$ into the expression of q_{4t}, which yields the rational functions on the variables a, ξ, η, θ and z. Since the variable a is large, the main impact will be played by the polynomial of a. The order of leading term of polynomials with respect to a to denominator of q_4 is 20, and its coefficient is the governing Equation (3.17). If the governing Equation (3.17) does not equal to zero, then the solution will tend to the background solution

$$
q_4 = \left[1 + O\left(\frac{1}{a}\right)\right]\mathrm{e}^{\mathrm{i}t}. \tag{3.18}
$$

If the governing Equation (3.17) equals to zero, we can obtain the following asymptotics

$$
q_4 = \left[1 - \frac{4(1 + 2\mathrm{i}\eta)}{1 + 4(\xi^2 + \eta^2)} + O\left(\frac{1}{a}\right)\right]\mathrm{e}^{\mathrm{i}t}. \tag{3.19}
$$

Combining them, we obtain the final asymptotics (3.16) in the whole (x, t) plane. ■

2) The fourth-order rogue waves with regular 7−gon shape. Similarly, by choosing the parameter $c = \left[\frac{\mathrm{i}}{2}(\lambda - \mathrm{i})\right]^3 a^7 \mathrm{e}^{7\mathrm{i}\theta}$, we have the following $f_{1\le i\le 4}$ and $g_{1\le i\le 4}$

$$\begin{aligned}
&f_1 = -2\chi, \qquad f_2 = 4\mathrm{i}t + \frac{4}{3}\chi^3 + 2\chi, \qquad f_3 = -8\mathrm{i}\chi^2 t - 8/3\chi^3 - \frac{4\chi^5}{15} - 4\mathrm{i}t,\\
&f_4 = -16\chi t^2 + 16\mathrm{i}\chi^2 t + \frac{8}{3}\mathrm{i}\chi^4 t + \frac{4}{5}\chi^5 + \frac{8\chi^7}{315} + \frac{4}{3}\chi^3 + a^7\mathrm{e}^{7\mathrm{i}\theta},\\
&g_1 = 1, \qquad g_2 = -2\chi^2, \qquad g_3 = 8\mathrm{i}\chi t + 2\chi^2 + \frac{2}{3}\chi^4,\\
&g_4 = 8t^2 - 8\mathrm{i}\chi t - \frac{16}{3}\mathrm{i}\chi^3 t - \frac{4}{3}\chi^4 - \frac{4\chi^6}{45}.
\end{aligned} \tag{3.20}$$

Inserting the above expressions (3.20) and $n = 4$ into the formula (3.11), we will obtain the fourth-order regular 7−gon shape rogue waves if the parameter a is large enough, in which there is a fundamental second-order rogue wave in the center.

Proposition 3. *The fourth-order rogue waves can be decomposed into the following form:*

$$q_4 = \left[1 + R_{2rw} - \sum_{i=1}^{7} \frac{4(1 + 2\mathrm{i}(t - t_i))}{1 + 4[(x - x_i)^2 + (t - t_i)^2]} + O\left(\frac{1}{a}\right)\right] \mathrm{e}^{\mathrm{i}t} \tag{3.21}$$

where $x_i = a\mathrm{Re}(z_i \mathrm{e}^{\mathrm{i}\theta})$ *and* $t_i = a\mathrm{Im}(z_i \mathrm{e}^{\mathrm{i}\theta})$, $z_i s$ *are the roots for the following characteristic equation:*

$$z^7 = \frac{1575}{128}. \tag{3.22}$$

This proposition can be proved similar to Proposition 1. By the numerical verification (Fig. 3 a), the locations of RW are predicted by the roots of governing Equation (3.22). The detailed analysis for this tool is proposed in the joint work of the first author with Zhang and Yan [73].

3.2 Hirota equation and W-shape soliton

For the Hirota equation $\gamma\delta \neq 0$, under the condition of modulational stability the periodic solution, rational W-shape soliton and periodic W-shape wave can be constructed by the formula of Bäcklund transformation [?, 52]. Consider the plane wave solution for the Hirota equation with $n = 1$ and $\gamma\delta \neq 0$

$$q_0 = a_1 \mathrm{e}^{\mathrm{i}\theta_1}, \qquad \theta_1 = b_1 x + c_1 t, \qquad c_1 = \frac{1}{2}\left(2a_1^2 - b_1^2\right)\gamma + \left(b_1^3 - 6a_1^2 b_1\right)\delta. \tag{3.23}$$

Then by the squared eigenfunction method, Equation (2.39) implies that the plane wave (3.23) is the modulational stable under the condition

$$b_1 = \frac{\gamma}{6\delta}, \tag{3.24}$$

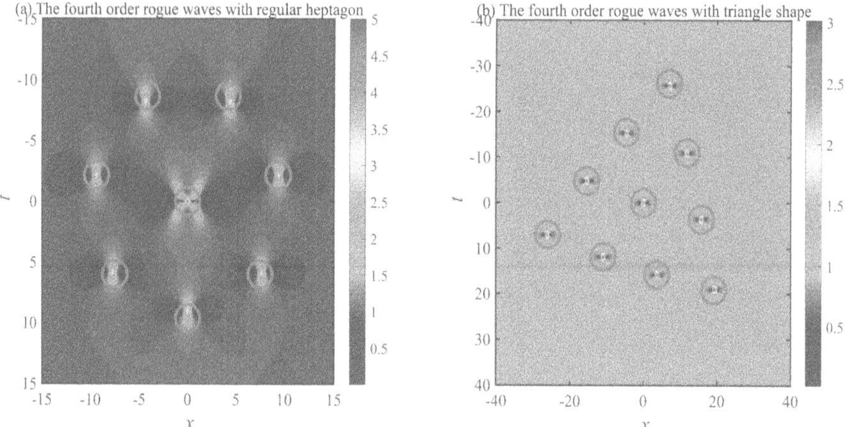

Figure 3. (a) The fourth order RW with regular heptagon shape: $a = 5$, (b) The triangle shape RW: $a = 15$, $\theta = 3\pi\mathrm{i}/4$.

which is given in the last section. It is interesting that we can construct the traveling wave and periodic traveling waves under the condition (3.24) by the Bäcklund transformation (2.11). These solutions with large amplitude perturbations on the plane wave cannot be interpreted by the linear stability analysis. The small one is consistent with the linear stability analysis.

To give the exact solutions for the Hirota equation, we just need to introduce the variables

$$A = 2\xi_1 \left[x + \left(\left(4\left(\lambda_1 - \frac{3}{4}b_1\right)^2 + \frac{3}{4}b_1^2 - 2a_1^2 \right)\delta + \gamma\left(\lambda_1 - b_1\right) \right) t \right], \tag{3.25}$$
$$\xi_1 = a_1 \cosh(\alpha_1 + \mathrm{i}\beta_1), \qquad \lambda_1 = a_1 \sinh(\alpha_1 + \mathrm{i}\beta_1).$$

By the formulas of Bäcklund transformation (2.11), we can obtain the general breather solutions [51]:

$$q^{[1]} = a_1 \left[\frac{\cosh(\alpha_1)\cosh(A_R + 2\mathrm{i}\beta_1) + \sin(\beta_1)\cos(A_I + 2\mathrm{i}\alpha_1)}{\cosh(\alpha_1)\cosh(A_R) - \sin(\beta_1)\cos(A_I)} \right] \mathrm{e}^{\mathrm{i}\theta_1}, \qquad A = A_R + \mathrm{i}A_I. \tag{3.26}$$

If $\alpha_1 = 0$ and $0 < \beta_1 < \frac{\pi}{2}$, we have

$$A_R = 0, \qquad A_I = a_1\cos(\beta_1)\left[2x + \left(8a_1^2\cos^2(\beta_1)\delta - \frac{1}{6\delta}\left(72a_1^2\delta^2 + \gamma^2\right) \right) t \right] \tag{3.27}$$

then the periodic solutions are given by

$$q^{[1]} = a_1 \left[-1 + \frac{\cos\left(2\,\beta\right) + 1}{1 - \sin\left(\beta\right)\cos(A_2)} \right] \mathrm{e}^{\mathrm{i}\theta_1}. \tag{3.28}$$

If β_1 is close to zero, we find the perturbations are small. The periodic wave propagates along the fixed direction without changing the shape, which provides an

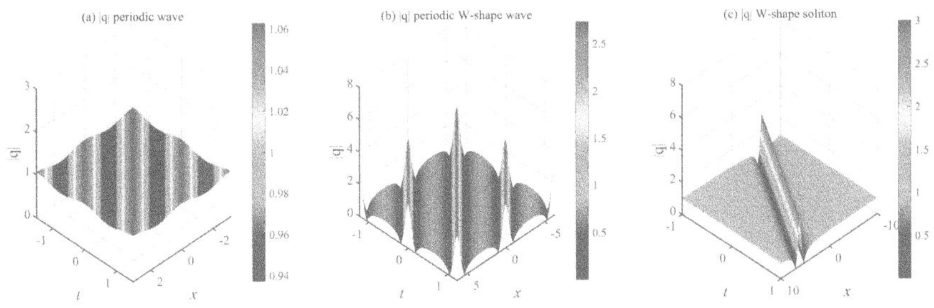

Figure 4. Parameters $\gamma = \delta = 1, a_1 = 1$, $b_1 = \frac{1}{6}$. (a) Periodic wave with small perturbations: $\alpha_1 = 0$, $\beta_1 = \frac{\pi}{100}$, (b) Periodic W-shape waves: $\alpha_1 = 0$, $\beta_1 = \frac{\pi}{3}$. (c) W-shape soliton.

evidence that the background wave is stable (Fig. 4 a). If $2\sin(\beta_1) > 1$, then the periodic wave will be the W-shape periodic wave, which also propagates stably but can not be interpreted by the theory of modulational instability (Fig. 4 b).

The rational solutions under the modulational stable background can turn to the W-shape soliton [52, 80]. At the branch point, we can obtain the rational W-shape solution, which can be readily derived as above by changing the factor $\xi(x+\lambda t) \to A$. The rational W-shape soliton is given by

$$q^{[1]} = a_1 \left[-1 + \frac{144\delta^2}{36\,\delta^2 + [12\delta\,x - (72\,\delta^2 + \gamma^2)t]^2} \right] \mathrm{e}^{\mathrm{i}\theta_1}. \tag{3.29}$$

By the computer graphic (Fig. 4 c), we find that the plotting exhibits the "W" shape. So we call it the rational W-shape soliton [52, 80, 81].

Actually, we can construct the multi-breather and other high-order solutions by the formula (2.11). These solutions have complicated dynamics, which needs further analysis to understand the dynamics.

4 Vector breathers and rogue waves

In the last section, we introduce the breathers, rogue waves and W-shape solitons for the scalar equation. The similar structures also exist for the vector integrable system. In this section, we consider the vector breathers and vector rogue waves. The vector W-shape soliton also can be obtained with an analogous manner for the vector Hirota equation [53].

4.1 Vector breathers

Departing from the plane wave background $q_i^{[0]} = a_i \mathrm{e}^{\mathrm{i}\theta_i}$, the Akhmediev breather solutions for the vector NLS equations can be constructed by the Bäcklund transformation (2.11), which is a type of spatial periodic and temporal localized solutions. To this purpose, we need to consider two parameters ξ_i and ξ_j with the same imaginary part for a fixed λ. In other words, the parameters ξ_i and ξ_j satisfy the following equations

$$2\lambda - \xi_{i/j} + \sum_{l=1}^{n} \frac{s_l a_l^2}{\xi_{i/j} + b_l} = 0, \tag{4.1}$$

which deduces the equation by subtracting the above two equations (4.1) each other:

$$1 + \sum_{l=1}^{n} \frac{s_l a_l^2}{(\xi_i + b_l)(\xi_j + b_l)} = 0. \tag{4.2}$$

Conversely, for arbitrary ξ_i, we can find a parameter λ such that

$$\lambda = \frac{1}{2}\left(\xi_i - \sum_{l=1}^{n} \frac{s_l a_l^2}{\xi_i + b_l}\right). \tag{4.3}$$

Moreover, for the fixed ξ_i, the parameter ξ_j satisfies the equation (4.2), then we have

$$2\lambda - \xi_j + \sum_{l=1}^{n} \frac{s_l a_l^2}{\xi_j + b_l} = 0. \tag{4.4}$$

Through the above analysis, we establish the equivalent relationship between Equations (4.1) and (4.2). Since the parameters ξ_i and ξ_j have the same imaginary part, we can set $\xi_j = \xi_i + \xi$, where $\xi \in \mathbb{R}$ will determines the temporal period of Akhmediev breather. Furthermore, the squared eigenfunction method discloses equivalent relationship between the dispersion relationship of modulational instability and existence conditions for the Akhmediev breathers, since they can be constructed from one special solution of Lax pair.

The exact form of Akhmediev breather for the vector NLS equations can be constructed from the Bäcklund transformation (2.11) with the above parameter setting:

$$q_i^{[1]} = a_i \left(\frac{H_i}{M}\right) \mathrm{e}^{\mathrm{i}\theta_i} \tag{4.5}$$

where

$$\begin{aligned}
H_i &= \frac{\xi_1^* + b_i}{\xi_1 + \xi + b_i} \frac{\mathrm{e}^{E_1}}{\xi_1 + \xi - \xi_1^*} + \frac{\xi_1^* + \xi + b_i}{\xi_1 + b_i} \frac{\mathrm{e}^{E_1^*}}{\xi_1 - \xi_1^* - \xi} \\
&+ \left(\frac{\xi_1^* + b_i}{\xi_1 + b_i} + \frac{\xi_1^* + \xi + b_i}{\xi_1 + \xi + b_i} \mathrm{e}^{E_1 + E_1^*}\right) \frac{1}{\xi_1 - \xi_1^*}, \\
M &= \frac{\mathrm{e}^{E_1}}{\xi_1 + \xi - \xi_1^*} + \frac{\mathrm{e}^{E_1^*}}{\xi_1 - \xi_1^* - \xi} + \frac{1 + \mathrm{e}^{E_1 + E_1^*}}{\xi_1 - \xi_1^*}, \\
E_1 &= \mathrm{i}\xi \left[x + (\xi_1 + \xi/2)t\right] + \ln(\alpha_1 + \mathrm{i}\beta_1).
\end{aligned} \tag{4.6}$$

We exhibit the explicit figure for the vector breathers by choosing the proper parameters (Fig. 5a,b). The period is 2π. And the absolute value of solutions $\mathbf{q}^{[1]}$

will attain the maximum by choosing the special parameters α_1 and β_1 as in the caption of Figure 5. For the component $q_1^{[1]}$ the extreme point is the peak, while the extreme point is the valley for the component $q_2^{[1]}$.

Now we continue to consider the multi-breather under the fixed period. By choosing the special parameters, we have the following multi-Akhmediev breather which is the nonlinear superposition for the single Akhmediev breathers:

$$q_i = a_i \left[\frac{\det(h_{j,k}^{[i]})}{\det(m_{j,k})} \right] \mathrm{e}^{\mathrm{i}\theta_i} \tag{4.7}$$

where

$$\begin{aligned} h_{j,k}^{[i]} &= \frac{\xi_j^* + b_i}{\xi_k + \xi + b_i} \frac{\mathrm{e}^{E_k}}{\xi_k + \xi - \xi_j^*} + \frac{\xi_j^* + \xi + b_i}{\xi_k + b_i} \frac{\mathrm{e}^{E_j^*}}{\xi_k - \xi_j^* - \xi} \\ &\quad + \left(\frac{\xi_j^* + b_i}{\xi_k + b_i} + \frac{\xi_j^* + \xi + b_i}{\xi_k + \xi + b_i} \mathrm{e}^{E_k + E_j^*} \right) \frac{1}{\xi_k - \xi_j^*}, \\ m_{j,k} &= \frac{\mathrm{e}^{E_k}}{\xi_k + \xi - \xi_j^*} + \frac{\mathrm{e}^{E_j^*}}{\xi_k - \xi_j^* - \xi} + \frac{1 + \mathrm{e}^{E_k + E_j^*}}{\xi_k - \xi_j^*}, \\ E_i &= \mathrm{i}\xi \left[x + (\xi_i + \xi/2)t \right] + \ln(\alpha_i + \mathrm{i}\beta_i). \end{aligned} \tag{4.8}$$

Especially, we consider the two-breather solutions, which involve the parameters α_is, β_is to control the locations of breather. Here we still take the parameters such that the $|\mathbf{q}|^2$ will attain the maximum at the origin (Fig. 5c,d). It is seen that the figures of both components exhibit anti-symmetric structures.

4.2 Vector rogue waves and high-order ones

To construct the vector rogue waves, we need to use the vector solutions Φ_s by the formulas (2.11). For the vector rogue waves we can rewrite the formulas (2.11) in a compact way based on the relationships:

Proposition 4. *Suppose there is a solution*

$$\begin{aligned} \Phi_l &= \frac{1}{p_l \epsilon_l^{g_l}} K \sum_{k=0}^{p_l - 1} L_k^{[l]} \mathrm{e}^{\vartheta_l^{[k]}}, \\ \vartheta_l^{[k]} &= \mathrm{i} \left[(\xi_l^{[k]} - \xi_l)x + (\eta_l^{[k]} - \eta_l)t + c_l^{[k]} \right] - \frac{2k g_l \pi \mathrm{i}}{p_l}, \\ L_k^{[l]} &= \left[1, \frac{a_1}{\xi_k^{[l]} + b_1}, \cdots, \frac{a_n}{\xi_k^{[l]} + b_n} \right]^{\mathrm{T}}, \end{aligned} \tag{4.9}$$

where $\xi_l^{[k]} = \xi_l^{[0]}(\mathrm{e}^{\frac{2k\pi\mathrm{i}}{p_l}} \epsilon_l)$, $\xi_l^{[0]} = \xi_l(\epsilon_l)$, $\eta_l^{[k]} = \eta_l^{[0]}(\mathrm{e}^{\frac{2k\pi\mathrm{i}}{p_l}} \epsilon_l)$, $\eta_l^{[0]} = \eta_l(\epsilon_l)$, $c_l^{[k]} =$

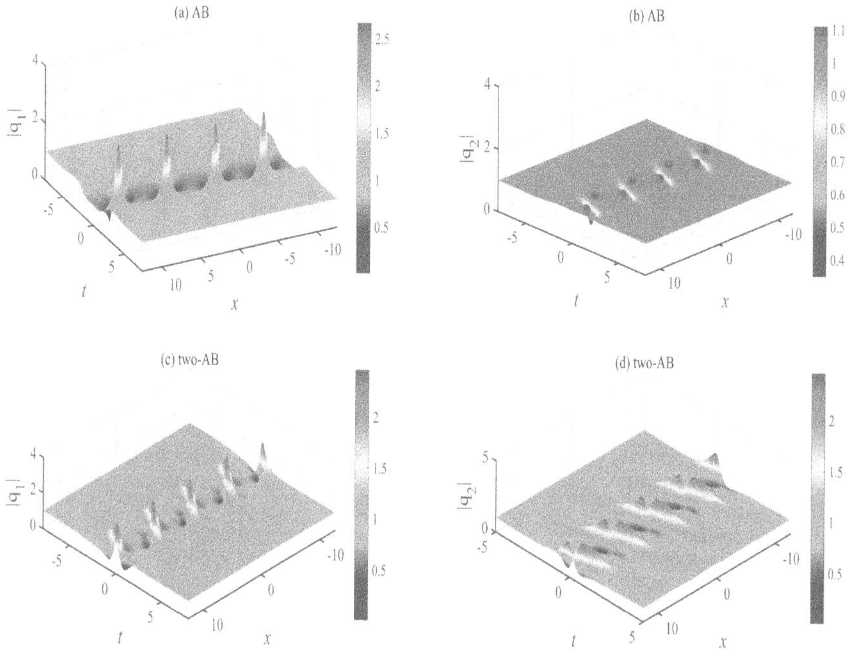

Figure 5. Parameters $a_1 = a_2 = 1$, $b_1 = -b_2 = 1$, and $\xi = 1$. (a,b) Akhmediev breather: $\xi_1 = -\frac{1}{2} - \frac{1}{4}\sqrt{2+2\sqrt{33}} - \frac{\mathrm{i}}{4}\sqrt{-2+2\sqrt{33}}$, $\alpha_1 = .8387281050, \beta_1 = .5445504250$, (c,d) Two-Akhmediev breather: $\xi_2 = -\frac{1}{2} + \frac{1}{4}\sqrt{2+2\sqrt{33}} + \frac{\mathrm{i}}{4}\sqrt{-2+2\sqrt{33}}$, $\alpha_1 = -0.770386699453238, \alpha_2 = -1.90823763640347, \beta_1 = -0.716323099896296, \beta_2 = -0.333845052373437$.

$\sum_{i=1}^{(r_l-1)\times p_l} \alpha_l^{[i]}(\mathrm{e}^{\frac{2k\pi\mathrm{i}}{p_l}}\epsilon_l)^i$, *then we obtain*

$$\begin{aligned}\frac{\Phi_l^\dagger\Omega\Phi_k}{\lambda_k-\lambda_l^*} &= \frac{\left[\mathrm{e}^{\vartheta_l^{[0]*}}, \ \cdots, \ \mathrm{e}^{\vartheta_l^{[r_l-1]*}}\right]}{p_l p_k \epsilon_l^{*g_l}\epsilon_k^{g_k}} \begin{bmatrix} \frac{1}{\xi_k^{[0]}-\xi_l^{[0]*}} & \cdots & \frac{1}{\xi_k^{[r_k-1]}-\xi_l^{[0]*}} \\ \vdots & \ddots & \vdots \\ \frac{1}{\xi_k^{[0]}-\xi_l^{[r_l-1]*}} & \cdots & \frac{1}{\xi_k^{[r_k-1]}-\xi_l^{[r_l-1]*}} \end{bmatrix} \begin{bmatrix} \mathrm{e}^{\vartheta_k^{[0]}} \\ \vdots \\ \mathrm{e}^{\vartheta_k^{[r_k-1]}} \end{bmatrix} \\ &= \frac{r_k r_l}{p_k p_l}\sum_{i=0}^{r_l-1}\sum_{j=0}^{r_k-1} \frac{\epsilon_l^{*ip_l}\epsilon_k^{jp_k}}{(ip_l+g_l)!(jp_k+g_k)!}\frac{d^{ip_l+jp_k+g_l+g_k}}{d\epsilon_l^{*(ip_l+g_l)}d\epsilon_k^{jp_k+g_k}}\left(\frac{\mathrm{e}^{\vartheta_l^{[0]*}+\vartheta_k^{[0]}}}{\xi_k-\xi_l^*}\right)\Big|_{\epsilon_l=0,\epsilon_k=0} \\ &\quad + O\left(\epsilon_l^{*r_l p_l}, \epsilon_k^{r_k p_k}\right),\end{aligned} \tag{4.10}$$

and

$$\begin{aligned}&\frac{\Phi_l^\dagger\Omega\Phi_k}{\lambda_k-\lambda_l^*} - \frac{\Phi_{l,1}^*\Phi_{k,j+1}}{a_j\mathrm{e}^{\theta_j}} \\ &= \frac{\left[\mathrm{e}^{\vartheta_l^{[0]*}}, \ \cdots, \ \mathrm{e}^{\vartheta_l^{[r_l-1]*}}\right]}{p_l p_k \epsilon_l^{*g_l}\epsilon_k^{g_k}} \begin{bmatrix} \frac{1}{\xi_k^{[0]}-\xi_l^{[0]*}}\frac{\xi_l^{[0]*}+b_j}{\xi_k^{[0]}+b_j} & \cdots & \frac{1}{\xi_k^{[r_k-1]}-\xi_l^{[0]*}}\frac{\xi_l^{[0]*}+b_j}{\xi_k^{[r_k-1]}+b_j} \\ \vdots & \ddots & \vdots \\ \frac{1}{\xi_k^{[0]}-\xi_l^{[r_l-1]*}}\frac{\xi_l^{[r_l-1]*}+b_j}{\xi_k^{[0]}+b_j} & \cdots & \frac{1}{\xi_k^{[r_k-1]}-\xi_l^{[r_l-1]*}}\frac{\xi_l^{[r_l-1]*}+b_j}{\xi_k^{[r_k-1]}+b_j} \end{bmatrix} \begin{bmatrix} \mathrm{e}^{\vartheta_k^{[0]}} \\ \vdots \\ \mathrm{e}^{\vartheta_k^{[r_k-1]}} \end{bmatrix} \\ &= \frac{r_k r_l}{p_k p_l}\sum_{i=0}^{r_l-1}\sum_{j=0}^{r_k-1} \frac{\epsilon_l^{*ip_l}\epsilon_k^{jp_k}}{(ip_l+g_l)!(jp_k+g_k)!}\frac{d^{ip_l+jp_k+g_l+g_k}}{d\epsilon_l^{*(ip_l+g_l)}d\epsilon_k^{jp_k+g_k}}\left(\frac{\xi_l^{[0]*}+b_j}{\xi_k^{[0]}+b_j}\frac{\mathrm{e}^{\vartheta_l^{[0]*}+\vartheta_k^{[0]}}}{\xi_k^{[0]}-\xi_l^{[0]*}}\right)\Big|_{\epsilon_l=0,\epsilon_k=0} \\ &\quad + O\left(\epsilon_l^{*r_l p_l}, \epsilon_k^{r_k p_k}\right),\end{aligned} \tag{4.11}$$

Based on the above Proposition 4, we will obtain the high-order rogue wave solutions for the vector NLS equations, which can be stated as the following theorem:

Theorem 2. *Suppose there are m double roots for the system in the upper half plane $\xi_1, \xi_2, \cdots, \xi_m$, and the multiple numbers are p_1, p_2, $\cdots$, p_m respectively, the corresponding orders are r_1, r_2, $\cdots$, r_m, then the general multi-high order rogue waves are given by*

$$q_j = a_j\left(\frac{\det(\mathbf{H}_j)}{\det(\mathbf{M})}\right)\mathrm{e}^{\theta_j}, \qquad |q|^2 = \sum_{i=1}^{n}|a_i|^2 + \ln_{xx}(\det(\mathbf{M})), \tag{4.12}$$

where

$$\mathbf{M} = \begin{bmatrix} M_{11} & M_{12} & \cdots & M_{1m} \\ M_{21} & M_{22} & \cdots & M_{2m} \\ \vdots & \vdots & \ddots & \vdots \\ M_{m1} & M_{m2} & \cdots & M_{mm} \end{bmatrix}, \qquad \mathbf{H}_j = \begin{bmatrix} H_{11}^{[j]} & H_{12}^{[j]} & \cdots & H_{1m}^{[j]} \\ H_{21}^{[j]} & H_{22}^{[j]} & \cdots & H_{2m}^{[j]} \\ \vdots & \vdots & \ddots & \vdots \\ H_{m1}^{[j]} & H_{m2}^{[j]} & \cdots & H_{mm}^{[j]} \end{bmatrix} \tag{4.13}$$

and

$$M_{lm} = \left(\frac{1}{(p_l j + g_l)!(p_m k + g_m)!} \frac{d^{p_l j + p_m k + g_l + g_m}}{d\epsilon_l^{*(p_l j + g_l)} d\epsilon_m^{p_m k + g_m}} \frac{e^{\vartheta_m + \vartheta_l^*}}{\xi_m - \xi_l^*} \Big|_{\epsilon_l^* = 0, \epsilon_m = 0} \right)_{1 \leq l \leq r_l, 1 \leq m \leq r_m},$$

$$H_{lm}^{[j]} = \left(\frac{1}{(p_l j + g_l)!(p_m k + g_m)!} \frac{d^{p_l j + p_m k}}{d\epsilon_l^{*p_l j} d\epsilon_m^{p_m k}} \frac{e^{\vartheta_m + \vartheta_l^*}}{\xi_m - \xi_l^*} \frac{\xi_l^* + b_j}{\xi_m + b_j} \Big|_{\epsilon_l^* = 0, \epsilon_m = 0} \right)_{1 \leq l \leq r_l, 1 \leq m \leq r_m}, \tag{4.14}$$

and $\vartheta_l = \mathrm{i}\left[(\xi_l(\epsilon_l) - \xi_l(0))x + (\eta_l(\epsilon_l) - \eta_l(0))t + \sum_{i=1}^{(r_l-1)\times p_l} \alpha_i^{[l]} \epsilon_l^i\right]$.

Through the above theorem on the high-order rogue waves, different types of rogue waves can be constructed, which includes the single high-order rogue waves and multi-high-order rogue waves. These solutions are closely related with the base-band modulational instability [8] or the resonant modulational instability [78]. For the fundamental rogue waves, they are well understood in the frame of modulational instability partially. However, for the high-order rogue waves, they are still puzzles for us.

Especially we consider the special example for the three component NLS equations. Choosing the parameters $n = 3$, the background solutions $a_1 = a_2 = a_3 = 1$, $b_1 = -b_3 = 1$, $b_2 = 0$, then we have three multiple roots $\xi_1 = \mathrm{i}$, $\xi_2 = \frac{1+\mathrm{i}}{\sqrt{2}}$, $\xi_3 = \frac{-1+\mathrm{i}}{\sqrt{2}}$ and the corresponding $\lambda_1 = \frac{3\mathrm{i}}{2}$, $\lambda_2 = \lambda_3 = \mathrm{i}\sqrt{2}$. Since they are double roots, then we can expand them in the neighborhood of branch points:

$$\begin{aligned} \lambda_1(\epsilon_1) &= \mathrm{i}\left(\frac{3}{2} - \frac{\epsilon_1^2}{4}\right), \\ \xi_1(\epsilon_1) &= \mathrm{i} + \epsilon_1 - \frac{\mathrm{i}}{2}\epsilon_1^2 + \frac{1}{8}\epsilon_1^3 + O(\epsilon_1^4), \end{aligned} \tag{4.15}$$

and

$$\begin{aligned} \lambda_2(\epsilon_2) &= \mathrm{i}\sqrt{2} - \frac{\epsilon_2^2}{\sqrt{2}}, \\ \xi_2(\epsilon_2) &= \frac{1+\mathrm{i}}{\sqrt{2}} + \epsilon_2 - \sqrt{2}\left(\frac{1+3\mathrm{i}}{4}\right)\epsilon_2^2 - \left(1 - \frac{9\mathrm{i}}{8}\right)\epsilon_2^3 + O(\epsilon_2^4), \end{aligned} \tag{4.16}$$

and

$$\begin{aligned} \lambda_3(\epsilon_3) &= \mathrm{i}\sqrt{2} + \frac{\epsilon_3^2}{\sqrt{2}}, \\ \xi_3(\epsilon_3) &= \frac{-1+\mathrm{i}}{\sqrt{2}} + \epsilon_3 + \sqrt{2}\left(\frac{1-3\mathrm{i}}{4}\right)\epsilon_3^2 - \left(1 + \frac{9\mathrm{i}}{8}\right)\epsilon_3^3 + O(\epsilon_3^4). \end{aligned} \tag{4.17}$$

And $\vartheta_i = \mathrm{i}(\xi_i(\epsilon_i) - \xi_i)\left[x + \frac{1}{2}(\xi_i(\epsilon_i) + \xi_i) + \sum_{k=1}^{r_i}(\alpha_i^{[k]} + \mathrm{i}\beta_i^{[k]})\epsilon_i^{2k-1}\right]$. From the formulas in the above theorem 2, we show the first-order rogue waves with $r_1 = 1, r_2 =$

$r_3 = 0$ and $\alpha_1^{[1]} = -\frac{1}{2}$, $\beta_1^{[1]} = 0$:

$$
\begin{aligned}
q_1 &= -\mathrm{i}\left(1 + R_1^{[1]}(x,t)\right)\mathrm{e}^{\mathrm{i}\theta_1}, & R_1^{[1]}(x,t) &= \frac{4\mathrm{i}(x-t)-2}{4t^2+4x^2+1} \\
q_2 &= -\left(1 + R_2^{[1]}(x,t)\right)\mathrm{e}^{\mathrm{i}\theta_2}, & R_2^{[1]}(x,t) &= -\frac{4+8\mathrm{i}t}{4t^2+4x^2+1} \\
q_3 &= \mathrm{i}\left(1 + R_3^{[1]}(x,t)\right)\mathrm{e}^{\mathrm{i}\theta_3}, & R_3^{[1]}(x,t) &= \frac{-4\mathrm{i}(x+t)-2}{4t^2+4x^2+1}.
\end{aligned}
\tag{4.18}
$$

The second group of the first-order rogue wave with parameters $r_1 = 0, r_2 = 1, r_3 = 0$, and $\alpha_2^{[1]} = -\frac{1}{\sqrt{2}}$, $\beta_2^{[1]} = 0$:

$$
\begin{aligned}
q_1 &= \frac{1-\mathrm{i}}{\sqrt{2}}\left[1 + R_1^{[2]}(x,t)\right]\mathrm{e}^{\mathrm{i}\theta_1}, & R_1^{[2]}(x,t) &= \frac{2\mathrm{i}x + (2\sqrt{2}-2)\mathrm{i}t - 2 + \sqrt{2}}{2\sqrt{2}tx + 2t^2 + 2x^2 + 1} \\
q_2 &= -\mathrm{i}\left[1 + R_2^{[2]}(x,t)\right]\mathrm{e}^{\mathrm{i}\theta_2}, & R_2^{[2]}(x,t) &= \frac{-2 + 2\mathrm{i}\sqrt{2}x}{2\sqrt{2}tx + 2t^2 + 2x^2 + 1} \\
q_3 &= \frac{-1+\mathrm{i}}{\sqrt{2}}\left[1 + R_3^{[2]}(x,t)\right]\mathrm{e}^{\mathrm{i}\theta_3}, & R_3^{[2]}(x,t) &= -\frac{2\mathrm{i}x + (2\sqrt{2}+2)\mathrm{i}t + 2 + \sqrt{2}}{2\sqrt{2}tx + 2t^2 + 2x^2 + 1}.
\end{aligned}
\tag{4.19}
$$

The third group of the first-order rogue waves with the parameters $r_1 = r_2 = 0, r_3 = 1$, and $\alpha_3^{[1]} = -\frac{1}{\sqrt{2}}$, $\beta_3^{[1]} = 0$:

$$
\begin{aligned}
q_1 &= \frac{-1-\mathrm{i}}{\sqrt{2}}\left[1 + R_1^{[3]}(x,t)\right]\mathrm{e}^{\mathrm{i}\theta_1}, & R_1^{[3]}(x,t) &= \frac{2\mathrm{i}x - (2\sqrt{2}+2)\mathrm{i}t - 2 - \sqrt{2}}{-2\sqrt{2}tx + 2t^2 + 2x^2 + 1} \\
q_2 &= \mathrm{i}\left[1 + R_2^{[3]}(x,t)\right]\mathrm{e}^{\mathrm{i}\theta_2}, & R_2^{[3]}(x,t) &= -\frac{2 + 2\mathrm{i}\sqrt{2}x}{-2\sqrt{2}tx + 2t^2 + 2x^2 + 1} \\
q_3 &= \frac{1+\mathrm{i}}{\sqrt{2}}\left[1 + R_3^{[3]}(x,t)\right]\mathrm{e}^{\mathrm{i}\theta_3}, & R_3^{[3]}(x,t) &= \frac{-2\mathrm{i}x + (2\sqrt{2}-2)\mathrm{i}t - 2 + \sqrt{2}}{-2\sqrt{2}tx + 2t^2 + 2x^2 + 1}.
\end{aligned}
\tag{4.20}
$$

Inserting the above expressions (4.15), (4.16) and (4.17) into the formula (4.12), we can obtain the triple-rogue waves by choosing $r_1 = r_2 = r_3 = 1$. If the parameters $\alpha_i^{[1]}$ and $\beta_i^{[1]}$ are large enough and the distances among locations $L_1 = (-\alpha_1^{[1]} - \frac{1}{2}, -\beta_1^{[1]})$, $L_2 = -(\alpha_1^{[1]} - \beta_1^{[1]} + 2 + \sqrt{2}, \sqrt{2}\beta_1^{[1]} - \frac{1}{2} - \sqrt{2})$, $L_3 = -(\alpha_1^{[1]} + \beta_1^{[1]} + 2 + \sqrt{2}, \sqrt{2}\beta_1^{[1]} + \frac{1}{2} + \sqrt{2})$ are also large enough, then by the analogous method the triple-rogue wave solutions can be decomposed into the following form:

$$
\begin{aligned}
q_i = \Bigg[& 1 + R_i^{[1]}(x + \alpha_1^{[1]} + \frac{1}{2}, t + \beta_1^{[1]}) + R_i^{[2]}(x + \alpha_1^{[1]} - \beta_1^{[1]} + 2 + \sqrt{2}, t + \sqrt{2}\beta_1^{[1]} \\
& - \frac{1}{2} - \sqrt{2}) + R_i^{[3]}(x + \alpha_1^{[1]} + \beta_1^{[1]} + 2 + \sqrt{2}, t + \sqrt{2}\beta_1^{[1]} + \frac{1}{2} + \sqrt{2}) \\
& + O(\frac{1}{\min(\alpha_i^{[1]}, \beta_i^{[1]})}) \Bigg] \mathrm{e}^{\mathrm{i}\theta_i + i\frac{\pi \mathrm{i}}{2}},
\end{aligned}
$$

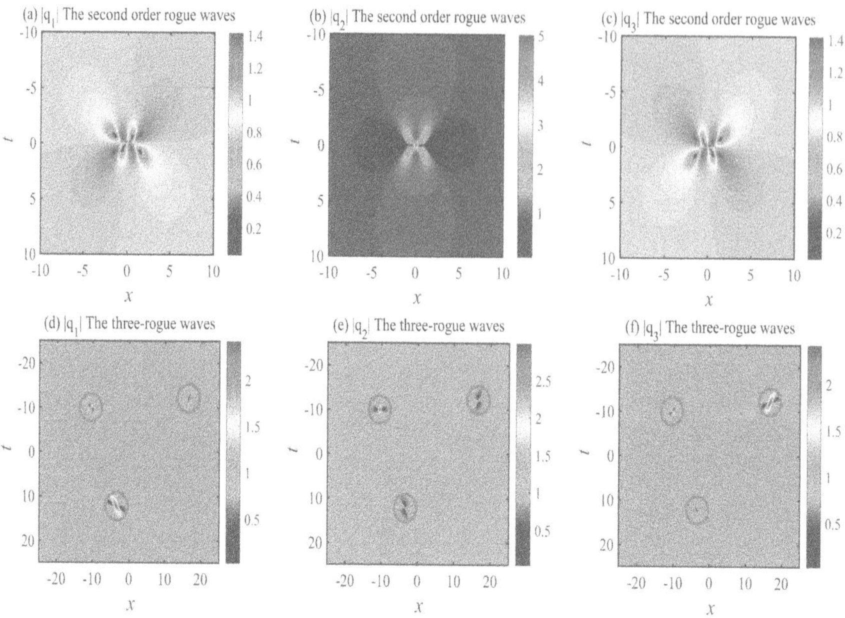

Figure 6. (a,b,c) The second order vector RWs: $\alpha_1^{[1]} = -1/2$, $\alpha_1^{[2]} = -\frac{1}{12}$, $\beta_1^{[1]} = \beta_1^{[2]} = 0$; (d,e,f) The multi-RW: $\alpha_1^{[1]} = 10, \alpha_2^{[1]} = -10, \alpha_3^{[1]} = 10, \beta_1^{[1]} = 10, \beta_2^{[1]} = 10, \beta_3^{[1]} = -10$.

(4.21)

which is shown in Fig. 6 (a,b,c). Noted that the triple-rogue wave is different from the second-order rogue wave of scalar NLS equation, although they both admit three fundamental rogue waves, since the multi-rogue wave can admit several different fundamental patterns (induced by different MI branches) and the position of each one can be more arbitrary than the scalar high-order rogue waves [13, 30, 32, 39, 45].

Taking the parameters $r_1 = 2, r_2 = r_3 = 0$, we will obtain the second-order rogue waves. By taking the parameters such that

$$\alpha_1^{[1]}, \alpha_1^{[2]}, \beta_1^{[1]}, \beta_1^{[2]} = \arg\max(|q|^2(0,0)) = 27 \tag{4.22}$$

we can obtain the rogue wave with maximal value at the origin. The parameters can be solved exactly: $\alpha_1^{[1]} = -\frac{1}{2}$, $\alpha_1^{[2]} = -\frac{1}{12}$, $\beta_1^{[1]} = \beta_1^{[2]} = 0$. Under this parameters setting, we can plot the figures for the second-order vector rogue waves (Fig. 6(d,e,f)). We can see that the high-order vector rogue wave admits more abundant patterns than the scalar ones [13, 26, 30, 32, 45, 63, 73].

5 Discussions and Conclusions

In this work, the exact solutions of NLS equations and Hirota equations on the plane wave background are constructed by the Bäcklund transformations. The linear stability analysis for the plane wave solutions is considered by the squared eigenfunction methods. Along this way, we know that the solutions of Lax pair not only can be used to construct the nontrivial solutions, but also to construct the solutions of linearized spectral problems.

Furthermore, the localized waves solutions of scalar NLS equations and Hirota equations are analyzed in detail. For the single breather solutions, we give the complete classifications by the dynamics. An oscillating breather is demonstrated for the first time. For the rogue waves, we provide a way to decompose high-order rogue waves into several lower order rogue waves by the special ansatz method. The decomposition method is meaningful for rogue wave pattern studies.

For the vector NLS equations, we construct the Akhmediev breather solutions and multi-Akhmdediev breathers by the formulas of Bäcklund transformation. The rogue wave solutions of vector NLS equations are constructed by the Bäcklund transformation. Especially, we consider the high-order rogue waves and multi-rogue wave for the three-component NLS equations. The multi- or high-order rogue waves can be decomposed into several fundamental rogue waves. For the vector NLS system, the dispersion relation of perturbations possesses much more MI branches, which means that there are multi-Akhmediev breathers and multi-rogue waves [50, 77], in contrast to high-order rogue waves. The paralleled works are analyzed for the vector NLS equations. Similar discussion can be extended to other nonlinear integrable systems [9, 31, 42, 47, 65, 66, 67, 68, 70, 75], high dimensional systems [17, 28, 76], and even nonlocal nonlinear systems [18, 19, 34, 35, 61, 71].

Acknowledgements

Liming Ling is supported by the National Natural Science Foundation of China (Grant No. 11771151), the Guangzhou Science and Technology Program of China (Grant No. 201904010362), and the Fundamental Research Funds for the Central Universities of China (Grant No. 2019MS110). Li-Chen Zhao is supported by National Natural Science Foundation of China (Grant No. 11775176, 12022513), Basic Research Program of Natural Science of Shaanxi Province (Grant No. 2018KJXX-094)

References

[1] Ablowitz M J and Segur H, *Solitons and the Inverse Scattering Transform*, SIAM, Philadelphia, 1985.

[2] Akhmediev N, Ankiewicz A and Soto-Crespo J M, *Rogue waves and rational solutions of the nonlinear Schrödinger equation*, Phys. Rev. E, **80**, 026601, 2009.

[3] Akhmediev N, Ankiewicz A and Taki M, *Waves that appear from nowhere and disappear without a trace*, Phys. Lett. A, **373**, 675-678, 2009.

[4] Akhmediev N, Dudley J M, Solli D R and Turitsyn S K, *Recent progress in investigating optical rogue waves*, J. Opt., **15**, 060201, 2013.

[5] Akhmediev N and Korneev V I, *Modulation instability and periodic solutions of the nonlinear Schrödinger equation,* Theor. Math. Phys., **69**, 1089-1093, 1986.

[6] Akhmediev N, Soto-Crespo J M and Ankiewicz A, *Extreme waves that appear from nowhere: on the nature of rogue waves,* Phys. Lett. A, **373**, 2137-2145, 2009.

[7] Bailung H, Sharma S K, and Nakamura Y, *Observation of Peregrine Solitons in a Multicomponent Plasma with Negative Ions,* Phys. Rev. Lett., **107**, 255005, 2011.

[8] Baronio F, Conforti M, Degasperis A, Lombardo S, Onorato M, and Wabnitz S, *Vector rogue waves and baseband modulation instability in the defocusing regime,* Phys. Rev. Lett., **113,** 034101, 2014.

[9] Baronio F, Conforti M, Degasperis A and Lombardo S, *Rogue waves emerging from the resonant interaction of three waves,* Phys. Rev. Lett., **111,** 114101, 2013.

[10] Baronio F, Degasperis A, Conforti M and Wabnitz S, *Solutions of the vector nonlinear Schrödinger equations: evidence for deterministic rogue waves,* Phys. Rev. Lett., **109,** 044102, 2012.

[11] Bilman D and Miller P D, *A robust inverse scattering transform for the focusing nonlinear Schrödinger equation,* Comm. Pure Appl. Math., **72,** 1722-1805, 2019.

[12] Bilman D and Buckingham R, *Large-order asymptotics for multiple-pole solitons of the focusing nonlinear Schrödinger equation,* J. Nonlinear Sci., **29,** 2185-2229, 2019.

[13] Bilman D, Ling L and Miller P D, *Extreme superposition: rogue waves of infinite order and the Painlevé-III hierarchy,* Duke Math. J., **169,** 671-760, 2020.

[14] Biondini G and Kovacic G, *Inverse scattering transform for the focusing nonlinear Schrödinger equation with nonzero boundary conditions,* J. Math. Phys., **55,** 031506, 2014.

[15] Chabchoub A, Hoffmann N P and Akhmediev N, *Rogue wave observation in a water wave tank,* Phys. Rev. Lett., **106,** 204502, 2011.

[16] Chabchoub A, Hoffmann N, M. Onorato, A. Slunyaev, A. Sergeeva, E. Pelinovsky, and N. Akhmediev, *Observation of a hierarchy of up to fifth-order rogue waves in a water tank,* Phys. Rev. E, **86,** 056601, 2012.

[17] Chen J, Chen Y, Feng B-F and Maruno K, *Rational solutions to two- and one-dimensional multicomponent Yajima-Oikawa systems,* Phys. Lett. A, **379,** 1510-1519, 2015.

[18] Chen K, Deng X, Lou S, Zhang D-J, *Solutions of nonlocal equations reduced from the AKNS hierarchy,* Stud. Appl. Math., **141,** 113-141, 2018.

[19] Dai C-Q and Huang W-H. *Multi-rogue wave and multi-breather solutions in PT-symmetric coupled waveguides,* Appl. Math. Lett., **32,** 35-40, 2014.

[20] Degasperis A and Lombardo S, *Integrability in action: solitons, instability and rogue waves[M]//Rogue and Shock Waves in Nonlinear Dispersive Media.* Springer, Cham, 23-53, 2016.

[21] Degasperis A, Lombardo S and Sommacal M, *Coupled nonlinear Schrödinger equations: spectra and instabilities of plane waves, Nonlinear Systems and Their Remarkable Mathematical Structures,* 206-248, 2019.

[22] Degasperis A, Lombardo S and Sommacal M, *Rogue wave type solutions and spectra of coupled nonlinear Schrödinger equations,* Fluids, **4,** 57, 2019.

[23] Doktorov E V and Leble S B, *A Dressing Method in Mathematical Physics,* Springer-Verlag, Berlin, 2007.

[24] Duan L, Yang Z Y, Gao P and Yang W L, *Excitation conditions of several fundamental nonlinear waves on continuous-wave background,* Phys. Rev. E, **99,** 012216, 2019.

[25] Duan L, Zhao L-C, Xu H X, Liu C, Yang ZY and Yang W L, *Soliton excitations on a continuous-wave background in the modulational instability regime with fourth-order effects,* Phys. Rev. E, **95,** 042212, 2017.

[26] Dubard P, Gaillard P, Klein C and Matveev V B, *On multi-rogue wave solutions of the NLS equation and positon solutions of the KdV equation,* The European Physical Journal Special Topics, **185,** 247-258, 2010.

[27] Dudley J M, Dias F, Erkintalo M and Genty G, *Instabilities, breathers and rogue waves in optics,* Nature Photonics, **8,** 755-764, 2014.

[28] Geng X, Shen J and Xue B, *A Hermitian symmetric space Fokas-Lenells equation: Solitons, breathers, rogue waves,* Annals of Physics, **404,** 115-131, 2019.

[29] Guo B and Ling L, *Rogue wave, breathers and bright-dark-rogue solutions for the coupled Schrödinger equations,* Chin. Phys. Lett., **28,** 110202-110202, 2011.

[30] Guo B, Ling L and Liu Q P, *Nonlinear Schrödinger equation: Generalized Darboux transformation and rogue wave solutions,* Phys. Rev. E, **85,** 026607, 2012.

[31] Guo B, Ling L and Liu Q P, *High-order solutions and generalized Darboux transformations of derivative nonlinear Schrödinger equations,* Stud. Appl. Math., **130,** 317-344, 2013.

[32] He J S, Zhang H R, Wang L H, Porsezian K and Fokas A S, *Generating mechanism for higher-order rogue waves,* Phys. Rev. E, **87,** 052914, 2013.

[33] Hirota R, *The Direct Method in Soliton Theory,* Cambridge: Cambridge University Press, 2004.

[34] Huang X and Ling L, *Soliton solutions for the nonlocal nonlinear Schrödinger equation,* The European Physical Journal Plus, **131,** 148, 2016.

[35] Ji J, Xu Z and Zhu Z-N, *Nonintegrable spatial discrete nonlocal nonlinear Schrödinger equation,* Chaos, **29,** 103129, 2019.

[36] Kaup D J and J Yang. *The inverse scattering transform and squared eigenfunctions for a degenerate 3×3 operator,* Inverse Problems, **25,** 105010, 2009.

[37] Kaup D J, Reiman A and Bers A, *Space-time evolution of nonlinear three-wave interactions: I. Interaction in a homogeneous medium,* Rev. Mod. Phys. **51,** 275-310, 1979.

[38] Kedziora D J, Ankiewicz A and Akhmediev N, *Circular rogue wave clusters,* Phys. Rev. E, **84,** 056611, 2011.

[39] Kedziora D J, Ankiewicz A, and Akhmediev N, *Classifying the hierarchy of nonlinear-Schrödinger-equation rogue-wave solutions,* Phys. Rev. E, **88,** 013207, 2013.

[40] Kibler B, Fatome J, Finot C, Millot G, Dias F, Genty G, Akhmediev N and Dudley J M, *The Peregrine soliton in nonlinear fibre optics,* Nature Physics, **6,** 790-795, 2010.

[41] Kuznetsov E, *Solitons in a parametrically unstable plasma,* Sov. Phys. Dokl. **22,** 507-508 (1977).

[42] Ling L, *The algebraic representation for high order solution of Sasa-Satsuma equation,* Discrete Contin. Dyn. Syst. Ser. S, **9,** 1975-2010, 2016.

[43] Ling L, Guo B and Zhao L C, *High-order rogue waves in vector nonlinear Schrödinger equations,* Phys. Rev. E, **89,** 041201, 2014.

[44] Ling L, Feng B F and Zhu Z, *Multi-soliton, multi-breather and higher order rogue wave solutions to the complex short pulse equation,* Physica D: Nonlinear Phenomena, **327,** 13-29, 2016.

[45] Ling L and Zhao L C, *Simple determinant representation for rogue waves of the nonlinear Schrödinger equation,* Phys. Rev. E, **88,** 043201, 2013.

[46] Ling L and Zhao L C, *Modulational instability and homoclinic orbit solutions in vector nonlinear Schrödinger equation,* Commun. Nonlinear Sci. Numer. Simul., **72,** 449-471, 2019.

[47] Ling L and Zhao L C, *Integrable pair-transition-coupled nonlinear Schrödinger equations,* Phys. Rev. E, **92,** 022924, 2015.

[48] Ling L, Zhao L C and Guo B, *Darboux transformation and multi-dark soliton for N-component nonlinear Schrödinger equations,* Nonlinearity, **28,** 3243-3261, 2015.

[49] Ling L, Zhao L C and Guo B, *Darboux transformation and classification of solution for mixed coupled nonlinear Schrödinger equations,* Commun. Nonlinear Sci. Numer. Simul., **32,** 285-304, 2016.

[50] Ling L, Zhao L C, Yang Z Y and Guo B, *Generation mechanisms of fundamental rogue wave spatial-temporal structure,* Phys. Rev. E, **96,** 022211, 2017.

[51] Liu C, Yang Z Y, Zhao L C, Duan L, Yang G and Yang W L, *Symmetric and asymmetric optical multipeak solitons on a continuous wave background in the femtosecond regime,* Phys. Rev. E, **94,** 042221, 2016.

[52] Liu C, Yang Z Y, Zhao L C and Yang W L, *State transition induced by higher-order effects and background frequency,* Phys. Rev. E, 91, 022904, 2015.

[53] Liu C, Yang Z Y, Zhao L C and Yang W L, *Transition, coexistence, and interaction of vector localized waves arising from higher-order effects,* Annals of Physics, **362,** 130-138, 2015.

[54] Ma Y C, *The perturbed plane-wave solutions of the cubic Schrödinger equation,* Stud. Appl. Math., **60,** 43-58, 1979.

[55] Manakov S V, *On the theory of two-dimensional stationary self-focusing of electromagnetic waves,* Soviet Physics-JETP, **38,** 248-253, 1974.

[56] Matveev V B and Salle M A, *Darboux Transformation and Solitons,* Springer-Verlag, Berlin, 1991.

[57] Mu G, Qin Z and Grimshaw R, Dynamics of *Rogue Waves on a Multisoliton Background in a Vector Nonlinear Schrödinger Equation,* SIAM J. Appl. Math., **75,** 1-20, 2015.

[58] Ohta Y and Yang J, *General high-order rogue waves and their dynamics in the nonlinear Schrödinger equation,* Proc. R. Soc. A, **468,** 1716-1740, 2012.

[59] Onorato M, Residori S, Bortolozzo U, Montina A and Arecchi F T, *Rogue waves and their generating mechanisms in different physical contexts,* Phys. Rep., **528,** 47-89, 2013.

[60] Peregrine D H, *Water waves, nonlinear Schrödinger equations and their solutions,* J. Aust. Math. Soc. Ser. B, **25,** 16-43, 1983.

[61] Rao J, Cheng Y, Porsezian K, Mihalache D and He J, *PT-symmetric nonlocal Davey-Stewartson I equation: soliton solutions with nonzero background,* Phys. D, **401,** 132180, 2020.

[62] Soto-Crespo J M, Ankiewicz A, Devine N, and Akhmediev N, *Modulation instability, Cherenkov radiation, and Fermi-Pasta-Ulam recurrence,* J. Opt. Soc. Am. B, **29,** 1930-1936, 2012.

[63] Wang L H, Porsezian K and He J S, *Breather and rogue wave solutions of a generalized nonlinear Schrödinger equation,* Phys. Rev. E, **87,** 053202, 2013.

[64] Wang L, Yang C, Wang J, He J S, *The height of an nth-order fundamental rogue wave for the nonlinear Schrödinger equation,* Phys. Lett. A, **381,** 1714-1718, 2017.

[65] Wang X, Li Y, Huang F and Chen Y, *Rogue wave solutions of AB system,* Commun. Nonlinear Sci. Numer. Simul., **20,** 434-442, 2015.

[66] Wang X, Wei J, Wang L, and Zhang J, *Baseband modulation instability, rogue waves and state transitions in a deformed Fokas-Lenells equation,* Nonlinear Dynamics, **97,** 343-353, 2019.

[67] Wang X, Wei J and Geng X, *Rational solutions for a (3+1)-dimensional nonlinear evolution equation,* Commun. Nonlinear Sci. Numer. Simul., **83,** 105116, 2020.

[68] Wen X, Yang Y and Yan Z, *Generalized perturbation (n, M)-fold Darboux transformations and multi-rogue-wave structures for the modified self-steepening nonlinear Schrödinger equation,* Phys. Rev. E, **92,** 012917, 2015.

[69] Yan Z, *Vector financial rogue waves,* Phys. Lett. A, **375,** 4274-4279, 2011.

[70] Yang B, Chen J and Yang J, *Rogue waves in the generalized derivative nonlinear Schrödinger equations,* J. Nonlinear Sci., **30,** 3027-3056, 2020.

[71] Yang B and Yang J, *Rogue waves in the nonlocal $\mathcal{PT}$-symmetric nonlinear Schrödinger equation,* Lett. Math. Phys., **109,** 945-973, 2019.

[72] Zakharov V E and Ostrovsky L A, *Modulation instability: the beginning,* Phys. D, **238,** 540-548, 2009.

[73] Zhang G, Ling L and Yan Z, *Higher-order vector Peregrine solitons and asymptotic estimates for the multi-component nonlinear Schrödinger equations,* arXiv:2012.15603, 2020.

[74] Zhang G and Yan Z, *Three-component nonlinear Schrödinger equations: Modulational instability, Nth-order vector rational and semi-rational rogue waves, and dynamics,* Commun. Nonlinear Sci. Numer. Simul., **62,** 117-133, 2018.

[75] Zhang G, Yan Z and Wen X, *Modulational instability, beak-shaped rogue waves, multi-dark-dark solitons and dynamics in pair-transition-coupled nonlinear Schrödinger equations,* Proc. R. Soc. A, **473,** 2203, 2017.

[76] Zhang G, Yan Z and Wen X, *Three-wave resonant interactions: Multi-dark-dark-dark solitons, breathers, rogue waves, and their interactions and dynamics,* Physica D: Nonlinear Phenomena, **366,** 27-42, 2017.

[77] Zhao L-C, Guo B and Ling L, *High-order rogue wave solutions for the coupled nonlinear Schrödinger equations-II,* J. Math. Phys., **57,** 043508, 2016.

[78] Zhao L-C and Ling L, *Quantitative relations between modulational instability and several well-known nonlinear excitations,* JOSA B, **33,** 850-856, 2016.

[79] Zhao L-C, Ling L and Yang Z-Y, *Mechanism of Kuznetsov-Ma breathers,* Phys. Rev. E, **97,** 022218, 2018.

[80] Zhao L-C, Li S-C and Ling L, *Rational W-shaped solitons on a continuous-wave background in the Sasa-Satsuma equation,* Phys. Rev. E, **89,** 023210, 2014.

[81] Zhao L-C, Li S-C and Ling L, *W-shaped solitons generated from a weak modulation in the Sasa-Satsuma equation,* Phys. Rev. E, **93,** 032215, 2016.

[82] Zhao L-C, and Liu J, *Localized nonlinear waves in a two-mode nonlinear fiber,* JOSA B, **29,** 3119-3127, 2012.

[83] Zhao L-C and Liu J, *Rogue-wave solutions of a three-component coupled nonlinear Schrödinger equation,* Phys. Rev. E, **87,** 013201, 2013.

[84] Zhao L-C, Xin G G and Yang Z Y, *Rogue-wave pattern transition induced by relative frequency,* Phys. Rev. E, **90,** 022918, 2014.

$B5$. Algebro-geometric solutions to the modified Blaszak-Marciniak lattice hierarchy

Wei Liu[a,b], *Xianguo Geng*[a] *and Bo Xue*[a,1]

[a] *School of Mathematics and Statistics, Zhengzhou University, 100 Kexue Road, Zhengzhou, Henan 450001, P.R.China*

[b] *Department of Mathematics and Physics, Shijiazhuang Tiedao University, 17 Northeast, Second Inner Ring, Shijiazhuang, Hebei 050043, P.R.China*

Abstract

The modified Blaszak-Marciniak (mBM) lattice hierarchy related to a 3×3 matrix problem is proposed with the help of the zero-curvature equation and the Lenard recursion equations. Based on the characteristic polynomial of Lax matrix for the mBM lattice hierarchy, we introduce a trigonal curve $\mathcal{K}_{m-1}$ of arithmetic genus $m-1$ and construct the related Baker-Akhiezer function, meromorphic function. By using asymptotic expansion of the meromorphic function and its theta function representation, we obtain the algebro-geometric solutions to the mBM lattice hierarchy.

1 Introduction

Integrable lattice equations, a special kind of differential-difference equations, are always used to model many nonlinear phenomena that occur in nature. Many scholars have paid much attention to seeking new integrable equations and the corresponding spectral problems [1, 5, 8, 9, 17, 20, 25]. Among the numerous equations, the Blaszak-Marciniak (BM) lattice equation [4]

$$u_t = w^+ - w^-, \quad v_t = u^- w^- - uw, \quad w_t = w(v - v^+), \tag{1.1}$$

as an application of r-matrix formalism to the algebra of shift operators, was originally introduced by Blaszak and Marciniak in 1994. After that, it attracted extensive research efforts and remarkable integrable properties have been achieved, such as the Hamiltonian structures, master symmetries, conserved densities, generalized symmetries, and so on [18, 23, 24, 31]. The combination of the homotopy perturbation method and Padé techniques [29] was used to derive soliton solutions to Eq. (1.1). Algebro-geometric solutions to the BM lattice hierarchy [14] were given according to the Lenard recursion sequences and the trigonal curve. The BM lattice hierarchy with self-consistent sources was obtained through the discrete zero curvature equation [30]. Explicit solutions of Eq. (1.1) are introduced by means of Darboux transformation [32].

Algebro-geometric (or quasi-periodic, or finite-band) solutions of soliton equations are explicit solutions closely related to the inverse spectral theory, which

[1] Corresponding author. *E-mail address:* xuebo@zzu.edu.cn.

show the quasi-periodic behavior of nonlinear phenomenon and characteristic for Liouville integrability. The algebro-geometric technique is a systematic method to obtain quasi-periodic solutions of nonlinear integrable equations, which was developed by pioneers such as Ablowitz, Dubrovin, Its, Novikov, Matveev, McKean and co-authors (see, e.g., Refs. [2, 3, 11, 19, 21] and the references therein). Refs. [6, 7] proposed a general approach to yield all algebro-geometric solutions of the entire Boussinesq hierarchy related to a third-order differential operator. In Refs. [12, 15, 16], Geng et al. successfully introduced the trigonal curve utilizing the characteristic polynomial of the Lax matrix and constructed algebro-geometric solutions to the 3×3 matrix spectral problem, including the modified Boussinesq flows, the coupled mKdV hierarchy, the Manakov hierarchy, the Blaszak-Marciniak lattice hierarchy, the four-component Toda lattice hierarchy and others [13, 26, 27, 28].

In this chapter, we derive a hierarchy of modified Blaszak-Marciniak (mBM) lattice equations, in which the first nontrivial member can be reduced to the following two equations:

$$\begin{aligned} u_t &= uw(w^+ - w^-) + w^- v - w^+ v^+, \\ v_t &= v(uw - u^- w^-) + v(ww^+ - w^{--} w^-), \\ w_t &= w(u^+ w^+ - u^- w^-) + w(w^+ w^{++} - w^{--} w^-) + w(v - v^+), \end{aligned} \tag{1.2}$$

and

$$u_t = \frac{w^+}{v^{++}} - \frac{w^-}{v^-}, \quad v_t = \frac{u^- w^-}{v^-} - \frac{uw}{v^+}, \quad w_t = \frac{w}{v} - \frac{w}{v^+}. \tag{1.3}$$

The main aim of the present chapter is to construct the algebro-geometric solutions for the mBM hierarchy. In Section 2, we derive the mBM lattice equations associated with a 3×3 matrix spectral problem by virtue of the zero-curvature equation and Lenard recurrence relations. In Section 3, a trigonal curve $\mathcal{K}_{m-1}$ of arithmetic genus $m-1$ is defined with the help of the characteristic polynomial of the Lax matrix for the stationary mBM lattice hierarchy. And the associated stationary Baker-Akhiezer function and meromorphic function are shown. In Section 4, we discuss the asymptotic expansions of the meromorphic function ϕ and the Baker-Akhiezer function ψ_1 by using the Riccati-type equation. Then the explicit theta representations of ϕ, ψ_1 and the potentials u, v, w are given resorting to the third-kind Abelian differential. The last section is an extension of Secs. 3 and 4. The properties of the Baker-Akhiezer function and the meromorphic function in Section 4 are applicable to the time-dependent case. Moreover, we obtain the explicit theta function representations for the mBM lattice hierarchy. The mBM flows are straightened out under the Abel map.

2 mBM hierarchy

Throughout this chapter, we assume that u, v and w satisfy the conditions

$$\begin{aligned} &u(\cdot,t),\ v(\cdot,t),\ w(\cdot,t) \in \mathbb{C}^{\mathbb{Z}}, \quad t \in \mathbb{R}, \\ &u(n,\cdot),\ v(n,\cdot),\ w(n,\cdot) \in C^1(\mathbb{R}), \quad n \in \mathbb{Z}, \\ &v(n,t) \neq 0, \quad (n,t) \in \mathbb{Z} \times \mathbb{R}, \end{aligned}$$

where $\mathbb{C}^{\mathbb{Z}}$ denotes the set of all complex-valued sequences indexed by $\mathbb{Z}$. Let us define the shift operators $E^{\pm}$ and difference operator Δ by

$$Ef(n) = f(n+1), \quad E^{-1}f(n) = f(n-1), \quad \Delta f(n) = (E-1)f(n), \quad n \in \mathbb{Z}.$$

For the sake of simplicity, we usually denote

$$f(n) = f, \quad E^{\pm}f = f^{\pm}, \quad f(n+k) = E^k f, \quad n, k \in \mathbb{Z}.$$

In this section, we shall derive the mBM hierarchy associated with a 3×3 matrix spectral problem

$$E\psi = U\psi, \quad \psi = \begin{pmatrix} \psi_1 \\ \psi_2 \\ \psi_3 \end{pmatrix}, \quad U = \begin{pmatrix} 0 & 1 & 0 \\ \lambda + v & u & 1 \\ \lambda w & 0 & 0 \end{pmatrix}, \tag{2.1}$$

where u, v, w are three potentials, and λ is a constant spectral parameter. Now, we introduce the following Lenard recurrence relations

$$K\hat{g}_{j,+} = J\hat{g}_{j+1,+}, \quad \hat{g}_{j,+} = (\hat{a}_{j,+}, \hat{b}_{j,+}, \hat{c}_{j,+})^{\mathrm{T}}, \quad j \geq 0, \tag{2.2a}$$

$$K\check{g}_{j,+} = J\check{g}_{j+1,+}, \quad \check{g}_{j,+} = (\check{a}_{j,+}, \check{b}_{j,+}, \check{c}_{j,+})^{\mathrm{T}}, \quad j \geq 0, \tag{2.2b}$$

$$J\hat{g}_{j,-} = K\hat{g}_{j+1,-}, \quad \hat{g}_{j,-} = (\hat{a}_{j,-}, \hat{b}_{j,-}, \hat{c}_{j,-})^{\mathrm{T}}, \quad j \geq 0, \tag{2.3a}$$

$$J\check{g}_{j,-} = K\check{g}_{j+1,-}, \quad \check{g}_{j,-} = (\check{a}_{j,-}, \check{b}_{j,-}, \check{c}_{j,-})^{\mathrm{T}}, \quad j \geq 0, \tag{2.3b}$$

with the starting points

$$\hat{g}_{0,+} = (-w^-, 1, w^{--}w^- + u^-w^-)^{\mathrm{T}}, \tag{2.4a}$$

$$\check{g}_{0,+} = (1, 0, -(E+1)^{-1}(u + w^-))^{\mathrm{T}}, \tag{2.4b}$$

$$\hat{g}_{0,-} = (0, \frac{1}{v}, 1)^{\mathrm{T}}, \tag{2.4c}$$

$$\check{g}_{0,-} = (\check{a}_{0,-}, \frac{\check{c}_{0,-}}{v}, \check{c}_{0,-})^{\mathrm{T}}, \tag{2.4d}$$

where

$$\check{c}_{0,-}^2 + v\check{a}_{0,-}\check{a}_{0,-}^+ = 1, \quad \check{c}_{0,-}^+ + \check{c}_{0,-} + u\check{a}_{0,-}^+ = 0, \quad \check{a}_{0,-} \neq 0,$$

and we define two difference operators K and J as

$$K = \begin{pmatrix} EvE - v + u\Delta uE & 0 & u\Delta E \\ v\Delta uE & 0 & v(E^2-1) \\ wE\Delta uE & -w\Delta v & w(E^3-1) \end{pmatrix},$$

$$J = \begin{pmatrix} 1 - E^2 & E^{-1}w - EwE & 0 \\ E^{-1}w - wE^2 - \Delta uE & uwE - E^{-1}uw & 1 - E^2 \\ 0 & w\Delta & 0 \end{pmatrix}.$$

To ensure $\hat{g}_{j,\pm}$ and $\check{g}_{j,\pm}$ can be uniquely determined by the recursion relations (2.2) and (2.3), the summation constants are set zero. For example, the first member reads as

$$\hat{g}_{1,+} = \begin{pmatrix} E^{-1}(w\hat{v} - w^2\hat{w} - w^2u - u^-w^-w - u^+ww^+ - ww^+w^{++} - w^{--}w^-w) \\ ww^+ + w^-w + w^{--}w^- + uw + u^-w^- - v \\ \hat{c}_{1,+} \end{pmatrix},$$

where

$$\begin{aligned}
&\hat{v} = v + v^+,\\
&\hat{w} = w^+ + w^-,\\
&\hat{c}_{1,+} = E^{-1}[u^-(w^-)^2w + uw^-w^2 + uw^2w^+ + uw^2w^- + u^2w^2 + u^+w^-ww^+\\
&\qquad + uw^{--}w^-w + u^-uw^-w + uu^+ww^+ - w^-wv^- - w^-wv - uvw - w^-wv^+\\
&\qquad - uwv^+ + (w^-w)^2 + (w^-)^2w^{--}w + w^2w^-w^+ + w^-ww^+w^{++} + uww^+w^{++}\\
&\qquad + w^{---}w^{--}w^-w + u^{--}w^{--}w^-w].
\end{aligned}$$

To generate the mBM hierarchy associated with the discrete spectral problem (2.1), we solve the stationary zero-curvature equation

$$(EV)U - UV = 0, \quad V = (V_{ij})_{3\times 3} = \begin{pmatrix} V_{11} & V_{12} & V_{13} \\ V_{21} & V_{22} & V_{23} \\ \lambda V_{31} & \lambda V_{32} & V_{33} \end{pmatrix}, \tag{2.5}$$

which is equivalent to

$$\begin{aligned}
&(\lambda + v)V_{12}^+ + \lambda wV_{13}^+ - V_{21} = 0,\\
&V_{11}^+ + uV_{12}^+ - V_{22} = 0,\\
&V_{12}^+ - V_{23} = 0,\\
&(\lambda + v)(V_{22}^+ - V_{11}) + \lambda wV_{23}^+ - uV_{21} - \lambda V_{31} = 0,\\
&V_{21}^+ + u(V_{22}^+ - V_{22}) - (\lambda + v)V_{12} - \lambda V_{32} = 0,\\
&V_{22}^+ - (\lambda + v)V_{13} - uV_{23} - V_{33} = 0,\\
&(\lambda + v)V_{32}^+ + wV_{33}^+ - wV_{11} = 0,\\
&V_{31}^+ + uV_{32}^+ - wV_{12} = 0,\\
&V_{32}^+ - wV_{13} = 0.
\end{aligned} \tag{2.6}$$

$$\begin{aligned}
&V_{11} = c, && V_{12} = a, && V_{13} = b,\\
&V_{21} = (\lambda + v)a^+ + \lambda wb^+, && V_{22} = c^+ + ua^+, && V_{23} = a^+,\\
&V_{31} = w^-a^- - u^-w^-b^-, && V_{32} = w^-b^-, && V_{33} = c^{++} + \Delta ua^+ - (\lambda + v)b.
\end{aligned} \tag{2.7}$$

Substituting (2.7) into (2.6), we get the Lenard equation

$$KG = \lambda JG, \quad G = (a, b, c)^{\mathrm{T}}. \tag{2.8}$$

The function G is expanded into a Laurent series in λ:

$$G = \sum_{j\geq 0} G_{j,+}\lambda^{-j}, \tag{2.9}$$

where $G_{j,+} = (a_{j,+}, b_{j,+}, c_{j,+})^{\mathrm{T}}$. Then we arrive at the following recursion relations

$$KG_{j,+} = JG_{j+1,+}, \quad JG_{0,+} = 0, \quad j \geq 0. \tag{2.10}$$

Since $JG_{0,+} = 0$ has a solution

$$G_{0,+} = \alpha_{0,+}\hat{g}_{0,+} + \beta_{0,+}\check{g}_{0,+}, \tag{2.11}$$

$G_{j,+}$ can be expressed as

$$G_{j,+} = \alpha_{0,+}\hat{g}_{j,+} + \beta_{0,+}\check{g}_{j,+} + \cdots + \alpha_{j,+}\hat{g}_{0,+} + \beta_{j,+}\check{g}_{0,+}, \quad j \geq 0, \tag{2.12}$$

where $\alpha_{j,+}, \beta_{j,+}$ are arbitrary constants.

On the other hand, the function G is expanded into a Laurent series in λ:

$$G = \sum_{j\geq 0} G_{j,-}\lambda^{j}, \tag{2.13}$$

where $G_{j,-} = (a_{j,-}, b_{j,-}, c_{j,-})^{\mathrm{T}}$, then (2.8) is equivalent to the following Lenard recursion equations:

$$JG_{j,-} = KG_{j+1,-}, \quad KG_{0,-} = 0, \quad j \geq 0. \tag{2.14}$$

Since $KG_{0,-} = 0$ has a solution

$$G_{0,-} = \alpha_{0,-}\hat{g}_{0,-} + \beta_{0,-}\check{g}_{0,-}, \tag{2.15}$$

then we have

$$G_{j,-} = \alpha_{0,-}\hat{g}_{j,-} + \beta_{0,-}\check{g}_{j,-} + \cdots + \alpha_{j,-}\hat{g}_{0,-} + \beta_{j,-}\check{g}_{0,-}, \quad j \geq 0, \tag{2.16}$$

where $\alpha_{j,-}, \beta_{j,-}$ are arbitrary constants.

Let ψ satisfy the spectral problem (2.1) and the auxiliary problem

$$\psi_{t_{\underline{r}}} = \widetilde{V}^{(\underline{r})}\psi, \quad \widetilde{V}^{(\underline{r})} = (\widetilde{V}^{(\underline{r})}_{ij})_{3\times 3}, \quad \underline{r} = (r_1, r_2), \tag{2.17}$$

with each entry $\widetilde{V}^{(\underline{r})}_{ij} = V_{ij}(\tilde{a}^{(\underline{r})}, \tilde{b}^{(\underline{r})}, \tilde{c}^{(\underline{r})})$,

$$\tilde{a}^{(\underline{r})} = \sum_{j=0}^{r_1} \tilde{a}_{j,+}\lambda^{r_1-j} + \sum_{j=0}^{r_2-1} \tilde{a}_{j,-}\lambda^{-r_2+j}, \tag{2.18a}$$

$$\tilde{b}^{(\underline{r})} = \sum_{j=0}^{r_1} \tilde{b}_{j,+}\lambda^{r_1-j} + \sum_{j=0}^{r_2-1} \tilde{b}_{j,-}\lambda^{-r_2+j}, \tag{2.18b}$$

$$\tilde{c}^{(\underline{r})} = \sum_{j=0}^{r_1} \tilde{c}_{j,+}\lambda^{r_1-j} + \sum_{j=0}^{r_2-1} \tilde{c}_{j,-}\lambda^{-r_2+j}, \tag{2.18c}$$

where $\tilde{a}_{j,\pm}, \tilde{b}_{j,\pm}$ and $\tilde{c}_{j,\pm}$ are determined by $\widetilde{G}_j = (\tilde{a}_j, \tilde{b}_j, \tilde{c}_j)^{\mathrm{T}}$, and

$$\widetilde{G}_{j,\pm} = \sum_{l=0}^{j} \tilde{\alpha}_{l,\pm} \hat{g}_{j-l,\pm} \quad j \geq 0, \tag{2.19}$$

and $\{\tilde{\alpha}_{j,\pm}\}$ are constants independent of $\{\alpha_{j,\pm}\}$. Then the compatibility condition of (2.1) and (2.17) yields the zero-curvature equation, $U_{t_{\underline{r}}} - (E\widetilde{V}^{(\underline{r})})U + U\widetilde{V}^{(\underline{r})} = 0$, which is equivalent to the mBM lattice hierarchy

$$(u_{t_{\underline{r}}}, v_{t_{\underline{r}}}, w_{t_{\underline{r}}})^{\mathrm{T}} = \widetilde{X}_{\underline{r}} = K\widetilde{G}_{r_1,+} - J\widetilde{G}_{r_2-1,-}, \tag{2.20}$$

where $\underline{r} = (r_1, r_2)$, $r_1 \geq 0$, $r_2 \geq 1$. As $\underline{r} = (0,1)$, the first nontrivial member of hierarchy (2.20) is as follows

$$\begin{aligned} u_{t_{(0,1)}} =& \tilde{\alpha}_{0,+}\left[uw(w^+ - w^-) + w^- v - w^+ v^+\right] + \tilde{\alpha}_{0,-}\left(\frac{w^+}{v^{++}} - \frac{w^-}{v^-}\right), \\ v_{t_{(0,1)}} =& \tilde{\alpha}_{0,+}\left[v(uw - u^- w^-) + v(ww^+ - w^{--}w^-)\right] + \tilde{\alpha}_{0,-}\left(\frac{u^- w^-}{v^-} - \frac{uw}{v^+}\right), \\ w_{t_{(0,1)}} =& \tilde{\alpha}_{0,+}\left[w(u^+ w^+ - u^- w^-) + w(w^+ w^{++} - w^{--}w^-) + w(v - v^+)\right] \\ &+ \tilde{\alpha}_{0,-}\left(\frac{w}{v} - \frac{w}{v^+}\right), \end{aligned} \tag{2.21}$$

which is reduced to (1.2) and (1.3) for $\tilde{\alpha}_{0,+} = 1, \tilde{\alpha}_{0,-} = 0, t_{(0,1)} = t$ and $\tilde{\alpha}_{0,+} = 0, \tilde{\alpha}_{0,-} = 1, t_{(0,1)} = t$, respectively.

3 Stationary Baker-Akhiezer function

In this section, we shall introduce the stationary Baker-Akhiezer function, a trigonal curve $\mathcal{K}_{m-1}$ of degree m and the associated meromorphic function. Let us consider the stationary mBM lattice hierarchy, which is equivalent to the stationary zero-curvature equation

$$(EV^{(\underline{q})})U - UV^{(\underline{q})} = 0, \quad V^{(\underline{q})} = (V_{ij}^{(\underline{q})})_{3\times 3}, \quad \underline{q} = (q_1, q_2), \tag{3.1}$$

with the elements $V_{ij}^{(\underline{q})} = V_{ij}(a^{(\underline{q})}, b^{(\underline{q})}, c^{(\underline{q})})$,

$$a^{(\underline{q})} = \sum_{j=0}^{q_1} a_{j,+}\lambda^{q_1-j} + \sum_{j=0}^{q_2-1} a_{j,-}\lambda^{-q_2+j} \quad j \geq 0, \tag{3.2a}$$

$$b^{(\underline{q})} = \sum_{j=0}^{q_1} b_{j,+}\lambda^{q_1-j} + \sum_{j=0}^{q_2-1} b_{j,-}\lambda^{-q_2+j} \quad j \geq 0, \tag{3.2b}$$

$$c^{(\underline{q})} = \sum_{j=0}^{q_1} c_{j,+}\lambda^{q_1-j} + \sum_{j=0}^{q_2-1} c_{j,-}\lambda^{-q_2+j} \quad j \geq 0, \tag{3.2c}$$

and $a_{j,\pm}, b_{j,\pm}, c_{j,\pm}$ are determined by (2.12) and (2.16). A direct calculation shows that $yI - \lambda^{q_2}V^{(\underline{q})}$ also satisfies the zero-curvature equation. Then the characteristic polynomial of the Lax matrix $\lambda^{q_2}V^{(\underline{q})}$ is a constant independent of variable n with the expansion

$$\det\left(yI - \lambda^{q_2}V^{(\underline{q})}\right) = y^3 - y^2R_m(\lambda) + yS_m(\lambda) - T_m(\lambda), \tag{3.3}$$

where $R_m(\lambda)$, $S_m(\lambda)$ and $T_m(\lambda)$ are polynomials with constant coefficients of λ

$$\begin{aligned}
R_m &= \lambda^{q_2}\left(V_{11}^{(\underline{q})} + V_{22}^{(\underline{q})} + V_{33}^{(\underline{q})}\right) = -\alpha_{0,+}\lambda^{q+1} - \alpha_{1,+}\lambda^{q} + \cdots + 2\alpha_{0,-},\\
S_m &= \lambda^{2q_2}\left(\begin{vmatrix} V_{11}^{(\underline{q})} & V_{12}^{(\underline{q})}\\ V_{21}^{(\underline{q})} & V_{22}^{(\underline{q})}\end{vmatrix} + \begin{vmatrix} V_{11}^{(\underline{q})} & V_{13}^{(\underline{q})}\\ \lambda V_{31}^{(\underline{q})} & V_{33}^{(\underline{q})}\end{vmatrix} + \begin{vmatrix} V_{22}^{(\underline{q})} & V_{23}^{(\underline{q})}\\ \lambda V_{32}^{(\underline{q})} & V_{33}^{(\underline{q})}\end{vmatrix}\right)\\
&= -\beta_{0,+}^2\lambda^{2q+1} - 2\beta_{0,+}\beta_{1,+}\lambda^{2q} + \cdots + (\alpha_{0,-} + \beta_{0,-})(\alpha_{0,-} - \beta_{0,-}),\\
T_m &= \lambda^{3q_2}\begin{vmatrix} V_{11}^{(\underline{q})} & V_{12}^{(\underline{q})} & V_{13}^{(\underline{q})}\\ V_{21}^{(\underline{q})} & V_{22}^{(\underline{q})} & V_{23}^{(\underline{q})}\\ \lambda V_{31}^{(\underline{q})} & \lambda V_{32}^{(\underline{q})} & V_{33}^{(\underline{q})}\end{vmatrix}\\
&= \lambda\left(\alpha_{0,+}\beta_{0,+}^2\lambda^{3q+1} + \beta_{0,+}(\alpha_{1,+}\beta_{0,+} + 2\alpha_{0,+}\beta_{1,+})\lambda^{3q} + \cdots\right),
\end{aligned} \tag{3.4}$$

and $q = q_1 + q_2$. Then one naturally leads to the trigonal curve $\mathcal{K}_{m-1}$ of degree m by

$$\mathcal{K}_{m-1}: \quad \mathscr{F}_{m-1}(\lambda, y) = y^3 - y^2R_m + yS_m(\lambda) - T_m(\lambda) = 0, \tag{3.5}$$

where $m = 3q + 2$ for $\alpha_{0,+}\beta_{0,+} \neq 0$. So we assume that $\alpha_{0,+}\beta_{0,+} \neq 0, \alpha_{0,-} \neq \pm\beta_{0,-}$. Adding the points P_{∞_1} and P_{∞_2} on $\mathcal{K}_{m-1}$, one of which is a double branch point and the other is not a branch point. Without loss of generality, we let P_{∞_2} be the double branch point. For the sake of convenience, the compactification of the curve $\mathcal{K}_{m-1}$ is still denoted by the same symbol. The discriminant of (3.5) is

$$\Delta(\lambda) = 4S_m^3 - R_m^2S_m^2 + 4R_m^3T_m - 18R_mS_mT_m + 27T_m^2 = -4\alpha_{0,+}^4\beta_{0,+}^2\lambda^{6q+5} + \cdots, \tag{3.6}$$

The Riemann-Hurwitz formula shows that the arithmetic genus of $\mathcal{K}_{m-1}$ is $3q + 1$. Therefore, $\mathcal{K}_{m-1}$ becomes a three-sheeted Riemann surface of arithmetic genus $m - 1$ if it is nonsingular or smooth. Here, the meaning of nonsingularity is that $(\frac{\partial \mathscr{F}_{m-1}}{\partial \lambda}, \frac{\partial \mathscr{F}_{m-1}}{\partial y})\,|_{(\lambda,y)=(\lambda_0,y_0)} \neq 0$ at each point $P_0 = (\lambda_0, y_0) \in \mathcal{K}_{m-1}$.

Now we introduce the stationary Baker-Akhiezer function by

$$E\psi(P, n, n_0) = U(u(n), v(n), w(n); \lambda(P))\psi(P, n, n_0), \tag{3.7a}$$

$$\lambda^{q_2}V^{(\underline{q})}(u(n), v(n), w(n); \lambda(P))\psi(P, n, n_0) = y(P)\psi(P, n, n_0), \tag{3.7b}$$

$$\psi_1(P, n_0, n_0) = 1, \quad P \in \mathcal{K}_{m-1} \setminus \{P_{\infty_1}, P_{\infty_2}\}, \quad n, n_0 \in \mathbb{Z}. \tag{3.7c}$$

The meromorphic function $\phi(P, n)$ on $\mathcal{K}_{m-1}$ defined by

$$\phi(P, n) = \frac{\psi_1^+(P, n, n_0)}{\psi_1(P, n, n_0)}, \quad P \in \mathcal{K}_{m-1},\ n \in \mathbb{Z}, \tag{3.8}$$

such that

$$\psi_1(P,n,n_0) = \begin{cases} \prod_{n'=n_0}^{n-1} \phi(P,n'), & n \geq n_0+1, \\ 1, & n = n_0, \\ \prod_{n'=n}^{n_0-1} \phi(P,n')^{-1}, & n \leq n_0 - 1. \end{cases} \tag{3.9}$$

From (3.7) after tedious calculations, we obtain

$$\begin{aligned} \phi =& \frac{yV_{23}^{(q)} + C_m}{yV_{13}^{(q)} + A_m} = \frac{y^2 V_{13}^{(q)} - y(A_m + V_{13}^{(q)} R_m) + B_m}{E_{m-1}} \\ =& \frac{F_{m-1}}{y^2 V_{23}^{(q)} - y(C_m + V_{23}^{(q)} R_m) + D_m}, \end{aligned} \tag{3.10}$$

where

$$A_m = \lambda^{q_2}(V_{12}^{(q)} V_{23}^{(q)} - V_{13}^{(q)} V_{22}^{(q)}), \tag{3.11a}$$
$$B_m = \lambda^{2q_2}(V_{13}^{(q)}(V_{11}^{(q)} V_{33}^{(q)} - \lambda V_{13}^{(q)} V_{31}^{(q)}) + V_{12}^{(q)}(V_{11}^{(q)} V_{23}^{(q)} - V_{13}^{(q)} V_{21}^{(q)})), \tag{3.11b}$$
$$C_m = \lambda^{q_2}(V_{13}^{(q)} V_{21}^{(q)} - V_{11}^{(q)} V_{23}^{(q)}), \tag{3.11c}$$
$$D_m = \lambda^{2q_2}(V_{23}^{(q)}(V_{22}^{(q)} V_{33}^{(q)} - \lambda V_{23}^{(q)} V_{32}^{(q)}) + V_{21}^{(q)}(V_{13}^{(q)} V_{22}^{(q)} - V_{12}^{(q)} V_{23}^{(q)})), \tag{3.11d}$$

$$E_{m-1} = \lambda^{2q_2}(\lambda(V_{13}^{(q)})^2 V_{32}^{(q)} + V_{12}^{(q)} V_{13}^{(q)}(V_{22}^{(q)} - V_{33}^{(q)}) - (V_{12}^{(q)})^2 V_{23}^{(q)}), \tag{3.12a}$$
$$F_{m-1} = \lambda^{2q_2}(\lambda(V_{23}^{(q)})^2 V_{31}^{(q)} + V_{21}^{(q)} V_{23}^{(q)}(V_{11}^{(q)} - V_{33}^{(q)}) - (V_{21}^{(q)})^2 V_{13}^{(q)}). \tag{3.12b}$$

For later use, we also introduce

$$\mathcal{A}_m = \lambda^{q_2}(\lambda V_{13}^{(q)} V_{32}^{(q)} - V_{12}^{(q)} V_{23}^{(q)}), \tag{3.13a}$$
$$\mathcal{B}_m = \lambda^{2q_2}(V_{12}^{(q)}(V_{11}^{(q)} V_{22}^{(q)} - V_{12}^{(q)} V_{21}^{(q)}) + \lambda V_{13}^{(q)}(V_{11}^{(q)} V_{32}^{(q)} - V_{12}^{(q)} V_{31}^{(q)})). \tag{3.13b}$$

We can easily show that there exist various relations among the polynomials A_m, B_m, C_m, D_m, E_{m-1}, F_{m-1}, $\mathcal{A}_m$, $\mathcal{B}_m$, R_m, S_m and T_m, some of which are listed below:

$$V_{23}^{(q)} E_{m-1} = V_{13}^{(q)} B_m - (V_{13}^{(q)})^2 S_m - A_m^2 - V_{13}^{(q)} R_m A_m, \tag{3.14a}$$
$$C_m E_{m-1} = (V_{13}^{(q)})^2 T_m + A_m B_m, \tag{3.14b}$$

$$V_{13}^{(q)} F_{m-1} = V_{23}^{(q)} D_m - (V_{23}^{(q)})^2 S_m - C_m^2 - V_{23}^{(q)} R_m C_m, \tag{3.15a}$$
$$A_m F_{m-1} = (V_{23}^{(q)})^2 T_m + C_m D_m, \tag{3.15b}$$

$$E_{m-1} = \lambda^{q_2}(V_{13}^{(q)}\mathcal{A}_m - V_{12}^{(q)}A_m), \qquad F_{m-1} = -\lambda w E_{m-1}^+, \tag{3.16a}$$
$$\mathcal{A}_m^+ = C_m, \qquad \mathcal{B}_m^+ = D_m. \tag{3.16b}$$

From (3.1), (3.2), (3.12) and (3.16), E_{m-1}, F_{m-1} can be written as follows

$$E_{m-1}(\lambda, n) = \lambda^{-q_2}\alpha_{0,+}^2\beta_{0,+}\prod_{j=1}^{m-1}(\lambda - \mu_j(n)), \tag{3.17a}$$

$$F_{m-1}(\lambda, n) = -\lambda^{-q_2+1}w\alpha_{0,+}^2\beta_{0,+}\prod_{j=1}^{m-1}(\lambda - \mu_j^+(n)). \tag{3.17b}$$

Defining $\{\hat{\mu}_j(n)\}_{j=1,\ldots,m-1} \subset \mathcal{K}_{m-1}$ and $\{\hat{\mu}_j^+(n)\}_{j=1,\ldots,m-1} \subset \mathcal{K}_{m-1}$ by

$$\hat{\mu}_j(n) = (\mu_j(n), y(\hat{\mu}_j(n))) = \left(\mu_j(n), -\frac{A_m(\mu_j(n), n)}{V_{13}^{(q)}(\mu_j(n), n)}\right) \in \mathcal{K}_{m-1}, \tag{3.18a}$$

$$\hat{\mu}_j^+(n) = \left(\mu_j^+(n), y(\hat{\mu}_j^+(n))\right) = \left(\mu_j^+(n), -\frac{C_m(\mu_j^+(n), n)}{V_{23}^{(q)}(\mu_j^+(n), n)}\right) \in \mathcal{K}_{m-1}. \tag{3.18b}$$

4 Algebro-geometric solutions to the stationary mBM hierarchy

In this section, we will derive the explicit Riemann theta function representations for the meromorphic function ϕ, the Baker-Akhiezer function ψ_1, and the potentials u, v, w of the stationary mBM hierarchy. A direct calculation shows that $\phi(P, n)$ satisfies the Riccati-type equation

$$\begin{aligned} &\phi^+(P,n)\phi(P,n)\phi^-(P,n) - u(n)\phi(P,n)\phi^-(P,n) - v(n)\phi^-(P,n) \\ &\quad = \lambda w^-(n) + \lambda\phi^-(P,n). \end{aligned} \tag{4.1}$$

***Lemma* 4.1.** (i) Near $P_{\infty_1} \in \mathcal{K}_{m-1}$, introducing the local coordinate $\lambda = \zeta^{-1}$, one has

$$\phi(P,n) \underset{\zeta\to 0}{=} -w + w(v^+ - w^+w^{++} - w^+u^+)\zeta + O(\zeta^2), \quad \textit{as} \quad P \to P_{\infty_1}. \tag{4.2}$$

(ii) Near $P_{\infty_2} \in \mathcal{K}_{m-1}$, introducing the local coordinate $\lambda = \xi^{-2}$, one has

$$\phi(P,n) \underset{\xi\to 0}{=} \xi^{-1} + (E+1)^{-1}(u + w^-) + O(\xi), \quad \textit{as} \quad P \to P_{\infty_2}. \tag{4.3}$$

(iii) Near $P_{0_s} \in \mathcal{K}_{m-1}, s = 1, 2, 3$, introducing the local coordinate $\lambda = \varrho$, one has

$$\phi(P,n) \underset{\varrho\to 0}{=} \begin{cases} -\dfrac{w}{v^+}\varrho + O(\varrho^2), & \textit{as}\ P \to P_{0_1}, \\ \dfrac{1 - \check{c}_{0,-}}{\check{a}_{0,-}} + O(\rho), & \textit{as}\ P \to P_{0_2}, \\ \dfrac{-1 - \check{c}_{0,-}}{\check{a}_{0,-}} + O(\varrho), & \textit{as}\ P \to P_{0_3}. \end{cases} \tag{4.4}$$

Proof. We can insert the three sets of ansatzs

$$
\begin{aligned}
&(1)\ \phi \underset{\zeta\to 0}{=} \phi_0+\phi_1\zeta+O(\zeta^2), && P\to P_{\infty_1},\\
&(2)\ \phi \underset{\xi\to 0}{=} \kappa_{-1}\xi^{-1}+\kappa_0+\kappa_1\xi+O(\xi^2), && P\to P_{\infty_2},\\
&(3)\ \phi \underset{\varrho\to 0}{=} \chi_{s,0}+\chi_{s,1}\varrho+\chi_{s,2}\varrho^2+O(\varrho^3), && P\to P_{0_s}, \quad s=1,2,3,
\end{aligned}
$$

into the Riccati-type Equation (4.1). A comparison of the same powers of ζ,ξ,ϱ, respectively, then the Lemma can be proved. ■

***Lemma* 4.2.** (i) Near $P_{\infty_1}, P_{\infty_2}\in\mathcal{K}_{m-1}$ introducing the local coordinate $\lambda=\zeta^{-1}$ and $\lambda=\xi^{-2}$, respectively, one has

$$\psi_1(P,n,n_0) \underset{\zeta\to 0}{=} \Gamma_0(n,n_0)(1+O(\zeta)), \qquad as\ \ P\to P_{\infty_1}, \quad \lambda=\zeta^{-1}, \tag{4.5}$$

$$\psi_1(P,n,n_0) \underset{\xi\to 0}{=} \xi^{n_0-n}(1+O(\xi)), \qquad as\ \ P\to P_{\infty_2}, \quad \lambda=\xi^{-2}, \tag{4.6}$$

$$\Gamma_0(n,n_0) \underset{\zeta\to 0}{=} \begin{cases} \prod_{n'=n_0}^{n-1}(-w(n')), & n\geq n_0+1,\\ 1, & n=n_0,\\ \prod_{n'=n}^{n_0-1}(-w(n'))^{-1}, & n\leq n_0-1.\end{cases}$$

(ii) Near $P_{0_s}\in\mathcal{K}_{m-1}, s=1,2,3$ introducing the local coordinate $\lambda=\varrho$, one has

$$\psi_1(P,n,n_0) \underset{\zeta\to 0}{=} \begin{cases} \Gamma_1(n,n_0)\varrho^{n-n_0}(1+O(\varrho)), & as\ \ P\to P_{0_1},\\ \Gamma_2(n,n_0)(1+O(\varrho)), & as\ \ P\to P_{0_2},\\ \Gamma_3(n,n_0)(1+O(\varrho)), & as\ \ P\to P_{0_3},\end{cases} \quad \lambda=\varrho, \tag{4.7}$$

$$\Gamma_1(n,n_0) \underset{\varrho\to 0}{=} \begin{cases} \prod_{n'=n_0}^{n-1}\left(-\dfrac{w(n')}{v^+(n')}\right), & n\geq n_0+1,\\ 1, & n=n_0,\\ \prod_{n'=n}^{n_0-1}\left(-\dfrac{w(n')}{v^+(n')}\right)^{-1}, & n\leq n_0-1,\end{cases}$$

$$\Gamma_s(n,n_0) \underset{\varrho\to 0}{=} \begin{cases} \prod_{n'=n_0}^{n-1}\chi_{s,0}(n'), & n\geq n_0+1,\\ 1, & n=n_0, \qquad s=2,3.\\ \prod_{n'=n}^{n_0-1}(\chi_{s,0}(n'))^{-1}, & n\leq n_0-1,\end{cases}$$

Observing (4.2)-(4.7) and (3.10), we obtain the divisors of $\phi(P,n)$ and $\psi_1(P,n,n_0)$

$$(\phi(P,n)) = \mathscr{D}_{P_{0_1},\hat{\mu}_1^+(n),\ldots,\hat{\mu}_{m-1}^+(n)}(P)-\mathscr{D}_{P_{\infty_2},\hat{\mu}_1(n),\ldots,\hat{\mu}_{m-1}(n)}(P), \tag{4.8a}$$

$$(\psi_1(P,n,n_0)) = \mathscr{D}_{\hat{\mu}_1(n),\ldots,\hat{\mu}_{m-1}(n)}(P)-\mathscr{D}_{\hat{\mu}_1(n_0),\ldots,\hat{\mu}_{m-1}(n_0)}(P)+(n-n_0)(P_{0_1}-P_{\infty_2}). \tag{4.8b}$$

Equip the Riemann surface $\mathcal{K}_{m-1}$ with homology basis $\{\mathbb{a}_j, \mathbb{b}_j\}_{j=1}^{m-1}$, which are independent and have intersection numbers as follows

$$\mathbb{a}_j \circ \mathbb{b}_k = \delta_{jk}, \quad \mathbb{a}_j \circ \mathbb{a}_k = 0, \quad \mathbb{b}_j \circ \mathbb{b}_k = 0, \quad j,k = 1,\ldots,m-1.$$

For the present, we will choose our basis as the following set

$$\tilde{\omega}_l(P) \;=\; \frac{1}{3y^2 - 2yR_m + S_m}\begin{cases} \lambda^{l-1}d\lambda, & 1 \le l \le 2q+1, \\ (y - \frac{1}{3}R_m)\lambda^{l-2q-2}d\lambda, & 2q+2 \le l \le m-1, \end{cases} \tag{4.9}$$

which are $m-1$ linearly independent holomorphic differentials on $\mathcal{K}_{m-1}$. By using the homology basis $\{\mathbb{a}_j\}_{j=1}^{m-1}$ and $\{\mathbb{b}_j\}_{j=1}^{m-1}$, the period matrices $A = (A_{jk})$ and $B = (B_{jk})$ can be constructed from

$$A_{jk} = \int_{\mathbb{a}_k} \tilde{\omega}_j, \qquad B_{jk} = \int_{\mathbb{b}_k} \tilde{\omega}_j. \tag{4.10}$$

It is possible to show that the matrices A and B are invertible [10, 22]. Now we define the matrices C and τ by $C = A^{-1}, \tau = A^{-1}B$. The matrix τ can be shown to be symmetric ($\tau_{jk} = \tau_{kj}$) and has a positive-definite imaginary part (Im $\tau > 0$). If we normalized $\tilde{\omega}_l(P)$ into new basis $\underline{\omega} = (\omega_1, \ldots, \omega_{m-1})$,

$$\omega_j = \sum_{l=1}^{m-1} C_{jl}\tilde{\omega}_l, \tag{4.11}$$

then we have $\int_{\mathbb{a}_k} \omega_j = \delta_{jk}, \quad \int_{\mathbb{b}_k} \omega_j = \tau_{jk}, \quad j,k = 1,\ldots,m-1.$

Let $\omega^{(3)}_{Q_1,Q_2}$ denote the normalized Abelian differential of the third kind holomorphic on $\mathcal{K}_{m-1} \setminus \{Q_1, Q_2\}$ with simple poles at Q_l with residues $(-1)^{l+1}$, $l = 1,2$, then

$$\int_{\mathbb{a}_k} \omega^{(3)}_{Q_1,Q_2} = 0, \quad \int_{\mathbb{b}_k} \omega^{(3)}_{Q_1,Q_2} = 2\pi i \int_{Q_2}^{Q_1} \omega_k, \quad k = 1,\ldots,m-1. \tag{4.12}$$

Especially, one obtains for $\omega^{(3)}_{P_{\infty_2},P_{0_1}}(P)$

$$\begin{aligned}
&\omega^{(3)}_{P_{\infty_2},P_{0_1}}(P) \underset{\zeta\to 0}{=} O(1)d\zeta, && as \quad P \to P_{\infty_1}, \quad \zeta = \lambda^{-1}, \\
&\omega^{(3)}_{P_{\infty_2},P_{0_1}}(P) \underset{\xi\to 0}{=} (\xi^{-1} + \omega_0^{\infty_2} + O(\xi))d\xi, && as \quad P \to P_{\infty_2}, \quad \xi = \lambda^{-\frac{1}{2}}, \\
&\omega^{(3)}_{P_{\infty_2},P_{0_1}}(P) \underset{\varrho\to 0}{=} \begin{cases} (-\varrho^{-1} + \omega_0^{0_1} + O(\varrho))d\varrho, & as \quad P \to P_{0_1}, \\ O(1)d\varrho, & as \quad P \to P_{0_2}, \\ O(1)d\varrho, & as \quad P \to P_{0_3}, \end{cases} && \varrho = \lambda.
\end{aligned} \tag{4.13}$$

Then

$$\int_{Q_0}^{P} \omega^{(3)}_{P_{\infty_2},P_{0_1}}(P) \underset{\zeta\to 0}{=} e_1(Q_0)+O(\zeta), \qquad as \quad P\to P_{\infty_1}, \quad \zeta=\lambda^{-1},$$

$$\int_{Q_0}^{P} \omega^{(3)}_{P_{\infty_2},P_{0_1}}(P) \underset{\xi\to 0}{=} \ln\xi + e_2(Q_0)+\omega_0^{\infty_2}\xi+O(\xi^2), \qquad as \quad P\to P_{\infty_2}, \quad \xi=\lambda^{-\frac{1}{2}},$$

$$\int_{Q_0}^{P} \omega^{(3)}_{P_{\infty_2},P_{0_1}}(P) \underset{\varrho\to 0}{=} \begin{cases} -\ln\varrho + e_{0_1}(Q_0)+\omega_0^{0_1}\varrho+O(\varrho^2), & as \quad P\to P_{0_1}, \\ e_{0_2}(Q_0)+O(\varrho), & as \quad P\to P_{0_2}, \qquad \varrho=\lambda, \\ e_{0_3}(Q_0)+O(\varrho), & as \quad P\to P_{0_3}, \end{cases} \tag{4.14}$$

where Q_0 is an appropriately chosen base point on $\mathcal{K}_{m-1}\setminus\{P_{\infty_1},P_{\infty_2},P_{0_1},P_{0_2},P_{0_3}\}$, and $e_1(Q_0), e_2(Q_0), e_{0_s}(Q_0), s=1,2,3$ are integration constants.

Let $\mathcal{T}_{m-1}$ be the period lattice $\{\underline{z}\in\mathbb{C}^{m-1}|\underline{z}=\underline{N}+\underline{M}\tau, \quad \underline{N},\underline{M}\in\mathbb{Z}^{m-1}\}$. The complex torus $\mathcal{J}_{m-1}=\mathbb{C}^{m-1}/\mathcal{T}_{m-1}$ is called the Jacobian variety of $\mathcal{K}_{m-1}$. An Abel map $\underline{\mathcal{A}}:\mathcal{K}_{m-1}\to\mathcal{J}_{m-1}$ is defined by

$$\underline{\mathcal{A}}(P)=\Big(\int_{Q_0}^{P}\omega_1,\ldots,\int_{Q_0}^{P}\omega_{m-1}\Big) \quad (\mathrm{mod}\ \mathcal{T}_{m-1}) \tag{4.15}$$

with the natural linear extension to the factor group $\mathrm{Div}(\mathcal{K}_{m-1})$

$$\underline{\mathcal{A}}(\sum l_kP_k)=\sum l_k\underline{\mathcal{A}}(P_k). \tag{4.16}$$

Define

$$\underline{\rho}(n)=\underline{\mathcal{A}}(\sum_{k=1}^{m-1}\hat{\mu}_k(n))=\sum_{k=1}^{m-1}\underline{\mathcal{A}}(\hat{\mu}_k(n))=\sum_{k=1}^{m-1}\int_{Q_0}^{\hat{\mu}_k(n)}\underline{\omega},$$

where $\underline{\rho}(n)=(\rho_1(n),\ldots,\rho_{m-1}(n))$ and $\underline{\omega}=(\omega_1,\ldots,\omega_{m-1})$.

Let $\theta(\underline{z})$ denote the Riemann theta function associated with $\mathcal{K}_{m-1}$:

$$\begin{aligned} &\theta(\underline{z})=\sum_{\underline{N}\in\mathbb{Z}^{m-1}}\exp\big\{\pi i\langle\underline{N}\tau,\underline{N}\rangle+2\pi i\langle\underline{N},\underline{z}\rangle\big\}, \\ &\underline{z}=(z_1,\ldots,z_{m-1})\in\mathbb{C}^{m-1}, \quad \langle\underline{N},\underline{z}\rangle=\sum_{i=1}^{m-1}N_iz_i, \quad \langle\underline{N}\tau,\underline{N}\rangle=\sum_{i,j=1}^{m-1}\tau_{ij}N_iN_j. \end{aligned} \tag{4.17}$$

For brevity, we use the abbreviation

$$\begin{aligned} &\theta(\underline{z}(P,\underline{\hat{\mu}}(n)))=\theta(\underline{\Lambda}-\underline{\mathcal{A}}(P)+\underline{\rho}(n)), \quad P\in\mathcal{K}_{m-1}, \\ &\underline{\hat{\mu}}(n)=\{\hat{\mu}_1(n),\ldots,\hat{\mu}_{m-1}(n)\}\in\sigma^{m-1}\mathcal{K}_{m-1}, \end{aligned}$$

where $\sigma^{m-1}\mathcal{K}_{m-1}$ denotes the $(m-1)$-th symmetric power of $\mathcal{K}_{m-1}$ and $\underline{\Lambda} = (\Lambda_1, \ldots, \Lambda_{m-1})$ is the vector of Riemann constant depending on the base point Q_0 by the following expression

$$\Lambda_j = \frac{1}{2}(1+\tau_{jj}) - \sum_{\substack{l=1\\ l\neq j}}^{m-1} \int_{\mathrm{a}_l} \omega_l(P) \int_{Q_0}^{P} \omega_j, \quad j = 1, \ldots, m-1.$$

***Theorem* 4.3.** We assume that the curve $\mathcal{K}_{m-1}$ is nonsingular. Let $P = (\lambda, y) \in \mathcal{K}_{m-1} \setminus \{P_{\infty_1}, P_{\infty_2}\}$, $(n, n_0) \in \mathbb{Z}^2$. If $\mathcal{D}_{\underline{\hat{\mu}}(n)}$ is nonspecial for $n \in \mathbb{Z}$, then

$$\phi(P,n) = \frac{\theta(\underline{z}(P_{\infty_2}, \underline{\hat{\mu}}(n)))}{\theta(\underline{z}(P_{\infty_2}, \underline{\hat{\mu}}^+(n)))} \frac{\theta(\underline{z}(P, \underline{\hat{\mu}}^+(n)))}{\theta(\underline{z}(P, \underline{\hat{\mu}}(n)))} \exp\left(e_2(Q_0) - \int_{Q_0}^{P} \omega^{(3)}_{P_{\infty_2}, P_{0_1}} \right), \tag{4.18}$$

and

$$\psi_1(P,n,n_0) = \frac{\theta(\underline{z}(P_{\infty_2}, \underline{\hat{\mu}}(n_0)))}{\theta(\underline{z}(P_{\infty_2}, \underline{\hat{\mu}}(n)))} \frac{\theta(\underline{z}(P, \underline{\hat{\mu}}(n)))}{\theta(\underline{z}(P, \underline{\hat{\mu}}(n_0)))} \exp\left((n-n_0)\left(e_2(Q_0) - \int_{Q_0}^{P} \omega^{(3)}_{P_{\infty_2}, P_{0_1}} \right)\right). \tag{4.19}$$

The Abel map linearizes that auxiliary divisor $\mathcal{D}_{\underline{\hat{\mu}}(n)}$ in the sense that

$$\underline{\rho}(n) = \underline{\rho}(n_0) + (n-n_0)(\underline{A}(P_{\infty_2}) - \underline{A}(P_{0_1})). \tag{4.20}$$

***Proof*.** Applying Abel's theorem to (4.8b) yields (4.20). It follows from (4.14) that

$$\begin{aligned}
&\exp\left(e_2(Q_0) - \int_{Q_0}^{P} \omega^{(3)}_{P_{\infty_2}, P_{0_1}} \right) \underset{\zeta\to 0}{=} \exp\left(e_2(Q_0) - e_1(Q_0)\right)(1+O(\zeta)), \quad as \quad P \to P_{\infty_1},\\
&\exp\left(e_2(Q_0) - \int_{Q_0}^{P} \omega^{(3)}_{P_{\infty_2}, P_{0_1}} \right) \underset{\xi\to 0}{=} \xi^{-1}(1 - \omega_0^{\infty_2}\xi + O(\xi^2)), \quad as \quad P \to P_{\infty_2},\\
&\exp\left(e_2(Q_0) - \int_{Q_0}^{P} \omega^{(3)}_{P_{\infty_2}, P_{0_1}} \right) \underset{\varrho\to 0}{=} \varrho \cdot \exp\left(e_2(Q_0) - e_{0_1}(Q_0)\right) + O(\varrho^2), \quad as \quad P \to P_{0_1}.
\end{aligned} \tag{4.21}$$

We let Φ denote the right-hand side of (4.18). From the expression for Φ, (4.21) and (4.4), we know that Φ and ϕ share the same local zeros and poles and identical essential singularities at $P_{\infty_1}, P_{\infty_2}$. By the Riemann-Roch theorem, we conclude that $\Phi/\phi = \gamma$, where γ is a constant. Using (4.3) and (4.21), we compute

$$\frac{\Phi}{\phi} \underset{\xi\to 0}{=} \frac{(1+O(\xi))(\xi^{-1}+O(1))}{\xi^{-1}+O(1)} \underset{\xi\to 0}{=} 1 + O(\xi), \quad P \to P_{\infty_2}, \tag{4.22}$$

which shows that $\gamma = 1$. This proves (4.18). We obtain representation (4.19) using (3.9) and (4.18). ∎

***Theorem* 4.4.** We assume that $\mathcal{K}_{m-1}$ is nonsingular. Suppose that $\mathcal{D}_{\underline{\hat{\mu}}(n)}$ is nonspecial for $n \in \mathbb{Z}$, then

$$\begin{aligned} u(n) =& \frac{\theta(\underline{z}(P_{\infty_2}, \underline{\hat{\mu}}^-(n)))}{\theta(\underline{z}(P_{\infty_2}, \underline{\hat{\mu}}(n)))} \frac{\theta(\underline{z}(P_{\infty_1}, \underline{\hat{\mu}}(n)))}{\theta(\underline{z}(P_{\infty_1}, \underline{\hat{\mu}}^-(n)))} \exp\left(e_2(Q_0) - e_1(Q_0)\right) \\ & - 2\omega_0^{\infty_2} - \sum_{j=0}^{m-1} d_{j,0}^{(\infty_2)} \frac{\partial}{\partial z_j} \ln \frac{\theta(\underline{z}(P_{0_1}, \underline{\hat{\mu}}^+(n)))}{\theta(\underline{z}(P_{0_1}, \underline{\hat{\mu}}^-(n)))}, \end{aligned} \tag{4.23}$$

$$v(n) = \frac{\theta(\underline{z}(P_{\infty_1}, \underline{\hat{\mu}}(n)))}{\theta(\underline{z}(P_{\infty_1}, \underline{\hat{\mu}}^-(n)))} \frac{\theta(\underline{z}(P_{0_1}, \underline{\hat{\mu}}^-(n)))}{\theta(\underline{z}(P_{0_1}, \underline{\hat{\mu}}(n)))} \exp\left(e_{0_1}(Q_0) - e_1(Q_0)\right), \tag{4.24}$$

$$w(n) = - \frac{\theta(\underline{z}(P_{\infty_2}, \underline{\hat{\mu}}(n)))}{\theta(\underline{z}(P_{\infty_2}, \underline{\hat{\mu}}^+(n)))} \frac{\theta(\underline{z}(P_{\infty_1}, \underline{\hat{\mu}}^+(n)))}{\theta(\underline{z}(P_{\infty_1}, \underline{\hat{\mu}}(n)))} \exp\left(e_2(Q_0) - e_1(Q_0)\right). \tag{4.25}$$

Proof. Applying Abel's theorem to (4.4), we obtain

$$\underline{\rho}^+(n) + \underline{A}(P_{0_1}) = \underline{\rho}(n) + \underline{A}(P_{\infty_2}), \tag{4.26}$$

and we conclude that

$$\theta(\underline{z}(P_{\infty_2}, \underline{\hat{\mu}}^+(n))) = \theta(\underline{z}(P_{0_1}, \underline{\hat{\mu}}(n))). \tag{4.27}$$

Because $y = V_{11}^{(q)} + V_{12}^{(q)} \phi + V_{13}^{(q)} \frac{\lambda w^-}{\phi^-}$, we obtain

$$y \underset{\xi \to 0}{=} \xi^{-2q-1}(\beta_{0,+} + \beta_{1,+}\xi^2 + O(\xi^4)), \tag{4.28}$$

with the local coordinates $\lambda = \xi^{-2}$ near P_{∞_2}. From (4.8a) and (4.9), we obtain

$$\begin{aligned} \omega_j &= \sum_{l=1}^{m-1} C_{jl} \tilde{\omega}_l \\ &= \sum_{l=1}^{2q+1} C_{jl} \frac{\lambda^{l-1} d\lambda}{3y^2 - 2yR_m + S_m} + \sum_{l=2q+2}^{m-1} C_{jl} \frac{(y - R_m/3)\lambda^{l-2q-2} d\lambda}{3y^2 - 2yR_m + S_m}, \end{aligned} \tag{4.29}$$

where $j = 1, \ldots, m-1$. By directly computing, we obtain the asymptotic expansion

$$\omega_j \underset{\xi \to 0}{=} (d_{j,0}^{(\infty_2)} + O(\xi)) d\xi \quad P \to P_{\infty_2}, \tag{4.30}$$

where

$$d_{j,0}^{(\infty_2)} = - \frac{1}{\alpha_{0,+}\beta_{0,+}} C_{j,2q+1} - \frac{1}{3\beta_{0,+}} C_{j,m-1}.$$

Expanding the ratios of the Riemann theta function in (4.18), we obtain

$$\frac{\theta(\underline{z}(P,\underline{\hat{\mu}}^+(n)))}{\theta(\underline{z}(P,\underline{\hat{\mu}}(n)))} = \frac{\theta(\underline{\Lambda}-\underline{A}(P)+\underline{\rho}^+(n))}{\theta(\underline{\Lambda}-\underline{A}(P)+\underline{\rho}(n))} = \frac{\theta\left(\underline{\Lambda}-\underline{A}(P_{\infty_2})+\underline{\rho}^+(n)+\int_P^{P_{\infty_2}}\underline{\omega}\right)}{\theta\left(\underline{\Lambda}-\underline{A}(P_{\infty_2})+\underline{\rho}(n)+\int_P^{P_{\infty_2}}\underline{\omega}\right)}$$

$$\underset{\xi\to 0}{=} \frac{\theta(\ldots,\Lambda_j-A_j(P_{\infty_2})+\rho_j^+(n)-d_{j,0}^{(\infty_2)}\xi+O(\xi^2),\ldots)}{\theta(\ldots,\Lambda_j-A_j(P_{\infty_2})+\rho_j(n)-d_{j,0}^{(\infty_2)}\xi+O(\xi^2),\ldots)}$$

$$\underset{\xi\to 0}{=} \frac{\theta_2^+ - \sum_{j=1}^{m-1} d_{j,0}^{(\infty_2)}\frac{\partial}{\partial z_j}\theta_2^+\xi+O(\xi^2)}{\theta_2 - \sum_{j=1}^{m-1} d_{j,0}^{(\infty_2)}\frac{\partial}{\partial z_j}\theta_2\xi+O(\xi^2)}$$

$$\underset{\xi\to 0}{=} \frac{\theta_2^+}{\theta_2}\left(1-\sum_{j=1}^{m-1} d_{j,0}^{(\infty_2)}\frac{\partial}{\partial z_j}\ln\frac{\theta_2^+}{\theta_2}\xi+O(\xi^2)\right), \quad P\to P_{\infty_2}, \tag{4.31}$$

where $\theta_2=\theta(\underline{z}(P_{\infty_2},\underline{\hat{\mu}}(n))), \theta_2^+=\theta(\underline{z}(P_{\infty_2},\underline{\hat{\mu}}^+(n)))$. Therefore,

$$\begin{aligned}
\phi(P,n) &\underset{\zeta\to 0}{=} \frac{\theta_2}{\theta_2^+}\frac{\theta(\underline{z}(P_{\infty_1},\underline{\hat{\mu}}^+(n)))}{\theta(\underline{z}(P_{\infty_1},\underline{\hat{\mu}}(n)))}\exp(e_2(Q_0)-e_1(Q_0))(1+O(\zeta)), && P\to P_{\infty_1},\\
\phi(P,n) &\underset{\xi\to 0}{=} \xi^{-1}(1-\omega_0^{\infty_2}\xi+O(\xi^2))\left(1-\sum_{j=1}^{m-1} d_{j,0}^{(\infty_2)}\frac{\partial}{\partial z_j}\ln\frac{\theta_2^+}{\theta_2}\xi+O(\xi^2)\right), && P\to P_{\infty_2},\\
\phi(P,n) &\underset{\varrho\to 0}{=} \frac{\theta_2}{\theta_2^+}\frac{\theta(\underline{z}(P_{0_1},\underline{\hat{\mu}}^+(n)))}{\theta(\underline{z}(P_{0_1},\underline{\hat{\mu}}(n)))}\exp(e_2(Q_0)-e_{0_1}(Q_0))(\varrho+O(\varrho^2)), && P\to P_{0_1}.
\end{aligned} \tag{4.32}$$

On the other hand, from (4.2), (4.3) and (4.4), we have

$$\phi(P,n) \underset{\zeta\to 0}{=} -w+w(v^+-w^+w^{++}-w^+u^+)\zeta+O(\zeta^2), \quad P\to P_{\infty_1}, \tag{4.33a}$$

$$\phi(P,n) \underset{\xi\to 0}{=} \xi^{-1}+(E+1)^{-1}(u+w^-)+O(\xi), \quad P\to P_{\infty_2}, \tag{4.33b}$$

$$\phi(P,n) \underset{\varrho\to 0}{=} -\frac{w}{v^+}\varrho+O(\varrho^2), \quad P\to P_{0_1}. \tag{4.33c}$$

Comparing (4.32) and (4.33), we obtain representations (4.23)-(4.25). ■

The $\mathbb{b}$-period of the differential $\omega^{(3)}_{P_{\infty_2},P_{0_1}}$ is denoted by

$$\underline{U}^{(3)}=(U_1^{(3)},\ldots,U_{m-1}^{(3)}), \quad U_j^{(3)}=\frac{1}{2\pi i}\int_{\mathbb{b}_j}\omega^{(3)}_{P_{\infty_2},P_{0_1}}, \quad j=1,\ldots,m-1. \tag{4.34}$$

Combining (4.12), (4.20), (4.34) and (4.23)-(4.25) shows the remarkable linearity of the theta function representations for $u(n), v(n), w(n)$ with respect to $n \in \mathbb{Z}$. In fact, one can rewritten (4.23)-(4.25) as

$$\begin{aligned} u(n) &= -2\omega_0^{\infty_2} + \Big(\frac{\theta(\underline{K}_2 - \underline{U}^{(3)} + \underline{U}^{(3)}n)}{\theta(\underline{K}_2 + \underline{U}^{(3)}n)} \frac{\theta(\underline{K}_1 + \underline{U}^{(3)}n)}{\theta(\underline{K}_1 - \underline{U}^{(3)} + \underline{U}^{(3)}n)} \\ &\quad \times e^{e_2(Q_0)-e_1(Q_0)}\Big) - \sum_{j=0}^{m-1} d_{j,0}^{(\infty_2)} \frac{\partial}{\partial z_j} \ln \frac{\theta(\underline{K}_0 + \underline{U}^{(3)} + \underline{U}^{(3)}n)}{\theta(\underline{K}_0 - \underline{U}^{(3)} + \underline{U}^{(3)}n)}, \\ v(n) &= \frac{\theta(\underline{K}_1 + \underline{U}^{(3)}n)}{\theta(\underline{K}_1 - \underline{U}^{(3)} + \underline{U}^{(3)}n)} \frac{\theta(\underline{K}_0 - \underline{U}^{(3)} + \underline{U}^{(3)}n)}{\theta(\underline{K}_0 + \underline{U}^{(3)}n)} e^{e_{0_1}(Q_0)-e_1(Q_0)}, \\ w(n) &= -\frac{\theta(\underline{K}_2 + \underline{U}^{(3)}n)}{\theta(\underline{K}_2 + \underline{U}^{(3)} + \underline{U}^{(3)}n)} \frac{\theta(\underline{K}_1 + \underline{U}^{(3)} + \underline{U}^{(3)}n)}{\theta(\underline{K}_1 + \underline{U}^{(3)}n)} e^{e_2(Q_0)-e_1(Q_0)}, \end{aligned} \tag{4.35}$$

where

$$\underline{K}_0 = \underline{\Lambda} - \underline{A}(P_{0_1}) + \underline{\rho}(n_0) - n_0\underline{U}^{(3)},$$

$$\underline{K}_1 = \underline{\Lambda} - \underline{A}(P_{\infty_1}) + \underline{\rho}(n_0) - n_0\underline{U}^{(3)},$$

$$\underline{K}_2 = \underline{\Lambda} - \underline{A}(P_{\infty_2}) + \underline{\rho}(n_0) - n_0\underline{U}^{(3)}.$$

5 Algebro-geometric solutions of the time-dependent mBM hierarchy

In this section, we will extend the results of Secs. 3 and 4 to the time-dependent case. By analogy with (3.7), we introduce the time-dependent Baker-Akhiezer function

$$\begin{aligned} &E\psi(P,n,n_0,t_{\underline{r}},t_{0,\underline{r}}) = U(u(n,t_{\underline{r}}),v(n,t_{\underline{r}}),w(n,t_{\underline{r}});\lambda(P))\psi(P,n,n_0,t_{\underline{r}},t_{0,\underline{r}}), \\ &\psi_{t_{\underline{r}}}(P,n,n_0,t_{\underline{r}},t_{0,\underline{r}}) = \widetilde{V}^{(\underline{r})}(u(n,t_{\underline{r}}),v(n,t_{\underline{r}}),w(n,t_{\underline{r}});\lambda(P))\psi(P,n,n_0,t_{\underline{r}},t_{0,\underline{r}}), \\ &\lambda^{q_2}V^{(\underline{q})}(u(n,t_{\underline{r}}),v(n,t_{\underline{r}}),w(n,t_{\underline{r}});\lambda(P))\psi(P,n,n_0,t_{\underline{r}},t_{0,\underline{r}}) = y(P)\psi(P,n,n_0,t_{\underline{r}},t_{0,\underline{r}}), \\ &\psi_1(P,n_0,n_0,t_{0,\underline{r}},t_{0,\underline{r}}) = 1, \quad P \in \mathcal{K}_{m-1} \setminus \{P_{\infty_1},P_{\infty_2}\}, \quad (n,t_{\underline{r}}),(n_0,t_{0,\underline{r}}) \in \mathbb{Z}\times\mathbb{R}. \end{aligned} \tag{5.1}$$

The compatibility conditions of the first three expressions in (5.1) yield that

$$U_{t_{\underline{r}}} - (E\widetilde{V}^{(\underline{r})})U + U\widetilde{V}^{(\underline{r})} = 0, \tag{5.2a}$$

$$(EV^{(\underline{q})})U - UV^{(\underline{q})} = 0, \tag{5.2b}$$

$$V^{(\underline{q})}_{t_{\underline{r}}} - [\widetilde{V}^{(\underline{r})}, V^{(\underline{q})}] = 0. \tag{5.2c}$$

A direct calculation shows that $yI - \lambda^{q_2}V^{(\underline{q})}$ satisfies (5.2b) and (5.2c). The characteristic polynomial of Lax matrix $\lambda^{q_2}V^{(\underline{q})}$ is a constant independent of variables n and $t_{\underline{r}}$ with the expansion

$$\det(yI - \lambda^{q_2}V^{(\underline{q})}) = y^3 - y^2R_m(\lambda) + yS_m(\lambda) - T_m(\lambda),$$

where $R_m(\lambda), S_m(\lambda)$ and $T_m(\lambda)$ are defined as in (3.4). Then the time-dependent mBM lattice curve $\mathcal{K}_{m-1}$ is determined by

$$\mathcal{K}_{m-1}: \quad \mathscr{F}_{m-1}(\lambda, y) = y^3 - y^2 R_m(\lambda) + y S_m(\lambda) - T_m(\lambda) = 0.$$

The meromorphic function $\phi(P, n, t_{\underline{r}})$ on $\mathcal{K}_{m-1}$ defined by

$$\phi(P, n, t_{\underline{r}}) = \frac{\psi_1^+(P, n, n_0, t_{\underline{r}}, t_{0,\underline{r}})}{\psi_1(P, n, n_0, t_{\underline{r}}, t_{0,\underline{r}})}, \quad P \in \mathcal{K}_{m-1}, \ (n, t_{\underline{r}}) \in \mathbb{Z} \times \mathbb{R}. \tag{5.3}$$

By (5.1), (5.3) implies that

$$\begin{aligned}\phi(P, n, t_{\underline{r}}) &= \frac{y V_{23}^{(q)}(\lambda, n, t_{\underline{r}}) + C_m(\lambda, n, t_{\underline{r}})}{y V_{13}^{(q)}(\lambda, n, t_{\underline{r}}) + A_m(\lambda, n, t_{\underline{r}})} \\ &= \frac{F_{m-1}(\lambda, n, t_{\underline{r}})}{y^2 V_{23}^{(q)}(\lambda, n, t_{\underline{r}}) - y(C_m(\lambda, n, t_{\underline{r}}) + V_{23}^{(q)}(\lambda, n, t_{\underline{r}}) R_m(\lambda)) + D_m(\lambda, n, t_{\underline{r}})} \\ &= \frac{y^2 V_{13}^{(q)}(\lambda, n, t_{\underline{r}}) - y(A_m(\lambda, n, t_{\underline{r}}) + V_{13}^{(q)}(\lambda, n, t_{\underline{r}}) R_m(\lambda)) + B_m(\lambda, n, t_{\underline{r}})}{E_{m-1}(\lambda, n, t_{\underline{r}})},\end{aligned} \tag{5.4}$$

where $P = (\lambda, y) \in \mathcal{K}_{m-1}, (n, t_{\underline{r}}) \in \mathbb{Z}\times\mathbb{R}$, and $A_m(\lambda, n, t_{\underline{r}}), B_m(\lambda, n, t_{\underline{r}}), C_m(\lambda, n, t_{\underline{r}})$, $D_m(\lambda, n, t_{\underline{r}}), E_{m-1}(\lambda, n, t_{\underline{r}}), F_{m-1}(\lambda, n, t_{\underline{r}})$ are defined as in (3.11) and (3.12). $\mathcal{A}_m(\lambda, n, t_{\underline{r}}$ and $\mathcal{B}_m(\lambda, n, t_{\underline{r}})$ are introduced as in (3.13). Therefore (3.14)-(3.16) also hold in the present context. Similarly, we write

$$E_{m-1}(\lambda, n, t_{\underline{r}}) = \lambda^{-q_2} \alpha_{0,+}^2 \beta_{0,+} \prod_{j=1}^{m-1} (\lambda - \mu_j(n, t_{\underline{r}})), \tag{5.5a}$$

$$F_{m-1}(\lambda, n, t_{\underline{r}}) = -\lambda^{-q_2+1} \alpha_{0,+}^2 \beta_{0,+} w \prod_{j=1}^{m-1} (\lambda - \mu_j^+(n, t_{\underline{r}})). \tag{5.5b}$$

Defining $\{\hat{\mu}_j(n, t_{\underline{r}})\}_{j=1,\ldots,m-1} \subset \mathcal{K}_{m-1}$ and $\{\hat{\mu}_j^+(n, t_{\underline{r}})\}_{j=1,\ldots,m-1} \subset \mathcal{K}_{m-1}$ by

$$\hat{\mu}_j(n, t_{\underline{r}}) = \Big(\mu_j(n, t_{\underline{r}}), y(\hat{\mu}_j(n, t_{\underline{r}}))\Big) = \Big(\mu_j(n, t_{\underline{r}}), -\frac{A_m(\mu_j(n, t_{\underline{r}}), n, t_{\underline{r}})}{V_{13}^{(q)}(\mu_j(n, t_{\underline{r}}), n, t_{\underline{r}})}\Big), \tag{5.6a}$$

$$\hat{\mu}_j^+(n, t_{\underline{r}}) = \Big(\mu_j^+(n, t_{\underline{r}}), y(\hat{\mu}_j^+(n, t_{\underline{r}}))\Big) = \Big(\mu_j^+(n, t_{\underline{r}}), -\frac{C_m(\mu_j^+(n, t_{\underline{r}}), n, t_{\underline{r}})}{V_{23}^{(q)}(\mu_j^+(n, t_{\underline{r}}), n, t_{\underline{r}})}\Big). \tag{5.6b}$$

Observing (5.4), we obtain the divisor of $\phi(P, n, t_{\underline{r}})$ as follows

$$(\phi(P, n, t_{\underline{r}})) = \mathscr{D}_{P_{0_1}, \hat{\mu}_1^+(n,t_{\underline{r}}), \ldots, \hat{\mu}_{m-1}^+(n,t_{\underline{r}})}(P) - \mathscr{D}_{P_{\infty_2}, \hat{\mu}_1(n,t_{\underline{r}}), \ldots, \hat{\mu}_{m-1}(n,t_{\underline{r}})}(P), \tag{5.7}$$

that is, $P_{0_1}, \hat{\mu}_1^+(n,t_{\underline{r}}), \ldots, \hat{\mu}_{m-1}^+(n,t_{\underline{r}})$ are the m simple zeros of $\phi(P,n,t_{\underline{r}})$ and P_{∞_2}, $\hat{\mu}_1(n,t_{\underline{r}}), \ldots, \hat{\mu}_{m-1}(n,t_{\underline{r}})$ are its m simple poles.

Similarly, $\phi(P,n,t_{\underline{r}})$ satisfies the Riccati-type equation

$$\begin{aligned} \phi^+(P,n,t_{\underline{r}})\phi(P,n,t_{\underline{r}})\phi^-(P,n,t_{\underline{r}}) - u(n,t_{\underline{r}})\phi(P,n,t_{\underline{r}})\phi^-(P,n,t_{\underline{r}}) \\ -v(n,t_{\underline{r}})\phi^-(P,n,t_{\underline{r}}) = \lambda w^-(n,t_{\underline{r}}) + \lambda\phi^-(P,n,t_{\underline{r}}). \end{aligned} \tag{5.8}$$

Differentiating (5.3) with respect to $t_{\underline{r}}$ and using (5.1), we obtain

$$\begin{aligned} \phi_{t_{\underline{r}}} &= \left(\frac{\psi_1^+}{\psi_1}\right)_{t_{\underline{r}}} = \frac{\psi_1^+}{\psi_1}\left(\frac{\psi_{1,t_{\underline{r}}}^+}{\psi_1^+} - \frac{\psi_{1,t_{\underline{r}}}}{\psi_1}\right) = \phi\Delta\left(\frac{\psi_{1,t_{\underline{r}}}}{\psi_1}\right) \\ &= \phi\Delta\left(\widetilde{V}_{11}^{(\underline{r})} + \widetilde{V}_{12}^{(\underline{r})}\phi + \widetilde{V}_{13}^{(\underline{r})}\frac{\lambda w^-}{\phi^-}\right). \end{aligned} \tag{5.9}$$

Hence,

$$\frac{\phi(P,n,t_{\underline{r}})_{t_{\underline{r}}}}{\phi(P,n,t_{\underline{r}})} = \Delta\left(\widetilde{V}_{11}^{(\underline{r})}(\lambda,n,t_{\underline{r}}) + \widetilde{V}_{12}^{(\underline{r})}(\lambda,n,t_{\underline{r}})\phi(P,n,t_{\underline{r}}) + \widetilde{V}_{13}^{(\underline{r})}(\lambda,n,t_{\underline{r}})\frac{\lambda w^-}{\phi^-(P,n,t_{\underline{r}})}\right). \tag{5.10}$$

We can describe the dynamics of the zeros $\{\mu_j(n,t_{\underline{r}})\}_{j=1,\ldots,m-1}$ of $\lambda^{q_2}E_{m-1}(\lambda,n,t_{\underline{r}})$ by the following Dubrovin-type equations.

***Lemma* 5.1.** Suppose that the zeros $\{\mu_j(x,t_{\underline{r}})\}_{j=1,\ldots,m-1}$ of $\lambda^{q_2}E_{m-1}(\lambda,x,t_{\underline{r}})$ remain distinct for $(n,t_{\underline{r}}) \in \mathbb{Z}\times\mathbb{R}$, then $\{\mu_j(x,t_{\underline{r}})\}_{j=1,\ldots,m-1}$ satisfy the system of differential equations

$$\begin{aligned} &\mu_{j,t_{\underline{r}}}(n,t_{\underline{r}}) \\ &= \left[\widetilde{V}_{13}^{(\underline{r})}(\mu_j(n,t_{\underline{r}}),n,t_{\underline{r}})V_{12}^{(\underline{q})}(\mu_j(n,t_{\underline{r}}),n,t_{\underline{r}}) - \widetilde{V}_{12}^{(\underline{r})}(\mu_j(n,t_{\underline{r}}),n,t_{\underline{r}})V_{13}^{(\underline{q})}(\mu_j(n,t_{\underline{r}}),n,t_{\underline{r}})\right] \\ &\times \frac{[3y^2(\hat{\mu}_j(n,t_{\underline{r}})) - 2y(\hat{\mu}_j(n,t_{\underline{r}}))R_m(\mu_j(n,t_{\underline{r}})) + S_m(\mu_j(n,t_{\underline{r}}))]}{\mu_j^{-q_2}(n,t_{\underline{r}})\alpha_{0,+}^2\beta_{0,+}\prod_{\substack{k=1\\k\neq j}}^{m-1}(\mu_j(n,t_{\underline{r}}) - \mu_k(n,t_{\underline{r}}))}, \quad 1\le j\le m-1. \end{aligned} \tag{5.11}$$

Proof. Using (3.11)-(3.13) and (5.2c), we obtain

$$\begin{aligned} E_{m-1,t_{\underline{r}}}(\lambda,n,t_{\underline{r}}) =& \lambda^{2q_2}\left(\lambda(V_{13}^{(\underline{q})})^2V_{32}^{(\underline{q})} + V_{12}^{(\underline{q})}V_{13}^{(\underline{q})}(V_{22}^{(\underline{q})} - V_{33}^{(\underline{q})}) - (V_{12}^{(\underline{q})})^2V_{23}^{(\underline{q})}\right)_{t_{\underline{r}}} \\ =& \left(3\widetilde{V}_{11}^{(\underline{r})} - \widetilde{R}_m\right)E_{m-1} + \left(\widetilde{V}_{12}^{(\underline{r})}V_{13}^{(\underline{q})} - \widetilde{V}_{13}^{(\underline{r})}V_{12}^{(\underline{q})}\right)S_m \\ &+2\left(\widetilde{V}_{12}^{(\underline{r})}A_m - \widetilde{V}_{13}^{(\underline{r})}A_m\right)R_m + 3\widetilde{V}_{12}^{(\underline{r})}\left(V_{23}^{(\underline{q})}A_m - V_{22}^{(\underline{q})}A_m\right) \\ &- 3\widetilde{V}_{13}^{(\underline{r})}\left(\lambda V_{32}^{(\underline{q})}A_m - V_{33}^{(\underline{q})}A_m\right), \end{aligned} \tag{5.12}$$

where $\widetilde{R}_m = \widetilde{V}_{11}^{(\underline{r})} + \widetilde{V}_{22}^{(\underline{r})} + \widetilde{V}_{33}^{(\underline{r})}$. By virtue of (5.5a), (5.6a) and (3.16), we have

$$\left.\frac{A_m}{V_{13}^{(q)}}\right|_{\lambda=\mu_j(n,t_{\underline{r}})} = \left.\frac{\mathcal{A}_m}{V_{12}^{(q)}}\right|_{\lambda=\mu_j(n,t_{\underline{r}})} = -y(\hat{\mu}_j(n,t_{\underline{r}})). \tag{5.13}$$

Then

$$\begin{aligned}
&(\widetilde{V}_{12}^{(\underline{r})}A_m - \widetilde{V}_{13}^{(\underline{r})}\mathcal{A}_m)\big|_{\lambda=\mu_j(n,t_{\underline{r}})} = -y(\hat{\mu}_j(n,t_{\underline{r}}))(\widetilde{V}_{12}^{(\underline{r})}V_{13}^{(q)} - \widetilde{V}_{13}^{(\underline{r})}V_{12}^{(q)})\big|_{\lambda=\mu_j(n,t_{\underline{r}})},\\
&\widetilde{V}_{12}^{(\underline{r})}(V_{23}^{(q)}\mathcal{A}_m - V_{22}^{(q)}A_m)\big|_{\lambda=\mu_j(n,t_{\underline{r}})} = y^2(\hat{\mu}_j(n,t_{\underline{r}}))\widetilde{V}_{12}^{(\underline{r})}V_{13}^{(q)}\big|_{\lambda=\mu_j(n,t_{\underline{r}})},\\
&\widetilde{V}_{13}^{(\underline{r})}(\lambda V_{32}^{(q)}A_m - V_{33}^{(q)}\mathcal{A}_m)\big|_{\lambda=\mu_j(n,t_{\underline{r}})} = y^2(\hat{\mu}_j(n,t_{\underline{r}}))\widetilde{V}_{13}^{(\underline{r})}V_{12}^{(q)}\big|_{\lambda=\mu_j(n,t_{\underline{r}})}.
\end{aligned} \tag{5.14}$$

Hence,

$$\begin{aligned}
E_{m-1,t_{\underline{r}}}(\lambda,n,t_{\underline{r}})\big|_{\lambda=\mu_j(n,t_{\underline{r}})} &= (\widetilde{V}_{12}^{(\underline{r})}V_{13}^{(q)} - \widetilde{V}_{13}^{(\underline{r})}V_{12}^{(q)})\big|_{\lambda=\mu_j(n,t_{\underline{r}})}\\
&\times [3y^2(\hat{\mu}_j(n,t_{\underline{r}})) - 2y(\hat{\mu}_j(n,t_{\underline{r}}))R_m(\mu_j(n,t_{\underline{r}})) + S_m(\mu_j(n,t_{\underline{r}}))].
\end{aligned} \tag{5.15}$$

On the other hand, (5.5a) implies that

$$E_{m-1,t_{\underline{r}}}(\lambda,n,t_{\underline{r}})\big|_{\lambda=\mu_j(n,t_{\underline{r}})} = -\mu_j^{-q_2}(n,t_{\underline{r}})\mu_{j,t_{\underline{r}}}(n,t_{\underline{r}})\alpha_{0,+}^2\beta_{0,+}\prod_{\substack{k=1\\k\neq j}}^{m-1}(\mu_j(n,t_{\underline{r}})-\mu_k(n,t_{\underline{r}})), \tag{5.16}$$

which, together with (5.15), leads to (5.11). ■

From the first two expressions in (5.1), we obtain

$$\begin{aligned}
&\psi_1(P,n,n_0,t_{\underline{r}},t_{0,\underline{r}})\\
&= \exp\left(\int_{t_{0,\underline{r}}}^{t_{\underline{r}}}\left[\widetilde{V}_{11}^{(\underline{r})}(\lambda,n_0,t') + \widetilde{V}_{12}^{(\underline{r})}(\lambda,n_0,t')\phi(P,n_0,t') + \widetilde{V}_{13}^{(\underline{r})}(\lambda,n_0,t')\frac{\lambda w^-}{\phi^-(P,n_0,t')}\right]dt'\right)\\
&\times\begin{cases}\prod_{n'=n_0}^{n-1}\phi(P,n',t_{\underline{r}}), & n\geq n_0+1,\\ 1, & n=n_0,\\ \prod_{n'=n}^{n_0-1}\phi(P,n',t_{\underline{r}})^{-1}, & n\leq n_0-1.\end{cases}
\end{aligned} \tag{5.17}$$

We note that

$$\psi_1(P,n,n_0,t_{\underline{r}},t_{0,\underline{r}}) = \psi_1(P,n,n_0,t_{\underline{r}},t_{\underline{r}})\psi_1(P,n_0,n_0,t_{\underline{r}},t_{0,\underline{r}}), \tag{5.18}$$

where $P=(\lambda,y)\in\mathcal{K}_{m-1}\backslash\{P_{\infty_1},P_{\infty_2}\}$, $(n,t_{\underline{r}}),(n_0,t_{0,\underline{r}})\in\mathbb{Z}\times\mathbb{R}$.

Analyzing the integrand in (5.17), we introduce a function $I_{\underline{r}}(P,n,t_{\underline{r}})$ by

$$\begin{aligned} I_{\underline{r}}(P,n,t_{\underline{r}}) &= \widetilde{V}_{11}^{(\underline{r})}(\lambda,n,t_{\underline{r}}) + \widetilde{V}_{12}^{(\underline{r})}(\lambda,n,t_{\underline{r}})\phi(P,n,t_{\underline{r}}) + \widetilde{V}_{13}^{(\underline{r})}(\lambda,n,t_{\underline{r}})\frac{\lambda w^-(n,t_{\underline{r}})}{\phi^-(P,n,t_{\underline{r}})} \\ &= I_{r_1,+}(P,n,t_{\underline{r}}) + I_{r_2,-}(P,n,t_{\underline{r}}), \end{aligned} \tag{5.19}$$

where

$$I_{r_1,+} = \sum_{l=0}^{r_1} \tilde{\alpha}_{r_1-l,+}\hat{I}_{l,+}, \tag{5.20a}$$

$$\hat{I}_{r_1,+} = \hat{c}_+^{(r_1)} + \hat{a}_+^{(r_1)}\phi + \hat{b}_+^{(r_1)}\frac{\lambda w^-}{\phi^-}, \tag{5.20b}$$

$$I_{r_2,-} = \sum_{l=0}^{r_2} \tilde{\alpha}_{r_2-l,-}\hat{I}_{l,-}, \tag{5.20c}$$

$$\hat{I}_{r_2,-} = \hat{c}_-^{(r_2)} + \hat{a}_-^{(r_2)}\phi + \hat{b}_-^{(r_2)}\frac{\lambda w^-}{\phi^-}, \tag{5.20d}$$

$$\hat{I}_{0,-} = 0, \tag{5.20e}$$

and

$$\hat{g}_+^{(r_1)} = (\hat{a}_+^{(r_1)}, \hat{b}_+^{(r_1)}, \hat{c}_+^{(r_1)})^{\mathrm{T}} = \sum_{l=0}^{r_1} \hat{g}_{l,+}\lambda^{r_1-l},$$

$$\hat{g}_-^{(r_2)} = (\hat{a}_-^{(r_2)}, \hat{b}_-^{(r_2)}, \hat{c}_-^{(r_2)})^{\mathrm{T}} = \sum_{l=0}^{r_2-1} \hat{g}_{l,-}\lambda^{-r_2+l}.$$

***Lemma* 5.2.** (i) Assuming $(n,t_{\underline{r}}) \in \mathbb{Z}\times\mathbb{R}$, $\lambda=\zeta^{-1}$ denoting the local coordinate near P_{∞_1}, then we have

$$\left.\begin{aligned} \hat{I}_{r_1,+}(P,n,t_{\underline{r}}) &\underset{\zeta\to 0}{=} -\zeta^{-r_1-1} + \hat{b}_{r_1+1,+} + O(\zeta), \\ \hat{I}_{r_2,-}(P,n,t_{\underline{r}}) &\underset{\zeta\to 0}{=} -\hat{b}_{r_2-1,-} + O(\zeta), \end{aligned}\right. \quad P\to P_{\infty_1}. \tag{5.21}$$

(ii) Assuming $(n,t_{\underline{r}}) \in \mathbb{Z}\times\mathbb{R}$, $\lambda=\xi^{-2}$ denoting the local coordinate near P_{∞_2}, then we have

$$\left.\begin{aligned} \hat{I}_{r_1,+}(P,n,t_{\underline{r}}) &\underset{\xi\to 0}{=} -(\hat{a}_{r_1+1,+} + w^-\hat{b}_{r_1+1,+})\xi + O(\xi^2), \\ \hat{I}_{r_2,-}(P,n,t_{\underline{r}}) &\underset{\xi\to 0}{=} (\hat{a}_{r_2-1,-} + w^-\hat{b}_{r_2-1,-})\xi + O(\xi^2), \end{aligned}\right. \quad P\to P_{\infty_2}. \tag{5.22}$$

(iii) Assuming $(n,t_{\underline{r}}) \in \mathbb{Z}\times\mathbb{R}$, $\lambda=\varrho$ denoting the local coordinate near P_{0_j}, $j=1,2,3$, then we have

$$\hat{I}_{r_1,+}(P,n,t_{\underline{r}}) \underset{\varrho\to 0}{=} \begin{cases} \hat{c}_{r_1,+} - v\hat{b}_{r_1,+} + O(\varrho), & P\to P_{0_1}, \\ \hat{c}_{r_1,+} + \hat{a}_{r_1,+}\chi_{s,0} + O(\varrho), & P\to P_{0_s}, \quad s=2,3. \end{cases} \tag{5.23}$$

$$\hat{I}_{r_2,-}(P,n,t_{\underline{r}}) \underset{\varrho\to 0}{=} \begin{cases} -\hat{b}_{r_2-1,-} + u^+\hat{a}^{++}_{r_2,-} + \hat{c}^{++}_{r_2,-} + \hat{c}^{+}_{r_2,-} + O(\varrho), & P \to P_{0_1}, \\ \varrho^{-r_2} - (\hat{c}_{r_2,-} + \hat{a}_{r_2,-}\chi_{s,0}) + O(\varrho), & P \to P_{0_s}, \quad s = 2,3. \end{cases} \tag{5.24}$$

Proof. We use the inductive method to prove (5.21).

$$\begin{aligned} \hat{I}_{0,+}(P,n,t_{\underline{r}}) &= \hat{c}_{0,+} + \hat{a}_{0,+}\phi + \hat{b}_{0,+}\frac{\lambda w^-}{\phi^-} = -\zeta^{-1} + \hat{b}_{1,+} + O(\zeta), \\ \hat{I}_{1,-}(P,n,t_{\underline{r}}) &= \hat{c}_{0,-}\lambda^{-1} + \hat{a}_{0,-}\lambda^{-1}\phi + \hat{b}_{0,-}\lambda^{-1}\frac{\lambda w^-}{\phi^-} = -\hat{b}_{0,-} + O(\zeta), \end{aligned} \quad P \to P_{\infty_1}. \tag{5.25}$$

Therefore, (5.21) holds for $r_1 = 0, r_2 = 1$, respectively. We suppose that (5.21) has the expansions

$$\begin{aligned} \hat{I}_{r_1,+}(P,n,t_{\underline{r}}) &= -\zeta^{r_1-1} + \sum_{j=0}^{\infty} \sigma_j(n,t_{\underline{r}})\zeta^j, \\ \hat{I}_{r_2,-}(P,n,t_{\underline{r}}) &= \sum_{j=0}^{\infty} \delta_j(n,t_{\underline{r}})\zeta^j, \end{aligned} \quad P \to P_{\infty_1}. \tag{5.26}$$

for some coefficients $\{\sigma_j(n,t_{\underline{r}})\}_{j\in\mathbb{N}}$ and $\{\delta_j(n,t_{\underline{r}})\}_{j\in\mathbb{N}}$ to be determined. From (5.10), (5.19), (5.20) and (4.2), we obtain

$$\phi(P,n,t_{\underline{r}})_{t_{\underline{r}}} = \phi(P,n,t_{\underline{r}})\Delta\hat{I}_{\underline{r}}(P,n,t_{\underline{r}}), \tag{5.27a}$$

$$\phi_{j,t_{\underline{r}}} = \phi_0\Delta\sigma_j + \phi_1\Delta\sigma_{j-1} + \cdots + \phi_j\Delta\sigma_0, \quad j \geq 0, \quad r_2 = 1, \tag{5.27b}$$

$$\phi_{j,t_{\underline{r}}} = \phi_0\Delta\delta_j + \phi_1\Delta\delta_{j-1} + \cdots + \phi_j\Delta\delta_0, \quad j \geq 0, \quad r_1 = 0. \tag{5.27c}$$

Explicitly, it follows from (2.20) that

$$\begin{aligned} \Delta\sigma_0 &= \frac{w_{t_{\underline{r}}}}{w} = \Delta\hat{b}_{r_1+1,+}, \\ \Delta\sigma_1 &= \frac{\phi_{1,t_{\underline{r}}}}{\phi_0} - \frac{\phi_1}{\phi_0}\Delta\sigma_0 \\ &= \Delta\left(w\hat{a}_{r_1+1,+} + u^+\hat{a}^{++}_{r_1+1,+} - (w^+ + u)w\hat{b}_{r_1+1,+} + (E^2+E)\hat{c}_{r_1+1,+}\right), \\ \Delta\delta_0 &= \frac{w_{t_{\underline{r}}}}{w} = -\Delta\hat{b}_{r_2-1,-}, \\ \Delta\delta_1 &= \frac{\phi_{1,t_{\underline{r}}}}{\phi_0} - \frac{\phi_1}{\phi_0}\Delta\delta_0 \\ &= \Delta\left(-w\hat{a}_{r_2-1,-} - u^+\hat{a}^{++}_{r_2-1,-} + (w^+ + u)w\hat{b}_{r_2-1,-} - (E^2+E)\hat{c}_{r_2-1,-}\right). \end{aligned} \tag{5.28}$$

Hence, we have

$$\begin{aligned} \sigma_0 &= \hat{b}_{r_1+1,+}, \\ \sigma_1 &= w\hat{a}_{r_1+1,+} + u^+\hat{a}^{++}_{r_1+1,+} - (w^+ + u)w\hat{b}_{r_1+1,+} + (E^2+E)\hat{c}_{r_1+1,+}, \\ \delta_0 &= -\hat{b}_{r_2-1,-}, \\ \delta_1 &= -w\hat{a}_{r_2-1,-} - u^+\hat{a}^{++}_{r_2-1,-} + (w^+ + u)w\hat{b}_{r_2-1,-} - (E^2+E)\hat{c}_{r_2-1,-}, \end{aligned} \tag{5.29}$$

where the summation constants are set equal to zero if we take into account that there are no arbitrary constants in the expansions of $\phi(P,n,t_{\underline{r}})$ near P_{∞_1} nor in the homogeneous coefficients $\hat{a}_{r_1+1,+}, \hat{b}_{r_1+1,+}, \hat{c}_{r_1+1,+}, \hat{a}_{r_2-1,-}, \hat{b}_{r_2-1,-}, \hat{c}_{r_2-1,-}$ with the condition that $\Delta\Delta^{-1} = \Delta^{-1}\Delta = 1$. Therefore,

$$\begin{aligned}
\hat{I}_{r_1+1,+}(P,n,t_{\underline{r}}) &= \hat{c}_+^{(r_1+1)} + \hat{a}_+^{(r_1+1)}\phi + \hat{b}_+^{(r_1+1)}\frac{\lambda w^-}{\phi^-} \\
&= \zeta^{-1}\hat{I}_{r_1,+} + \hat{c}_{r_1+1,+} + \hat{a}_{r_1+1,+}\phi + \hat{b}_{r_1+1,+}\frac{\lambda w^-}{\phi^-} \\
&= -\zeta^{-r_1-2} + \hat{b}_{r_1+2,+} + O(\zeta), \qquad P \to P_{\infty_1}, \\
\hat{I}_{r_2+1,-}(P,n,t_{\underline{r}}) &= \hat{c}_-^{(r_2+1)} + \hat{a}_-^{(r_2+1)}\phi + \hat{b}_-^{(r_2+1)}\frac{\lambda w^-}{\phi^-} \\
&= \zeta\hat{I}_{r_2,-} + \zeta\hat{c}_{r_2,-} + \zeta\hat{a}_{r_2,-}\phi + \zeta\hat{b}_{r_2,-}\frac{w^-}{\zeta\phi^-} \\
&= -\hat{b}_{r_2,-} + O(\zeta), \qquad P \to P_{\infty_1}.
\end{aligned} \tag{5.30}$$

The proof of (5.21) is complete. Relations (5.22)-(5.24) can be proved similarly. ∎

From Lemma 5.2, (5.19) and (5.20), we arrive at

$$I_{\underline{r}}(P,n,t_{\underline{r}}) \underset{\zeta\to 0}{=} -\sum_{l=0}^{r_1} \tilde{\alpha}_{r_1-l,+}\zeta^{-l-1} - \tilde{\alpha}_{r_1+1,+} + \tilde{b}_{r_1+1,+} - \tilde{b}_{r_2-1,-} + O(\zeta), \quad P \to P_{\infty_1}. \tag{5.31}$$

$$I_{\underline{r}}(P,n,t_{\underline{r}}) \underset{\xi\to 0}{=} O(\xi), \quad P \to P_{\infty_2}. \tag{5.32}$$

$$I_{\underline{r}}(P,n,t_{\underline{r}}) \underset{\varrho\to 0}{=} \begin{cases} O(1), & P \to P_{0_1}, \\ \sum_{l=0}^{r_2} \tilde{\alpha}_{r_2-l,-}\varrho^{-l} + O(1), & P \to P_{0_s}, \quad s = 2,3. \end{cases} \tag{5.33}$$

Let $\omega^{(2)}_{P_{\infty_1},j}, \omega^{(2)}_{P_{0_s},j}, s = 2,3, j \geq 2$ denote the normalized Abelian differential of the second kind holomorphic on $\mathcal{K}_{m-1} \setminus \{P_{\infty_1}, P_{0_s}\}$ with a pole of order j at P_{∞_1} and $P_{0_s}, s = 2,3$,

$$\omega^{(2)}_{P_{\infty_1},j} \underset{\zeta\to 0}{=} (\zeta^{-j} + O(1))d\zeta, \quad \text{as} \quad P \to P_{\infty_1}, \quad \zeta = \lambda^{-1}, \tag{5.34a}$$

$$\omega^{(2)}_{P_{0_s},j} \underset{\zeta\to 0}{=} (\varrho^{-j} + O(1))d\zeta, \quad \text{as} \quad P \to P_{0_s}, \quad \varrho = \lambda, \quad s = 2,3, \tag{5.34b}$$

with vanishing a-periods: $\int_{\mathfrak{a}_k} \omega^{(2)}_{P_{\infty_1},j} = 0, \int_{\mathfrak{a}_k} \omega^{(2)}_{P_{0_s},j} = 0, s = 2,3, k = 1,\ldots,m-1.$

Moreover, we define

$$\widetilde{\Omega}^{(2)}_{\underline{r}} = \sum_{l=0}^{r_1}(l+1)\tilde{\alpha}_{r_1-l,+}\omega^{(2)}_{P_{\infty_1},l+2} - \sum_{l=1}^{r_2} l\tilde{\alpha}_{r_2-l,-}(\omega^{(2)}_{P_{0_2},l+1} + \omega^{(2)}_{P_{0_3},l+1}). \tag{5.35}$$

Then

$$\begin{aligned}
&\int_{Q_0}^{P} \tilde{\Omega}_{\underline{r}}^{(2)} \underset{\zeta\to 0}{=} -\sum_{l=0}^{r_1} \tilde{\alpha}_{r_1-l,+}\zeta^{-l-1} + \tilde{e}_1^{(2)}(Q_0) + O(\zeta), \quad as \quad P \to P_{\infty_1},\\
&\int_{Q_0}^{P} \tilde{\Omega}_{\underline{r}}^{(2)} \underset{\xi\to 0}{=} \tilde{e}_2^{(2)}(Q_0) + O(\xi), \quad as \quad P \to P_{\infty_2},\\
&\int_{Q_0}^{P} \tilde{\Omega}_{\underline{r}}^{(2)} \underset{\varrho\to 0}{=} \begin{cases} \tilde{e}_{0_1}^{(2)}(Q_0) + O(\varrho), & as \quad P \to P_{0_1},\\ \sum\limits_{l=1}^{r_2} \tilde{\alpha}_{r_2-l,-}\varrho^{-l} + \tilde{e}_{0_2}^{(2)}(Q_0) + O(\varrho), & as \quad P \to P_{0_2},\\ \sum\limits_{l=1}^{r_2} \tilde{\alpha}_{r_2-l,-}\varrho^{-l} + \tilde{e}_{0_3}^{(2)}(Q_0) + O(\varrho), & as \quad P \to P_{0_3}, \end{cases}
\end{aligned} \tag{5.36}$$

where $\tilde{e}_1^{(2)}(Q_0), \tilde{e}_2^{(2)}(Q_0), \tilde{e}_{0_s}^{(2)}(Q_0), s = 1, 2, 3$ are integration constants.

Using these preparatory results, we write the representations of $\phi(P, n, t_{\underline{r}})$ and $\psi_1(P, n, n_0, t_{\underline{r}}, t_{0,\underline{r}})$ in terms of theta functions in the following forms.

***Theorem* 5.3.** We assume that the curve $\mathcal{K}_{m-1}$ is nonsingular. Let $P = (\lambda, y) \in \mathcal{K}_{m-1} \setminus \{P_{\infty_1}, P_{\infty_2}\}$, $(n, n_0, t_{\underline{r}}, t_{0,\underline{r}}) \in \mathbb{Z}^2 \times \mathbb{R}^2$. If $\mathcal{D}_{\underline{\hat{\mu}}(n,t_{\underline{r}})}$ is nonspecial for $(n, t_{\underline{r}}) \in \mathbb{Z} \times \mathbb{R}$, then

$$\phi(P, n, t_{\underline{r}}) = \frac{\theta(\underline{z}(P_{\infty_2}, \underline{\hat{\mu}}(n, t_{\underline{r}})))}{\theta(\underline{z}(P_{\infty_2}, \underline{\hat{\mu}}^+(n, t_{\underline{r}})))} \frac{\theta(\underline{z}(P, \underline{\hat{\mu}}^+(n, t_{\underline{r}})))}{\theta(\underline{z}(P, \underline{\hat{\mu}}(n, t_{\underline{r}})))} \exp\left(e_2(Q_0) - \int_{Q_0}^{P} \omega_{P_{\infty_2}, P_{0_1}}^{(3)} \right), \tag{5.37}$$

and

$$\begin{aligned}
\psi_1(P, n, n_0, t_{\underline{r}}, t_{0,\underline{r}}) = {} & \frac{\theta(\underline{z}(P_{\infty_2}, \underline{\hat{\mu}}(n_0, t_{0,\underline{r}})))}{\theta(\underline{z}(P_{\infty_2}, \underline{\hat{\mu}}(n, t_{\underline{r}})))} \frac{\theta(\underline{z}(P, \underline{\hat{\mu}}(n, t_{\underline{r}})))}{\theta(\underline{z}(P, \underline{\hat{\mu}}(n_0, t_{0,\underline{r}})))}\\
& \times \exp\left((n - n_0)\left(e_2(Q_0) - \int_{Q_0}^{P} \omega_{P_{\infty_2}, P_{0_1}}^{(3)} \right) + (t_{\underline{r}} - t_{0,\underline{r}}) \left(\int_{Q_0}^{P} \tilde{\Omega}_{\underline{r}}^{(2)} - \tilde{e}_2^{(2)}(Q_0) \right) \right).
\end{aligned} \tag{5.38}$$

Proof. As in Theorem 4.3, we conclude that $\phi(P, n, t_{\underline{r}})$ has the form (5.37). For $t_{0,\underline{r}} = t_{\underline{r}}$, $\psi_1(P, n, n_0, t_{\underline{r}}, t_{0,\underline{r}})$ turns into

$$\begin{aligned}
\psi_1(P, n, n_0, t_{\underline{r}}, t_{\underline{r}}) = {} & \frac{\theta(\underline{z}(P_{\infty_2}, \underline{\hat{\mu}}(n_0, t_{\underline{r}})))}{\theta(\underline{z}(P_{\infty_2}, \underline{\hat{\mu}}(n, t_{\underline{r}})))} \frac{\theta(\underline{z}(P, \underline{\hat{\mu}}(n, t_{\underline{r}})))}{\theta(\underline{z}(P, \underline{\hat{\mu}}(n_0, t_{\underline{r}})))}\\
& \times \exp\left((n - n_0)\left(e_2(Q_0) - \int_{Q_0}^{P} \omega_{P_{\infty_2}, P_{0_1}}^{(3)} \right) \right).
\end{aligned} \tag{5.39}$$

It remains to use (5.17) to investigate

$$\psi_1(P,n_0,n_0,t_{\underline{r}},t_{0,\underline{r}}) = \exp\left(\int_{t_{0,\underline{r}}}^{t_{\underline{r}}} I_{\underline{r}}(P,n_0,t')dt'\right). \tag{5.40}$$

Letting $\Psi_1(P,n_0,n_0,t_{\underline{r}},t_{0,\underline{r}})$ denote the right-hand side of (5.38) for $n=n_0$, that is,

$$\begin{aligned}\Psi_1(P,n_0,n_0,t_{\underline{r}},t_{0,\underline{r}}) =& \frac{\theta(\underline{z}(P_{\infty_2},\hat{\underline{\mu}}(n_0,t_{0,\underline{r}})))}{\theta(\underline{z}(P_{\infty_2},\hat{\underline{\mu}}(n_0,t_{\underline{r}})))}\frac{\theta(\underline{z}(P,\hat{\underline{\mu}}(n_0,t_{\underline{r}})))}{\theta(\underline{z}(P,\hat{\underline{\mu}}(n_0,t_{0,\underline{r}})))}\\ &\times\exp\left((t_{\underline{r}}-t_{0,\underline{r}})\left(\int_{Q_0}^{P}\widetilde{\Omega}_{\underline{r}}^{(2)}-\tilde{e}_2^{(2)}(Q_0)\right)\right).\end{aligned} \tag{5.41}$$

Next we will prove that

$$\psi_1(P,n_0,n_0,t_{\underline{r}},t_{0,\underline{r}}) = \Psi_1(P,n_0,n_0,t_{\underline{r}},t_{0,\underline{r}}). \tag{5.42}$$

Using (3.16), (5.4), (5.19) and Lemma 5.1, we obtain

$$\begin{aligned}I_{\underline{r}}(P,n,t_{\underline{r}}) &= \widetilde{V}_{11}^{(\underline{r})}(\lambda,n,t_{\underline{r}}) + \widetilde{V}_{12}^{(\underline{r})}(\lambda,n,t_{\underline{r}})\phi(P,n,t_{\underline{r}}) + \widetilde{V}_{13}^{(\underline{r})}(\lambda,n,t_{\underline{r}})\frac{\lambda w^-(n,t_{\underline{r}})}{\phi^-(P,n,t_{\underline{r}})}\\ &= \widetilde{V}_{11}^{(\underline{r})} + \widetilde{V}_{12}^{(\underline{r})}\frac{y^2V_{13}^{(\underline{q})} - y(A_m + V_{13}^{(\underline{q})}R_m) + B_m}{E_{m-1}}\\ &\quad + \widetilde{V}_{13}^{(\underline{r})}\frac{\lambda w^-[y^2(V_{23}^{(\underline{q})})^- - y(C_m^- + (V_{23}^{(\underline{q})})^- R_m) + D_m^-]}{F_{m-1}^-}\\ &= \frac{1}{E_{m-1}}\Big[\frac{1}{3}E_{m-1,t_{\underline{r}}} + \frac{1}{3}\widetilde{R}_m E_{m-1} + (\widetilde{V}_{12}^{(\underline{r})}V_{13}^{(\underline{q})} - \widetilde{V}_{13}^{(\underline{r})}V_{12}^{(\underline{q})})\\ &\quad\times(y^2 - yR_m + \tfrac{2}{3}S_m) - (\widetilde{V}_{12}^{(\underline{r})}A_m - \widetilde{V}_{13}^{(\underline{r})}\mathcal{A}_m)(y - \tfrac{1}{3}R_m)\Big]\\ &\underset{\lambda\to\mu_j(n,t_{\underline{r}})}{=} -\frac{\mu_{j,t_{\underline{r}}}}{\lambda - \mu_j(n,t_{\underline{r}})} + O(1)\\ &\underset{\lambda\to\mu_j(n,t_{\underline{r}})}{=} \partial_{t_{\underline{r}}}\ln(\lambda - \mu_j(n,t_{\underline{r}})) + O(1),\end{aligned} \tag{5.43}$$

then

$$\begin{aligned}\psi_1(P,n_0,n_0,t_{\underline{r}},t_{0,\underline{r}}) &= \exp\left(\int_{t_{0,\underline{r}}}^{t_{\underline{r}}} \partial_{t'}\ln(\lambda - \mu_j(n_0,t'))dt'\right)\\ &= \frac{\lambda - \mu_j(n_0,t_{\underline{r}})}{\lambda - \mu_j(n_0,t_{0,\underline{r}})}\cdot O(1)\\ &= \begin{cases}(\lambda - \mu_j(n_0,t_{\underline{r}}))O(1), & P \text{ near } \hat{\mu}_j(n_0,t_{\underline{r}}) \neq \hat{\mu}_j(n_0,t_{0,\underline{r}}),\\ O(1), & P \text{ near } \hat{\mu}_j(n_0,t_{\underline{r}}) = \hat{\mu}_j(n_0,t_{0,\underline{r}}),\\ (\lambda - \mu_j(n_0,t_{0,\underline{r}}))^{-1}O(1), & P \text{ near } \hat{\mu}_j(n_0,t_{0,\underline{r}}) \neq \hat{\mu}_j(n_0,t_{\underline{r}}),\end{cases}\end{aligned} \tag{5.44}$$

where $O(1) \neq 0$. Hence, all zeros and poles of $\psi_1(P, n_0, n_0, t_{\underline{r}}, t_{0,\underline{r}})$ and $\Psi_1(P, n_0, n_0, t_{\underline{r}}, t_{0,\underline{r}})$ on $\mathcal{K}_{m-1} \setminus \{P_{\infty_1}, P_{\infty_2}\}$ are simple and coincident. Moreover, using (5.31)-(5.33) and (5.36), it is easy to see that the essential singularities of $\psi_1(P, n_0, n_0, t_{\underline{r}}, t_{0,\underline{r}})$ and $\Psi_1(P, n_0, n_0, t_{\underline{r}}, t_{0,\underline{r}})$ at P_{∞_1} and P_{∞_2} are identical. Therefore, we can apply the uniqueness result for Baker-Akhiezer function to conclude that (5.42) holds, because $\mathcal{D}_{\underline{\hat{\mu}}(n,t_{\underline{r}})}$ is nonspecial for all $(n, t_{\underline{r}}) \in \mathbb{Z} \times \mathbb{R}$. Finally we obtain (5.38). ■

The $\mathbb{b}$-period of the differential $\widetilde{\Omega}_{\underline{r}}^{(2)}$ is denoted by

$$\underline{\widetilde{U}}_{\underline{r}}^{(2)} = \left(\widetilde{U}_{\underline{r},1}^{(2)}, \ldots, \widetilde{U}_{\underline{r},m-1}^{(2)}\right), \quad \widetilde{U}_{\underline{r},j}^{(2)} = \frac{1}{2\pi i} \int_{\mathbb{b}_j} \widetilde{\Omega}_{\underline{r}}^{(2)}, \quad j = 1, \ldots, m-1. \tag{5.45}$$

***Theorem* 5.4.** We have the relation

$$\underline{\rho}(n, t_{\underline{r}}) = \underline{\rho}(n_0, t_{0,\underline{r}}) + \underline{U}^{(3)}(n - n_0) - \underline{\widetilde{U}}_{\underline{r}}^{(2)}(t_{\underline{r}} - t_{0,\underline{r}}) \pmod{\mathcal{T}_{m-1}}. \tag{5.46}$$

Proof. We introduce a meromorphic differential

$$\Omega(n, n_0, t_{\underline{r}}, t_{0,\underline{r}}) = \frac{\partial}{\partial \lambda} \ln \left(\psi_1(P, n, n_0, t_{\underline{r}}, t_{0,\underline{r}})\right) d\lambda. \tag{5.47}$$

From representation (5.38), we obtain

$$\Omega(n, n_0, t_{\underline{r}}, t_{0,\underline{r}}) = (t_{\underline{r}} - t_{0,\underline{r}}) \widetilde{\Omega}_{\underline{r}}^{(2)} - (n - n_0) \omega_{P_{\infty_2}, P_{0_1}}^{(3)} + \sum_{j=1}^{m-1} \omega_{\mu_j(n,t_{\underline{r}}), \mu_j(n_0, t_{0,\underline{r}})}^{(3)} + \overline{\omega}, \tag{5.48}$$

where $\overline{\omega}$ denotes a holomorphic differential on $\mathcal{K}_{m-1}$, i.e., $\overline{\omega} = \sum_{j=1}^{m-1} \overline{e}_j \omega_j$ for some $\overline{e}_j \in \mathbb{C}, j = 1, \ldots, m-1$. Because $\psi_1(P, n, n_0, t_{\underline{r}}, t_{0,\underline{r}})$ is single-valued on $\mathcal{K}_{m-1}$, all $\mathbb{a}$- and $\mathbb{b}$- period of Ω are integer multiples of $2\pi i$, and hence

$$2\pi i M_k = \int_{\mathbb{a}_k} \Omega(n, n_0, t_{\underline{r}}, t_{0,\underline{r}}) = \int_{\mathbb{a}_k} \overline{\omega} = \overline{e}_k, \quad k = 1, \ldots, m-1, \tag{5.49}$$

for some $M_k \in \mathbb{Z}$.

Similarly, for some $N_k \in \mathbb{Z}$,

$$\begin{aligned} 2\pi i N_k &= \int_{\mathbb{b}_k} \Omega(n, n_0, t_{\underline{r}}, t_{0,\underline{r}}) \\ &= (t_{\underline{r}} - t_{0,\underline{r}}) \int_{\mathbb{b}_k} \widetilde{\Omega}_{\underline{r}}^{(2)} - (n - n_0) \int_{\mathbb{b}_k} \omega_{P_{\infty_2}, P_{0_1}}^{(3)} + \sum_{j=1}^{m-1} \int_{\mathbb{b}_k} \omega_{\hat{\mu}_j(n,t_{\underline{r}}), \hat{\mu}_j(n_0, t_{0,\underline{r}})}^{(3)} + \int_{\mathbb{b}_k} \overline{\omega} \\ &= 2\pi i (t_{\underline{r}} - t_{0,\underline{r}}) \widetilde{U}_{\underline{r},k}^{(2)} - 2\pi i (n - n_0) U_k^{(3)} + 2\pi i \sum_{j=1}^{m-1} \int_{\hat{\mu}_j(n_0, t_{0,\underline{r}})}^{\hat{\mu}_j(n,t_{\underline{r}})} \omega_k + 2\pi i \sum_{j=1}^{m-1} M_j \tau_{jk}. \end{aligned} \tag{5.50}$$

Therefore, we have

$$\underline{N} = (t_{\underline{r}} - t_{0,\underline{r}})\underline{\widetilde{U}}_{\underline{r}}^{(2)} - (n - n_0)\underline{U}^{(3)} + \sum_{j=1}^{m-1} \int_{Q_0}^{\hat{\mu}_j(n,t_{\underline{r}})} \underline{\omega} - \sum_{j=1}^{m-1} \int_{Q_0}^{\hat{\mu}_j(n_0,t_{0,\underline{r}})} \underline{\omega} + \underline{M}\tau, \tag{5.51}$$

where $\underline{N} = (N_1, \ldots, N_{m-1})$ and $\underline{M} = (M_1, \ldots, M_{m-1}) \in \mathbb{Z}^{m-1}$. So (5.51) is equivalent to (5.46). ■

***Theorem* 5.5.** We assume that $\mathcal{K}_{m-1}$ is nonsingular. Suppose that $\mathcal{D}_{\underline{\hat{\mu}}(n,t_{\underline{r}})}$ is nonspecial for $(n, t_{\underline{r}}) \in \mathbb{Z} \times \mathbb{R}$, then

$$\begin{aligned}
u(n,t_{\underline{r}}) =& \frac{\theta(\underline{z}(P_{\infty_2}, \underline{\hat{\mu}}^-(n,t_{\underline{r}})))}{\theta(\underline{z}(P_{\infty_2}, \underline{\hat{\mu}}(n,t_{\underline{r}})))} \frac{\theta(\underline{z}(P_{\infty_1}, \underline{\hat{\mu}}(n,t_{\underline{r}})))}{\theta(\underline{z}(P_{\infty_1}, \underline{\hat{\mu}}^-(n,t_{\underline{r}})))} \exp\left(e_2(Q_0) - e_1(Q_0)\right) \\
&- 2\omega_0^{\infty_2} - \sum_{j=0}^{m-1} d_{j,0}^{(\infty_2)} \frac{\partial}{\partial z_j} \ln \frac{\theta(\underline{z}(P_{0_1}, \underline{\hat{\mu}}^+(n,t_{\underline{r}})))}{\theta(\underline{z}(P_{0_1}, \underline{\hat{\mu}}^-(n,t_{\underline{r}})))}, \\
v(n,t_{\underline{r}}) =& \frac{\theta(\underline{z}(P_{\infty_1}, \underline{\hat{\mu}}(n,t_{\underline{r}})))}{\theta(\underline{z}(P_{\infty_1}, \underline{\hat{\mu}}^-(n,t_{\underline{r}})))} \frac{\theta(\underline{z}(P_{0_1}, \underline{\hat{\mu}}^-(n,t_{\underline{r}})))}{\theta(\underline{z}(P_{0_1}, \underline{\hat{\mu}}(n,t_{\underline{r}})))} \exp\left(e_{0_1}(Q_0) - e_1(Q_0)\right), \\
w(n,t_{\underline{r}}) =& - \frac{\theta(\underline{z}(P_{\infty_2}, \underline{\hat{\mu}}(n,t_{\underline{r}})))}{\theta(\underline{z}(P_{\infty_2}, \underline{\hat{\mu}}^+(n,t_{\underline{r}})))} \frac{\theta(\underline{z}(P_{\infty_1}, \underline{\hat{\mu}}^+(n,t_{\underline{r}})))}{\theta(\underline{z}(P_{\infty_1}, \underline{\hat{\mu}}(n,t_{\underline{r}})))} \exp\left(e_2(Q_0) - e_1(Q_0)\right).
\end{aligned} \tag{5.52}$$

From Theorem 5.4, we can rewrite (5.52) as

$$\begin{aligned}
u(n,t_{\underline{r}}) =& \frac{\theta(\underline{\widetilde{K}}_2 + \underline{U}^{(3)}n - \underline{U}^{(3)} - \underline{\widetilde{U}}_{\underline{r}}^{(2)} t_{\underline{r}})}{\theta(\underline{\widetilde{K}}_2 + \underline{U}^{(3)}n - \underline{\widetilde{U}}_{\underline{r}}^{(2)} t_{\underline{r}})} \frac{\theta(\underline{\widetilde{K}}_1 + \underline{U}^{(3)}n - \underline{\widetilde{U}}_{\underline{r}}^{(2)} t_{\underline{r}})}{\theta(\underline{\widetilde{K}}_1 + \underline{U}^{(3)}n - \underline{U}^{(3)} - \underline{\widetilde{U}}_{\underline{r}}^{(2)} t_{\underline{r}})} \cdot e^{e_2(Q_0)-e_1(Q_0)} \\
&- 2\omega_0^{\infty_2} - \sum_{j=0}^{m-1} d_{j,0}^{(\infty_2)} \frac{\partial}{\partial z_j} \ln \frac{\theta(\underline{\widetilde{K}}_0 + \underline{U}^{(3)}n + \underline{U}^{(3)} - \underline{\widetilde{U}}_{\underline{r}}^{(2)} t_{\underline{r}})}{\theta(\underline{\widetilde{K}}_0 + \underline{U}^{(3)}n - \underline{U}^{(3)} - \underline{\widetilde{U}}_{\underline{r}}^{(2)} t_{\underline{r}})}, \\
v(n,t_{\underline{r}}) =& \frac{\theta(\underline{\widetilde{K}}_1 + \underline{U}^{(3)}n - \underline{\widetilde{U}}_{\underline{r}}^{(2)} t_{\underline{r}})}{\theta(\underline{\widetilde{K}}_1 + \underline{U}^{(3)}n - \underline{U}^{(3)} - \underline{\widetilde{U}}_{\underline{r}}^{(2)} t_{\underline{r}})} \frac{\theta(\underline{\widetilde{K}}_0 + \underline{U}^{(3)}n - \underline{U}^{(3)} - \underline{\widetilde{U}}_{\underline{r}}^{(2)} t_{\underline{r}})}{\theta(\underline{\widetilde{K}}_0 + \underline{U}^{(3)}n - \underline{\widetilde{U}}_{\underline{r}}^{(2)} t_{\underline{r}})} \cdot e^{e_{0_1}(Q_0)-e_1(Q_0)}, \\
w(n,t_{\underline{r}}) =& \frac{-\theta(\underline{\widetilde{K}}_2 + \underline{U}^{(3)}n - \underline{\widetilde{U}}_{\underline{r}}^{(2)} t_{\underline{r}})}{\theta(\underline{\widetilde{K}}_2 + \underline{U}^{(3)}n + \underline{U}^{(3)} - \underline{\widetilde{U}}_{\underline{r}}^{(2)} t_{\underline{r}})} \frac{\theta(\underline{\widetilde{K}}_1 + \underline{U}^{(3)}n + \underline{U}^{(3)} - \underline{\widetilde{U}}_{\underline{r}}^{(2)} t_{\underline{r}})}{\theta(\underline{\widetilde{K}}_1 + \underline{U}^{(3)}n - \underline{\widetilde{U}}_{\underline{r}}^{(2)} t_{\underline{r}})} \cdot e^{e_2(Q_0)-e_1(Q_0)},
\end{aligned} \tag{5.53}$$

where

$$\begin{aligned}
\underline{\widetilde{K}}_0 &= \underline{\Lambda} - \underline{\mathcal{A}}(P_{0_1}) + \underline{\rho}(n_0, t_{0,\underline{r}}) - \underline{U}^{(3)}n_0 + \underline{\widetilde{U}}_{\underline{r}}^{(2)} t_{0,\underline{r}}, \\
\underline{\widetilde{K}}_s &= \underline{\Lambda} - \underline{\mathcal{A}}(P_{\infty_s}) + \underline{\rho}(n_0, t_{0,\underline{r}}) - \underline{U}^{(3)}n_0 + \underline{\widetilde{U}}_{\underline{r}}^{(2)} t_{0,\underline{r}}, \quad s = 1, 2.
\end{aligned}$$

Acknowledgements

This work is supported by the National Natural Science Foundation of China (Grant Nos. 11971442, 11871440 and 11931017), the Natural Science Foundation of Hebei Province (Grant No. A2020210005), the Foundation of Hebei Education Department (Grant No. QN2018050).

References

[1] Ablowitz M J and Clarkson P A, *Solitons, Nonlinear Evolution Equation and Inverse Scattering*, Cambridge University Press, Cambridge, 1991.

[2] Ablowitz M J, Kaup D J and Newell A C, The inverse scattering transform-fourier analysis for nonlinear problems, *Stud. Appl. Math.* **53**, 249–315, 1974.

[3] Belokolos E D and Bobenko A I, *Algebro-Geometric Approach to Nonlinear Integrable Equations*, Springer-Verlag, Berlin, 1994.

[4] Blaszak M and Marciniak K, R-matrix approach to lattice integrable systems, *J. Math. Phys.* **35**, 4661–4682, 1994.

[5] Dubrovin B A, Theta functions and non-linear equations, *Russian Math. Surveys* **36**, 11–92, 1981.

[6] Dickson R, Gesztesy F and Unterkofler K, A new approach to the Boussinesq hierarchy, *Math. Nachr.* **198**, 51–108, 1999.

[7] Dickson R, Gesztesy F and Unterkofler K, Algebro-geometric solutions of the Boussinesq hierarchy, *Rev. Math. Phys.* **11**, 823–879, 1999.

[8] Geng X G, Dai H H and Zhu J Y, Decomposition of the discrete Ablowitz-Ladik hierarchy, *Stud. Appl. Math.* **118**, 281–312, 2007.

[9] Gesztesy F, Holden H, Michor J and Teschl G, *Soliton Equations and Their Algebro-Geometric Solutions. Volume II: (1+1)-Dimensional Discrete Models*, Cambridge University Press, Cambridge, 2008.

[10] Griffiths P and Harris J, *Principles of Algebraic Geometry*, Wiley, New York, 1994.

[11] Gesztesy F and Ratnaseelan R, An alternative approach to algebro-geometric solutions of the AKNS hierarchy, *Rev. Math. Phys.* **10**, 345–391, 1998.

[12] Geng X G, Wu L H and He G L, Algebro-geometric constructions of the modified Boussinesq flows and quasi-periodic solutions, *Phys. D* **240**, 1262–1288, 2011.

[13] Geng X G and Wang H, Algebro-geometric constructions of quasi-periodic flows of the Newell hierarchy and applications, *IMA J. Appl. Math.* **82**, 97–130, 2017.

[14] Geng X G and Zeng X, Application of the trigonal curve to the Blaszak-Marciniak lattice hierarchy, *Teoret. Mat. Fiz.* **190**, 21–47, 2017.

[15] Geng X G, Zhai Y Y and Dai H H, Algebro-geometric solutions of the coupled modified Korteweg-de Vries hierarchy, *Adv. Math.* **263**, 123–153, 2014.

[16] He G L, Wu L H and Geng X G, Finite genus solutions to the mixed Boussinesq equation, *Sci. Sin. Math.* **42**, 711–734, 2012.

[17] Hu X B, Wang D L, Tam H.W and Xue W M, Soliton solutions to the Jimbo-Miwa equations and the Fordy-Gibbons-Jimbo-Miwa equation, *Phys. Lett. A* **262**, 310–320, 1999.

[18] Hu X B, Wang D L and Tam H W, Integrable extended Blaszak-Marciniak lattice and another extended lattice with their Lax pairs, *Theor. Math. Phys.* **127**, 738–743, 2001.

[19] Its A R and Matveev V B, Schrödinger operators with finite-gap spectrum and N-soliton solutions of the Korteweg-de Vries equation, *Theoret. Math. Phys.* **23**, 343–355, 1975.

[20] Lundmark H and Szmigielski J, An inverse spectral problem related to the Geng-Xue two-component peakon equation, *Mem. Am. Math. Soc.* **244**, 1–87, 2016.

[21] McKean H P, Integrable Systems and Algebraic Curves, in Global analysis, *Springer Berlin Heidelberg*, 83–200, 1979.

[22] Mumford D, *Tata Lectures on Theta II*, Birkhäuser, Boston, Mass, 1984.

[23] Sahadevan R and Khousalya S, Similarity reduction, generalized symmetries and integrability of Belov-Chaltikian and Blaszak-Marciniak lattice equations, *J. Math. Phys.* **42**, 3854–3870, 2001.

[24] Sahadevan R and Khousalya S, Master symmetries for Volterra equation, Belov-Chaltikian and Blaszak-Marciniak lattice equations, *J. Math. Anal. Appl.* **280**, 241–251, 2003.

[25] Toda M, *Theory of nonlinear lattices*, Springer-Verlag, Berlin, 1981.

[26] Wang H and Geng G X, Algebro-geometric solutions to a new hierarchy of soliton equations, *Zh. Mat. Fiz. Anal. Geom.* **11** 359–398, 2015.

[27] Wu L H, Geng X G and He G L, Algebro-geometric solutions to the Manakov hierarchy, *Appl. Anal.* **95**, 769–800, 2016.

[28] Wei J, Geng X G and Zeng X, Quasi-periodic solutions to the hierarchy of four-component Toda lattices, *J. Geom. Phys.* **106**, 26–41, 2016.

[29] Yildirim A and Kocak H, An efficient technique for solving the Blaszak-Marciniak lattice by combining the Homotopy perturbation and Padé techniques, *Int. J. Comput. Math.* **9**, 1240024, 2012.

[30] Yu F J and Li L, A Blaszak-Marciniak lattice hierarchy with self-consistent sources, *Int. J. Mod. Phys. B* **25**, 3371–3379, 2011.

[31] Zhu Z N, Wu X N, Xue W M and Ding Q, Infinitely many conservation laws of two Blaszak-Marciniak three-field lattice hierarchies, *Phys. Lett. A* **297**, 387–395, 2002.

[32] Zhao H Q, Zhu Z N and Zhang J L, Darboux transformations and new explicit solutions for a Blaszak-Marciniak three-field lattice equation, *Commun. Theor. Phys.* **56**, 23–30, 2011.

B6. Long-time asymptotic behavior of the modified Schrödinger equation via $\bar{\partial}$-steepest descent method

Yiling Yang and Engui Fan

School of Mathematical Sciences, Fudan University, Shanghai 200433, P.R. China.

Abstract

We consider the Cauchy problem for the modified NLS equation

$$iu_t + u_{xx} + 2\rho|u|^2u + i(|u|^2u)_x = 0,$$
$$u(x,0) = u_0(x) \in H^{2,2}(R),$$

where $H^{2,2}(R)$ is a weighted Sobolev space. Using nonlinear steepest descent method and combining the $\bar{\partial}$-analysis, we show that inside any fixed cone

$$C(x_1, x_2, v_1, v_2) = \left\{(x,t) \in R^2 | x = x_0 + vt, x_0 \in [x_1, x_2], v \in [v_1 + 4\rho, v_2 + 4\rho]\right\},$$

the long-time asymptotics of the modified NLS equation can be characterized with an $N(I)$-soliton on discrete spectrum and leading order term $O(t^{-1/2})$ on continuous spectrum up to a residual error order $O(t^{-3/4})$.

1 Introduction

In this paper, we study the long-time asymptotic behavior for the initial value problem of the modified NLS equation

$$iu_t + u_{xx} + 2\rho|u|^2u + i(|u|^2u)_x = 0, \qquad u(x,0) = u_0(x), \tag{1.1}$$

where $\rho \in R$ and the initial data $u_0(x)$ belongs to the weighted Sobolev space

$$H^{2,2}(R) = \left\{f \in L^2(R); x^2 f, f'' \in L^2(R)\right\}.$$

The modified NLS Equation (1.1) was proposed to describe the nonlinear propagation of the Alfvèn waves, the femtosecond optical pulses in a nonlinear single-mode optical fiber and the deep-water gravity waves [26, 29]. The term $i(|u|^2u)_x$ in Equation (1.1) is called the self-steepening term, which causes an optical pulse to become asymmetric and steepen upward at the trailing edge [1, 36]. Equation (1.1) also describes the short pulses that propagate in a long optical fiber characterized by a nonlinear refractive index [27, 31]. Brizhik et al. showed that the modified NLS Equation (1.1), unlike the classical NLS equation

$$iu_t + u_{xx} + 2|u|^2u = 0, \tag{1.2}$$

possesses static localized solutions when the effective nonlinearity parameter is larger than a certain critical value [2]. In the 1970s, Wadati et al. showed that

Equation (1.1) is completely integrable by inverse scattering transformation [32]. In recent years, various exact solutions for Equation (1.1) also have been extensively discussed by analytical method, Hirota bilinear method and Darboux transformation respectively [14, 21, 25, 33]. The Hamiltonian structure from mathematical structures for Equation (1.1) was given [18]. The inverse transformation and dressing method were used to construct N-soliton solutions of the modified NLS Equation (1.1) with zero boundary conditions considered [5, 6, 11]. Recently, we presented inverse transformation for the modified NLS Equation (1.1) with nonzero boundary conditions by using the Riemann-Hilbert method [37]. From the determinant expressions of N-soliton solutions of the modified NLS Equation (1.1), the asymptotic behavior of the N-soliton solutions in the case of $t \to \infty$ was directly derived [7]. Kitaev and Vartanian[19] applied Deift-Zhou method to obtain long-time asymptotic solution of Equation (1.1) with decaying initial value. They derived an explicit functional form for the next-to-leading-order $O(t^{-1/2})$ term, that is [19],

$$u(x,t) = \frac{c}{\sqrt{t}} + O(t^{-1}\text{log}t), \tag{1.3}$$

where c is related to initial value and phase point.

The study on the long-time behavior of nonlinear wave equations solvable by the inverse scattering method was first carried out by Manakov [22]. Zakharov and Manakov give the first result for large-time asymptotic of solutions for the NLS Equation (1.2) with decaying initial value in 1976 [39]. The inverse scattering method also worked for long-time behavior of integrable systems such as KdV, Landau-Lifshitz and the reduced Maxwell-Bloch system [3, 12, 28]. In 1993, Deift and Zhou developed a nonlinear steepest descent method to rigorously obtain the long-time asymptotics behavior of the solution for the MKdV equation by deforming contours to reduce the original Riemann-Hilbert problem to a model one whose solution is calculated in terms of parabolic cylinder functions [42]. Later, this method was applied to the focusing NLS equation, KdV equation, Fokas-Lenells equation and derivative NLS equation, etc. [10, 13, 34, 35, 43].

In recent years, McLaughlin and Miller further presented a $\bar{\partial}$ steepest descent method to analyze the asymptotic of orthogonal polynomials with non analytical weights. This method combines steepest descent with $\bar{\partial}$-problem rather than the asymptotic analysis of singular integrals on contours [23, 24]. When it is applied to integrable systems, the $\bar{\partial}$ steepest descent method also has advantages, such as avoiding delicate estimates involving L^p estimates of Cauchy projection operators, and leading the non-analyticity in the RHP reductions to a $\bar{\partial}$-problem in some sectors of the complex plane which can be solved by being recast into an integral equation and by using Neumann series. Dieng and McLaughin use it to study the defocusing NLS equation under essentially minimal regularity assumptions on finite mass initial data [9]. Cussagna and Jenkins study the defocusing NLS equation with finite density initial data [8]. They were also successfully applied to prove asymptotic stability of N-soliton solutions to focusing NLS [4] which has been conjectured for a long-time [38]. Jenkins and Liu study the derivative nonlinear Schrödinger equation for generic initial data in a weighted Sobolev space [15]. Their work decomposes the solution into the sum of a finite number of separated solitons and a radiative part as when $t \to \infty$. And the dispersive part contains two components, one coming from the continuous spectrum and another from the interaction of the discrete and continuous spectrum.

In our chapter, we obtain the long-time asymptotic behavior of the solution of modified NLS Equation (1.1) with initial data $u_0 \in H^{2,2}$ by using the steepest descent method and $\bar{\partial}$ steepest descent method. In recent work on the focusing NLS equation, Borghese, Jenkins and McLaughlin showed how to treat a problem with discrete and continuous spectral data [4]. The work on derivative NLS equation due to Jenkins and Liu is the special case of modified NLS with $\rho = 0$ [15].

This chapter is arranged as follows. Following the idea in [15], we reduce the **RHP 1** in the following context into two parts, one describes the asymptotic behavior of solitons, and another model computes the contributions due to the interactions of solitons and radiation. In Section 1, we first make gauge transformation

$$u = qe^{-i\int_{-\infty}^{x}|q|^2dy}, \qquad u_0 = q_0e^{-i\int_{-\infty}^{x}|q_0|^2dy},$$

and change the modified NLS Equation (1.1) into a new equation which is more convenient to be dealt with. We describe the forward scattering transform step and necessary results, and establish the inverse scattering transform with a vector **RHP 1**. In Section 2, we define a new row-vector RHP $M^{(1)}$ by (2.17) which deforms the contour R such that the jump matrix (2.18) approaches the identity exponentially fast away from the critical point z_0 (see Figure 2). In Section 3, we deform **RHP 2** on a new contour $\Sigma^{(2)}$ whose jump matrices approach the identity exponentially fast away from the critical point z_0 by defining a new unknown $M^{(2)}$(3.5) which solves a mixed $\bar{\partial}$-Riemann-Hilbert problem–**RHP 3**. In Section 4, we decompose $M^{(2)}$ into a model Riemann-Hilbert problem–**RHP 4** with solution M^{RHP} and a pure **$\bar{\partial}$-problem 5** with solution $M^{(3)}$. To solve M^{RHP}, we divide it into an outer model $M^{(out)}$ for the soliton components in Section 5, and an inner model $M^{(z_0)}$ for the stationary phase point z_0 which is constructed by M^{pc} by parabolic cylinder functions in Section 6. The outer and inner models together with error $E(z)$ build M^{RHP} in (4.12), where $E(z)$ is the solution of a small-norm Riemann-Hilbert problem in Section 7. Then we solve the **$\bar{\partial}$-problem 5** for $M^{(3)}$ in Section 8. Thus, previous result we obtain

$$M(z) = M^{(3)}(z)M^{RHP}(z)R^{(2)}(z)^{-1}T(z)^{\sigma_3},$$

for brevity we omit the dependence of (x,t). Then from the asymptotic behavior as $|t| \to \infty$ of every function and the reconstruction formula, we get (9.4). But our ultimate purpose is to get the long-time asymptotic behavior of $u(x,t)$, so we need to establish the asymptotic formula of $e^{-i\int_{-\infty}^{x}|q|^2dy}$. Unlike in reference [15], which is our special case with $\rho = 0$, we need to calculate $M_+(\rho)$ and study its long-time asymptotic behavior. At the point ρ, $M(z)$ does not have such simple nice properties like it at $\rho = 0$, for example, his $M(z)$ is continuous at 0 and its calculation is simple. We calculate the values of $M_+(\rho)$ the parts in the corresponding Section and combine all results to obtain a long-time estimation of the phase factor $e^{-i\int_{-\infty}^{x}|q|^2dy}$

in Section 9. Compared with the result (1.3), we get a more general result (9.17)-(9.18), which improves error estimate as a sharp error $O(t^{-3/4})$ for more general initial data $u_0 \in H^{2,2}(R)$.

The modified NLS Equation (1.1) admits the Lax pair [12]

$$\phi_x = L_0\phi, \quad \phi_t = M_0\phi, \tag{1.4}$$

where

$$L_0 = -i(k^2-\rho)\sigma_3 + kU,$$

$$M_0 = -2i(k^2-\rho)^2\sigma_3 + 2k(k^2-\rho)U - ik^2U^2\sigma_3 - ikU_x\sigma_3 + kU^3\sigma_3,$$

and

$$\sigma_3 = \begin{pmatrix} 1 & 0 \\ 0 & -1 \end{pmatrix}, \quad U = \begin{pmatrix} 0 & u \\ -\bar{u} & 0 \end{pmatrix}.$$

To avoid the imaginary axis becoming its boundary, we use a new Lax pair which is written in terms of $z = k^2$. And we discard the symmetry of the equation to avoid the factor $z^{1/2}$. So we make the following transformation

$$\phi = \begin{pmatrix} k & 0 \\ 0 & 1 \end{pmatrix}\psi,$$

and get a new Lax pair [6]

$$\psi_x = L\psi, \quad \psi_t = M\psi, \tag{1.5}$$

where

$$L = -i(z-\rho)\sigma_3 + \Lambda U,$$

$$M = -2i(z-\rho)^2\sigma_3 + 2(z-\rho)\Lambda U - izU^2\sigma_3 - i\Lambda U_x\sigma_3 + \Lambda U^3\sigma_3,$$

and

$$\Lambda = \begin{pmatrix} 1 & 0 \\ 0 & z \end{pmatrix}.$$

In order to have good asymptotics at $z \to \pm\infty$, because of the necessity of that the diagonal line of $\psi \to I$ as $z \to \pm\infty$, we first make the following transformation

$$u = qe^{-i\int_{-\infty}^{x}|q|^2dy}, \quad u_0 = q_0e^{-i\int_{-\infty}^{x}|q_0|^2dy}, \tag{1.6}$$

which defines an invertible mapping in $L^2(R) : u \to q$, mapping soliton solutions to soliton solutions, and dense open sets to dense open sets in weighted Sobolev spaces. It has inverse transformation

$$q = ue^{i\int_{-\infty}^{x}|u|^2dy}, \quad q_0 = u_0e^{i\int_{-\infty}^{x}|u_0|^2dy}. \tag{1.7}$$

Then Equation (1.1) is gauge-equivalent to the equation

$$iq_t + q_{xx} + 2\rho|q|^2 q + i(|q|^2 q)_x - 2i|q|^2 q_x = 0, \tag{1.8}$$

with Lax pair transformation

$$\psi = e^{-i/2\int_{-\infty}^{x}|q|^2 dy\sigma_3}\Psi, \tag{1.9}$$

which leads to the Lax pair of Equation (1.8)

$$\Psi_x = -i(z-\rho)\sigma_3\Psi + Q\Psi, \tag{1.10}$$

$$\Psi_t = -2i(z-\rho)^2\sigma_3\Psi + P\Psi, \tag{1.11}$$

where

$$Q = \Lambda Q_0 + \frac{i}{2}q^2\sigma_3, \quad Q_0 = \begin{pmatrix} 0 & q \\ -\bar{q} & 0 \end{pmatrix},$$

$$P = 2(z-\rho)\Lambda Q_0 - izQ_0^2\sigma_3 - i\Lambda(Q_0)_x\sigma_3 + \Lambda Q_0^3\sigma_3.$$

We first consider the situation that q is only dependent on $x \in R$. Consider the Jost solutions of (1.10), which are restricted by the boundary conditions that

$$\Psi_\pm \sim e^{-i(z-\rho)\sigma_3 x}, \quad x \to \pm\infty. \tag{1.12}$$

Denote

$$\Psi_\pm = \begin{pmatrix} \Psi_\pm^1 & \hat{\Psi}_\pm^1 \\ \Psi_\pm^2 & \hat{\Psi}_\pm^2 \end{pmatrix},$$

then the first and second columns of Equation (1.10) can be written respectively as

$$\begin{pmatrix} \Psi_\pm^1 \\ \Psi_\pm^2 \end{pmatrix}_x = -i(z-\rho)\sigma_3\begin{pmatrix} \Psi_\pm^1 \\ \Psi_\pm^2 \end{pmatrix} + Q\begin{pmatrix} \Psi_\pm^1 \\ \Psi_\pm^2 \end{pmatrix}, \tag{1.13}$$

$$\begin{pmatrix} \hat{\Psi}_\pm^1 \\ \hat{\Psi}_\pm^2 \end{pmatrix}_x = -i(z-\rho)\sigma_3\begin{pmatrix} \hat{\Psi}_\pm^1 \\ \hat{\Psi}_\pm^2 \end{pmatrix} + Q\begin{pmatrix} \hat{\Psi}_\pm^1 \\ \hat{\Psi}_\pm^2 \end{pmatrix}, \tag{1.14}$$

from which we can obtain

$$(\Psi_\pm^2)_x = i(z-\rho)\Psi_\pm^2 - z\bar{q}\Psi_\pm^1 - \frac{i}{2}|q|^2\Psi_\pm^2, \tag{1.15}$$

$$(\Psi_\pm^1)_x = -i(z-\rho)\Psi_\pm^1 + q\Psi_\pm^2 + \frac{i}{2}|q|^2\frac{i}{2}|q|^2\Psi_\pm^1, \tag{1.16}$$

$$(\hat{\Psi}_\pm^1)_x = -i(z-\rho)\hat{\Psi}_\pm^1 + q\hat{\Psi}_\pm^2 + \frac{i}{2}|q|^2\hat{\Psi}_\pm^1, \tag{1.17}$$

$$(\hat{\Psi}_\pm^2)_x = i(z-\rho)\hat{\Psi}_\pm^2 - z\bar{q}\hat{\Psi}_\pm^1 - \frac{i}{2}|q|^2\hat{\Psi}_\pm^2. \tag{1.18}$$

By simple calculation, we obtain

$$\overline{\left(-\frac{1}{z}\Psi_{\pm}^{2}\right)}_{x} = -i(\bar{z}-\rho)\overline{\left(-\frac{1}{z}\Psi_{\pm}^{2}\right)} + q\overline{\Psi_{\pm}^{1}} + \frac{i}{2}|q|^{2}\overline{\left(-\frac{1}{z}\Psi_{\pm}^{2}\right)}, \tag{1.19}$$

$$\overline{(\Psi_{\pm}^{1})_{x}} = i(\bar{z}-\rho)\overline{\Psi_{\pm}^{1}} - \bar{z}\bar{q}\overline{\left(-\frac{1}{z}\Psi_{\pm}^{2}\right)} - \frac{i}{2}|q|^{2}\overline{\Psi_{\pm}^{1}}. \tag{1.20}$$

Comparing their asymptotic condition (1.12) and equation, respectively, we get the symmetry of Ψ as follow

$$\overline{\Psi_{\pm}^{1}(\bar{z})} = \hat{\Psi}_{\pm}^{2}, \quad \overline{-\frac{1}{z}\Psi_{\pm}^{2}(\bar{z})} = \hat{\Psi}_{\pm}^{1}. \tag{1.21}$$

These solutions can be expressed as Volterra type integrals

$$\begin{pmatrix} \Psi_{\pm}^{1} \\ \Psi_{\pm}^{2} \end{pmatrix} = e^{-i(z-\rho)\sigma_{3}} + \int_{x}^{\pm\infty} e^{i(z-\rho)\hat{\sigma}_{3}} Q \begin{pmatrix} \Psi_{\pm}^{1} \\ \Psi_{\pm}^{2} \end{pmatrix} dy. \tag{1.22}$$

Then we obtain that Ψ_{+}^{1} and Ψ_{+}^{2} are analysis in C^{-}. In the same way we obtain that Ψ_{-}^{1} and Ψ_{-}^{2} are analysis in C^{+}.

By making the transformation

$$\varphi_{\pm} = \Psi_{\pm} e^{i(z-\rho)\sigma_{3}}, \tag{1.23}$$

we then have

$$\varphi_{\pm} \sim I, \quad x \to \pm\infty,$$

and $\varphi_{\pm}$ satisfy an equivalent Lax pair

$$(\varphi_{\pm})_{x} = -i(z-\rho)[\sigma_{3}, \varphi_{\pm}] + Q\varphi_{\pm}, \tag{1.24}$$

$$(\varphi_{\pm})_{t} = -2i(z-\rho)^{2}[\sigma_{3}, \varphi_{\pm}] + P\varphi_{\pm}. \tag{1.25}$$

Since $\operatorname{tr}(-i(z-\rho)\sigma_{3} + Q) = 0$ in (1.10) and (1.11), by using the Able formula, we have

$$(\det \Psi_{\pm})_{x} = 0. \tag{1.26}$$

Again by using the relation

$$\det(\varphi_{\pm}) = \det(\Psi_{\pm} e^{i(z-\rho)\sigma_{3}}) = \det(\Psi_{\pm}),$$

we get $(\det \varphi_{\pm})_{x} = 0$, which means that $\det(\varphi_{\pm})$ is independent of x. So we obtain that

$$\det \varphi_{\pm} = \lim_{x \to \pm\infty} \det(\varphi_{\pm}) = \det I = 1, \tag{1.27}$$

which implies that $\varphi_{\pm}$ are inverse matrices.

Since $\Psi_\pm$ are two fundamental matrix solutions of the Lax pair (1.10) and (1.11), there exists a linear relation between Ψ_+ and Ψ_-, namely

$$\Psi_+(x,z) = \Psi_-(x,z)S(z), \qquad z \in R, \tag{1.28}$$

where $S(z)$ is called scattering matrix which only depend on z, and $\det S(z) = 1$ by using (1.27). From the symmetry of $\Psi_\pm$ (1.21), $S(z)$ can be written as

$$S(z) = \begin{pmatrix} a(z) & b(z) \\ -z\overline{b(z)} & \overline{a(z)} \end{pmatrix},$$

then we have

$$|a(z)|^2 + z|b(z)|^2 = 1. \tag{1.29}$$

We introduce the reflection coefficient

$$r(z) = \frac{\overline{b(z)}}{a(z)}. \tag{1.30}$$

Then we immediately have $1 + z|r(z)|^2 = |a(z)|^{-2}$. Calculating individual elements of the matrix Equation (1.28), then we obtain that the function a(z) and b(z) may be computed via the Wronskian formula:

$$a(z) = \Psi_+^1 \hat{\Psi}_-^2 - \Psi_+^2 \hat{\Psi}_-^1, \qquad b(z) = \hat{\Psi}_+^1 \hat{\Psi}_-^2 - \hat{\Psi}_+^2 \hat{\Psi}_-^1. \tag{1.31}$$

Thus a(z) has an analytic continuation to C^+, but b(z) has no analyticity.

To get the Riemann-Hilbert problem, it is necessary to discuss the asymptotic behaviors of the Jost solutions and scattering matrix as $z \to \infty$. For convenience, we denote two elements of the first column of $\varphi_\pm$ as $\varphi_\pm^1$ and $\varphi_\pm^2$, respectively. We consider the following asymptotic expansions

$$\varphi_\pm^j = \varphi_\pm^{j,(0)} + \frac{\varphi_\pm^{j,(1)}}{z} + \frac{\varphi_\pm^{j,(2)}}{z^2} + O(z^{-3}), \qquad \text{as } z \to \infty, \qquad j = 1, 2, \tag{1.32}$$

where $\varphi_\pm^{j,(k)}$ are independent of z, $k = 0, 1, 2, \ldots$, $j = 1, 2$.

Substituting (1.32) into the Lax pair (1.24) and comparing the coefficients, we obtain

$$\begin{aligned} &2i\varphi_\pm^{2,(0)} - \bar{q}\varphi_\pm^{1,(0)} = 0, \\ &(\varphi_\pm^{1,(0)})_x = \frac{i}{2}|q|^2\varphi_\pm^{1,(0)} + q\varphi_\pm^{2,(0)}, \end{aligned}$$

from which we obtain $(\varphi_\pm^{1,(0)})_x = 0$, which means

$$\varphi_\pm^1 \to 1, \quad \varphi_\pm^2 \to -\frac{i}{2}\bar{q}, \qquad \text{as } z \to \infty.$$

Therefore, from (1.31) we have

$$a(z) \to 1, \quad \text{as } z \to \infty.$$

When z is real, we have

$$b(z) = O(z), \quad r(z) = O(z), \quad \text{as } z \to \infty.$$

Zeros of a on R are known to occur and they correspond to spectral singularities [40]. They are excluded from our analysis in this chapter. To deal with our following task, we assume our initial data satisfy this assumption.

Assumption 1. *The initial data $u_0 \in H^{2,2}(R)$ and it generates generic scattering data which satisfy that*

1. *$a(z)$ has no zeros on R.*
2. *$a(z)$ only has finite number of simple zeros.*
3. *$a(z)$ and $r(z)$ belong $H^{2,2}(R)$.*

In the absence of spectral singularities (real zeros of $a(z)$), there also exist $c \in (0,1)$ such that $c < |a(z)| < 1/c$ for $z \in R$, which implies $1 + z|r(z)|^2 > c^2 > 0$ for $z \in R$.

We assume that $a(z)$ has N simple zeros $z_n \in C^+$, n = 1, 2,.., N, $a(z_n) = 0$. Denote $Z=\{z_n\}_{n=1}^N$ which is the set of the zeros of $a(z)$. A standard result of the scattering theory has the following trace formula for the transmission coefficient:

$$a(z) = \prod_{k=1}^{N} \frac{z - z_k}{z - \bar{z}_k} \exp\left(i \int_R \frac{k(s)ds}{s - z} \right), \tag{1.33}$$

where

$$k(s) = -\frac{1}{2\pi} \log(1 + s|r(s)|^2). \tag{1.34}$$

Combining two transformations (1.23) and (1.28) gives

$$\varphi_+ = \varphi_- e^{-(z-\rho)x\hat{\sigma}_3} S(z), \tag{1.35}$$

from which we have

$$a(z) = \lim_{x \to -\infty} \varphi_+^1. \tag{1.36}$$

From (1.24), we have

$$(\varphi_+)_x = (z-\rho)\left(\varphi_+ \begin{pmatrix} i & 0 \\ 0 & -i \end{pmatrix} - \begin{pmatrix} i & 0 \\ \bar{q} & -i \end{pmatrix} \varphi_+\right) + \begin{pmatrix} \frac{i}{2}|q|^2 & q \\ -\rho\bar{q} & \frac{-i}{2}|q|^2 \end{pmatrix} \varphi_+. \tag{1.37}$$

Note that $\lim_{x \to +\infty} \varphi_+ = I$, as $z = \rho$ is a regular point of this system of equations, we have

$$\begin{aligned} \varphi_+(x,t,\rho) =& e^{\frac{-i\sigma_3}{2} \int_x^{+\infty} |q(y,t)|^2 dy} \\ & \begin{pmatrix} 1 & -\int_x^{+\infty} q(y,t) e^{i\int_y^{+\infty} |q(s,t)|^2 ds} dy \\ \int_x^{+\infty} \bar{q}(y,t) e^{-i\int_y^{+\infty} |q(s,t)|^2 ds} dy & 1 \end{pmatrix}. \end{aligned} \tag{1.38}$$

Then we take $x \to \infty$ and obtain that

$$a(\rho) = e^{-\frac{i}{2}\int_{-\infty}^{+\infty}|q(y,t)|^2dy} = \prod_{k=1}^{N}\frac{\rho - z_k}{\rho - \bar{z}_k}\exp\left(i\int_R \frac{k(s)ds}{s-\rho}\right). \tag{1.39}$$

Similarly we have

$$(\varphi_-)_x = (z-\rho)\left(\varphi_-\begin{pmatrix} i & 0 \\ 0 & -i\end{pmatrix} - \begin{pmatrix} i & 0 \\ \bar{q} & -i\end{pmatrix}\varphi_-\right) + \begin{pmatrix} \frac{i}{2}|q|^2 & q \\ -\rho\bar{q} & \frac{-i}{2}|q|^2\end{pmatrix}\varphi_-, \tag{1.40}$$

and

$$\begin{aligned}\varphi_-(x,t,\rho) =& e^{\frac{i\sigma_3}{2}\int_{-\infty}^{x}|q(y,t)|^2dy} \\ &\begin{pmatrix} 1 & -\int_{-\infty}^{x} q(y,t)e^{-i\int_{-\infty}^{y}|q(s,t)|^2ds}dy \\ \int_{-\infty}^{x}\bar{q}(y,t)e^{i\int_{-\infty}^{y}|q(s,t)|^2ds}dy & 1\end{pmatrix}.\end{aligned} \tag{1.41}$$

Then we begin to calculate residue conditions. At any z_n, (Ψ_+^1, Ψ_+^2) and (Ψ_-^1, Ψ_-^2) are linearly dependent. Specifically, there exists a constant b_k such that:

$$(\Psi_+^1, \Psi_+^2) = b_k(\Psi_-^1, \Psi_-^2).$$

Denote *norming constant* $c_k = b_k/a'(z_k)$, for initial data q_0, the collection $D = \left\{r(z), \{z_k, c_k\}_{k=1}^N\right\}$ is called the *scattering data* for q_0. It is an elementary calculation to show that the Sectionally meromorphic matrices defined as follow

$$N(x,z) = \begin{cases} \begin{pmatrix} a(z)^{-1}\varphi_+^1 & \hat{\varphi}_-^1 \\ a(z)^{-1}\varphi_+^2 & \hat{\varphi}_-^2 \end{pmatrix}, & \text{as } z \in D^+, \\ \begin{pmatrix} \varphi_-^1 & \overline{a(\bar{z})}^{-1}\hat{\varphi}_+^1 \\ \varphi_-^2 & \overline{a(\bar{z})}^{-1}\hat{\varphi}_+^2 \end{pmatrix}, & \text{as } z \in D^-, \end{cases} \tag{1.42}$$

which solves the following Riemann-Hilbert problem:

Riemann-Hilbert problem 0. Find a matrix-valued function z$\in C \to N(z;x)$ which satisfies:

- Analyticity: $N(x,z)$ is meromorphic in $\mathbb{C}\setminus R$ and has single poles;
- Jump condition: N has continuous boundary values $N_\pm$ on R and

$$N^+(x,z) = N^-(x,z)V(z), \quad z \in R, \tag{1.43}$$

where

$$V(z) = \begin{pmatrix} 1 + z|r(z)|^2 & -e^{-2i(z-\rho)x}r(z) \\ -e^{2i(z-\rho)x}z\overline{r(z)} & 1\end{pmatrix}; \tag{1.44}$$

- Asymptotic behaviors:

$$N(x,z) = \begin{pmatrix} 1 & 0 \\ -\frac{i}{2}\bar{q} & 1\end{pmatrix} + O(z^{-1}), \quad z \to \infty; \tag{1.45}$$

- Residue conditions: N has simple poles at each point in $Z \bigcup \bar{Z}$ with:

$$\mathrm{Res}_{z=z_n} N(z) = \lim_{z\to z_n} N(z) \begin{pmatrix} 0 & 0 \\ c_n e^{2i(z_n-\rho)x} & 0 \end{pmatrix}, \tag{1.46}$$

$$\mathrm{Res}_{z=\bar{z}_n} N(z) = \lim_{z\to \bar{z}_n} N(z) \begin{pmatrix} 0 & -z_n^{-1}\bar{c}_n e^{-2i(\bar{z}_n-\rho)x} \\ 0 & 0 \end{pmatrix}. \tag{1.47}$$

From the asymptotic behavior of the functions $\varphi_\pm$, we have the following reconstruction formula:

$$q(x) = 2i \lim_{z\to\infty} [zN(z;x)]_{12}. \tag{1.48}$$

Now we are going to take into account the time. If q also depends on t (i.e., $q = q(x,t)$), we can obtain the functions a and b as above for all times t $\in R$. Taking account of (1.11), we have

$$(a(z;t))_t = 0, \quad (b(z;t))_t = -4i(z-\rho)^2 b(z;t). \tag{1.49}$$

Then we can obtain time dependences of scattering data which can be expressed as the following replacement

$$c(z_n) \to c(t, z_n) = c(0, z_n)e^{-4i(z_n-\rho)^2 t}, \tag{1.50}$$

$$r(z) \to r(t,z) = r(0,z)e^{-4i(z-\rho)^2 t} \tag{1.51}$$

In particular, if at time $t = 0$ the initial function $q(x,0)$ produces N simple zeros $z_1,\ldots,z_N$ of $a(z;0)$ and if q evolves accordingly to the (1.8), then $q(x,t)$ will produce exactly the same N simple zeros at any other time $t \in R$. Altogether the scattering data of a function q(x, t), which is a solution of (1.8), is given at time t by

$$\left\{ e^{-4i(z-\rho)^2 t} r(z), \left\{ z_k, e^{-4i(z_n-\rho)^2 t} c_k \right\}_{k=1}^{N} \right\},$$

where $\left\{ r(z), \{z_k, c_k\}_{k=1}^N \right\}$ are obtained from the initial data $q(x,0) = q_0(x)$. Let us introduce the phase function:

$$\theta(z) = (z-\rho)\frac{x}{t} + 2(z-\rho)^2. \tag{1.52}$$

For convenience we denote $\theta_n = \theta(z_n)$. The (time-dependent) inverse spectral problem is defined by the following Riemann-Hilbert problem (RHP modified NLS) and reconstruction formula:

RHP of modified NLS. Find a matrix-valued function z$\in C \to N(z;x,t)$ which satisfies:

- Analyticity: $N(z;x,t)$ is meromorphic in $\mathbb{C} \setminus R$ and has single poles;
- Jump condition: M has continuous boundary values $N_\pm$ on R and

$$N^+(z;x,t) = N^-(z;x,t)V(z), \quad z \in R, \tag{1.53}$$

where

$$V(z) = \begin{pmatrix} 1+z|r(z)|^2 & -e^{-2it\theta}r(z) \\ -e^{2it\theta}z\overline{r(z)} & 1 \end{pmatrix}; \tag{1.54}$$

- Asymptotic behaviors:

$$N(z;x,t) \sim \begin{pmatrix} 1 & 0 \\ -\frac{i}{2}\bar{q} & 1 \end{pmatrix} + O(z^{-1}), \qquad z \to \infty; \tag{1.55}$$

- Residue conditions: N has simple poles at each point in $Z \bigcup \bar{Z}$ with:

$$\mathrm{Res}_{z=z_n} N(z) = \lim_{z\to z_n} N(z) \begin{pmatrix} 0 & 0 \\ c_n e^{2i\theta_n t} & 0 \end{pmatrix}, \tag{1.56}$$

$$\mathrm{Res}_{z=\bar{z}_n} N(z) = \lim_{z\to \bar{z}_n} N(z) \begin{pmatrix} 0 & -z_n^{-1}\bar{c}_n e^{-2i\bar{\theta}_n t} \\ 0 & 0 \end{pmatrix}. \tag{1.57}$$

Reconstruction formula is

$$q(x,t) = 2i \lim_{z\to\infty} [zN(z;x,t)]_{12}. \tag{1.58}$$

Inserting the time dependence into (1.46) and (1.47), we end up exactly with (1.56) and (1.57). We summarize the method of (inverse) scattering works as follows in Figure 1:

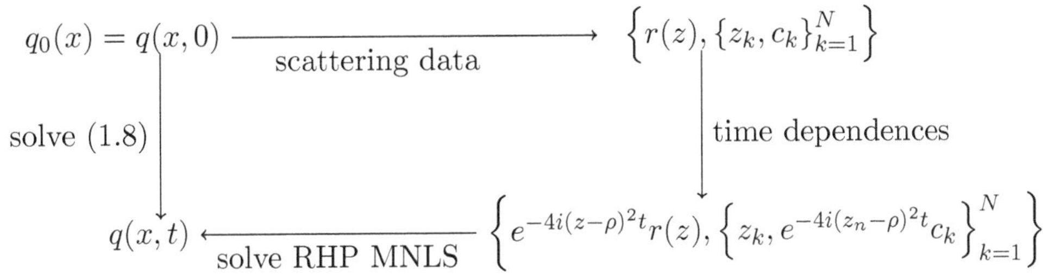

$$\begin{array}{ccc} q_0(x) = q(x,0) & \xrightarrow{\text{scattering data}} & \left\{r(z), \{z_k, c_k\}_{k=1}^N\right\} \\ \text{solve (1.8)} \downarrow & & \downarrow \text{time dependences} \\ q(x,t) & \xleftarrow{\text{solve RHP MNLS}} & \left\{e^{-4i(z-\rho)^2 t}r(z), \left\{z_k, e^{-4i(z_n-\rho)^2 t}c_k\right\}_{k=1}^N\right\} \end{array}$$

Figure 1

Because RHP of the modified NLS is not properly normalized, we only consider the first row of $N(z;x,t)$ which is denoted as $M(z;x,t)$.

Riemann-Hilbert Problem 1 (RHP1). Find a row vector-valued function z$\in C \to M(z;x,t)$ which satisfies:

- Analyticity: $M(z;x,t)$ is meromorphic in $\mathbb{C}\setminus R$ and has single poles;
- Jump condition: M has continuous boundary values $M_\pm$ on R and

$$M^+(z;x,t) = M^-(z;x,t)V(z), \qquad z \in R, \tag{1.59}$$

where

$$V(z) = \begin{pmatrix} 1+z|r(z)|^2 & -e^{-2i\theta t}r(z) \\ -e^{2it\theta t}z\overline{r(z)} & 1 \end{pmatrix}; \tag{1.60}$$

- Asymptotic behaviors: There exists p independent of z that

$$M(x,z) = \begin{pmatrix} 1 & 0 \end{pmatrix} + O(z^{-1}), \quad z \to \infty; \tag{1.61}$$

- Residue conditions: M has simple poles at each point in $Z \bigcup \bar{Z}$ with:

$$\mathrm{Res}_{z=z_n} M(z) = \lim_{z\to z_n} M(z) \begin{pmatrix} 0 & 0 \\ c_n e^{2i\theta_n t} & 0 \end{pmatrix}, \tag{1.62}$$

$$\mathrm{Res}_{z=\bar{z}_n} M(z) = \lim_{z\to \bar{z}_n} M(z) \begin{pmatrix} 0 & -z_n^{-1}\bar{c}_n e^{-2i\bar{\theta}_n t} \\ 0 & 0 \end{pmatrix}. \tag{1.63}$$

2 Conjugation

The long-time asymptotic analysis of RHP1 is determined by the growth and decay of the exponential function $e^{2it\theta}$ appearing in both the jump relation and the residue conditions. In this Section, we describe a new transform: $M(z) \to M^{(1)}(z)$, from which we make that the RHP1 is well behaved as $|t| \to \infty$ along any characteristic line. Let $z_0 = -\frac{x}{4t} + \rho$ be the (unique) critical point of the phase function $\theta(z)$. Then we have

$$\theta = (z-\rho)(2z - 4z_0 + 2\rho), \quad \mathrm{Re}(2it\theta) = -8t\mathrm{Imz}(\mathrm{Re}z - z_0). \tag{2.1}$$

The partition $\Delta^{\pm}_{z_0,\eta}$ of $\{1,...,N\}$ for $z_0 \in R$, $\eta = \mathrm{sgn}(t)$ is defined as follows:

$$\Delta^{+}_{z_0,\eta} = \{k \in \{1,...,N\} \,|\mathrm{Re}(z_k) > z_0\},$$
$$\Delta^{-}_{z_0,\eta} = \{k \in \{1,...,N\} \,|\mathrm{Re}(z_k) < z_0\}.$$

This partition splits the residue coefficients c_n in two sets which is shown in Figure 2.

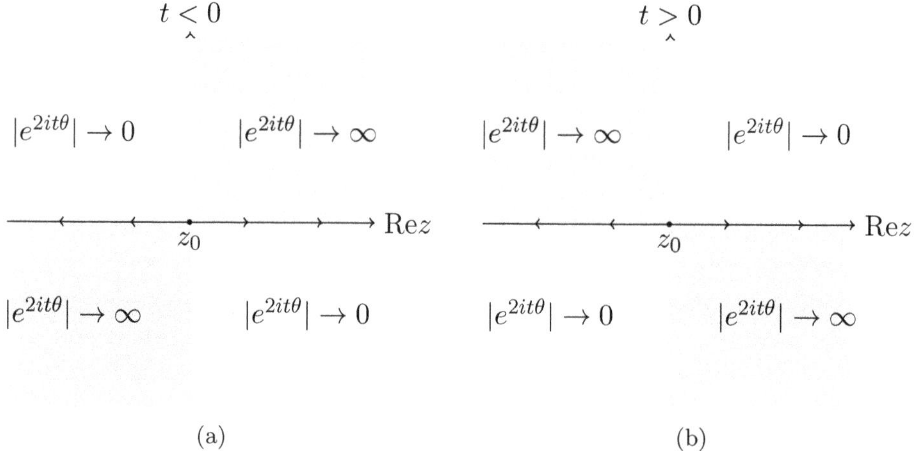

Figure 2: In the yellow region, $|e^{2it\theta}| \to \infty$ when $t \to \pm\infty$, respectively. And in the white region, $|e^{2it\theta}| \to 0$ when $t \to \pm\infty$, respectively.

To introduce a transformation which renormalizes **RHP1** with well-conditioned for $|t| \to \infty$ and fixed z_0, we denote following functions:

$$I_{z_0}^{\eta} = \{s \in R| -\infty < \eta s \le \eta z_0\}, \quad \delta(z) = \delta(z, z_0, \eta) = \exp\left(i \int_{I_{z_0}^{\eta}} \frac{k(s)ds}{s-z} \right) \tag{2.2}$$

$$T(z) = T(z, z_0, \eta) = \prod_{k \in \Delta_{z_0,\eta}^{-}} \frac{z - \bar{z}_k}{z - z_k} \delta(z), \tag{2.3}$$

$$\beta(z, z_0, \eta) = -\eta k(z_0) \log(\eta(z - z_0 + 1)) + \int_{I_{z_0}^{\eta}} \frac{k(s) - X(s)k(z_0)}{s - z} ds, \tag{2.4}$$

$$T_0(z_0) = T_0(z_0, \eta) = \prod_{k \in \Delta_{z_0,\eta}^{-}} \frac{z_0 - \bar{z}_k}{z_0 - z_k} e^{i\beta(z_0, z_0, \eta)}, \tag{2.5}$$

where $k(s)$ is defined in (1.34), and $X(s)$ is the characteristic function of the interval $\eta z_0 - 1 < \eta s < \eta z_0$. In all of the above formulas, we choose the principal branch of power and logarithm functions. From (1.33) we find that the function $T(z, z_0)$ is a partial transmission coefficient which approaches $a(z)^{-1}$ as $z_0 \to \eta$.

Proposition 1. *The function defined by (2.3) has the following properties:*
(a) T is meromorphic in $C \setminus I_{z_0}^{\eta}$, for each $n \in \Delta_{z_0,\eta}^{-}$, $T(z)$ has a simple pole at z_n and a simple zero at $\bar{z}_n$;
(b) For $z \in C \setminus I_{z_0}^{\eta}$, $\overline{T(\bar{z})}T(z) = 1$;
(c) For $z \in I_{z_0}^{\eta}$, as z approaches the real axis from above and below, T has boundary values $T_{\pm}$, which satisfy:

$$T_+(z) = (1 + z|r(z)|^2)T_-(z), \quad z \in I_{z_0}^{\eta}; \tag{2.6}$$

(d) As $|z| \to \infty$ with $|arg(z)| \le c < \pi$,

$$T(z) = 1 + \frac{i}{z}\left[2 \sum_{k \in \Delta_{z_0,\eta}^{-}} Im(z_k) - \int_{I_{z_0}^{\eta}} k(s)ds \right]; \tag{2.7}$$

(e) As $z \to z_0$, along $z = z_0 + e^{i\psi}l$, $l > 0$, $|\psi| \le c < \pi$,

$$|T(z, z_0, \eta) - T_0(z_0, \eta)(\eta(z - z_0))^{i\eta k(z_0)}| \le C|z - z_0|^{1/2}. \tag{2.8}$$

Proof. Parts (a) and (b) are elementary consequences of the Definition (2.3). And for Part (c), as it definition we only need to consider $\int_{I_{z_0}^{\eta}} \frac{k(s)ds}{s-z}$, and via the residue theorem we obtain the consequence. For Part (d), we expand the product term and the factor $(s-z)^{-1}$ for large z

$$\prod_{k\in\Delta^-_{z_0,\eta}} \frac{z-\bar{z}_k}{z-z_k} = \prod_{k\in\Delta^-_{z_0,\eta}} \left[1+\frac{2i}{z}Im(z_k)+O(z^{-2})\right], \tag{2.9}$$

$$\delta(z) = exp\left(-\frac{i}{z}\int_{I^\eta_{z_0}} k(s)ds + O(z^{-2})\right) = 1-\frac{i}{z}\int_{I^\eta_{z_0}} k(s)ds + O(z^{-2}). \tag{2.10}$$

Noting the fact that $\| k \|_{L^1(R)} \leq (2\pi)^{-1} \| r \|_{H^{2,2}(R)}$, the integration in (2.10) also makes sense. Combining (2.9) and (2.10), we obtain (2.7) immediately. For Part (e), we rewrite T:

$$T(z,z_0) = \prod_{k\in\Delta^-_{z_0,\eta}} \frac{z-\bar{z}_k}{z-z_k}(\eta(z-z_0))^{i\eta k(z_0)} \exp(i\beta(z,z_0,\eta)).$$

Thus

$$\begin{aligned}&|T(z,z_0,\eta)-T_0(z_0,\eta)(\eta(z-z_0))^{i\eta k(z_0)}| = \\ &|(\eta(z-z_0))^{i\eta k(z_0)}||\prod_{k\in\Delta^-_{z_0,\eta}} \frac{z-\bar{z}_k}{z-z_k}e^{i\beta(z,z_0,\eta)} - \prod_{k\in\Delta^-_{z_0,\eta}} \frac{z_0-\bar{z}_k}{z_0-z_k}e^{i\beta(z_0,z_0,\eta)}|.\end{aligned} \tag{2.11}$$

We consider the case that $\eta=1$, the other case $\eta=-1$ is similarly. Note the fact that

$$\begin{aligned}|(\eta(z-z_0))^{ik(z_0)}| &= |l^{ik(z_0)} \exp\left(\frac{\psi}{2\pi}\log(1+z_0|r(z_0|^2)\right)| \\ &\leq \exp\left(\frac{1}{2}\log(1+z_0|r(z_0|^2)\right) \\ &= \sqrt{1+z_0|r(z_0|^2},\end{aligned} \tag{2.12}$$

and

$$|\prod_{k\in\Delta^-_{z_0,\eta}} \frac{z-\bar{z}_k}{z-z_k} - \prod_{k\in\Delta^-_{z_0,\eta}} \frac{z_0-\bar{z}_k}{z_0-z_k}| = O(l), \qquad \text{as } l\to 0, \tag{2.13}$$

with constants depending on ψ. And for function $\beta(z,z_0,1)$ we have

$$\begin{aligned}&|\beta(z,z_0,1)-\beta(z_0,z_0,1)| = \\ &|-k(z_0)\log(le^{i\psi}+1) + \int_{-\infty}^{z_0}\left(\frac{1}{s-z}-\frac{z}{s-z_0}\right)(k(s)-X(s)k(z_0))\,ds| \\ &|\int_{-\infty}^{z_0-1}\left(\frac{1}{s-z}-\frac{z}{s-z_0}\right)k(s)ds| \\ &+|\int_{z_0-1}^{z_0}\left(\frac{1}{s-z}-\frac{z}{s-z_0}\right)(k(s)-X(s)k(z_0))\,ds| + O(t),\end{aligned} \tag{2.14}$$

where

$$\int_{-\infty}^{z_0-1}\left(\frac{1}{s-z}-\frac{z}{s-z_0}\right)k(s)ds=\int_{-\infty}^{z_0-1}\frac{le^{i\psi}}{(s-z)(s-z_0)}k(s)ds=O(l), \tag{2.15}$$

and

$$|\int_{z_0-1}^{z_0}\left(\frac{1}{s-z}-\frac{z}{s-z_0}\right)(k(s)-X(s)k(z_0))\,ds|=$$
$$|\int_{z_0-1}^{z_0}\frac{le^{i\psi}}{s-z_0+le^{i\psi}}k'(z')ds|\le l\left(\int_{z_0-1}^{z_0}|\frac{1}{s-z_0+le^{i\psi}}|^2ds\right)^{1/2}\parallel k'\parallel_{L^2}$$
$$\le cl^{\frac{1}{2}}, \tag{2.16}$$

with c depending on $\parallel k \parallel_{H^2}$. Combining the above equations, we finally get the result. ■

We now use T to define a new unknown function $M^{(1)}$:

$$M^{(1)}(z)=M(z)T(z)^{-\sigma_3}. \tag{2.17}$$

Note that if $M(z)$ is solution of **RHP1**, then $M^{(1)}(z)$ satisfies the following **RHP**.

Riemann-Hilbert Problem 2 (RHP2). Find a row vector valued function $M^{(1)}(z;x,t)$ which satisfies:

- Analyticity: $M^{(1)}(z;x,t)$ is meromorphic in $\mathbb{C}\setminus R$ and has single poles;
- Jump condition: $M^{(1)}$ has continuous boundary values $M_\pm^{(1)}$ on R and

$$M_+^{(1)}(z;x,t)=M_-^{(1)}(z;x,t)V^{(1)}(z),\qquad z\in R, \tag{2.18}$$

where

$$\text{as } z\in I_{z_0}^{\eta}, V^{(1)}(z)=\begin{pmatrix}1 & -r(z)T(z)^2e^{-2it\theta}\\ 0 & 1\end{pmatrix}\begin{pmatrix}1 & 0\\ -z\overline{r(z)}T(z)^{-2}e^{2it\theta} & 1\end{pmatrix}, \tag{2.19}$$

$$\text{as } z\in R\setminus I_{z_0}^{\eta}, V^{(1)}(z)=\begin{pmatrix}1 & 0\\ -\frac{z\overline{r(z)}T_-(z)^{-2}}{1+z|r(z)|^2}e^{2it\theta} & 1\end{pmatrix}\begin{pmatrix}1 & -\frac{r(z)T_+(z)^2}{1+z|r(z)|^2}e^{-2it\theta}\\ 0 & 1\end{pmatrix}; \tag{2.20}$$

- Asymptotic behaviors:

$$M^{(1)}(z;x,t)=\begin{pmatrix}1 & 0\end{pmatrix}+O(z^{-1}),\qquad z\to\infty; \tag{2.21}$$

- Residue conditions: M has simple poles at each point in $Z\bigcup\bar{Z}$ with:

1. When $n \in \Delta^+_{z_0,\eta}$,

$$\operatorname*{Res}_{z=z_n} M^{(1)}(z) = \lim_{z\to z_n} M^{(1)}(z) \begin{pmatrix} 0 & 0 \\ c_n T(z_n)^{-2} e^{2i\theta_n t} & 0 \end{pmatrix}, \tag{2.22}$$

$$\operatorname*{Res}_{z=\bar{z}_n} M^{(1)}(z) = \lim_{z\to \bar{z}_n} M^{(1)}(z) \begin{pmatrix} 0 & -z_n^{-1} T(\bar{z}_n)^2 \bar{c}_n e^{-2i\bar{\theta}_n t} \\ 0 & 0 \end{pmatrix}. \tag{2.23}$$

2. When $n \in \Delta^-_{z_0,\eta}$,

$$\operatorname*{Res}_{z=z_n} M^{(1)}(z) = \lim_{z\to z_n} M^{(1)}(z) \begin{pmatrix} 0 & (c_n (1/T)'(z_n)^2 e^{2i\theta_n t})^{-1} \\ 0 & 0 \end{pmatrix}, \tag{2.24}$$

$$\operatorname*{Res}_{z=\bar{z}_n} M^{(1)}(z) = \lim_{z\to \bar{z}_n} M^{(1)}(z) \begin{pmatrix} 0 & 0 \\ (-z_n T'(\bar{z}_k)^2 \bar{c}_n e^{-2i\bar{\theta}_n t})^{-1} & 0 \end{pmatrix}. \tag{2.25}$$

Proof. The analyticity, jump condition and asymptotic behaviors of $M^{(1)}(z)$ is directly from its definition, Proposition 1 and the properties of M. As for residues, because $T(z)$ is analytic at each z_n and $\bar{z}_n$ for $n \in \Delta^+_{z_0,\eta}$, from (1.57), (1.62) and (2.17) we obtain residue conditions at these points immediately. For $n \in \Delta^-_{z_0,\eta}$, we denote $M(z) = (M_1(z), M_2(z))$, then $M^{(1)}(z) = \left(M_1^{(1)}(z), M_2^{(1)}(z)\right) = \left(M_1(z)T^{-1}(z), M_2(z)T(z)\right)$. $T(z)$ has a simple zero at $\bar{z}_n$ and a pole at z_n, so z_n is no longer the pole of $M_1^{(1)}(z)$ with $\bar{z}_n$ becoming the pole of it. And $M_2^{(1)}(z)$ has the opposite situation. It has a pole at z_n and a removable singularity at $\bar{z}_n$. We consider the residue condition of $M_2^{(1)}(z)$ at pole z_n,

$$\begin{aligned} M_1^{(1)}(z_n) &= \lim_{z\to z_n} M_1(z) T^{-1}(z) = \operatorname*{Res}_{z=z_n} M_1(z) (1/T)'(z_n) \\ &= c_n (1/T)'(z_n) e^{2i\theta_n t} M_2(z_n), \\ \operatorname*{Res}_{z=z_n} M_2^{(1)}(z) &= \operatorname*{Res}_{z=z_n} M_2(z) T(z) = M_2(z_k) \left[(1/T)'(z_k)\right]^{-1} \\ &= (c_n (1/T)'(z_n)^2 e^{2i\theta_n t})^{-1} M_1^{(1)}(z_n). \end{aligned} \tag{2.26}$$

Then we have (2.24), and (2.25) is similar. ■

3 Mixed $\bar{\partial}$-Riemann-Hilbert Problem

Next we introduce transformations of the jump matrix which deform the contours to other new contours defined as follows:

$$\Sigma_k = z_0 + e^{(2k-1)i\pi/4} R_+, \qquad k = 1, 2, 3, 4, \tag{3.1}$$

$$\Sigma^{(2)} = \Sigma_1 \cup \Sigma_2 \cup \Sigma_3 \cup \Sigma_4. \tag{3.2}$$

Along $\Sigma^{(2)}$, the jumps are decaying. But the price we pay for this non-analytic transformation is that it shows new unknown nonzero $\bar{\partial}$ derivatives inside the regions in which the extensions are introduced and satisfies a mixed $\bar{\partial}$-Riemann-Hilbert

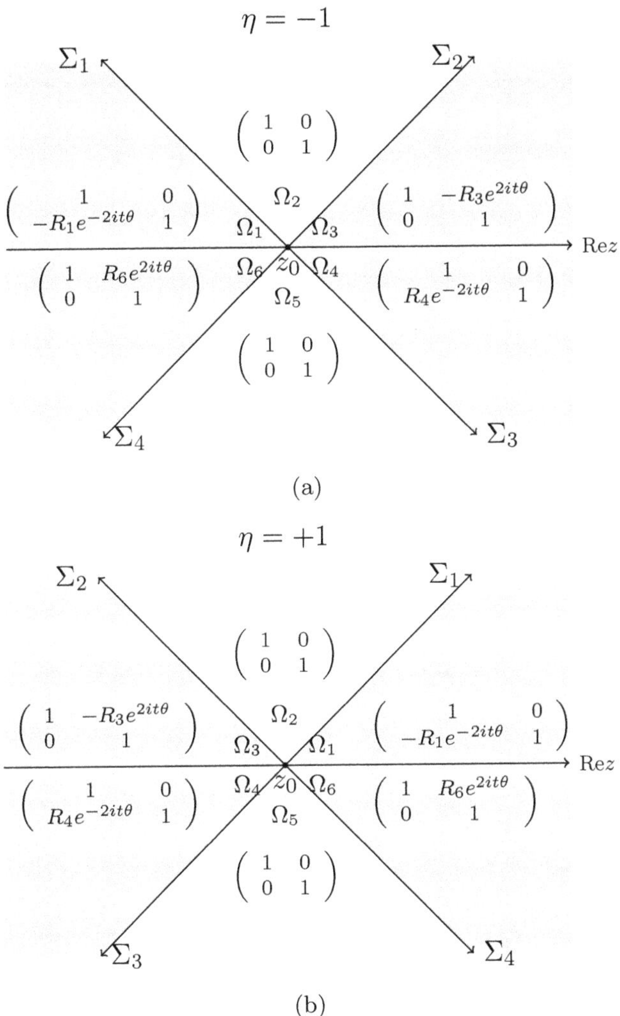

Figure 3: In the yellow region, $|e^{2it\theta}| \to \infty$ when $t \to \pm\infty$, respectively. And in white region, $|e^{2it\theta}| \to 0$ when $t \to \pm\infty$, respectively.

problem. R and $\Sigma^{(2)}$ separate C into six open sectors. We denote these sectors by Ω_k, $k = 1, ..., 6$, starting with sector Ω_1 between $I_{z_0}^{\eta}$ and Σ_1 and numbered consecutively continuing counterclockwise for $\eta = 1$ (η = -1 is similarly), as shown in Figure 3.

Additionally, let

$$\mu = \frac{1}{2} \min_{\lambda, \gamma \in Z \cup \bar{Z} \text{ and } \lambda \neq \gamma} |\lambda - \gamma|. \tag{3.3}$$

As we assuming there is no pole on the real axis, we obtain $\mu <$dist(Z, R). Then we define $X_Z \in C_0^{\infty}(C, [0, 1])$ which only supported on the neighborhood of $Z \cup \bar{Z}$,

$$X_Z(z) = \begin{cases} 1 & \text{dist}(z, Z \cup \bar{Z}) < \mu/3 \\ 0 & \text{dist}(z, Z \cup \bar{Z}) > 2\mu/3. \end{cases} \tag{3.4}$$

In order to deform the contour R to the contour $\Sigma^{(2)}$, we introduce a new unknown function $M^{(2)}$.

$$M^{(2)}(z) = M^{(1)}(z)R^{(2)}(z). \tag{3.5}$$

And we need $R^{(2)}(z)$ to satisfy the following conditions: First, $M^{(2)}$ has no jump on the real axis, so we choose the boundary values of $R^{(2)}(z)$ through the factorization of $V^{(1)}(z)$ in (2.18) where the new jumps on $\Sigma^{(2)}$ match a well-known model RHP. Second, we need to control the norm of $R^{(2)}(z)$, so that the $\bar{\partial}$-contribution to the long-time asymptotics of $q(x,t)$ can be ignored. Third the residues are unaffected by the transformation. So we choose $R^{(2)}(z)$ as shown in Figure 3, where the function R_j, $j = 1, 3, 4, 6$ is defined as follows.

Proposition 2. *R_j: $\bar{\Omega}_j \to C$, $j = 1, 3, 4, 6$ have boundary values as follows:*

$$R_1(z) = \begin{cases} z\bar{r}(z)T(z)^{-2} & z \in I^{\eta}_{z_0} \\ z_0\bar{r}(z_0)T_0(z_0)^{-2}(\eta(z-z_0))^{-2i\eta k(z_0)}(1 - X_Z(z)) & z \in \Sigma_1 \end{cases}, \tag{3.6}$$

$$R_3(z) = \begin{cases} \dfrac{r(z_0)T_0(z_0)^2}{1+z_0|r(z_0)|^2}(\eta(z-z_0))^{2i\eta k(z_0)}(1 - X_Z(z)) & z \in \Sigma_2 \\ \dfrac{r(z)T_+(z)^2}{1+z|r(z)|^2} & z \in R \setminus (I^{\eta}_{z_0} \cup \{z_0\}) \end{cases}, \tag{3.7}$$

$$R_4(z) = \begin{cases} -\dfrac{z\bar{r}(z)T_-(z)^{-2}}{1+z|r(z)|^2} & z \in R \setminus (I^{\eta}_{z_0} \cup \{z_0\}) \\ -\dfrac{z_0\bar{r}(z_0)T_0(z_0)^{-2}}{1+z_0|r(z_0)|^2}(\eta(z-z_0))^{-2i\eta k(z_0)}(1 - X_Z(z)) & z \in \Sigma_3 \end{cases}, \tag{3.8}$$

$$R_6(z) = \begin{cases} -r(z_0)T_0(z_0)^2(\eta(z-z_0))^{2i\eta k(z_0)}(1 - X_Z(z)) & z \in \Sigma_4 \\ -r(z)T(z)^2 & z \in I^{\eta}_{z_0} \end{cases}. \tag{3.9}$$

then we can find a fixed constant $c_1{=}c_1(q_0)$, we have

$$|R_j(z)| \le c_1\left(\sin^2(\arg(z-z_0)) + \langle Re(z)\rangle^{-1/2}\right), \tag{3.10}$$

$$|\bar{\partial}R_j(z)| \le c_1\left(|\bar{\partial}X_Z(z)| + |z^{m_k}p'_k(Rez)| + |z-z_0|^{-1/2}\right), \tag{3.11}$$

$$\bar{\partial}R_j(z) = 0, \quad \text{if } z \in \Omega_2 \cup \Omega_5 \text{ or } dist(z, Z \cup \bar{Z}) < \mu/3, \tag{3.12}$$

where $m_1 = m_4 = 1$, $m_3 = m_6 = 0$, and

$$p_1(z) = \bar{r}(z), \qquad p_3(z) = \frac{r(z)}{1+z|r(z)|^2}, \tag{3.13}$$

$$p_4(z) = -\frac{zr(z)}{1+z|r(z)|^2}, \qquad p_6(z) = -r(z). \tag{3.14}$$

Proof. Define the functions

$$f_1(z) = z_0 p_1(z_0) T(z)^2 T_0(z_0)^{-2} (\eta(z - z_0))^{-2i\eta k(z_0)}, \qquad z \in \bar{\Omega}_1 \tag{3.15}$$

$$f_3(z) = p_3(z_0) T(z)^{-2} T_0(z_0)^2 (\eta(z - z_0))^{2i\eta k(z_0)}, \qquad z \in \bar{\Omega}_3. \tag{3.16}$$

Let $z - z_0 = le^{i\psi}$, then using the above function we can give the construction for $R_1(z)$ and $R_3(z)$

$$R_1(z) = [f_1(z) + \eta(\mathrm{Re}z p_1(\mathrm{Re}z) - f_1(z)) cos(2\psi)] \, T(z)^{-2}(1 - X_Z(z)), \tag{3.17}$$

$$R_3(z) = [f_3(z) + \eta(p_3(\mathrm{Re}z) - f_3(z)) cos(2\psi)] \, T(z)^2 (1 - X_Z(z)). \tag{3.18}$$

The construction of $R_4(z)$ and $R_6(z)$ are similar. We proof this proposition for example of $R_1(z)$, the case of the others is the same. And from the definition of $R_1(z)$ we can easily obtain that $\bar{\partial} R_j(z) = 0$, if $z \in \Omega_2 \cup \Omega_5$ or $\mathrm{dist}(z, Z \cup \bar{Z}) < \mu/3$. Then we estimate $|R_1(z)|$,

$$\begin{aligned} |R_1(z)| \leq & |f_1(z) T(z)^{-2} (1 - X_Z(z))(1 - \eta cos(2\psi))| \\ & + |\mathrm{Re}z p_1(\mathrm{Re}z) T(z)^{-2} (1 - X_Z(z)) cos(2\psi)|. \end{aligned} \tag{3.19}$$

Note that $(1 - X_Z(z))$ is zero in $(z, Z \cup \bar{Z}) < \mu/3$, so $|T(z)^{-2}(1 - X_Z(z))|$ is bounded. And from $r \in H^2(R)$, which means $p_1 \in H^2(R)$ we have $|p_1(u)| \lesssim \langle u \rangle^{-1/2}$. Together with (2.12) we have (3.10). Since

$$\bar{\partial} = \frac{1}{2} (\partial_x + i\partial_y) = \frac{e^{i\psi}}{2} \left(\partial_l + il^{-1}\partial_\psi\right),$$

we have

$$\begin{aligned} \bar{\partial} R_1(z) = & - [f_1(z) + \eta(\mathrm{Re}z p_1(\mathrm{Re}z) - f_1(z)) cos(2\psi)] \, T(z)^{-2} \bar{\partial} X_Z(z) \\ & + \eta \frac{\mathrm{Re}z}{2} p_1'(\mathrm{Re}z) cos(2\psi) T(z)^{-2} (1 - X_Z(z)) \\ & - \frac{ie^{i\psi}\eta}{|z - z_0|} (\mathrm{Re}z p_1(\mathrm{Re}z) - f_1(z)) sin(2\psi) T(z)^{-2} (1 - X_Z(z)). \end{aligned} \tag{3.20}$$

Because $\bar{\partial} X_Z(z)$ is supported in $\mathrm{dist}(z, Z \cup \bar{Z}) \in [\mu/3, 2\mu/3]$, similarly, we have the first two terms are bounded. For the last term, we have

$$\begin{aligned} |\mathrm{Re}z p_1(\mathrm{Re}z) - f_1(z)| \leq & |\mathrm{Re}z p_1(\mathrm{Re}z) - z_0 p_1(z_0)| \\ & + |z_0 p_1(z_0)| |1 - T(z)^2 T_0(z_0)^{-2} (\eta(z - z_0))^{-2i\eta k(z_0)}|. \end{aligned} \tag{3.21}$$

For the first term using Cauchy-Schwarz, we have

$$\begin{aligned} |Rez p_1(Rez) - z_0 p_1(z_0)| & \leq | \int_{z_0}^{\mathrm{Re}z} s p'(s) + p(s) ds| \\ & \leq |z - z_0|^{1/2} \parallel s p'(s) + p(s) \parallel_{L^2(R)} \\ & \leq 2|z - z_0|^{1/2} \parallel p(s) \parallel_{H^2(R)} . \end{aligned} \tag{3.22}$$

And together with **Proposition 1**, which also implies $|T_0(z_0)|$ and $|(\eta(z-z_0))^{-2i\eta k(z_0)}|$ are bounded in a neighborhood of z_0, we come to the consequence for the estimation of $|\bar{\partial}R_1|$. ■

Then we use $R^{(2)}$ to define a new unknown function

$$M^{(2)} = M^{(1)}R^{(2)}, \tag{3.23}$$

which satisfies the following mixed $\bar{\partial}$-Riemann-Hilbert problem.

Mixed $\bar{\partial}$-Riemann-Hilbert problem 3 (RHP3). Find a row vector valued function z$\in C \to M^{(2)}(z;x,t)$ with the following properties:

• Analyticity: $M^{(2)}(z;x,t)$ is continuous with Sectionally continuous first partial derivatives in $\mathbb{C}\setminus(\Sigma^{(2)}\cup Z\cup\bar{Z})$ and meromorphic in $\Omega_2\cup\Omega_5$;

• Jump condition: $M^{(2)}$ has continuous boundary values $M_\pm^{(2)}$ on R and

$$M_+^{(2)}(z;x,t) = M_-^{(2)}(z;x,t)V^{(2)}(z), \qquad z\in R, \tag{3.24}$$

where

$$V^{(2)}(z) = I + (1-X_Z(z))V_0^{(2)}, \tag{3.25}$$

$$V_0^{(2)} = \begin{cases} \begin{pmatrix} 0 & 0 \\ -z\overline{r(z)}T_0(z_0)^{-2}(\eta(z-z_0))^{-2i\eta k(z_0)}e^{2it\theta} & 0 \end{pmatrix}, & z\in\Sigma_1; \\ \begin{pmatrix} 0 & -\dfrac{r(z)T_0(z_0)^2}{1+z|r(z)|^2}(\eta(z-z_0))^{2i\eta k(z_0)}e^{-2it\theta} \\ 0 & 0 \end{pmatrix}, & z\in\Sigma_2; \\ \begin{pmatrix} 0 & 0 \\ -\dfrac{z\overline{r(z)}T_0(z_0)^{-2}}{1+z|r(z)|^2}(\eta(z-z_0))^{-2i\eta k(z_0)}e^{2it\theta} & 0 \end{pmatrix}, & z\in\Sigma_3; \\ \begin{pmatrix} 0 & -r(z)T_0(z_0)^2(\eta(z-z_0))^{2i\eta k(z_0)}e^{-2it\theta} \\ 0 & 0 \end{pmatrix}, & z\in\Sigma_4, \end{cases} \tag{3.26}$$

• Asymptotic behaviors:

$$M^{(2)}(z;x,t) \sim \begin{pmatrix} 1 & 0 \end{pmatrix} + O(z^{-1}), \qquad z\to\infty. \tag{3.27}$$

- $\bar{\partial}$-Derivative: For $\mathbb{C}\setminus(\Sigma^{(2)}\cup Z\cup\bar{Z})$ we have $\bar{\partial}M^{(2)} = M^{(1)}\bar{\partial}R^{(2)}$, where

$$\bar{\partial}R^{(2)} = \begin{cases} \begin{pmatrix} 0 & 0 \\ -\bar{\partial}R_1 e^{2it\theta} & 0 \end{pmatrix}, & z \in \Omega_1; \\ \begin{pmatrix} 0 & -\bar{\partial}R_3 e^{-2it\theta} \\ 0 & 0 \end{pmatrix}, & z \in \Omega_3; \\ \begin{pmatrix} 0 & 0 \\ \bar{\partial}R_4 e^{2it\theta} & 0 \end{pmatrix}, & z \in \Omega_4; \\ \begin{pmatrix} 0 & \bar{\partial}R_6 e^{-2it\theta} \\ 0 & 0 \end{pmatrix}, & z \in \Omega_6; \\ 0, & z \in \Omega_2 \cup \Omega_5. \end{cases} \tag{3.28}$$

- Residue conditions: M has simple poles at each point in $Z\bigcup\bar{Z}$ with:
  1. When $n \in \Delta^+_{z_0,\eta}$,

$$\operatorname*{Res}_{z=z_n} M^{(2)}(z) = \lim_{z\to z_n} M^{(2)}(z)\begin{pmatrix} 0 & 0 \\ c_n T(z_n)^{-2} e^{2i\theta_n t} & 0 \end{pmatrix}, \tag{3.29}$$

$$\operatorname*{Res}_{z=\bar{z}_n} M^{(2)}(z) = \lim_{z\to \bar{z}_n} M^{(2)}(z)\begin{pmatrix} 0 & -z_n^{-1} T(\bar{z}_n)^{2} \bar{c}_n e^{-2i\bar{\theta}_n t} \\ 0 & 0 \end{pmatrix}. \tag{3.30}$$

  2. When $n \in \Delta^-_{z_0,\eta}$,

$$\operatorname*{Res}_{z=z_n} M^{(2)}(z) = \lim_{z\to z_n} M^{(2)}(z)\begin{pmatrix} 0 & (c_n(1/T)'(z_n)^2 e^{2i\theta_n t})^{-1} \\ 0 & 0 \end{pmatrix}, \tag{3.31}$$

$$\operatorname*{Res}_{z=\bar{z}_n} M^{(2)}(z) = \lim_{z\to \bar{z}_n} M^{(2)}(z)\begin{pmatrix} 0 & 0 \\ (-z_n T'(\bar{z}_k)^2 \bar{c}_n e^{-2i\bar{\theta}_n t})^{-1} & 0 \end{pmatrix}. \tag{3.32}$$

4 Model RHP and pure $\bar{\partial}$-problem

To solve **RHP3**, we decompose it into a model Riemann-Hilbert Problem and a pure $\bar{\partial}$-Problem. First we build a solution M^{RHP} to a model RH problem as following

Riemann-Hilbert problem 4 (RHP4). Find a matrix-valued function $M^{RHP}(z;x,t)$ with the following properties:

- Analyticity: $M^{RHP}(z;x,t)$ is meromorphic in $\mathbb{C}\setminus(\Sigma^{(2)}\cup Z\cup\bar{Z})$;
- Jump condition: M^{RHP} has continuous boundary values $M^{RHP}_\pm$ on R:

$$M^{RHP}_+(z;x,t) = M^{RHP}_-(z;x,t)V^{(2)}(z), \quad z \in R; \tag{4.1}$$

- Symmetry: $M^{RHP}_{22}(z) = \overline{M^{RHP}_{11}(\bar{z})}$, $M^{RHP}_{21}(z) = -z\overline{M^{RHP}_{12}(\bar{z})}$;

• Asymptotic behaviors:

$$M^{RHP}(z;x,t) \sim \begin{pmatrix} 1 & 0 \\ \alpha & 1 \end{pmatrix} + O(z^{-1}), \quad z \to \infty \tag{4.2}$$

for a constant α determined by the symmetry condition above;

• Residue conditions: M^{RHP} has simple poles at each point in $Z \bigcup \bar{Z}$ with:

1. When $n \in \Delta^{+}_{z_0,\eta}$,

$$\operatorname*{Res}_{z=z_n} M^{RHP}(z) = \lim_{z\to z_n} M^{RHP}(z) \begin{pmatrix} 0 & 0 \\ c_n T(z_n)^{-2} e^{2i\theta_n t} & 0 \end{pmatrix}, \tag{4.3}$$

$$\operatorname*{Res}_{z=\bar{z}_n} M^{RHP}(z) = \lim_{z\to \bar{z}_n} M^{RHP}(z) \begin{pmatrix} 0 & -z_n^{-1} T(\bar{z}_n)^2 \bar{c}_n e^{-2i\bar{\theta}_n t} \\ 0 & 0 \end{pmatrix}. \tag{4.4}$$

2. When $n \in \Delta^{-}_{z_0,\eta}$,

$$\operatorname*{Res}_{z=z_n} M^{RHP}(z) = \lim_{z\to z_n} M^{RHP}(z) \begin{pmatrix} 0 & (c_n (1/T)'(z_n)^2 e^{2i\theta_n t})^{-1} \\ 0 & 0 \end{pmatrix}, \tag{4.5}$$

$$\operatorname*{Res}_{z=\bar{z}_n} M^{RHP}(z) = \lim_{z\to \bar{z}_n} M^{RHP}(z) \begin{pmatrix} 0 & 0 \\ (-z_n T'(\bar{z}_k)^2 \bar{c}_n e^{-2i\bar{\theta}_n t})^{-1} & 0 \end{pmatrix}. \tag{4.6}$$

And we will prove the existence of M^{RHP} and construct its asymptotic expansion for $t \to \infty$ later. Before this, we first consider to use M^{RHP} to get the solution of following pure $\bar{\partial}$-problem

$$M^{(3)}(z) = M^{(2)}(z) M^{RHP}(z)^{-1}. \tag{4.7}$$

$\bar{\partial}$-problem 5. Find a vector-valued function z∈ $C \to M^{(3)}(z;x,t)$ with the following properties:

• Analyticity: $M^{(3)}(z;x,t)$ is continuous with Sectionally continuous first partial derivatives in $\mathbb{C} \setminus (\Sigma^{(2)} \cup Z \cup \bar{Z})$ and meromorphic in $\Omega_2 \cup \Omega_5$.

• Asymptotic behavior:

$$M^{(3)}(z;x,t) \sim \begin{pmatrix} 1 & 0 \end{pmatrix} + O(z^{-1}), \quad z \to \infty; \tag{4.8}$$

• $\bar{\partial}$-Derivative: For $\mathbb{C} \setminus (\Sigma^{(2)} \cup Z \cup \bar{Z})$ we have $\bar{\partial} M^{(3)} = M^{(3)} W^{(3)}$, where

$$W^{(3)} = M^{RHP}(z) \bar{\partial} R^{(2)}(z) M^{RHP}(z)^{-1}. \tag{4.9}$$

Proof. For the property of solutions $M^{(2)}$ and M^{RHP} of **mixed-$\bar{\partial}$-RHP** and **RHP 4**, respectively, the analyticity and asymptotic behavior come immediately. Since $M^{(2)}$ and M^{RHP} have same jump matrix, we have

$$\begin{aligned} M_-^{(3)}(z)^{-1} M_+^{(3)}(z) &= M_-^{RHP}(z) M_-^{(2)}(z)^{-1} M_+^{(2)}(z) M_+^{RHP}(z)^{-1} \\ &= M_-^{RHP}(z) V^{(2)}(z) \left(M_-^{RHP}(z) V^{(2)}(z) \right)^{-1} \\ &= I, \end{aligned}$$

which means $M^{(3)}$ has no jumps and is everywhere continuous. Then we proof that $M^{(3)}$ has no pole. For instance, if $\lambda \in Z \cup \bar{Z}$ and let N denote the nilpotent matrix which appears in the left side of the corresponding residue condition of **mixed-$\bar{\partial}$-RHP** and **RHP 4**, we have the Laurent expansions in $z-\lambda$

$$M^{(2)}(z) = a(\lambda)\left[\frac{N}{z-\lambda}+I\right]+O(z-\lambda), \quad M^{RHP}(z) = A(\lambda)\left[\frac{N}{z-\lambda}+I\right]+O(z-\lambda), \tag{4.10}$$

where $a(\lambda)$ and $A(\lambda)$ are the constant row vector and matrix in their respective expansions. And from $M^{RHP}(z)^{-1} = \sigma_2 M^{RHP}(z)^T \sigma_2$, we have

$$\begin{aligned} M^{(3)}(z) &= \left\{a(\lambda)\left[\frac{N}{z-\lambda}+I\right]\right\}\left\{\left[\frac{-N}{z-\lambda}+I\right]\sigma_2 A(\lambda)^T\sigma_2\right\} + O(z-\lambda) \\ &= O(1), \end{aligned} \tag{4.11}$$

which means $M^{(3)}$ has removable singularities at λ. And the $\bar{\partial}$-derivative of $M^{(3)}$ is from the $\bar{\partial}$-derivative of $M^{(3)}$ and the analyticity of M^{RHP}. ■

Then we begin to prove the existence of M^{RHP} and explain its characteristics. We construct the solution M^{RHP} as follows:

$$M^{RHP} = \begin{cases} E(z)M^{(out)} & z \notin U_{z_0} \\ E(z)M^{(z_0)} & z \in U_{z_0} \end{cases}, \tag{4.12}$$

where U_{z_0} is the neighborhood of z_0

$$U_{z_0} = \{z : |z-z_0| \le \mu/3\}. \tag{4.13}$$

From the definition we can easily find that M^{RHP} is pole free. This decomposes M^{RHP} into two part: $M^{(out)}$ solves the pure RHP obtained by ignoring the jump conditions of **RHP 4**, which is shown in Section 5; $M^{(z_0)}$ uses parabolic cylinder functions to build a matrix whose jumps exactly match those of $M^{(2)}$ in a neighborhood of the critical point z_0, which is shown in Section 6. And $E(z)$ is the error function, which is a solution of a small-norm Riemann-Hilbert problem shown in Section 8.

5 Outer model Riemann-Hilbert problem

In this Section we build an outer model Riemann-Hilbert problem and research its property. From the previous Section we find that the matrix function M^{RHP} is meromorphic away from the contour $\Sigma^{(2)}$, and on the $\Sigma^{(2)}$, the boundary value satisfies $M_+^{RHP}(z;x,t) = M_-^{RHP}(z;x,t)V^{(2)}(z)$ which have the following proposition.

Proposition 3. *For $V^{(2)}$ which is the jump matrix of M^{RHP}, we have*

$$\| V^{(2)} - I \|_{L^\infty(\Sigma^{(2)})} = O(e^{-4t|z-z_0|^2}). \tag{5.1}$$

Proof. We proof it for example in Σ_1. The other case can be proven in the same way. For $z \in \Sigma_1$, we have

$$\| V^{(2)} - I \|_{L^\infty(\Sigma^{(2)})} = \| R_1(z)e^{2it\theta} \|_{L^\infty(\Sigma^{(2)})} . \tag{5.2}$$

And from **Proposition 1**, **Proposition 2** and (2.1) we have

$$|R_1(z)e^{2it\theta}| \lesssim e^{-Im(2t\theta)} + \langle \mathrm{Re}(z)\rangle^{-1/2} e^{-Im(2t\theta)} \lesssim \left(1 + \langle \mathrm{Re}(z)\rangle^{-1/2}\right) e^{-4t|z-z_0|^2}. \tag{5.3}$$

So we come to the consequence. ■

This proposition means that the jump $V^{(2)}$ is uniformly near identity. So outside the U_{z_0} there is only exponentially small error (in t) by completely ignoring the jump condition of M^{RHP}. Then we can introduce the following outer model problem and obtain existence and uniqueness of its solution by **Proposition 4** .

Riemann-Hilbert problem 6. Find a matrix-valued function z$\in C \to M^{(out)}(z; x, t$ with the following properties:

- Analyticity: $M^{(out)}(z; x, t)$ is meromorphic in $\mathbb{C} \setminus (\Sigma^{(2)} \cup Z \cup \bar{Z})$;
- Symmetry: $M^{(out)}_{22}(z) = \overline{M^{(out)}_{11}(\bar{z})}$, $M^{(out)}_{21}(z) = -z\overline{M^{(out)}_{12}(\bar{z})}$;
- Asymptotic behaviors:

$$M^{(out)}(z; x, t) \sim \begin{pmatrix} 1 & 0 \\ \alpha & 1 \end{pmatrix} + O(z^{-1}), \quad z \to \infty \tag{5.4}$$

for a constant α determined by the symmetry condition above;

- Residue conditions: $M^{(out)}$ has simple poles at each point in $Z \bigcup \bar{Z}$ satisfying the same residue relations of M^{RHP}.

Proposition 4. *Fhe following Riemann-Hilbert problem, which is the reflectionless case $r \equiv 0$ of **RHP MNLS**, has unique solution.*

Problem 1. *Given discrete data $\sigma_d = \{(z_k, c_k)\}_{k=1}^N$, and $Z = \{z_k\}_{k=1}^N$. Find a matrix-valued function z$\in C \to m(z; x, t|\sigma_d)$ with the following properties:*

- *Analyticity: $m(z; x, t|\sigma_d)$ is meromorphic in $\mathbb{C} \setminus (\Sigma^{(2)} \cup Z \cup \bar{Z})$;*
- *Symmetry: $m_{22}(z; x, t|\sigma_d) = \overline{m_{11}(\bar{z}; x, t|\sigma_d)}$, $m_{21}(z; x, t|\sigma_d) = -z\overline{m_{12}(\bar{z}; x, t|\sigma_d)}$, which means $m(z; x, t|\sigma_d) = z^{\sigma_3/2}\sigma_2\overline{m(\bar{z}; x, t|\sigma_d)}\sigma_2^{-1}z^{-\sigma_3/2}$;*
- *Asymptotic behaviors:*

$$m(z; x, t|\sigma_d) \sim \begin{pmatrix} 1 & 0 \\ \alpha & 1 \end{pmatrix} + O(z^{-1}), \quad z \to \infty \tag{5.5}$$

for a constant α determined by the symmetry condition above;

- *Residue conditions: $m(z; x, t|\sigma_d)$ has simple poles at each point in $Z \bigcup \bar{Z}$ satisfying*

$$\operatorname*{Res}_{z=z_n} m(z; x, t|\sigma_d) = \lim_{z \to z_n} m(z; x, t|\sigma_d)\tau_k, \tag{5.6}$$

$$\operatorname*{Res}_{z=\bar{z}_n} m(z; x, t|\sigma_d) = \lim_{z \to \bar{z}_n} m(z; x, t|\sigma_d)\hat{\tau}_k, \tag{5.7}$$

where τ_k is a nilpotent matrix satisfies

$$\tau_k = \begin{pmatrix} 0 & 0 \\ \gamma_k & 0 \end{pmatrix}, \quad \hat{\tau}_k = z_k^{\sigma_3/2} \sigma_2 \overline{\tau_k} \sigma_2^{-1} z_k^{-\sigma_3/2}, \quad \gamma_k = c_k e^{2i[(z-\rho)x + 2(z-\rho)^2 t]}. \tag{5.8}$$

Moreover, the solution satisfies

$$\| m(z; x, t|\sigma_d)^{-1} \|_{L^\infty(C \setminus (Z \cup \bar{Z}))} \lesssim 1. \tag{5.9}$$

Proof. The uniqueness of the solution follows from Liouville's Theorem. And the symmetries in **Problem 1** means that its solution must admit a partial fraction expansion of the following form

$$m(z; x, t|\sigma_d) = \begin{pmatrix} 1 & 0 \\ \alpha & 1 \end{pmatrix} + \sum_{k=1}^{N} \left[\frac{1}{z - z_k} \begin{pmatrix} \nu_k(x,t) & 0 \\ \beta_k(x,t) & 0 \end{pmatrix} + \frac{1}{z - \bar{z}_k} \begin{pmatrix} 0 & -z_k^{-1} \overline{\beta_k(x,t)} \\ 0 & \overline{\nu_k(x,t)} \end{pmatrix} \right]. \tag{5.10}$$

Following the proof in theorem 4.3 of [16] we similarly have the existence of a solution. Since det$(m(z; x, t|\sigma_d))$=1, we only need to consider $\| m(z; x, t|\sigma_d) \|_{L^\infty(C \setminus (Z \cup \bar{Z}))}$. And from (5.10) we simply obtain the consequence. ■

From the Trace formula we have

$$a(z) = \exp\left[-\frac{1}{2\pi i} \int_R \frac{\log[1 + \zeta |r(\zeta)|^2]}{\zeta - z} d\zeta \right] \prod_{n=1}^{N} \frac{z - z_n}{z - \bar{z}_n}, \qquad z \in C^+. \tag{5.11}$$

Let $\triangle \subseteqq \{1, 2, ..., N\}$, and define

$$a_\triangle(z) = \prod_{k \in \triangle} \frac{z - z_k}{z - \bar{z}_k} \exp\left[-\frac{1}{2\pi i} \int_R \frac{\log[1 + \zeta |r(\zeta)|^2]}{\zeta - z} d\zeta \right]. \tag{5.12}$$

The renormalization

$$m^\triangle(z|\sigma_d) = m(z|\sigma_d) a^\triangle(z)^{\sigma_3}, \tag{5.13}$$

splits the poles according to the choice of $\triangle$, and it satisfies the following modified discrete Riemann-Hilbert problem.

Problem 2. Given discrete data $\sigma_d = \{(z_k, c_k)\}_{k=1}^N$, and $\triangle \subseteqq \{1, 2, ..., N\}$, find a matrix-valued function z$\in C \to m^\triangle(z; x, t|\sigma_d)$ with following properties:

- Analyticity: $m^\triangle(z; x, t|\sigma_d)$ is meromorphic in $\mathbb{C} \setminus (\Sigma^{(2)} \cup Z \cup \bar{Z})$;
- Symmetry: $m_{22}^\triangle(z; x, t|\sigma_d) = \overline{m_{11}^\triangle(\bar{z}; x, t|\sigma_d)}$, $m_{21}^\triangle(z; x, t|\sigma_d) = -z\overline{m_{12}^\triangle(\bar{z}; x, t|\sigma_d)}$, which means $m^\triangle(z; x, t|\sigma_d) = z^{\sigma_3/2} \sigma_2 \overline{m^\triangle(\bar{z}; x, t|\sigma_d)} \sigma_2^{-1} z^{-\sigma_3/2}$;
- Asymptotic behaviors:

$$m^\triangle(z; x, t|\sigma_d) \sim \begin{pmatrix} 1 & 0 \\ \alpha & 1 \end{pmatrix} + O(z^{-1}), \qquad z \to \infty \tag{5.14}$$

for a constant α determined by the symmetry condition above;

• Residue conditions: $m(z;x,t|\sigma_d)$ has simple poles at each point in $Z\bigcup \bar{Z}$ satisfying

$$\operatorname*{Res}_{z=z_n} m^{\triangle}(z;x,t|\sigma_d) = \lim_{z\to z_n} m^{\triangle}(z;x,t|\sigma_d)\tau_k^{\triangle}, \tag{5.15}$$

$$\operatorname*{Res}_{z=\bar{z}_n} m^{\triangle}(z;x,t|\sigma_d) = \lim_{z\to \bar{z}_n} m^{\triangle}(z;x,t|\sigma_d)\hat{\tau}_k^{\triangle}, \tag{5.16}$$

where τ_k is a nilpotent matrix satisfies

$$\tau_k^{\triangle} = \begin{cases} \begin{pmatrix} 0 & 0 \\ \gamma_k a^{\triangle}(z)^2 & 0 \end{pmatrix} & k \notin \triangle \\ \begin{pmatrix} 0 & \gamma_k^{-1} a'^{\triangle}(z)^{-2} \\ 0 & 0 \end{pmatrix} & k \in \triangle \end{cases}, \qquad \hat{\tau}_k^{\triangle} = z_k^{\sigma_3/2}\sigma_2\overline{\tau}_k^{\triangle}\sigma_2^{-1}z_k^{-\sigma_3/2},$$
$$\gamma_k = c_k e^{2i[(z-\rho)x+2(z-\rho)^2t]}. \tag{5.17}$$

Since $m^{\triangle}(z;x,t|\sigma_d)$ is an explicit transformation of $m(z;x,t|\sigma_d)$, from **Proposition** 4 we obtain the existence and uniqueness of the solution of **Problem 2**. If $q_{sol}(x,t) = q_{sol}(x,t;\sigma_d)$ denotes the N-soliton solution of (1.8) encoded by **Problem 1**, we also have the reconstruction formula that

$$q_{sol}(x,t) = \lim_{z\to\infty} 2iz(m^{\triangle}(z;x,t|\sigma_d))_{12}, \tag{5.18}$$

which shows that each normalization encodes $q_{sol}(x,t)$ in the same way. If we choose $\triangle$ appropriately, the asymptotic limits $|t|\to\infty$ with $z_0 = -x/4t+\rho$ bounded are under better asymptotic control. Then we consider the long-time behavior of soliton solutions.

Give pairs points $x_1 \le x_2 \in R$ and velocities $v_1 \le v_2 \in R$, and define the cone

$$C(x_1,x_2,v_1,v_2) = \left\{(x,t)\in R^2 | x = x_0+vt \text{ ,with } x_0\in[x_1,x_2],\ v\in[v_1+4\rho, v_2+4\rho]\right\}. \tag{5.19}$$

Denote $I = [-v_2/4, -v_1/4]$, and let

$$Z(I) = \{z_k\in Z|\mathrm{Re}z_k \in I\}, \qquad N(I) = |Z(I)|,$$
$$Z^-(I) = \{z_k\in Z|\mathrm{Re}z_k < -v_2/4+\rho\}, \quad Z^+(I) = \{z_k\in Z|\mathrm{Re}z_k > -v_1/4+\rho\},$$
$$c_k(I) = c_k \prod_{\mathrm{Re}z_n\in I_{z_0}^{\eta}\setminus I} \left(\frac{z_k-z_n}{z_k-\bar{z}_n}\right)^2 \exp\left[-\frac{1}{\pi i}\int_{I_{z_0}^{\eta}} \frac{\log[1+\zeta|r(\zeta)|^2]}{\zeta-z}d\zeta\right]. \tag{5.20}$$

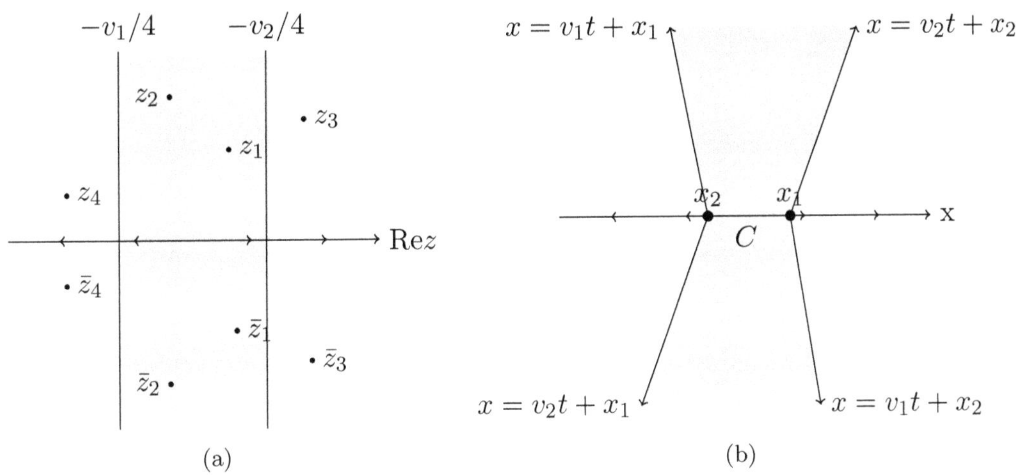

Figure 4: (a) In the example here, the original data has four pairs zero points of discrete spectrum, but insidecone C only three pairs points with $Z(I) = \{z_1, z_2\}$; (b) the cone $C(x_1, x_2, v_1, v_2)$.

Then we have a modified discrete scattering data $D(I) = \{(z_k, c_k(I))|z_k \in Z(I)\}$ (see Figure 4) and obtain following lemma:

Lemma 1. *Fix reflection-less data* $D = \{(z_k, c_k)\}_{k=1}^N$, $D(I) = \{(z_k, c_k(I))|z_k \in Z(I)\}$. *Then as* $|t| \to \infty$ *with* $(x,t) \in C(x_1, x_2, v_1, v_2)$, *we have*

$$m(z; x, t|D) = \left(I + O(e^{-8\mu(I)|t|})\right) m(z; x, t|D(I)), \tag{5.21}$$

where $\mu(I) = \min_{z_K \in Z\setminus Z(I)} \{Im(z_k) dist(Re(z_k), I)\}$.

Proof. Let $\triangle^+(I) = \{k|\text{Re}(z_k) < -v_2/4\}$, $\triangle^-(I) = \{k|\text{Re}(z_k) > -v_1/4\}$. Then we consider the case that $\triangle$ in **Problem 1** is $\triangle^\eta(I)$, where $\eta = \text{sgn}(t)$. For $z \in Z\setminus Z(I)$ and $(x,t) \in C(x_1, x_2, v_1, v_2)$, denote $x = x_0 + (v_0 + 4\rho)t$, where $x_0 \in [x_1, x_2]$ and $v_0 \in [v_1, v_2]$. Note that the residue coefficients (5.17) have that

$$|\tau_k^{\triangle^\pm(I)}| = |c_k e^{-2x_0 Imz}||e^{-8tImz(\text{Re}z + v_0/4)}|. \tag{5.22}$$

So they have following asymptotic property:

$$\| \tau_k^{\triangle^\pm(I)} \| = O(e^{-8\mu(I)|t|}), \qquad t \to \pm\infty. \tag{5.23}$$

D_k is a small disks centred in each $z_k \in Z\setminus Z(I)$ with radius smaller than μ. Denote ∂D_k as the boundary of D_k. Then we can introduce a new transformation which can remove the poles $z_k \in Z \setminus Z(I)$ and these residues change to near-identity jumps.

$$\tilde{m}^{\triangle^\eta(I)}(z; x, t|D) = \begin{cases} m^{\triangle^\eta(I)}(z; x, t|D) \left(I - \frac{\tau_k^{\triangle^\eta(I)}}{z - z_k}\right) & z \in D_k \\ m^{\triangle^\eta(I)}(z; x, t|D) \left(I - \frac{\hat{\tau}_k^{\triangle^\eta(I)}}{z - \bar{z}_k}\right) & z \in \bar{D}_k \\ m^{\triangle^\eta(I)}(z; x, t|D) & elsewhere \end{cases} \tag{5.24}$$

Comparing with $m^{\triangle^\eta(I)}(z|D)$, the new function has new jump in each ∂D_k which is denoted by $\tilde{V}(Z)$. Then using (5.23) we have

$$\| \tilde{V}(Z) - I \|_{L^\infty(\tilde{\Sigma})} = O(e^{-8\mu(I)|t|}), \quad \tilde{\Sigma} = \bigcup_{z_k \in Z \setminus Z(I)} (\partial D_k \cup \partial \bar{D}_k) . \tag{5.25}$$

After transformation, $\tilde{m}^{\triangle^\eta(I)}(z|D)$ has same poles and residue conditions as $m(z;x,t|D(I$ So we denote $m_0(z) = \tilde{m}^{\triangle^\eta(I)}(z|D)m(z|D(I))^{-1}$, which has no poles. And it has jump matrix for $z \in \tilde{\Sigma}$,

$$m_0^+(z) = m_0^-(z)V_{m_0}(z), \quad V_{m_0}(z) = m(z|D(I))\tilde{V}(Z)m(z|D(I))^{-1}. \tag{5.26}$$

From (5.25) and (5.9) applied to $m(z|D(I))$ the theory of small norm **RHPs** [30],[41], we have $m_0(z)$ exists and $m_0(z) = I + O(e^{-8\mu(I)|t|})$ for $t \to \pm\infty$. Then we have the consequence. ■

Using reconstruction formula to $m(z;x,t|D)$, we immediately have the following corollary.

Corollary 1. Let $q_{sol}(x,t;D)$ and $q_{sol}(x,t;D(I))$ denote the N-soliton solution of (1.8) corresponding to discrete scattering data D and D(I), respectively. And I, $C(x_1,x_2,v_1,v_2)$, $D(I)$ is given above. As $|t| \to \infty$ with $(x,t) \in C(x_1,x_2,v_1,v_2)$, we have

$$\lim_{z\to\infty} 2iz(m(z;x,t|\sigma_d))_{12} = q_{sol}(x,t;D) = q_{sol}(x,t;D(I)) + O(e^{-4\mu(I)|t|}). \tag{5.27}$$

Then we go back to the outer model and obtain the following corollary.

Corollary 2. There exists a unique solution $M^{(out)}$ of **Riemann-Hilbert problem 6** with

$$M^{(out)}(z) = m^{\triangle_{\bar{z}_0,\eta}}(z|D^{(out)}) \tag{5.28}$$

$$= M(z;x,t|D(I)) \prod_{\mathrm{Re} z_n \in I^\eta_{z_0} \setminus I} \left(\frac{z - z_n}{z - \bar{z}_n}\right)^{\sigma_3} + O(e^{-4\mu(I)|t|}), \tag{5.29}$$

where $D^{(out)} = \{z_k, c_k(z_0)\}_{k=1}^N$, $M(z;x,t|D(I))$ is the solution of **Problem 1**, with $D(I) = \{(z_k, c_k(I))|z_k \in Z(I)\}$ and

$$c_k(z_0) = c_k \exp\left[-\frac{1}{\pi i}\int_{I^\eta_{z_0}} \frac{\log[1+\zeta|r(\zeta)|^2]}{\zeta - z} d\zeta\right].$$

Then substituting (5.29) into (5.9), we immediately have

$$\| M^{(out)}(z)^{-1} \|_{L^\infty(C\setminus(Z\cup\bar{Z}))} \lesssim 1. \tag{5.30}$$

Moreover, we have reconstruction formula

$$\lim_{z\to\infty} 2iz(M^{(out)})_{12} = q_{sol}(x,t;D^{(out)}), \tag{5.31}$$

where the $q_{sol}(x,t;D^{(out)})$ is the N-soliton solution of (1.8) corresponding to discrete scattering data $D^{(out)}$. And

$$q_{sol}(x,t;D^{(out)}) = q_{sol}(x,t;D(I)) + O(e^{-4\mu(I)|t|}), \quad \text{for } t \to \pm\infty. \tag{5.32}$$

As we can find that $M(z;x,t|D(I))$ is the solution with reflectionless scattering data case $\{r \equiv 0; \{(z_k, c_k(I))|z_k \in Z(I)\}\}$, combining (1.38), (1.41) and (1.39), we have the following proposition.

Proposition 5. *The unique solution $M(z;x,t|D(I))$ of **Problem 1** has that*

$$[M(\rho;x,t|D(I))_+]_{11} = e^{i/2\int_{-\infty}^{x}|q_{sol}(x,t|D(I))|^2dy}, \tag{5.33}$$

$$[M(\rho;x,t|D(I))_+]_{12} = -e^{i/2\int_{-\infty}^{x}|q_{sol}(x,t|D(I))|^2dy} \int_{-\infty}^{x} q_{sol}(y,t|D(I))e^{-i\int_{-\infty}^{y}|q_{sol}(s,t|D(I))|^2ds}dy, \tag{5.34}$$

with symmetry $[M_+(\rho;x,t|D(I))]_{22} = \overline{[M_+(\rho;x,t|D(I))]_{11}}$, $[M_+(\rho;x,t|D(I))]_{21} = -\rho\overline{[M_+(\rho;x,t|D(I))]_{12}}$.

6 Local model RHP near z_0

From **Proposition** 3 we find that $V^{(2)} - I$ does not have a uniformly small jump for a large time. So in the neighborhood U_{z_0} of z_0, we establish a local model to arrive at a uniformly small jump Riemann-Hilbert problem for function $E(Z)$. Notice that $\forall p \in Z \cup \bar{Z}$, $p \notin U_{z_0}$. Let $\xi = \xi(z)$ denote the local variable

$$\xi(z) = \sqrt{|8t|}(z - z_0). \tag{6.1}$$

Then we have

$$2t\theta = \xi^2/2 - 4t(z_0-\rho)^2, \quad (\eta(z-z_0))^{2i\eta k(z_0)} = (\eta\xi)^{2i\eta k(z_0)}e^{-i\eta k(z_0)log|8t|}.$$

Let

$$r_{z_0} = -r(z_0)T_0(z_0)^2e^{-i\eta k(z_0)log|8t|}e^{4ti(z_0-\rho)^2}, \quad s_{z_0} = z\bar{r}_{z_0}, \tag{6.2}$$

then we have $1 + r_{z_0}s_{z_0} = 1 + z|r(z_0)|^2 \neq 0$. And since $1 - X_Z(z) = 1$ for $z \in U_{z_0}$, the jump matrix $V^{(2)}$ limit in U_{z_0} denoted as $V^{(pc)}$ has become

$$V^{(pc)} = \begin{cases} \begin{pmatrix} 1 & 0 \\ s_{z_0}(\eta\xi)^{-2i\eta k(z_0)}e^{\xi^2 i/2} & 1 \end{pmatrix} & z \in \Sigma_1 \\ \begin{pmatrix} 1 & \dfrac{r_{z_0}}{1+r_{z_0}s_{z_0}}(\eta\xi)^{2i\eta k(z_0)}e^{-\xi^2 i/2} \\ 0 & 1 \end{pmatrix} & z \in \Sigma_2 \\ \begin{pmatrix} 1 & 0 \\ \dfrac{s_{z_0}}{1+r_{z_0}s_{z_0}}(\eta\xi)^{-2i\eta k(z_0)}e^{\xi^2 i/2} & 1 \end{pmatrix} & z \in \Sigma_3 \\ \begin{pmatrix} 1 & r_{z_0}(\eta\xi)^{2i\eta k(z_0)}e^{-\xi^2 i/2} \\ 0 & 1 \end{pmatrix} & z \in \Sigma_4 \end{cases}. \tag{6.3}$$

Then we have the following **RHP6** which does not possess the symmetry condition shared by **RHP4**, because we only use this model for bounded values z.

Parabolic Cylinder Model Riemann-Hilbert problem 6 Find an analytic function $M^{(pc)}(z;z_0,\eta)$: $C\setminus\Sigma^{(2)}\to SL_2(C)$ such that

- Analyticity: $M^{pc}(z;z_0,\eta)$ is meromorphic in $\mathbb{C}\setminus(\Sigma^{(2)}\cup Z\cup\bar{Z})$;
- Jump condition: M^{pc} has continuous boundary values $M^{pc}_{\pm}$ on R and

$$M^{pc}_{+}(z;z_0,\eta)=M^{pc}_{-}(z;z_0,\eta)V^{(pc)}(z;z_0,\eta),\qquad z\in\Sigma^{(2)}; \tag{6.4}$$

- Asymptotic behaviors:

$$M^{pc}(z;z_0,\eta)\sim I+O(z^{-1}),\qquad z\to\infty. \tag{6.5}$$

The precise details of the construction for this solution, differ only slightly from the construction for KdV [42]. In fact,this type of model problem is typical in integrable systems whenever there is a phase function. Here in our system is θ, which has a quadratic critical point along the real line. Here we only give the necessary details, and to arrive at our consequence, we also need its boundedness property shown in the following **Lemma** which proof can be find in [20], appendix D.

Proposition 6. *The solution of **RHP 6** is given as follow:*

$$\begin{aligned}M^{pc}(z;z_0,+)&=F(\xi(z);r_{z_0},s_(z_0)),\\ M^{pc}(z;z_0,-)&=\sigma_2F(-\xi(z);r_{z_0},s_(z_0))\sigma_2,\end{aligned} \tag{6.6}$$

where $\kappa=k(z_0)$

$$F(\xi;r,s)=\Phi(\xi;r,s)P(\xi;r,s)e^{i\xi^2\sigma_3/4}\xi^{-i\kappa\sigma_3} \tag{6.7}$$

$$P(\xi;r,s)=\begin{cases}\begin{pmatrix}1&0\\ s&1\end{pmatrix} & \xi\in\Omega_1\\ \begin{pmatrix}1&\dfrac{r}{1+rs}\\ 0&1\end{pmatrix} & \xi\in\Omega_3\\ \begin{pmatrix}1&0\\ \dfrac{-s}{1+rs}&1\end{pmatrix} & \xi\in\Omega_4\\ \begin{pmatrix}1&-r\\ 0&1\end{pmatrix} & \xi\in\Omega_6\\ I & z\in\Omega_2\cup\Omega_5\end{cases}. \tag{6.8}$$

And when $\xi\in C^+$

$$\Phi(\xi;r,s)=\begin{pmatrix}e^{-3\pi\kappa/4}D_{i\kappa}(\xi e^{-3i\pi/4}) & -i\beta_{12}e^{\pi(\kappa-i)/4}D{-i\kappa-1}(\xi e^{-i\pi/4})\\ i\beta_{21}e^{-3\pi(\kappa+i)/4}D{i\kappa-1}(\xi e^{-3i\pi/4}) & e^{\pi\kappa/4}D_{-i\kappa}(\xi e^{-i\pi/4})\end{pmatrix}; \tag{6.9}$$

when $\xi\in C^-$

$$\Phi(\xi;r,s)=\begin{pmatrix}e^{\pi\kappa/4}D_{i\kappa}(\xi e^{i\pi/4}) & -i\beta_{12}e^{-3\pi(\kappa-i)/4}D{-i\kappa-1}(\xi e^{3i\pi/4})\\ i\beta_{21}e^{\pi(\kappa+i)/4}D{i\kappa-1}(\xi e^{i\pi/4}) & e^{-3\pi\kappa/4}D_{-i\kappa}(\xi e^{3i\pi/4})\end{pmatrix}.$$

$$(6.10)$$

Here $D_a(a)$ denotes the parabolic cylinder functions, and β_{21} and β_{12} are complex constants

$$\beta_{12} = \beta_{12}(r,s) = \frac{\sqrt{2\pi}e^{-\kappa\pi/2}e^{i\pi/4}}{s\Gamma(-i\kappa)}, \quad \beta_{21} = \beta_{21}(r,s) = \frac{-\sqrt{2\pi}e^{-\kappa\pi/2}e^{-i\pi/4}}{r\Gamma(i\kappa)}. \quad (6.11)$$

Then we consider the asymptotic behavior of the solution. When $\xi \to \infty$,

$$F(\xi; r, s) = I + \frac{1}{\xi}\begin{pmatrix} 0 & -i\beta_{12}(r,s) \\ i\beta_{21}(r,s) & 0 \end{pmatrix} + O(\xi^{-2}), \quad (6.12)$$

from which we have for $z \in \partial U_{z_0}$

$$M^{pc}(z; z_0, \eta) = I + \frac{|8t|^{-1/2}}{z - z_0}\begin{pmatrix} 0 & -iA_{12}(z_0,\eta) \\ iA_{21}(z_0,\eta) & 0 \end{pmatrix} + O(|t|^{-1}), \quad (6.13)$$

with

$$\begin{aligned} A_{12}(z_0,+) &= \beta_{12}(r_{z_0}, s_{z_0}), \quad A_{21}(z_0,+) = \beta_{21}(r_{z_0}, s_{z_0}), \\ A_{12}(z_0,-) &= -\beta_{21}(r_{z_0}, s_{z_0}), \quad A_{12}(z_0,-) = -\beta_{12}(r_{z_0}, s_{z_0}), \end{aligned} \quad (6.14)$$

satisfying the following conditions

$$\begin{aligned} |A_{12}(z_0,\eta)|^2 &= \frac{k(z_0)}{z_0}, \quad A_{12}(z_0,\eta) = z_0\overline{A_{12}(z_0,\eta)}, \\ arg\,(A_{12}(z_0,+)) &= \frac{\pi}{4} + arg\Gamma(ik(z_0)) - arg(-z_0\overline{r(z_0)}) - k(z_0)log|8t| \\ &+ \frac{1}{\pi}\int_{-\infty}^{z_0} log|s - z_0|dslog(1 + z|r(z)|^2) + 4tz_0^2, \\ arg\,(A_{12}(z_0,-)) &= \frac{\pi}{4} - arg\Gamma(ik(z_0)) - arg(-z_0\overline{r(z_0)}) + k(z_0)log|8t| \\ &+ \frac{1}{\pi}\int_{z_0}^{\infty} log|s - z_0|dslog(1 + z|r(z)|^2) + 4tz_0^2. \end{aligned}$$

And the ∞-norm of M^{pc} has

$$\| M^{pc}(z; z_0, \eta) \|_\infty \lesssim 1, \quad \| M^{pc}(z; z_0, \eta)^{-1} \|_\infty \lesssim 1. \quad (6.15)$$

Using (6.6) we can define the local model

$$M^{(z_0)}(z) = M^{(out)}(z)M^{pc}(z; z_0, \eta), \quad z \in U_{z_0}, \quad (6.16)$$

which have the same jump matrix $V^{(2)}$ as M^{RHP} from M^{pc} and the same residue conditions as M^{RHP} from $M^{(out)}$. So we have that it is a bounded function in U_{z_0}.

7 Small norm RHP for E(z)

In this Section we consider the error $E(z)$. From its definition (4.12) and the analyticity of $M^{(out)}$ and $M^{(z_0)}$, we can obtain that $E(z)$ is analytic in $C \setminus \Sigma^{(E)}$, where

$$\Sigma^{(E)} = \partial U_{z_0} \cup (\Sigma^{(2)} \setminus U_{z_0}).$$

We will prove that for large times, the error $E(z)$ solves following small norm Riemann-Hilbert problem which we can expand asymptotically.

Riemann-Hilbert problem 7. Find a matrix-valued function $E(z)$: z$\in C \to E(z)$ with the following properties:

• Analyticity: $E(z)$ is meromorphic in $\mathbb{C} \setminus (\Sigma^{(E)} \cup Z \cup \bar{Z})$;

• Symmetry: $E_{22}(z) = \overline{E_{11}(\bar{z})}$, $E_{21}(z) = -z\overline{E_{12}(\bar{z})}$, which means $E(z) = z^{\sigma_3/2}\sigma_2\overline{E(\bar{z})}\sigma_2^{-1}z^{-\sigma_3/2}$;

• Asymptotic behaviors:

$$E(z) \sim \begin{pmatrix} 1 & 0 \\ \alpha_E & 1 \end{pmatrix} + O(z^{-1}), \quad |z| \to \infty \tag{7.1}$$

for a constant α_E determined by the symmetry condition above;

• Jump condition: E has continuous boundary values $E_\pm$ on $\Sigma^{(E)}$ satisfy $E_+(z) = E_-(z)V^{(E)}$, where

$$V^{(E)} = \begin{cases} M^{(out)}(z)V^{(2)}(z)M^{(out)}(z)^{-1} & \xi \in \Sigma^{(2)} \setminus U_{z_0} \\ M^{(out)}(z)M^{pc}(z)^{-1}M^{(out)}(z)^{-1} & \xi \in \partial U_{z_0} \end{cases}. \tag{7.2}$$

From **Proposition 3**, (6.13), (5.30), the jump matrix have

$$|V^{(E)} - I| = \begin{cases} e^{-4|t(z-z_0)^2|} & \xi \in \Sigma^{(2)} \setminus U_{z_0} \\ |t|^{-1/2} & \xi \in \partial U_{z_0} \end{cases}, \tag{7.3}$$

which implies that for $k \in N$, $p \leq 1$

$$\parallel V^{(E)} - I \parallel_{L^{p,k}(R) \cup L^\infty(R)} = O(|t|^{-1/2}). \tag{7.4}$$

This uniformly approaches to zero, as $|t| \to \infty$ of $V^{(E)} - I$ make **Riemann-Hilbert problem 7** a small-norm Riemann-Hilbert problem, for which there is a well-known existence and uniqueness theorem[10, 41, 43]. Then we have the following Lemma.

Lemma 2. *There exists a unique solution $E(z)$ of **RHP 7** stratifies*

$$\parallel E - \begin{pmatrix} 1 & 0 \\ \alpha_E & 1 \end{pmatrix} \parallel_{L^\infty(C \setminus \Sigma^{(E)})} \lesssim |t|^{-1/2}. \tag{7.5}$$

And when $z \to \infty$,

$$E(z) \sim \begin{pmatrix} 1 & 0 \\ \alpha_E & 1 \end{pmatrix} + \frac{E_1}{z} + O(z^{-2}), \tag{7.6}$$

$\bar{\alpha}_E = (E_1)_{12}$ *and it has*

$$2i(E_1)_{12} = |t|^{-1/2} f(x,t) + O(|t|^{-1}), \tag{7.7}$$

where

$$f(x,t) = 2^{-1/2}\left[A_{12}(z_0,\eta) M_{11}^{(out)}(z_0)^2 + A_{21}(z_0,\eta) M_{12}^{(out)}(z_0)^2\right]. \tag{7.8}$$

Proof. Because **RHP 7** is not the standard Riemann-Hilbert problem, we need to construct the solution $E(z)$ row-by-row. For the first row, we denote $e^1 = (E_{11}, E_{12})$, which have the following property:

1. $e^1 \sim (1, 0)$ as $z \to \infty$;
2. e^1 has continuous boundary values $e^1_\pm$ on $\Sigma^{(E)}$ satisfy $e^1_+(z) = e^1_-(z) V^{(E)}$.

Then by Plemelj formula we have

$$e^1(z) = (1,0) + \frac{1}{2\pi i}\int_{\Sigma^{(E)}} \frac{((1,0)+\mu_1(s))\left(V^{(E)} - I\right)}{s-z} ds, \tag{7.9}$$

where the $\mu_1 \in L^2(\Sigma^{(E)})$ is the unique solution of the following equation:

$$(1 - C_E)\mu_1 = C_E\left((1,0)\right). \tag{7.10}$$

C_E is an integral operator defined as

$$C_E(f)(z) = C_-\left(f(V^{(E)} - I)\right), \tag{7.11}$$

where the C_- is the usual Cauchy projection operator on $\Sigma^{(E)}$

$$C_-(f)(s) = \lim_{z \to \Sigma_-^{(E)}} \frac{1}{2\pi i}\int_{\Sigma^{(E)}} \frac{f(s)}{s-z} ds. \tag{7.12}$$

Then by (7.4) we have

$$\| C_E \| \leq \| C_- \| \| V^{(E)} - I \|_\infty \lesssim O(t^{-1/2}), \tag{7.13}$$

which means for sufficiently large t, $\| C_E \| < 1$. Then we have 1-C_E is invertible, so the μ_1 has unique existence. Moreover,

$$\| \mu_1 \|_{L^2(\Sigma^{(E)})} \lesssim \frac{\| C_E \|}{1 - \| C_E \|} \lesssim |t|^{-1/2}. \tag{7.14}$$

Then we can prove that $e^1 \sim (1,0)$.

$$|e^1(z) - (1,0)| \leq \left|\frac{1}{2\pi i}\int_{\Sigma^{(E)}} \frac{(1,0)(V^{(E)} - I)}{s-z} ds\right| + \left|\frac{1}{2\pi i}\int_{\Sigma^{(E)}} \frac{\mu_1(s)(V^{(E)} - I)}{s-z} ds\right|. \tag{7.15}$$

When there exists a constant $d > 0$ such that $\inf_{z\in\Sigma^{(E)}} |s - z| > d$, we have

$$|e^1(z)-(1,0)| \leq \frac{1}{2\pi d}\left(\| V^{(E)} - I \|_{L^1} + \| \mu_1 \|_{L^2}\| V^{(E)} - I \|_{L^2}\right) \lesssim |t|^{-1/2}. \quad (7.16)$$

And for z approaching to $\Sigma^{(E)}$, because the jump matrix on the contours $\Sigma^{(E)}$ are locally analytic, we can make a invertible transformation $e^1 \to \tilde{e}^1$(for example, inversion transformation of a circle center at z_0 with radius of $\mu/6$), which transforms $\Sigma^{(E)}$ to a new contour $\tilde{\Sigma}^{(E)}$ with different points of self-interSection. Similarly we have $|\tilde{e}^1(z) - (1,0)|$ is bounded on $\Sigma^{(E)}$, then we obtain the boundedness of e^1.

Then we consider the second row $e^2 = (E_{21}, E_{22})$; similarly it has the following property:

1. $e^2 \sim (1, \alpha_E)$ as $z \to \infty$;
2. e^2 has continuous boundary values $e^1_\pm$ on $\Sigma^{(E)}$ satisfy $e^2_+(z) = e^2_-(z)V^{(E)}$,

where from the symmetry $E_{21}(z) = -z\overline{E_{12}(\bar{z})}$ and (7.9) making $z \to \infty$ we obtain

$$\alpha_E = -\left[\frac{1}{2\pi i}\int_{\Sigma^{(E)}} ((1,0) + \mu_1(s))\,(V^{(E)} - I)ds\right]_2, \quad (7.17)$$

where the subscript 2 means the second element. By (7.4) and (7.14) we have

$$|\alpha_E| \lesssim |t^{-1/2}|. \quad (7.18)$$

In the same way we obtain

$$e^2(z) = (\alpha_E, 1) + \frac{1}{2\pi i}\int_{\Sigma^{(E)}} \frac{((\alpha_E, 1) + \mu_2(s))\,(V^{(E)} - I)}{s - z}ds, \quad (7.19)$$

where the $\mu_2 \in L^2(\Sigma^{(E)})$ is the unique solution of the following equation:

$$(1 - C_E)\mu_1 = C_E\left((\alpha_E, 1)\right). \quad (7.20)$$

Also we have $\| \mu_2 \|_{L^2(\Sigma^{(E)})} \lesssim |t|^{-1/2}$ and $|e^2(z) - (\alpha_E, 1)| \lesssim |t|^{-1/2}$. Now we denote $E = (e^1, e^2)^T$, $\mu = (\mu^1, \mu^2)^T$, then E has expansion for $z \to \infty$

$$E = \begin{pmatrix} 1 & 0 \\ \alpha_E & 0 \end{pmatrix} + \frac{E_1}{z} + O(z^{-2}), \quad (7.21)$$

where

$$E_1 = \frac{-1}{2\pi i}\int_{\Sigma^{(E)}} \left(\begin{pmatrix} 1 & 0 \\ \alpha_E & 0 \end{pmatrix} + \mu(s)\right)(V^{(E)} - I)ds. \quad (7.22)$$

Using (7.18), (7.4), (7.2) (6.13) and (7.14) we have

$$\begin{aligned} E_1 =& \frac{-1}{2\pi i}\int_{\partial U_{z_0}} (V^{(E)} - I)ds + O(|t^{-1}|) \\ =& M^{(out)}(Z_0)A(z_0, \eta)M^{(out)}(Z_0)^{-1}|8t|^{-1/2} + O(|t^{-1}|). \end{aligned} \quad (7.23)$$

■

When estimating the solution $u(x,t)$ of the MNLS Equation (1.1), we need the following result which provides the large-time behavior of the error term $E(\rho)$.

Proposition 7. *When $|t| \to \infty$, the unique solution $E(z)$ of* ***RHP 7*** *described by above Lemma satisfies:*

1. when $\rho \in U_{z_0}$,

$$E_{11}(\rho) = 1 - \sum_{s=z_0,\rho} \frac{|t|^{-1/2}}{4\sqrt{2\pi}} \left[A_{21}(z_0,\eta) M_{12}^{(out)}(s) \overline{M_{11}^{(out)}(s)} - s A_{12}(z_0,\eta) M_{11}^{(out)}(s) \overline{M_{12}^{(out)}(s)} \right] + O(|t|^{-1}), \tag{7.24}$$

$$E_{12}(\rho) = \sum_{s=z_0,\rho} \frac{|t|^{-1/2}}{4\sqrt{2\pi}} \left[A_{21}(z_0,\eta) \overline{M_{11}^{(out)}(s)}^{2} + s^2 A_{12}(z_0,\eta) \overline{M_{12}^{(out)}(s)}^{2} \right] + O(|t|^{-1}); \tag{7.25}$$

2. when $\rho \notin U_{z_0}$,

$$E_{11}(\rho) = 1 - \frac{|t|^{-1/2}}{4\sqrt{2\pi}} \left[A_{21}(z_0,\eta) M_{12}^{(out)}(z_0) \overline{M_{11}^{(out)}(z_0)} - z_0 A_{12}(z_0,\eta) M_{11}^{(out)}(z_0) \overline{M_{12}^{(out)}(z_0)} \right] + O(|t|^{-1}), \tag{7.26}$$

$$E_{12}(\rho) = \frac{|t|^{-1/2}}{4\sqrt{2\pi}} \left[A_{21}(z_0,\eta) \overline{M_{11}^{(out)}(z_0)}^{2} + z_0^2 A_{12}(z_0,\eta) \overline{M_{12}^{(out)}(z_0)}^{2} \right] + O(|t|^{-1}). \tag{7.27}$$

$A_{12}(z_0,\eta)$ and $A_{12}(z_0,\eta)$ are given in ***Proposition 6****. Together with $E_{22}(\rho) = \overline{E_{11}(\rho)}$, $E_{21}(\rho) = -z\overline{E_{12}(\rho)}$ we obtain the long-time behavior of $E(\rho)$.*

Proof. We only calculate $e_1(\rho)$, because $e_2(\rho)$ can obtain by symmetry of $E(z)$. From (7.9) and (7.2) we have

$$\begin{aligned} e_1(\rho) - (1,0) &= \frac{1}{2\pi i} \int_{\Sigma^{(E)}} \frac{((1,0) + \mu_1(s)) (V^{(E)} - I)}{s - \rho} ds \\ &= \frac{1}{2\pi i} \int_{\Sigma^{(2)} \setminus U_{z_0}} \frac{((1,0) + \mu_1(s))}{s - \rho} M^{(out)} (V^{(2)} - I) [M^{(out)}]^{-1} ds \\ &+ \frac{1}{2\pi i} \int_{\partial U_{z_0}} \frac{((1,0) + \mu_1(s))}{s - \rho} M^{(out)} (M^{pc} - I) [M^{(out)}]^{-1} ds. \end{aligned} \tag{7.28}$$

For the first integral, we calculate on $\Sigma_1 \setminus U_{z_0}$. The others are similar. Let $s = z_0 + le^{i\pi/4}$, $l \in [\mu/3, +\infty]$, and note that $|s - \rho| > \sqrt{2}\mu/6$. Then together with (7.3) and (7.14) we obtain

$$\begin{aligned}
&\| \int\limits_{\Sigma^{(2)}\setminus U_{z_0}} \frac{((1,0)+\mu_1(s))\,(V^{(E)}-I)}{s-\rho} ds \,\|_\infty \\
&\lesssim \int\limits_{\mu/3}^{+\infty} e^{-4|tl^2|} dl + \int\limits_{\mu/3}^{+\infty} |\mu_1(l) e^{-4|tl^2|} dl \\
&\lesssim |t|^{-1} + |t|^{-1/2} \| \mu_1 \|_{L^2} \lesssim |t|^{-1}.
\end{aligned} \tag{7.29}$$

For the second integral, from (6.13) and (7.14) we have

$$\begin{aligned}
&\frac{1}{2\pi i} \int\limits_{\partial U_{z_0}} \frac{((1,0)+\mu_1(s))}{s-\rho} M^{(out)}(M^{pc}-I)[M^{(out)}]^{-1} ds \\
&= -\frac{|t|^{-1/2}}{4\sqrt{2}\pi i} \int\limits_{\partial U_{z_0}} \frac{(1,0)}{(s-\rho)(s-z_0)} M^{(out)} A(z_0,\eta)[M^{(out)}]^{-1} ds + O(|t|^{-1}),
\end{aligned} \tag{7.30}$$

where

$$\begin{aligned}
&(1,0)M^{(out)}A(z_0,\eta)[M^{(out)}]^{-1} \\
&= i\left(A_{21}M^{(out)}_{12}M^{(out)}_{22} + A_{12}M^{(out)}_{11}M^{(out)}_{21}, A_{21}[M^{(out)}_{22}]^2 + A_{12}[M^{(out)}_{21}]^2\right).
\end{aligned} \tag{7.31}$$

Then by residue theorem and the symmetry, we obtain the result. ■

Now combine the above Lemmas and proposition about the boundedness of $M^{(out)}$, E, M^{pc} we have

Proposition 8. *M^{RHP} is the unique solution of **RHP 4**, which is in $L^\infty\left(C \setminus (\Sigma^{(2)} \cap supp(1-X_Z))\right)$ and*

$$\| M^{RHP}(z)^{\pm 1} \|_{L^\infty(\Sigma^{(2)} \cap supp(1-X_Z))} \lesssim 1. \tag{7.32}$$

And to estimate the gauge factor of the solution $u(x,t)$ of the MNLS equation (1.1), we also need the large-time behavior of $M_+^{RHP}(\rho)$.

Proposition 9. *M^{RHP} is the unique solution of **RHP 4**, then as $|t| \to \infty$, $M_+^{RHP}(\rho)$ have*

***1.** when $\rho \in U_{z_0}$,*

$$\begin{aligned}
[M_+^{RHP}(\rho)]_{11} &= [M_+(\rho;x,t|D(I))]_{11} \prod_{Rez_n \in I^\eta_{z_0}\setminus I} \left(\frac{\rho - z_n}{\rho - \bar{z}_n}\right) \\
&\quad + |t|^{-1/2} G_1(\rho, z_0; x, t|D(I) + O(|t|^{-1});
\end{aligned} \tag{7.33}$$

$$\begin{aligned}
[M_+^{RHP}(\rho)]_{12} &= [M_+(\rho;x,t|D(I))]_{12} \prod_{Rez_n \in I^\eta_{z_0}\setminus I} \left(\frac{\rho - \bar{z}_n}{\rho - z_n}\right) \\
&\quad + |t|^{-1/2} G_2(\rho, z_0; x, t|D(I)) + O(|t|^{-1}),
\end{aligned} \tag{7.34}$$

where

$$
\begin{aligned}
&G_1(\rho, z_0; x, t|D(I)) = \\
&- [M_+(\rho; x, t|D(I))]_{11} \prod_{Rez_n \in I_{z_0}^{\eta} \setminus I} \left(\frac{\rho - z_n}{\rho - \bar{z}_n}\right) G_1^{(1)}(\rho, z_0; x, t|D(I)) \\
&- [M_+(\rho; x, t|D(I))]_{11} \prod_{Rez_n \in I_{z_0}^{\eta} \setminus I} \left(\frac{\rho - z_n}{\rho - \bar{z}_n}\right) G_1^{(2)}(\rho, z_0; x, t|D(I)) \\
&- \rho\overline{[M_+(\rho; x, t|D(I))]_{12}} \prod_{Rez_n \in I_{z_0}^{\eta} \setminus I} \left(\frac{\rho - z_n}{\rho - \bar{z}_n}\right) G_1^{(3)}(\rho, z_0; x, t|D(I)) \\
&- \rho\overline{[M_+(\rho; x, t|D(I))]_{12}} \prod_{Rez_n \in I_{z_0}^{\eta} \setminus I} \left(\frac{\rho - z_n}{\rho - \bar{z}_n}\right) G_1^{(4)}(\rho, z_0; x, t|D(I)); \qquad (7.35)
\end{aligned}
$$

and

$$
\begin{aligned}
&G_2(\rho, z_0; x, t|D(I)) = \\
&- [M_+(\rho; x, t|D(I))]_{12} \prod_{Rez_n \in I_{z_0}^{\eta} \setminus I} \left(\frac{\rho - \bar{z}_n}{\rho - z_n}\right) G^{(1)}(\rho, z_0; x, t|D(I)) \\
&+ [M_+(\rho; x, t|D(I))]_{12} \prod_{Rez_n \in I_{z_0}^{\eta} \setminus I} \left(\frac{\rho - \bar{z}_n}{\rho - z_n}\right) G^{(2)}(\rho, z_0; x, t|D(I)) \\
&+ \overline{[M_+(\rho; x, t|D(I))]_{11}} \prod_{Rez_n \in I_{z_0}^{\eta} \setminus I} \left(\frac{\rho - z_n}{\rho - \bar{z}_n}\right) G^{(3)}(\rho, z_0; x, t|D(I)) \\
&+ \overline{[M_+(\rho; x, t|D(I))]_{11}} \prod_{Rez_n \in I_{z_0}^{\eta} \setminus I} \left(\frac{\rho - z_n}{\rho - \bar{z}_n}\right) G^{(4)}(\rho, z_0; x, t|D(I)). \qquad (7.36)
\end{aligned}
$$

Here,

$$
\begin{aligned}
G^{(1)}(\rho, z_0; x, t|D(I)) =& \frac{1}{4\sqrt{2}\pi} A_{21}(z_0, \eta) \sum_{s = z_0, \rho} M_{12}(s; x, t|D(I)) \\
& \overline{M_{11}(s; x, t|D(I))} \prod_{Rez_n \in I_{z_0}^{\eta} \setminus I} \left(\frac{s - \bar{z}_n}{s - z_n}\right)^2, \qquad (7.37)
\end{aligned}
$$

$$
\begin{aligned}
G^{(2)}(\rho, z_0; x, t|D(I)) =& \frac{1}{4\sqrt{2}\pi} A_{12}(z_0, \eta) \sum_{s = z_0, \rho} M_{11}(s; x, t|D(I)) \\
& \overline{sM_{12}(s; x, t|D(I))} \prod_{Rez_n \in I_{z_0}^{\eta} \setminus I} \left(\frac{s - z_n}{s - \bar{z}_n}\right)^2, \qquad (7.38)
\end{aligned}
$$

$$G^{(3)}(\rho, z_0; x, t|D(I)) = \frac{1}{4\sqrt{2}\pi} A_{21}(z_0,\eta) \sum_{s=z_0,\rho} \overline{M_{11}(s;x,t|D(I))}^2 \prod_{Rez_n \in I^{\eta}_{z_0}\setminus I} \left(\frac{s-\bar{z}_n}{s-z_n}\right)^2, \tag{7.39}$$

$$G^{(4)}(\rho, z_0; x, t|D(I)) = \frac{1}{4\sqrt{2}\pi} A_{12}(z_0,\eta) \sum_{s=z_0,\rho} s^2\overline{M_{12}(s;x,t|D(I))}^2 \prod_{Rez_n \in I^{\eta}_{z_0}\setminus I} \left(\frac{s-z_n}{s-\bar{z}_n}\right)^2, \tag{7.40}$$

2. when $\rho \notin U_{z_0}$,

$$[M_+^{RHP}(\rho)]_{11} = [M_+(\rho;x,t|D(I))]_{11} \prod_{Rez_n \in I^{\eta}_{z_0}\setminus I} \left(\frac{\rho-z_n}{\rho-\bar{z}_n}\right) + |t|^{-1/2} H_1(\rho,z_0;x,t|D(I) + O(|t|^{-1}); \tag{7.41}$$

$$[M_+^{RHP}(\rho)]_{12} = [M_+(\rho;x,t|D(I))]_{12} \prod_{Rez_n \in I^{\eta}_{z_0}\setminus I} \left(\frac{\rho-\bar{z}_n}{\rho-z_n}\right) + |t|^{-1/2} H_2(\rho,z_0;x,t|D(I)) + O(|t|^{-1}), \tag{7.42}$$

where

$$\begin{aligned}
&H_1(\rho, z_0; x, t|D(I)) = \\
&- |[M_+(\rho;x,t|D(I))]_{11}|^2 \prod_{Rez_n \in I^{\eta}_{z_0}\setminus I} \left(\frac{\rho-\bar{z}_n}{\rho-z_n}\right) \frac{1}{4\sqrt{2}\pi} A_{21}(z_0,\eta)[M_+(\rho;x,t|D(I))]_{12} \\
&- [M_+(\rho;x,t|D(I))]_{11}^2 \prod_{Rez_n \in I^{\eta}_{z_0}\setminus I} \left(\frac{\rho-z_n}{\rho-\bar{z}_n}\right)^3 \frac{1}{4\sqrt{2}\pi} \rho A_{12}(z_0,\eta)\overline{M_{12}(\rho;x,t|D(I))} \\
&- \frac{1}{4\sqrt{2}\pi} A_{21}(z_0,\eta)\overline{M_{11}(\rho;x,t|D(I))}^2 \rho \overline{M_{12}(\rho;x,t|D(I))} \prod_{Rez_n \in I^{\eta}_{z_0}\setminus I} \left(\frac{\rho-\bar{z}_n}{\rho-z_n}\right) \\
&- \frac{1}{4\sqrt{2}\pi} \rho^3 A_{12}(z_0,\eta)\overline{M_{12}(\rho;x,t|D(I))}^3 \prod_{Rez_n \in I^{\eta}_{z_0}\setminus I} \left(\frac{\rho-z_n}{\rho-\bar{z}_n}\right)^3 \\
&+ \frac{iA_{21}(z_0,\eta)}{2\sqrt{2}(\rho-z_0)} \prod_{Rez_n \in I^{\eta}_{z_0}\setminus I} \left(\frac{\rho-\bar{z}_n}{\rho-z_n}\right) \left[M_{12}(\rho;x,t|D(I)) + \overline{M_{11}(\rho;x,t|D(I))}\right];
\end{aligned} \tag{7.43}$$

$$
\begin{aligned}
&H_2(\rho, z_0; x, t|D(I)) = \\
&-\frac{iA_{12}(z_0,\eta)}{2\sqrt{2}(\rho - z_0)} \prod_{Rez_n \in I^{\eta}_{z_0} \setminus I} \left(\frac{\rho - \bar{z}_n}{\rho - z_n}\right) \left[M_{12}(\rho; x, t|D(I)) + \overline{M_{11}(\rho; x, t|D(I))}\right] \\
&- [M_+(\rho; x, t|D(I))]^2_{12} \prod_{Rez_n \in I^{\eta}_{z_0} \setminus I} \left(\frac{\rho - \bar{z}_n}{\rho - z_n}\right)^3 \frac{1}{4\sqrt{2}\pi} A_{21}(z_0,\eta) \overline{[M_+(\rho; x, t|D(I))]_{11}} \\
&+ |[M_+(\rho; x, t|D(I))]_{12}|^2 \frac{1}{4\sqrt{2}\pi} \rho A_{12}(z_0,\eta)[M_+(\rho; x, t|D(I))]_{11} \prod_{Rez_n \in I^{\eta}_{z_0} \setminus I} \left(\frac{\rho - z_n}{\rho - \bar{z}_n}\right) \\
&+ \frac{1}{4\sqrt{2}\pi} A_{21}(z_0,\eta) \overline{[M_+(\rho; x, t|D(I))]_{11}}^3 \prod_{Rez_n \in I^{\eta}_{z_0} \setminus I} \left(\frac{\rho - \bar{z}_n}{\rho - z_n}\right)^3 \\
&+ \overline{[M_+(\rho; x, t|D(I))]_{11}} \frac{1}{4\sqrt{2}\pi} \rho^2 A_{12}(z_0,\eta) \overline{[M_+(\rho; x, t|D(I))]_{12}}^2 \prod_{Rez_n \in I^{\eta}_{z_0} \setminus I} \left(\frac{\rho - z_n}{\rho - \bar{z}_n}\right).
\end{aligned}
\tag{7.44}
$$

And from the symmetry of M^{RHP}*:* $M^{RHP}_{22}(z) = \overline{M^{RHP}_{11}(\bar{z})}$, $M^{RHP}_{21}(z) = -z\overline{M^{RHP}_{12}(\bar{z})}$, *we obtain the whole result of* $M^{RHP}_+(\rho)$.

Proof. 1. When $\rho \in U_{z_0}$, from (4.12), we have

$$M^{RHP}_+(\rho) = E(\rho) M^{(out)}_+(\rho), \tag{7.45}$$

from which we have

$$[M^{RHP}_+(\rho)]_{11} = E_{11}(\rho)[M^{(out)}_+(\rho)]_{11} + E_{12}(\rho)[M^{(out)}_+(\rho)]_{21}, \tag{7.46}$$

$$[M^{RHP}_+(\rho)]_{12} = E_{11}(\rho)[M^{(out)}_+(\rho)]_{12} + E_{12}(\rho)[M^{(out)}_+(\rho)]_{22}. \tag{7.47}$$

Combining with (5.29) and **Proposition 7** we come to the result.

2. When $\rho \notin U_{z_0}$, from (4.12) we have

$$M^{RHP}_+(\rho) = E(\rho) M^{(out)}_+(\rho) M^{pc}_+(\rho; z_0, \eta), \tag{7.48}$$

from which we have

$$
\begin{aligned}
[M^{RHP}_+(\rho)]_{11} =& E_{11}(\rho)[M^{(out)}_+(\rho)]_{11}[M^{PC}_+(\rho)]_{11} + E_{12}(\rho)[M^{(out)}_+(\rho)]_{21}[M^{PC}_+(\rho)]_{11} \\
&+ E_{11}(\rho)[M^{(out)}_+(\rho)]_{12}[M^{PC}_+(\rho)]_{21} + E_{12}(\rho)[M^{(out)}_+(\rho)]_{22}[M^{PC}_+(\rho)]_{21},
\end{aligned}
\tag{7.49}
$$

$$
\begin{aligned}
[M^{RHP}_+(\rho)]_{12} =& E_{11}(\rho)[M^{(out)}_+(\rho)]_{11}[M^{PC}_+(\rho)]_{12} + E_{12}(\rho)[M^{(out)}_+(\rho)]_{21}[M^{PC}_+(\rho)]_{12} \\
&+ E_{11}(\rho)[M^{(out)}_+(\rho)]_{12}[M^{PC}_+(\rho)]_{22} + E_{12}(\rho)[M^{(out)}_+(\rho)]_{22}[M^{PC}_+(\rho)]_{22}.
\end{aligned}
\tag{7.50}
$$

Combining with (5.29), (6.13) and **Proposition 7**, we come to the result. ■

8 $\bar{\partial}$ Problem

$\bar{\partial}$-Problem 5 of $M^{(3)}$ is equivalent to the integral equation

$$M^{(3)}(z) = (1,0)+\frac{1}{\pi}\int_C \frac{\bar{\partial}M^{(3)}(s)}{z-s}dm(s) = (1,0)+\frac{1}{\pi}\int_C \frac{M^{(3)}(s)W^{(3)}(s)}{z-s}dm(s), \quad (8.1)$$

where $W^{(3)}(s) = M^{RHP}(s)\bar{\partial}R^{(2)}(s)M^{RHP}(s)^{-1}$, and $m(s)$ is the Lebesgue measure on the C. If we denote C_z as the left Cauchy-Green integral operator,

$$fC_z(z) = \frac{1}{\pi}\int_C \frac{f(s)W^{(3)}(s)}{z-s}dm(s),$$

then

$$M^{(3)}(z) = (1,0)\left(I - C_z\right)^{-1}. \quad (8.2)$$

To prove the existence of operator $\left(I - C_z\right)^{-1}$, we have the following Lemma.

Lemma 3. *There exists a constant C such that the operator C_z satisfies that*

$$\| C_z \|_{L^\infty\to L^\infty}\leq C|t|^{-1/4}. \quad (8.3)$$

Proof. For any $f \in L^\infty$,

$$\| fC_z \|_{L^\infty}\leq\| f \|_{L^\infty} \frac{1}{\pi}\int_C \frac{|W^{(3)}(s)|}{|z-s|}dm(s), \quad (8.4)$$

where

$$|W^{(3)}(s)| \leq\| M^{RHP} \|_{L^\infty} |\bar{\partial}R^{(2)}(s)| \, \| M^{RHP} \|^{-1}_{L^\infty}\lesssim |\bar{\partial}R^{(2)}(s)|.$$

So we only need to estimate

$$\frac{1}{\pi}\int_C \frac{|\bar{\partial}R^{(2)}(s)|}{|z-s|}dm(s).$$

We only prove the case $\eta = +1$; the case $\eta = -1$ can be proven in the same way. For $\bar{\partial}R^{(2)}(s)$ is a piece-wise function; we detail the case in the region Ω_1, the other regions are similar. From (3.11) and (3.28) we have

$$\| C_z \|_{L^\infty\to L^\infty}\leq C(I_1 + I_2 + I_3), \quad (8.5)$$

where for $s = u + vi$,

$$I_1 = \iint_{\Omega_1} \frac{|\bar{\partial}X_Z(s)e^{-8tv(u-z_0)}|}{|z-s|}dudv, \qquad I_2 = \iint_{\Omega_1} \frac{|sp_1'(u)e^{-8tv(u-z_0)}|}{|z-s|}dudv,$$

$$I_3 = \iint_{\Omega_1} \frac{|s-z_0|^{-1/2}e^{-8tv(u-z_0)}}{|z-s|}dudv. \quad (8.6)$$

First we bound I_1, For $z=\alpha+\beta i$, note that

$$\| (s-z)^{-1} \|_{L^2(v+z_0,+\infty)}^2 = \int_{v+z_0}^{\infty} \frac{1}{v-\beta}\left[\left(\frac{u-\alpha}{v-\beta}\right)^2+1\right]^{-1} d\left(\frac{u-\alpha}{v-\beta}\right) \leq \frac{\pi}{v-\beta}, \quad (8.7)$$

then we have

$$\begin{aligned} I_1 &= \int_0^{\infty}\int_{v+z_0}^{\infty} \frac{|\bar{\partial}X_Z(s)e^{-8tv(u-z_0)}|}{|z-s|}dudv \\ &\leq \int_0^{\infty} \| \bar{\partial}X_Z(s) \|_{L_u^2(v+z_0,\infty)} \| (s-z)^{-1} \|_{L^2(v+z_0,+\infty)} e^{-8tv^2}dv \\ &\lesssim \int_0^{\infty} \frac{e^{-8tv^2}}{|v-\beta|^{1/2}}dv \leq t^{-1/4}\int_R \frac{e^{-8(\sqrt{t}\beta+w)^2}}{|w|^{1/2}}dw \leq C_1 t^{-1/4}. \end{aligned} \quad (8.8)$$

And for I_2,

$$\begin{aligned} I_2 &\leq \int_0^{\infty} \| (u^2+v^2)^{1/2}p_1'(u) \|_{L_u^2(v+z_0,\infty)} \| (s-z)^{-1} \|_{L^2(v+z_0,+\infty)} e^{-8tv^2}dudv \\ &\lesssim \int_0^{\infty} \| p_1(u) \|_{H_u^2(R)} \frac{(1+v)e^{-8tv^2}}{|v-\beta|^{1/2}}dv. \end{aligned} \quad (8.9)$$

And using $e^{-m} \leq m^{-1/4}$ for $m \geq 0$ we obtain

$$\begin{aligned} \int_0^{\beta} \beta^{-1/2}\frac{ve^{-8t\beta^2(v/\beta)^2}}{\beta|v/\beta-1|^{1/2}}d(v/\beta) &= \int_0^{1} \beta^{-1/2}\frac{we^{-8t\beta^2w^2}}{|w-1|^{1/2}}dw \\ &\lesssim t^{-1/4}\int_0^1 w(1-w)^{-1/2}dw \leq Ct^{-1/4}. \end{aligned} \quad (8.10)$$

Together with

$$\begin{aligned} &\int_{\beta}^{\infty} ve^{-8tv^2}(v-\beta)^{-1/2}dv \\ &= \int_0^{\infty} t^{-3/4}e^{-8[\sqrt{t}(w+\beta)]^2}(\sqrt{t}w)^{1/2} + \beta t^{-1/4}e^{-8[\sqrt{t}(w+\beta)]}(\sqrt{t}w)^{-1/2}d\sqrt{t}w \\ &\lesssim t^{-1/4}, \end{aligned} \quad (8.11)$$

we have that there exists a constant $C_2 > 0$ such that $I_2 \leq C_2 t^{-1/4}$. For I_3, we choose $p > 2$ and q Hölder conjugate to p, and notice that

$$\begin{aligned}\| (s-z)^{-1} \|^q_{L^q_u(v+z_0,\infty)} &= \int\limits_{v+z_0}^{\infty} \left[1+\left(\frac{u-\alpha}{v-\beta}\right)^2\right]^{-q/2} |v-\beta|^{-q+1} d\left(\frac{u-\alpha}{v-\beta}\right) \\ &\leq C_q |v-\beta|^{-q+1}, \\ \| (s-z_0)^{-1/2} \|^p_{L^p_u(v+z_0,\infty)} &= \int\limits_{v+z_0}^{\infty} \left[1+\left(\frac{u-z_0}{v}\right)^2\right]^{-p/4} |v|^{-p/2+1} d\left(\frac{u-z_0}{v}\right) \\ &\leq C_p |v|^{-p/2+1},\end{aligned}$$

by Hölder inequality we have

$$\begin{aligned} I_3 &\leq \int\limits_0^{\infty} e^{-8tv^2} \| (s-z_0)^{-1/2} \|_{L^p_u(v+z_0,\infty)} \| (s-z)^{-1} \|_{L^q_u(v+z_0,\infty)} dv \\ &\leq max[C_p, C_q] \int\limits_0^{\infty} e^{-8tv^2} v^{1/p-1/2} |v-\beta|^{1/q-1} dv. \end{aligned} \tag{8.12}$$

And using the same way of estimating I_2, we obtain a constant $C_3 > 0$ such that $I_3 \leq C_3 t^{-1/4}$. Finally we come to the result by combining the above equations. ∎

From this Lemma, we obtain that for sufficiently large t, $\| C_z \|_{L^\infty \to L^\infty} < 1$. So the operator $(I - C_z)^{-1}$ exists, which means $M^{(3)}$ uniquely exists with property

$$\| M^{(3)} \|_\infty \lesssim 1. \tag{8.13}$$

To recover the long-time asymptotic behavior of $q(x,t)$ by reconstruction formula, we need to consider the asymptotic behavior of $M_1^{(3)}$, where $M_1^{(3)}$ is given by the Laurent-expansion of $M^{(3)}$ as $z \to \infty$

$$M^{(3)}(z) = (1,0) + \frac{M_1^{(3)}(x,t)}{z} + \frac{1}{z\pi} \int\limits_C \frac{sM^{(3)}(s)W^{(3)}(s)}{z-s} dm(s), \tag{8.14}$$

and

$$M_1^{(3)}(x,t) = \frac{1}{\pi} \int\limits_C M^{(3)}(s)W^{(3)}(s) dm(s). \tag{8.15}$$

Then we start to estimate $M_1^{(3)}$.

Lemma 4. *For $z = iy$, $y \in R$ and $y \to +\infty$, we have*

$$|M_1^{(3)}(x,t)| \lesssim |t|^{-3/4}. \tag{8.16}$$

Proof. From **Lemma 3** and (8.2), we have $\| M^{(3)} \|_\infty \lesssim 1$. And we only estimate the integral on Ω_1, since the other estimates are similar. As in the above Lemma, by (3.11) and (3.28) we obtain

$$|\frac{1}{\pi}\int_{\Omega_1} M^{(3)}(s)W^{(3)}(s)dm(s)| \lesssim \frac{1}{\pi}\int_{\Omega_1} |W^{(3)}(s)|dm(s) \lesssim I_4 + I_5 + I_6, \tag{8.17}$$

where for $s - z_0 = u + vi$

$$I_4 = \iint_{\Omega_1} |\bar{\partial} X_Z(s)| e^{-8tvu} dudv, \quad I_5 = \iint_{\Omega_1} |sp_1'(u)| e^{-8tvu} dudv,$$
$$I_6 = \iint_{\Omega_1} |s - z_0|^{-1/2} e^{-8tvu} dudv. \tag{8.18}$$

By Cauchy-Schwarz inequality we have

$$\begin{aligned}|I_4| \leq & \int_0^\infty \| \bar{\partial} X_Z(s) \|_{L^2_u(v+z_0,\infty)} \left(\int_v^\infty e^{-16tvu} du\right)^{1/2} dv \\ \lesssim & \int_0^\infty t^{-1/2} v^{-1/2} e^{-8tv^2} dv = \int_0^\infty t^{-3/4} (\sqrt{t}v)^{-1/2} e^{-8(\sqrt{t}v)^2} d(\sqrt{t}v) \\ \leq & C_4 t^{-3/4}.\end{aligned} \tag{8.19}$$

And for I_5,

$$\begin{aligned}|I_5| \leq & \int_0^\infty \| \left((u+z_0)^2 + v^2\right)^{1/2} p_1' \|_{L^2_u(v+z_0,\infty)} \left(\int_v^\infty e^{-16tvu} du\right)^{1/2} dv \\ \lesssim & \int_0^\infty v^{-1/2} t^{-1/2} \| p_1(u) \|_{H^2_u(R)} (1+v) e^{-8tv^2} dv \\ \lesssim & \int_0^\infty t^{-3/4} (\sqrt{t}v)^{-1/2} e^{-8(\sqrt{t}v)^2} + t^{-5/4} (\sqrt{t}v)^{1/2} e^{-8(\sqrt{t}v)^2} d(\sqrt{t}v) \\ \leq & C_5 t^{-3/4}.\end{aligned} \tag{8.20}$$

Finally we consider I_6 in the same way we do with I_3 by Hölder inequality with $2 < p < 4$

$$\begin{aligned}|I_6| \leq & \int_0^\infty \left(\int_v^\infty e^{-8qtvu} du\right)^{1/q} v^{1/p-1/2} dv \lesssim \int_0^\infty t^{-1/q} v^{2/p-3/2} e^{-8tv^2} dv \\ = & \int_0^\infty t^{-3/4} (\sqrt{t}v)^{2/p-3/2} e^{-8(\sqrt{t}v)^2} d(\sqrt{t}v) \\ \leq & C_6 t^{-3/4}.\end{aligned} \tag{8.21}$$

These estimates together show the consequence. ■

But our eventual aim is to obtain the long-time asymptotic behavior of $u(x,t)$, so we need the following estimation about $M^{(3)}(\rho)$.

Proposition 10. *The unique solution $M^{(3)}$ of $\bar{\partial}$-Problem 5 satisfies*

$$M^{(3)}(\rho) = (1,0) + O(t^{-3/4}), \tag{8.22}$$

for sufficiently large times $|t| > 0$, where the implied constant is independent of t.

Proof.

$$M^{(3)}(\rho) = (1,0) + \frac{1}{\pi}\int_C \frac{M^{(3)}(s)W^{(3)}(s)}{\rho - s}dm(s), \tag{8.23}$$

where

$$M^{(3)}(s)W^{(3)}(s) = \\ \left(M_1^{(3)}(s)W_{11}^{(3)}(s) + M_2^{(3)}(s)W_{21}^{(3)}(s), M_1^{(3)}(s)W_{12}^{(3)}(s) + M_2^{(3)}(s)W_{22}^{(3)}(s)\right). \tag{8.24}$$

Note that $\bar{\partial}R^{(2)}$ has zeros on its diagonal, so together with $W^{(3)} = M^{RHP}\bar{\partial}R^{(2)}[M^{RHP}]^{-}$ we obtain

$$\begin{aligned} W_{11}^{(3)} &= \bar{\partial}R_{21}^{(2)}M_{12}^{RHP}M_{22}^{RHP} - \bar{\partial}R_{12}^{(2)}M_{11}^{RHP}M_{21}^{RHP}, &(8.25)\\ W_{12}^{(3)} &= -\bar{\partial}R_{21}^{(2)}[M_{12}^{RHP}]^2 + \bar{\partial}R_{12}^{(2)}[M_{11}^{RHP}]^2, &(8.26)\\ W_{21}^{(3)} &= \bar{\partial}R_{21}^{(2)}[M_{22}^{RHP}]^2 - \bar{\partial}R_{12}^{(2)}[M_{21}^{RHP}]^2, &(8.27)\\ W_{22}^{(3)} &= -\bar{\partial}R_{21}^{(2)}M_{12}^{RHP}M_{22}^{RHP} + \bar{\partial}R_{12}^{(2)}M_{11}^{RHP}M_{21}^{RHP}. &(8.28) \end{aligned}$$

Then using (8.13) and (7.32) to control the size of each term in the integral, and the symmetry of M^{RHP}, we have

$$|M_1^{(3)}(\rho) - 1| \lesssim \int_C |\frac{s}{\rho - s}\bar{\partial}R_{21}^{(2)}| + |\frac{\bar{\partial}R_{12}^{(2)}}{\rho - s}|dm(s) = O(|t|^{-3/4}), \tag{8.29}$$

$$|M_0^{(3)}(\rho)| \lesssim \int_C |\frac{s^2}{\rho - s}\bar{\partial}R_{21}^{(2)}| + |\frac{s}{\rho - s}\bar{\partial}R_{21}^{(2)}| + |\frac{\bar{\partial}R_{12}^{(2)}}{\rho - s}|dm(s) = O(|t|^{-3/4}) \tag{8.30}$$

where the last equality of each estimation we use in the same way which used to bound $\int_C |W^{(3)}(z)|dm(z)$ in the above Lemma to establish the result. ■

9 Long-time asymptotics for modified NLS equation

Now we begin to consider the long-time asymptotics of $q(x,t)$ at first, which is the solution of (1.8). Inverting the sequence of transformations (2.17), (3.5), (4.7) and (4.12), we have

$$\begin{aligned} M(z) =& M^{(3)}(z)M^{RHP}(z)R^{(2)}(z)^{-1}T(z)^{\sigma_3} \\ =& M^{(3)}(z)E(z)M^{(out)}(z)R^{(2)}(z)^{-1}T(z)^{\sigma_3}, \quad \text{when } z\in C\setminus U_{z_0} \end{aligned} \tag{9.1}$$

To reconstruct the solution $q(x,t)$, we take $z\to\infty$ along the straight line z_0+R^+i. Then we have that eventually $z\in\Omega_2$, which means $R^{(2)}(z)=I$. From (2.7), (5.31), (7.6) and (8.14), we have

$$M=\left(I+\frac{M_1^{(3)}}{z}+...\right)\left(I+\frac{E_1}{z}+...\right)\left(I+\frac{M_1^{(out)}}{z}+...\right)\left(I+\frac{T_1^{\sigma_3}}{z}+...\right), \tag{9.2}$$

which means the coefficient of the z^{-1} in the Laurent expansion of M is

$$M_1=M_1^{(3)}+E_1+M_1^{(out)}+T_1^{\sigma_3}. \tag{9.3}$$

So from (1.58), (5.32), (7.7) and (8.16) we have

$$q(x,t)=q_{sol}(x,t;D(I))+|t|^{-1/2}f(x,t)+O(|t|^{-3/4}), \tag{9.4}$$

where $f(x,t)$ is given in (7.8).

Now we begin to construct the solution $u(x,t)$ of (1.1) with initial data u_0 by the transformation

$$u(x,t)=q(x,t)e^{-i\int_{-\infty}^{x}|q(y,t)|^2dy}. \tag{9.5}$$

Because we already have the long-time asymptotics of $q(x,t)$, we only need to consider $e^{-i\int_{-\infty}^{x}|q(y,t)|^2dy}$. From (1.38)and (1.39), we have

$$\begin{aligned} e^{-i\int_{-\infty}^{x}|q(y,t)|^2dy} &= \left(\frac{a(\rho)}{\varphi_+^1(\rho)}\right)^2 = M_1^+(\rho)^{-2} \\ &= \left[M^{(3)}(\rho)M_+^{RHP}(\rho)R_+^{(2)}(\rho)^{-1}T(\rho)^{\sigma_3}\right]_1^{-2} \\ &= \left[M^{(3)}(\rho)M_+^{RHP}(\rho)R_+^{(2)}(\rho)^{-1}\right]_1^{-2}T(\rho)^{-2} \end{aligned} \tag{9.6}$$

From the definition of $R^{(2)}$ in Figure 3, we have following situation:

- when $x>0$, which means when $\eta=+1$, $\rho<z_0$ or when $\eta=-1$, $\rho>z_0$, $R^{(2)}$ is an upper triangular matrix $\begin{pmatrix} 1 & * \\ 0 & 1 \end{pmatrix}$. So

$$\begin{aligned} \left[M^{(3)}(\rho)M_+^{RHP}(\rho)R_+^{(2)}(\rho)^{-1}\right]_1 =& [M^{(3)}(\rho)]_1[M_+^{RHP}(\rho)]_{11}+[M^{(3)}(\rho)]_2[M_+^{RHP}(\rho)]_{21} \\ =& [M_+^{RHP}(\rho)]_{11}+O(|t|^{-3/4}). \end{aligned} \tag{9.7}$$

From Proposition 9, when $\rho \notin U_{z_0}$ we have

$$\left[M^{(3)}(\rho)M_+^{RHP}(\rho)R_+^{(2)}(\rho)^{-1}\right]_1 = [M_+(\rho;x,t|D(I))]_{11} \prod_{\mathrm{Re}z_n \in I_{z_0}^{\eta}\setminus I} \left(\frac{\rho - z_n}{\rho - \bar{z}_n}\right) + |t|^{-1/2} H_1(\rho, z_0; x, t|D(I) + O(|t|^{-3/4}), \tag{9.8}$$

where H_1 is given in (7.43). And when $\rho \in U_{z_0}$,

$$\left[M^{(3)}(\rho)M_+^{RHP}(\rho)R_+^{(2)}(\rho)^{-1}\right]_1 = [M_+(\rho;x,t|D(I))]_{11} \prod_{\mathrm{Re}z_n \in I_{z_0}^{\eta}\setminus I} \left(\frac{\rho - z_n}{\rho - \bar{z}_n}\right) + |t|^{-1/2} G_1(\rho, z_0; x, t|D(I) + O(|t|^{-3/4}), \tag{9.9}$$

where G_1 is given in (7.35).

- when $x < 0$, which means when $\eta = +1$, $\rho > z_0$ or when $\eta = -1$, $\rho < z_0$, $R_+^{(2)}(\rho)$ is a lower triangular matrix $\begin{pmatrix} 1 & 0 \\ -R_1(\rho) & 1 \end{pmatrix}$, where we note that $\theta(\rho) = 0$. So

$$\begin{aligned}
&\left[M^{(3)}(\rho)M_+^{RHP}(\rho)R_+^{(2)}(\rho)^{-1}\right]_1 \\
=&[M^{(3)}(\rho)]_1[M_+^{RHP}(\rho)]_{11} + [M^{(3)}(\rho)]_2[M_+^{RHP}(\rho)]_{21} \\
&- R_1(\rho)\left\{[M^{(3)}(\rho)]_1[M_+^{RHP}(\rho)]_{12} + [M^{(3)}(\rho)]_2[M_+^{RHP}(\rho)]_{22}\right\} \\
=&[M^{(3)}(\rho)]_1[M_+^{RHP}(\rho)]_{11} - R_1(\rho)[M^{(3)}(\rho)]_1[M_+^{RHP}(\rho)]_{12} \\
&+ O(|t|^{-3/4}).
\end{aligned} \tag{9.10}$$

From Proposition 9, when $\rho \notin U_{z_0}$ we have

$$\begin{aligned}
&\left[M^{(3)}(\rho)M_+^{RHP}(\rho)R_+^{(2)}(\rho)^{-1}\right]_1 = \\
&[M_+(\rho;x,t|D(I))]_{11} \prod_{\mathrm{Re}z_n \in I_{z_0}^{\eta}\setminus I} \left(\frac{\rho - z_n}{\rho - \bar{z}_n}\right) \\
&- R_1(\rho)[M_+(\rho;x,t|D(I))]_{12} \prod_{\mathrm{Re}z_n \in I_{z_0}^{\eta}\setminus I} \left(\frac{\rho - \bar{z}_n}{\rho - z_n}\right) \\
&+ |t|^{-1/2}\left[H_1(\rho, z_0; x, t|D(I) - R_1(\rho)H_2(\rho, z_0; x, t|D(I))\right] + O(|t|^{-3/4}),
\end{aligned} \tag{9.11}$$

where H_1 and H_2 is given in (7.43) and (7.44) respectively. And when $\rho \in U_{z_0}$,

$$\begin{aligned}
&\left[M^{(3)}(\rho)M_+^{RHP}(\rho)R_+^{(2)}(\rho)^{-1}\right]_1 = \\
&[M_+(\rho;x,t|D(I))]_{11} \prod_{\mathrm{Re}z_n \in I_{z_0}^{\eta}\setminus I} \left(\frac{\rho - z_n}{\rho - \bar{z}_n}\right) \\
&- R_1(\rho)[M_+(\rho;x,t|D(I))]_{12} \prod_{\mathrm{Re}z_n \in I_{z_0}^{\eta}\setminus I} \left(\frac{\rho - \bar{z}_n}{\rho - z_n}\right) \\
&+ |t|^{-1/2}\left[G_1(\rho, z_0; x, t|D(I) - R_1(\rho)G_2(\rho, z_0; x, t|D(I))\right] + O(|t|^{-3/4}),
\end{aligned} \tag{9.12}$$

where G_1 and G_2 are given in (7.35) and (7.36), respectively.

Combining the above results and **Proposition 5**, we have following result.

Theorem 1. *Let $u(x,t)$ be the solution of (1.1) with initial data $u_0 = u(x,t=0) \in H^{2,2}(R)$ which has corresponding scattering data $\left\{r, \{z_k, c_k\}_{k=1}^N\right\}$. Fixed $x_1, x_2, v_1, v_2 \in R$ with $x_1 \leq x_2$ and $v_1 \leq v_2$. Let $I = [-v_2/4, -v_1/4]$, and $z_0 = -x/4t + \rho$. Denote $u_{sol}(x,t)$ as the $N(I)$ soliton corresponding to reflection-less scattering data $\left\{r \equiv 0, \{z_k, c_k(I)\}_{k=1}^N\right\}$ which is given in (5.20). As $|t| \to \infty$ with $(x,t) \in C(x_1, x_2, v_1, v_2)$, we have*

- *when $x > 0$, from which we have when $\eta = +1$, $\rho < z_0$ or when $\eta = -1$, $\rho > z_0$.*

If $\rho \notin U_{z_0}$, we obtain

$$u(x,t) = u_{sol}(x,t)S(\rho)\left(1 + F_1(x,t)|t|^{-1/2}\right) + O(|t|^{-3/4}), \tag{9.13}$$

and if $\rho \in U_{z_0}$, we obtain

$$u(x,t) = u_{sol}(x,t)S(\rho)\left(1 + F_2(x,t)|t|^{-1/2}\right) + O(|t|^{-3/4}), \tag{9.14}$$

where $S(\rho) = \prod_{Rez_n \in I_{z_0}^\eta \setminus I} \left(\frac{\rho - \bar{z}_n}{\rho - z_n}\right)^2 T(\rho)^2$ is a constant depending on ρ, η, z_0 and two sets of scattering data,

$$F_1(x,t) = \frac{f(x,t)}{q_{sol}(x,t)} - 2exp\left(-\frac{i}{2}\int_{-\infty}^{x} |q_{sol}(y,t)|^2 dy\right) \prod_{Rez_n \in I_{z_0}^\eta \setminus I} \left(\frac{\rho - \bar{z}_n}{\rho - z_n}\right) H_1, \tag{9.15}$$

$$F_2(x,t) = \frac{f(x,t)}{q_{sol}(x,t)} - 2exp\left(-\frac{i}{2}\int_{-\infty}^{x} |q_{sol}(y,t)|^2 dy\right) \prod_{Rez_n \in I_{z_0}^\eta \setminus I} \left(\frac{\rho - \bar{z}_n}{\rho - z_n}\right) G_1. \tag{9.16}$$

- *when $x < 0$, from which we have when $\eta = +1$, $\rho > z_0$ or when $\eta = -1$, $\rho < z_0$.*

If $\rho \notin U_{z_0}$, we obtain

$$u(x,t) = u_{sol}(x,t)B(x,t)^2\left(1 + F_3(x,t)|t|^{-1/2}\right) + O(|t|^{-3/4}), \tag{9.17}$$

and if $\rho \in U_{z_0}$, we obtain

$$u(x,t) = u_{sol}(x,t)B(x,t)^2\left(1 + F_4(x,t)|t|^{-1/2}\right) + O(|t|^{-3/4}), \tag{9.18}$$

where

$$B(x,t) = \left(\prod_{Rez_n \in I_{z_0}^{\eta} \setminus I} \left(\frac{\rho - \bar{z}_n}{\rho - z_n} \right) + \prod_{Rez_n \in I_{z_0}^{\eta} \setminus I} \left(\frac{\rho - z_n}{\rho - \bar{z}_n} \right) R_1(\rho) \int_x^{+\infty} u_{sol}(y,t) dy \right)^{-1}, \tag{9.19}$$

$$F_3(x,t) = \frac{f(x,t)}{q_{sol}(x,t)} - 2B(x,t) \left(H_1 - R_1(\rho) H_2 \right), \tag{9.20}$$

$$F_4(x,t) = \frac{f(x,t)}{q_{sol}(x,t)} - 2B(x,t) \left(G_1 - R_1(\rho) G_2 \right). \tag{9.21}$$

Acknowledgements

This work is supported by the National Science Foundation of China (Grant Nos. 11671095, 51879045).

References

[1] Agrawal G P, Nonlinear Fiber Optics, 4th ed., Academic Press, Boston, 2007.

[2] Brizhik L, Eremko A and Piette B, Solutions of a D-dimensional modified nonlinear Schrödinger equation, *Nonlinearity*, **16.4**, 1481-1497, 2003.

[3] Bikbaev R F, Asymptotic-behavior as t-infinity of the solution to the cauchy-problem for the Landau-Lifshitz equation, *Theor. Math. Phys*, **77**, 1117-1123, 1988.

[4] Borghese M, Jenkins R and McLaughlin K T R, Long-time asymptotic behavior of the focusing nonlinear Schrödinger equation, *Ann. I. H. Poincaré Anal*, **35**, 887-920, 2018.

[5] Chen Z Y and Huang N N, Explicit N-soliton solution of the modified nonlinear Schrödinger equation, *Phys. Rev. A*, **41**, 4066-4069, 1990.

[6] Chen Z Y, An inverse scattering transformation for the modified nonlinear Schrödinger equation, *Commun. Theor. Phys.*, **15**, 271-276, 1991.

[7] Chen Z Y, Asymptotic behaviors of multi-soliton solutions of the cubic and the modified nonlinear Schrodinger equation *Commun. Theor. Phys.*, **13**, 299-306, 1990.

[8] Cuccagna S and Jenkins R, On asymptotic stability of N-solitons of the defocusing nonlinear Schrödinger equation, *Comm. Math. Phys*, **343**, 921-969, 2016.

[9] Dieng M and McLaughlin K D T, Dispersive asymptotics for linear and integrable equations by the Dbar steepest descent method, Nonlinear dispersive partial differential equations and inverse scattering, 253-291, Fields Inst. Commun., 83, Springer, New York, 2019.

[10] Deift P and Zhou X, Long-time asymptotics for solutions of the NLS equation with initial data in a weighted Sobolev space, *Comm. Pure Appl. Math.*, **56**, 1029-1077, 2003.

[11] Doktorov E V and Leble S B, A Dressing Method in Mathematical Physics, Springer, The Netherlands, 2007.

[12] Fokas A S and Its A R, Soliton generation for initial-boundary-value problems, *Phys. Rev. Lett.*, **68**, 3117-3120, 1992.

[13] Grunert K and Teschl G, Long-time asymptotics for the Korteweg de Vries equation via nonlinear steepest descent. *Math. Phys. Anal. Geom.*, **12**, 287-324, 2009.

[14] He J, Xu S and Cheng Y, The rational solutions of the mixed nonlinear Schrödinger equation, *AIP Advances*, **5**, 017105, 2015.

[15] Jenkins R, Liu J Q, Perry P and Sulem C, Global well-posedness for the derivative nonlinear Schrödinger equation, *Commun. Partial Diff. Eqs.*, **43**, 1151-1195, 2018.

[16] Jenkins R, Liu J Q, Perry P and Sulem C, The derivative nonlinear Schrödinger equation: Global well-posedness and soliton resolution, *Quarterly Appl Math.* **1**, 33-73, 2020.

[17] Jenkins R, Liu J Q, Perry P and Sulem C, Soliton resolution for the derivative nonlinear Schrödinger equation *Commun. Math. Phys.*, **363**, 1003-1049, 2018.

[18] Karsten T, Hamiltonian form of the modified nonlinear Schrödinger equation for gravity waves on arbitrary depth, *J. Fluid. Mech.*, **670**, 404-426, 2011.

[19] Kitaev A V and Vartanian A H, Leading-order temporal asymptotics of the modified nonlinear Schrödinger equation: solitonless sector, *Inverse Prooblems,* **13**, 1311-1339, 1997.

[20] Liu J Q, Perry P and Sulem C, Long-time behavior of solutions to the derivative nonlinear Schrödinger equation for soliton-free initial data *Ann. Inst. Henri Poincaré Anal. Non Linéaire*, **35**, 217-265, 2018.

[21] Liu L S and Wang W Z, Exact N-soliton solution of the modified nonlinear Schrödinger equation, *Phys. Rev. E*, **48**, 3054-3059, 1993.

[22] Manakov S V, Nonlinear Fraunhofer diffraction, *Sov. Phys. JETP,* **38**, 693-696, 1974.

[23] McLaughlin K T R and Miller P D, The $\bar{\partial}$ steepest descent method and the asymptotic behavior of polynomials orthogonal on the unit circle with fixed and exponentially varying non-analytic weights, *Int. Math. Res. Not.*, 2006 , Art. ID 48673.

[24] McLaughlin K T R and Miller P D, The $\bar{\partial}$ steepest descent method for orthogonal polynomials on the real line with varying weights, *Int. Math. Res. Not.*, IMRN 2008 , Art. ID 075.

[25] Mihalache D, Truta N, Panoiu N and Baboiu D, Analytic method for solving the modified nonlinear Schrödinger equation describing soliton propagation along optical fibers, *Phys. Rev. A*, **47**, 3190-3194, 1993.

[26] Mio K, Ogino T, Minami K and Takeda S, Modified nonlinear Schrödinger equation for Alfvèn waves propagating along the magnetic field in cold plasmas, *J. Phys. Soc. Jpn.*, **41**, 265-271, 1976.

[27] Nakatsuka H, Grischkowsky D and Balant A C, Nonlinear picosecond-pulse propagation through optical fibers with positive group velocity dispersion, *Phys. Rev. Lett.*, **47**, 910-913, 1981.

[28] Schuur P C, Asymptotic analysis of soliton products, *Lecture Notes in Mathematics,* **1232**, 1986.

[29] Stiassnie M, Note on the modified nonlinear Schrödinger equation for deep water waves, *Wave Motion*, **6**, 431-433, 1984.

[30] Trogdon T and Olver S, Riemann-Hilbert problems, their numerical solution, and the computation of nonlinear special functions, *SIAM*, Philadelphia, PA, 2016.

[31] Tzoar N and Jain M, Self-phase modulation in long-geometry optical waveguides, *Phys. Rev. A*, **23**, 1266-1270, 1981.

[32] Wadati M, Konno K and Ichikawa Y H, A generalization of inverse scattering method. *J. Phys. Soc. Jpn.*, **46**, 1965-1966, 1979.

[33] Wen X Y, Yang Y Q and Yan Z Y, Generalized perturbation (n,M)-fold Darboux transformations and multi-rogue-wave structures for the modified self-steepening nonlinear Schrödinger equation, *Phys. Rev. E*, **92**, 012917, 2015.

[34] Xu J and Fan EG, Long-time asymptotics for the Fokas-Lenells equation with decaying initial value problem: Without solitons, *J. Differential Equations,* **259**, 1098-1148, 2015.

[35] Xu J, Fan E G and Chen Y, Long-time asymptotic for the derivative nonlinear Schrodinger equation with step-like initial value, *Math. Phys. Anal. Geom.*, **16**, 253-288, 2013.

[36] Yang J K, Nonlinear Waves in Integrable and Nonintegrable Systems, SIAM, Philadelphia, 2010.

[37] Yang Y L and Fan E G, Riemann-Hilbert approach to the modified nonlinear Schrödinger equation with non-vanishing asymptotic boundary conditions, *Physica D.,* **417**, 132811, 2021.

[38] Zakharov V E and Shabat A B, Exact theory of two-dimensional self-focusing and one-dimensional self-modulation of waves in nonlinear media, *Sov. Phys. JETP,* **34**, 62-69, 1972.

[39] Zakharov V E and Manakov S V, Asymptotic behavior of nonlinear wave systems integrated by the inverse scattering method, *Sov. Phys. JETP,* **44**, 106-112, 1976.

[40] Zhou X, Direct and inverse scattering transforms with arbitrary spectral singularities, *Commun. Math. Phys.*, **42**, 895-938, 1989.

[41] Zhou X, The Riemann-Hilbert problem and inverse scattering, *SIAM J. Math. Anal.*, **20**, 966-986, 1989.

[42] Zhou X and Deift P, A steepest descent method for oscillatory Riemann-Hilbert problems. *Ann. Math.*, **137**, 295-368, 1993.

[43] Zhou X and Deift P, Long-time behavior of the non-focusing nonlinear Schrödinger equation–a case study, *Lectures in Mathematical Sciences*, Graduate School of Mathematical Sciences, University of Tokyo, 1994.

B7. Two hierarchies of multiple solitons and soliton molecules of (2+1)-dimensional Sawada-Kotera type equation

Ruoxia Yao[a], *Wei Wang*[a,b] *and Yan Li*[a]

[a]*School of Computer Science, Shaanxi Normal University, Xi'an, 710119, P.R.China*

[b]*Information and Education Technology Center, Xi'an University of Finance and Economics, Xi'an, 710062, P.R.China*

Abstract

Two transformations $v = p\,(\ln f)_{xx} (p = 2, 4)$ that produce a trilinear and a quintic linear equation solved by two pairs of Hirota's bilinear equations respectively and the corresponding bilinearizations of a (2+1)-dimensional Sawada-Kotera (2DSK) type equation are reported. The two pairs of Hirota's bilinear equations, one is the normal (2+1)-dimensional bilinear SK equation and the other is a variant bilinear Kadomtsev-Petviashvili (KP) equation, are both obtained from the SK and variant KP (vKP) equations by $v = 2\,(\ln f)_{xx}$. Two hierarchies of multiple solitons and soliton molecules are discussed starting from them.

1 Introduction

In the last two decades, an increasing number of researchers have taken more attention to the study of various kinds of solitons and soliton molecules. The simplified Hirota bilinear method [1–6] could construct multiple soliton solutions effectively. Up to now, we know that an integrable nonlinear partial differential equation (PDE) usually possesses only one set of multiple soliton solutions under certain dependent variable transformation while it admits the PDE to pass Hirota three-soliton condition [7–9]; however, we find that (2+1)-dimensional Sawada-Kotera (2DSK) equation possesses two hierarchies of multiple soliton solutions [16]. This very special but intrinsic property is novel and reported recently by us. The two hierarchies of the multiple soliton solutions of the 2DSK equation differ significantly with respect to their underlying mathematical structures and the evolution behavior of the solutions, of which actually there exists an interesting relation that will be given systematically in the future.

In this chapter, we concentrate on the following equation:

$$\begin{aligned} &v_t + v_{5x} + 15v_x v_{xx} + 15vv_{3x} + 45v^2 v_x + 5v_{xxy} + 15vv_y + 15v_x \textstyle\int v_y \mathrm{d}x \\ &-5\textstyle\int v_{yy}\mathrm{d}x + 30av_y + 30av_{3x} + 180avv_x + 180a^2 v_x = 0, \end{aligned} \tag{1.1}$$

where $v = v(x, y, t)$, and when $a = 0$, that is the (2+1)-dimensional Sawada-Kotera equation [11, 14] proposed firstly in [12]. Equation (1.1) also falls in the category of 2DSK equations and has multiple solitons, soliton molecules and resonant soliton structures with nonzero background.

2 New bilinearizations of 2DSK equation (1.1)

Introducing a potential variable transformation $v = u_x$, $u = u(x, y, t)$, Eq. (1.1) reduces to

$$\begin{aligned} &u_{xt} + u_{6x} + 5u_{3xy} - 5u_{yy} + 15u_{xx}u_{3x} + 15u_xu_{4x} + 30au_{4x} + 15u_xu_{xy} \\ &+30au_{xy} + 15u_{xx}u_y + 45u_x^2u_{xx} + 180au_xu_{2x} + 180a^2u_{2x} = 0. \end{aligned} \tag{2.1}$$

Here, we mainly consider Eq. (2.1) for its possible multiple soliton and soliton molecule solutions. Beginning with an auxiliary dependent variable transformation

$$u = p\,(\ln f)_x, \quad f \equiv f_1 = 1 + \exp(\xi_1), \tag{2.2}$$

where $f = f(x, y, t)$, $\xi_1 = k_1x + l_1y + w_1t$, we derive

$$p = 2, \quad w_1 = -\frac{30\,ak_1^4 + 30\,ak_1l_1 + 180\,a^2k_1^2 - 5\,l_1^2 + 5\,k_1^3l_1 + k_1^6}{k_1}, \tag{2.3}$$

$$p = 4, \quad l_1 = -k_1^3 - 6ak_1, \quad w_1 = 9(k_1^5 + 10ak_1^3 + 20a^2k_1), \tag{2.4}$$

and then the trilinear and quintic linear equations of Eq. (2.1) later corresponding to (2.3) and (2.4), respectively. It is easy to check that (2.3) is not equivalent to (2.4). So before we go further, we should report the following fact. One can notice that in (2.3), l_1 is free, but in (2.4), $l_1 = -k_1^3 - 6ak_1$. Substituting $l_1 = -k_1^3 - 6ak_1$ into w_1 of (2.3) gives $w_1 = 9(k_1^5 + 10ak_1^3 + 20a^2k_1)$, which means that

$$p = 2, \quad l_1 = -k_1^3 - 6ak_1, \quad w_1 = 9(k_1^5 + 10ak_1^3 + 20a^2k_1) \tag{2.5}$$

is a subcase of (2.3). However, (2.4) is not a subcase of (2.3) apparently. Although the solution $u = 4(\ln f_1)_x$ with f_1 given in (2.2), and $\xi_1 = k_1x + l_1y + w_1t$ with the same dispersion relations $\{l_1 = -k_1^3 - 6ak_1,\ w_1 = 9(k_1^5 + 10ak_1^3 + 20a^2k_1)\}$ is indeed a solution of the 2DSK equation, it does not mean that it is a subcase of $p = 2$.

In fact, we know that for some function $f_1(\xi), \xi_1 = k_1x + l_1y + w_1t$,

$$4\ln\sqrt{f_1} = 2\ln f_1 \neq 4\ln f_1,$$

which indicates the solutions with the parameter selections (2.4) and (2.5) respectively (same dispersion relations, however different p values) are different. In other words, we can not find a function f_1 such that $\sqrt{f_1} = f_1$.

We further point out that even for the same dispersion relations

$$l_1 = -k_1^3 - 6ak_1, \ w_1 = 9(k_1^5 + 10ak_1^3 + 20a^2k_1),$$

we get only these two solutions with $p = 2$ and $p = 4$, respectively, of which the amplitudes are different. Amplitude (power) of a nonlinear wave is an important physical quantity for a physical and mathematical model, which can affect the evolution behavior of waves, such as the evolution velocity of a wave. Actually, the amplitude of the solution of case $p = 4$ is a double size of that of case $p = 2$. It is say that cases (2.3) and (2.4) do not overlap each other, which indicates an important phenomenon that one can get solutions with different amplitudes under the same dispersion relations. We would like to emphasize that just because of this fact and based on (2.3) and (2.4), especially (2.4), we obtain two hierarchies of multiple soliton solutions respectively for Eq. (2.1). Actually, in 2004, we have investigated some (1+1)–dimensional 7th–order KdV equations in [17], such as the 7th–order Lax and 7th-order SK-Ito equations, of which the 7th–order Lax equation possesses only one cluster of multiple soliton solutions as usual, but the 7th–order SK–Ito equation possesses two one-soliton solutions with different amplitudes under the cases $p = 42, 84$, respectively. However, the 7th–order SK–Ito equation does not possess two hierarchies of higher order multiple soliton solutions. As we have seen above, the solutions of cases $p = 2$ and $p = 4$ are different; however, they have some connections such as the solutions are different and do not overlap each other even with the same dispersion relations. For what it is worth, we will study deeply and widely further. Next, we have the following discussion.

2.1 Bilinearization of case $u = 2\,(\ln f)_x$

Substituting $u = 2\,(\ln f)_x$ first into Eq. (2.1) yields the following trilinear equation

$$\begin{aligned} P_1 \;\equiv\;& f^2(f_{7x} + 5f_{4xy} + f_{2xt} - 5f_{x2y} + 180a^2 f_{3x} + 30af_{2xy}) \\ &+f\,(-150af_{4x}f_x - 20f_{3xy}f_x - 5f_{4x}f_y - f_{2x}f_t - 7f_{6x}f_x - 2f_{xt}f_x - 5f_{4x}f_{3x} \\ &+60f_{3x}f_{2x} - 60af_xf_{xy} - 540a^2 f_{2x}f_x + 10f_{xy}f_y + 5f_xf_{2y} + 10f_{3x}f_{xy} \\ &+9f_{5x}f_{2x} - 30af_{2x}f_y) - 30f_{4x}f_xf_{2x} + 30f_{2xy}f_x^2 + 240af_{3x}f_x^2 \\ &+60af_x^2f_y - 180af_{2x}^2f_x - 10f_xf_y^2 + 20f_{3x}^2f_x + 10f_{3x}f_xf_y + 360a^2f_x^3 \\ &+2f_x^2f_t - 30f_{2x}f_xf_{xy} + 12f_{5x}f_x^2 = 0. \end{aligned} \tag{2.6}$$

By the definition of the Hirota bilinear operator, Eq. (2.6) can be decomposed slickly and directly into two parts

$$\begin{aligned} P_1 \;\equiv\;& \frac{1}{f^2}\frac{\partial}{\partial x}[30a(D_xD_y + 6aD_x^2 + D_x^4) + D_x^6 + D_xD_t + 5D_x^3D_y - 5D_y^2]f\cdot f \\ &-\frac{2f_x}{f^3}[30a(D_xD_y + 6aD_x^2 + D_x^4) + D_x^6 + D_xD_t + 5D_x^3D_y - 5D_y^2]f\cdot f, \\ =\;& 0. \end{aligned} \tag{2.7}$$

We deduce that

$$\Delta_{\mathrm{vKP}} \equiv (D_xD_y + 6aD_x^2 + D_x^4)f\cdot f \tag{2.8}$$

is a Hirota's bilinear equation of a variant KP equation of the form

$$u_{4x} + u_{xy} + 6u_x^2 + 6uu_{2x} + 3au_{2x} = 0 \tag{2.9}$$

under an independent variable transformation $u = 2\,(\ln f)_{xx}$, and

$$\Delta_{\mathrm{SK}} \equiv (D_x^6 + D_xD_t + 5D_x^3D_y - 5D_y^2)f\cdot f \tag{2.10}$$

is the standard (2+1)-dimensional bilinear SK equation obtained from $u = 2\,(\ln f)_x$. Hence, if $f = f(x, y, t)$ solves Eqs. (2.8) and (2.10), it solves the trilinear Eq. (2.6).

2.2 Bilinearization of case $u = 4(\ln f)_x$

If taking the following transformation alternatively

$$u = 4(\ln f)_x, \tag{2.11}$$

we obtain a quintic linear equation of Eq. (2.1) with the form

$$\begin{aligned}
P_2 \equiv\ & f^4(f_{7x} + 5f_{4xy} + f_{2xt} - 5f_{x2y} + 180a^2 f_{3x} + 30af_{2xy} + 30af_{5x}) \\
& + f^3(-150af_{4x}f_x - 20f_{3xy}f_x - 5f_{4x}f_y - f_{2x}f_t - 7f_{6x}f_x - 2f_{xt}f_x \\
& + 25f_{4x}f_{3x} - 60af_x f_{xy} + 420af_{3x}f_{2x} - 540a^2 f_{2x}f_x + 10f_{xy}f_y \\
& + 5f_x f_{2y} + 40f_{3x}f_{xy} + 39f_{5x}f_{2x} - 30af_{2x}f_y + 30f_{2x}f_{2xy}) \\
& + f^2(-270f_{4x}f_x f_{2x} + 60af_y f_x^2 - 10f_x f_y^2 - 120af_x^2 f_{3x} - 100f_{3x}^2 f_x \\
& - 30f_{2x}^2 f_y + 2f_x^2 f_t + 150f_{3x}f_{2x}^2 + 360a^2 f_x^3 - 180f_{2x}f_x f_{xy} \\
& - 1260af_{2x}^2 f_x - 20f_{3x}f_x f_y - 18f_{5x}f_x^2) + f(180f_{2x}f_x^2 f_y \\
& + 210f_{4x}f_x^3 + 540f_{3x}f_x^2 f_{2x} + 1800af_{2x}f_x^3 + 120f_x^3 f_{xy} - 450f_{2x}f_x) \\
& - 120f_x^4 f_y - 480f_{3x}f_x^4 + 360f_{2x}f_x^3 - 720af_x^5 = 0.
\end{aligned} \tag{2.12}$$

Similarly, Eq. (2.12) can be written as

$$\begin{aligned}
P_2 \equiv\ & \left[\frac{15}{f^4}\frac{\partial}{\partial x}[D_x^2(D_xD_y + 6aD_x^2 + D_x^4)] - \frac{60}{f^5}f_x D_x^2(D_xD_y + 6aD_x^2 + D_x^4)\right] f\cdot f \\
& + \frac{1}{f^2}\frac{\partial}{\partial x}[30a(D_xD_y + 6aD_x^2 + D_x^4) + D_x^6 + D_xD_t + 5D_x^3D_y - 5D_y^2]f\cdot f \\
& - \frac{2}{f^3}f_x[30a(D_xD_y + 6aD_x^2 + D_x^4) + D_x^6 + D_xD_t + 5D_x^3D_y - 5D_y^2]f\cdot f \\
=\ & P_1 + \left(\frac{15}{f^4}\frac{\partial}{\partial x}Q - \frac{60}{f^5}f_x Q\right) f\cdot f \\
=\ & 0,
\end{aligned} \tag{2.13}$$

where $Q = D_x^2(D_xD_y + 6aD_x^2 + D_x^4)$ and P_1 is the trilinear Equation (2.7), which is also governed by the variant KP bilinear Equation (2.8), and the standard (2+1)-dimensional bilinear SK Equation (2.10). Hence, if $f = f(x, y, t)$ solves Eqs. (2.8) and (2.10), it also solves the quintic linear Eq. (2.13).

It shows that the obtained trilinear and quintic linear equations both are solved by the pair of bilinear equations which usually serves as a role to construct multiple soliton solutions. It is well known that Eq. (2.10) ensures the N–soliton solutions of the 2DSK equation starting from $v = 2(\ln f)_{xx}$, while Eq. (2.8) ensures the N–soliton solutions of the variant KP equation. Hence for the 2DSK type equation (2.1), an explicit connection between the standard 2DSK equation and a variant KP equation is established.

3 Multiple soliton and soliton molecule solutions

To construct multiple soliton and soliton molecule solutions of the 2DSK Eq. (2.1), let us first recall the simplified direct Hirota method [10]. It starts with a dependent variable transformation shown as before and with the following formulae:

$$f_1 = 1+\exp(\xi_1), \quad (3.1)$$

$$f_2 = 1+\exp(\xi_1)+\exp(\xi_2)+h_{1,2}\exp(\xi_1+\xi_2), \quad (3.2)$$

$$\begin{aligned} f_3 = {} & 1+\exp(\xi_1)+\exp(\xi_2)+\exp(\xi_3) \\ & +h_{1,2}\exp(\xi_1+\xi_2)+h_{1,3}\exp(\xi_1+\xi_3)+h_{2,3}\exp(\xi_2+\xi_3) \\ & +h_{1,2}h_{1,3}h_{2,3}\exp(\xi_1+\xi_2+\xi_3), \end{aligned} \quad (3.3)$$

$$\cdots\cdots$$

$$f_n = \sum_{\mu\in 0,1}\exp\left(\sum_{i=1}^{n}\mu_i\xi_i+\sum\mu_i\mu_j H_{ij}\right), \quad (3.4)$$

where $H_{i,j} = \ln h_{i,j}$, $\xi_i = \xi_i(x,y,t) = k_i x + l_i y + w_i t + \phi_i, (i = 1,\ldots,n)$, and k_i, l_i, w_i, ϕ_i are arbitrary constants for all i.

3.1 One-soliton solutions

From Eq. (2.2) along with Eqs. (2.3) and (2.4), we obtain two one-soliton solutions of Eq. (1.1)

$$\begin{aligned} & v = 4(\ln f_1)_{xx},\ f_1 = 1+\exp[k_1x-(k_1^3+6ak_1)y+9(k_1^5+10ak_1^3+20a^2k_1)t+\phi_1], \\ & v = 2(\ln f_1)_{xx},\ f_1 = 1+\exp\left[k_1x+l_1y+w_1t+\phi_1\right], \end{aligned} \quad (3.5)$$

where w_1 is given in (2.3) and $k_1 \neq 0, l_1 \neq 0$, ϕ_1 is free and denotes position parameter. They also can be rewritten in terms of hyperbolic cosine functions [13].

$$\begin{aligned} & v = 4\ln\left[\cosh\left(\tfrac{k_1}{2}x-\tfrac{k_1^3+6ak_1}{2}y+\tfrac{9(k_1^5+10ak_1^3+20a^2k_1)}{2}t+\phi_1\right)\right]_{xx}, \\ & v = 2\ln\left[\cosh\left(\tfrac{k_1}{2}x+\tfrac{l_1}{2}y+\tfrac{w_1}{2}t+\phi_1\right)\right]_{xx}. \end{aligned} \quad (3.6)$$

It has been indicated in Section 2 that two cases in (3.6) not only do not overlap each other, but also the first solution is not a subcase of the second one.

3.2 Two-soliton and soliton molecule solutions

Starting from $u = 2(\ln f_2)_x$ and $u = 4(\ln f_2)_x$ respectively, we obtain two types of two-soliton solutions of Eq. (2.1), then solutions of Eq. (1.1) via $v = u_x$. Specifically, under the case of $u = 2(\ln f_2)_x$, we obtain a soliton molecule solution.

Case I $u = 2(\ln f_2)_x$

Subcase I-1 Two-soliton solution

To find the soliton molecule solution, one can give the soliton ones. Starting from $u = 2(\ln f_2)_x$ for Eq. (2.1) we obtain a two-soliton solution with f_2 being the form of (3.2), with

$$w_i = -\frac{30\,ak_i^4+30\,ak_il_i+180\,a^2k_i^2-5\,l_i^2+5\,k_i^3l_i+k_i^6}{k_i}, \qquad h_{1,2} = \frac{C_1}{C_2}, \quad (3.7)$$

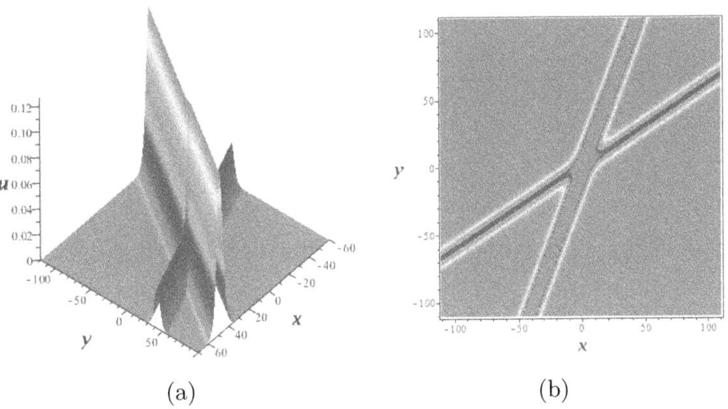

(a) (b)

Figure 1. Depiction of the evolution of the two-soliton solution (3.9) for Eq. (1.1) with (3.8).

and

$$C_1 = -36\, ak_2^3k_1^3 + 2\, l_1k_2^2k_1^3 - 3\, l_2k_2^2k_1^3 + 18\, ak_2^2k_1^4 + 18\, ak_2^4k_1^2 - 3\, l_1k_2^3k_1^2 + 2\, l_2k_2^3k_1^2 + l_2k_2k_1^4 + l_1k_1k_2^4 - 2\, l_2l_1k_2k_1 + l_2^2k_1^2 + l_1^2k_2^2 + 4\, k_2^4k_1^4 - 3\, k_2^3k_1^5 + k_2^2k_1^6 + k_2^6k_1^2 - 3\, k_2^5k_1^3,$$

$$C_2 = -2\, l_2l_1k_2k_1 + k_2^6k_1^2 + k_2^2k_1^6 + 4\, k_2^4k_1^4 + 3\, k_2^3k_1^5 + l_2^2k_1^2 + 3\, k_2^5k_1^3 + l_1^2k_2^2 + l_2k_2k_1^4 + 3\, l_2k_2^2k_1^3 + 2\, l_2k_2^3k_1^2 + 2\, l_1k_2^2k_1^3 + 3\, l_1k_2^3k_1^2 + l_1k_1k_2^4 + 18\, ak_2^2k_1^4 + 36\, ak_2^3k_1^3 + 18\, ak_2^4k_1^2,$$

where $i = 1,\ 2$, and $k_i \neq 0, l_i$ are free parameters.

Choosing $a = 1/2,\ k_1 = -1/2,\ k_2 = 1/5,\ l_1 = 1/5,\ l_2 = -2/5$ in (3.7), we have

$$a = \frac{1}{2}, \quad k_1 = -\frac{1}{2}, \quad k_2 = \frac{1}{4}, \quad l_1 = \frac{1}{5}, \quad l_2 = -\frac{2}{5},$$
$$w_1 = \frac{3321}{160}, \quad w_2 = -\frac{11061}{5120}, \quad h_{1,2} = \frac{3679}{1379}. \tag{3.8}$$

Then the corresponding two-soliton solution of Eq. (1.1) is

$$v = \tfrac{1379}{8}\, \tfrac{5516\exp(z_1)+16090\exp(z_2)+3679\exp(z_3)+1379\exp(z_4)+14716\exp(z_5)}{(1379+1379\exp(z_1)+1379\exp(z_4)+3679\exp(z_2))^2},$$
$$z_1 = -\tfrac{1}{2}\,x + \tfrac{1}{5}\,y + \tfrac{3321}{160}\,t + \phi_1, \quad z_2 = -\tfrac{1}{4}\,x - \tfrac{1}{5}\,y + \tfrac{95211}{5120}\,t + \phi_1 + \phi_2,$$
$$z_3 = -\tfrac{3}{4}\,x + \tfrac{201483}{5120}\,t + 2\phi_1 + \phi_2, \quad z_4 = \tfrac{1}{4}\,x - \tfrac{2}{5}\,y - \tfrac{11061}{5120}\,t + \phi_2, \tag{3.9}$$
$$z_5 = -\tfrac{3}{5}\,y + \tfrac{8415}{512}\,t + \phi_1 + 2\phi_2.$$

The structure of two-soliton solution with $\phi_1 = 0,\ \phi_2 = 25$ is shown in Fig. 1.

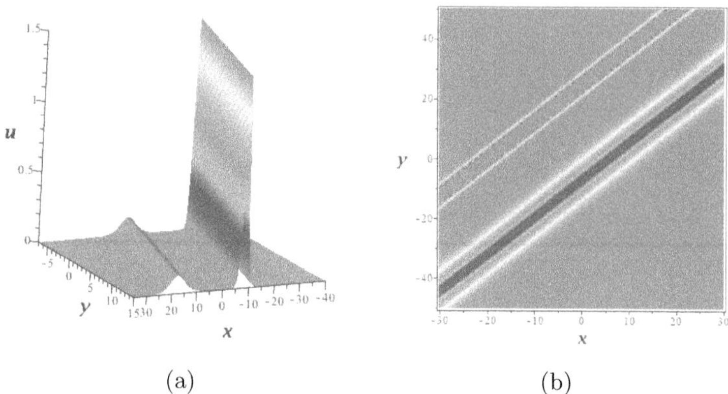

Figure 2. Depiction of the evolution of the soliton molecule solution of Eq. (1.1) with (3.11).

Subcase I-2 Soliton molecule solution

To obtain soliton molecule solution, one can use the velocity resonant scheme proposed in [15]

$$\frac{k_2}{k_1} = \frac{w_2}{w_1} = \frac{l_2}{l_1},$$

which is combined with (3.7) to give

$$k_2 = \pm\sqrt{-\frac{k_1^3 + 30\,ak_1 + 5\,l_1}{k_1}}, \quad l_2 = \pm\frac{l_1}{k_1}\sqrt{-\frac{k_1^3 + 30\,ak_1 + 5\,l_1}{k_1}}. \tag{3.10}$$

It shows that we indeed obtain soliton molecule under this circumstance. Figure 2 depicts its structure with the parameter selections

$$k_1 = -\frac{1}{2}, \quad k_2 = \sqrt{3}, \quad l_1 = \frac{2}{5}, \quad l_2 = -\frac{4}{5}\sqrt{3}, \quad w_1 = -\frac{71}{32}, \quad w_2 = -\frac{359}{5}\sqrt{3}$$

$$a = \frac{1}{40}, \quad h_{1,2} = -\frac{229 + 117\sqrt{3}}{-229 + 117\sqrt{3}}, \quad \phi_1 = 0, \quad \phi_2 = 30. \tag{3.11}$$

Case II $u = 4(\ln f_2)_x$

Beginning with $u = 4(\ln f_2)_x$, we obtain a two-soliton solution of Eq. (1.1) with f_2 being the form of (3.2), and

$$l_i = -k_i^3 - 6ak_i, \quad w_i = 9k_i^5 + 10ak_i^3 + 20a^2k_i, \quad h_{1,2} = \left(\frac{k_1 - k_2}{k_1 + k_2}\right)^2, \tag{3.12}$$

where $i = 1, 2$, a and k_1, k_2 $(k_1 \neq \pm k_2)$ are free parameters.

In (3.12), setting

$$a = \frac{1}{2}, \quad h_{1,2} = \frac{1}{9}, \quad k_1 = -\frac{1}{6}, \quad k_2 = -\frac{1}{3} \tag{3.13}$$

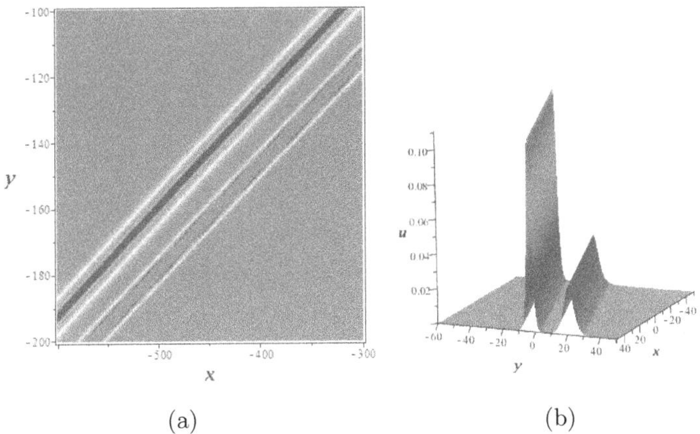

(a) (b)

Figure 3. Depiction of the evolution of the two-soliton solution of Eq. (1.1) when (3.12) along with (3.13) and (3.14) hold.

gives

$$l_1 = \frac{109}{216}, \quad l_2 = \frac{28}{27}, \quad w_1 = -\frac{666}{864}, \quad w_2 = -\frac{451}{27}, \quad h_{1,2} = \frac{1}{9}. \tag{3.14}$$

Then we have a two-soliton solution, of which the evolution 3D graph and the density plot are demonstrated in Fig. 3 with $\phi_1 = 0, \phi_2 = 20$. We find that the two-soliton solution of this case shows a celebrated characteristic of resonant behavior directly, which will be verified again by the derived soliton molecules while constructing three-soliton solutions below.

3.3 Three-soliton and soliton molecule solutions

Likewise, starting from $u = 2(\ln f)_x$ and $u = 4(\ln f)_x$, we obtain two types of three-soliton solutions of Eq. (1.1), respectively. Specifically, under the case of $u = 2(\ln f_2)_x$, two types of three-soliton solutions are presented and based on one of them, a soliton molecule solution is reported.

Case I $u = 2(\ln f)_x$

Starting from $u = 2(\ln f)_x$, two types of three-soliton solutions of (2.1) are given, then solutions of (1.1) via $v = u_x$.

Subcase I-1 The conditions of the first three-soliton solution are

$$\begin{aligned} h_{i,j} &= \left(\tfrac{k_i-k_j}{k_i+k_j}\right)^2 \quad (k_i \neq \pm k_j,\ 1 \leq i < j \leq 3), \\ l_i &= \tfrac{k_i\left(k_i^2 k_3 - k_3^3 - l_3\right)}{k_3} \quad (i = 1, 2), \\ w_i &= -\tfrac{180\,a^2 k_i^2 - 5\,l_i^2 + 5\,k_i^3 l_i + 30\,a k_i l_i + 30\,a k_i^4 + k_i^6}{k_i} \quad (i = 1, \ldots, 3), \end{aligned} \tag{3.15}$$

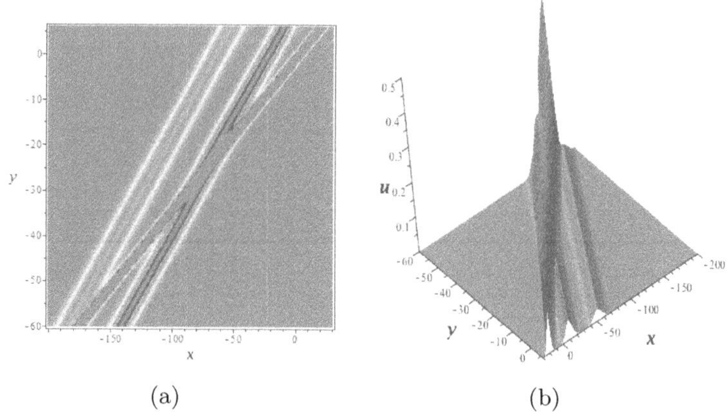

(a) (b)

Figure 4. Depiction of the evolution of the resonance soliton solution for Eq. (1.1) with (3.17) and (3.18).

or w_i is written as

$$\begin{aligned} w_i &= -\frac{k_i\left(180\,a^2k_3^2-9\,k_i^4k_3^2+15\,k_i^2k_3^4+15\,k_i^2k_3l_3-5\,k_3^6-10\,k_3^3l_3-5\,l_3^2+30\,ak_3^4+30\,ak_3l_3\right)}{k_3^2},\\ w_3 &= -\frac{180\,a^2k_3^2-5\,l_3^2+5\,k_3^3l_3+30\,ak_3l_3+30\,ak_3^4+k_3^6}{k_3}, \end{aligned} \tag{3.16}$$

where $i = 1, 2$ and a, l_3 are arbitrary constants.

In (3.15), setting

$$a = 2, \quad k_1 = -\frac{1}{3}, \quad k_2 = \frac{1}{7}, \quad k_3 = \frac{1}{4}, \quad l_3 = -\frac{1}{6} \tag{3.17}$$

gives

$$\begin{aligned} &\frac{103}{432}, \quad l_2 = -\frac{1469}{16464}, \quad w_1 = \frac{1568579}{6912}, \quad w_2 = -\frac{3771299299}{38723328},\\ &w_3 = -\frac{1569769}{9216}, \quad h_{1,2} = \frac{25}{4}, \quad h_{1,3} = 49, \quad h_{2,3} = \frac{9}{121}. \end{aligned} \tag{3.18}$$

The graphs of the three-soliton solution with $\phi_1 = 0, \phi_2 = 10, \phi_3 = 15$ are shown in Fig. 4. No soliton molecule could be found under the consideration of the velocity resonant scheme. However, we observe that this three-soliton solution exhibits resonance behavior between a single soliton and a two-soliton molecule.

Subcase I-2(a) The second three-soliton solution

The conditions of the second three-soliton solution are

$$\begin{aligned}
&l_1 = -\tfrac{k_1\left(k_1^2k_3-k_3^3-l_3\right)}{k_3}, \quad l_2 = -\tfrac{1}{2}\,\tfrac{k_2\left(3\,k_1^2k_3-2\,k_3^3-2\,l_3-k_2^2k_3\right)}{k_3},\\
&w_1 = -\tfrac{k_1\left(-9\,k_1^4k_3^2+30\,ak_3^4+30\,ak_3l_3+15\,k_1^2k_3^4+15\,k_3k_1^2l_3+180\,a^2k_3^2-5\,k_3^6-10\,k_3^3l_3-5\,l_3^2\right)}{k_3^2},\\
&w_2 = -\tfrac{1}{4}\,\tfrac{C_1}{k_3^2}, \quad w_3 = -\tfrac{180\,a^2k_3^2-5\,l_3^2+5\,k_3^3l_3+30\,ak_3l_3+30\,ak_3^4+k_3^6}{k_3},\\
&h_{i,j} = \left(\tfrac{k_i-k_j}{k_i+k_j}\right)^2 \ (i=1,\, j=2,3), \quad h_{2,3} = \tfrac{C_2}{C_3},
\end{aligned} \tag{3.19}$$

where

$$\begin{aligned}
C_1 &= k_2\left(9\,k_2^4k_3^2 + 720\,a^2k_3^2 - 45\,k_1^4k_3^2 + 60\,k_1^2k_3^4 + 60\,k_3k_1^2l_3 - 20\,k_3^6 - 40\,k_3^3l_3\right.\\
&\quad\left. -20\,l_3^2 - 180\,ak_1^2k_3^2 + 120\,ak_3^4 + 120\,ak_3l_3 + 180\,ak_2^2k_3^2\right),\\
C_2 &= 24\,ak_3^3 - 6\,k_1^2k_3^3 + 3\,k_1^4k_3 + 4\,l_3k_3^2 + 4\,l_3k_2^2 - 8\,k_2k_3^4 - 6\,k_2^3k_3^2 + 10\,{k_2}^2k_3^3\\
&\quad +3\,k_2^4k_3 + 4\,k_3^5 + 6\,k_2k_3^2k_1^2 - 8\,k_2k_3l_3 - 48\,ak_3^2k_2 - 6\,k_3k_1^2k_2^2 + 24\,ak_3k_2^2,\\
C_3 &= 4\,k_3^5 + 3\,k_2^4k_3 + 3\,k_1^4k_3 + 4\,l_3k_3^2 - 6\,k_1^2k_3^3 + 6\,k_2^3{k_3}^2 + 24\,ak_3^3 + 10\,k_2^2k_3^3\\
&\quad +4\,l_3k_2^2 + 8\,k_2k_3^4 + 8\,k_2k_3l_3 - 6\,k_3k_1^2k_2^2 - 6\,k_2k_3^2k_1^2 + 24\,ak_3k_2^2 + 48\,ak_3^2k_2.
\end{aligned}$$

In (3.19), choosing

$$a=2, \quad k_1=-1, \quad k_2=\frac{1}{3}, \quad k_3=\frac{1}{4}, \quad l_3=-\frac{1}{6}, \tag{3.20}$$

gives

$$\begin{aligned}
&l_1=\frac{77}{48}, \quad l_2=-\frac{295}{432}, \quad w_1=\frac{1529539}{2304}, \quad w_2=-\frac{1340099}{6912},\\
&w_3=-\frac{1569769}{9216}, \quad h_{1,2}=4, \quad h_{1,3}=\frac{25}{9}, \quad h_{2,3}=\frac{967}{5911}.
\end{aligned} \tag{3.21}$$

Figures 5 (a) and (b) display its structures when $\phi_1=0$, $\phi_2=20$, $\phi_3=30$.

Subcase I-2(b) Soliton molecule solution

Under the subcase I-2(a) of this section, there exists a soliton molecule solution. Similarly, using the velocity resonance scheme

$$\frac{k_3}{k_2}=\frac{w_3}{w_2}=\frac{l_3}{l_2},$$

we have

$$k_3=\pm\sqrt{\frac{3}{2}\,k_1^2-\frac{1}{2}\,k_2^2}, \quad l_3=\pm\frac{1}{10}\sqrt{\frac{3}{2}\,k_1^2-\frac{1}{2}\,k_2^2}\,\left(3\,k_1^2+k_2^2+60\,a\right). \tag{3.22}$$

For (3.22), choosing

$$a=2, \quad k_1=-1, \quad k_2=\frac{1}{3}, \quad k_3=\frac{1}{3}\sqrt{13}, \quad l_3=-\frac{554}{135}\sqrt{13} \tag{3.23}$$

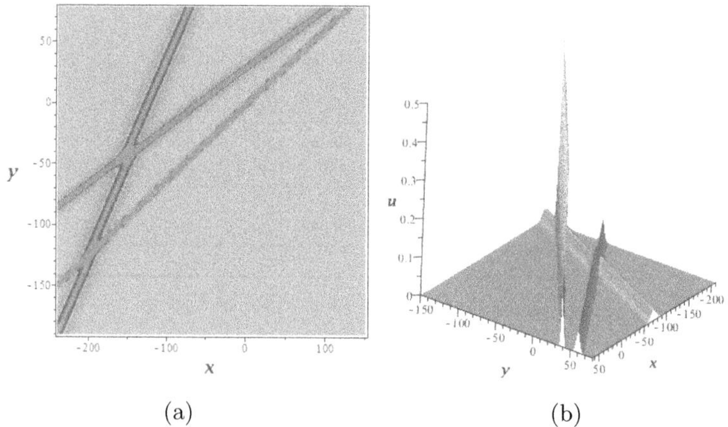

Figure 5. Depiction of the interaction of three-soliton solutions of Eq. (1.1) described by (3.19) with the parameter selections (3.20) and (3.21).

gives birth to

$$\begin{aligned} &l_1 = \tfrac{178}{15}, \quad l_2 = -\tfrac{554}{135}, \quad w_1 = -\tfrac{31249}{45}, \quad w_2 = \tfrac{34949}{135}, \\ &w_3 = \tfrac{34949}{135}\sqrt{13}, \quad h_{1,2} = 4, \quad h_{1,3} = \frac{\frac{2}{3}\sqrt{13}+\frac{22}{9}}{\left(-1+\frac{1}{3}\sqrt{13}\right)^2}, \quad h_{2,3} = \frac{29\sqrt{13}-91}{29\sqrt{13}+91}. \end{aligned} \tag{3.24}$$

Figures 6(a) and (b) show the structures of this soliton molecule solution of Eq. (1.1) with parameter selections (3.24) and $\phi_1 = 0$, $\phi_2 = 30$, $\phi_3 = 60$.

Subcase I-3 Three soliton solution of the variant KP equation

As an unexpected result, in Section 2, we know that Eq. (2.8) is the Hirota bilinear form of the variant KP Equation (2.9). We know that Eq. (2.9) could pass the three-soliton solution condition. The three-soliton solution of Eq. (2.9) is obtained from $u = 2(\ln f)_{xx}$, where $f = f(x, y, t) = f(\xi)$ is of the form (3.3), and $\xi = k_i x + l_i y + w_i t + \phi_i$ with

$$l_i = -k_i^3 - 3ak_i, \quad h_{i,j} = \left(\frac{k_i - k_j}{k_i + k_j}\right)^2 \tag{3.25}$$

and k_i, k_j $(k_i \neq \pm k_j,\ 1 \leq i < j \leq 3)$, a, w_i $(i = 1, \ldots, 3)$ are free parameters.

Figure 7 shows the structure of this three-soliton solution with the following parameter selections in (3.26)

$$\begin{aligned} &a = \tfrac{1}{2}, \quad k_1 = -\tfrac{1}{2}, \quad k_2 = \tfrac{1}{5}, \quad k_3 = \tfrac{1}{4}, \quad l_1 = \tfrac{7}{8}, \quad l_2 = -\tfrac{77}{250}, \\ &l_3 = -\tfrac{25}{64}, \quad w_1 = -1, \quad w_2 = \tfrac{1}{2}, \quad w_3 = 2, \quad h_{1,2} = \tfrac{49}{9}, \\ &h_{1,3} = 9, \quad h_{2,3} = \tfrac{1}{81}, \quad \phi_1 = 0, \quad \phi_2 = 8, \quad \phi_3 = 20. \end{aligned} \tag{3.26}$$

Fixing $y = -100$ and t varying from $-15, -11, -10, 15, 25, 35$, we present some evolution graphs shown in Figures 8 and 9.

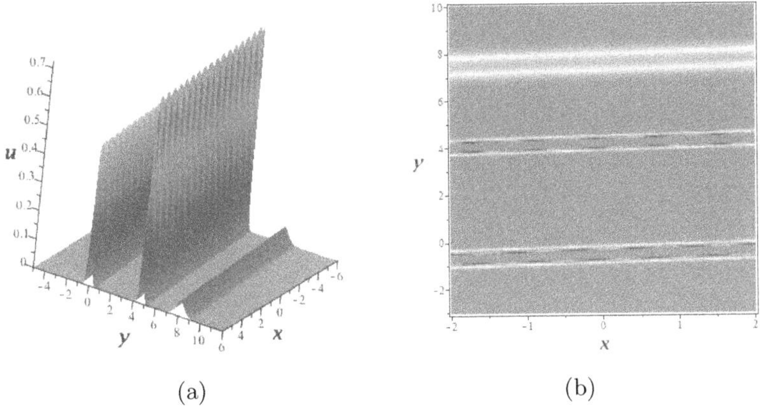

Figure 6. Depict of the evolution of the resonance soliton solution for Eq. (1.1) when (3.22) along with (3.24) hold.

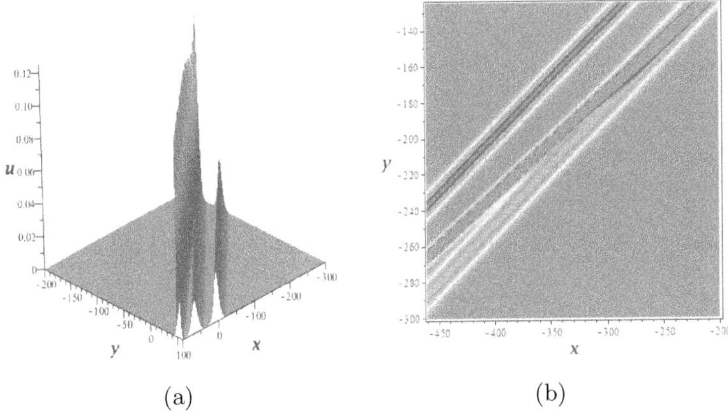

Figure 7. Depiction of the evolution of the three soliton solution of Eq. (2.9) when (3.25) along with (3.26) hold.

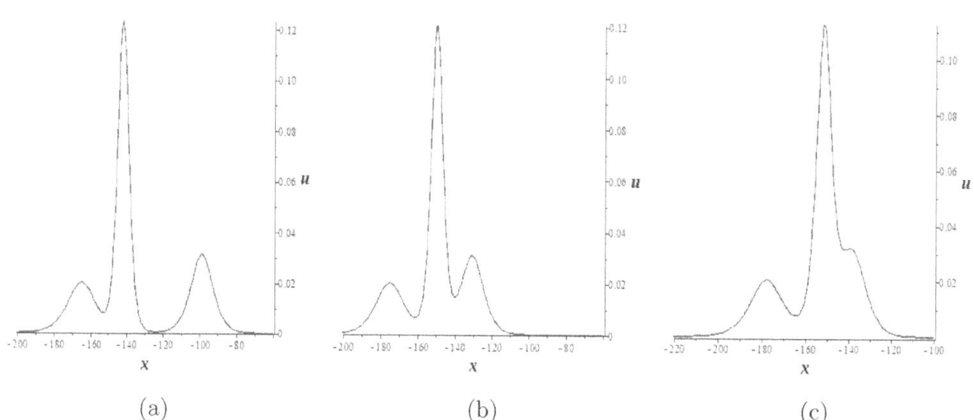

Figure 8. Depiction of the evolution of the three-soliton solution of Eq. (2.9) with (3.25) and (3.26).

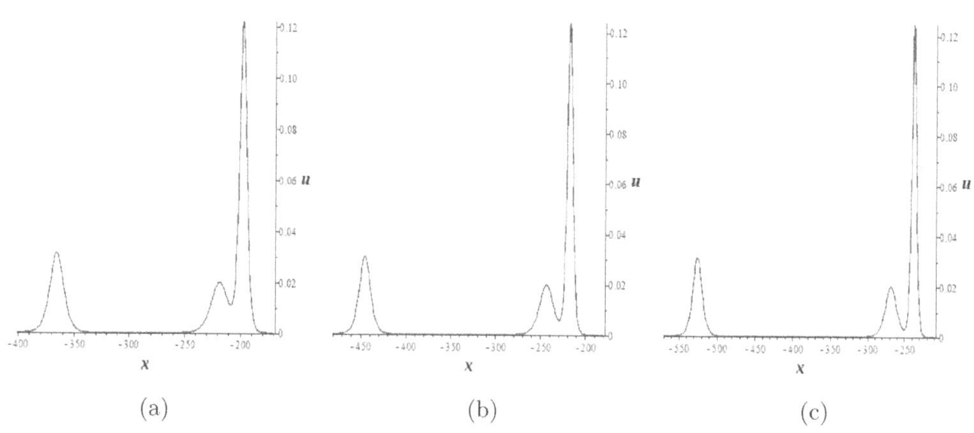

Figure 9. Depiction of the evolution of the three-soliton solution of Eq. (2.9) when (3.25) along with (3.26) hold.

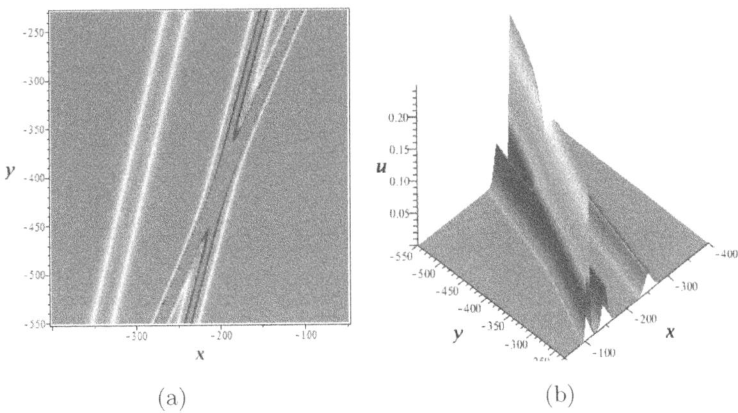

(a) (b)

Figure 10. Depiction of the evolution of the three-soliton solution for Eq. (1.1) when (3.15) along with (3.28) hold.

Case II Starting from $u = 4(\ln f)_x$, we obtain a three-soliton solution of Eq. (2.1), then a solution $v = 4(\ln f)_{xx}$ of Eq. (1.1). The conditions of the three-soliton solution are

$$l_i = -k_i^3 - 6ak_i, \quad w_i = 9k_i(k_i^4 + 10ak_i^2 + 20a^2), \quad h_{i,j} = \left(\frac{k_i - k_j}{k_i + k_j}\right)^2, \tag{3.27}$$

where $k_i, k_j\,(k_i \neq \pm k_j,\ i = 1, \ldots, 3)$ are free parameters. Figure 10 shows the structure of the three-soliton solution of Eq. (1.1) under the case of $v = 4(\ln f)_{xx}$ with the parameter selections in (3.28)

$$\begin{aligned} &a = \tfrac{1}{24},\ k_1 = -\tfrac{1}{2},\ k_2 = \tfrac{1}{4},\ k_3 = \tfrac{1}{6},\ l_1 = \tfrac{1}{4},\ l_2 = -\tfrac{5}{64},\ l_3 = -\tfrac{5}{108},\\ &w_1 = -\tfrac{29}{32},\ w_2 = \tfrac{149}{1024},\ w_3 = \tfrac{61}{864},\ h_{1,2} = 9,\ h_{1,3} = 4,\ h_{2,3} = \tfrac{1}{25},\\ &\phi_1 = 0,\ \phi_2 = 20,\ \phi_3 = 30. \end{aligned} \tag{3.28}$$

It is easy to see this three-soliton solution also exhibits resonance behavior between a one-soliton and a two-soliton molecule. However, it is different with that of the soliton molecule derived by carrying out some resonant conditions under case $p = 2$. It is a hint of predicting some specific property of this kind of 2DSK equation that is exhibiting inherent resonant behavior during the evolving process of nonlinear wave.

3.4 N–soliton solutions of (1.1)

The N–soliton solution $v = 2\,(\ln f_N)_{xx}$ of Eq. (1.1) is

$$
\begin{aligned}
h_{i,j} &= \left(\tfrac{k_i-k_j}{k_i+k_j}\right)^2 \quad (1 \le i < j \le n),\\
l_i &= \tfrac{k_i\left(k_i^2 k_3 - k_3^3 - l_n\right)}{k_3} \quad (i = 1, \ldots, n-1),\\
w_i &= -\tfrac{180\,a^2 k_i^2 - 5\,l_i^2 + 5\,k_i^3 l_i + 30\,a k_i l_i + 30\,a k_i^4 + k_i^6}{k_i} \quad (i = 1, \ldots, n).
\end{aligned}
\tag{3.29}
$$

The N–soliton solution $v = 4\,(\ln f_N)_{xx}$ of Eq. (1.1) is

$$
\begin{aligned}
h_{i,j} &= \left(\tfrac{k_i-k_j}{k_i+k_j}\right)^2 \quad (1 \le i < j \le n)\\
l_i &= -k_i^3 - 6ak_i, \quad w_i = 9\,k_i(k_i^4 + 10ak_i^2 + 20a^2) \quad (i = 1, \ldots, n).
\end{aligned}
\tag{3.30}
$$

In (3.29) and (3.30), $k_i, k_j\,(k_i \ne \pm k_j, i = 1, \ldots, n)$ are free parameters.

4 Concluding remarks

In short, this chapter is devoted to exploring new multiple soliton and soliton molecule solutions under two dependent variable transformations that could give trilinear and quintic linear equations that both can be solved by a pair of Hirota's bilinear equations relating to obtaining two hierarchies of multiple soliton and molecule soliton solutions for the discussed 2DSK type equation. Up to now, the 2DSK-type equation is the only nonlinear PDE that possesses two hierarchies of multiple soliton solutions. The new transformation proposed by us could not give birth to Hirota's bilinear form directly by using the classical Hirota bilinear method; however, we construct another set of multiple soliton solutions to the 2DSK equation. By observation, we notice that the solutions of case $p = 4$ exhibit inherent resonant behavior during the evolving process of the nonlinear waves. It is a hint of predicting some specific property of this kind of 2DSK equation and is worthy of being studied in the future.

Acknowledgements

The work is sponsored by the National Natural Science Foundations of China (Grant Nos. 11975131, 11435005), the Natural Science Basic Research Program of Shaanxi Province (Grant No. 2021JZ-21), the Fundamental Research Funds for the Central Universities (Grant No. 2020CBLY013) and K. C. Wong Magna Fund in Ningbo University.

References

[1] Hirota R and Satsuma J, Exact solutions of the Korteweg-de Vries equation for multiple collisions of solitons, *Phys. Rev. Lett.* **27**, 1192-1194, 1971.

[2] Hirota R and Satsuma J, N-soliton solutions of model equations for shallow water waves, *J. Phys. Soc. Japan* **40**, 611-612, 1976.

[3] Hirota R and Ito M, Resonance of solitons in one dimension, *J. Phys. Soc. Japan* **52**, 744-748, 1983.

[4] Hereman W, Adams P J, Eklund H L, Hickman M S and Herbst B M, Direct Methods and Symbolic Software for Conservation Laws of Nonlinear Equations, *In: Advances in Nonlinear Waves and Symbolic Computation,* Ed.:Yan ZY, Nova Science publishers, New York, 2008.

[5] Hereman W and Nuseir A, Symbolic methods to construct exact solutions of nonlinear partial differential equations, *Math. Comput. Simul.* **43**, 13-27, 1997.

[6] Hirota R and Satsuma J, An extension of nonlinear evolution equations of the KdV (mKdV) type to higher order, *J. Phys. Soc. Japan* **49**, 771-778, 1980.

[7] Hietarinta J, A search of bilinear equations passing Hirota's three-soliton condition: I. KdV type bilinear equations, *J. Math. Phys.* **28**, 1732-1742, 1987.

[8] Hietarinta J, A search of bilinear equations passing Hirota's three-soliton condition: II. mKdV-type bilinear equations, *J. Math. Phys.* **28**, 2094-2101, 1987.

[9] Hietarinta J, A search of bilinear equations passing Hirota's three-soliton condition: IV. complex bilinear equations, *J. Math. Phys.* **29**, 628-635, 1989.

[10] Hirota R, The Direct Method in Soliton Theory, Camb. Univ. Press, Cambridge, 2004.

[11] Konopelchenko B and Dubrovsky V, Some new integrable nonlinear evolution equations in $2+1$ dimensions, *Phys. Lett. A* **102**, 15-17, 1984.

[12] Kadomtsev B B and Petviashvili V I, On the stability of solitary waves in weakly dispersive media, *Sov. Phys. Doklady.* **15**, 539-541, 1970.

[13] Lou S Y, $\hat{P}-\hat{T}-\hat{C}$ symmetry invariant and symmetry breaking soliton solutions, *J. Math. Phys.* **591**, No.083507 (20pp), 2018.

[14] Nucci M C, Painlevé property and pseudopotentials for non-linear evolution equations, *J. Phys. A: Math. Gen.* **22**, 2897-2913, 1989.

[15] Yan Z W and Lou S Y, Soliton molecules in Sharma–Tasso–Olver–Burgers equation, *Appl. Math. Lett.* **104**, No.106271 (7pp), 2020.

[16] Yao R X, Li Y and Lou S Y, A new set and new relations of multiple soliton solutions of (2+1)-dimensional Sawada–Kotera equation, *Commun. Nonlinear Sci. Numer. Simulat.* **99**, No.105820 (11pp), 2021.

[17] Yao R X, Xu G Q and Li Z B, Conservation laws and soliton solutions for generalized seventh order KdV equation, *Commun. Theor. Phys.* **41**, 487-492, 2004.

B8. Dressing the boundary: exact solutions of soliton equations on the half-line

Cheng Zhang

Department of Mathematics, Shanghai University
99 Shangda Road, Shanghai 200444, P. R. China

Abstract

This chapter reviews the approach called "dressing the boundary" in view of solving initial-boundary value problems for a class of integrable partial differential equations (PDEs) on the half-line. It is based on the notions of integrable boundary conditions developed by Sklyanin. By dressing the boundary, we dress simultaneously the bulk and boundary integrable structures. We take the nonlinear Schrödinger equation on the half-line under the Robin boundary conditions as our primary example, and compute multi-soliton solutions. This method admits an inverse scattering transform interpretation, and can be applied to a wide range of soliton models.

1 Introduction

Initial-boundary value problems for integrable nonlinear partial differential equations (PDEs) represent one of the very basic problems in integrable systems. The crucial aspect is that boundary conditions are inherent in the definition of integrability. For instance, in the inverse scattering transform to integrate the Korteweg-de Vries (KdV) equation defined on the whole line, the vanishing boundary conditions, *i.e.*, the KdV field and its derivatives vanish rapidly as the space variable tends to infinity, should be explicitly imposed [22]. In the Hamiltonian formulation of integrable field theory, the vanishing boundary conditions are also needed in order to ensure the existence of infinitely many conserved quantities [21, 33]. Indeed, one could argue that integrable PDEs are said to be integrable only if they obey certain special boundary behaviors. The typical choices are either the vanishing boundary conditions for the system defined on the whole line, or periodic boundary conditions for the system defined on a circle. This aspect is well summarized by Krichever and Novikov in their review paper [26]:

> Let us point out that the KdV system, as well as other nontrivial completely integrable by IST (inverse scattering transform) partial differential equation (PDE) systems, are indeed completely integrable in any reasonable sense for rapidly decreasing or periodic (quasi-periodic) boundary conditions only.

As to *open boundary problems* for integrable systems, *i.e.*, the systems are defined on a closed interval; one of the systematic approaches to characterizing integrability was due to Sklyanin [29, 30] in the framework of integrable Hamiltonian field theory.

This gave rise to the notions of integrable boundary conditions. In Sklyanin's approach, the central objects are the so-called classical reflection equations involving both the classical r matrix and the boundary K matrix. For an integrable PDE defined on an interval, the boundary conditions at the two ends of the interval are encoded into two boundary K matrices subject to some algebraic constraints. Then integrability in the presence of boundaries is defined as the existence of infinitely many conserved quantities in involution.

The inverse scattering transform is an analytic approach to solving initial value problems for integrable PDEs [1]. In order to treat initial-boundary value problems for integrable PDEs on the half-line, or more generally on an interval, with generic boundary conditions, the inverse scattering transform has been remarkably generalized by Fokas to the unified transform method [16, 17, 18]. The key idea of the unified transform method is to treat simultaneously the initial and boundary data in the direct scattering process. Then, both the initial and boundary data are regarded at the same footing, and the associated scattering data can be put into certain functional constraints usually formulated as certain Riemann-Hilbert problems in the inverse space.

Although Fokas's unified transform method has been successfully applied to a wide range of integrable PDEs with boundaries, in general, it is a difficult task to solve inversely the Riemann-Hilbert problems. Moreover, in contrast to Sklyanin's approach, there is no clear definition of integrable boundary conditions in the unified transform method. Note that a special class of boundary conditions, called *linearizable* boundary conditions, do exist in Fokas's approach [17, 18]. The linearizable boundary conditions reflect certain symmetry of the scattering data that can be used to reduce the Riemann-Hilbert problems. For certain models, they coincide with the integrable boundary conditions, cf. [17, 29]. However, the linearizable boundary conditions are, a priori, not equivalent to integrable boundary conditions. As an important application of the unified transform method, asymptotic solutions of integrable PDEs at large time, mostly accompanied with the linearizable boundary conditions, can be derived [13, 17, 18].

In this chapter, we review the approach called "dressing the boundary", recently developed by the author, to solve initial-boundary value problems for certain integrable PDEs on the half-line equipped with integrable boundary conditions [38]. We concentrate on the focusing nonlinear Schrödinger (NLS) equation as our primary example. The NLS equation is an important model in mathematical physics (see for instance [2, 14, 34] for its integrable aspects). It can be solved by the inverse scattering transform under the vanishing boundary conditions [34], which leads to multi-soliton solutions. The NLS model with boundaries has also been extensively studied in the literature. Sklyanin was the first to derive integrable boundary conditions of the model defined on an interval. His work was then followed by other important contributions to characterize the analytic and integrable aspects of the half-line NLS model [4, 15, 25, 31]. The unified transform method has also been applied to the half-line NLS model, and asymptotic soliton solutions at large time were derived using the nonlinear steepest descent method [17, 20]. The qualitative behaviors of initial-boundary value problems for the NLS equation on the half-line under the homogeneous and non-homogeneous Robin boundary conditions can also be found, for instance, in [9, 13, 19].

In spite of a good understanding of the model, exact soliton solutions of the NLS equation on the half-line subject to Robin boundary conditions were only obtained rather recently in [8] where a nonlinear mirror-image technique was implemented. The method consists in extending the half-line space domain to the whole line. This reflects the space inversion symmetry of NLS and allows us to obtain soliton solutions of the model using the inverse scattering transform by uniquely looking at the positive semi-axis. This technique was also successfully applied to the vector NLS model [11] and was used to obtain boundary-bound solitons [7] that are static solitons subject to the boundary conditions.

Our approach to solving the NLS equation on the half-line mainly follows Sklyanin's formalism of integrable boundary conditions. It is well known that soliton solutions of integrable PDEs can be obtained in an algebraic fashion using the Darboux-dressing transformations, cf. [35, 36]. The essential idea of our method is to *dress* simultaneously the *bulk* and *boundary* integrable structures, which are respectively represented by the Lax pair and an integrable boundary constraint. Using the zero seed solutions, soliton solutions of NLS on the half-line can be directly computed in the boundary-dressing process. One of the particular advantages of our method is that it does not require extension of the space domain to the whole line. It also admits a natural inverse scattering transform interpretation. Although the main content of the chapter is focusing on a particular model, the method we present here can be applied to a wide range of classical integrable PDEs on the half-line equipped with integrable boundary conditions, cf. [23, 24, 32, 39, 40] .

The chapter is organized as follows: we first provide a simple argument in Section 2 to characterize integrable boundary conditions for the NLS equation; in Section 3, we present the procedure of dressing the boundary and derive multi-soliton solutions of the NLS equation on the half-line; we give an the inverse scattering transform interpretation of the boundary-dressing method in Section 4.

2 Notions of integrable boundary conditions

Taking the focusing NLS equation as our primary example, we briefly review how to determine the appropriate boundary conditions for the model to be integrable in the cases with and without boundaries. For the case without boundary where the space variable is defined on a circle, the quasi-periodic boundary conditions are the typical conditions. For the case with boundaries where the space variable is defined on an interval, Sklyanin's formalism for integrable boundary conditions is presented. As a consequence, the integrable half-line NLS model is also derived.

2.1 NLS with quasi-periodic boundary conditions

Consider the focusing NLS equation

$$iq_t + q_{xx} + 2|q|^2 q = 0\,, \tag{2.1}$$

where $q := q(x,t)$ is a complex field, and x and t are respectively the space and time variables. The equation is the result of the compatibility between the following linear differential equations

$$U\,\phi = \phi_x\,, \quad V\,\phi = \phi_t\,. \tag{2.2}$$

Here, the matrix-valued functions U and V, known as the Lax pair of the NLS equation, depend on x, t and also on a spectral parameter λ. They are in the forms (we drop the x, t and λ dependence for conciseness unless there is ambiguity)

$$U = -i\lambda\sigma_3 + Q\,, \quad V = -2i\lambda^2\sigma_3 + 2\lambda\,Q - iQ_x\,\sigma_3 - iQ^2\,\sigma_3\,, \tag{2.3}$$

where

$$\sigma_3 = \begin{pmatrix} 1 & 0 \\ 0 & -1 \end{pmatrix}, \quad Q = \begin{pmatrix} 0 & q \\ -\bar{q} & 0 \end{pmatrix}. \tag{2.4}$$

We will call U and V the *x-part* and *t-part* of the Lax pair, respectively. We briefly show how the integrability of the NLS equation under the quasi-periodic boundary conditions is obtained by deriving a generating function for conserved quantities, *cf.* [14]. The quasi-periodic boundary conditions read

$$q(x,t) = q(x+L,t)e^{i\theta}\,, \tag{2.5}$$

where $0 \le \theta < 2\pi$ and θ is independent of t. This also implies

$$Q(x,t) = \Theta_+ Q(x+L,t)\Theta_-\,, \quad \Theta_\pm = \exp(\pm i\frac{\theta}{2}\sigma_3)\,. \tag{2.6}$$

By restricting the space domain to $[0,L)$, one can compute the *monodromy matrix* $T(\lambda)$ by integrating the x-part of the Lax pair

$$T(t;\lambda) = \overset{\curvearrowleft}{\exp}\int_0^L U(\xi,t,\lambda)\,d\xi\,, \tag{2.7}$$

where $\overset{\curvearrowleft}{\exp}$ denotes the path-ordered exponential. The monodromy matrix encodes the spectral properties in the direct scattering transform for a given NLS field. Having the quasi-periodic boundary conditions and the monodromy matrix, one can prove the following result.

Lemma 1. *For the Lax pair* (2.2) *with q satisfying the quasi-periodic boundary conditions* (2.5), *one has*

$$\frac{d}{dt}tr(\Theta_- T(t;\lambda)) = 0\,. \tag{2.8}$$

Proof: By differentiating the monodromy matrix with respect to t, one gets

$$\frac{d}{dt}T(t;\lambda) = V(L,t;\lambda)T(t;\lambda) - T(t;\lambda)V(0,t;\lambda)\,. \tag{2.9}$$

Then, one has

$$\frac{d}{dt}\operatorname{tr}(\Theta_- T(t;\lambda)) = \operatorname{tr}(\Theta_- V(L,t;\lambda)T(t;\lambda) - \Theta_- T(t;\lambda)V(0,t;\lambda)). \tag{2.10}$$

Because of the quasi-periodic boundary conditions (2.5), $V(L,t;\lambda)$ and $V(0,t;\lambda)$ are related by

$$V(L,t;\lambda) = \Theta_+ V(0,t;\lambda)\Theta_- . \tag{2.11}$$

Replacing this relation into (2.10) completes the proof. ■

Following the above simple considerations, the quantity $\operatorname{tr}(\Theta_- T(t;\lambda))$ can be considered as the generating function for infinitely many conserved quantities by taking $\operatorname{tr}(\Theta_- T(t;\lambda))$ as a series expansion in λ. This indicates that the NLS model under the quasi-periodic boundary conditions is integrable. More systematic treatments of the model as an integrable Hamiltonian system are based on the r matrix formalism [14]. Using the following canonical Poisson bracket

$$\{q(x), \bar{q}(y)\} = -i\delta(x-y), \tag{2.12}$$

one can prove that the quantity $\Theta_- T$ obeys the Poisson relation

$$\{\Theta_- T(\lambda) \underset{,}{\otimes} \Theta_- T(\eta)\} = [r(\lambda-\eta), (\Theta_- T(\lambda)) \otimes (\Theta_- T(\eta))], \tag{2.13}$$

where $\otimes$ is the standard tensor product for matrices, and the operation $\{\cdot \underset{,}{\otimes} \cdot\}$ stands for the Poisson bracket in tensor-product form. The quantity $r(\lambda)$ is called the classical r matrix satisfying the classical Yang-Baxter equation

$$[r_{12}(\lambda-\eta), r_{13}(\lambda) + r_{23}(\eta)] + [r_{13}(\lambda), r_{23}(\eta)] = 0. \tag{2.14}$$

It is in the form

$$r(\lambda) = \frac{1}{2\lambda}\mathcal{P}, \quad \mathcal{P} = \begin{pmatrix} 1 & 0 & 0 & 0 \\ 0 & 0 & 1 & 0 \\ 0 & 1 & 0 & 0 \\ 0 & 0 & 0 & 1 \end{pmatrix}, \tag{2.15}$$

where $\mathcal{P}$ is the permutation matrix acting on $\mathbb{C}^2 \otimes \mathbb{C}^2$. By taking the trace of (2.13), one has

$$\{\operatorname{tr}(\Theta_- T(\lambda)), \operatorname{tr}(\Theta_- T(\eta))\} = 0, \tag{2.16}$$

which implies there are infinitely many quantities in involution. Note that the classical r matrix formalism represents a universal integrable structure characterizing a wide class of classical integrable models, cf. [5, 14].

Remark 2.1. *For $x \in \mathbb{R}$, the case of the NLS field vanishes rapidly as $x \to \pm\infty$, can be treated similarly by using* $\operatorname{tr}(T(t;\lambda))$ *as the generating function with*

$$T(t;\lambda) = \overset{\curvearrowleft}{\exp} \int_{-\infty}^{\infty} U(\xi,t,\lambda)\, d\xi . \tag{2.17}$$

2.2 NLS on an interval and integrable boundary conditions

We proceed to Sklyanin's formalism for integrable boundary conditions [29, 30]. Consider the NLS equation (2.28) on an interval by restricting the space domain to $[0, L]$. The key idea is to formulate a *double-row monodromy matrix* which takes the boundary behaviors into account (see Figure 1 for a pictorial illustration). The integrable boundary conditions are encoded into the so-called *boundary matrices*, which are subject to certain algebraic constraints. The integrability of the model can be showed by using the trace of the double-row monodromy matrix as a generating function for the conserved quantities.

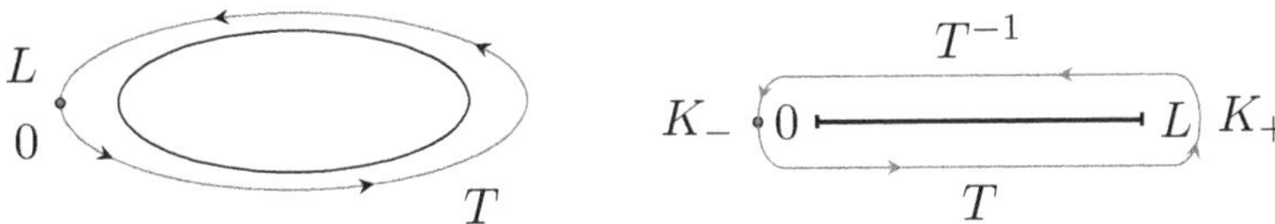

Figure 1. Illustrations of the monodromy matrices under the quasi-periodic boundary conditions (left) and the double-row monodromy matrix for NLS on an interval (right).

For the NLS equation on an interval, one defines the double-row monodromy matrix (we follow the path as indicated in Figure 1) as

$$\mathcal{T}(t;\lambda) = K_-(\lambda)T^{-1}(t;-\lambda)K_+(\lambda)T(t;\lambda)\,, \tag{2.18}$$

where $K_\pm$ are the boundary matrices, and $T(t;\lambda)$ is the usual monodromy matrix as defined in (2.7). We assume that $K_\pm$ are nondegenerate and t-independent. Then, one has the following results.

Lemma 2. *Consider the double-row monodromy matrix $\mathcal{T}$. If the boundary matrices $K_\pm(\lambda)$ satisfy*

$$K_+(\lambda)\,V(L,t;\lambda) = V(L,t;-\lambda)\,K_+(\lambda)\,, \tag{2.19a}$$

$$K_-(\lambda)\,V(0,t;-\lambda) = V(0,t;\lambda)\,K_-(\lambda)\,. \tag{2.19b}$$

then,

$$\frac{d}{dt}\,\mathrm{tr}(\mathcal{T}(t;\lambda)) = 0\,. \tag{2.20}$$

The proof is similar to that of Lemma 1. The integrable boundary conditions at $x = 0$ and $x = L$ are determined respectively by the boundary constraints (2.19a) and (2.19b). Clearly, the boundary constraints (2.19) provide a sufficient condition under which the quantity $\mathrm{tr}(\mathcal{T}(t;\lambda))$ can be possibly considered as a generating function for infinitely many conserved quantities. Moreover, it can be shown that if the boundary matrices $K_\pm$ satisfy the following set of equations

$$\begin{aligned}[][r(\lambda-\eta)\,, K_\pm(\lambda)\otimes K_\pm(\eta)] &= (I\otimes K_\pm(\eta))\; r(\lambda+\eta)\; (K_\pm(\lambda)\otimes I)\\ &\quad - (K_\pm(\lambda)\otimes I)\; r(\lambda+\eta)\; (I\otimes K_\pm(\eta))\,,\end{aligned} \tag{2.21}$$

known as the classical reflection equations, then

$$\{\operatorname{tr}\mathcal{T}(\lambda)\,,\,\operatorname{tr}\mathcal{T}(\eta)\} = 0\,. \tag{2.22}$$

Therefore, $\operatorname{tr}\mathcal{T}(\lambda)$ can be considered as the generating function in the Hamiltonian sense.

There are a number of points we would like to clarify regarding Sklyanin's formalism to integrable boundary conditions:

- the formalism has a quantum analog; in particular, the classical reflection equation can be seen as the classical limit of the quantum reflection equations which give rise to integrable boundary conditions for quantum models [12, 30];
- the formalism requires that the x-part of the Lax pair satisfies the so-called *ultra-local* r matrix structure [3, 14, 29]; however, it is not clear how to implement this formalism to models, such as the KdV equation, which do not possess the ultra-local r matrix structure;
- the requirement that the boundary matrices are t-independent is not essential; we could have t-dependent boundary matrices which lead to dynamical boundary conditions [3, 37] (see Section 2.3 for an example);
- although the NLS equation defined on an interval is integrable, it is still not clear how to solve such model by performing an inverse scattering transform to the double-row monodromy matrix.

2.3 NLS on the half-line

Following Sklyanin's formalism, the boundary matrices $K_\pm(\lambda)$ can be treated separately. The integrable NLS equation on the half-line can be obtained by taking $L \to \infty$ with the NLS field vanishes rapidly as $x \to \infty$. In this case, the boundary matrix K_+ can be chosen to be some matrix proportional to the identity matrix, which is a trivial solution to the boundary constraint (2.19b) and the classical reflection equations (2.21). The boundary conditions at $x = 0$ can be determined by (2.19a). For conciseness, we use K instead of K_-; then one needs to solve the boundary constraint

$$K(\lambda)\,V(-\lambda)\big|_{x=0} = V(\lambda)\big|_{x=0}\,K(\lambda)\,, \tag{2.23}$$

with $K(\lambda)$ also satisfying (2.21).

A class of solutions to the boundary constraint (2.23), given by Sklyanin [29], is the Robin boundary conditions

$$q_x\big|_{x=0} = \alpha\, q\big|_{x=0}\,, \quad \alpha \in \mathbb{R}\,, \tag{2.24}$$

where the boundary matrix $K(\lambda)$ is in the form

$$K(\lambda) = \begin{pmatrix} f_\alpha(\lambda) & 0 \\ 0 & 1 \end{pmatrix}, \quad f_\alpha(\lambda) = \frac{i\,\alpha + 2\lambda}{i\,\alpha - 2\lambda}\,. \tag{2.25}$$

One can easily check that $K^{-1}(\lambda) = K(-\lambda)$, and $K(\lambda)$ satisfies the classical reflection equations (2.21). The real parameter α controls the boundary behaviors. The limiting cases of α give rise either to Dirichlet boundary conditions

$$q\big|_{x=0} = 0\,, \quad \alpha \to \infty\,, \quad K = I\,, \tag{2.26}$$

or to the Neumann boundary conditions

$$q_x\big|_{x=0} = 0\,, \quad \alpha = 0\,, \quad K = -\sigma_3\,. \tag{2.27}$$

We also exclude the cases $\lambda = \pm\frac{i\alpha}{2}$ due to the nondegeneracy of $K(\lambda)$.

Remark 2.2. *The boundary constraint* (2.23) *and the associated Robin boundary conditions* (2.24) *for the NLS equation on the half-line have been found following various perspectives: in [4, 25, 31], the boundary constraint* (2.23) *was interpreted as a Bäcklund transformation exploring the space inversion symmetry $x \to -x$ of the NLS model; in the Fokas uniform transform, the boundary constraint* (2.23) *was known as the linearizable boundary conditions which were used in the reduction of the scattering data [17, 18, 20].*

The presence of boundary conditions adds some freedom to the NLS equation at the boundary. As pointed out by Fokas, cf. [13], the half-line NLS model under the Robin boundary conditions (2.24) is equivalent to the Gross-Pitaevskii equation on the whole line

$$iq_t + q_{xx} + 2|q|^2 q + u(x)q = 0\,, \tag{2.28}$$

with even initial data $q(x, t_0) = q(-x, t_0)$, equipped with a delta-potential $u(x) = -2\alpha\delta(x)$, where $\delta(x)$ is the Dirac-delta function. This is due to the fact that the delta function introduces a jump in the derivative at $x = 0$. After integration, one gets $q_x(0+) - q_x(0-) - 2\alpha q(0) = 0$, which leads to the Robin boundary conditions.

Sklyanin's formalism could also contain dynamical versions of boundary matrices and integrable boundary conditions [3]. Zambon [37] gave dynamical boundary conditions at $x = 0$

$$q_x = \frac{iq_t}{2\Omega} - \frac{\Omega\, q}{2} + \frac{q|q|^2}{2} - \frac{\alpha^2 q}{2\Omega}\,, \tag{2.29}$$

where $\Omega = \sqrt{\beta^2 - |q|^2}$ with α and β being real parameters. They are integrable boundary conditions for the half-line NLS model associated with a dynamical boundary matrix K. Note that exact solutions of this model were recently obtained [24, 32] using the approach of dressing the boundary which will be the main content of next Section.

3 Dressing the boundary

Dressing transformations, also know as Darboux transformations, are algebraic approaches to constructing soliton solutions for integrable PDEs [35, 36] (see also the monographs [5, 27]. By dressing the boundary, we adapt the dressing procedure in view of solving the initial-boundary value problem for integrable PDEs on the half-line equipped with integrable boundary conditions. The key idea is to dress simultaneously the Lax pair and the boundary constraint.

3.1 Dressing transformations and soliton solutions

We first briefly review the dressing transformations and multi-soliton solutions for the NLS equation on the whole line under the rapidly vanishing boundary conditions.

Dressing transformations are special gauge transformations which preserve the forms of the Lax pair. We use $U[0]$, $V[0]$ and $\phi[0]$ to denote the *undressed* Lax system (2.2). Suppose there is a gauge transformation

$$\phi[1] = D[1]\,\phi[0]\,, \tag{3.1}$$

where $\phi[1]$ satisfies the *newly* transformed Lax pair

$$U[1]\,\phi[1] = \phi[1]_x\,, \quad V[1]\,\phi[1] = \phi[1]_t\,. \tag{3.2}$$

$U[1], V[1]$ and $U[0], V[0]$ are related by

$$D[1]_x = D[1]\,U[0] - U[1]\,D[1]\,, \quad D[1]_t = D[1]\,V[0] - V[1]\,D[1]\,. \tag{3.3}$$

We call $D[1]$ the dressing matrix. There are several equivalent representations of $D[1]$, and here we adopt the polynomial (in λ) form:

$$D[1] = (\lambda - \bar{\lambda}_1) + (\bar{\lambda}_1 - \lambda_1)\,P[1]\,, \quad P[1] = \frac{\psi_1\psi_1^\dagger}{\psi_1^\dagger\psi_1}\,. \tag{3.4}$$

The 2-vector $\psi_1 = (\mu_1, \nu_1)^\intercal$ is a special solution of the undressed Lax pair at $\lambda = \lambda_1$, and $\psi_1^\dagger$ denotes the transpose conjugate of ψ_1. $D[1]$ defines a one-step dressing transformation. It adds a pair of zeros $\{\lambda_1, \bar{\lambda}_1\}$ to $\phi[0]$ as

$$\det D[1] = (\lambda - \lambda_1)\,(\lambda - \bar{\lambda}_1)\,. \tag{3.5}$$

Putting $D[1]$ into (3.3) gives $Q[1]$ in terms of $Q[0]$ and $P[1]$:

$$Q[1] = Q[0] - i(\lambda_1 - \bar{\lambda}_1)[\sigma_3, P[1]]\,. \tag{3.6}$$

This is the *reconstruction formula*, from which a new solution $q[1]$ can be easily extracted. We commonly refer to $q[0]$ as *seed solutions.* Since we are focusing on soliton solutions of the NLS equation, the zero seed solutions $q[0] = 0$ are used in this section, which correspond to solutions vanishing rapidly as $x \to \pm\infty$.

N-step dressing transformations can be constructed by iteration. Assume that there exist N linearly-independent special solutions $\psi_j = (\mu_j, \nu_j)^\intercal$ of the undressed Lax pair (2.2) evaluated respectively at λ_j, $j = 1 \ldots N$, then the N-step dressing matrix $D[N]$ is in the form

$$D[N] = \left((\lambda - \bar{\lambda}_N) + (\bar{\lambda}_N - \lambda_N)\,P[N]\right) \cdots \left((\lambda - \bar{\lambda}_1) + (\bar{\lambda}_1 - \lambda_1)\,P[1]\right)\,, \tag{3.7}$$

where

$$P[j] = \frac{\psi_j[j-1]\,\psi_j^\dagger[j-1]}{\psi_j^\dagger[j-1]\,\psi_j[j-1]}, \quad \psi_j[j-1] = D[j-1]\big|_{\lambda=\lambda_j}\,\psi_j\,. \tag{3.8}$$

A series expansion of $D[N]$ in λ leads to

$$D[N] = \lambda^N + \lambda^{N-1}\,\Sigma_1 + \lambda^{N-2}\,\Sigma_2 \cdots + \Sigma_N\,, \tag{3.9}$$

with the matrices Σ_j, $j = 1 \ldots N$ to be determined. Using

$$U[N]\,\phi[N] = \phi[N]_x\,, \quad V[N]\,\phi[N] = \phi[N]_t\,, \tag{3.10}$$

where $\phi[N] = D[N]\,\phi[0]$, one has the reconstruction formula for N-soliton solutions

$$q[N] = q[0] + 2i\,\Sigma_1^{(1,2)}\,, \tag{3.11}$$

where $\Sigma_1^{(1,2)}$ is the $(1,2)$ entry of Σ_1. It can be put into the compact form [27]:

$$q[N] = q[0] + 2i\frac{\Delta_1}{\Delta_2}\,, \tag{3.12}$$

where

$$\Delta_1 = \begin{vmatrix} -\lambda_1^N\mu_1 & \cdots & -\lambda_N^N\mu_N & \bar\lambda_1^N\bar\nu_1 & \cdots & \bar\lambda_N^N\bar\nu_N \\ \lambda_1^{N-2}\nu_1 & \cdots & \lambda_N^{N-2}\nu_N & \bar\lambda_1^{N-2}\bar\mu_1 & \cdots & \bar\lambda_N^{N-2}\bar\mu_N \\ \cdots & \cdots & \cdots & \cdots & \cdots & \cdots \\ \nu_1 & \cdots & \nu_N & \bar\mu_1 & \cdots & \bar\mu_N \\ \lambda_1^{N-1}\mu_1 & \cdots & \lambda_N^{N-1}\mu_N & -\bar\lambda_1^{N-1}\bar\nu_1 & \cdots & -\bar\lambda_N^{N-1}\bar\nu_N \\ \cdots & \cdots & \cdots & \cdots & \cdots & \cdots \\ \mu_1 & \cdots & \mu_N & -\bar\nu_1 & \cdots & -\bar\nu_N \end{vmatrix}\,, \tag{3.13}$$

$$\Delta_2 = \begin{vmatrix} \lambda_1^{N-1}\nu_1 & \cdots & \lambda_N^{N-1}\nu_N & \bar\lambda_1^{N-1}\bar\mu_1 & \cdots & \bar\lambda_N^{N-1}\bar\mu_N \\ \cdots & \cdots & \cdots & \cdots & \cdots & \cdots \\ \nu_1 & \cdots & \nu_N & \bar\mu_1 & \cdots & \bar\mu_N \\ \lambda_1^{N-1}\mu_1 & \cdots & \lambda_N^{N-1}\mu_N & -\bar\lambda_1^{N-1}\bar\nu_1 & \cdots & -\bar\lambda_N^{N-1}\bar\nu_N \\ \cdots & \cdots & \cdots & \cdots & \cdots & \cdots \\ \mu_1 & \cdots & \mu_N & -\bar\nu_1 & \cdots & -\bar\nu_N \end{vmatrix}\,. \tag{3.14}$$

This formula will be used later in constructing half-line NLS solitons.

3.2 Dressing the boundary and soliton solutions on the half-line

We consider the half-line NLS model subject to Robin boundary conditions as described in Section 2.3 by restricting x to $[0,\infty)$. Let a seed solution $q[0]$ be given, and let $\psi_1(\lambda_1)$ be a special solution of the undressed Lax pair $U[0]$ and $V[0]$. If there exists a *paired* special solution $\psi_2(\lambda_2)$ such that

$$\psi_2(0,t,\lambda_2) = K(\lambda_2)\,\psi_1(0,t,\lambda_1)\,, \quad \lambda_2 = -\lambda_1\,, \quad \bar\lambda_1 \neq -\lambda_1\,, \tag{3.15}$$

where $K(\lambda)$ is the boundary matrix defined in (2.23), then $q[0]$ satisfies the boundary conditions imposed by the boundary constraint (2.23). This can be easily understood by using the t-part of the Lax pair (2.2). Since ψ_j, $j=1,2$, satisfies

$$V[0](0,t,\lambda_j)\,\psi_j(0,t,\lambda_j)=\psi_{j\,t}(0,t,\lambda_j)\,, \tag{3.16}$$

using the relation (3.15), one gets

$$K(\lambda_j)\,V[0](-\lambda_j)\,K(-\lambda_j)\,\psi_j=V[0](\lambda_j)\,\psi_j=\psi_{j\,t}(\lambda_j)\,, \tag{3.17}$$

at $x=0$. This implies $V[0]$ is subject to the boundary constraint (2.23). We exclude for the moment the case λ_1 is a pure imaginary because of the condition $\bar{\lambda}_1\neq-\lambda_1$. In fact, the existence of (3.15) is a strong condition allowing us to construct soliton solutions satisfying the boundary conditions imposed by (2.23).

Proposition 1 (Dressing the boundary: one soliton case). *Consider the half-line NLS model. Assume that there exist paired special solutions* $\{\psi_1,\widehat{\psi}_1\}$ *of the undressed Lax pair* (2.2) *associated with the parameters* $\{\lambda_1,\widehat{\lambda}_1\}$ *such that*

$$\widehat{\psi}_1(0,t,\widehat{\lambda}_1)=K(\widehat{\lambda}_1)\,\psi_1(0,t,\lambda_1)\,,\quad \widehat{\lambda}_1=-\lambda_1\,,\quad \bar{\lambda}_1\neq-\lambda_1\,, \tag{3.18}$$

where $K(\lambda)$ *is given in* (2.25), *then a two-step dressing transformation using such pair leads to a* $V[2]$ *satisfying*

$$K(\lambda)\,V[2](-\lambda)|_{x=0}=V[2](\lambda)\,K(\lambda)|_{x=0}\,. \tag{3.19}$$

The so-constructed $q[2]$ *satisfies the Robin boundary conditions* (2.24). *We use* $\widehat{q}[1]$ *to denote such* $q[2]$.

Proof: In order that $\widehat{q}[1]$ satisfies the Robin boundary conditions (2.24), we need to prove the relation (3.19). Let $D[2](\lambda)$ be the dressing matrix constructed from $\{\psi_1,\widehat{\psi}_1\}$. One knows that $V[2](\lambda)$ and $V[0](\lambda)$ are connected by

$$V[2](\lambda)=D[2]_t(\lambda)\,D[2]^{-1}(\lambda)+D[2](\lambda)\,V[0](\lambda)\,D[2]^{-1}(\lambda)\,. \tag{3.20}$$

One can easily verify that if $D[2](\lambda)$ satisfies

$$D[2](\lambda)\,K(\lambda)|_{x=0}=K(\lambda)\,D[2](-\lambda)|_{x=0}\,, \tag{3.21}$$

then $V[2](\lambda)$ satisfies (3.19). We multiply both sides of (3.21) by an irrelevant factor $(i\,\alpha-2\lambda)$. Since $D[2]$ can be expressed as a matrix polynomial of degree 2 in λ and that $(i\alpha-2\lambda)K(\lambda)=i\,\alpha I+2\lambda\sigma_3$, the left-hand side and right-hand side of (3.21) are now matrix polynomials of degree 3. We use $L(\lambda)$ and $R(\lambda)$ to denote them

$$L(\lambda)=D[2](\lambda)\,K(\lambda)=\lambda^3L_0+\lambda^2L_1+\lambda^2L_2+L_3\,, \tag{3.22a}$$
$$R(\lambda)=K(\lambda)D[2](-\lambda)=\lambda^3R_0+\lambda^2R_1+\lambda^2R_2+R_3\,. \tag{3.22b}$$

Clearly, $L_0=R_0$, $L_3=R_3$, and L_1,L_2,R_1,R_2 can be determined by zeros and the associated kernel vectors of $L(\lambda),R(\lambda)$. Knowing the relation (3.18), one has

$$R(\lambda)|_{\lambda=-\lambda_1}\psi_1=0\,,\ L(\lambda)|_{\lambda=-\lambda_1}\psi_1=0\,,\ R(\lambda)|_{\lambda=\lambda_1}\widehat{\psi}_1=0\,,\ L(\lambda)|_{\lambda=\lambda_1}\widehat{\psi}_1=0\,, \tag{3.23}$$

evaluated at $x = 0$. Moreover, let $\varphi_1 = \sigma_2 \bar{\psi}_1$, $\sigma_2 = \begin{pmatrix} 0 & -i \\ i & 0 \end{pmatrix}$, one can also verify

$$R(\lambda)|_{\lambda=-\bar{\lambda}_1}\varphi_1 = 0\,, \; L(\lambda)|_{\lambda=-\bar{\lambda}_1}\varphi_1 = 0\,, \; R(\lambda)|_{\lambda=\bar{\lambda}_1}\widehat{\varphi}_1 = 0\,, \; L(\lambda)|_{\lambda=\bar{\lambda}_1}\widehat{\varphi}_1 = 0\,, \quad (3.24)$$

at $x = 0$, where $\widehat{\varphi}_1$ is defined in the same way as φ_1. The above formulae show that $L(\lambda)$ and $R(\lambda)$ has the same zeros and the associated kernel vectors, which, in turn, implies that $L(\lambda) = R(\lambda)$. This completes the proof. ■

The above construction is the realization of dressing the boundary. The existence of the paired special solutions $\{\psi_1(\lambda_1), \widehat{\psi}_1(\widehat{\lambda}_1)\}$ satisfying the conditions (3.15) implies that the t-part of undressed Lax pair $V[0]$ satisfies the boundary constraint (2.23). Thus the seed solution $q[0]$ is presumed to be subject to the Robin boundary conditions (2.24). Dressing the Lax pair using the pair $\{\psi_1(\lambda_1), \widehat{\psi}_1(\widehat{\lambda}_1)\}$ preserves the boundary constraint (2.23), and the so-constructed solution $\widehat{q}[1]$ represents a one-soliton solution on the half-line subject to the Robin boundary conditions (2.24).

One can repeatedly dress the boundary using N paired special solutions to derive N-soliton solutions on the half-line. This requires the existence of N paired special solutions $\{\psi_j(\lambda_j), \widehat{\psi}_j(\widehat{\lambda}_j)\}$, $j = 1, \ldots, N$, of the undressed Lax pair (2.2) obeying

$$\widehat{\psi}_j(0,t,\widehat{\lambda}_j) = K(\widehat{\lambda}_j)\,\psi_j(0,t,\lambda_j)\,, \quad \widehat{\lambda}_j = -\lambda_j\,, \quad \bar{\lambda}_j \neq -\lambda_j\,, \quad \widehat{\lambda}_k \neq \lambda_j\,. \quad (3.25)$$

Then, the so-constructed $q[2N]$ corresponds to an N-soliton solution of NLS on the half-line satisfying the Robin boundary conditions (2.24). Again, we use $\widehat{q}[N]$ to denote such $q[2N]$. The requirement $\widehat{\lambda}_k \neq \lambda_j$ ensures that all the special solutions are independent. By construction, the boundary constraint (2.23) are preserved at each step of the dressing process.

It remains to find such pairs $\{\psi_j(\lambda_j), \widehat{\psi}_j(\widehat{\lambda}_j)\}$. For $x \geq 0$, using the zero seed solution $q[0] = 0$, one has special solutions $\psi_j(\lambda_j)$, $j = 1, \ldots, N$ in the forms

$$\psi_j(\lambda_j) = \begin{pmatrix} \mu_i \\ \nu_i \end{pmatrix} = e^{(-i\lambda_j x - 2i\lambda_j^2 t)\,\sigma_3} \begin{pmatrix} u_j \\ v_j \end{pmatrix}, \quad (3.26)$$

where λ_j is a complex parameter, and u_j, v_j are complex constants. Clearly, with each ψ_j, one can associate a paired $\widehat{\psi}_j$ in the form

$$\widehat{\psi}_j(\widehat{\lambda}_j) = K(-\lambda_j)\,\psi_j(-\lambda_j)\,. \quad (3.27)$$

Now having the data $\{\psi_j(\lambda_j), \widehat{\psi}_j(\widehat{\lambda}_j)\}$, we are ready to compute the N-soliton solutions of the NLS equation on the half-line. Two solitons interacting with the boundary are illustrated in Figures 2 and 3.

3.3 Boundary-bound solitons

Static soliton solutions arise as the special solutions $\psi_j(\lambda_j)$ having pure imaginary parameters λ_j. Since $\bar{\lambda}_j = -\lambda_j$, each ψ_j needs to be paired with itself to eventually make the boundary constraint preserved (2.23) in the boundary-dressing process. This generates solitons bounded to the boundary.

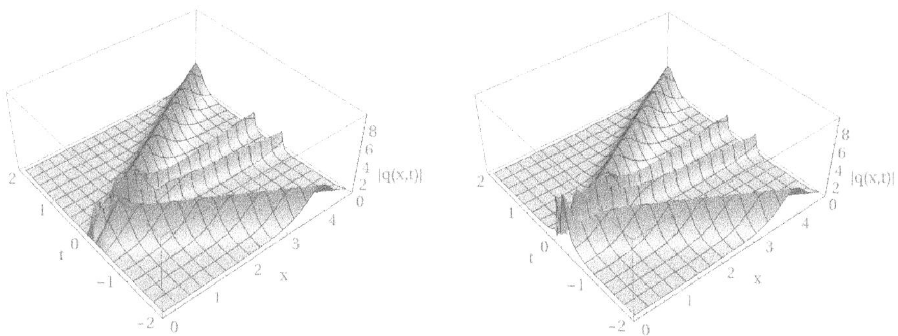

Figure 2. Two-soliton solution satisfying the Dirichlet boundary conditions (left) and Neumann boundary conditions (right).

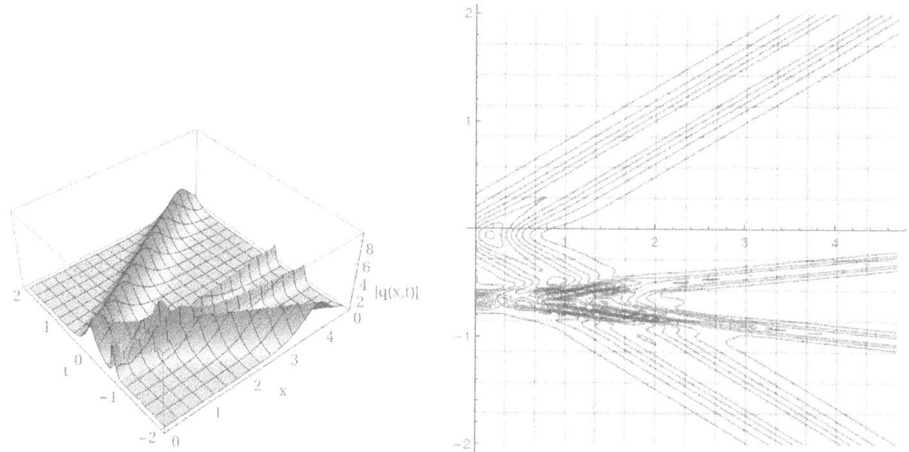

Figure 3. Two-soliton solution satisfying the Robin boundary conditions ($\alpha = 3$).

Proposition 2 (Boundary-bound solitons). *Assume that, associated with N distinct pure imaginary parameters $\lambda_j = i\kappa_j$, $\kappa_j \in \mathbb{R}$, there exist N special solutions $\psi_j(\lambda_j)$, $j = 1, \ldots, N$, of the undressed Lax system* (2.2) *with zero seed solution $q[0] = 0$, in the forms*

$$\psi_j(\lambda_j) = \psi_j(i\kappa_j) = \begin{pmatrix} \mu_i \\ \nu_i \end{pmatrix} = e^{\left(\kappa_j x + 2i\kappa_j^2 t\right)\sigma_3} \begin{pmatrix} u_j \\ 1 \end{pmatrix}, \quad x \geq 0, \tag{3.28}$$

such that κ_j satisfies

$$f_\alpha(i\kappa_j) = \frac{\alpha + 2\kappa_j}{\alpha - 2\kappa_j} < 0, \quad \kappa_j \neq \pm\frac{\alpha}{2}, \tag{3.29}$$

with α being a real parameter and u_j being defined as

$$u_j = \sqrt{-f_\alpha(i\kappa_j)}^{(-1)^N}, \tag{3.30}$$

then the so-construction $q[N]$ corresponds to static N-soliton solutions satisfying the Robin boundary conditions (2.24).

Proof: The proof is split into two cases: when N is odd and when N is even. Moreover, for simplicity, we only consider the cases where $N = 1$ and $N = 2$ since the odd N and the even N cases can be understood as generalizations.

Case $N = 1$: since the zero seed solution is imposed, one has $V[0] = -2i\lambda^2\sigma_3$. The one-step dressing transformation involves a dressing matrix $D[1]$ constructed from a special solution $\psi_1(\lambda_1)$. Using the identity $(iaI \pm 2\lambda\sigma_3) = (ia \mp 2\lambda)K(\pm\lambda)$, one can show if $D[1]$ obeys

$$(iaI - 2\lambda\sigma_3)D[1](\lambda)|_{x=0} = D[1](-\lambda)(iaI + 2\lambda\sigma_3)|_{x=0}\,, \tag{3.31}$$

then the dressed $V[1]$ satisfies the boundary constraints (2.23). It remains to prove the relation (3.31). Knowing that $D[1](\lambda)$ is a matrix polynomial of degree 1 in λ, the left-hand side and right-hand side of (3.31) are thus polynomials of degree 2. Denote them by

$$L(\lambda) =(iaI - 2\lambda\sigma_3)D[1](\lambda) = \lambda^2 L_0 + \lambda^1 L_1 + L_2\,, \tag{3.32}$$
$$R(\lambda) =D[1](-\lambda)(iaI + 2\lambda\sigma_3) = \lambda^2 R_0 + \lambda^1 R_1 + R_2\,. \tag{3.33}$$

Clearly, $L_0 = R_0$, $L_2 = R_2$. This explains the presumed form (3.31): $D[1](\lambda)$ is of odd degree in λ, thus (3.31) ensures that the leading and zero-degree terms of both sides are equal. The equality (3.31) holds, if

$$K(\lambda_1)\psi_1|_{x=0} = \sigma_2\bar{\psi}_1|_{x=0}\,. \tag{3.34}$$

In fact, let $\varphi_1 = \sigma_2\bar{\psi}_1$, $D[1](-\lambda_1)\varphi_1 = 0$. One can show that

$$L(\lambda)|_{\lambda=\lambda_1}\psi_1 = 0\,,\; R(\lambda)|_{\lambda=\lambda_1}\psi_1 = 0\,,\; L(\lambda)|_{\lambda=-\lambda_1}\varphi_1 = 0\,,\; R(\lambda)|_{\lambda=-\lambda_1}\varphi_1 = 0\,, \tag{3.35}$$

meaning that $L(\lambda)$ and $R(\lambda)$ share the same zeros and the associated kernel vectors, thus $L(\lambda) = R(\lambda)$. Moreover, the constraint (3.34) imposes

$$f_\alpha(i\kappa_1)u_1 = -i\bar{v}_1\,, \quad v_1 = i\bar{u}_1\,, \tag{3.36}$$

where u_1, v_1 are elements of the constant vector appearing in the expression of ψ_j (3.26). The above constraints lead to

$$f_\alpha(i\kappa_1)|u_1|^2 = -|v_1|^2\,. \tag{3.37}$$

This implies the requirement that $f_\alpha(i\kappa_1) < 0$. Without loss of generality, letting $|v_1|^2 = 1$ and u_1 be real, then $u_1 = 1/\sqrt{-f_\alpha(i\kappa_1)}$. One recovers the statements (3.30) for $N = 1$. Note that the assumption (3.31) is needed for any odd N, this imposes similar constraints on u_j, v_j as shown in (3.37).

Case $N = 2$: similarly, one needs the following identity

$$(iaI - 2\lambda\sigma_3)D[2](\lambda)|_{x=0} = D[2](-\lambda)(iaI - 2\lambda\sigma_3)|_{x=0}\,. \tag{3.38}$$

Then the dressed $V[2]$ satisfies the boundary constraints (2.23). As previously explained, this identity ensures the equality of the leading and zero-degree terms of both sides of (3.38). The relation holds if

$$K(-\lambda_j)\psi_j|_{x=0} = \sigma_2\bar{\psi}_j|_{x=0}\,, \quad j = 1, 2\,. \tag{3.39}$$

In components, it reads

$$f_\alpha(-i\kappa_j)|u_j|^2 = \frac{1}{f_\alpha(i\kappa_j)}|u_j|^2 = -|v_j|^2, \quad j = 1, 2. \tag{3.40}$$

Again let $|v_j|^2 = 1$ and u_j be real, one obtains (3.30) for $N = 2$. This condition is true for any even N. ∎

Note that the requirement $f_\alpha(i\kappa_j) < 0$ excludes the case of Dirichlet boundary conditions (2.26) for the boundary-bound solitons. Following the above proposition, one can easily compute static solitons bounded to the boundary under the Robin boundary conditions. When there are multi-static solitons bounded to the boundary, the interference phenomena take place (see Figure 4). One can dress the boundary using both the moving and static soliton data, since the boundary constraint (2.23) is preserved at each step of the dressing process. This generates interactions between moving solitons and boundary-bounded solitons (see Figure 5). Similar results were obtained in [7] using the mirror-image technique.

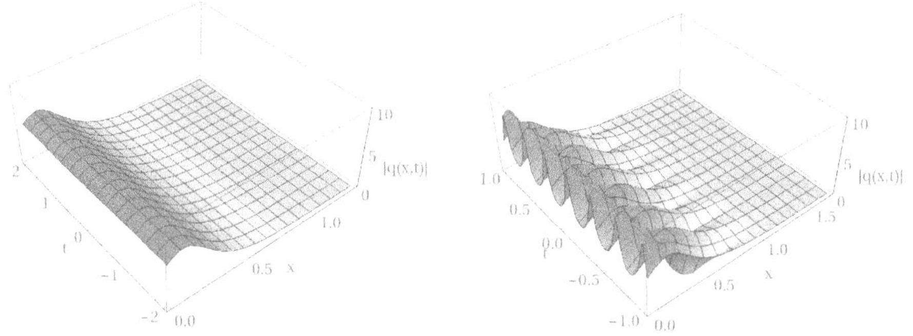

Figure 4. Boundary-bound solitons under the Robin boundary conditions ($\alpha = 3$): the magnitude (norm) is constant along the boundary for the one-soliton case (left); an interference phenomenon takes place for a doubly-boundary-bound soliton (right).

4 Dressing the boundary as an inverse scattering transform

In this section, we present an inverse scattering transform interpretation of the process of dressing the boundary. This relies on switching the roles of initial and boundary data.

The conventional inverse scattering transform is an analytic approach to solving initial value problems for integrable PDEs typically in the case of vanishing boundary conditions for $x \in \mathbb{R}, t \geq 0$. It is essentially made of three steps: (1) the direct scattering process where the initial conditions at $t = 0$ are transformed into scattering data using the x-part of the Lax pair; (2) the time-evolution process where the scattering data evolve linearly in time according to the t-part of the Lax pair; (3) the inverse problem where the scattering data are put into the reconstruction formula to recover solutions of the integrable PDEs.

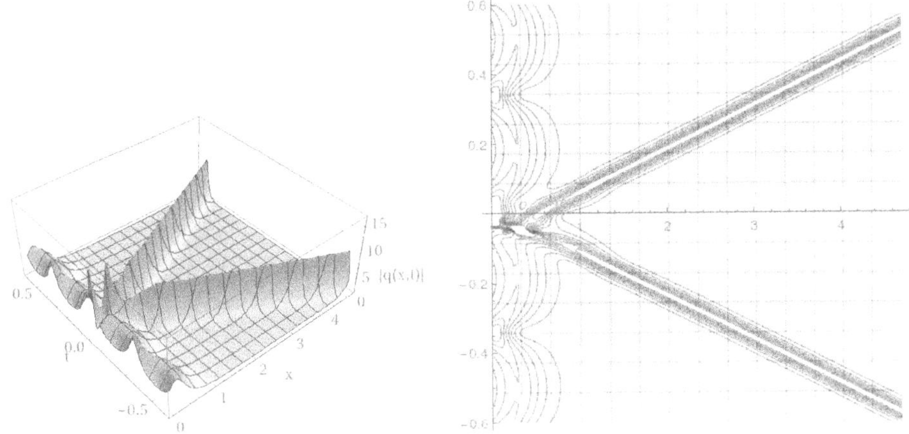

Figure 5. Interaction between a (moving) soliton and a doubly-boundary-bound soliton under the Robin boundary conditions ($\alpha = 3$).

In order to fit the approach of dressing the boundary into an inverse scattering transform scheme, one needs to fix the space-time domain to be $x \geq 0$, $t \in \mathbb{R}$. In contrast to the usual inverse scattering transform, the roles of space and time variables are switched, and we are dealing with a boundary value problem where an *initial* boundary profile is imposed by the boundary conditions at $x = 0$. It turns out that the boundary value problem can be solved using a *space-evolution* process where the scattering data are determined by the t-part of the Lax pair at $x = 0$ and evolve linearly in x for $x \geq 0$.

To make the statements precise, we first perform the direct scattering for the t-part of the Lax pair at $x = 0$. We only consider the vanishing asymptotic conditions where the NLS field q and its x derivatives are of of Schwartz type, *i.e.*, vanish rapidly as $t \to \pm\infty$. This leads to the Jost solutions

$$\lim_{t\to\pm\infty} \phi_\pm(0,t;\lambda) = e^{-2i\lambda^2 t\sigma_3}\,. \tag{4.1}$$

By integrating the t-part of the Lax pair (2.2), due to the λ^2-dependence of the spectral parameter, the analytic domain of Jost solutions can be naturally split into four quadrants, which leads to a monodromy matrix $S(k)$ in the form

$$S(\lambda) = \widehat{\exp} \int_{-\infty}^{\infty} V(0,\tau,\lambda)\, d\tau = \begin{pmatrix} a^{(24)}(\lambda) & \bar{b}^{(13)}(\lambda) \\ -b^{(24)}(\bar{\lambda}) & \bar{a}^{(13)}(\bar{\lambda}) \end{pmatrix}. \tag{4.2}$$

Here the subscript of $a^{(24)}(\lambda)$ means that the scattering function $a^{(24)}(\lambda)$ can be analytically continued to the union of the quadrants (2) and (4) (see Figure 6 for the distribution of the quadrants). This also applies to other scattering functions.

Apparently, the direct scattering of V at $x = 0$ differs from the usual inverse scattering transform only by a switch between the x-part and the t-part of the Lax pair (2.2). This also changes the roles of initial and boundary conditions: instead

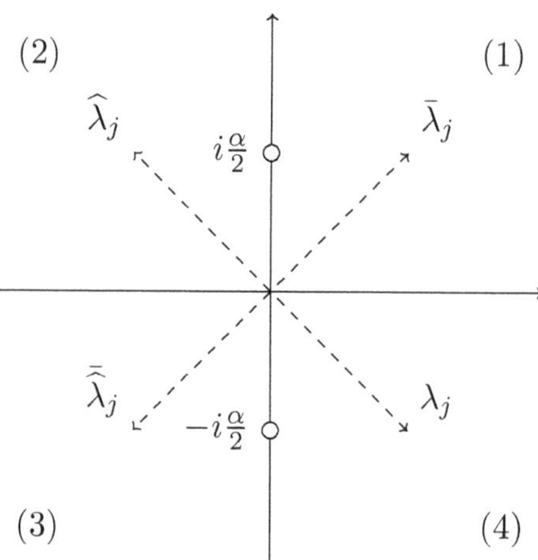

Figure 6. Zeros of the scattering function $a^{(24)}(\lambda)$ in the presence of Robin boundary conditions in the λ-plane, $\lambda \neq \pm i\frac{\alpha}{2}$ due to the nondegeneracy of the boundary matrix K.

of characterizing initial profile at $t = 0$ following the usual direct scattering, the boundary conditions are considered and encoded into a monodromy matrix $S(k)$ at $x = 0$. Then following the asymptotic conditions imposed to the NLS field q as $t \to \pm\infty$, one can easily show that the scattering data evolve linearly in $x \geq 0$ as $S(\lambda)$ obeys

$$\frac{\partial S(k)}{\partial x} = -i\lambda[\sigma_3, S(k)]\,. \tag{4.3}$$

The Jost solutions can be put into certain (x,t)-dependent Riemann-Hilbert problems, which eventually lead to soliton solutions of NLS with zeros of $a^{(24)}(\lambda)$ appearing in the union of the quadrants (2) and (4).

Having the above space-evolution process in mind, we are ready to implement the Robin boundary conditions into the system. Since V is required to obey the boundary constraint (2.23), this imposes an additional constraint on S

$$S(-\lambda) = K(-\lambda)\, S(\lambda)\, K(\lambda)\,, \tag{4.4}$$

with the boundary matrix $K(\lambda)$ defined in (2.25). Consequently, the Robin boundary conditions imply that if λ_j is a zero of $a^{(24)}(\lambda)$, so does $-\lambda_j$, since

$$a^{(24)}(-\lambda) = a^{(24)}(\lambda)\,. \tag{4.5}$$

The associated norming constants (they can be understood as ratio of u_j and v_j appearing in the special solutions in the dressing transformations) are related by

$$b^{(24)}(-\lambda_j) = f_a(-\lambda) b^{(24)}(\lambda_j)\,. \tag{4.6}$$

Therefore, the paired zeros of $a^{(24)}(\lambda)$ (see Figure 6) and the relation between the paired norming constants give the underlying reason of the paired special solutions $\{\psi_j(\lambda_j), \widehat{\psi}_j(\widehat{\lambda}_j)\}$ in the process of dressing the boundary.

Remark 4.1. *The relation* (4.4) *is in contrast to the mirror-image technique where the pairing of zeros of the scattering function is related by spectra* $\{\lambda_j, -\bar{\lambda}_j\}$, *cf. [8].*

5 Concluding remarks

In this chapter, we review the approach called "dressing the boundary" in view of solving initial-boundary value problems for the NLS equation on the half-line, cf. [38]. By dressing the boundary, we dress simultaneously both the Lax pair and integrable boundary conditions, which allows us to compute directly soliton solutions of the NLS model on the half-line subject to the integrable Robin boundary conditions. One particular advantage of our method is that it does not require extension of the space domain to the whole line as did the nonlinear mirror-image technique [7, 8, 11]. It also admits a natural inverse scattering transform interpretation based on the space-evolution process, in which the initial boundary profile at $x = 0$ is encoded into scattering data that evolve linearly on the half-line. Then, the half-line solutions can be obtained through the inverse process.

Although we only consider the focusing NLS model on the half-line, this method can be applied to a wide range of classical integrable PDEs on the half-line equipped with integrable boundary conditions. For instance, soliton solutions of the sine-Gordon equation and the vector NLS equation on the half-line were obtained [39, 40]. This method has also been recently adapted to more complicated cases such as defect problems for the NLS equation [23], and the NLS equation on the half-line with dynamical integrable boundary conditions [24, 32].

Some particularly important extensions of the method are to dress the boundary of integrable models on a star-graph [10], and to solve integrable PDEs on an interval where the algebraic-geometric integration technique is needed [6, 26, 28].

Acknowledgements

This project is supported by NSFC No. 11875040.

References

[1] Ablowitz M J, Kaup D J, Newell A C and Segur H, The inverse scattering transform-Fourier analysis for nonlinear problems, *Stud. Appl. Math.* **53**, 249–315, 1974.

[2] Ablowitz M J, Prinari B and Trubatch A D, *Discrete and continuous nonlinear Schrödinger systems*, Cambridge University Press, 2004.

[3] Avan J, Caudrelier V and Crampé N, From Hamiltonian to zero curvature formulation for classical integrable boundary conditions, *J Phys. A: Math. Theor.* **51**, 30LT01, 2018.

[4] Bikbaev R F and Tarasov V O, Initial boundary value problem for the nonlinear Schrödinger equation, *J. Phys. A: Math. Theor.* **24**, 2507–2516, 1991.

[5] Babelon O, Bernard D and Talon M, *Introduction to Classical Integrable Systems*, Cambridge University Press, 2003.

[6] Belokolos E D (Ed), *Algebro-Geometric Approach to Nonlinear Integrable Equations*, Springer, Berlin, 1994.

[7] Biondini G and Bui A, On the nonlinear Schrödinger equation on the half line with homogeneous Robin boundary conditions, *Stud. Appl. Math.* **129**, 249–271, 2012.

[8] Biondini G and Hwang G, Solitons, boundary value problems and a nonlinear method of images, *J. Phys. A: Math. Theor.* **42**, 205–217, 2009.

[9] Bona J L, Sun S M and Zhang B Y, Nonhomogeneous boundary-value problems for one-dimensional nonlinear Schrödinger equations, *J. Math. Pures Appl.* **109**, 1–66, 2018.

[10] Caudrelier V, On the inverse scattering method for integrable PDEs on a star graph, *Comm. Math. Phys.* **338**, 893–917, 2015.

[11] Caudrelier V and Zhang C, The vector nonlinear Schrödinger equation on the half-line, *J. Phys. A: Math. Theor.* **45**, 105201, 2012.

[12] Cherednik I V, Factorizing particles on a half-line and root systems, *Theor. Math. Phys.* **61**, 977–983, 1984.

[13] Deift P and Park J, Long-time asymptotics for solutions of the NLS equation with a delta potential and even initial data, *Int. Math. Res. Not.* **2011**, 5505–5624, 2011.

[14] Faddeev L D and Takhtajan L A, *Hamiltonian Methods in the Theory of Solitons*, Springer, Berlin, 2007.

[15] Fokas A S, An initial-boundary value problem for the nonlinear Schrödinger equation, *Physica D* **35**, 167–185, 1989.

[16] Fokas A S, A unified transform method for solving linear and certain nonlinear PDEs, *Proc. R. Soc. Lond. A* **453**, 1411–1443, 1997.

[17] Fokas A S, Integrable nonlinear evolution equations on the half-line, *Comm. Math. Phys.* **230**, 1–39, 2002.

[18] Fokas A S, *A Unified Approach to Boundary Value Problems*, SIAM, Philadelphia, 2008.

[19] Fokas A S, Himonas A and Mantzavinos D, The nonlinear Schrödinger equation on the half-line, *Trans. Am. Math. Soc.* **369**, 681–709, 2017.

[20] Fokas A S, Its A R and Sung L Y, The nonlinear Schrödinger equation on the half-line, *Nonlinearity*, **18**, 1771, 2005.

[21] Gardner C S, Korteweg-de Vries Equation and generalizations. IV. the Korteweg-de Vries Equation as a Hamiltonian System, *J. Math. Phys.* **12**, 1548–1551, 1971.

[22] Gardner C S, Greene J M, Kruskal M D and Miura R M, Method for solving the Korteweg-de Vries equation, *Phys. Rev. Lett.* **19**, 1095, 1967.

[23] Gruner K T, Soliton solutions of the nonlinear Schrödinger equation with defect conditions, *arXiv:1908.05101*, 2019.

[24] Gruner K T, Dressing a new integrable boundary of the nonlinear Schrödinger equation, *arXiv:2008.03272*, 2020.

[25] Habibullin I T, The Bäcklund transformation and integrable initial boundary value problems, *Math. Notes. Acad. Sci. USSR.* **49**, 130–137, 1991.

[26] Krichever I and Novikov S P, Periodic and almost-periodic potentials in inverse problems, *Inverse Probl.* **15**, R117, 1999.

[27] Matveev V B and Salle M A, *Darboux transformations and solitons*, Springer-Verlag, Berlin, 1991.

[28] Novikov S P, The periodic problem for the Korteweg-de Vries equation, *Funct. Anal. Appl.* **8**, 236–246, 1974.

[29] Sklyanin E K, Boundary conditions for integrable equation, *Funct. Anal. Appl.* **21**, 164–166, 1987.

[30] Sklyanin E K, Boundary conditions for integrable quantum systems, *J. Phys. A: Math. Gen.* **21**, 2375, 1988.

[31] Tarasov V O, The integrable initial-boundary value problem on a semiline: nonlinear Schrodinger and sine-Gordon equations, *Inverse Probl.* **7**, 435, 1991.

[32] Xia B, On the nonlinear Schrödinger equation with a time-dependent boundary condition, *arXiv:2008.03955*, 2020.

[33] Zakharov V E, Faddeev L D, Korteweg-de Vries equation: A completely integrable Hamiltonian system, *Funct. Anal. Appl.* **5** 280–287, 1971.

[34] Zakharov V E and Shabat A B, Exact theory of two-dimensional self-focusing and one-dimensional self-modulation of waves in non-linear media, *Soviet Phys. JETP.* **34**, 62–69, 1972.

[35] Zakharov V E and Shabat A B, A scheme for integrating the nonlinear equations of mathematical physics by the method of the inverse scattering problem. I, *Funct. Anal. Appl.* **8**, 226–235, 1974.

[36] Zakharov V E and Shabat A B, Integration of nonlinear equations of mathematical physics by the method of inverse scattering. II, *Funct. Anal. Appl.* **13**, 166–174, 1979.

[37] Zambon C, The classical nonlinear Schrödinger model with a new integrable boundary, *J. High Energy Phys.* **2014**, 36, 2014.

[38] Zhang C, Dressing the boundary: on soliton solutions of the nonlinear Schrödinger equation on the half-line, *Stud. Appl. Math.* **142**, 190–212, 2019.

[39] Zhang C, Cheng Q and Zhang D-J, Soliton solutions of the sine-Gordon equation on the half-line, *Appl. Math. Lett.* **86**, 64–69, 2018.

[40] Zhang C and Zhang D-J, Vector NLS solitons interacting with a boundary, *arXiv:1903.01791*, 2019.

B9. From integrable spatial discrete hierarchy to integrable nonlinear PDE hierarchy

Hai-Qiong Zhao [a] *and Zuo-Nong Zhu* [b]

[a] *Department of Applied Mathematics, Shanghai University of International Business and Economics, Shanghai 201620, P.R. China*

[b] *School of Mathematical Sciences, Shanghai Jiao Tong University, 800 Dongchuan Road, Shanghai, 200240, P. R. China*

Abstract

We review the connection between integrability of integrable spatial discrete hierarchy and integrable nonlinear PDE hierarchy. We take two main integrability results as examples. The first one is the integrability theory of the Kac-Moerbeke hierarchy including infinitely many commuting vector fields, the Lax pairs and the bi-Hamiltonian structure systematically yields the corresponding theory of the KdV hierarchy in the continuous limit. The second one is the continuous limit theory of integrable spatially discrete Hirota equation including the Lax pairs, the Darboux transformation and soliton solutions and the integrability of Hirota equation.

1 Introduction

It is well known that integrable systems include integrable nonlinear PDEs, integrable nonlinear ODEs, integrable nonlinear differential-difference equations (e.g., spatial discrete), and integrable difference equations. It is an interesting and important topic to study the connection among these different integrable systems. In this paper, we focus on the topic that from integrable spatial discrete hierarchy to integrable nonlinear PDE hierarchy.

The more deep relations between the Kac-Moerbeke hierarchy (discrete KdV hierarchy) and KdV hierarchy; generalized N-fields Volterra hierarchy and the KdV-type hierarchy associated to the simple Lie algebra $sp(N)$; N-fields Kupershmidt hierarchy and the Gelfand-Dickey hierarchy corresponding to the Lie algebra $sl(N+1)$; Toda hierarchy and KdV hierarchy have been developed [1]-[5], [7]-[9],[11]-[13].

• The Lax spectral equation associated Kac-Moerbeke hierarchy:

$$L^{KM}(a)\phi := (aE + E^{-1}a)\phi = \lambda_d\phi, \tag{1.1}$$

relates to the Lax spectral equation for KdV hierarchy

$$L^{KdV}(u)\phi := (\partial^2 + u)\phi = \lambda_c\phi. \tag{1.2}$$

• The Lax spectral equation associated generalized N-fields Volterra hierarchy:

$$L^{\epsilon}(a)\phi := \frac{1}{2}(\sum_{j=1}^{N}(a_j E^{(2j-1)\epsilon} + a_{j[(-2j+1)\epsilon]}E^{(-2j+1)\epsilon}))\phi = \lambda_d\phi, \tag{1.3}$$

relates to the Lax spectral equation for KdV-type hierarchy associated the Lie algebra $sp(N)$

$$L^{sp(N)}(u)\phi := (\partial^{2N} + \frac{1}{2}\sum_{l=1}^{N}(u_l\partial^{2N-2l} + \partial^{2N-2l}u_l)\phi = \lambda_c\phi. \tag{1.4}$$

• The Lax spectral equation associated N-fields Kupershmidt hierarchy:

$$L^{\epsilon}(\alpha)\phi := (E^{\epsilon} + \sum_{j=1}^{N}\alpha_{2j-1}E^{(-2j+1)\epsilon})\phi = \lambda_d\phi, \tag{1.5}$$

relates to the Lax spectral equation for Gelfand-Dickey hierarchy associated the Lie algebra $sl(N+1)$

$$L(u)\phi := (\partial^{N+1} + \sum_{k=1}^{N}u_k\partial^{N-k})\phi = \lambda_c\phi. \tag{1.6}$$

In this paper, we will give a review on the connection between Kac-Moerbeke hierarchy and KdV hierarchy due to the work of Morosi and Pizzocchero. The KdV theory including the infinitely many commuting vector fields, the Lax pairs and the bi-Hamiltonian structure is recovered systematically through the continuous limit of the corresponding integrable spatial discrete KdV hierarchy. We also give a connection between integrable spatial discrete Hirota equation and the Hirota equation [10]. The theory of Hirota equation including the Lax pair, the Darboux transformation and explicit solutions are recovered from the corresponding theory of the integrable spatial discrete Hirota equation in the continuous limit.

2 From integrable spatial discrete KdV hierarchy to KdV hierarchy

In this section, we review the theory of the discrete integrable KdV equation including infinitely many commuting vector fields, the Lax pairs and the bi-Hamiltonian structure yield the corresponding theory of the KdV equation in the continuous limit.

2.1. The construction of discrete KdV hierarchy

Consider the Lax spectral equation

$$L(a)\phi = (aE + E^{-1}a)\phi = \lambda\phi \tag{2.1}$$

$$\frac{d\phi}{dt_s} = A_s\phi, \qquad A_s = \left(L^{2s}\right)_{skew}, \tag{2.2}$$

with $a = a(n,t)$. Then the discrete KdV hierarchy can be described by

$$\frac{dL}{dt_s} = [A_s, L], \tag{2.3}$$

where the operators $A_s = \left(L^{2s}\right)_{skew}$ are constructed by the requirement that keeping the structures of L^{2s} and A_s are skew-symmetric operators. So, A_s is given by

$$A_s = \left(L^{2s}\right)_{skew} = \sum_{r=1}^{r_{max}} (g_r(a)E^r - E^{-r}g_r(a)). \tag{2.4}$$

For example, we have

$$\begin{aligned} &A_1 = \left(L^2\right)_{skew} = aa_1E^2 - E^{-2}aa_1, \\ &A_2 = \left(L^4\right)_{skew} = \alpha E^4 - E^{-4}\alpha + \beta E^2 - E^{-2}\beta, \\ &\alpha = aa_1a_2a_3, \quad \beta = aa_1(a_{-1}^2 + a^2 + a_1^2 + a_2^2). \end{aligned} \tag{2.5}$$

By the structures of the operators A_s and L, we have

$$[A_s, L] = X_s(a)E + E^{-1}X_s(a). \tag{2.6}$$

e.g.,

$$\begin{aligned} &X_1(a) = a(E^2 - 1)a_{-1}^2, \\ &X_2(a) = a(E^2 - 1)(a_{-1}^4 + a^2a_{-1}^2 + a_{-2}^2a_{-1}^2), \\ &X_3(a) = a(E^2 - 1)(a_{-1}^6 + a^4a_{-1}^2 + a_{-1}^2a_{-2}^4 + 2a^2a_{-1}^4 + 2a_{-1}^4a_{-2}^2 \\ &+a^2a_{-1}^2a_{-2}^2 + a_{-1}^2a^2a_1^2 + a_{-1}^2a_{-2}^2a_{-3}^2). \end{aligned}$$

In fact,$X_j(a)$ can be recurrently given by

$$X_{j+l}(a) = R^lX_j(a), j, l = 1, 2, 3, ... \tag{2.7}$$

where recurrent operator R is given by

$$R = a(E + 1)(a^2E - E^{-1}a^2)(E - 1)^{-1}a^{-1}.$$

So, the Lax equation hierarchy can be written as

$$\frac{dL}{dt_s} = \frac{da}{dt_s}E + E^{-1}\frac{da}{dt_s} = X_s(a)E + E^{-1}X_s(a). \tag{2.8}$$

Thus, the Kac-Moerbeke (KM) hierarchy (2.8)—-a discrete KdV hierarchy can be given by a recurrent form:

$$\frac{da}{dt_s} = R^{s-1}X_1(a). \tag{2.9}$$

The first several members of the hierarchy are written as

$$\begin{aligned} &\frac{da}{dt_1} = X_1(a) = a(E^2 - 1)a_{-1}^2, \\ &\frac{da}{dt_2} = X_2(a) = a(E^2 - 1)(a_{-1}^4 + a^2a_{-1}^2 + a_{-2}^2a_{-1}^2), \\ &\frac{da}{dt_3} = X_3(a) = a(E^2 - 1)(a_{-1}^6 + a^4a_{-1}^2 + a_{-1}^2a_{-2}^4 + 2a^2a_{-1}^4 + 2a_{-1}^4a_{-2}^2 \\ &+a^2a_{-1}^2a_{-2}^2 + a_{-1}^2a^2a_1^2 + a_{-1}^2a_{-2}^2a_{-3}^2), \end{aligned}$$

....

We remark here that the Lax spectral operator

$$L(a) = (aE + E^{-1}a)$$

already was considered on discrete inverse scattering problems by Case and Kac [2]. The original idea of the Lax spectral operator comes from the discretization of Lax operator of KdV equation. In fact, let ue consider the Schrödinger scattering problem:

$$\psi_{xx} + (\lambda_c - q)\psi = 0. \tag{2.10}$$

A discretization of the last equation is

$$\frac{\psi_{n+1} + \psi_{n-1} - 2\psi_n}{h^2} + (\lambda_c - q_n)\psi_n = 0. \tag{2.11}$$

Using the substitution $v_n = g_n\psi_n, g_n = exp(\frac{h^2 q_n}{2})$ yields

$$g_n^{-1}g_{n+1}^{-1}v_{n+1} + g_{n-1}^{-1}g_n^{-1}v_{n-1} - (exp(h^2 q_n) + exp(-h^2\lambda_c))exp(-h^2 q_n)v_n = 0. \tag{2.12}$$

Set $a_n = exp(-\frac{h^2}{2}(q_{n+1} + q_n))$, $\lambda = 1 + exp(-h^2\lambda_c)$. It follows that

$$a_n v_{n+1} + a_{n-1}v_{n-1} = \lambda v_n. \tag{2.13}$$

2.2 From Lax pair of discrete KdV hierarchy to Lax pair of KdV hierarchy
It is well known that KdV hierarchy is constructed by considering the following Lax spectral equation:

$$L^{KdV}\phi := (\partial_{xx} + u)\phi = \lambda\phi, \tag{2.14}$$

$$\frac{d\phi}{d\tau_s} = B_s^{KdV}\phi, \tag{2.15}$$

where

$$B_s^{KdV} = 4^{s-1}((L^{KdV})^{s-\frac{1}{2}})_+, \tag{2.16}$$

where $+$ denotes the projection on the nonnegative powers of ∂_x in the algebra of formal pseudo differential operators. So, KdV hierarchy is written as

$$\frac{dL^{KdV}}{d\tau_s} = [B_s^{KdV}, L^{KdV}].$$

The KdV hierarchy can also be given currently by

$$u_{\tau_s} = \Phi^s u_x, \qquad s = 1, 2, 3.... \tag{2.17}$$

where the current operator Φ is given by

$$\Phi = \partial^2 + 4u + 2u_x\partial^{-1}. \tag{2.18}$$

How to get B_s^{KdV}? Note that

$$\begin{aligned}(L^{KdV})^{\frac{1}{2}} &= \partial + \tfrac{1}{2}u\partial^{-1} - \tfrac{1}{4}u_x\partial^{-2} + \tfrac{1}{8}(u_{xx} - u^2)\partial^{-3} \\ &\quad +(\tfrac{3}{8}uu_x - \tfrac{1}{16}u_{xxx})\partial^{-4} + \ldots..\end{aligned} \tag{2.19}$$

One has

$$\begin{aligned}B_2^{KdV} &= 4(L^{KdV})^{\frac{3}{2}}|_+ = 4\partial_{xxx} + 6u\partial_x + 3u_x, \\ B_3^{KdV} &= 4^2(L^{KdV})^{\frac{5}{2}}|_+ = 16\partial^5 + 40u\partial^3 + 60u_x\partial^2 \\ &+(30u^2 + 50u_{xx})\partial + 30uu_x + 15u_{xxx}, \\ B_4^{KdV} &= 4^3(L^{KdV})^{\frac{7}{2}}|_+ = \ldots\ldots\end{aligned}$$

Let us show that how to obtain the Lax pair of KdV hierarchy from Lax pair of discrete KdV hierarchy. Consider the transformation

$$a(n\epsilon, t) \to 2 + \epsilon^2 u(x, \tau). \tag{2.20}$$

One can show that

$$\lim_{\epsilon\to 0} \frac{L^\epsilon(u) - 4}{2\epsilon^2} = L^{KdV}(u) = \partial_{xx} + u. \tag{2.21}$$

In fact, set $a_n = 2 + \epsilon^2 u_n$, then

$$\frac{(L-4)\phi_n}{2\epsilon^2} = \frac{\phi_{n+1} + \phi_{n-1} - 2\phi_n}{\epsilon^2} + \frac{u_n\phi_{n+1} + u_{n-1}\phi_{n-1}}{2}.$$

So

$$\lim_{\epsilon\to 0} \frac{(L-4)\phi_n}{2\epsilon^2} = (\partial_{xx} + u)\phi.$$

Let us see that how to derive the KdV equation and a fifth-order KdV equation. Note that

$$\begin{aligned}&X_1(a)|_{a=2+\epsilon^2 u} \to X_1^\epsilon(u) = 16\epsilon^3 u_x + \frac{8}{3}\epsilon^5(u_{xxx} + 6uu_x) \\ &+\frac{2}{15}\epsilon^7(u_{5x} + 20uu_{xxx} + 30u_xu_{xx} + 30u^2u_x) + O(\epsilon^9), \\ &X_2(a)|_{a=2+\epsilon^2 u} \to X_2^\epsilon(u) = 384\epsilon^3 u_x + 128\epsilon^5(u_{xxx} + 6uu_x) \\ &+\frac{96}{5}\epsilon^7(u_{5x} + \frac{40}{3}uu_{xxx} + \frac{70}{3}u_xu_{xx} + 30u^2u_x) + O(\epsilon^9), \\ &X_3(a)|_{a=2+\epsilon^2 u} \to X_3^\epsilon(u) = 7680\epsilon^3 u_x + 3840\epsilon^5(u_{xxx} + 6uu_x) \\ &+960\epsilon^7(u_{5x} + 12uu_{xxx} + 22u_xu_{xx} + 30u^2u_x) + O(\epsilon^9).\end{aligned}$$

Set $X_2(u) = X_2^\epsilon - 24X_1^\epsilon$, and $X_3(u) = X_3^\epsilon - 40X_2^\epsilon + 480X_1^\epsilon$, then

$$X_2(u) = 64\epsilon^5(u_{xxx} + 6uu_x) + O(\epsilon^7) \tag{2.22}$$

$$X_3(u) = 256\epsilon^7(u_{5x} + 10uu_{xxx} + 20u_xu_{xx} + 30u^2u_x) + O(\epsilon^9) \tag{2.23}$$

Thus, spatial discrete integrable equations

$$\frac{da}{dt} = X_2(a) - 24X_1(a), \tag{2.24}$$

$$\frac{da}{dt} = X_3(a) - 40X_2(a) + 480X_1(a), \tag{2.25}$$

lead to the KdV and a fifth-order KdV, respectively, under the transformations:

$$\tau = 64\epsilon^3 t, \qquad \tau = 256\epsilon^5 t. \tag{2.26}$$

In general case, how to get the vector fields $X_s(u), s = 2, 3, 4, ...$? By the formula

$$B_s^\epsilon(u) = A_s^\epsilon(u) + \sum_{j=1}^{s-1} c_{sj} A_j^\epsilon(u), \tag{2.27}$$

where c_{sj} is defined by

$$c_{sj} = c_{s-1,j-1} - 16c_{s-1,j}, \qquad j = 1, 2, ..., s-1$$

$$c_{ss} = 1, \quad c_{s0} = -4^s \begin{pmatrix} 2s \\ s \end{pmatrix} - \sum_{i=1}^{s-1} 4^i \begin{pmatrix} 2i \\ i \end{pmatrix} c_{si}, s = 1, 2, ...$$

So, we have

$$B_2^\epsilon(u) = A_2^\epsilon(u) - 24A_1^\epsilon(u),$$
$$B_3^\epsilon(u) = A_3^\epsilon(u) + 480A_1^\epsilon(u) - 40A_2^\epsilon(u),$$
$$B_4^\epsilon(u) = A_4^\epsilon(u) - 8960A_1^\epsilon(u) + 1120A_2^\epsilon(u) - 56A_3^\epsilon(u),$$
$$.......$$

and

$$X_s(u) = X_s^\epsilon(u) + \sum_{j=1}^{s-1} c_{sj} X_j^\epsilon(u). \tag{2.28}$$

Further, we obtain

$$X_s(u) = 4^{s+1}\epsilon^{2s+1}\Phi^{s-1}u_x. \tag{2.29}$$

In general case, we can prove

$$\lim_{\epsilon\to 0} \frac{B_s^\epsilon(u)}{4^{s+1}\epsilon^{2s-1}} = B_s^{KdV}(u). \tag{2.30}$$

But, the proof of eigenfunction for the evolution of time is very difficult (see Ref.[7] for details). Let us give two examples of the proof of eigenfunction with the evolution of time.

Example 1 Prove

$$\lim_{\epsilon\to 0} \frac{B_2^\epsilon(u)}{4^3\epsilon^3} = B_2^{KdV}(u) = 4\partial_{xxx} + 6u\partial_x + 3u_x. \tag{2.31}$$

Note that

$$B_2^\epsilon(u) = (A_2 - 24A_1)|_{a=2+\epsilon^2 u(x,\tau)}.$$

Using Taylor's formula

$$f(x + r\epsilon) = \sum_{j=0}^{\infty} \frac{1}{j!}(r\epsilon)^j \frac{d^j f}{dx^j}(x), \tag{2.32}$$

and after a careful calculation, we have

$$A_1|_{a=2+\epsilon^2 u(x,\tau)}\phi = 16\epsilon\phi_x + \epsilon^3(\frac{32}{3}\phi_{xxx} + 16u\phi_x + 8u_x\phi) + O(\epsilon^5), \tag{2.33}$$

$$A_2|_{a=2+\epsilon^2 u(x,\tau)}\phi = 384\epsilon\phi_x + \epsilon^3(512\phi_{xxx} + 768u\phi_x + 384u_x\phi) + O(\epsilon^5). \tag{2.34}$$

We thus have

$$\lim_{\epsilon\to 0} \frac{B_2^\epsilon(u)}{4^3\epsilon^3} = B_2^{KdV} = 4\partial_{xxx} + 6u\partial_x + 3u_x. \tag{2.35}$$

Example 2 Prove

$$\lim_{\epsilon\to 0} \frac{B_3^\epsilon(u)}{4^4\epsilon^5} = B_3^{KdV}(u). \tag{2.36}$$

Note that

$$\begin{aligned} A_1|_{a=2+\epsilon^2 u(x,\tau)}\phi &= 16\epsilon\phi_x + \epsilon^3(\frac{32}{3}\phi_{xxx} + 16u\phi_x + 8u_x\phi) \\ &+\frac{2\epsilon^5}{15}(16\phi_{5x} + 80u\phi_{xxx} + 25u_{xxx}\phi \\ &+120u_x\phi_{xx} + 90u_{xx}\phi_x + 30u^2\phi_x + 30uu_x\phi) + O(\epsilon^7), \end{aligned}$$

$$\begin{aligned} A_2|_{a=2+\epsilon^2 u(x,\tau)}\phi &= 384\epsilon\phi_x + \epsilon^3(512\phi_{xxx} + 768u\phi_x + 384u_x\phi) \\ &+\frac{32\epsilon^5}{5}(48\phi_{5x} + 160u\phi_{xxx} + 55u_{xxx}\phi \\ &+240u_x\phi_{xx} + 190u_{xx}\phi_x + 90u^2\phi_x + 90uu_x\phi) + O(\epsilon^7), \end{aligned}$$

$$\begin{aligned} A_3|_{a=2+\epsilon^2 u(x,\tau)}\phi &= 7680\epsilon\phi_x + \epsilon^3 3840(4\phi_{xxx} + 6u\phi_x + 3u_x\phi) \\ &+960\epsilon^5(16\phi_{5x} + 48u\phi_{xxx} + 17u_{xxx}\phi \\ &+72u_x\phi_{xx} + 58u_{xx}\phi_x + 30u^2\phi_x + 30uu_x\phi) + O(\epsilon^7). \end{aligned}$$

So, one has

$$B_3^\epsilon(u) = (A_3 - 40A_2 + 480A_1)|_{a=2+\epsilon^2 u(x,\tau)} = 256\epsilon^5 B_3^{KdV}(u) + O(\epsilon^7). \tag{2.37}$$

This leads to

$$\lim_{\epsilon\to 0} \frac{B_3^\epsilon(u)}{4^4\epsilon^5} = B_3^{KdV}(u). \tag{2.38}$$

Similarly, one can show

$$\lim_{\epsilon\to 0}\frac{B_4^{\epsilon}(u)}{4^5\epsilon^7}=B_4^{KdV}(u), \tag{2.39}$$

where

$$B_4^{\epsilon}(u)=(A_4-8960A_1+1120A_2-56A_3)|_{a=2+\epsilon^2u(x,\tau)}. \tag{2.40}$$

Thus, the Lax pairs of discrete KdV hierarchy yields the ones of KdV hierarchy, and discrete KdV hierarchy yields the KdV hierarchy.

2.3. From the bi-Hamiltonian structure of discrete KdV hierarchy to the bi-Hamiltonian structure of KdV hierarchy

The KM hierarchy possesses the bi-Hamiltonian structure:

$$\frac{da}{dt_s}=R^{s-1}X_1(a)=P_2\frac{\delta}{\delta a_n}f_{s-1}=P_1\frac{\delta}{\delta a_n}f_s, \tag{2.41}$$

where $\frac{\delta}{\delta u_i}$ stands for the discrete variational derivative defined by

$$\frac{\delta}{\delta u_i}f=\sum_{k\in Z}E^{-k}\frac{\partial}{\partial u_i^{(k)}}f,\qquad u_i^{(k)}=E^ku_i. \tag{2.42}$$

P_1 and P_2 are two Poisson tensor operators,

$$P_1=\frac{a}{2}\left(E-E^{-1}\right)a, \tag{2.43}$$

$$P_2=RP_1=\frac{a}{2}(E+1)(a^2E-E^{-1}a^2)(1+E^{-1})a, \tag{2.44}$$

and f_s, $s=0,1,2,...$ are the Hamiltonians of the KM hierarchy given by

$$f_0=\sum_{n=-\infty}^{+\infty}\log a_n, \tag{2.45a}$$

and

$$f_s=\frac{1}{2s}Trace\left(L^{KM}(a)\right)^{2s},\qquad s=1,2,3... \tag{2.45b}$$

where the $Trace$ is defined by

$$f_s=\frac{1}{2s}\sum_{n=-\infty}^{+\infty}g_{n,s},\qquad L^{2s}=\sum_{j=1}^{s}\alpha_jE^{2j}+E^{-2j}\alpha_j+g_{n,s} \tag{2.46}$$

e.g.,

$$f_1=\sum_{n=-\infty}^{+\infty}a_n^2,\qquad f_2=\sum_{n=-\infty}^{+\infty}(a_n^2a_{n+1}^2+\frac{1}{2}a_n^4).$$

It is well known that KdV hierarchy possesses the bi-Hamiltonian structure:

$$\frac{d}{d\tau_s}u = S_2\frac{\delta}{\delta u}h_s^{KdV} = S_1\frac{\delta}{\delta u}h_{s+1}^{KdV}, \quad s = 1, 2, 3, ... \tag{2.47}$$

where $S_1 = \partial_x, S_2 = \Phi S_1$, $\frac{\delta}{\delta u}$ stands for the variational derivative and Hamiltonians h_s can be characterized by

$$h_s^{KdV} = \frac{4^s}{2s+1}Trace(L^{KdV})^{s+\frac{1}{2}}, \tag{2.48}$$

where $Trace$ is defined by

$$Trace(G) = \int_{-\infty}^{+\infty} g_{-1}dx, \qquad G = \sum_{j=-\infty}^{m} g_j\partial^j, \tag{2.49}$$

e.g.,

$$h_0^{KdV} = \frac{1}{2}\int_{-\infty}^{+\infty} udx, \qquad h_1^{KdV} = \frac{1}{2}\int_{-\infty}^{+\infty} u^2dx,$$

$$h_2^{KdV} = \int_{-\infty}^{+\infty} (u^3 - \frac{1}{2}u_x^2)dx.$$

Considering the expansion of two Poisson tensors, which can be written as,

$$\begin{aligned} P_1 &= 4\epsilon\partial_x + \frac{2}{3}\epsilon^3(\partial_{xxx} + 6u\partial_x + 3u_x) + o(\epsilon^3), \\ P_2 &= 64\epsilon\partial_x + \frac{80}{3}\epsilon^3(\partial_{xxx} + \frac{24}{5}u\partial_x + \frac{12}{5}u_x) + o(\epsilon^3). \end{aligned} \tag{2.50}$$

Further, one has

$$\begin{aligned} \mathfrak{P}_1 &:= P_1 = 4\epsilon S_1 + o(\epsilon), \\ \mathfrak{P}_2 &:= P_2 - 16P_1 = 16\epsilon^3 S_2 + o(\epsilon^3). \end{aligned} \tag{2.51}$$

In order to give an appropriate connection between the bi-Hamiltonian structures for discrete KdV hierarchy and KdV hierarchy, we should write re-combination Kac-Moerbeke vector fields $Z_s(a)$ as

$$Z_s(a) := A_s(a) + \sum_{j=1}^{s-1} c_{sj}A_j(a) = \mathfrak{P}_1\frac{\delta}{\delta a_n}h_s = \mathfrak{P}_2\frac{\delta}{\delta a_n}h_{s-1} \tag{2.52}$$

Thus, one can check that

$$\begin{aligned} \lim_{\epsilon\longrightarrow 0}\frac{\mathfrak{P}_1\frac{\delta}{\delta a_n}h_s}{4^{s+1}\epsilon^{2s+1}} &= S_1\frac{\delta}{\delta u}h_s^{KdV}, \\ \lim_{\epsilon\longrightarrow 0}\frac{\mathfrak{P}_2\frac{\delta}{\delta a_n}h_{s-1}}{4^{s+1}\epsilon^{2s+1}} &= S_2\frac{\delta}{\delta u}h_{s-1}^{KdV}, \end{aligned} \tag{2.53}$$

e.g., note that

$$h_1 = f_1 - 8f_0 - c_1,$$
$$h_2 = f_2 - 24f_1 + 96f_0 - c_2,$$

we have

$$\frac{\delta}{\delta a_n} h_1 = 2a_n - \frac{8}{a_n} = 4u\epsilon^2 + o(\epsilon^2),$$
$$\frac{\delta}{\delta a_n} h_2 = 2a_n^3 + 2a_n a_{n+1}^2 + 2a_{n-1}^2 a_n - 48a_n + \frac{96}{a_n}$$
$$= 16(3u^2 + u_{xx})\epsilon^4 + o(\epsilon^4).$$

Thus, we obtain

$$\lim_{\epsilon \longrightarrow 0} \frac{\mathfrak{P}_1 \frac{\delta}{\delta a_n} h_2}{4^3 \epsilon^5} = S_1(3u^2 + u_{xx}) = S_1 \frac{\delta}{\delta u} h_2^{KdV},$$
$$\lim_{\epsilon \longrightarrow 0} \frac{\mathfrak{P}_2 \frac{\delta}{\delta a_n} h_1}{4^3 \epsilon^5} = S_2 u = S_2 \frac{\delta}{\delta u} h_1^{KdV}. \tag{2.54}$$

This shows that the bi-Hamiltonian representation of re-combination KM hierarchy yields the ones of the KdV hierarchy.

3 From integrability of spatial discrete Hirota equation to integrability of Hirota equation

In this section, we will focus on the topic that from integrability of spatial discrete Hirota equation to integrability of Hirota equation (see [10] and there references). The Hirota equation,

$$iv_\tau = \gamma(v_{xx} + 2|v|^2 v) + i\alpha(v_{xxx} + 6|v|^2 v_x), \tag{3.1}$$

was introduced by Hirota and studied on the integrability, soliton solution, rogue wave, gauge-equivalence, and physical applications in a number of papers. Spatial discrete Hirota equation has received a great deal of attention due to its wide range of physical applications. An integrable spatial discrete Hirota equation is

$$i\frac{d\psi_n}{dt} = \gamma[\psi_{n+1} - 2\psi_n + \psi_{n-1} + |\psi_n|^2(\psi_{n+1} + \psi_{n-1})]$$
$$+\alpha i[(1 + |\psi_n|^2)(\psi_{n+1} - \psi_{n-1})]. \tag{3.2}$$

Its soliton solutions and rogue waves were obtained. However, the continuum limit of these solutions, whether solitons or rogue waves, does not yield the corresponding solutions to the Hirota equation.

Aiming to construct such a spatial discrete Hirota equation that from the integrability of the spatial discrete Hirota equation one can obtain the integrability of the Hirota equation, we introduce and investigate the following integrable spatially discrete Hirota equation [10]:

$$\begin{aligned}\tfrac{d}{dt}u_n = \ & \alpha(1+|u_n|^2)[u_{n+2}-u_{n-2}+2u_{n-1}-2u_{n+1}+u_n^*(u_{n+1}^2-u_{n-1}^2)\\ & -|u_{n-1}|^2u_{n-2}+|u_{n+1}|^2u_{n+2}+u_n(u_{n-1}^*u_{n+1}-u_{n+1}^*u_{n-1})]\\ & -\beta i(1+|u_n|^2)(u_{n+1}+u_{n-1})+2\beta i u_n.\end{aligned} \tag{3.3}$$

Under the transformation

$$u_n=\delta v(n\delta,2\delta^3 t)\triangleq\delta v(x,\tau),\qquad \beta=2\gamma\delta, \tag{3.4}$$

the spatially discrete Hirota equation (3.3) yields the Hirota equation (3.1). We have constructed the Darboux transformation and soliton solutions to equation (3.3). We have shown that the continuum limit for the spatially discrete Hirota equation including the Lax pair, the Darboux transformation and soliton solutions yields the corresponding results of the Hirota equation as $\delta\to 0$. Here we present our main results (see Ref. [10] for details).

3.1 Darboux transformation and soliton solution of the spatially discrete Hirota equation

We first give the Lax pair of spatially discrete Hirota equation (3.3), and then we construct its Darboux transformation. Furthermore, by using the Darboux transformation, we obtain multi-soliton solutions to spatially discrete Hirota equation (3.3).

Lax pair. The spatially discrete Hirota equation (3.3) has the following Lax pair:

$$\psi_{n+1}=L_n\psi_n,\qquad \frac{d\psi_n}{dt}=M_n\psi_n, \tag{3.5}$$

where the matrices L_n and M_n have the forms:

$$\begin{aligned}L_n &= \begin{pmatrix}\lambda & u_n\\ -u_n^* & \lambda^{-1}\end{pmatrix},\\ M_n &= \alpha\begin{pmatrix}A_n(\lambda,\lambda^{-1},u_n) & B_n(\lambda,\lambda^{-1},u_n)\\ -B_n(\lambda^{-1},\lambda,u_n^*) & A_n(\lambda^{-1},\lambda,u_n^*)\end{pmatrix}\\ &+\beta i\begin{pmatrix}C_n(\lambda,\lambda^{-1},u_n) & D_n(\lambda,\lambda^{-1},u_n)\\ D_n(\lambda^{-1},\lambda,u_n^*) & -C_n(\lambda^{-1},\lambda,u_n^*)\end{pmatrix},\end{aligned}$$

where

$$\begin{aligned}A_n(\lambda,\lambda^{-1},u_n) &= \frac{\lambda^4-\lambda^{-4}}{2}-\lambda^2+\lambda^{-2}+\lambda^2u_nu_{n-1}^*-\lambda^{-2}u_{n-1}u_n^*\\ &-2u_nu_{n-1}^*+(u_nu_{n-1}^*)^2+(1+|u_{n-1}|^2)u_nu_{n-2}^*+(1+|u_n|^2)u_{n+1}u_{n-1}^*,\\ B_n(\lambda,\lambda^{-1},u_n) &= \lambda^3u_n+\lambda^{-3}u_{n-1}+\lambda[(1+|u_n|^2)u_{n+1}+u_n^2u_{n-1}^*\\ &-2u_n]+\lambda^{-1}[(1+|u_{n-1}|^2)u_{n-2}+u_{n-1}^2u_n^*-2u_{n-1}],\\ C_n(\lambda,\lambda^{-1},u_n) &= -u_nu_{n-1}^*-\frac{(\lambda-\lambda^{-1})^2}{2},\\ D_n(\lambda,\lambda^{-1},u_n) &= -u_n\lambda+u_{n-1}\lambda^{-1}.\end{aligned}$$

Darboux transformation. N-times Darboux transformation can be constructed by

$$\psi_n[N] = T_n[N]\psi_n, \tag{3.6}$$

with

$$T_n[N] = \begin{pmatrix} \lambda^N + \sum\limits_{k=1}^{N} f_n^{(N-2k)}\lambda^{N-2k} & \sum\limits_{k=1}^{N} g_n^{(N-2k+1)}\lambda^{N-2k+1} \\ (-1)^{N+1}\sum\limits_{k=1}^{N} g_n^{*(N-2k+1)}\lambda^{-N+2k-1} & (-1)^N(\lambda^{-N} + \sum\limits_{k=1}^{N} f_n^{*(N-2k)}\lambda^{-N+2k}) \end{pmatrix}, \tag{3.7}$$

where $f_n^{(k)}$, $k = -N, -N+2, ..., N-2$ and $g_n^{(k)}$, $k = -N+1, -N+3, ..., N-1$ are determined as follows

$$\begin{array}{c} (\lambda_j^N + \sum\limits_{k=1}^{N} f_n^{(N-2k)}\lambda_j^{N-2k})\psi_{n,1}^{(j)} + (\sum\limits_{k=1}^{N} g_n^{(N-2k+1)}\lambda_j^{N-2k+1})\psi_{n,2}^{(j)} = 0 \\ ((\lambda_j^*)^{-N} + \sum\limits_{k=1}^{N} f_n^{(N-2k)}(\lambda_j^*)^{-N+2k})\psi_{n,2}^{*(j)} - (\sum\limits_{k=1}^{N} g_n^{(N-2k+1)}(\lambda_j^*)^{-N+2k-1})\psi_{n,1}^{*(j)} = 0 \\ j = 1, 2, ..., N \end{array}, \tag{3.8}$$

and

$$u_n[N] = -u_n f_{n+1}^{(-N)} - g_{n+1}^{(-N+1)}, \tag{3.9}$$

where

$$f_n^{(-N)} = -\frac{\Delta_1}{\Delta}, \qquad g_n^{(-N+1)} = -\frac{\Delta_2}{\Delta}, \tag{3.10}$$

$$\Delta = \begin{vmatrix} \lambda_1^{-N}\psi_{n,1}^{(1)} & \lambda_1^{-N+1}\psi_{n,2}^{(1)} & \cdots & \lambda_1^{N-2}\psi_{n,1}^{(1)} & \lambda_1^{N-1}\psi_{n,2}^{(1)} \\ \lambda_2^{-N}\psi_{n,1}^{(2)} & \lambda_2^{-N+1}\psi_{n,2}^{(2)} & \cdots & \lambda_2^{N-2}\psi_{n,1}^{(2)} & \lambda_2^{N-1}\psi_{n,2}^{(2)} \\ \vdots & \vdots & \vdots & \vdots & \vdots \\ \lambda_N^{-N}\psi_{n,1}^{(N)} & \lambda_N^{-N+1}\psi_{n,2}^{(N)} & \cdots & \lambda_N^{N-2}\psi_{n,1}^{(N)} & \lambda_N^{N-1}\psi_{n,2}^{(N)} \\ (\lambda_1^*)^N\psi_{n,2}^{*(1)} & -(\lambda_1^*)^{N-1}\psi_{n,1}^{*(1)} & \cdots & (\lambda_1^*)^{-N+2}\psi_{n,2}^{*(1)} & -(\lambda_1^*)^{-N+1}\psi_{n,1}^{*(1)} \\ (\lambda_2^*)^N\psi_{n,2}^{*(2)} & -(\lambda_2^*)^{N-1}\psi_{n,1}^{*(2)} & \cdots & (\lambda_2^*)^{-N+2}\psi_{n,2}^{*(2)} & -(\lambda_2^*)^{-N+1}\psi_{n,1}^{*(2)} \\ \vdots & \vdots & \vdots & \vdots & \vdots \\ (\lambda_N^*)^N\psi_{n,2}^{*(N)} & -(\lambda_N^*)^{N-1}\psi_{n,1}^{*(N)} & \cdots & (\lambda_N^*)^{-N+2}\psi_{n,2}^{*(N)} & -(\lambda_N^*)^{-N+1}\psi_{n,1}^{*(N)} \end{vmatrix}$$

Δ_1 and Δ_2 are described respectively by Δ, but the first column and the second column in the Δ are changed respectively to
$(\lambda_1^N\psi_{n,1}^{(1)}, \lambda_2^N\psi_{n,1}^{(2)}, ..., \lambda_N^N\psi_{n,1}^{(N)}, (\lambda_1^*)^{-N}\psi_{n,2}^{*(1)}, (\lambda_2^*)^{-N}\psi_{n,2}^{*(2)}, ..., (\lambda_N^*)^{-N}\psi_{n,2}^{*(N)})^T$.
Example 1 Taking $N = 1$, we have

$$\psi_n[1] = T_n^{(1)}(\lambda)\psi_n, \tag{3.11}$$

where the Darboux transformation matrix $T_n^{(1)}(\lambda)$ is written as

$$T_n^{(1)}(\lambda) = \begin{pmatrix} \lambda + \lambda^{-1}T_{n,1}^{(1)} & T_{n,2}^{(1)} \\ T_{n,2}^{(1)*} & -\lambda^{-1} - \lambda T_{n,1}^{(1)*} \end{pmatrix}, \tag{3.12}$$

where

$$T_{n,1}^{(1)} = \frac{-(\lambda_1|\psi_{n,1}^{(1)}|^2 + (\lambda_1^*)^{-1}|\psi_{n,2}^{(1)}|^2)}{\Delta_1},$$

$$T_{n,2}^{(1)} = \frac{(|\lambda_1|^{-2} - |\lambda_1|^2)\psi_{n,1}^{(1)}\psi_{n,2}^{(1)*}}{\Delta_1}, \qquad \Delta_1 = \lambda_1^{-1}|\psi_{n,1}^{(1)}|^2 + \lambda_1^*|\psi_{n,2}^{(1)}|^2.$$

$$u_n[1] = -u_n T_{n+1,1}^{(1)} - T_{n+1,2}^{(1)}. \tag{3.13}$$

The new solution $u_n[1]$ can also be written as

$$u_n[1] = \lambda_1^2 u_n \left[\frac{|\lambda_1|^{-2}|\psi_{n,1}^{(1)}|^2 + |\psi_{n,2}^{(1)}|^2}{|\lambda_1|^2|\psi_{n,1}^{(1)}|^2 + |\psi_{n,2}^{(1)}|^2}\right] + \left(\frac{\lambda_1^3 - \lambda_1(\lambda_1^*)^{-2}}{|\lambda_1|^2|\psi_{n,1}^{(1)}|^2 + |\psi_{n,2}^{(1)}|^2}\right)\psi_{n,1}^{(1)}\psi_{n,2}^{(1)*}. \tag{3.14}$$

Soliton solutions. Taking a seed solution $u_n = 0$, and solving the corresponding spectral equation gives

$$\psi_{n,1}^{(j)} = c_1^{(j)} e^{Z(\lambda_j)}, \qquad \psi_{n,2}^{(j)} = c_2^{(j)} e^{-Z(\lambda_j)}, \tag{3.15}$$

where $Z(\lambda_j) = n \ln \lambda_j + \chi(\lambda_j)t$, $\lambda_j = e^{a_j + b_j i}$, $\chi(\lambda_j) = \alpha\left(\frac{\lambda_j^4 - \lambda_j^{-4}}{2} - \lambda_j^2 + \lambda_j^{-2}\right) - \frac{\beta i}{2}(\lambda_j - \lambda_j^{-1})^2$, and $c_j, j = 1, 2$ are arbitrary complex parameters.
Example 2 Using Darboux transformation, we get

$$u_n[1] = \frac{c_1^{(1)} c_2^{(1)*}\lambda_1(\lambda_1^*)^{-2}(|\lambda_1|^4 - 1)e^{2iZ_I(\lambda_1)}}{|c_2^{(1)}|^2 e^{-2Z_R(\lambda_1)} + |c_1^{(1)}\lambda_1|^2 e^{2Z_R(\lambda_1)}}, \tag{3.16}$$

where $Z_R(\lambda_1)$ and $Z_I(\lambda_1)$ are the real and imaginary parts of $Z(\lambda_1)$ respectively,

$$\begin{aligned} Z_R(\lambda_1) &= a_1 n + \{\alpha[\cos(4b_1)\sinh(4a_1) - 2\cos(2b_1)\sinh(2a_1)] \\ &\quad + \beta\sin(2b_1)\sinh(2a_1)\}t, \\ Z_I(\lambda_1) &= b_1 n + \{\alpha[\sin(4b_1)cosh(4a_1) - 2\sin(2b_1)\cosh(2a_1)] \\ &\quad - \beta\cos(2b_1)\cosh(2a_1) + \beta\}t. \end{aligned}$$

We thus obtain a one-soliton solution:

$$u_n[1] = \frac{c_1^{(1)} c_2^{(1)*}\lambda_1(|\lambda_1|^4 - 1)e^{2iZ_I(\lambda_1)}}{2\bar{\lambda}_1^2|c_1^{(1)}c_2^{(1)}\lambda_1|} sech(2Z_R(\lambda_1) + \epsilon), \quad e^\epsilon = |\frac{c_1^{(1)}\lambda_1}{c_2^{(1)}}|. \tag{3.17}$$

Example 3 Using 2-times Darboux transformation, we obtain a two-soliton solution

$$u_n[2] = g_{n+1}^{(-1)}, \tag{3.18}$$

where

$$g_n^{(-1)} = \frac{\Delta_2}{\Delta}, \tag{3.19}$$

with

$$\Delta = \begin{vmatrix} \lambda_1^{-2}\psi_{n,1}^{(1)} & \lambda_1^{-1}\psi_{n,2}^{(1)} & \psi_{n,1}^{(1)} & \lambda_1\psi_{n,2}^{(1)} \\ \lambda_2^{-2}\psi_{n,1}^{(2)} & \lambda_2^{-1}\psi_{n,2}^{(2)} & \psi_{n,1}^{(2)} & \lambda_2\psi_{n,2}^{(2)} \\ (\lambda_1^*)^2\psi_{n,2}^{*(1)} & -(\lambda_1^*)\psi_{n,1}^{*(1)} & \psi_{n,2}^{*(1)} & -(\lambda_1^*)^{-1}\psi_{n,1}^{*(1)} \\ (\lambda_2^*)^2\psi_{n,2}^{*(2)} & -(\lambda_2^*)\psi_{n,1}^{*(2)} & \psi_{n,2}^{*(2)} & -(\lambda_2^*)^{-1}\psi_{n,1}^{*(2)} \end{vmatrix},$$

and

$$\Delta_2 = \begin{vmatrix} \lambda_1^{-2}\psi_{n,1}^{(1)} & \lambda_1^{2}\psi_{n,1}^{(1)} & \psi_{n,1}^{(1)} & \lambda_1\psi_{n,2}^{(1)} \\ \lambda_2^{-2}\psi_{n,1}^{(2)} & \lambda_2^{2}\psi_{n,1}^{(2)} & \psi_{n,1}^{(2)} & \lambda_2\psi_{n,2}^{(2)} \\ (\lambda_1^*)^2\psi_{n,2}^{*(1)} & (\lambda_1^*)^{-2}\psi_{n,2}^{*(1)} & \psi_{n,2}^{*(1)} & -(\lambda_1^*)^{-1}\psi_{n,1}^{*(1)} \\ (\lambda_2^*)^2\psi_{n,2}^{*(2)} & (\lambda_2^*)^{-2}\psi_{n,2}^{*(2)} & \psi_{n,2}^{*(2)} & -(\lambda_2^*)^{-1}\psi_{n,1}^{*(2)} \end{vmatrix}.$$

3.2 From integrability of the spatially discrete Hirota equation to the integrability of Hirota equation

We will show that continuum limit for the spatially discrete integrable Hirota equation (3.3) including the Lax pair, the Darboux transformations, and soliton solutions can yield the corresponding results of the Hirota equation. We set

$$\psi_n = \delta\phi + o(\delta), \qquad \lambda = e^{z\delta} \tag{3.20}$$

where ϕ and z are eigenfunction and eigenvalue of spectral problem of Hirota equation, respectively.

Continuum limit for the Lax pair. Under the transformations (3.4) and (3.20) , we see that the connection between the discrete linear eigenvalue problem for discrete Hirota equation and linear eigenvalue problem for Hirota equation can be given by

$$\begin{aligned} &E\psi_n - L_n\psi_n = (\phi_x - U\phi)\delta^2 + o(\delta^2), \\ &\frac{d\psi_n}{dt} - M_n\psi_n = 2(\phi_\tau - V\phi)\delta^4 + o(\delta^4), \end{aligned} \tag{3.21a}$$

where

$$U = \begin{pmatrix} z & v \\ -v^* & -z \end{pmatrix}, \qquad V = \alpha\begin{pmatrix} V_1^{(1)} & V_2^{(1)} \\ V_3^{(1)} & -V_1^{(1)} \end{pmatrix} + i\gamma\begin{pmatrix} V_1^{(2)} & V_2^{(2)} \\ V_3^{(2)} & -V_1^{(2)} \end{pmatrix} \tag{3.21b}$$

with

$$\begin{aligned} V_1^{(1)} &= 4z^3 + 2z|v|^2 - vv_x^* + v^*v_x, & V_1^{(2)} &= -|v|^2 - 2z^2, \\ V_2^{(1)} &= 4z^2v + 2zv_x + v_{xx} + 2|v|^2v, & V_2^{(2)} &= -2zv - v_x, \\ V_3^{(1)} &= -4z^2v^* + 2zv_x^* - v_{xx}^* - 2|v|^2v^*, & V_3^{(2)} &= -v_x^* + 2zv^*. \end{aligned} \tag{3.21c}$$

One can check that the zero-curvature equation $U_\tau - V_x + UV - VU = 0$ gives rise to the Hirota equation. This means that the Lax pair of spatially discrete Hirota equation yields that of the Hirota equation in the continuum limit. Let us assume $\phi^{(j)} = (\phi_1^{(j)}, \phi_2^{(j)})^T$ satisfy

$$\phi_x^{(j)} = U(z_j)\phi^{(j)}, \qquad \phi_\tau^{(j)} = V(z_j)\phi^{(j)}, \qquad j = 1, 2, 3..., N. \tag{3.22}$$

To assure consistency with (3.20), we define the relation between $\psi_{n,i}^{(j)}(t)$ and $\phi_i^{(j)}(x,\tau)$ as follows

$$\psi_{n,i}^{(j)}(t) = \delta\phi_i^{(j)}(x,\tau) + o(\delta), \qquad i = 1, 2. \tag{3.23}$$

Continuum limit for the Darboux transformation. Notice that (3.12) have the expansions

$$T_{n,2}^{(1)} = 2D_2^{(1)}\delta + o(\delta), \qquad T_{n,1}^{(1)} = -1 + 2D_1^{(1)}\delta + o(\delta), \tag{3.24}$$

where

$$D_1^{(1)} = \frac{z_1^*|\phi_2^{(1)}|^2 - z_j|\phi_1^{(1)}|^2}{|\phi_1^{(1)}|^2 + |\phi_2^{(1)}|^2}, \qquad D_2^{(1)} = \frac{-2Re(z_1)\phi_1^{(1)*}\phi_2^{(1)}}{|\phi_1^{(1)}|^2 + |\phi_2^{(1)}|^2},$$

then the expansion of the Darboux matrix is

$$T_n^{(1)}(\lambda) = 2\delta D^{(1)}(z) + o(\delta), \tag{3.25}$$

where

$$D^{(1)}(z) = \begin{pmatrix} z + D_1^{(1)} & D_2^{(1)} \\ D_2^{(1)*} & z - D_1^{(1)*} \end{pmatrix}.$$

Thus, we have

$$\lim_{\delta\to 0}\frac{\psi_n[1]}{2\delta^2} = \lim_{\delta\to 0}\frac{T_n^{(1)}(\lambda)\psi_n(\lambda)}{2\delta^2} = \phi[1], \tag{3.26a}$$

$$\lim_{\delta\to 0}\frac{u_n[1]}{\delta} = \lim_{\delta\to 0}\frac{-u_nT_{n+1,1}^{(1)} - T_{n+1,2}^{(1)}}{\delta} = v[1], \tag{3.26b}$$

where

$$\phi[1] = D^{(1)}(z)\phi, \qquad v[1] = v - 2D_2^{(1)}. \tag{3.26c}$$

One can examine that (3.26c) satisfies the linear equation

$$\phi[1]_x = U[1]\phi[1], \qquad \phi[1]_\tau = V[1]\phi[1]. \tag{3.27}$$

where $U[1]$ and $V[1]$ are defined by U and V, respectively, but the potential v is replaced by $v[1]$.

Further, considering the N-times Darboux transformation, we have

$$f_{n+1}^{(-N)} = -1 + O(\delta), \qquad g_{n+1}^{(-N+1)} = -2\delta\frac{\nabla_2}{\nabla} + o(\delta), \tag{3.28}$$

where

$$\nabla = \begin{vmatrix} \phi_1^{(1)} & \phi_2^{(1)} & z_1\phi_1^{(1)} & z_1\phi_2^{(1)} & \cdots & z_1^{N-1}\phi_1^{(1)} & z_1^{N-1}\phi_2^{(1)} \\ \phi_1^{(2)} & \phi_2^{(2)} & z_2\phi_1^{(2)} & z_2\phi_2^{(2)} & \cdots & z_2^{N-1}\phi_1^{(2)} & z_2^{N-1}\phi_2^{(2)} \\ \vdots & \vdots & \vdots & \vdots & \vdots & \vdots & \vdots \\ \phi_1^{(N)} & \phi_2^{(N)} & z_N\phi_1^{(N)} & z_N\phi_2^{(N)} & \cdots & z_N^{N-1}\phi_1^{(N)} & z_N^{N-1}\phi_2^{(N)} \\ \phi_2^{*(1)} & -\phi_1^{*(1)} & z_1^*\phi_2^{*(1)} & -z_1^*\phi_1^{*(1)} & \cdots & (z_1^*)^{N-1}\phi_2^{*(1)} & -(z_1^*)^{N-1}\phi_1^{*(1)} \\ \phi_2^{*(2)} & -\phi_1^{*(2)} & z_2^*\phi_2^{*(2)} & -z_2^*\phi_1^{*(2)} & \cdots & (z_2^*)^{N-1}\phi_2^{*(2)} & -(z_2^*)^{N-1}\phi_1^{*(2)} \\ \vdots & \vdots & \vdots & \vdots & \vdots & \vdots & \vdots \\ \phi_2^{*(N)} & -\phi_1^{*(N)} & z_N^*\phi_2^{*(N)} & -z_N^*\phi_1^{*(N)} & \cdots & (z_N^*)^{N-1}\phi_2^{*(N)} & -(z_N^*)^{N-1}\phi_1^{*(N)} \end{vmatrix}$$

and ∇_2 is described by ∇, but the last second column in the ∇ is changed to $(z_1^N\phi_1^{(1)}, z_2^N\phi_1^{(2)}, ..., z_N^N\phi_1^{(N)}, -(z_1^*)^N\phi_2^{*(1)}, -(z_2^*)^N\phi_2^{*(2)}, ..., -(z_N^*)^N\phi_2^{*(N)})^T$. The determinant representation of N-times Darboux transformation leads to

$$\lim_{\delta\longrightarrow 0}\frac{u_n[N]}{\delta} = v[N]. \tag{3.29}$$

Here $v[N]$ is the solution to Hirota equation via N-times Darboux transformation given by

$$v[N] = v + 2\frac{\nabla_2}{\nabla}. \tag{3.30}$$

We thus reach the conclusion that the continuum limits of N-times Darboux transformation and explicit solutions of the discrete Hirota equation (3.3) yield the corresponding results of the Hirota equation (3.1).

Continuum limit for the soliton solutions. Setting $\lambda_j = e^{z_j\delta}$, $c_i^{(j)} = \delta c_i^{(j)}$, and noting $\beta = 2\gamma\delta$, we have

$$\begin{aligned} Z(\lambda_j) &= n\ln(\lambda_j) + \chi(\lambda_j)t = z_j n\delta + (4\alpha z_j^3 - 2\gamma i z_j^2)2\delta^3 t + o(\delta^3) \\ &= X(z_j) + o(\delta^3) \end{aligned} \tag{3.31}$$

with $X(z_j) = z_j x + (4\alpha z_j^3 - 2\gamma i z_j^2)\tau$. Then, the relation between the solution of discrete eigenfunction (3.15) and continuous counterpart is consistent with the definition of (3.23), which is given by

$$\psi_{n,i}^{(j)} = \delta\phi_i^{(j)} + o(\delta), \qquad i = 1, 2 \tag{3.32a}$$

where

$$\phi_1^{(j)} = c_1^{(j)} e^{X(z_j)}, \qquad \phi_2^{(j)} = c_2^{(j)} e^{-X(z_j)} \tag{3.32b}$$

It should be remarked that $\phi^{(1)}(z_1) = (\phi_1^{(1)}(z_1), \phi_2^{(1)}(z_1))^T$ is just the eigenfunction of the Hirota equation (3.1) corresponding to the seed solution $v = 0$.

Example 4 The relation between the one-soliton solution (3.16) of spatially discrete Hirota equation and the one-soliton solution of the Hirota equation is given by

$$u_n[1] = \delta v[1] + o(\delta), \tag{3.33a}$$

where

$$v[1] = \frac{2c_1^{(1)}c_2^{(1)*}z_{1,R}}{|c_1^{(1)}c_2^{(1)}|}e^{2iX_I(z_1)}sech(2X_R(z_1)+\epsilon), \quad e^{\epsilon} = |\frac{c_1^{(1)}}{c_2^{(1)}}| \tag{3.33b}$$

Example 5 This example shows that continuum limit of the two-soliton solution to spatially discrete Hirota equation (3.3) leads to the two-soliton solution to Hirota equation (3.1). Note that

$$\begin{aligned}\Delta \ &=\delta^4\begin{vmatrix}(1-2z_1\delta+2z_1^2\delta^2)\phi_1^{(1)} & (1-z_1\delta+\frac{z_1^2}{2}\delta^2)\phi_2^{(1)} & \phi_1^{(1)} & (1+z_1\delta+\frac{z_1^2}{2}\delta^2)\phi_2^{(1)}\\ (1-2z_2\delta+2z_2^2\delta^2)\phi_1^{(2)} & (1-z_2\delta+\frac{z_2^2}{2}\delta^2)\phi_2^{(2)} & \phi_1^{(2)} & (1+z_2\delta+\frac{z_2^2}{2}\delta^2)\phi_2^{(2)}\\ (1+2z_1^*\delta+2z_1^{*2}\delta^2)\phi_2^{*(1)} & -(1+z_1^*\delta+\frac{z_1^{*2}}{2}\delta^2)\phi_1^{*(1)} & \phi_2^{*(1)} & -(1-z_1^*\delta+\frac{z_1^{*2}}{2}\delta^2)\phi_1^{*(1)}\\ (1+2z_2^*\delta+2z_2^{*2}\delta^2)\phi_2^{*(2)} & -(1+z_2^*\delta+\frac{z_2^{*2}}{2}\delta^2)\phi_1^{*(2)} & \phi_2^{*(2)} & -(1-z_2^*\delta+\frac{z_2^{*2}}{2}\delta^2)\phi_1^{*(2)}\end{vmatrix}+o(\delta^6)\\ &=4\delta^6G+o(\delta^6)\end{aligned}$$

and

$$\begin{aligned}\Delta_2 \ &=\delta^4\begin{vmatrix}K_1 & (1+2z_1\delta+2z_1^2\delta^2+\frac{4}{3}z_1^3\delta^3)\phi_1^{(1)} & \phi_1^{(1)} & (1+z_1\delta+\frac{z_1^2}{2}\delta^2+\frac{1}{6}z_1^3\delta^3)\phi_2^{(1)}\\ K_2 & (1+2z_2\delta+2z_2^2\delta^2+\frac{4}{3}z_2^3\delta^3)\phi_1^{(2)} & \phi_1^{(2)} & (1+z_2\delta+\frac{z_2^2}{2}\delta^2+\frac{1}{6}z_2^3\delta^3)\phi_2^{(2)}\\ H_1 & (1-2z_1^*\delta+2z_1^{*2}\delta^2-\frac{4}{3}z_1^{*3}\delta^3)\phi_2^{*(1)} & \phi_2^{*(1)} & (z_1^*\delta-\frac{z_1^{*2}}{2}\delta^2+\frac{1}{6}z_1^{*3}\delta^3-1)\phi_1^{*(1)}\\ H_2 & (1-2z_2^*\delta+2z_2^{*2}\delta^2-\frac{4}{3}z_2^{*3}\delta^3)\phi_2^{*(2)} & \phi_2^{*(2)} & (z_2^*\delta-\frac{z_2^{*2}}{2}\delta^2+\frac{1}{6}z_2^{*3}\delta^3-1)\phi_1^{*(2)}\end{vmatrix}+o(\delta^7)\\ &=8\delta^7F+o(\delta^7),\end{aligned}$$

where

$$\begin{aligned}K_j &= (1-2z_j\delta+2z_j^2\delta^2-\tfrac{4}{3}z_j^3\delta^3)\phi_1^{(j)},\\ H_j &= (1+2z_j^*\delta+2z_j^{*2}\delta^2+\tfrac{4}{3}z_j^{*3}\delta^3)\phi_2^{*(j)},\end{aligned} \qquad j=1,2$$

$$F = \begin{vmatrix} z_1^2\phi_1^{(1)} & z_1\phi_1^{(1)} & \phi_1^{(1)} & \phi_2^{(1)}\\ z_2^2\phi_1^{(2)} & z_2\phi_1^{(2)} & \phi_1^{(2)} & \phi_2^{(2)}\\ z_1^{*2}\phi_1^{*(1)} & -z_1^*\phi_2^{*(1)} & \phi_2^{*(1)} & -\phi_1^{*(1)}\\ z_2^{*2}\phi_2^{*(2)} & -z_2^*\phi_2^{*(2)} & \phi_2^{*(2)} & -\phi_1^{*(2)}\end{vmatrix},$$

$$G = \begin{vmatrix} -z_1\phi_1^{(1)} & \phi_2^{(1)} & \phi_1^{(1)} & z_1\phi_2^{(1)}\\ -z_2\phi_1^{(2)} & \phi_2^{(2)} & \phi_1^{(2)} & z_2\phi_2^{(2)}\\ z_1^*\phi_2^{*(1)} & -\phi_1^{*(1)} & \phi_2^{*(1)} & z_1^*\phi_1^{*(1)}\\ z_2^*\phi_2^{*(2)} & -\phi_1^{*(2)} & \phi_2^{*(2)} & z_2^*\phi_1^{*(2)}\end{vmatrix}.$$

Thus, the connection between two-soliton solution of spatially discrete Hirota equation (3.3) and Hirota equation is given by

$$u_n[2] = \delta v[2] + o(\delta), \tag{3.34}$$

where

$$v[2] = \frac{2F}{G}. \tag{3.35}$$

4 Conclusions

In this paper, we focus on the topic that the connection between the integrability of integrable spatial discrete hierarchy and integrable nonlinear PDE hierarchy, mainly including infinitely many commuting vector fields, the Lax pairs, the bi-Hamiltonian structure and soliton solutions. We take integrable spatial discrete KdV hierarchy and KdV hierarchy, integrable spatial discrete Hirota equation and Hirota equation as two examples. As a matter of fact, it is also very interesting to consider the continuous limit theory of other topics, such as the Toda hierarchy and KdV hierarchy, symmetries [6],[14], physically important solutions (double-hump soliton, rogue wave, breather) [10],[15]-[20], geometric properties, etc.

Acknowledgements

The work of Zhu is supported by National Natural Science Foundation of China (Grant Nos. 11671255 and 12071286), and by the Ministry of Economy and Competitiveness of Spain under contract MTM2016-80276-P (AEI/FEDER,EU), and Zhao is partially supported by Natural Science Foundation of Shanghai (Grant Nos. 20ZR1421900, 17ZR1411600).

References

[1] Case K M and Kac M, A discrete version of the inverse scattering problem, *J. Math. Phys.* **14**, 594-603, 1973.

[2] Case K M, On discrete inverse scattering problems II, *J. Math. Phys.* **14**, 916-920, 1973.

[3] Gieseker D, The Toda hierarchy and the KdV hierarchy, *Commun. Math. Phys.* **181**, 587-603, 1996.

[4] Kupershmidt B A, Discrete Lax equations and differential-difference calculus, Astérisque No.123, Soc. Math. France, Paris, 1985.

[5] Kac M and Moerbeke P, On an explicitly soluble system of nonlinear differential equations related to certain Toda lattices, *Adv. Math.***16**, 160-169, 1975.

[6] Levi D and Rodriguez M A, Symmetry group of partial differential equations and of differential difference equations: the Toda lattice versus the Korteweg-de Vries equation, *J. Phys. A: Math. Gen.***25**, L975-L979, 1992.

[7] Morosi C and Pizzocchero L, On the continuous limit of integrable lattices I. The Kac-Moerbeke system and KdV theory, *Commun. Math. Phys.* **180**, 505-528, 1996.

[8] Morosi C and Pizzocchero L, On the continuous limit of integrable lattices II. Volterra system and sp(N) theories, *Rev. Math. Phys.* **2**, 235-270, 1998.

[9] Morosi C and Pizzocchero L, On the continuous limit of integrable lattices III. Kupershmidt systems and sl(N+1) KdV theories, *J. Phys. A: Math. Gen.* **31**, 2727-2746, 1998.

[10] Pickering A, Zhao H Q and Zhu Z N, On the continuum limit for a semidiscrete Hirota equation, *Proc. R. Soc. A* **472**, 20160628, 2016.

[11] Schwarz M, Korteweg-de Vries and nonlinear equations related to the Toda lattice, *Adv. Math.* **44**, 132-154, 1982.

[12] Toda M and Wadati M, A soliton and two solitons in an exponential lattice and related equations, *J. Phys. Soc. Japan* **34**, 18-25, 1973.

[13] Zeng Y B, Lin R L and Cao X, The relation between the Toda hierarchy and the KdV hierarchy, *Phys. Lett. A* **251** 177-183, 1999.

[14] Zhang D J, Chen S T, Symmetries for the Ablowitz-Ladik hierarchy: Part II. Integrable discrete nonlinear Schrödinger equations and discrete AKNS hierarchy, *Stud. Appl. Math.***125**, 419-443, 2010.

[15] Zhao H Q, On a new semi-discrete integrable combination of Burgers and Sharma-Tasso-Olver equation, *Chaos* **27** 023102, 2017.

[16] Zhao H Q and Zhu Z N, Multi-soliton, multi-positon, multi-negaton, and multi-periodic solutions of a coupled Volterra lattice system and their continuous limits, *J. Math. Phys.* **52**, 023512, 2011.

[17] Zhao H Q and Yuan J Y, A semi-discrete integrable multi-component coherently coupled nonlinear Schrödinger system, *J. Phys. A: Math. Theor.***49**, 275204, 2016.

[18] Zhao H Q and Yu G F, Discrete rational and breather solution in the spatial discrete complex modified Korteweg-de Vries equation and continuous counterparts, *Chaos* **27**, 043113, 2017.

[19] Zhao H Q, Yuan J Y and Zhu Z N, Integrable semi-discrete Kundu-Eckhaus equation: Darboux transformation, breather, rogue wave and continuous limit theory, *J. Nonlinear Sci.* **28**, 43-68, 2018.

[20] Zhu Z N, Zhao H Q and Wu X N, On the continuous limits and integrability of a new coupled semidiscrete mKdV system, *J. Math. Phys.***52**, 043508, 2011.

Milton Keynes UK
Ingram Content Group UK Ltd.
UKHW051905071024
449327UK00025B/2104

9 780367 541125